Signal Processing First

James H. McClellan
Georgia Institute of Technology

Ronald W. Schafer
Georgia Institute of Technology

Mark A. Yoder
Rose-Hulman Institute of Technology

PEARSON
Prentice Hall

Pearson Education, Inc.
Upper Saddle River, NJ 07458

Library of Congress Cataloging-in-Publication Data Available

Vice President and Editorial Director, ECS: *Marcia Horton*
Publisher: *Tom Robbins*
Editorial Assistant: *Eric Van Ostenbridge*
Vice President and Director of Production and Manufacturing, ESM: *David W. Riccardi*
Executive Managing Editor: *Vince O'Brien*
Managing Editor: *David A. George*
Production Editor: *Rose Kernan*
Director of Creative Services: *Paul Belfanti*
Creative Director: *Carole Anson*
Art Director: *Jayne Conte*
Art Editor: *Gregory Dulles*
Cover Designer: *Bruce Kenselaar*
Manufacturing Manager: *Trudy Pisciotti*
Manufacturing Buyer: *Lisa McDowell*
Marketing Manager: *Holly Stark*
Cat: *Percy*

Printed in the United States of America

2 2022

ISBN 0-13-090999-8

Pearson Education Ltd., *London*
Pearson Education Australia Pty. Ltd., *Sydney*
Pearson Education Singapore, Pte. Ltd.
Pearson Education North Asia Ltd., *Hong Kong*
Pearson Education Canada, Inc., *Toronto*
Pearson Educación de Mexico, S.A. de C.V.
Pearson Education—Japan, *Tokyo*
Pearson Education Malaysia, Pte. Ltd.
Pearson Education, Inc., *Upper Saddle River, New Jersey*

Contents

13 Computing the Spectrum 389

A Complex Numbers 427

Preface

This book and its accompanying CD-ROM are the result of almost 10 years of work that originated from, and was guided by, the premise that signal processing is the best starting point for the study of both electrical engineering and computer engineering. In the summer of 1993, two of us (J. H. McC and R. W. S) began to develop a one-quarter course that was to become the first course for Georgia Tech computer engineering students, who were at that time following an overlapping, but separate, curriculum track from the electrical engineering students in the School of ECE. We argued that the subject of digital signal processing (DSP) had everything we wanted in a first course for computer engineers: it introduced the students to the use of mathematics as a language for thinking about engineering problems; it laid useful groundwork for subsequent courses; it made a strong connection to digital computation as a means for implementing systems; and it offered the possibility of interesting applications to motivate beginning engineers to do the hard work of connecting mathematics and computation to problem solving.

We were not the first to have this idea. In particular, two books by Professor Ken Steiglitz of Princeton University had a major impact on our thinking.[1] The major reasons that it was feasible in 1993 to experiment with what came to be known at Georgia Tech as the "DSP First" approach were: (1) the easy accessibility of increasingly powerful personal computers and (2) the availability of MATLAB, a powerful and easy-to-use software environment for numerical computation. Indeed, Steiglitz's 1972 book was well ahead of its time, since DSP had few practical applications, and even simple simulations on then-available batch processing computers required significant programming effort. By the early 1990s, however, DSP applications such as CD audio, high-speed modems, and cell phones were widespread due to the availability of low-cost "DSP chips" that could perform extensive computation in

[1] *An Introduction to Discrete Systems*, John Wiley & Sons, 1972, and *A Digital Signal Processing Primer: With Applications to Computer Music*, Addison-Wesley Publishing Company, 1996.

"real time." Thus, integrated circuit technology was the driving force that simultaneously provided the wherewithal both for a convenient PC-based laboratory experience for learning DSP and for creating a climate of applications that provided motivation for that study.

From the beginning, we believed that "hands-on" experience with real signals was crucial. This is provided by a "laboratory" based on MATLAB running on PCs. In the laboratory assignments, students gain direct reinforcement from hearing and seeing the effects of filtering operations that they have implemented on sound and image signals. They synthesize music from sinusoids, and they see that those same sinusoids are the basis for the data modems that they use routinely to access the Internet. We also found that MATLAB made it possible to quickly develop demonstration programs for visualizing and clarifying complicated mathematical concepts. By 1995, we had written notes covering the topics in our course, and we had amassed a large amount of computer-based supporting material. Mark Yoder, while on sabbatical leave from Rose-Hulman, had the idea to put all of this material in a form that other teachers (and students) could access easily. That idea led to a CD-ROM that captured the entire contents of our course web site. It included demonstrations and animations used in classes, laboratory assignments, and solved homework problems. As teachers, this material has changed the way we present ideas, because it offers new ways to visualize a concept "beyond the equations." Over the years, our web site has continued to grow. We anticipate that this growth will continue, and that users of this material will see new ideas take shape in the form of additional demos and labs. In 1998, all of this material was packaged together in a textbook/CD-ROM, and we gave it the descriptive title *DSP First: A Multimedia Approach*.

No sooner had we finished *DSP First*, then Georgia Tech switched from a quarterly system to semesters, and our expanded course became "Signal Processing First," the first course for computer engineers and electrical engineers. However, we found ourselves with a textbook that only covered two-thirds of the material that we needed to include for the semester-long, required signals and systems core course in our semester curriculum.[2] This led to another four years of development that included four new chapters on continuous-time signal processing and the Fourier transform; many new laboratory assignments in areas such as filtering, Fourier series, and analog and digital communications; many new demos and visualizations; hundreds of new homework problems and solutions; and updates of many of our original computer demos.

The present text is our effort to implement an expanded version of our basic philosophy. It is a conventional book, although, as our title *Signal Processing First* suggests, the distinguishing feature of the text (and the accompanying CD-ROM) is that it presents signal processing at a level consistent with an introductory ECE course, i.e., the sophomore level in a typical U.S. university. The list of topics in the book is not surprising, but since we must combine signal processing concepts with some introductory ideas, the progression of topics may strike some teachers as unconventional. Part of the reason for this is that in the electrical engineering curriculum, signals and systems and DSP typically have been treated as junior- and senior-level courses, for which a traditional background of linear circuits and linear systems is assumed. We continue to believe strongly that there are compelling reasons for turning this order around, since the early study of signal processing affords a perfect opportunity to show electrical and computer engineering students that mathematics and digital computation can be the key to understanding familiar engineering applications. Furthermore, this approach makes the subject much more accessible to students in other

[2] *DSP First*, which remains in print, is still appropriate for a quarter-length course or for a "signal processing lite" course in non-ECE fields. This book is currently under revision along the lines of changes in Chapters 1–8 of *Signal Processing First*.

majors such as computer science and other engineering fields. This point is increasingly important because non-specialists are beginning to use DSP techniques routinely in many areas of science and technology.

Signal Processing First is organized to move from simple continuous-time sinusoidal signals, to discrete-time signals and systems, then back to continuous-time, and finally, the discrete and continuous are mixed together as they often are in real engineering systems. A look at the table of contents shows that the book begins very simply (Chapter 2) with a detailed discussion of continuous-time sinusoidal signals and their representation by complex exponentials. This is a topic traditionally introduced in a linear circuits course. We then proceed to introduce the spectrum concept (Chapter 3) by considering sums of sinusoidal signals with a brief introduction to Fourier series. At this point we make the transition to discrete-time signals by considering sampled sinusoidal signals (Chapter 4). This allows us to introduce the important issues in sampling without the additional complexity of Fourier transforms. Up to this point in the text, we have only relied on the simple mathematics of sine and cosine functions. The basic linear system concepts are then introduced with simple FIR filters (Chapter 5). The key concept of frequency response is derived and interpreted for FIR filters (Chapter 6), and then we introduce z-transforms (Chapter 7) and IIR systems (Chapter 8). The first eight chapters are very similar to the those of *DSP First*. At this point, we return to continuous-time signals and systems with the introduction of convolution integrals (Chapter 9) and then frequency response for continuous-time systems (Chapter 10). This leads naturally to a discussion of the Fourier transform as a general representation of continuous-time signals (Chapter 11). The last two chapters of the book cap off the text with discussions of applications of the concepts discussed in the early chapters. At this stage, a student who has faithfully read the text, worked homework problems, and done the laboratory assignments related to the early chapters will be rewarded with the ability to understand applications involving linear filtering, amplitude modulation, the sampling theorem and discrete-time filtering, and spectrum analysis.

At Georgia Tech, our sophomore-level, 15-week course covers most of the content of Chapters 2–12 in a format involving two one-hour lectures, one 1.5 hour recitation, and one 1.5 hour laboratory period per week. As mentioned previously, we place considerable emphasis on the lab because we believe that it is essential for motivating our students to learn the mathematics of signal processing, and because it introduces our students to the use of powerful software in engineering analysis and design. At Rose-Hulman, we use *Signal Processing First* in a junior-level, 10-week course that covers Chapters 4–13. The Rose format is four one-hour lectures per week. The students use MATLAB throughout the course, but do not have a separate laboratory period.

As can be seen from the previous discussion, *Signal Processing First* is not a conventional signals and systems book. One difference is the inclusion of a significant amount of material on sinusoids and complex phasor representations. In a traditional electrical engineering curriculum, these basic notions are covered under the umbrella of linear circuits taken before studying signals and systems. Indeed, our choice of title for this book and its ancestor is designed to emphasize this departure from tradition. An important point is that teaching signal processing first also opens up new approaches to teaching linear circuits, since there is much to build upon that will allow redirected emphasis in the circuits course. At Georgia Tech, we use the fact that students have already seen phasors and sinusoidal steady-state response to move more quickly from resistive circuits to AC circuits. Furthermore, students have also seen the important concepts of frequency response and poles and zeros before studying linear circuits. This allows more emphasis on circuits as linear systems. For example, the

Laplace transform is used in the circuits course as a tool for solving the particular systems problems associated with linear circuits. This has resulted in a new textbook with accompanying CD-ROM co-authored by Professors Russell Mersereau and Joel Jackson.[3]

A second difference from conventional signals and systems texts is that *Signal Processing First* emphasizes topics that rely on "frequency domain" concepts. This means that topics like Laplace transforms, state space, and feedback control, are absent. At Georgia Tech, these topics are covered in the required linear circuits course and in a junior-level "tier two" course on control systems. Although our text has clearly been shaped by a specific point of view, this does not mean that it and the associated CD-ROM can only be used in the way that they are used at Georgia Tech. For example, at Rose-Hulman the material on sinusoids and phasors is skipped in a junior-level course because students have already had this material in a circuits course. This allows us to cover the latter part of the text in one quarter. Indeed, by appropriate selection of topics, our text can be used for either a one-quarter or one-semester signals and systems course that emphasizes communications and signal processing applications from the frequency domain point of view. For most electrical engineering curricula, the control-oriented topics would have to be covered elsewhere. In other curricula, such as computer science and computer engineering, *Signal Processing First* emphasizes those topics that are most relevant to multimedia computing, and the control-oriented topics are generally not a required part of the curriculum. This is also likely to be true in other engineering fields where data acquisition and frequency domain analysis is assuming a prominent role in engineering analysis and design.[4]

[3] R. M. Mersereau and J. R. Jackson, *Circuits: A Systems Perspective*, to be published by Pearson Prentice Hall, Pearson Education, Inc.

[4] Note that the latter chapters of *Signal Processing First* require a calculus background. On the other hand, *DSP First* does not.

The CD-ROM that accompanies the present text contains all of the material that we currently use in teaching our one-semester first course for sophomore electrical and computer engineering students. This type of material has become a common supplement for lecturing in an age where "computers in education" is the buzz word. These new forms of computer-based media provide powerful ways to express ideas that motivate our students in their engineering education. As authors, we continue to experiment with different modes of presentation, such as the narrations and movies on the accompanying CD-ROM, along with the huge archive of solved problems. In the original *DSP First* CD-ROM we noticed that finding material was a challenge, so we have provided a search engine on this CD-ROM in order to make is easy to find relevant material from keywords searches. Now, for example, if you want to know why `firfilt.m` is in the *SP-First Toolbox*, you can just search for `firfilt.m` and see all the labs and homework that use it.

This text and its associated CD-ROM represents an untold amount of work by the three authors and many students and colleagues. Fortunately, we have been able to motivate a number of extremely talented students to contribute to this project. Of the many participants, five students who served as award-winning teaching assistants over many terms provided essential material to the CD-ROM. Jeff Schodorf developed the original aliasing and reconstruction demos for Chapter 4, and did much of the early organization of all the *DSP First* CD-ROM demos along with Mark Yoder. David Anderson apprenticed with Jeff and then took over the course as its primary TA. David contributed new labs and redesigned the *DSP First* lab format so that the CD-ROM version would be easy to use. Jordan Rosenthal developed a consistent way to write GUIs that has now been used in all of our demonstrations. Many other students have benefited from his extraordinary MATLAB expertise.

Greg Krudysz wrote the `CON2DIS` demo and has now taken over the primary role in developing GUIs.

In addition, many undergraduates have implemented MATLAB programs, graphical user interfaces (GUIs), and demos that are an important part of this CD-ROM. Most notably, Craig Ulmer developed `PeZ` as a multi-year undergraduate research project and contributed some of the other GUIs used in the labs. Koon Kong overhauled `PeZ` for later versions of MATLAB. Joseph Stanley made our first movie, the tuning fork movie. Amer Abufadel developed the image filtering demo for Chapter 6. Emily Eaton wrote the Music GUI and provided many of the musical scores and piano performances needed for the songs in the labs. Rajbabu Velmurugan improved the Music GUI and provided last minute updates for all the GUIs labs. Janak Patel wrote most of help files for the GUIs. Greg Slabaugh wrote the Fourier series demo as a JAVA applet, and Mustayeen Nayeem converted it into the MATLAB Fourier series demo. Budyanto Junus wrote the first LTI demo. Mehdi Javaramand developed parts of the `Phasor Races` GUI. Sam Li has participated in the development of many parts of the labs. He, Arthur Hinson, and Ghassan Al-Regib also developed many questions for the pre-labs and warm-ups in the labs. Kathy Harrington created lists of keywords for searching homework problems and edited an extensive set of frequently asked questions for the labs. Bob Paterno recorded a large number of tutorial movies about MATLAB.

During the past few years many professors have participated in the sophomore course ECE-2025 at Georgia Tech as lecturers and recitation instructors. Many of them have also written problem solutions that are included on this CD-ROM. We an indebted to the following for permitting us to include their solutions: Randy Abler, Yucel Altunbasak, John Bordelon, Giorgio Casinovi, Russ Callen, Kate Cummings, Richard Dansereau, Steve DeWeerth, Michael Fan, Bruno Frazier, Faramarz Fekri, Elias Glytsis, Monty Hayes, Bonnie Heck, Mary Ann Ingram, Paul Hasler, Chuanyi Ji, Aaron Lanterman, Russell Mersereau, Geofferey Li, Steve McLaughlin, Mohamed Moad, Bill Sayle, Mark Smith, Whit Smith, David Taylor, Erik Verriest, Doug Williams, Tony Yezzi, and Tong Zhou.

We are also indebted to Wayne Padgett and Bruce Black, who have taught ECE-280 at Rose-Hulman and have contributed many good ideas.

We also want to acknowledge the contributions or our Publisher, Tom Robbins at Pearson Prentice Hall. Very early on, he bought into the concept of *DSP First* and supported and encouraged us at every step in that project and its continuation. He also arranged for reviews of the text and the CD-ROM by some very thoughtful and careful reviewers, including Filson Glantz, S. Hossein Mousavinezhad, Geoffrey Orsak, Mitch Wilkes, Robert Strum, James Kaiser, Victor DeBrunner, Timothy Schultz, and Anna Baraniecki.

Finally, we want to recognize the understanding and support of our wives (Carolyn McClellan, Dorothy Schafer, and Sarah Yoder). Carolyn's photo of the cat Percy appears on the cover after undergoing some DSP. They have patiently supported us as this seemingly never-ending project continued to consume energy and time. Indeed, this project will continue on beyond the present text and CD-ROM since there are just too many ideas yet to explore. That is the appeal of the computer-based and Web-based approach. It can easily grow to incorporate the innovative visualizations and experiments that others will provide.

J. H. McC
R.W.S.
M.A.Y.

CHAPTER 1

Introduction

This is a book about signals and systems. In this age of multimedia computers, audio and video entertainment systems, and digital communication systems, it is almost certain that you, the reader of this text, have formed some impression of the meaning of the terms ***signal*** and ***system***, and you probably use the terms often in daily conversation.

It is likely that your usage and understanding of the terms are correct within some rather broad definitions. For example, you may think of a signal as "something" that carries information. Usually, that something is a pattern of variations of a physical quantity that can be manipulated, stored, or transmitted by physical processes. Examples include speech signals, audio signals, video or image signals, biomedical signals, radar signals, and seismic signals, to name just a few. An important point is that signals can take many equivalent forms or representations. For example, a speech signal is produced as an acoustic signal, but it can be converted to an electrical signal by a microphone, or a pattern of magnetization on a magnetic tape, or even a string of numbers as in digital audio recording.

The term ***system*** may be somewhat more ambiguous and subject to interpretation. For example, we often use "system" to refer to a large organization that administers or implements some process, such as the "Social Security system" or the "airline transportation system." However, we will be interested in a much narrower definition that is very closely linked to signals. More specifically, a system, for our purposes, is something that can manipulate, change, record, or transmit signals. For example, an audio compact disk (CD) recording stores or represents a music signal as sequence of numbers. A CD player is a system for converting the numbers stored on the disk (i.e., the numerical representation of the signal) to an acoustic signal that we can hear. In general, systems *operate* on signals to produce new signals or new signal representations.

Our goal in this text is to develop a framework wherein it is possible to make precise statements about both signals and systems. Specifically, we want to show that mathematics is an appropriate language for describing and understanding signals and systems. We also want to show that the representation of signals and systems by mathematical equations allows us to understand how signals and systems interact and how we can design and implement systems that achieve a prescribed purpose.

1-1 Mathematical Representation of Signals

Signals are patterns of variations that represent or encode information. They have a central role in measurement, in probing other physical systems, in medical technology, and in telecommunication, to name just a few areas.

Many signals are naturally thought of as a pattern of variations in time. A good example is a speech signal, which initially arises as a pattern of changing air pressure in the vocal tract. This pattern, of course, evolves with time, creating what we often call a ***time waveform***. Figure 1-1 shows a plot of a speech waveform. In this plot, the vertical axis represents air pressure or microphone voltage and the horizontal axis represents time. Notice that there are four plots in the figure corresponding to four contiguous time segments of the speech waveform. The second plot is a continuation of the first, and so on, with each graph corresponding to a time interval of 50 milliseconds (msec).

The speech signal in Fig. 1-1 is an example of a one-dimensional ***continuous-time signal***. Such signals can be represented mathematically as a function of a single independent variable, which is normally denoted t. Although in this particular case we cannot write a simple

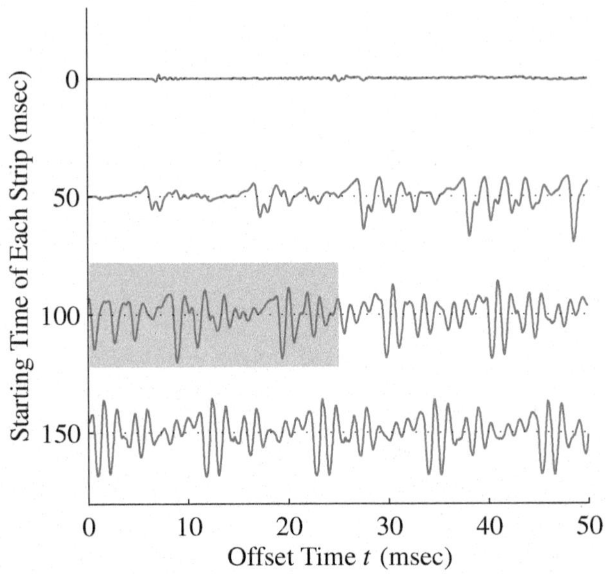

Figure 1-1: Plot of part of a speech signal. This signal can be represented as a function of a single (time) variable, $s(t)$. The shaded region is shown in more detail in Fig. 1-2.

equation that describes the graph of Fig. 1-1 in terms of familiar mathematical functions, we can nevertheless associate a function $s(t)$ with the graph. Indeed, the graph itself can be taken as a definition of the function that assigns a number $s(t)$ to each instant of time (each value of t).

Many, if not most, signals originate as continuous-time signals. However, for a variety of reasons that will become increasingly obvious as we progress through this text, it is often desirable to obtain a discrete-time representation of a signal. This can be done by ***sampling*** a continuous-time signal at isolated, equally spaced points in time. The result is a sequence of numbers that can be represented as a function of an index variable that takes on only discrete values. This can be represented mathematically as $s[n] = s(nT_s)$, where n is an integer;

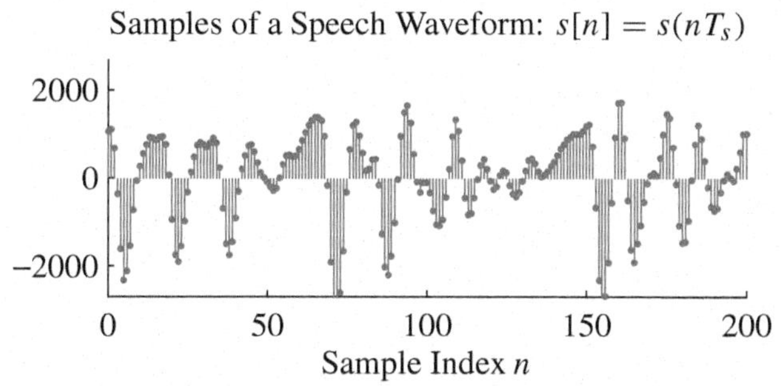

Figure 1-2: Example of a discrete-time signal that can be represented by a one-dimensional sequence or function of a discrete variable. Signal samples are taken from the shaded region of Fig. 1-1.

Figure 1-3: Example of a signal that can be represented by a function of two spatial variables.

i.e., $\{\ldots, -2, -1, 0, 1, 2, \ldots\}$, and T_s is the ***sampling period***.[1] This is, of course, exactly what we do when we plot values of a function on graph paper or on a computer screen. We cannot evaluate the function at every possible value of a continuous variable, only at a set of discrete points. Intuitively, we know that the closer the spacing of the points, the more the sequence retains the shape of the original continuous-variable function. Figure 1-2 shows an example of a short segment of a discrete-time signal that was derived by sampling the speech waveform of Fig. 1-1 with a sampling period of $T_s = 1/8$ msec. In this case, a vertical line with a dot at the end shows the location and the size of each of the isolated sequence values.

While many signals can be thought of as evolving patterns in time, many other signals are not time-varying patterns. For example, an image formed by focusing light through a lens is a spatial pattern, and thus is appropriately represented mathematically as a function of two spatial variables. Such a signal would be considered, in general, as a function of two independent variables; i.e., a picture might be denoted $p(x, y)$. A photograph is another example, such as the "gray-scale image" shown in Fig. 1-3. In this case, the value $p(x_0, y_0)$ represents the shade of gray at position (x_0, y_0) in the image.

Images such as that in Fig. 1-3 are generally considered to be two-dimensional continuous-variable signals, since we normally consider space to be a continuum. However, sampling can likewise be used to obtain a discrete-variable two-dimensional signal from a continuous-variable two-dimensional signal. Such a two-dimensional discrete-variable signal would be represented by a two-dimensional sequence or an array of numbers, and would be denoted $p[m, n] = p(m\Delta_x, n\Delta_y)$, where both m and n would take on only integer values, and Δ_x and Δ_y are the horizontal and vertical sampling periods, respectively.

Two-dimensional functions are appropriate mathematical representations of still images that do not change with time; on the other hand, videos are time-varying

[1] Note that our convention will be to use parentheses ( ) to enclose the independent variable of a continuous-variable function, and square brackets [ ] to enclose the independent variable of a discrete-variable function (sequence).

images that would require a third independent variable for time; i.e., $v(x, y, t)$. Video signals are intrinsically three-dimensional, and, depending on the type of video system, either two or all three variables may be discrete.

Our purpose in this section has been simply to introduce the idea that signals can be represented by mathematical functions. Although we will soon see that many familiar functions are quite valuable in the study of signals and systems, we have not even attempted to demonstrate that fact. Our sole concern is to make the connection between functions and signals, and, at this point, functions simply serve as abstract symbols for signals. Thus, for example, now we can refer to "the speech signal $s(t)$" or "the sampled image $p[m, n]$." Although this may not seem highly significant, we will see in the next section that it is indeed a very important step toward our goal of using mathematics to describe signals and systems in a systematic way.

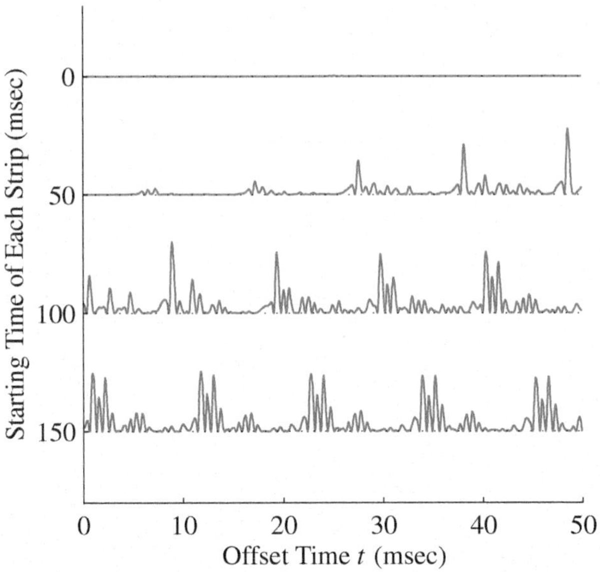

Figure 1-4: Output of a squarer system for the speech signal input of Fig. 1-1. The squarer system is defined by the equation $y(t) = [x(t)]^2$

1-2 Mathematical Representation of Systems

As we have already suggested, a system is something that transforms signals into new signals or different signal representations. This is a rather abstract definition, but it is useful as a starting point. To be more specific, we say that a one-dimensional continuous-time system takes an input signal $x(t)$ and produces a corresponding output signal $y(t)$. This can be represented mathematically by

$$y(t) = \mathcal{T}\{x(t)\} \tag{1.1}$$

which means that the input signal (waveform, image, etc.) is operated on by the system (symbolized by the operator $\mathcal{T}$) to produce the output $y(t)$. While this sounds very abstract at first, a simple example will show that this need not be mysterious. Consider a system such that the output signal is the square of the input signal. The mathematical description of this system is simply

$$y(t) = [x(t)]^2 \tag{1.2}$$

which says that at each time instant the value of the output is equal to the square of the input signal value at that same time. Such a system would logically be termed a "squarer system." Figure 1-4 shows the output of the squarer for the input of Fig. 1-1. As would be expected from the properties of the squaring operation, we see that the output signal is always nonnegative and the large signal values are emphasized relative to the small signal values.

The squarer system defined by (1.2) is an example of a ***continuous-time system***; i.e., a system whose input and output are continuous-time signals. Can we build a physical system that acts like the squarer system? The

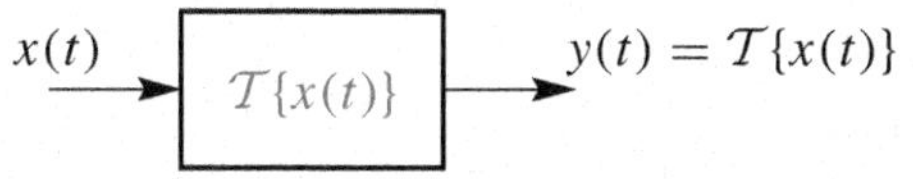

Figure 1-5: Block diagram representation of a continuous-time system.

answer is that the system of (1.2) can be approximated through appropriate connections of electronic circuits. On the other hand, if the input and output of the system are both discrete-time signals (sequences of numbers) related by

$$y[n] = (x[n])^2 \tag{1.3}$$

then the system would be a ***discrete-time system***. The implementation of the discrete-time squarer system would be trivial; one simply multiplies each discrete signal value by itself.

In thinking and writing about systems, it is often useful to have a visual representation of the system. For this purpose, engineers use ***block diagrams*** to represent operations performed in an implementation of a system and to show the interrelations among the many signals that may exist in an implementation of a complex system. An example of the general form of a block diagram is shown in Fig. 1-5. What this diagram shows is simply that the signal $y(t)$ is obtained from the signal $x(t)$ by the operation $\mathcal{T}\{\ \}$.

Another example of a system was suggested earlier when we discussed the sampling relationship between continuous-time signals and discrete-time signals. Therefore, we would define a ***sampler*** as a system whose input is a continuous-time signal $x(t)$ and whose output is the corresponding sequence of samples, defined by the equation

$$x[n] = x(nT_s) \tag{1.4}$$

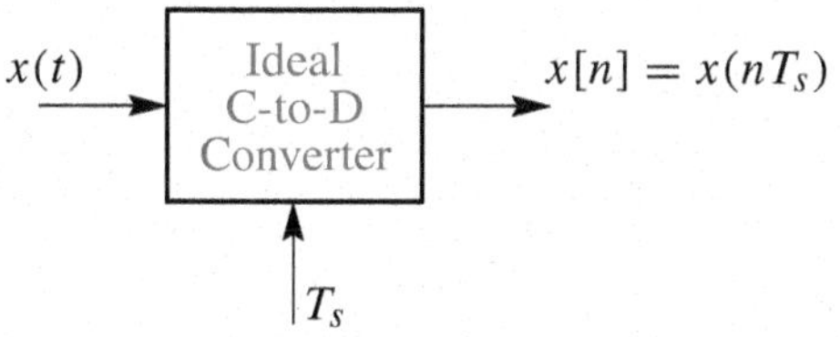

Figure 1-6: Block diagram representation of a sampler.

which simply states that the sampler "takes an instantaneous snapshot" of the continuous-time input signal once every T_s seconds. Thus, the operation of sampling fits our definition of a system, and it can be represented by the block diagram in Fig. 1-6. Often we will refer to the sampler system as an "ideal continuous-to-discrete converter" or ***ideal C-to-D converter***. In this case, as in the case of the squarer, the name that we give to the system is really just a description of what the system does.

1-3 Thinking About Systems

Block diagrams are useful for representing complex systems in terms of simpler systems, which are more easily understood. For example, Fig. 1-7 shows a block diagram representation of the process of recording and playback of an audio CD. This block diagram breaks the operation down into four subsystems. The first operation is A-to-D (analog-to-digital) conversion, which is a physical approximation to the ideal C-to-D converter defined in (1.4). An A-to-D converter produces finite-precision numbers as samples of the input signal (quantized to a limited number of bits), while the ideal C-to-D converter produces samples with infinite precision. For the high-accuracy A-to-D converters used in precision audio systems, the difference between an A-to-D converter and our idealized C-to-D converter is

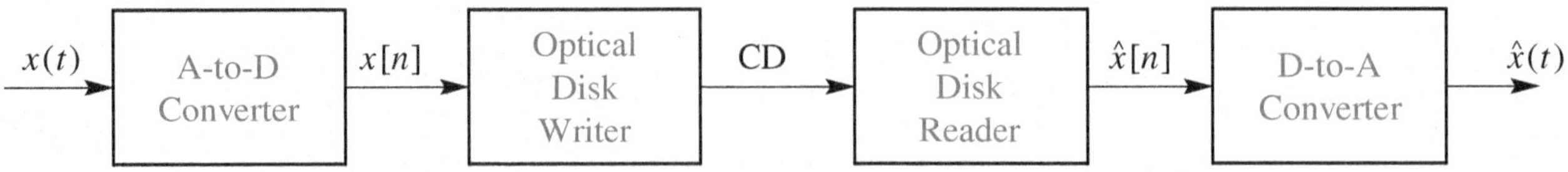

Figure 1-7: Simplified block diagram for recording and playback of an audio CD.

slight, but the distinction is very important. Only finite-precision quantized sample values can be stored in digital memory of finite size!

Figure 1-7 shows that the output of the A-to-D converter is the input to a system that writes the numbers $x[n]$ onto the optical disc. This is a complex process, but for our purposes it is sufficient to simply show it as a single operation. Likewise, the complex mechanical/optical system for reading the numbers off the optical disk is shown as a single operation. Finally, the conversion of the signal from discrete-time form to continuous-time (acoustic) form is shown as a system called a D-to-A (digital-to-analog) converter. This system takes finite precision binary numbers in sequence and fills in a continuous-time function between the samples. The resulting continuous-time signal could then be fed to other systems, such as amplifiers, loudspeakers, and headphones, for conversion to sound.

Systems like the CD audio system are all around us. Most of the time we do not need to think about how such systems work, but this example illustrates the value of thinking about a complex system in a hierarchical form. In this way, we can first understand the individual parts, then the relationship among the parts, and finally the whole system. By looking at the CD audio system in this manner, we see that a very important issue is the conversion from continuous-time to discrete-time and back to continuous-time, and we see that it is possible to consider these operations separately from the other parts of the system. The effect of connecting the parts is then relatively easy to understand. Details of some parts can be left to experts in other fields who, for example, can develop more detailed breakdowns of the optical disk subsystems.

1-4 The Next Step

The CD audio system is a good example of a discrete-time system. Buried inside the blocks of Fig. 1-7 are many discrete-time subsystems and signals. While we do not promise to explain all the details of CD players or any other complex system, we do hope to establish the foundations for the understanding of discrete- and continuous-time signals and systems so that this knowledge can be applied to understanding components of more complicated systems. In Chapter 2, we will start at a basic mathematical level and show how the well-known sine and cosine functions from trigonometry play a fundamental role in signal and system theory. Next, we will show how complex numbers can simplify the algebra of trigonometric functions. Subsequent chapters will introduce the concept of the frequency spectrum of a signal and the concept of filtering with a linear time-invariant system. By the end of the book, the diligent reader who has worked the problems, experienced the demonstrations, and done the laboratory exercises will be rewarded with a solid understanding of many of the key concepts underlying the digital multimedia information systems that are rapidly becoming commonplace.

CHAPTER 2

Sinusoids

We begin our discussion by introducing a general class of signals that are commonly called *cosine signals* or, equivalently, *sine signals*.[1] Collectively, such signals are called ***sinusoidal signals*** or, more concisely, ***sinusoids***. Although sinusoidal signals have simple mathematical representations, they are the most basic signals in the theory of signals and systems, and it is important to become familiar with their properties. The most general mathematical formula for a cosine signal is

$$x(t) = A\cos(\omega_0 t + \phi) \tag{2.1}$$

where $\cos(\cdot)$ denotes the cosine function that is familiar from the study of trigonometry. When defining a continuous-time signal, we typically use a function whose independent variable is t, a continuous variable that represents time. From (2.1) it follows that $x(t)$ is a mathematical function in which the angle of the cosine function is, in turn, a function of the variable t. The parameters A, ω_0, and ϕ are fixed numbers for a particular cosine signal. Specifically, A is called the ***amplitude***, ω_0 the ***radian frequency***, and ϕ the ***phase shift*** of the cosine signal.

Figure 2-1 shows a plot of the continuous-time cosine signal

$$x(t) = 10\cos(2\pi(440)t - 0.4\pi)$$

i.e., $A = 10$, $\omega_0 = 2\pi(440)$, and $\phi = -0.4\pi$ in (2.1). Note that $x(t)$ oscillates between A and $-A$, and that it repeats the same pattern of oscillations every $1/440 = 0.00227$ sec (approximately). This time interval is called the ***period*** of the sinusoid. We will show later in this

[1] It is also common to refer to cosine or sine signals as cosine or sine *waves*, particularly when referring to acoustic or electrical signals.

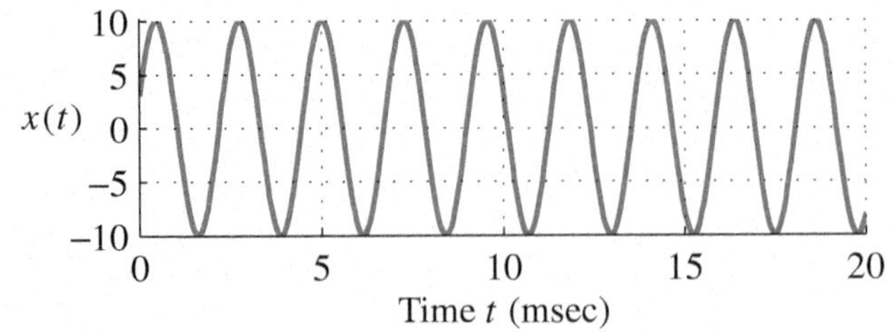

Figure 2-1: Sinusoidal signal generated from the formula: $x(t) = 10\cos(2\pi(440)t - 0.4\pi)$.

chapter that most features of the sinusoidal waveform are directly dependent on the choice of the parameters A, ω_0, and ϕ.

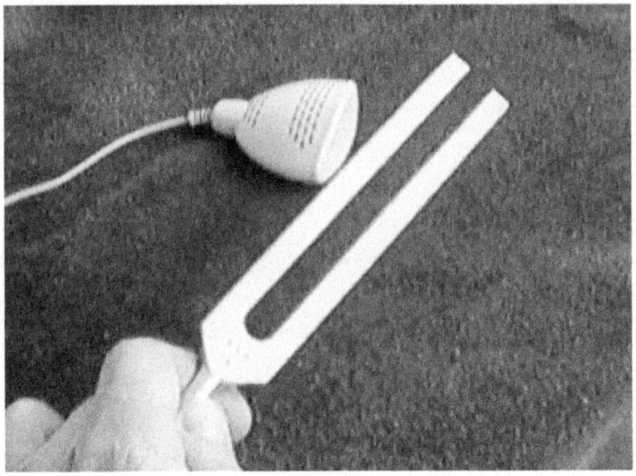

Figure 2-2: Picture of a tuning fork and a microphone.

2-1 Tuning Fork Experiment

One of the reasons that cosine waves are so important is that many physical systems generate signals that can be modeled (i.e., represented mathematically) as sine or cosine functions versus time. Among the most prominent of these are signals that are audible to humans. The tones or notes produced by musical instruments are perceived as different pitches. Although it is an oversimplification to equate notes to sinusoids and pitch to frequency, the mathematics of sinusoids is an essential first step to understanding complicated sound signals.

To provide some motivation for our study of sinusoids, we will begin by considering a very simple and familiar system for generating a sinusoidal signal. This system is a *tuning fork*, an example of which is shown in Fig. 2-2. When struck sharply, the tines of the tuning fork vibrate and emit a "pure" tone. This tone is at a single frequency, which is usually stamped on the tuning fork. It is common to find "A–440" tuning forks, because 440 hertz (Hz) is the frequency of A above middle C on a musical scale, and is often used as the reference note for tuning a piano and other musical instruments. If you can obtain a tuning fork, perform the following experiment:

> Strike the tuning fork against your knee, and then hold it close to your ear. You should hear a distinct "hum" at the frequency designated for the tuning fork. The sound will persist for a rather long time if you have struck the tuning fork properly; however, it is easy to do this experiment incorrectly. If you hit the tuning fork sharply on a hard surface such as a table, you will hear a high pitched metallic "ting" sound. This is *not* the characteristic sound that you are seeking. If you hold the tuning fork close to your ear, you will hear two tones: The higher-frequency "ting" will die away rapidly, and then the lower-frequency "hum" will be heard.

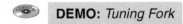

DEMO: *Tuning Fork*

With a microphone and a computer equipped with an A-to-D converter, we can make a digital recording of the signal produced by the tuning fork. The microphone converts the sound into an electrical signal, which in turn is converted to a sequence of numbers stored in the computer. Then these numbers can be plotted on the computer screen. A typical plot is shown in Fig. 2-3 for an A–440 tuning fork. In this case, the A-to-D converter sampled the output of the microphone at a rate of 5563.6 samples/sec.[2] The upper plot was constructed by connecting the sample values by straight lines. It appears that the signal generated by the tuning fork is very much like the cosine signal of Fig. 2-1. It oscillates between symmetric limits of amplitude and it also repeats periodically with a period of about 2.27 msec (0.00227 sec). As we will see in Section 2-3.1, this period is proportional to the reciprocal of ω_0; i.e., $2\pi/(2\pi(440)) \approx 0.00227$.

This experiment shows that common physical systems produce signals whose graphical representations look very much like cosine signals; i.e., they look very much like the graphical plots of the mathematical functions defined in (2.1). Later, in Section 2-7, we will add further credence to the sinusoidal model for the tuning fork sound by showing that cosine functions arise as solutions to the differential equation that (through the laws of physics) describes the motion of the tuning fork's tines. Before looking at the physics of the tuning fork, however, we should become more familiar with sinusoids and sinusoidal signals.

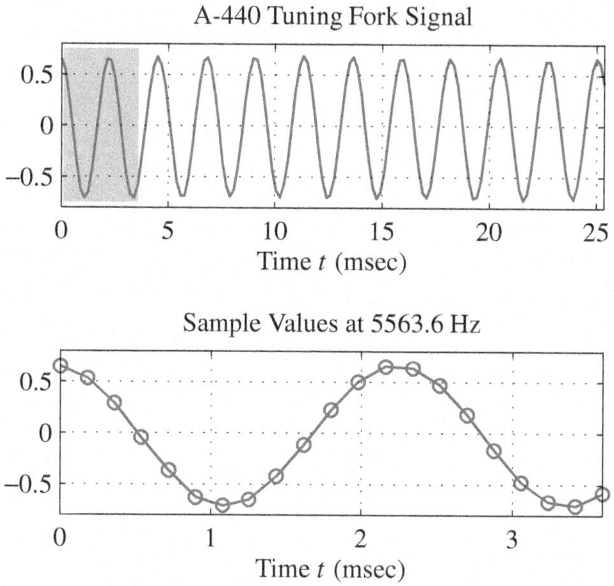

Figure 2-3: Recording of an A–440 tuning fork signal sampled at a sampling rate of 5563.6 samples/sec. The bottom plot, which consists of the first 3.6 msec taken from the top plot, shows the individual sample values (as circles).

2-2 Review of Sine and Cosine Functions

Sinusoidal signals are defined in terms of the familiar sine and cosine functions of trigonometry. A brief review of the properties of these basic trigonometric functions is useful, since these properties determine the properties of sinusoidal signals.

The sine and cosine functions are often introduced and defined through a diagram like Fig. 2-4. The trigonometric functions sine and cosine take an angle as their argument. We often think of angles in degrees, but where sine and cosine functions are concerned, angles must be dimensionless. Angles are therefore specified in radians. If the angle θ is in the first quadrant ($0 \leq \theta < \pi/2$ rad), then the sine of θ is the length y of the side of the triangle opposite the angle θ divided by the length r of the hypotenuse of the right triangle. Similarly, the cosine of θ is the ratio of the length of the adjacent side x to the length of the hypotenuse.

Note that as θ increases from 0 to $\pi/2$, $\cos\theta$ decreases from 1 to 0 and $\sin\theta$ increases from 0 to 1. When the

[2] This rate is one-quarter of the A-to-D conversion rate on a Macintosh computer.

Table 2-1: Basic properties of the sine and cosine functions.

Property	Equation
Equivalence	$\sin\theta = \cos(\theta - \pi/2)$ or $\cos(\theta) = \sin(\theta + \pi/2)$
Periodicity	$\cos(\theta + 2\pi k) = \cos\theta$, when k is an integer
Evenness of cosine	$\cos(-\theta) = \cos\theta$
Oddness of sine	$\sin(-\theta) = -\sin\theta$
Zeros of sine	$\sin(\pi k) = 0$, when k is an integer
Ones of cosine	$\cos(2\pi k) = 1$, when k is an integer.
Minus ones of cosine	$\cos[2\pi(k + \frac{1}{2})] = -1$, when k is an integer.

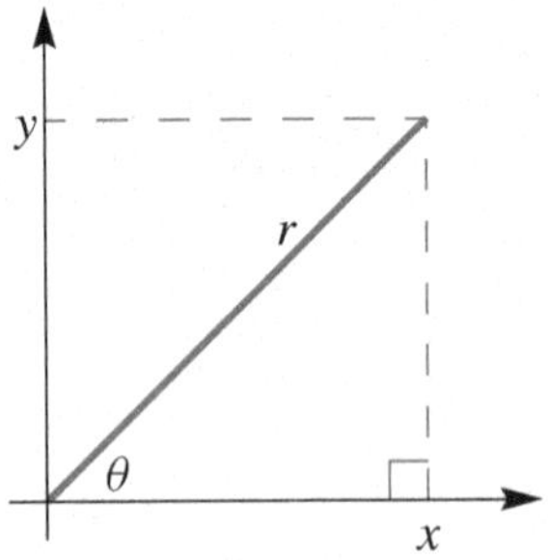

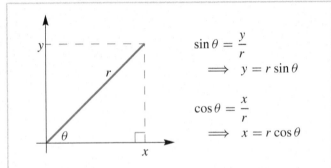

$$\sin\theta = \frac{y}{r} \implies y = r\sin\theta$$

$$\cos\theta = \frac{x}{r} \implies x = r\cos\theta$$

Figure 2-4: Definition of sine and cosine of an angle θ within a right triangle.

angle is greater than $\pi/2$ radians, the algebraic signs of x and y come into play, x being negative in the second and third quadrants and y being negative in the third and fourth quadrants. This is most easily shown by plotting the values of $\sin\theta$ and $\cos\theta$ as a function of θ, as in Fig. 2-5.[3] Several features of these plots are worthy of comment. The two functions have exactly the same shape. Indeed, the sine function is just a cosine function that is shifted to the right by $\pi/2$; i.e., $\sin\theta = \cos(\theta - \pi/2)$. Both functions oscillate between $+1$ and -1, and they repeat the same pattern periodically with period 2π. Furthermore, the sine function is an odd function of its argument, and the cosine is an even function. A summary of these and other properties is presented in Table 2-1.

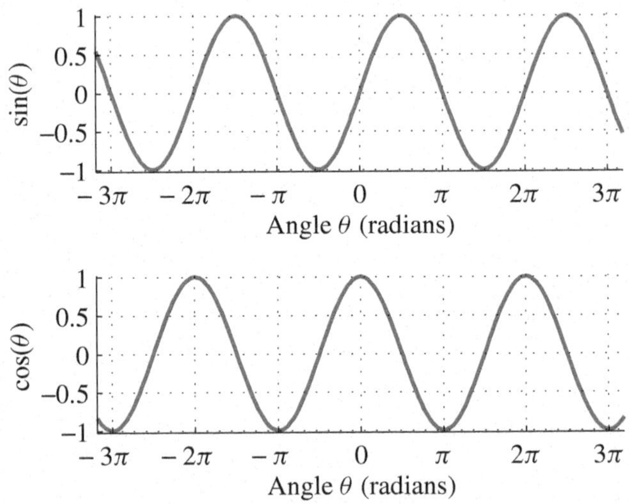

Figure 2-5: Sine and cosine functions plotted versus angle θ. Both functions have a period of 2π.

[3] It is a good idea to memorize the form of these plots, and to be able to sketch them accurately.

Table 2-2: Some basic trigonometric identities.

Number	Equation
1	$\sin^2\theta + \cos^2\theta = 1$
2	$\cos 2\theta = \cos^2\theta - \sin^2\theta$
3	$\sin 2\theta = 2\sin\theta\cos\theta$
4	$\sin(\alpha \pm \beta) = \sin\alpha\cos\beta \pm \cos\alpha\sin\beta$
5	$\cos(\alpha \pm \beta) = \cos\alpha\cos\beta \mp \sin\alpha\sin\beta$

Clearly, the sine and cosine functions are very closely related. This often leads to opportunities for simplification of expressions involving both sine and cosine functions. In calculus, we have the interesting property that the sine and cosine functions are derivatives of each other:

$$\frac{d\sin\theta}{d\theta} = \cos\theta \quad \text{and} \quad \frac{d\cos\theta}{d\theta} = -\sin\theta$$

That is, the cosine function gives the slope of the sine function, and the sine function is the negative of the slope of the cosine function. In trigonometry, there are many identities that can be used in simplifying expressions involving combinations of sinusoidal signals. Table 2-2 gives a brief table of *trigonometric identities* that will be useful. Recall from your study of trigonometry that these identities are not independent; e.g., identity 3 can be obtained from identity 4 by substituting $\alpha = \beta = \theta$. Also, these identities can be combined to derive other identities. For example, combining identity 1 with identity 2 leads to the identities

$$\cos^2\theta = \tfrac{1}{2}(1 + \cos 2\theta)$$
$$\sin^2\theta = \tfrac{1}{2}(1 - \cos 2\theta)$$

A more extensive table of trigonometric identities may be found in any book on trigonometry or in a book of mathematical tables.

EXERCISE 2.1: Use trigonometric identity #5 to derive an expression for $\cos 8\theta$ in terms of $\cos 9\theta$, $\cos 7\theta$, and $\cos\theta$.

NOTE: *Solutions to all Exercises are on the CD*

2-3 Sinusoidal Signals

The most general mathematical formula for a sinusoidal time signal is obtained by making the argument (i.e., the angle) of the cosine function be a function of t. The following equation gives two equivalent forms:

$$x(t) = A\cos(\omega_0 t + \phi) = A\cos(2\pi f_0 t + \phi) \qquad (2.2)$$

The two forms are related by defining $\omega_0 = 2\pi f_0$. In either form given in (2.2), there are three independent parameters. The names and interpretations of these parameters are as follows:

(a) A is called the ***amplitude***. The amplitude is a scaling factor that determines how large the cosine signal will be. Since the function $\cos\theta$ oscillates between $+1$ and -1, the signal $x(t)$ in (2.2) oscillates between $+A$ and $-A$.

(b) ϕ is called the ***phase shift***. The units of phase shift must be radians, since the argument of the cosine must be in radians. We will generally prefer to use the cosine function when defining the phase shift. If we happen to have a formula containing sine, e.g., $x(t) = A\sin(\omega_0 t + \phi')$, then we can rewrite it in terms of cosine if we use the equivalence property in Table 2-1. The result is:

$$x(t) = A\sin(\omega_0 t + \phi') = A\cos(\omega_0 t + \phi' - \pi/2)$$

so we define the phase shift to be $\phi = \phi' - \pi/2$ in (2.2). For simplicity and to prevent confusion, we often avoid using the sine function.

(c) ω_0 is called the ***radian frequency***. Since the argument of the cosine function must be in radians, which is dimensionless, the quantity $\omega_0 t$ must likewise be dimensionless. Thus, ω_0 must have units of rad/sec if t has units of sec. Similarly, $f_0 = \omega_0/2\pi$ is called the ***cyclic frequency***, and f_0 must have units of $\sec^{-1}$.

Example 2-1: Plotting Sinusoids

Figure 2-6 shows a plot of the signal

$$x(t) = 20\cos(2\pi(40)t - 0.4\pi) \tag{2.3}$$

In terms of our definitions, the signal parameters are $A = 20$, $\omega_0 = 2\pi(40)$, $f_0 = 40$, and $\phi = -0.4\pi$. The dependence of the signal on the amplitude parameter A is obvious; its maximum and minimum values are $+20$ and -20, respectively. The maxima occur at

$$t = \ldots, -0.02, 0.005, 0.03, \ldots$$

and the minima at

$$\ldots, -0.0325, -0.0075, 0.0175, \ldots$$

The time interval between successive maxima of the signal is $1/f_0 = 0.025$ sec. To understand why the signal has these properties, we will need to do a bit of analysis. ■

DEMO: *Sinusoids*

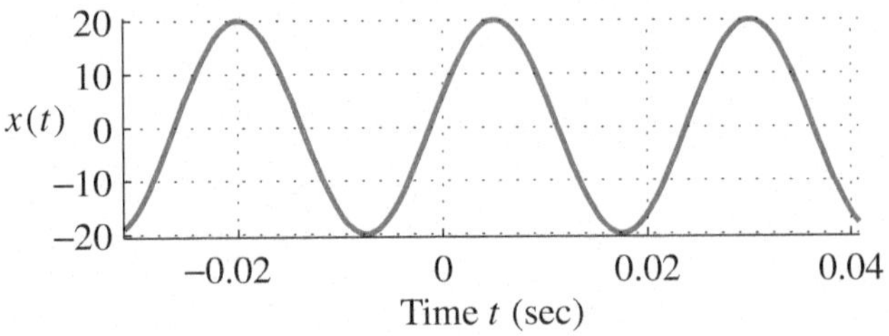

Figure 2-6: Sinusoidal signal with parameters $A = 20$, $\omega_0 = 2\pi(40)$, $f_0 = 40$, and $\phi = -0.4\pi$.

2-3.1 Relation of Frequency to Period

The sinusoid in Fig. 2-6 is clearly a periodic signal. The ***period*** of the sinusoid, denoted by T_0, is the length of one cycle of the sinusoid. In general, the frequency of the sinusoid determines its period, and the relationship can be found by examining the following equations:

$$\begin{aligned} x(t + T_0) &= x(t) \\ A\cos(\omega_0(t + T_0) + \phi) &= A\cos(\omega_0 t + \phi) \\ \cos(\omega_0 t + \omega_0 T_0 + \phi) &= \cos(\omega_0 t + \phi) \end{aligned}$$

Since the cosine function has a period of 2π, the equality above holds for all values of t if

$$\begin{aligned} \omega_0 T_0 = 2\pi \quad &\Longrightarrow \quad T_0 = \frac{2\pi}{\omega_0} \\ (2\pi f_0)T_0 = 2\pi \quad &\Longrightarrow \quad T_0 = \frac{1}{f_0} \end{aligned} \tag{2.4}$$

Since T_0 is the period of the signal, $f_0 = 1/T_0$ is the number of periods (cycles) per second. Therefore, cycles per second is an appropriate unit for f_0, and it was in general use until the 1960s.[4] When dealing with ω_0, the unit of radian frequency is rad/sec. The units of f_0 are often more convenient when describing the sinusoid, because cycles per second naturally define the period.

DEMO: *Sine Drill*

[4]The unit hertz (abbreviated Hz) was adopted in 1933 by the Electrotechnical Commission in honor of Heinrich Hertz, who first demonstrated the existence of radio waves.

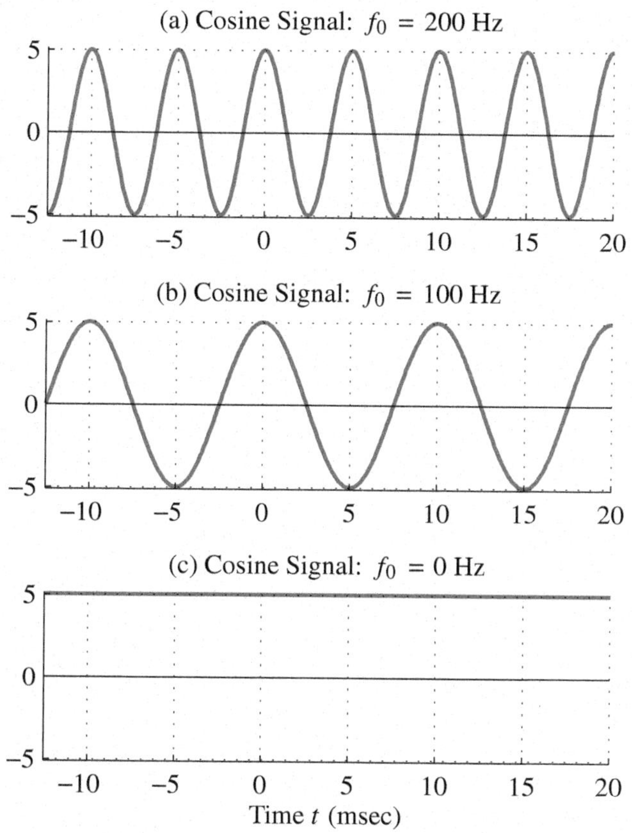

Figure 2-7: Cosine signals $x(t) = 5\cos(2\pi f_0 t)$ for several values of f_0: (a) $f_0 = 200$ Hz; (b) $f_0 = 100$ Hz; (c) $f_0 = 0$.

It is very important to understand the effect of the frequency parameter. Figure 2-7 shows this effect for several choices of f_0 in the signal

$$x(t) = 5\cos(2\pi f_0 t)$$

The two plots in Fig. 2-7(a,b) show the effect of changing f_0. As we expect, the waveform shape is similar for both values of frequency. However, for the higher frequency, the signal varies more rapidly with time; i.e., the cycle length is a shorter time interval. We have already seen in (2.4) that this is true because the period of a cosine signal is the reciprocal of the frequency. Note that when the frequency doubles ($100 \to 200$), the period is halved. This is an illustration of the general principle that the higher the frequency, the more rapid the signal waveform changes with time. Throughout this book we will see more examples of the inverse relationship between time and frequency.

Finally, notice that $f_0 = 0$ is a perfectly acceptable value, and when this value is used, the resulting signal is constant (Fig. 2-7(c)), since $5\cos(2\pi \cdot 0 \cdot t) = 5$ for all values of t. Thus, the constant signal, often called DC,[5] is, in fact, a sinusoid of zero frequency.

2-3.2 Phase Shift and Time Shift

The phase shift parameter ϕ (together with the frequency) determines the time locations of the maxima and minima of a cosine wave. To be specific, notice that the sinusoid (2.2) with $\phi = 0$ has a positive peak at $t = 0$. When $\phi \neq 0$, the phase shift determines how much the maximum of the cosine signal is shifted away from $t = 0$.

Before we examine this point in detail for sinusoids, it is useful to become familiar with the general concept of ***time-shifting*** a signal. Suppose that a signal $s(t)$ is defined by a known formula or graph. A simple example is the following triangularly shaped function:

$$s(t) = \begin{cases} 2t & 0 \le t \le \frac{1}{2} \\ \frac{1}{3}(4 - 2t) & \frac{1}{2} \le t \le 2 \\ 0 & \text{elsewhere} \end{cases} \tag{2.5}$$

This simple function has a slope of 2 for $0 \le t < \frac{1}{2}$ and a negative slope of $-\frac{2}{3}$ for $\frac{1}{2} < t \le 2$. Now consider the

[5]Electrical engineers use the abbreviation DC standing for direct current, which is a constant current.

function $x_1(t) = s(t-2)$. From the definition of $s(t)$, it follows that $x_1(t)$ is nonzero for

$$0 \leq (t-2) \leq 2 \qquad \Longrightarrow \qquad 2 \leq t \leq 4.$$

Within the time interval [2, 4], the formula for the shifted signal is:

$$x_1(t) = \begin{cases} 2(t-2) & 2 \leq t \leq 2\frac{1}{2} \\ \frac{1}{3}(8-2t) & 2\frac{1}{2} \leq t \leq 4 \\ 0 & \text{elsewhere} \end{cases} \tag{2.6}$$

In other words, $x_1(t)$ is simply the $s(t)$ function with its origin shifted to the *right* by 2 seconds. Similarly, $x_2(t) = s(t+1)$ is the $s(t)$ function shifted to the *left* by 1 second; its nonzero portion is located in the interval $-1 \leq t \leq 1$. The three signals $x(t) = s(t)$, $x_1(t) = s(t-2)$, and $x_2(t) = s(t+1)$ are all shown in Fig. 2-8.

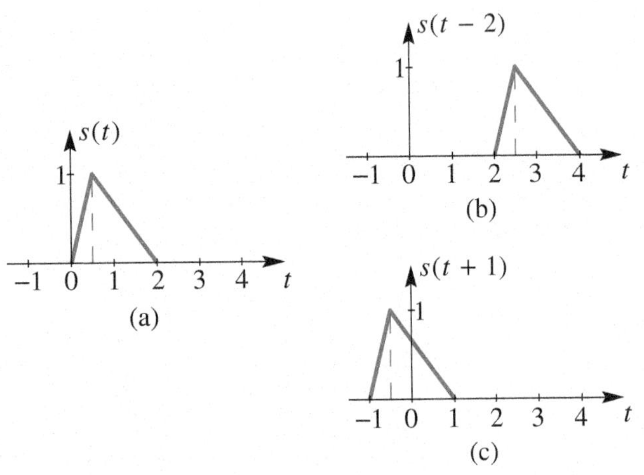

Figure 2-8: Illustration of time-shifting: (a) the triangular signal $s(t)$; (b) shifted to the right by 2 secs, $s(t-2)$; (c) shifted to the left by 1 sec, $s(t+1)$.

EXERCISE 2.2: Derive the equations for the shifted signal $x_2(t) = s(t+1)$.

We will have many occasions to consider time-shifted signals. Whenever a signal can be expressed in the form $x_1(t) = s(t - t_1)$, we say that $x_1(t)$ is a time-shifted version of $s(t)$. If t_1 is a positive number, then the shift is to the right, and we say that the signal $s(t)$ has been ***delayed*** in time. When t_1 is a negative number, then the shift is to the left, and we say that the signal $s(t)$ was ***advanced*** in time. In summary, time shift is essentially a redefinition of the time origin of the signal. In general, any function of the form $s(t - t_1)$ has its origin moved to the location $t = t_1$.

One way to determine the time shift for a cosine signal would be to find the positive peak of the sinusoid that is closest to $t = 0$. In the plot of Fig. 2-6, the time where this positive peak occurs is $t_1 = 0.005$ sec. Since the peak in this case occurs at a positive time (to the right of $t = 0$), we say that the time shift is a delay of the zero-phase cosine signal. Let $x_0(t) = A\cos(\omega_0 t)$ denote a cosine signal with zero phase shift. A delay of $x_0(t)$ can be converted to a phase shift ϕ by making the following comparison:

$$x_0(t - t_1) = A\cos\left(\omega_0(t - t_1)\right) = A\cos(\omega_0 t + \phi)$$
$$\cos(\omega_0 t - \omega_0 t_1) = \cos(\omega_0 t + \phi)$$

Since this equation must hold for all t, we must have $-\omega_0 t_1 = \phi$, which leads to

$$t_1 = -\frac{\phi}{\omega_0} = -\frac{\phi}{2\pi f_0}$$

Notice that the phase shift is negative when the time shift is positive (a delay). In terms of the period ($T_0 = 1/f_0$) we get the more intuitive formula

$$\phi = -2\pi f_0 t_1 = -2\pi\left(\frac{t_1}{T_0}\right) \tag{2.7}$$

which states that the phase shift is 2π times the fraction of a cycle given by the ratio of the time shift to the period.

DEMO: *Sine Drill*

Since the positive peak nearest to $t = 0$ must always lie within $|t_1| \leq T_0/2$, the phase shift can always be chosen to satisfy $-\pi < \phi \leq \pi$. However, the phase shift is also ambiguous because adding a multiple of 2π to the argument of a cosine function does not change the value of the cosine. This is a direct consequence of the fact that the cosine is periodic with period 2π. Each different multiple of 2π corresponds to picking a different peak of the periodic waveform. Thus, another way to compute the phase shift is to find any positive peak of the sinusoid and measure its corresponding time location. After the time location is converted to phase shift using (2.7), an integer multiple of 2π can be added to or subtracted from the phase shift to produce a final result between $-\pi$ and $+\pi$. This gives a final result identical to locating the peak that is within half a period of $t = 0$. The operation of adding or subtracting multiples of 2π is referred to as ***reducing modulo*** 2π, because it is similar to modulo reduction in mathematics, which amounts to dividing by 2π and taking the remainder. The value of phase shift that falls between $-\pi$ and $+\pi$ is called the ***principal value*** of the phase shift.

EXERCISE 2.3: In Fig. 2-6, it is possible to measure both a positive and a negative value of t_1 and then calculate the corresponding phase shifts. Which phase shift is within the range $-\pi < \phi \leq \pi$? Verify that the two phase shifts differ by 2π.

EXERCISE 2.4: Starting with the plot in Fig. 2-6, sketch a plot of $x(t - t_1)$ when $t_1 = 0.0075$. Repeat for $t_1 = -0.01$. Make sure that you shift in the correct direction. For each case, compute the phase shift of the resulting shifted sinusoid.

EXERCISE 2.5: For the signal in Fig. 2-6, $x(t) = 20\cos(2\pi(40)t - 0.4\pi)$, find G and t_1 so that the signal $y(t) = Gx(t - t_1)$ is equal to $5\cos(2\pi(40)t)$; i.e., obtain an expression for $y(t) = 5\cos(2\pi(40)t)$ in terms of $x(t)$.

2-4 Sampling and Plotting Sinusoids

All of the plots of sinusoids in this chapter were created using MATLAB. This had to be done with care, because MATLAB deals only with discrete signals represented by row or column matrices, but we are actually plotting a continuous function $x(t)$. For example, if we wish to plot a function such as

$$x(t) = 20\cos(2\pi(40)t - 0.4\pi)$$

which is shown in Fig. 2-6, we must evaluate $x(t)$ at a discrete set of times, $t_n = nT_s$, where n is an integer. If we do so, we obtain the sequence of samples

$$x(nT_s) = 20\cos(80\pi nT_s - 0.4\pi)$$

where T_s is called the *sample spacing* or ***sampling period***, and n is an integer. When plotting the function using the `plot` function in MATLAB, we must provide a pair of row or column vectors, one containing the time values and the other the computed function values to be plotted. For example, the MATLAB statements

```
n = -7:5;
Ts = 0.005;
tn = n*Ts;
xn = 20*cos(80*pi*tn - 0.4*pi);
plot(tn,xn)
```

would create a row vector `tn` of 13 numbers between -0.035 and 0.025 spaced by the sampling period 0.005, and a row vector `xn` of samples of $x(t)$. Then the MATLAB function `plot` draws the corresponding points, connecting them with straight line segments. Constructing the curve between sample points in this way is called *linear interpolation.* The solid gray curve in the upper plot of Fig. 2-9 shows the result of linear interpolation when the sample spacing is $T_s = 0.005$. Intuitively, we realize that if the points are very close together, we will see a smooth curve. The important question is, "How small must we make the sample spacing, so that the cosine signal can be accurately reconstructed between samples by linear interpolation?" A qualitative answer to this question is provided by Fig. 2-9, which shows plots produced by three different sampling periods.

Obviously, the sample spacing of $T_s = 0.005$ is not sufficiently close to create an accurate plot when the sample points are connected by straight lines. Note that sample points are shown as dots in the upper two plots.[6] With a spacing of $T_s = 0.0025$, the plot starts to approximate a cosine, but it is still possible to see places where it is clear that the points are connected by straight lines rather than the smooth cosine function. Only in the lower plot of Fig. 2-9, where the spacing is $T_s = 0.0001$, does the sampling spacing become so dense that our eye can easily see that the curve is a faithful representation of the cosine function.[7] A precise answer to the question posed above would require a mathematical definition of accuracy; our subjective judgment would be too prone to variability among different observers. However, we learn from this example that as the sampling period decreases, more samples are taken across one cycle of the periodic cosine signal. When $T_s = 0.005$, there are 5 samples per cycle; when $T_s = 0.0025$ there are 10 samples per cycle; and when $T_s = 0.0001$, there are 250 samples per cycle. It seems that 10 samples per cycle is not quite enough, and 250 samples per cycle is probably more than necessary, but, in general, the more samples per cycle, the smoother and more accurate is the linearly interpolated curve.

The choice of T_s also depends on the frequency of the cosine signal, because it is the number of samples

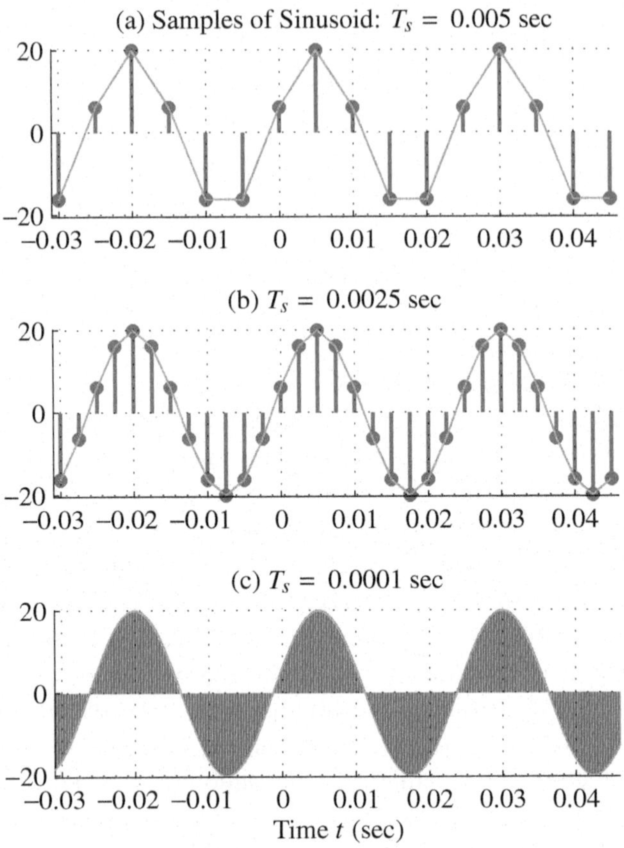

Figure 2-9: Sampled cosine signals $x(nT_s) = 20\cos(2\pi(40)nT_s - 0.4\pi)$ for several values of T_s: (a) $T_s = 0.005$ sec; (b) $T_s = 0.0025$ sec; (c) $T_s = 0.0001$ sec.

[6]This was achieved by adding a second plot using MATLAB's `hold` and `stem` functions.

[7]Here the points are so close together that we cannot show the discrete samples as individual large dots.

per cycle that matters in plotting. For example, if the frequency of the cosine signal were 2000 Hz instead of 40 Hz, then a sample spacing of $T_s = 0.0001$ would yield only 5 samples per cycle. The key to accurate reconstruction is to sample frequently enough so that the cosine signal does not change very much between sample points. This will depend directly on the frequency of the cosine signal.

The problem of plotting a cosine signal from a set of discrete samples depends on the interpolation method used. With MATLAB's built-in plotting function, linear interpolation is used to connect points by straight-line segments. An insightful question would be: "If a more sophisticated interpolation method can be used, how *large* can the sample spacing be such that the cosine signal can be reconstructed accurately from the samples?" Surprisingly, the theoretical answer to this question is that the cosine signal can be reconstructed *exactly* from its samples if the sample spacing is less than half the period, i.e., the average number of samples per cycle need be only slightly more than two! Linear interpolation certainly cannot achieve this result, but, in Chapter 4, where we examine the sampling process in more detail, we will illustrate how this remarkable result can be achieved. For now, our observation that smooth and accurate sinusoidal curves can be reconstructed from samples if the sampling period is "small enough" will be adequate for our purposes.

2-5 Complex Exponentials and Phasors

We have shown that cosine signals are useful mathematical representations for signals that arise in a practical setting, and that they are simple to define and interpret. However, it turns out that the analysis and manipulation of sinusoidal signals is often greatly simplified by dealing with related signals called *complex exponential signals.* Although the introduction of the unfamiliar and seemingly artificial concept of complex exponential signals may at first seem to be making the problem more difficult, we will soon see the value in this new representation. Before introducing the complex exponential signal, we will first review some basic concepts concerning complex numbers.[8]

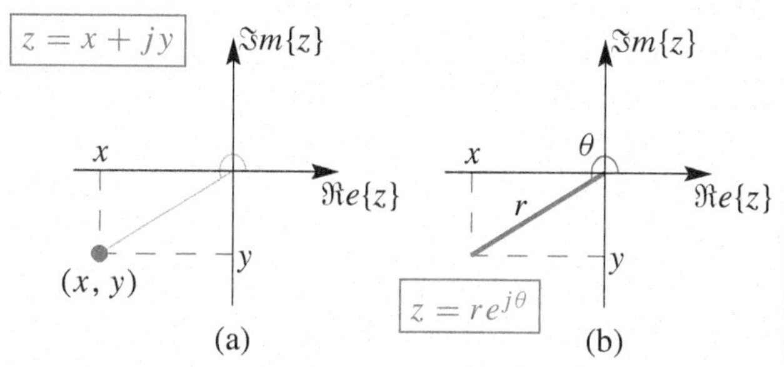

Figure 2-10: (a) Cartesian and (b) polar representations of complex numbers in the complex plane.

2-5.1 Review of Complex Numbers

A complex number z is an ordered pair of real numbers. Complex numbers may be represented by the notation $z = (x, y)$, where $x = \Re e\{z\}$ is the ***real part*** and $y = \Im m\{z\}$ is the ***imaginary part*** of z. Electrical engineers use the symbol j for $\sqrt{-1}$ instead of i, so we can also represent a complex number as $z = x + jy$. These two representations are called the ***Cartesian form*** of the complex number. Complex numbers are often represented as points in a ***complex plane***, where the real and imaginary parts are the horizontal and vertical coordinates, respectively, as shown in Fig. 2-10(a). With the Cartesian notation and the understanding that any

[8] Appendix A provides a more detailed review of the fundamentals of complex numbers. Readers who know the basics of complex numbers and how to manipulate them can skim Appendix A and skip to Section 2-5.2.

number multiplied by j is included in the imaginary part, the operations of complex addition, complex subtraction, complex multiplication, and complex division can be defined in terms of real operations on the real and imaginary parts. For example, the sum of two complex numbers is defined as the complex number whose real part is the sum of the real parts and whose imaginary part is the sum of the imaginary parts.

Since complex numbers can be represented as points in a plane, it follows that complex numbers are analogous to vectors in a two-dimensional space. This leads to a useful geometric interpretation of a complex number as a vector, shown in Fig. 2-10(b). Vectors have length and direction, so another way to represent a complex number is the ***polar form*** in which the complex number is represented by r, its vector length, together with θ, its angle with respect to the real axis. The length of the vector is also called the ***magnitude*** of z (denoted $|z|$), and the angle with the real axis is called the ***argument*** of z (denoted $\arg z$). This is indicated by the descriptive notation $z \longleftrightarrow r\angle\theta$, which is interpreted to mean that the vector representing z has length r and makes an angle θ with the real axis.

It is important to be able to convert between the Cartesian and polar forms of complex numbers. Figure 2-10(b) shows a complex number z and the quantities involved in both the Cartesian and polar representations. Using this figure, as well as some simple trigonometry and the Pythagorean theorem, we can derive a method for computing the Cartesian coordinates (x, y) from the polar variables $r\angle\theta$:

$$x = r\cos\theta \qquad \text{and} \qquad y = r\sin\theta \tag{2.8}$$

and, likewise, for going from Cartesian to polar form

$$r = \sqrt{x^2 + y^2} \qquad \text{and} \qquad \theta = \arctan\left(\frac{y}{x}\right) \tag{2.9}$$

Many calculators and computer programs have these two sets of equations built in, making the conversion between polar and Cartesian forms simple and convenient.

The $r\angle\theta$ notation is clumsy, and does not lend itself to ordinary algebraic rules. A much better polar form is given by using Euler's famous formula for the complex exponential

$$e^{j\theta} = \cos\theta + j\sin\theta \tag{2.10}$$

The Cartesian pair $(\cos\theta, \sin\theta)$ can represent any point on a circle of radius 1, so a slight generalization of (2.10) gives a representation valid for any complex number z

$$z = re^{j\theta} = r\cos\theta + jr\sin\theta \tag{2.11}$$

The complex exponential polar form of a complex number is most convenient when calculating a complex multiplication or division (see Appendix A for more details). It also serves as the basis for the complex exponential signal, which is introduced in the next section.

DEMO: *ZDrill*

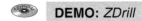

2-5.2 Complex Exponential Signals

The *complex exponential signal* is defined as

$$z(t) = A\,e^{j(\omega_0 t + \phi)} \tag{2.12}$$

Observe that the complex exponential signal is a complex-valued function of t, where the magnitude of $z(t)$ is $|z(t)| = A$ and the angle of $z(t)$ is $\arg z(t) = (\omega_0 t + \phi)$. Using Euler's formula (2.10), the complex exponential signal can be expressed in Cartesian form as

$$\begin{aligned} z(t) &= A\,e^{j(\omega_0 t + \phi)} \\ &= A\cos(\omega_0 t + \phi) + jA\sin(\omega_0 t + \phi) \end{aligned} \tag{2.13}$$

As with the real sinusoid, A is the *amplitude*, and should be a positive real number; ϕ is the *phase shift*; and ω_0 is the *frequency* in rad/sec. In (2.13) it is clear that the real part of the complex exponential signal is a real cosine

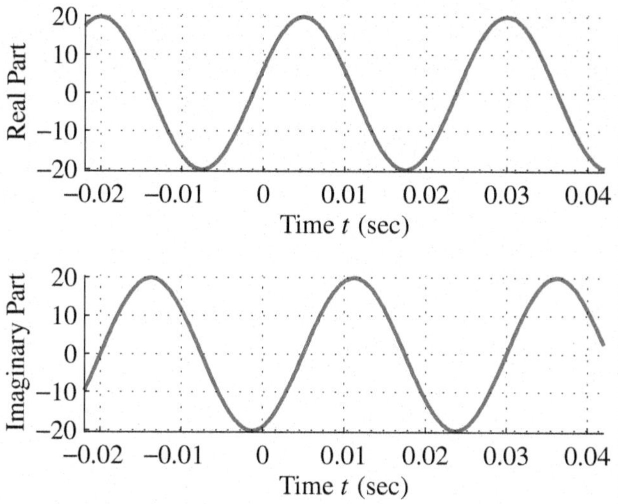

Figure 2-11: Real and imaginary parts of the complex exponential signal $z(t) = 20e^{j(2\pi(40)t-0.4\pi)}$. The phase difference between the two waves is 90° or $\pi/2$ rad.

signal as defined in (2.2), and that its imaginary part is a real sine signal. Figure 2-11 shows a plot of the following complex exponential signal:

$$\begin{aligned} z(t) &= 20\, e^{j(2\pi(40)t-0.4\pi)} \\ &= 20\, e^{j(80\pi t-0.4\pi)} \\ &= 20\cos(80\pi t - 0.4\pi) + j20\sin(80\pi t - 0.4\pi) \\ &= 20\cos(80\pi t - 0.4\pi) + j20\cos(80\pi t - 0.9\pi) \end{aligned}$$

Plotting a complex signal as a function of time requires two graphs, one for the real part and another for the imaginary part. Observe that the real and the imaginary parts of the complex exponential signal are both real sinusoidal signals, and they differ by only a phase shift of 0.5π rad.

The main reason that we are interested in the complex exponential signal is that it is an alternative *representation* for the real cosine signal. This is because we can always write

$$x(t) = \Re e\left\{Ae^{j(\omega_0 t+\phi)}\right\} = A\cos(\omega_0 t + \phi) \qquad (2.14)$$

In fact, the real part of the complex exponential signal shown in Fig. 2-11 is identical to the cosine signal plotted in Fig. 2-6. Although it may seem that we have complicated things by first introducing the imaginary part to obtain the complex exponential signal and then throwing it away by taking only the real part, we will see that many calculations are simplified by using properties of the exponents. It is possible, for example, to replace all trigonometric manipulations with algebraic operations on the exponents.

EXERCISE 2.6: Demonstrate that expanding the real part of $e^{j(\alpha+\beta)} = e^{j\alpha}e^{j\beta}$ will lead to identity #5 in Table 2-2. Also show that identity #4 is obtained from the imaginary part.

2-5.3 The Rotating Phasor Interpretation

When two complex numbers are multiplied, it is best to use the polar form for both numbers. To illustrate this, consider $z_3 = z_1 z_2$, where $z_1 = r_1 e^{j\theta_1}$ and $z_2 = r_2 e^{j\theta_2}$ and

$$z_3 = r_1 e^{j\theta_1} r_2 e^{j\theta_2} = r_1 r_2 e^{j\theta_1} e^{j\theta_2} = r_1 r_2 e^{j(\theta_1+\theta_2)}$$

We have used the law of exponents to combine the two complex exponentials. From this result we conclude that to multiply two complex numbers, we multiply the magnitudes and add the angles. If we consider one of the complex numbers to be represented by a fixed vector in the complex plane, then multiplication by a second complex number scales the length of the first vector by the magnitude of the second complex number and rotates

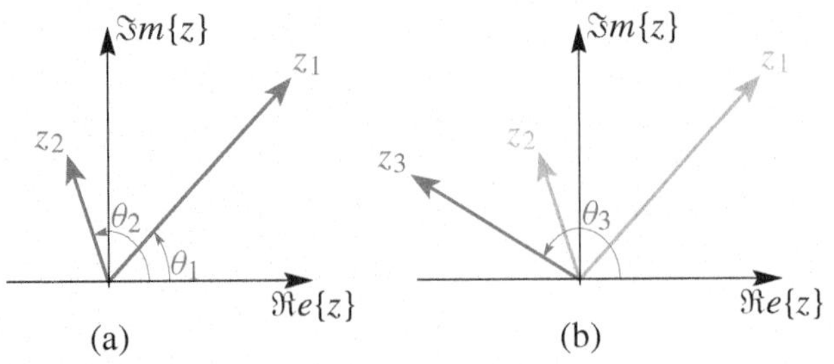

Figure 2-12: Geometric view of complex multiplication $z_3 = z_1 z_2$. The angles add: $\theta_3 = \theta_1 + \theta_2$

it by the angle of the second complex number. This is illustrated in Fig. 2-12 where it is assumed that $r_1 > 1$ so that $r_1 r_2 > r_2$.

This geometric view of complex multiplication leads to a useful interpretation of the complex exponential signal as a complex vector that rotates as time increases. If we define the complex number

$$X = A\,e^{j\phi} \tag{2.15}$$

then (2.12) can be expressed as

$$z(t) = X\,e^{j\omega_0 t} \tag{2.16}$$

i.e., $z(t)$ is the product of the complex number X and the complex-valued time function $e^{j\omega_0 t}$. The complex number X, which is aptly called the *complex amplitude*, is a polar representation created from the amplitude and the phase shift of the complex exponential signal. Taken together, the complex amplitude $X = Ae^{j\phi}$ and the frequency ω_0 are sufficient to represent $z(t)$, as well as the real cosine signal, $x(t) = A\cos(\omega_0 t + \phi)$, using (2.14). The complex amplitude is also called a *phasor.* Use of this terminology is common in electrical circuit theory, where complex exponential signals are used to greatly simplify the analysis and design of circuits. Since it is a complex number, X can be represented graphically as a vector in the complex plane, where the vector's magnitude ($|X| = A$) is the amplitude, and vector's angle ($\angle X = \phi$) is the phase shift of a complex exponential signal defined by (2.16). In the remainder of the text, the terms *phasor* and *complex amplitude* will be used interchangeably, because they refer to the same quantity defined in (2.15).

The complex exponential signal defined in (2.16) can also be written as

$$z(t) = X\,e^{j\omega_0 t} = Ae^{j\phi}e^{j\omega_0 t} = Ae^{j\theta(t)}$$

where

$$\theta(t) = \omega_0 t + \phi \qquad \text{(Radians)}$$

At a given instant in time, t, the value of the complex exponential signal, $z(t)$, is a complex number whose magnitude is A and whose argument is $\theta(t)$. Like any complex number, $z(t)$ can be represented as a vector in the complex plane. In this case, the tip of the vector always lies on the perimeter of a circle of radius A. Now, if t increases, the complex vector $z(t)$ will simply rotate at a constant rate, determined by the radian frequency ω_0. In other words, multiplying the phasor X by $e^{j\omega_0 t}$ as in (2.16) causes the fixed phasor X to rotate. (Since $|e^{j\omega_0 t}| = 1$, no scaling occurs.) Thus, another name for the complex exponential signal is *rotating phasor*.

If the frequency ω_0 is positive, the direction of rotation is counterclockwise, because $\theta(t)$ will increase with increasing time. Similarly, when ω_0 is negative, the angle $\theta(t)$ changes in the negative direction as time increases, so the complex phasor rotates clockwise. Thus, rotating phasors are said to have *positive frequency* if they rotate counterclockwise, and *negative frequency* if they rotate clockwise.

A rotating phasor makes one complete revolution every time the angle $\theta(t)$ changes by 2π radians. The time it takes to make one revolution is also equal to the period,

T_0, of the complex exponential signal, so

$$\omega_0 T_0 = (2\pi f_0)T_0 = 2\pi \qquad \Longrightarrow \qquad T_0 = \frac{1}{f_0}$$

Notice that the phase shift ϕ defines where the phasor is pointing when $t = 0$. For example, if $\phi = \pi/2$, then the phasor is pointing straight up when $t = 0$, whereas if $\phi = 0$, the phasor is pointing to the right when $t = 0$.

The plots in Fig. 2-13(a) illustrate the relationship between a single complex rotating phasor and the cosine signal waveform. The upper left plot shows the complex plane with two vectors. The vector at an angle in the third quadrant represents the signal

$$z(t) = e^{j(t-\pi/4)}$$

at the specific time $t = 1.5\pi$. The horizontal vector pointing to the left represents the real part of the vector $z(t)$ at the particular time $t = 1.5\pi$; i.e.,

$$x(1.5\pi) = \Re e\{z(1.5\pi)\} = \cos(1.5\pi - \pi/4) = -\frac{\sqrt{2}}{2}$$

As t increases, the rotating phasor $z(t)$ rotates in the counterclockwise direction, and its real part $x(t)$ oscillates left and right along the real axis. This is shown in the lower left plot, which shows how the real part of the rotating phasor has varied over one period, i.e., $0 \le t \le 2\pi$.

DEMO: *Rotating Phasors*

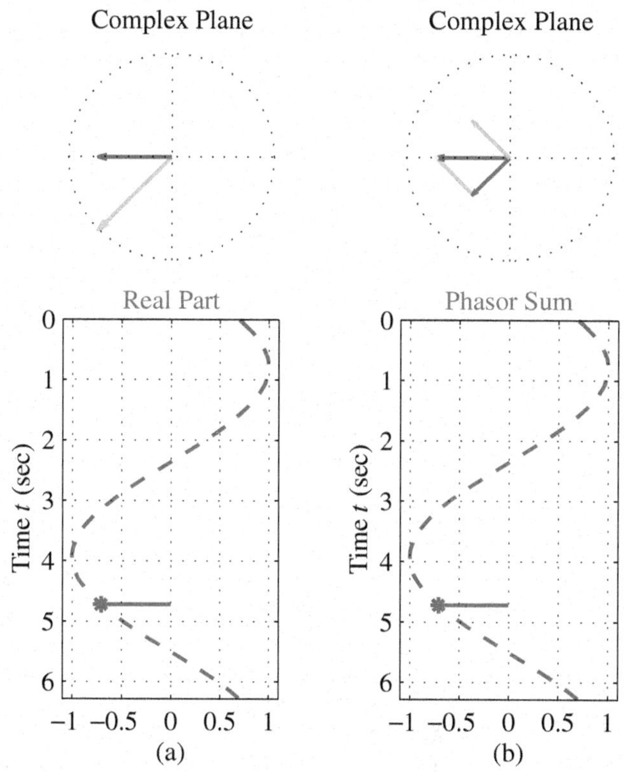

Figure 2-13: Rotating phasors: (a) single phasor rotating counterclockwise; (b) complex conjugate rotating phasors.

2-5.4 Inverse Euler Formulas

The inverse Euler formulas allow us to write the cosine function in terms of complex exponentials as

$$\cos\theta = \frac{e^{j\theta} + e^{-j\theta}}{2} \tag{2.17}$$

and also the sine function can be expressed as

$$\sin\theta = \frac{e^{j\theta} - e^{-j\theta}}{2j} \tag{2.18}$$

(See Appendix A for more details.)

Equation (2.17) can be used to express $\cos(\omega_0 t + \phi)$ in terms of a positive and a negative frequency complex exponential as follows:

$$\begin{aligned} A\cos(\omega_0 t + \phi) &= A\left(\frac{e^{j(\omega_0 t+\phi)} + e^{-j(\omega_0 t+\phi)}}{2}\right) \\ &= \tfrac{1}{2}Xe^{j\omega_0 t} + \tfrac{1}{2}X^* e^{-j\omega_0 t} \\ &= \tfrac{1}{2}z(t) + \tfrac{1}{2}z^*(t) \\ &= \Re e\{z(t)\} \end{aligned}$$

where * denotes complex conjugation.

This formula has an interesting interpretation. The real cosine signal with frequency ω_0 is actually composed of two complex exponential signals; one with positive frequency (ω_0) and the other with negative frequency ($-\omega_0$). The complex amplitude of the positive frequency complex exponential signal is $\frac{1}{2}X = \frac{1}{2}Ae^{j\phi}$, and the complex amplitude of the negative frequency complex exponential is $\frac{1}{2}X^* = \frac{1}{2}Ae^{-j\phi}$. In other words, the real cosine signal can be represented as the sum of two complex rotating phasors that are complex conjugates of each other.

DEMO: *Rotating Phasors*

Figure 2-13(b) illustrates how the sum of the two half-amplitude complex conjugate rotating phasors becomes the real cosine signal. In this case, the vector at an angle in the third quadrant is the complex rotating phasor $\frac{1}{2}z(t)$ at time $t = 1.5\pi$. As t increases after that time, the angle would increase in the counterclockwise direction. Similarly, the vector in the second quadrant (plotted with a light-orange line) is the complex rotating phasor $\frac{1}{2}z^*(t)$ at time $t = 1.5\pi$. As t increases after that time, the angle of $\frac{1}{2}z^*(t)$ will increase in the clockwise direction. The horizontal vector pointing to the right is the sum of these two complex conjugate rotating phasors. The result is the same as the real vector in the plot on the left, and therefore the real cosine wave traced out as a function of time is the same in both cases. The lower right shows the variation of the real values of $\cos(t - \pi/4)$ for $0 \le t \le 2\pi$.

This representation of real sinusoidal signals in terms of their positive and negative frequency components is a remarkably useful concept. The negative frequencies, which arise due to the complex exponential representation, turn out to lead to many simplifications in the analysis of signal and systems problems. We will develop this representation further in Chapter 3, where we introduce the idea of the spectrum of a signal.

EXERCISE 2.7: Show that the following representation can be derived for the real sine signal:

$$A\sin(\omega_0 t + \phi) = \tfrac{1}{2}Xe^{-j\pi/2}e^{j\omega_0 t} + \tfrac{1}{2}X^*e^{j\pi/2}e^{-j\omega_0 t}$$

where $X = Ae^{j\phi}$. In this case, the interpretation is that the sine signal is also composed of two complex exponentials with the same positive and negative frequencies, but the complex coefficients multiplying the terms are different from those of the cosine signal. Specifically, the sine signal requires additional phase shifts of $\mp\pi/2$ applied to the complex amplitude X and X^*, respectively.

2-6 Phasor Addition

There are many situations in which it is necessary to add two or more sinusoidal signals. When all signals have the same frequency, the problem simplifies. This problem arises in electrical circuit analysis, and it will arise again in Chapter 5, where we introduce the concept of discrete-time filtering. Thus it is useful to develop a mechanism for adding several sinusoids having the same frequency, *but with different amplitudes and phases.* Our goal is to prove that the following statement is true:

$$\sum_{k=1}^{N} A_k\cos(\omega_0 t + \phi_k) = A\cos(\omega_0 t + \phi) \qquad (2.19)$$

Equation (2.19) states that a sum of N cosine signals of differing amplitudes and phase shifts, but with the same frequency, can always be reduced to a single cosine signal of the same frequency. A proof of (2.19) can be accomplished by using trigonometric identities such as

$$\begin{aligned} A_k\cos(\omega_0 t + \phi_k) &= A_k\cos\phi_k\cos(\omega_0 t) \\ &\quad - A_k\sin\phi_k\sin(\omega_0 t) \end{aligned} \qquad (2.20)$$

We can expand the sum (2.19) into sines and cosines, collect terms involving $\cos(\omega_0 t)$ and those involving $\sin(\omega_0 t)$, and then finally use the same identity (2.20) in the reverse direction. However, this is exceedingly tedious to do as a numerical computation (see Exercise 2.8), and it leads to some very messy expressions if we wish to obtain a general formula. As we will see, a much simpler approach can be based on the complex exponential representation of the cosine signals.

EXERCISE 2.8: Use (2.20) to show that the sum

$$1.7\cos(20\pi t + 70\pi/180) + 1.9\cos(20\pi t + 200\pi/180)$$

reduces to $A\cos(20\pi t + \phi)$, where

$$\begin{aligned} A &= \{[1.7\cos(70\pi/180) + 1.9\cos(200\pi/180)]^2 \\ &\quad + [1.7\sin(70\pi/180) + 1.9\sin(200\pi/180)]^2\}^{1/2} \\ &= 1.532 \end{aligned}$$

and

$$\begin{aligned} \phi &= \tan^{-1}\left\{\frac{1.7\sin(70\pi/180) + 1.9\sin(200\pi/180)}{1.7\cos(70\pi/180) + 1.9\cos(200\pi/180)}\right\} \\ &= 141.79\pi/180 = 2.475 \text{ rads} \end{aligned}$$

The value of ϕ, given in radians, corresponds to 141.79°.

2-6.1 Addition of Complex Numbers

When two complex numbers are added, it is necessary to use the Cartesian form. If $z_1 = x_1 + jy_1$ and $z_2 = x_2 + jy_2$, then $z_3 = z_1 + z_2 = (x_1 + x_2) + j(y_1 + y_2)$; i.e., the real and imaginary parts of the sum are the sum of the real and imaginary parts, respectively. Using the vector interpretation of complex numbers, where both z_1 and z_2 are viewed as vectors with their tails at the origin, the sum z_3 is the result of vector addition, and is constructed as follows:

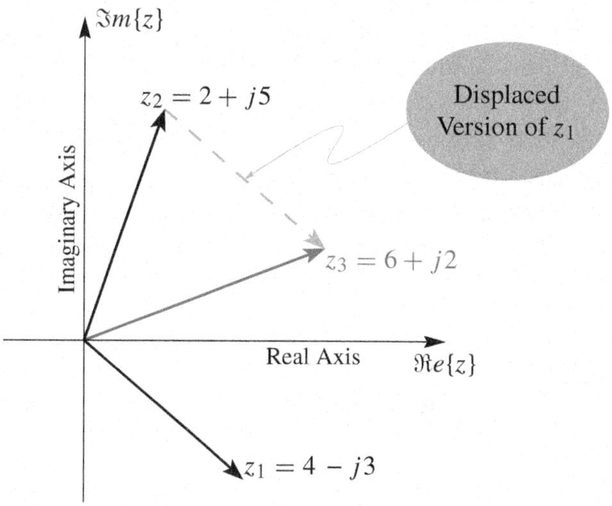

Figure 2-14: Graphical construction of complex number addition, $z_3 = z_1 + z_2$, shows that the process is the same as vector addition.

(a) Draw a copy of z_1 with its tail at the head of z_2. Call this displaced vector $\hat{z}_1$.

(b) Draw the vector from the origin to the head of $\hat{z}_1$. This is the sum z_3.

This process is depicted in Fig. 2-14 for the case $z_1 = 4 - j3$ and $z_2 = 2 + j5$.

2-6.2 Phasor Addition Rule

The phasor representation of cosine signals can be used to show the following result:

$$\begin{aligned} x(t) &= \sum_{k=1}^{N} A_k \cos(\omega_0 t + \phi_k) \\ &= A\cos(\omega_0 t + \phi) \end{aligned} \tag{2.21}$$

where N is any integer. That is, the sum of two or more cosine signals each having the same frequency, but having different amplitudes and phase shifts, can be expressed as a single equivalent cosine signal. The resulting amplitude (A) and phase (ϕ) of the term on the right-hand side of (2.21) can be computed from the individual amplitudes (A_k) and phases (ϕ_k) on the left-hand side by doing the following addition of complex numbers:

$$\sum_{k=1}^{N} A_k e^{j\phi_k} = Ae^{j\phi} \tag{2.22}$$

Equation (2.22) is the essence of the phasor addition rule. Proof of the phasor rule requires the following two pieces of information:

(a) Any sinusoid can be written in the form:

$$A\cos(\omega_0 t + \phi) = \Re e\{Ae^{j(\omega_0 t+\phi)}\} = \Re e\{Ae^{j\phi}\, e^{j\omega_0 t}\}$$

(b) For any set of complex numbers $\{X_k\}$ the sum of the real parts is equal to the real part of the sum, so we have

$$\Re e\left\{\sum_{k=1}^{N} X_k\right\} = \sum_{k=1}^{N} \Re e\{X_k\}$$

Proof of the phasor addition rule involves the following algebraic manipulations:

$$\begin{aligned}
\sum_{k=1}^{N} A_k \cos(\omega_0 t + \phi_k) &= \sum_{k=1}^{N} \Re e\left\{A_k e^{j(\omega_0 t+\phi_k)}\right\} \\
&= \Re e\left\{\sum_{k=1}^{N} A_k e^{j\phi_k} e^{j\omega_0 t}\right\} \\
&= \Re e\left\{\left(\sum_{k=1}^{N} A_k e^{j\phi_k}\right) e^{j\omega_0 t}\right\} \\
&= \Re e\left\{\left(Ae^{j\phi}\right)e^{j\omega_0 t}\right\} \\
&= \Re e\left\{Ae^{j(\omega_0 t+\phi)}\right\} \\
&= A\cos(\omega_0 t + \phi)
\end{aligned}$$

Two steps (shown in color) are important in this proof. In the third line, the complex exponential $e^{j\omega_0 t}$ is factored out of the summation because all the sinusoids have the same frequency. In going from the third line to the fourth, the crucial step is replacing the summation term in parentheses with a single complex number, $Ae^{j\phi}$, as defined in (2.22).

2-6.3 Phasor Addition Rule: Example

We now return to the example of Exercise 2.8, where

$$x_1(t) = 1.7\cos(20\pi t + 70\pi/180)$$
$$x_2(t) = 1.9\cos(20\pi t + 200\pi/180)$$

and the sum was found to be

$$\begin{aligned}
x_3(t) &= x_1(t) + x_2(t) \\
&= 1.532\cos(20\pi t + 141.79\pi/180)
\end{aligned}$$

The frequency of both sinusoids is 10 Hz, so the period is $T_0 = 0.1$ sec. The waveforms of the three signals are shown Fig. 2-15(b) and the phasors used to solve the problem are shown on the left in Fig. 2-15(a). Notice that the times where the maximum of each cosine signal occurs can be derived from the phase through the formula

$$t_m = -\frac{\phi T_0}{2\pi}$$

which gives

$$t_{m1} = -0.0194, \quad t_{m2} = -0.0556, \quad t_{m3} = -0.0394 \text{ sec.}$$

These times are marked with vertical dashed lines in the corresponding waveform plots in Fig. 2-15(b). The phasor addition of the two signals is computed in four steps.

(a) Represent $x_1(t)$ and $x_2(t)$ by the phasors:

$$X_1 = A_1 e^{j\phi_1} = 1.7e^{j70\pi/180}$$
$$X_2 = A_2 e^{j\phi_2} = 1.9e^{j200\pi/180}$$

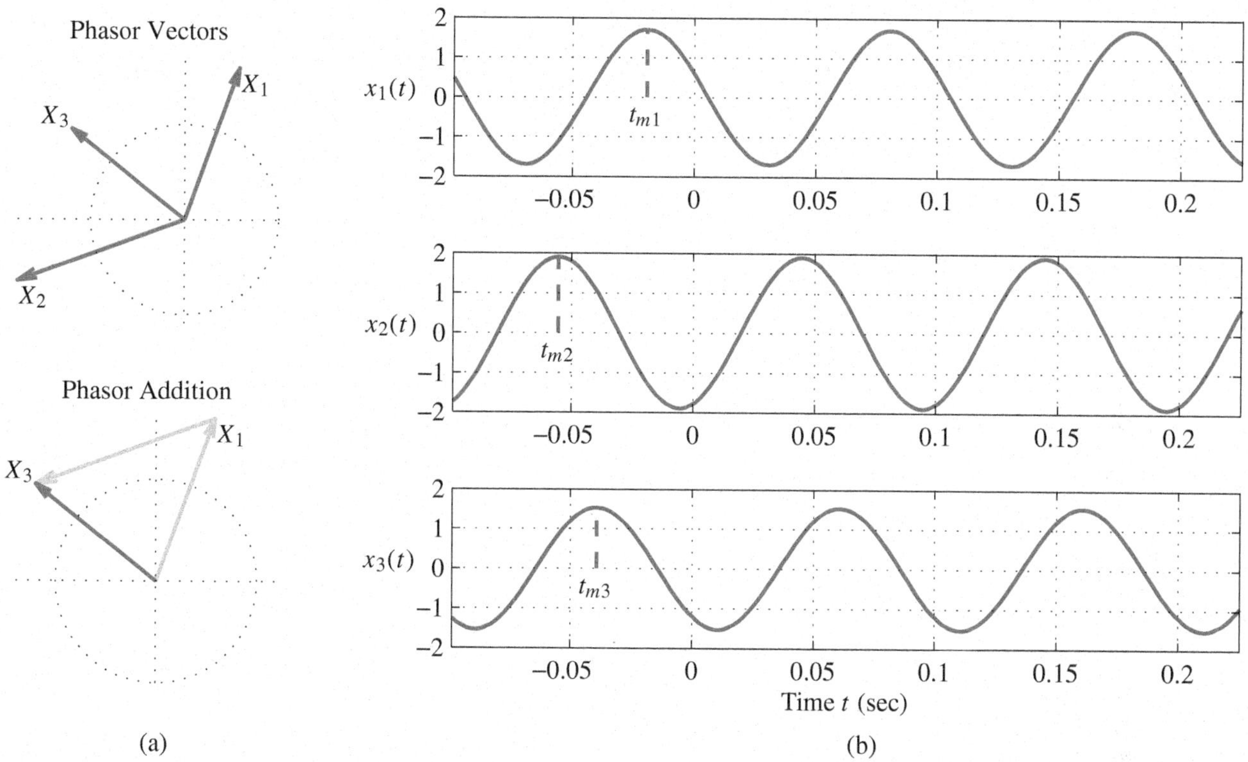

Figure 2-15: (a) Adding sinusoids by doing a phasor addition, which is actually a graphical vector sum. (b) The time of the signal maximum is marked on each $x_i(t)$ plot.

(b) Convert both phasors to rectangular form:

$$X_1 = 0.5814 + j1.597$$
$$X_2 = -1.785 - j0.6498$$

(c) Add the real parts and the imaginary parts:

$$\begin{aligned} X_3 &= X_1 + X_2 \\ &= (0.5814 + j1.597) + (-1.785 - j0.6498) \\ &= -1.204 + j0.9476 \end{aligned}$$

(d) Convert back to polar form, obtaining

$$X_3 = 1.532e^{j141.79\pi/180}$$

Therefore, the final formula for $x_3(t)$ is

$$x_3(t) = 1.532\cos(20\pi t + 141.79\pi/180)$$
$$\text{or} \quad x_3(t) = 1.532\cos(20\pi(t + 0.0394))$$

2-6.4 MATLAB Demo of Phasors

The process of phasor addition can be accomplished easily using MATLAB. The answer generated by MATLAB

Table 2-3: Phasor Addition Example

Z	=	X	+	jY	Magnitude	Phase	Ph/pi	Ph(deg)
Z1		0.5814		1.597	1.7	1.222	0.389	70.00
Z2		-1.785		-0.6498	1.9	-2.793	-0.889	-160.00
Z3		-1.204		0.9476	1.532	2.475	0.788	141.79

and printed with the special function `zprint` (provided on the CD-ROM) for this particular phasor addition is given in Table 2-3. Help on `zprint` gives

```
ZPRINT  print out complex # in rect
        and polar form
  usage:    zprint(z)
  z = vector of complex numbers
```

DEMO: *ZDrill*

The MATLAB code that generates Fig. 2-15 can be found on the CD and is also contained in the first lab (also on the CD). It uses the special MATLAB functions (provided on the CD-ROM) `zprint`, `zvect`, `zcat`, `ucplot`, and `zcoords` to make the vector plots.

LAB: *#2 Introduction to Complex Exponentials*

2-6.5 Summary of the Phasor Addition Rule

In this section, we have shown how a real cosine signal can be represented as the real part of a complex exponential signal (complex rotating phasor), and we have applied this representation to show how to simplify the process of adding several cosine signals of the same frequency.

$$x(t) = \sum_{k=1}^{N} A_k \cos(\omega_0 t + \phi_k) = A\cos(\omega_0 t + \phi)$$

In summary, all we have to do to get the cosine signal representation of the sum is:

(a) Obtain the phasor representation $X_k = A_k e^{j\phi_k}$ of each of the individual signals.

(b) Add the phasors of the individual signals to get $X = X_1 + X_2 + \cdots = Ae^{j\phi}$. This requires polar-to-Cartesian-to-polar format conversions.

(c) Multiply X by $e^{j\omega_0 t}$ to get $z(t) = Ae^{j\phi}e^{j\omega_0 t}$.

(d) Take the real part to get $x(t) = \Re e\{Ae^{j\phi}e^{j\omega_0 t}\} = A\cos(\omega_0 t + \phi) = x_1(t) + x_2(t) + \ldots$.

In other words, A and ϕ must be calculated by doing a vector sum of all the X_k phasors.

EXERCISE 2.9: Consider the two sinusoids,

$$x_1(t) = 5\cos(2\pi(100)t + \pi/3)$$
$$x_2(t) = 4\cos(2\pi(100)t - \pi/4)$$

Obtain the phasor representations of these two signals, add the phasors, plot the two phasors and their sum in the complex plane, and show that the sum of the two signals is

$$x_3(t) = 5.536\cos(2\pi(100)t + 0.2747)$$

In degrees the phase should be 15.74°. Examine the plots in Fig. 2-16 to see whether you can identify the cosine waves $x_1(t)$, $x_2(t)$, and $x_3(t) = x_1(t) + x_2(t)$.

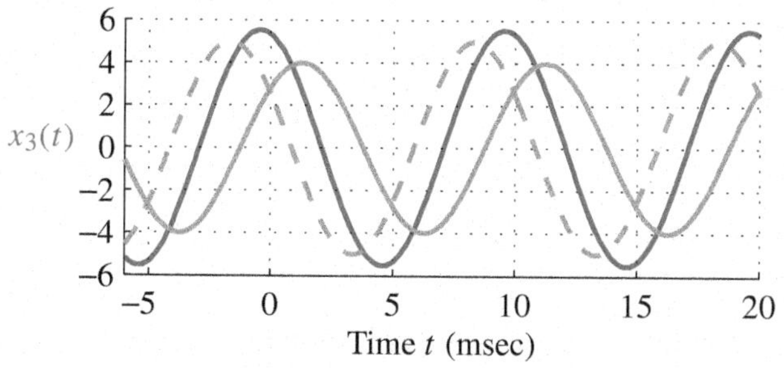

Figure 2-16: Two sinusoids and their sum, $x_3(t) = x_1(t) + x_2(t)$.

2-7 Physics of the Tuning Fork

In Section 2-1, we described a simple experiment in which a tuning fork was seen to generate a signal whose waveform looked very much like that of a sinusoidal signal. Now that we know a lot more about sinusoidal signals, it is worthwhile to take up this issue again. Is it a coincidence that the tuning-fork signal looks like a sinusoid, or is there a deeper connection between vibrations and sinusoids? In this section, we present a simple analysis of the tuning-fork system that shows that the tuning fork does indeed vibrate sinusoidally when given a displacement from its equilibrium position. The sinusoidal motion of the tuning-fork tines is transferred to the surrounding air particles, thereby producing the acoustic signal that we hear. This simple example illustrates how mathematical models of physical systems that are derived from fundamental physical principles can lead to concise mathematical descriptions of physical phenomena and of signals that result.

2-7.1 Equations from Laws of Physics

A simplified drawing of the tuning fork is shown in Fig. 2-17. As we have seen experimentally, when struck against a firm surface, the tines of the tuning fork vibrate and produce a "pure" tone. We are interested in deriving the equations that describe the physical behavior of the tuning fork so that we can understand the basic mechanism by which the sound is produced.[9] Newton's second law, $F = ma$, will lead to a differential equation whose solution is a sine or cosine function, or a complex exponential.

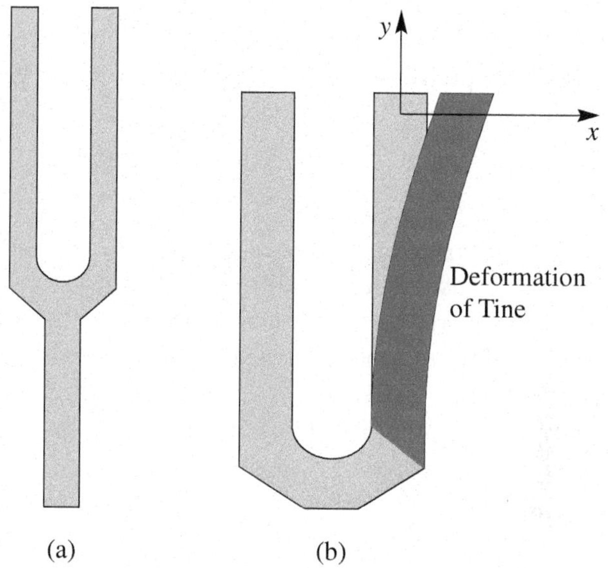

Figure 2-17: (a) Tuning fork. (b) The coordinate system needed to write equations for the vibration of the tine.

When the tuning fork is struck, one of the tines is deformed slightly from its rest position, as depicted in Fig. 2-17(b). We know from experience that unless the deformation was so large as to break or bend the metal, there would be a tendency for the tine to return to its original rest position. The physical law that governs this movement is Hooke's law. Although the tuning fork is made of a very stiff metal, we can think of it as an

[9]The generation and propagation of sound are treated in any general college physics text.

elastic material when the deformation is tiny. Hooke's law states that the restoring force is directly proportional to the amount of deformation. If we set up a coordinate system as in Fig. 2-17(b), the deformation is along the x-axis, and we can write

$$F = -kx$$

where the parameter k is the elastic constant of the material (i.e., its stiffness). The minus sign indicates that this restoring force acts in the negative direction when the displacement of the tine is in the positive x direction; i.e., it acts to pull the tine back toward the neutral position.

Now this restoring force due to stiffness produces an acceleration as dictated by Newton's second law; i.e.,

$$F = ma = m\,\frac{d^2x}{dt^2}$$

where m is the mass of the tine, and the second derivative with respect to time of position x is the acceleration of the mass along the x-axis. Since these two forces must balance each other (i.e., the sum of the forces is zero), we get a second-order differential equation that describes the motion $x(t)$ of the tine for all values of time t

$$m\,\frac{d^2x(t)}{dt^2} = -k\,x(t) \qquad (2.23)$$

This particular differential equation is rather easy to solve, because we can, in fact, guess the solution. From the derivative properties of sine and cosine functions, we are motivated to try as a solution the function

$$x(t) = \cos\omega_0 t$$

where the parameter ω_0 is a constant that must be determined. The second derivative of $x(t)$ is

$$\begin{aligned}\frac{d^2x(t)}{dt^2} &= \frac{d^2}{dt^2}(\cos\omega_0 t)\\ &= \frac{d}{dt}(-\omega_0\sin\omega_0 t)\\ &= -\omega_0^2\cos\omega_0 t\end{aligned}$$

Notice that the second derivative of the cosine function is the same cosine function multiplied by a constant $(-\omega_0^2)$. Therefore, when we substitute $x(t)$ into (2.23), we get

$$\begin{aligned}m\,\frac{d^2x(t)}{dt^2} &= -k\,x(t)\\ -m\,\omega_0^2\cos\omega_0 t &= -k\,\cos\omega_0 t\end{aligned}$$

Since this equation must be satisfied for all t, it follows that the coefficients of $\cos\omega_0 t$ on both sides of the equation must be equal, which leads to the following algebraic equation

$$-m\,\omega_0^2 = -k$$

This equation can be solved for ω_0, obtaining

$$\omega_0 = \pm\sqrt{\frac{k}{m}} \qquad (2.24)$$

Therefore, we conclude that one solution of the differential equation is

$$x(t) = \cos\left(\sqrt{\frac{k}{m}}\,t\right)$$

From our model, $x(t)$ describes the motion of the tuning-fork tine. Therefore we conclude that the tines oscillate sinusoidally. This motion is, in turn, transferred to the particles of air in the locality of the tines thereby producing the tiny variations in air pressure that make up an acoustic wave. The formula for the frequency lets us draw two conclusions:

(a) Of two tuning forks having the same mass, the stiffer one will produce a higher frequency. This is because the frequency is proportional to $\sqrt{k}$, which is in the numerator of (2.24).

(b) Of two tuning forks having the same stiffness, the heavier one will produce a lower frequency. This is because the frequency is inversely proportional to the square root of the mass, $\sqrt{m}$, which is in the denominator of (2.24).

DEMO: *Tuning Fork*

2-7.2 General Solution to the Differential Equation

There are many possible solutions to the tuning-fork differential equation (2.23). We can prove that the following function

$$x(t) = A\cos(\omega_0 t + \phi)$$

will satisfy the differential equation (2.23) by substituting back into (2.23), and taking derivatives. Once again the frequency must be $\omega_0 = \sqrt{k/m}$. Only the frequency ω_0 is constrained by our simple model; the specific values of the parameters A and ϕ are not important. From this we can conclude that any scaled or time-shifted sinusoid with the correct frequency will satisfy the differential equation that describes the motion of the tuning fork's tine. This implies that an infinite number of different sinusoidal waveforms can be produced in the tuning fork experiment. For any particular experiment, A and ϕ would be determined by the exact strength and timing of the sharp force that gave the tine its initial displacement. However, the frequency of all these sinusoids will be determined only by the mass and stiffness of the tuning-fork metal.

EXERCISE 2.10: Demonstrate that a complex exponential signal can also be a solution to the tuning-fork differential equation:

$$\frac{d^2x}{dt^2} = -\frac{k}{m}x(t)$$

By substituting $z(t)$ and $z^*(t)$ into both sides of the differential equation, show that the equation is satisfied for all t by both of the signals

$$z(t) = X\,e^{j\omega_0 t} \qquad \text{and} \qquad z^*(t) = X^*\,e^{-j\omega_0 t}$$

Determine the value of ω_0 for which the differential equation is satisfied.

2-7.3 Listening to Tones

The observer is an important part of any physical experiment. This is particularly true when the experiment involves listening to the sound produced. In the tuning-fork experiment, we perceive a tone with a certain pitch (related to the frequency) and loudness (related to the amplitude). The human ear and neural processing system respond to the frequency and amplitude of a sustained sound like that produced by the tuning fork, but the phase is not perceptible. This is because phase is really due to an arbitrary definition of the starting time of the sinusoid; i.e., a sustained tone sounds the same now as it did 5 minutes ago. On the other hand, we could pick up the sound with a microphone and sample or display the signal on an oscilloscope. In this case, it would be possible to make precise measurements of frequency and amplitude, but phase would be measured accurately only with respect to the time base of the sampler or the oscilloscope.

2-8 Time Signals: More Than Formulas

The purpose of this chapter has been to introduce the concept of a sinusoidal signal and to illustrate how sinusoidal signals can arise in real situations. Signals, as we have defined them, are varying patterns that convey or represent information, usually about the state or behavior of a physical system. We have seen by both theory and observation that a tuning fork generates a signal that can be represented mathematically as a sinusoidal signal. In the context of the tuning fork, the cosine wave conveys and represents information about the state of the tuning fork. Encoded in the sinusoidal waveform is information such as whether the tuning fork is vibrating or at rest, and, if it is vibrating, its frequency and the amplitude of its vibrations. This information can be extracted from the signal by human listeners, or it can be recorded for later processing by either humans or computers.

DEMO: *Clay Whistle*

Although the solution to the differential equation of the tuning fork (2.23) is a cosine function, the resulting mathematical formula is simply a model that results from an idealization of the tuning fork. It is important to recall that the signal is a separate entity from the formula. The actual waveform produced by a tuning fork is probably not exactly sinusoidal. The signal is *represented* by the mathematical formula $x(t) = A\cos(\omega_0 t + \phi)$, which can be derived from an idealized model based on physical principles. This model is a good approximation of reality, but an approximation nevertheless. Even so, this model is extremely useful since it leads directly to a useful mathematical representation of the signal produced by the tuning fork.

In the case of a complicated signal generated by a musical instrument, the signal cannot be so easily reduced to a mathematical formula. Figure 2-18 shows a short segment of a recording of orchestra music. Just the appearance of the waveform suggests a much more complex situation. Although it oscillates like the cosine wave, it clearly is not periodic (at least, as far as we can see from the given segment). Orchestra music consists of many instruments sounding different notes together. If each instrument produced a pure sinusoidal tone at the frequency of the note that is assigned to it, then the composite orchestra signal would be simply a sum

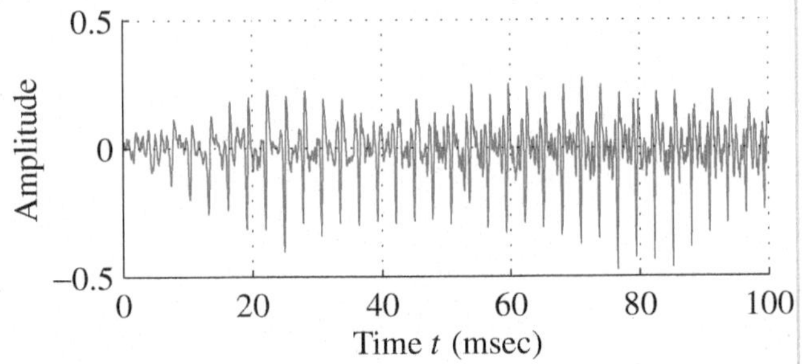

Figure 2-18: A short segment of an orchestra music signal.

of sinusoids with different frequencies, amplitudes, and phase shifts. While this is far from being a correct model for most instruments, it is actually a highly appropriate way to think about the orchestra signal. In fact, we will see very soon that sums of sinusoids of different frequencies, amplitudes, and phase shifts can result in an infinite variety of waveforms. Indeed, it is true that almost any signal can be represented as a sum of sinusoidal signals. When this concept was first introduced by Jean-Baptiste Joseph Fourier in 1807, it was received with great skepticism by the famous mathematicians of the world. Nowadays, this notion is commonplace (although no less remarkable). The mathematical and computational techniques of Fourier analysis underlie the frequency-domain analysis tools used extensively in electrical engineering and other areas of science and engineering.

2-9 Summary and Links

We have introduced sinusoidal signals in this chapter. We have attempted to show that they arise naturally as a result of simple physical processes and that they can be represented by familiar mathematical functions, and also by complex exponentials. The value of the mathematical representation of a signal is twofold. First, the mathematical representation provides a convenient formula to consistently describe the signal. For example, the cosine signal is completely described in terms of just three parameters. Second, by representing both signals and systems through mathematical expressions, we can make precise statements about the interaction between signals and systems.

LAB: *#1 Introduction to* MATLAB

In connection with this chapter, two laboratories are found on the CDROM. Lab #1 involves some introductory exercises on the basic elements of the MATLAB programming environment, and its use for manipulating complex numbers and plotting sinusoids. Appendix B is also available for a quick overview of essential ideas about MATLAB. Labs #2a and #2b deal with sinusoids and phasor addition. In these labs, students must re-create a phasor addition demonstration similar to Fig. 2-15. Copies of the lab write-ups are also found on the CD-ROM.

LAB: *#2 Introduction to Complex Exponentials*

On the CD-ROM, one can find the following resources:

(a) A tuning-fork movie that shows the experiment of striking the tuning fork and recording its sound. Several recorded sounds from different tuning forks are available, as well as sounds from clay whistles.

DEMO: *Links to Many Demos*

(b) A drill program written in MATLAB for working with the amplitude, phase, and frequency of sinusoidal plots.

(c) A set of movies showing rotating phasor(s) and how they generate sinusoids through the real part operator.

Finally, the CD-ROM also contains a wealth of solved homework problems that may be used for practice and self-study.

NOTE: *Hundreds of Solved Problems*

2-10 Problems

P-2.1 Define $x(t)$ as

$$x(t) = 3\cos(\omega_0 t - \pi/4)$$

For $\omega_0 = \pi/5$, make a plot of $x(t)$ that is valid over the range $-10 \le t \le 20$.

P-2.2 Figure P-2.2 is a plot of a sinusoidal wave. From the plot, determine values for the amplitude (A), phase (ϕ), and frequency (ω_0) needed in the representation:

$$x(t) = A\cos(\omega_0 t + \phi)$$

Give the answer as numerical values, *including the units* where applicable.

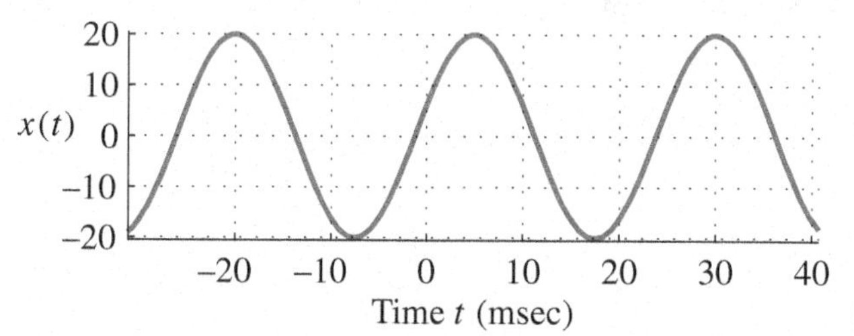

Figure P-2.2

P-2.3 Make a carefully labeled sketch for each of the following functions.

(a) Sketch $\cos\theta$ for values of θ in the range $0 \le \theta \le 6\pi$.

(b) Sketch $\cos(0.2\pi t)$ for values of t such that three periods of the function are shown.

(c) Sketch $\cos(2\pi t/T_0)$ for values of t such that three periods of the function are shown. Label the horizontal axis in terms of the parameter T_0.

(d) Sketch $\cos(2\pi t/T_0 + \pi/2)$ for values of t such that three periods of the function are shown.

P-2.4 Use the series expansions for e^x, $\cos(\theta)$, and $\sin(\theta)$ given here to verify Euler's formula.

$$e^x = 1 + x + \frac{x^2}{2!} + \frac{x^3}{3!} + \cdots$$

$$\cos(\theta) = 1 - \frac{\theta^2}{2!} + \frac{\theta^4}{4!} + \cdots$$

$$\sin(\theta) = \theta - \frac{\theta^3}{3!} + \frac{\theta^5}{5!} + \cdots$$

P-2.5 Use complex exponentials (i.e., phasors) to show the following trigonometric identities:

(a) $\cos(\theta_1 + \theta_2) = \cos(\theta_1)\cos(\theta_2) - \sin(\theta_1)\sin(\theta_2)$

(b) $\cos(\theta_1 - \theta_2) = \cos(\theta_1)\cos(\theta_2) + \sin(\theta_1)\sin(\theta_2)$

Hint: Write the left-hand side of each equation as the real part of a complex exponential.

P-2.6 Use Euler's formula for the complex exponential to prove DeMoivre's formula

$$(\cos\theta + j\sin\theta)^n = \cos n\theta + j\sin n\theta$$

Use it to evaluate $\left(\frac{3}{5} + j\frac{4}{5}\right)^{100}$.

P-2.7 Simplify the following expressions:

(a) $3e^{j\pi/3} + 4e^{-j\pi/6}$

(b) $(\sqrt{3} - j3)^{10}$

(c) $(\sqrt{3} - j3)^{-1}$

(d) $(\sqrt{3} - j3)^{1/3}$

(e) $\Re e\left\{je^{-j\pi/3}\right\}$

Give the answers in *both* Cartesian form $(x + jy)$ and polar form $(re^{j\theta})$.

P-2.8 Suppose that MATLAB is used to plot a sinusoidal signal. The following MATLAB code generates the signal and makes the plot. Derive a formula for the signal; then draw a sketch of the plot that will be done by MATLAB.

```
dt = 1/100;
tt = -1 : dt : 1;
Fo = 2;
zz = 300*exp(j*(2*pi*Fo*(tt - 0.75)));
xx = real( zz );
%
plot( tt, xx ), grid on
title( 'SECTION of a SINUSOID' )
xlabel('TIME   (sec)')
```

P-2.9 Define $x(t)$ as

$$x(t) = 2\sin(\omega_0 t + 45^\circ) + \cos(\omega_0 t)$$

(a) Express $x(t)$ in the form $x(t) = A\cos(\omega_0 t + \phi)$.

(b) Assume that $\omega_0 = 5\pi$. Make a plot of $x(t)$ over the range $-1 \le t \le 2$. How many periods are included in the plot?

(c) Find a complex-valued signal $z(t)$ such that $x(t) = \Re e\{z(t)\}$.

P-2.10 Define $x(t)$ as

$$x(t) = 5\cos(\omega t) + 5\cos(\omega t + 120°) + 5\cos(\omega t - 120°)$$

Simplify $x(t)$ into the standard sinusoidal form: $x(t) = A\cos(\omega t + \phi)$. Use phasors to do the algebra, but also provide a plot of the vectors representing each of the three phasors.

P-2.11 Solve the following equation for θ:

$$\Re e\{(1 + j)e^{j\theta}\} = -1$$

Give the answers in radians. Make sure that you find *all* possible answers.

P-2.12 Give two possible complex-valued solutions to the following differential equation:

$$\frac{d^2x(t)}{dt^2} = -100\,x(t)$$

P-2.13 Define the following complex exponential signal:

$$s(t) = 5e^{j\pi/3}e^{j10\pi t}$$

(a) Make a plot of $s_i(t) = \Im m\{s(t)\}$. Pick a range of values for t that will include exactly three periods of the signal.

(b) Make a plot of $q(t) = \Im m\{\dot{s}(t)\}$, where the dot means differentiation with respect to time t. Again plot three cycles of the signal.

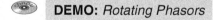

DEMO: *Rotating Phasors*

P-2.14 For the sinusoidal waveform shown in Fig. P-2.14, determine the complex phasor representation

$$X = A\,e^{j\phi}$$

i.e., find ω_0, ϕ, and A such that the waveform is represented by

$$x(t) = \Re e\{X\,e^{j\omega_0 t}\}$$

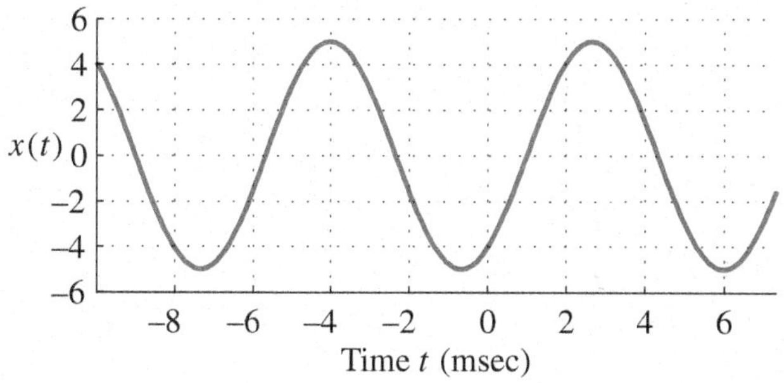

Figure P-2.14

P-2.15 Define $x(t)$ as

$$x(t) = 5\cos(\omega t + \tfrac{1}{3}\pi) + 7\cos(\omega t - \tfrac{5}{4}\pi) + 3\cos(\omega t)$$

Express $x(t)$ in the form $x(t) = A\cos(\omega t + \phi)$. Use complex phasor manipulations to obtain the answer. Explain your answer by giving a phasor diagram.

P-2.16 The phase of a sinusoid can be related to time shift as follows:

$$x(t) = A\cos(2\pi f_0 t + \phi) = A\cos(2\pi f_0(t - t_1))$$

In the following parts, assume that the period of the sinusoidal wave is $T_0 = 8$ sec.

(a) "When $t_1 = -2$ sec, the value of the phase is $\phi = \pi/2$." Explain whether this is True or False.

(b) "When $t_1 = 3$ sec, the value of the phase is $\phi = 3\pi/4$." Explain whether this is True or False.

(c) "When $t_1 = 7$ sec, the value of the phase is $\phi = \pi/4$." Explain whether this is True or False.

P-2.17 Define $x(t)$ as

$$x(t) = 5\cos(\omega t + \tfrac{3}{2}\pi) + 4\cos(\omega t + \tfrac{2}{3}\pi) + 4\cos(\omega t + \tfrac{1}{3}\pi)$$

(a) Express $x(t)$ in the form $x(t) = A\cos(\omega t + \phi)$ by finding the numerical values of A and ϕ.

(b) Plot all the phasors used to solve the problem in (a) in the complex plane.

P-2.18 Solve the following simultaneous equations by using the phasor method. Is the answer for A_1, A_2, ϕ_1, ϕ_2 unique? Provide a geometrical diagram to explain the answer.

$$\cos(\omega_0 t) = A_1\cos(\omega_0 t + \phi_1) + A_2\cos(\omega_0 t + \phi_2)$$
$$\sin(\omega_0 t) = 2A_1\cos(\omega_0 t + \phi_1) + A_2\cos(\omega_0 t + \phi_2)$$

P-2.19 Solve the following equation for M and ψ. Obtain *all* possible answers. Use the phasor method, and provide a geometrical diagram to explain the answer.

$$5\cos(\omega_0 t) = M\cos(\omega_0 t - \pi/6) + 5\cos(\omega_0 t + \psi)$$

Hint: Describe the figure in the z-plane given by the set $\{z : z = 5e^{j\psi} - 5\}$ where $0 \le \psi \le 2\pi$.

P-2.20 Let $x[n]$ be the complex exponential sequence

$$x[n] = 7e^{j(0.22\pi n - 0.25\pi)}$$

defined for $n = -\infty, \dots, -1, 0, 1, 2, \dots, \infty$. If we define a new sequence $y[n]$ to be the second difference

$$y[n] = x[n+1] - 2x[n] + x[n-1] \qquad \text{for all } n,$$

it is possible to express $y[n]$ in the form

$$y[n] = Ae^{j(\hat{\omega}_0 n + \phi)}$$

Determine the numerical values of A, ϕ, and $\hat{\omega}_0$.

P-2.21 In a mobile radio system (e.g., cell phones), there is one type of degradation that can be modeled easily with sinusoids. This is the case of *multipath fading* caused by reflections of the radio waves interfering destructively at some locations. Suppose that a transmitting tower sends a sinusoidal signal, and a mobile user receives not one but two copies of the transmitted signal: a direct-path transmission and a reflected-path signal (e.g., from a large building) as depicted in Fig. P-2.21.

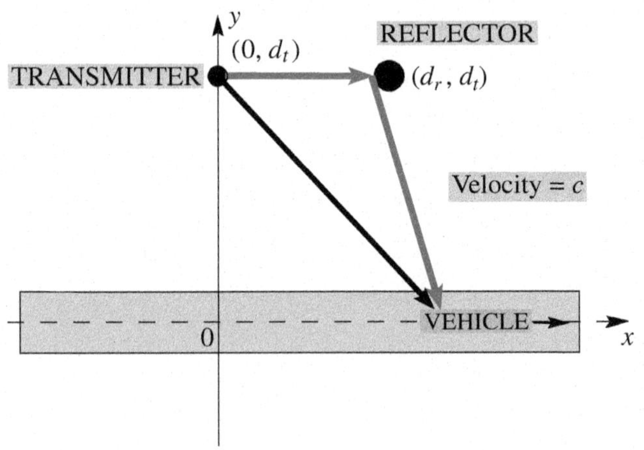

Figure P-2.21

The received signal is the sum of the two copies, and since they travel different distances they have different time delays. If the transmitted signal is $s(t)$, then the received signal[10] is

$$r(t) = s(t - t_1) + s(t - t_2)$$

In a mobile phone scenario, the distance between the mobile user and the transmitting tower is always

[10] For simplicity we are ignoring propagation losses: When a radio signal propagates over a distance R, its amplitude will be reduced by an amount that is proportional to $1/R^2$.

changing. Suppose that the direct-path distance is

$$d_1 = \sqrt{x^2 + 10^6} \quad \text{(meters)}$$

where x is the position of a mobile user who is moving along the x-axis. Assume that the reflected-path distance is

$$d_2 = \sqrt{(x - 55)^2 + 10^6} + 55 \quad \text{(meters)}$$

(a) The amount of the delay (in seconds) can be computed for both propagation paths, using the fact that the time delay is the distance divided by the speed of light (3×10^8 m/s). Determine t_1 and t_2 as a function of the mobile's position (x).

(b) Assume that the transmitted signal is

$$s(t) = \cos(300 \times 10^6 \pi t)$$

Determine the received signal when $x = 0$. Prove that the received signal is a sinusoid and find its amplitude, phase, and frequency when $x = 0$.

(c) The amplitude of the received signal is a measure of its strength. Show that as the mobile user moves, it is possible to find positions where the signal strength is zero. Find one such location.

(d) If you have access to MATLAB write a script that will plot signal strength versus position x, thus demonstrating that there are numerous locations where no signal is received. Use x in the range $-100 \le x \le 100$.

CHAPTER 3

Spectrum Representation

This chapter introduces the concept of the ***spectrum***, a compact representation of the frequency content of a signal that is composed of sinusoids. In Chapter 2, we learned about the properties of sinusoidal waveforms of the form

$$x(t) = A\cos(2\pi f_0 t + \phi) = \Re e\left\{Xe^{j2\pi f_0 t}\right\}$$

where $X = Ae^{j\phi}$ is a phasor, and we showed how phasors can simplify the addition of sinusoids of the *same* frequency. In this chapter, we will show how more complicated waveforms can be constructed out of sums of sinusoidal signals of *different* amplitudes, phases, and frequencies. As we will define it, the spectrum is simply the collection of amplitude, phase, and frequency information that allows us to express the signal in the form

$$x(t) = A_0 + \sum_{k=1}^{N} A_k \cos(2\pi f_k t + \phi_k)$$

$$= X_0 + \Re e\left\{\sum_{k=1}^{N} X_k\, e^{j2\pi f_k t}\right\}$$

where $X_0 = A_0$ is a real constant, and $X_k = A_k e^{j\phi_k}$ is the complex amplitude (i.e., phasor) for the complex exponential of frequency f_k. We will see that it is useful to show this information in a graphical representation. This visual form allows us to see interrelationships among the different frequency components and their relative amplitudes quickly and easily.

3-1 The Spectrum of a Sum of Sinusoids

One of the reasons that sinusoids are so important for our study is that they are the basic building blocks for making more complicated signals. Later in this chapter, we will show some extraordinarily complicated waveforms that can be constructed from rather simple combinations of basic cosine waves. The most general and powerful method for producing new signals from sinusoids is the *additive linear combination*, where a signal is created by adding together a constant and N sinusoids, each with a different frequency, amplitude, and phase. Mathematically, this signal may be represented by the equation

$$x(t) = A_0 + \sum_{k=1}^{N} A_k \cos(2\pi f_k t + \phi_k) \tag{3.1}$$

where each amplitude, phase, and frequency[1] may be chosen independently. Such a signal may also be represented in terms of the complex amplitude representations of the individual sinusoidal components; i.e.,

$$x(t) = X_0 + \sum_{k=1}^{N} \Re e\left\{X_k e^{j2\pi f_k t}\right\} \tag{3.2}$$

where $X_0 = A_0$ represents a real constant component, and each complex amplitude

$$X_k = A_k e^{j\phi_k}$$

represents the magnitude and phase of a rotating phasor whose frequency is f_k.

The *inverse Euler formula* gives a way to represent $x(t)$ in the alternative form

$$x(t) = X_0 + \sum_{k=1}^{N} \left\{ \frac{X_k}{2} e^{j2\pi f_k t} + \frac{X_k^*}{2} e^{-j2\pi f_k t} \right\} \tag{3.3}$$

As in the case of individual sinusoids, this form follows from the fact that the real part of a complex number is equal to one-half the sum of that number and its complex conjugate. Equation (3.3) also shows that each sinusoid in the sum decomposes into two rotating phasors, one with positive frequency, f_k, and the other with negative frequency, $-f_k$.

We define the ***two-sided spectrum*** of a signal composed of sinusoids as in (3.3) to be the set of $2N+1$ complex amplitudes and the $2N+1$ frequencies that specify the signal in the representation of (3.3). To be specific, although it is a somewhat awkward mathematical notation, our definition of the spectrum is just the set of pairs

$$\begin{aligned}\{ (0, X_0),\ \left(f_1, \tfrac{1}{2}X_1\right),\ \left(-f_1, \tfrac{1}{2}X_1^*\right), \ldots \\ \left(f_k, \tfrac{1}{2}X_k\right),\ \left(-f_k, \tfrac{1}{2}X_k^*\right), \ldots \}\end{aligned} \tag{3.4}$$

Each pair $(f_k, \frac{1}{2}X_k)$ indicates the size and relative phase of the sinusoidal component contributing at frequency f_k. It is common to refer to the spectrum as the ***frequency-domain representation*** of the signal. Instead of giving the time waveform itself (i.e., the time-domain representation), the frequency-domain representation simply gives the information required to synthesize it with (3.3)

Example 3-1: Two-Sided Spectrum

For example, consider the sum of a constant and two sinusoids:

$$\begin{aligned}x(t) =& 10 + 14\cos(200\pi t - \pi/3) \\ &+ 8\cos(500\pi t + \pi/2)\end{aligned}$$

When we apply the inverse Euler formula, we get the following five terms:

$$\begin{aligned}x(t) =& 10 + 7e^{-j\pi/3}e^{j2\pi(100)t} + 7e^{j\pi/3}e^{-j2\pi(100)t} \\ &+ 4e^{j\pi/2}e^{j2\pi(250)t} + 4e^{-j\pi/2}e^{-j2\pi(250)t}\end{aligned} \tag{3.5}$$

Note that the constant component of the signal, often called the ***DC component***, can be expressed as a complex

[1] For this chapter, we prefer cyclic frequency f_k to radian frequency $\omega_k = 2\pi f_k$, because it is easier to describe physical quantities such as musical notes in Hz.

exponential signal with zero frequency, i.e., $10e^{j0t} = 10$. Therefore, in the list form suggested in (3.4), the spectrum of this signal is the set of five rotating phasors represented by

$$\{(0,\ 10),\ (100,\ 7e^{-j\pi/3}),\ (-100,\ 7e^{j\pi/3}),\ (250,\ 4e^{j\pi/2}),\ (-250,\ 4e^{-j\pi/2})\} \quad \blacksquare$$

3-1.1 Notation Change

The relationship between X_k and the spectrum involves special cases because the factor of $\frac{1}{2}$ multiplies every X_k in the spectrum (3.4), except for X_0. The companion formulas such as (3.3) are cumbersome. Therefore, we introduce a_k as a new symbol for the complex amplitude in the spectrum, and define it as follows:

$$a_k = \begin{cases} A_0 & \text{for } k = 0 \\ \frac{1}{2}A_k e^{j\phi_k} & \text{for } k \neq 0 \end{cases} \tag{3.6}$$

This allows us to say that the spectrum is the set of $(f_k,\ a_k)$ pairs. The primary motivation for this notational change is to simplify the formulas for the Fourier series coefficients developed later in Section 3-4. For example, equation (3.3), which is similar to the Fourier series synthesis formula, can now be written as a single compact summation

$$x(t) = \sum_{k=-N}^{N} a_k e^{j2\pi f_k t} \tag{3.7}$$

where we have defined $f_0 = 0$.

3-1.2 Graphical Plot of the Spectrum

A plot of the spectrum is much more revealing than the list of $(f_k,\ a_k)$ pairs. Each frequency component can be represented by a vertical line at the appropriate frequency, and the length of the line can be drawn proportional to the magnitude, $|a_k|$. This is shown in Fig. 3-1 for the signal in (3.5). Each ***spectral line*** is labeled with the value of a_k to complete the information needed to define the spectrum. This simple but effective plot makes it easy to see two things: the relative location of the frequencies, and the relative amplitudes of the sinusoidal components. This is why the spectrum plot is widely used as a graphical representation of the signal. As we will see in Chapters 4 and 6, the frequency-domain representation is so useful because it is often very easy to see how systems affect a signal by determining what happens to the signal spectrum as it is transmitted through the system. This is why the spectrum is the key to understanding most complex processing systems such as radios, televisions, CD players, and the like.

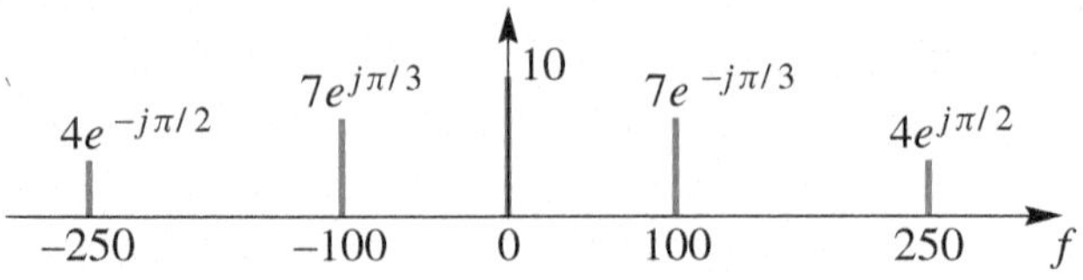

Figure 3-1: Spectrum plot for the signal $x(t) = 10 + 14\cos(200\pi t - \pi/3) + 8\cos(500\pi t + \pi/2)$. Units of frequency (f) are Hz. Negative frequency components should be included for completeness even though they are conjugates of the corresponding positive frequency components.

Notice that, for the example in Fig. 3-1, the complex amplitude of each negative frequency component is the complex conjugate of the complex amplitude at the corresponding positive frequency component. This is a general property of the spectrum whenever $x(t)$ is a real signal, because the complex rotating phasors with positive and negative frequency must combine to form a real signal (see Fig. 2-13(b) and the movie found on the CD-ROM).

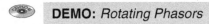

DEMO: *Rotating Phasors*

A general procedure for computing and plotting the spectrum for an arbitrarily chosen signal requires the study of Fourier analysis. However, by assuming that the signal of interest is the sum of a constant and one or more sinusoids, we can begin to explore the virtues of the spectrum representation. For a signal where we know the sinusoidal waveforms that comprise it, the procedure is straightforward. It is necessary only to express the cosines and sines as complex exponentials (by using the inverse Euler relation) and then to plot the complex amplitude of each of the positive and negative frequency components at the corresponding frequency. In other words, if it is known a priori that a signal is composed of a finite number of sinusoidal components, the process of analyzing that signal to find its ***spectral components*** involves writing an equation for the signal in the form of (3.3), and picking off the amplitude, phase, and frequency of each of its rotating phasor components.

In many other cases, spectrum analysis is not so simple, but it is nevertheless possible. For example, it is possible to represent any periodic waveform (even discontinuous signals) as a sum of complex exponential signals where the frequencies are all integer multiples of a common frequency, called the ***fundamental frequency***. Likewise, most (nonperiodic) signals can also be represented as a superposition of complex exponential signals. The mathematical tools for doing this analysis are called Fourier series (see Section 3-4) and Fourier transforms (Chapter 11). Before tackling the general case, we will show examples where the sum of just a few sinusoids can be used to produce audio signals that are interesting to hear, and we will relate the sounds to their spectra.

3-2 Beat Notes

When we multiply two sinusoids having different frequencies, we can create an interesting audio effect called a ***beat note***. The phenomenon, which may sound like a warble, is best heard by picking one of the frequencies to be very small (e.g., 10 Hz), and the other around 1 kHz. Some musical instruments naturally produce beating tones. Another use for multiplying sinusoids is modulation for radio broadcasting. AM radio stations use this method, which is called ***amplitude modulation***.

LAB: *#3 AM and FM Sinusoidal Signals*

3-2.1 Multiplication of Sinusoids

Although the signal is produced by multiplying two sinusoids, our spectrum representation demands that the signal be expressed as an *additive* linear combination of complex exponential signals. Other combinations of sinusoids would also have to be rewritten in the additive form in order to display their spectrum representation. This is illustrated by the following example.

Example 3-2: Spectrum of a Product

For the special case of a beat signal formed as the product of two sinusoids at 5 Hz and $\frac{1}{2}$ Hz

$$x(t) = \cos(\pi t)\sin(10\pi t) \tag{3.8}$$

it is necessary to rewrite $x(t)$ as a sum before its spectrum can be defined. The following technique for doing this relies on the inverse Euler formula

$$\begin{aligned} x(t) &= \left(\frac{e^{j\pi t} + e^{-j\pi t}}{2}\right)\left(\frac{e^{j10\pi t} - e^{-j10\pi t}}{2j}\right) \\ &= \tfrac{1}{4}e^{-j\pi/2}e^{j11\pi t} + \tfrac{1}{4}e^{-j\pi/2}e^{j9\pi t} \\ &\quad - \tfrac{1}{4}e^{-j\pi/2}e^{-j9\pi t} - \tfrac{1}{4}e^{-j\pi/2}e^{-j11\pi t} \\ &= \tfrac{1}{2}\cos(11\pi t - \pi/2) + \tfrac{1}{2}\cos(9\pi t - \pi/2) \end{aligned} \tag{3.9}$$

From the second and third lines of this derivation, it is obvious that there are four terms in the additive combination, and the four spectrum components are at frequencies $\pm 11\pi$ and $\pm 9\pi$ rad/sec which convert to 5.5, 4.5, −4.5, and −5.5 Hz. It is worth noting that neither of the original frequencies (5 Hz and $\frac{1}{2}$Hz) used to define $x(t)$ in (3.8) are in the spectrum. ■

EXERCISE 3.1: Let $x(t) = \sin^2(10\pi t)$. Find an additive combination in the form of (3.7) for $x(t)$, and then plot the spectrum. Count the number of frequency components in the spectrum. What is the highest frequency contained in $x(t)$? Use the inverse Euler formula rather than a trigonometric identity.

3-2.2 Beat Note Waveform

Beat notes are produced by adding two sinusoids with nearly identical frequencies, (e.g., by playing two neighboring piano keys). The example in (3.8) and (3.9) suggests that the product of two sinusoids is equivalent to a sum. Thus we can derive a general relationship between any beat signal, its spectrum, and the product form if we start with an additive combination of two closely spaced sinusoids:

$$x(t) = \cos(2\pi f_1 t) + \cos(2\pi f_2 t) \tag{3.10}$$

The two frequencies can be expressed as $f_1 = f_c - f_\Delta$ and $f_2 = f_c + f_\Delta$, where we have defined a *center frequency* $f_c = \frac{1}{2}(f_1 + f_2)$ and a *deviation frequency* $f_\Delta = \frac{1}{2}(f_2 - f_1)$, which we assume is much smaller than f_c. The spectrum of this beat signal is plotted in Fig. 3-2.

DEMO: *Spectrograms of Simple Sounds*

Using the complex exponential representation of the two cosines, we can rewrite $x(t)$ as a product of two cosines, and thereby have a form that is easier to plot in the time domain.The analysis proceeds as follows:

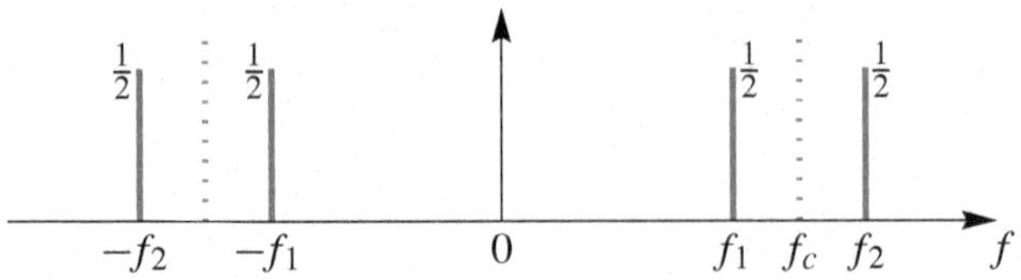

Figure 3-2: Spectrum of the beat signal in (3.10). The negative frequency lines are centered around $f = -f_c$.

$$\begin{aligned}
x(t) &= \cos(2\pi f_1 t) + \cos(2\pi f_2 t)\\
&= \Re e\left\{e^{j2\pi f_1 t}\right\} + \Re e\left\{e^{j2\pi f_2 t}\right\}\\
&= \Re e\left\{e^{j2\pi (f_c - f_\Delta)t} + e^{j2\pi (f_c + f_\Delta)t}\right\}\\
&= \Re e\left\{e^{j2\pi f_c t}\left(e^{-j2\pi f_\Delta t} + e^{j2\pi f_\Delta t}\right)\right\}\\
&= \Re e\left\{e^{j2\pi f_c t}\left(2\cos(2\pi f_\Delta t)\right)\right\}\\
&= 2\cos(2\pi f_\Delta t)\cos(2\pi f_c t)
\end{aligned} \tag{3.11}$$

Example 3-3: Time-Domain Plot of a Beat Note

For a numerical example, we take $f_c = 200$ and $f_\Delta = 20$ Hz so that

$$x(t) = 2\cos(2\pi(20)t)\cos(2\pi(200)t) \tag{3.12}$$

A *time-domain* plot of $x(t)$ is given in Fig. 3-3(b). ■

Figure 3-3(a) shows the two sinusoidal components, $2\cos(2\pi(20)t)$ and $\cos(2\pi(200)t)$, that make up the product in (3.12). The plot of the beat note is constructed by first drawing $2\cos(2\pi(20)t)$ and its negative version $-2\cos(2\pi(20)t)$ to define boundaries inside of which we then draw the higher frequency signal. These boundaries are called the signal's ***envelope***. The resulting beat note is plotted in Fig. 3-3(b), where it can be seen that the

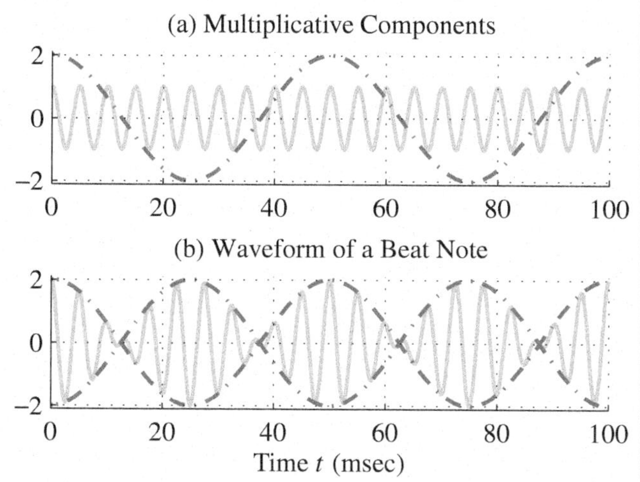

Figure 3-3: Multiplicative components of a beat note with $f_c = 200$ Hz and $f_\Delta = 20$ Hz. The time interval between nulls is $\frac{1}{2}(1/f_\Delta) = 25$ msec, which is dictated by the frequency difference.

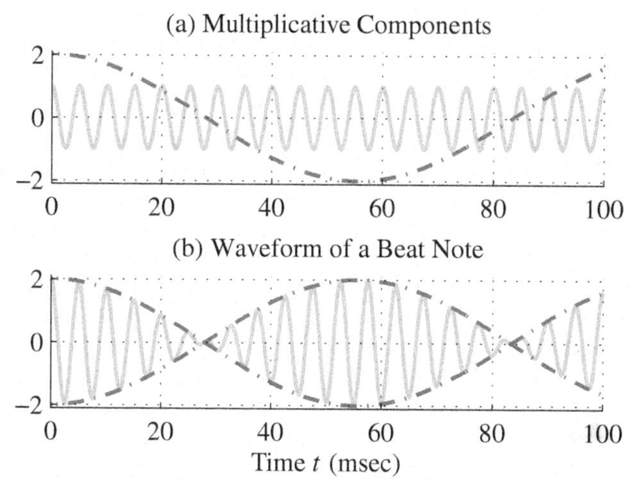

Figure 3-4: Beat note with $f_c = 200$ Hz and $f_\Delta = 9$ Hz. Nulls are now $\frac{1}{2}(1/f_\Delta) = 55.6$ msec apart.

effect of multiplying the higher-frequency sinusoid (200 Hz) by the lower-frequency sinusoid (at 20 Hz) is to change the envelope of the peaks of the higher-frequency waveform. If we listen to such an $x(t)$ we can hear that the f_Δ variation causes the signal to fade in and out because the signal envelope is rising and falling, as in Fig. 3-3(b). This is the phenomenon called "beating" of tones in music.

Example 3-4: Decreasing f_Δ

If f_Δ is decreased to 9 Hz, we see in Fig. 3-4(a,b) that the envelope of the 200 Hz tone changes much more slowly. The time interval between nulls of the envelope is $\frac{1}{2}(1/f_\Delta)$, so the more closely spaced the frequencies of the sinusoids in (3.10), the slower the envelope variation. These figures are simplified somewhat by using cosines for both terms in (3.10), but other phase relationships would give similar patterns. Finally, remember that the spectrum for $x(t)$ in Fig. 3-3 contains frequency components at ± 220 Hz and ± 180 Hz, while the spectrum for Fig. 3-4 has frequencies ± 209 Hz and ± 191 Hz. ■

Musicians use this beating phenomenon as an aid in tuning two instruments to the same pitch. When two notes are close but not identical in frequency, the beating phenomenon is heard. As one pitch is changed to become closer to the other, the effect disappears, and the two instruments are then "in tune."

3-2.3 Amplitude Modulation

LAB: *#3, AM and FM Sinusoidal Signals*

Multiplying sinusoids is also useful in modulation for communication systems. ***Amplitude modulation*** is the process of multiplying a low-frequency signal by a high-frequency sinusoid. It is the technique used to broadcast AM radio: In fact "AM" is just the abbreviation for

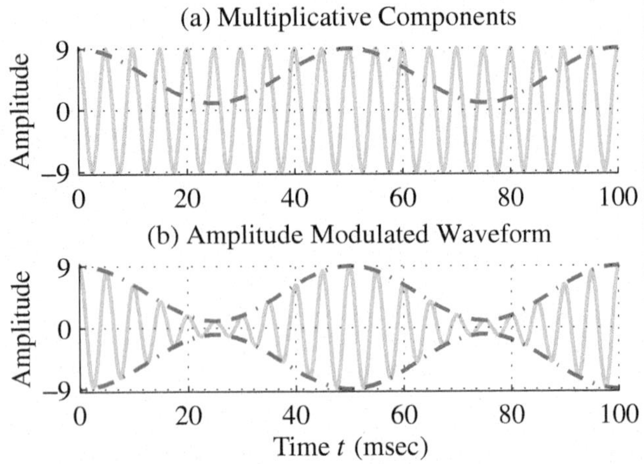

Figure 3-5: AM signal $f_c = 200$ Hz and $f_\Delta = 20$ Hz. The modulating signal has been superimposed as a light colored line to show its effect.

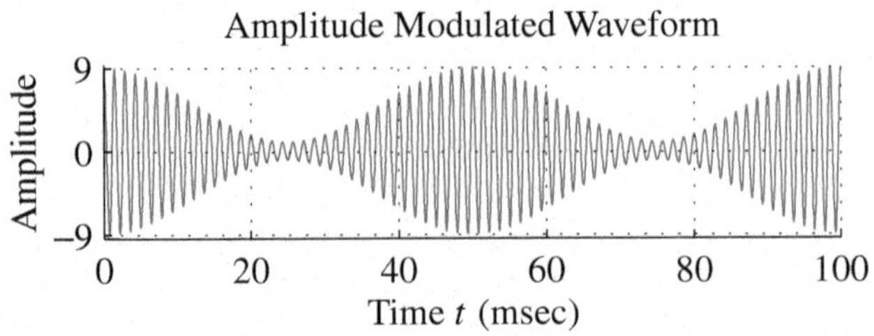

Figure 3-6: AM signal $f_c = 700$ Hz and $f_\Delta = 20$ Hz. The higher carrier frequency makes it possible to see the outline of the modulating cosine without drawing the envelope.

amplitude modulation. The AM signal is a product of the form

$$x(t) = v(t)\cos(2\pi f_c t) \tag{3.13}$$

where it is assumed that the frequency of the cosine term (f_c Hz) is much higher than any frequencies contained in the spectrum of $v(t)$, which represents the voice or music signal to be transmitted. The cosine wave $\cos(2\pi f_c t)$ in (3.13) is called the *carrier signal*, and its frequency is called the *carrier frequency*.

With our basic knowledge of the spectrum at this point, the form of $v(t)$ in (3.13) must be restricted to be a sum of sinusoids, but that is sufficient to understand how the modulation process works.

Example 3-5: Amplitude Modulation

If we let $v(t) = 5 + 4\cos(40\pi t)$ and $f_c = 200$ Hz, then the AM signal is a multiplication similar to the beat signal:

$$x(t) = [5 + 4\cos(40\pi t)]\cos(400\pi t) \tag{3.14}$$

A plot of this signal is given in Fig. 3-5, where it can be seen that the effect of multiplying the higher-frequency sinusoid (200 Hz) by the lower-frequency sinusoid (at 20 Hz) is to "modulate" (or change) the amplitude envelope of the carrier waveform—hence the name amplitude modulation for a signal like $x(t)$. ■

The primary difference between this AM signal and the beat signal is that the envelope never goes to zero. This is because the DC component (5) is greater than the amplitude (4) of the 20-Hz component. When the carrier frequency becomes very high compared to the frequencies in $v(t)$ as in Fig. 3-6, it is possible to see the outline of the modulating cosine without drawing the envelope signal explicitly. This characteristic simplifies the implementation of a detector circuit in AM broadcast receivers.

In the frequency domain, the AM signal spectrum is nearly the same as the beat signal, the only difference being a relatively large term at $f = f_c$. The spectrum can be derived by first breaking the time-domain signal into two terms

$$x(t) = 5\cos(400\pi t) + 4\cos(40\pi t)\cos(400\pi t) \tag{3.15}$$

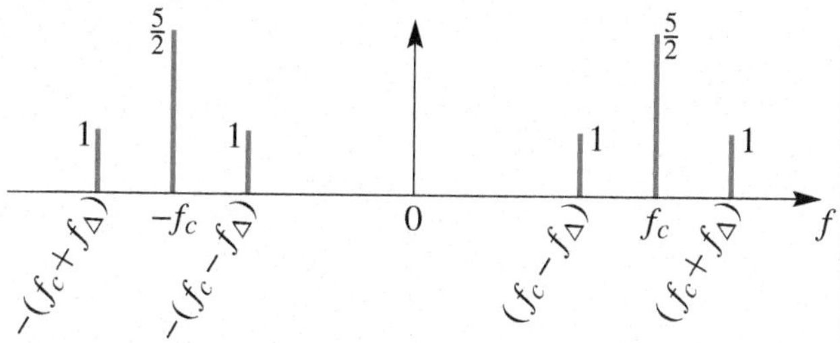

Figure 3-7: Spectrum of the AM signal in (3.16) and Fig. 3-5, where $f_c = 200$ Hz and $f_\Delta = 20$ Hz.

and then using our previous knowledge about the beat signal to get the following additive combination for the spectrum:

$$\begin{aligned} x(t) &= \tfrac{5}{2}e^{j400\pi t} + e^{j440\pi t} + e^{j360\pi t} \\ &\quad + \tfrac{5}{2}e^{-j400\pi t} + e^{-j440\pi t} + e^{-j360\pi t} \end{aligned} \qquad (3.16)$$

Thus there are six spectral components for $x(t)$ at the frequencies ± 220 Hz and ± 180 Hz, and also at the carrier frequency ± 200 Hz (Fig. 3-7). It is interesting to note that the spectrum for $x(t)$ contains two identical subsets, one centered at $f = f_c$ and the other at $f = -f_c$. These subsets each contain three spectral lines, and it is easy to show that each subset is just a frequency shifted version of the two-sided spectrum of $v(t)$. In Chapter 11 we will show that this is true for a much broader class of signals.

EXERCISE 3.2: Derive the spectrum of $v(t) = 5 + 2\cos(2\pi(20)t)$ and plot it as a function of frequency. Compare this result to the spectral plot for the AM signal in Fig. 3-7.

3-3 Periodic Waveforms

A periodic signal satisfies the condition that $x(t + T_0) = x(t)$ for all t, which states that the signal repeats its values every T_0 secs. The time interval T_0 is called the ***period*** of $x(t)$; if it is the smallest such repetition interval, it is called the ***fundamental period***. For example, the signal $x(t) = \cos^2(4\pi t)$ has a period of $\frac{1}{2}$ sec, but its fundamental period is $T_0 = \frac{1}{4}$ sec. In this section, we show that periodic signals can be synthesized by adding two or more cosine waves that have ***harmonically*** related frequencies; i.e., all frequencies are integer multiples of a frequency f_0. In other words, the signal would be synthesized as the sum of $N + 1$ cosine waves[2]

$$x(t) = A_0 + \sum_{k=1}^{N} A_k \cos(2\pi k f_0 t + \phi_k) \qquad (3.17)$$

where the frequency, f_k, of the kth cosine component in (3.17) is

$$f_k = k f_0 \qquad \text{(harmonic frequencies)}$$

The frequency f_k is called the kth ***harmonic*** of f_0 because it is an integer multiple of the basic frequency f_0, which is called the ***fundamental frequency***.

Example 3-6: Calculating f_0

The fundamental frequency is the *largest* f_0 such that $f_k = k f_0$. In mathematical terms, this is the *greatest common divisor,* so we can state

$$f_0 = \gcd\{f_k\}$$

For example, if the signal is the sum of sinusoids with frequencies 1.2, 2, and 6 Hz, then $f_0 = 0.4$ Hz, because 1.2 Hz is the 3rd harmonic, 2 Hz is the 5th harmonic, and 6 Hz is the 15th harmonic. ■

What is the period of $x(t)$? Since each cosine in (3.17) has a period of $1/f_0$, the sum must have exactly the same period and $x(t + 1/f_0) = x(t)$. Thus the period of $x(t)$ is

[2]The DC component is a cosine signal with zero frequency.

$T_0 = 1/f_0$, the reciprocal of the fundamental frequency. Since, $T_0 = 1/f_0$ is the smallest possible period, it is also the fundamental period.

Using the complex exponential representation of the cosines, we can write (3.17) as

$$\begin{aligned} x(t) &= \sum_{k=-N}^{N} a_k e^{j2\pi k f_0 t} \\ &= a_0 + 2\Re e\left\{\sum_{k=1}^{N} a_k e^{j2\pi k f_0 t}\right\} \end{aligned} \tag{3.18}$$

where a_k was defined by (3.6) in Section 3-1.1.

EXERCISE 3.3: Show that one possible period of the complex exponential signal $v_k(t) = e^{j2\pi k f_0 t}$ is $T_0 = 1/f_0$. Also show that the fundamental period is $1/(kf_0)$.

3-3.1 Synthetic Vowel

As an example of synthesizing a periodic signal, consider a case where the sum in (3.18) contains nonzero terms for $\{a_{\pm 2}, a_{\pm 4}, a_{\pm 5}, a_{\pm 16}, a_{\pm 17}\}$, and where the fundamental frequency is $f_0 = 100$ Hz. The numerical values of the complex amplitudes are listed in Table 3-1. This signal approximates the waveform produced by a man speaking the vowel sound "ah." The two-sided spectrum of this signal is plotted in Fig. 3-8. Note that all the frequencies are integer multiples of 100 Hz, even though there is no spectral component at 100 Hz itself. Also note that the negative frequency components have phase angles that are the negative of the phase angles of the corresponding positive frequency components, because the complex amplitudes for the negative frequencies are the complex conjugates of the complex amplitudes for the corresponding positive frequencies. Note that Fig. 3-8 shows the spectrum as two plots, one for the magnitudes and one for the phases. This is in contrast to Fig. 3-1,

Table 3-1: Complex amplitudes for the periodic signal that approximates the vowel "ah". The a_k coefficients are given for positive indices k, but the values for negative k are the conjugates, $a_{-k} = a_k^*$.

k	f_k (Hz)	a_k	Mag	Phase
1	100	0	0	0
2	200	$386 + j6101$	6 113	1.508
3	300	0	0	0
4	400	$-4433 + j14024$	14 708	1.877
5	500	$24000 - j4498$	24 418	−0.185
6	600	0	0	0
⋮	⋮	⋮	⋮	⋮
15	1500	0	0	0
16	1600	$828 - j6760$	6 811	−1.449
17	1700	$2362 + j0$	2 362	0

where it was convenient to make just one plot and label the spectral components with their complex amplitudes.

The synthetic vowel signal has ten spectral components, but only five sinusoidal terms when the real part is taken as in (3.18). It is interesting to examine the contribution of each real sinusoidal component separately. We can do this by successively plotting the waveforms corresponding to only one sinusoid, then two sinusoids, then three, etc. Figure 3-9 (top) shows a plot of the $k = 2$ sinusoidal term in Table 3-1 alone. Note that since the frequency of this component is $2f_0 = 200$ Hz, the waveform is periodic with period $1/200 = 5$ msec. The next panel of Fig. 3-9 shows shows $x_4(t)$ which is a plot of the sum of the $k = 2$ and $k = 4$ terms.

Now notice that the two frequencies are multiples of 200 Hz, so the period of $x_4(t)$ is still 5 msec. Figure 3-9 (middle) shows a plot of the sum of the first three terms, $x_5(t)$. Now we see that the period of the waveform is increased to 10 msecs. This is because the three frequencies, 200, 400, and 500 Hz are integer multiples

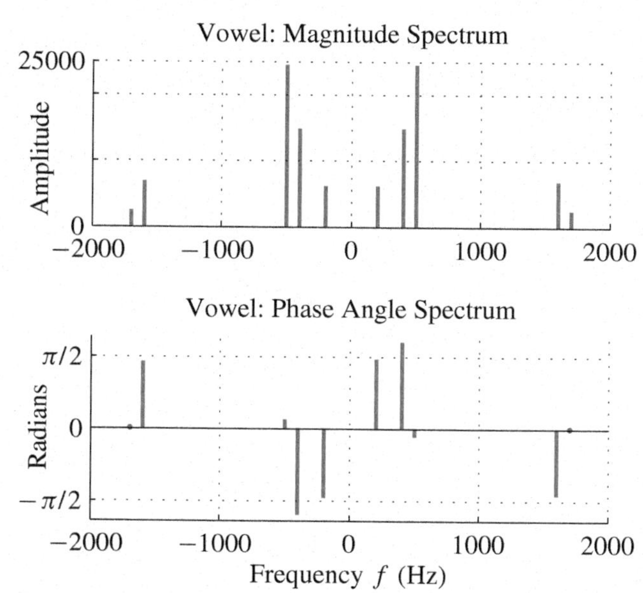

Figure 3-8: Spectrum of signal defined in Table 3-1. The magnitude is an even function with respect to $f = 0$; the phase is odd.

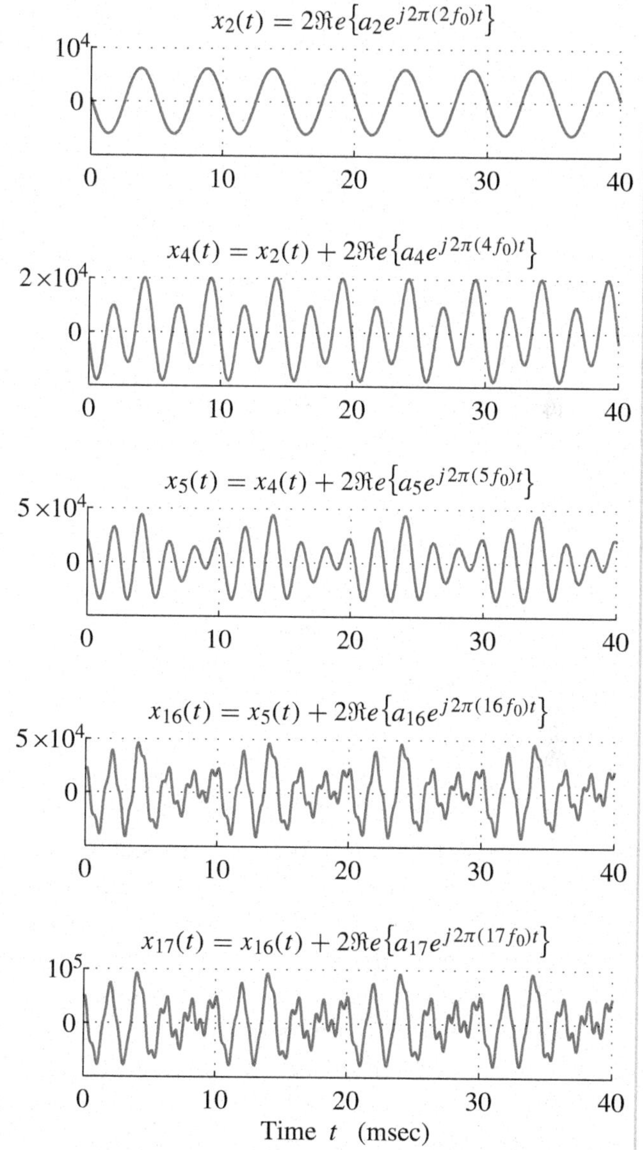

Figure 3-9: Sum of all five terms in Table 3-1. The 200-Hz term is shown in the top panel; additional terms are added one at a time until the entire vowel signal is created (bottom).

of 100 Hz; i.e., the fundamental frequency is now 100 Hz. Figure 3-9 (bottom) shows $x_{17}(t)$, the sum of all the terms in Table 3-1. Note that the period of $x_{17}(t)$ is $T_0 = 10$ msec, which equals $1/f_0$, even though there is no component with frequency f_0. The signal $x_{17}(t)$ is typical of waveforms for vowel sounds in speech. The high frequencies in the signal contribute the fine detail in the waveform. This is evident in Fig. 3-9 as the waveform becomes increasingly complicated and more rapidly varying as higher-frequency components such as the 16th and 17th harmonics are added.

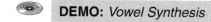

DEMO: *Vowel Synthesis*

3-3.2 Example of a Nonperiodic Signal

When we add harmonically related complex exponentials, we get a periodic result. What happens when

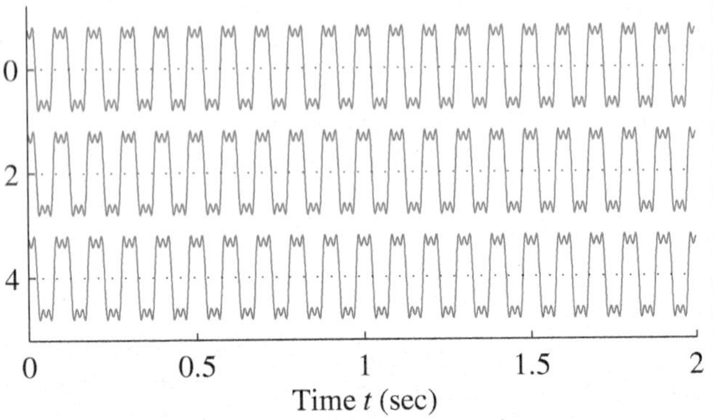

Figure 3-10: Sum of cosine waves of harmonic frequencies. The fundamental frequency of $x_h(t)$ is 10 Hz.

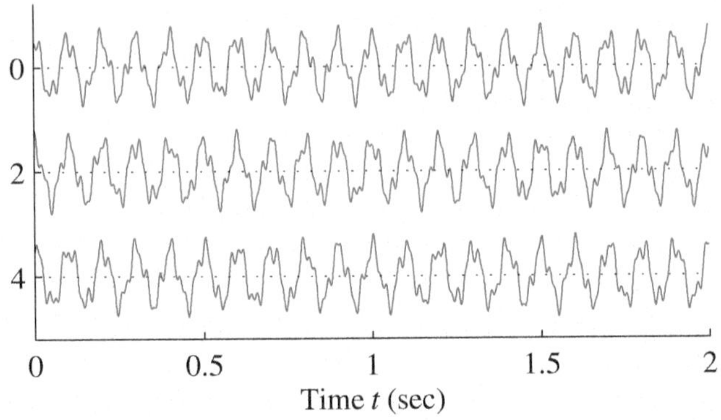

Figure 3-11: Sum of three cosine waves with nonharmonic frequencies, $x_2(t)$. No matter how hard you try, you cannot find an exact repetition in this signal.

the frequencies have no simple relation to one another? The sinusoidal synthesis formula

$$
\begin{aligned}
x(t) &= A_0 + \sum_{k=1}^{N} A_k \cos(2\pi f_k t + \phi_k) \\
&= A_0 + \sum_{k=1}^{N} \left(\tfrac{1}{2} A_k e^{j\phi_k} e^{j2\pi f_k t} + \tfrac{1}{2} A_k e^{-j\phi_k} e^{-j2\pi f_k t}\right)
\end{aligned}
$$

is still valid, but now we will make no assumptions about the individual frequencies f_k.

With a simple example, we want to demonstrate that periodicity is tied to harmonic frequencies. We can do this by taking a specific example. Consider the harmonic signal $x_h(t)$ made up from the first, third, and fifth harmonics of a square wave[3] with fundamental frequency $f_0 = 10$ Hz:

$$
x_h(t) = 2\cos(20\pi t) - \tfrac{2}{3}\cos(20\pi(3)t) + \tfrac{2}{5}\cos(20\pi(5)t)
$$

[3] We will discuss the square wave spectrum later in Sec. 3-6.1.

A plot of $x_h(t)$ is shown in Fig. 3-10 using a "strip chart" format. The plot consists of three lines, each one containing 2 sec of the signal. The first line starts at $t = 0$, the second at $t = 2$, and the third at $t = 4$. This lets us view a long section of the signal, which in this case is clearly periodic, with period equal to 1/10 sec.

DEMO: *Spectrograms: Simple Sounds: Square Wave*

Now we create a second signal that is just a slight perturbation from the first. Define $x_2(t)$ to be the sum of three sinusoids:

$$
\begin{aligned}
x_2(t) = &2\cos(20\pi t) - \tfrac{2}{3}\cos(20\pi\sqrt{8}t) \\
&+ \tfrac{2}{5}\cos(20\pi\sqrt{27}t)
\end{aligned}
$$

The amplitudes are the same, but the frequencies have been changed slightly. The plot of in Fig. 3-11 shows that $x_2(t)$ is not periodic.

The spectrum plots in Fig. 3-12 will help explain the difference between the signals in Figs. 3-11 and 3-10. In Fig. 3-12(a), the frequencies are integer multiples of a common frequency, $f_0 = 10$ Hz, so the waveform

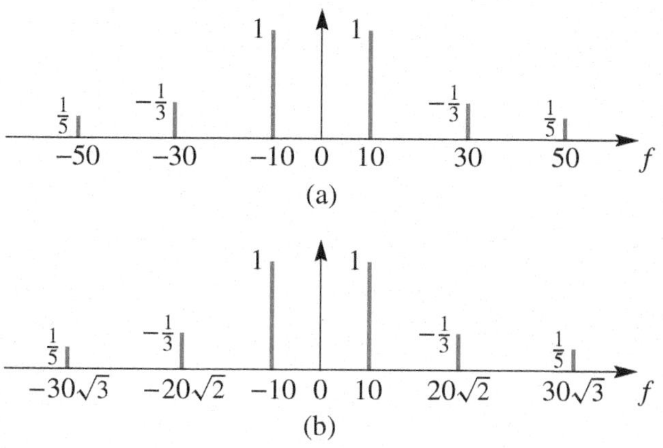

Figure 3-12: Spectrum of (a) the harmonic waveform in Fig. 3-10 which has a period of 0.1 sec, and (b) the nonharmonic waveform in Fig. 3-11 that is not periodic.

of Fig. 3-10 is periodic with period $T_0 = 1/10$ sec. The waveform of Fig. 3-11 is *nonperiodic*. We can justify this in the "frequency domain" by examining Fig. 3-12(b), which shows that the spectrum of the signal in Figs. 3-11 does not have a fundamental frequency, since the frequencies are not integer multiples of a common fundamental frequency. These ***spectrum*** plots show "how much" of each cosine wave is in the sum, and they are very similar. However, the frequencies are slightly different: $20\sqrt{2} = 28.28\cdots \approx 30$ and $30\sqrt{3} = 51.96\cdots \approx 50$. These slight shifts of frequency make a dramatic difference in the time waveform.

3-4 Fourier Series

The examples in Sec. 3-3 show that we can synthesize *periodic* waveforms by using a sum of *harmonically related* sinusoids. Now, we want to describe a general theory that shows how *any periodic signal can be synthesized with a sum of harmonically related sinusoids,* although the sum may need an infinite number of terms. This is the mathematical theory of ***Fourier series*** which uses the following representation:

$$x(t) = \sum_{k=-\infty}^{\infty} a_k\, e^{j(2\pi/T_0)kt} \tag{3.19}$$

where T_0 is the fundamental period of the periodic signal $x(t)$. The kth complex exponential in (3.19) has a frequency equal to $f_k = k/T_0$ Hz, so all the frequencies are integer multiples of the fundamental frequency $f_0 = 1/T_0$ Hz.[4]

There are two aspects of the Fourier theory: analysis and synthesis. Starting from $x(t)$ and calculating $\{a_k\}$ is called ***Fourier analysis***. The reverse process of starting from $\{a_k\}$ and generating $x(t)$ is called ***Fourier synthesis***.

The formula in (3.19) is the general synthesis formula. When the complex amplitudes are *conjugate-symmetric,* i.e., $a_{-k} = a_k^*$, the synthesis formula becomes a sum of sinusoids of the form

$$x(t) = A_0 + \sum_{k=1}^{\infty} A_k \cos((2\pi/T_0)kt + \phi_k) \tag{3.20}$$

where $A_0 = a_0$, and the amplitude and phase of the kth term comes from the polar form, $a_k = \frac{1}{2}A_k e^{j\phi_k}$. In other words, the condition $a_{-k} = a_k^*$ is sufficient for the synthesized waveform to be a *real* function of time.

By clever choice of the complex amplitudes a_k in (3.19), we can represent a number of interesting periodic waveforms, such as square waves, triangle waves, and so on. The fact that a discontinuous square wave can be represented with an infinite number of sinusoids was one of the amazing claims in Fourier's famous thesis of 1807.

[4]There are three ways to refer to the fundamental frequency: radian frequency ω_0 in rad/sec, cyclic frequency f_0 in Hz, or with the period T_0 in sec. Each one has its merits in certain situations. The relationship among these is $\omega_0 = 2\pi f_0 = 2\pi/T_0$.

It took many years before mathematicians were able to develop a rigorous convergence proof to verify Fourier's claim.

3-4.1 Fourier Series: Analysis

How do we derive the coefficients for the harmonic sum (3.19), i.e., how do we go from $x(t)$ to a_k? The answer is that we use the ***Fourier series integral*** to perform Fourier analysis. The complex amplitudes for any periodic signal can be calculated with the Fourier integral

$$a_k = \frac{1}{T_0}\int_0^{T_0} x(t)e^{-j(2\pi/T_0)kt}\,dt \qquad (3.21)$$

where T_0 is the fundamental period of $x(t)$. A special case of (3.21) is that the DC component is obtained by

$$a_0 = \frac{1}{T_0}\int_0^{T_0} x(t)\,dt \qquad (3.22)$$

A common interpretation of (3.22) is that a_0 is simply the average value of the signal over one period.

The Fourier integral (3.21) is convenient if we have a formula that defines $x(t)$ over one period. Two examples will be presented later to illustrate this point. On the other hand, if $x(t)$ is known only as a recording, then numerical methods such as those discussed in Chapter 13 will be needed.

3-4.2 Fourier Series Derivation

In this section, we present a derivation of the Fourier series integral formula (3.21). The derivation relies on a simple property of the complex exponential signal—the integral of a complex exponential over an integral number of periods is zero. In equation form,

$$\int_0^{T_0} e^{j(2\pi/T_0)kt}\,dt = 0 \qquad (3.23)$$

where T_0 is a period of the complex exponential whose frequency is $\omega_k = (2\pi/T_0)k$, and k is a nonzero integer. Here is the integration:

$$\begin{aligned}\int_0^{T_0} e^{j(2\pi/T_0)kt}\,dt &= \left.\frac{e^{j(2\pi/T_0)kt}}{j(2\pi/T_0)k}\right|_0^{T_0}\\ &= \frac{e^{j(2\pi/T_0)kT_0}-1}{j(2\pi/T_0)k} = 0\end{aligned}$$

The numerator is zero because $e^{j2\pi k} = 1$ for any integer k (positive or negative). This fact can also be justified if we use Euler's formula to separate the integral into its real and imaginary parts and then integrate cosine and sine separately—each one over k complete periods:

$$\begin{aligned}\int_0^{T_0} e^{j(2\pi/T_0)kt}\,dt &= \int_0^{T_0} \cos((2\pi/T_0)kt)\,dt\\ &\quad + j\int_0^{T_0} \sin((2\pi/T_0)kt)\,dt = 0\end{aligned}$$

A key ingredient in the infinite series representation (3.19) is the form of the complex exponentials, which all must repeat with the same period as the period of the signal $x(t)$, which is T_0. If we define $v_k(t)$ to be the complex exponential of frequency $\omega_k = (2\pi/T_0)k$, then

$$v_k(t) = e^{j(2\pi/T_0)kt} \qquad (3.24)$$

Even though the minimum duration period of $v_k(t)$ is smaller than T_0, the following shows that $v_k(t)$ still repeats with a period of T_0:

$$\begin{aligned} v_k(t+T_0) &= e^{j(2\pi/T_0)k(t+T_0)} \\ &= e^{j(2\pi/T_0)kt}\, e^{j(2\pi/T_0)kT_0} \\ &= e^{j(2\pi/T_0)kt}\, e^{j2\pi k} \\ &= e^{j(2\pi/T_0)kt} = v_k(t) \end{aligned}$$

where again we have used $e^{j2\pi k} = 1$ for any integer k (positive or negative).

Now we can generalize the zero-integral property (3.23) of the complex exponential to involve two signals:[5]

Orthogonality Property

$$\int_0^{T_0} v_k(t) v_\ell^*(t)\, dt = \begin{cases} 0 & \text{if } k \neq \ell \\ T_0 & \text{if } k = \ell \end{cases} \tag{3.25}$$

where the $*$ superscript in $v_\ell^*(t)$ denotes the complex conjugate.

Proof: Proving the orthogonality property is straightforward. We begin with

$$\begin{aligned} \int_0^{T_0} v_k(t) v_\ell^*(t)\, dt &= \int_0^{T_0} e^{j(2\pi/T_0)kt} e^{-j(2\pi/T_0)\ell t}\, dt \\ &= \int_0^{T_0} e^{j(2\pi/T_0)(k-\ell)t}\, dt \end{aligned}$$

There are two cases to consider for the last integral: when $k = \ell$ the exponent becomes zero, so the integral is

$$\begin{aligned} \int_0^{T_0} e^{j(2\pi/T_0)(k-\ell)t}\, dt &= \int_0^{T_0} e^{j0t}\, dt \\ &= \int_0^{T_0} 1\, dt = T_0 \end{aligned}$$

Otherwise, when $k \neq \ell$ the exponent is nonzero and we can invoke the zero-integral property in (3.23) to get

$$\int_0^{T_0} e^{j(2\pi/T_0)(k-\ell)t}\, dt = \int_0^{T_0} e^{j(2\pi/T_0)mt}\, dt = 0$$

where $m = k - \ell \neq 0$. ■

The orthogonality property of complex exponentials (3.25) simplifies the rest of the Fourier series derivation. If we assume that (3.19) is valid,

$$x(t) = \sum_{k=-\infty}^{\infty} a_k e^{j(2\pi/T_0)kt}$$

then we can multiply both sides by the complex exponential $v_\ell^*(t)$ and integrate over the period T_0.

$$\begin{aligned} &\int_0^{T_0} x(t) e^{-j(2\pi/T_0)\ell t}\, dt = \\ &\qquad \int_0^{T_0} \left(\sum_{k=-\infty}^{\infty} a_k e^{j(2\pi/T_0)kt} \right) e^{-j(2\pi/T_0)\ell t}\, dt \\ &\qquad = \sum_{k=-\infty}^{\infty} a_k \left(\int_0^{T_0} e^{j(2\pi/T_0)(k-\ell)t}\, dt \right) = a_\ell T_0 \end{aligned}$$

[5]The integral in (3.25) is called the "inner product" between $v_k(t)$ and $v_\ell(t)$, sometimes denoted as $\langle v_k(t), v_\ell(t) \rangle$.

Notice that we are able to isolate one of the complex amplitudes (a_ℓ) in the final step by applying the orthogonality property (3.25).

The crucial step above occurs when the order of the infinite summation and the integration is swapped. This is a delicate manipulation that depends on convergence properties of the infinite series expansion. It was also a topic of research that occupied mathematicians for a good part of the early 19th century. For our purposes, if we assume that $x(t)$ is a smooth function and has only a finite number of discontinuities within one period, then the swap is permissible.

The final analysis formula is obtained by dividing both sides of the equation by T_0 and writing a_ℓ on one side of the equation. Since ℓ could be any index, we can replace ℓ with k to obtain

Fourier Analysis Equation

$$a_k = \frac{1}{T_0}\int_0^{T_0} x(t)e^{-j(2\pi/T_0)kt}\,dt \qquad (3.26)$$

where $\omega_0 = 2\pi/T_0 = 2\pi f_0$ is the fundamental frequency of the periodic signal $x(t)$. This analysis formula goes hand in hand with the synthesis formula for periodic signals, which is

Fourier Synthesis Equation

$$x(t) = \sum_{k=-\infty}^{\infty} a_k e^{j(2\pi/T_0)kt} \qquad (3.27)$$

3-5 Spectrum of the Fourier Series

When we discussed the spectrum in Section 3-1, we described a graphical procedure for drawing the spectrum when $x(t)$ is composed of a sum of complex exponentials. By virtue of the synthesis formula (3.27), the Fourier series coefficients a_k are, in fact, the complex amplitudes that define the spectrum of $x(t)$. In order to illustrate this general connection between the Fourier series and the spectrum, we use the "sine-cubed" signal. First, we derive the a_k coefficients for $x(t) = \sin^3(3\pi t)$, and then we sketch its spectrum.

Example 3-7: Fourier Series without Integration

Determine the Fourier series coefficients of the signal:

$$x(t) = \sin^3(3\pi t)$$

Solution: There are two ways to get the a_k coefficients: plug $x(t)$ into the Fourier integral (3.26), or use the inverse Euler formula to expand $x(t)$ into a sum of complex exponentials. It is far easier to use the latter approach. Using the inverse Euler formula for $\sin(\cdot)$, we get the following expansion of the sine-cubed function:

$$\begin{aligned} x(t) &= \left(\frac{e^{j3\pi t} - e^{-j3\pi t}}{2j}\right)^3 \\ &= \tfrac{1}{-8j}\left(e^{j9\pi t} - 3e^{j6\pi t}e^{-j3\pi t} + 3e^{j3\pi t}e^{-j6\pi t} - e^{-j9\pi t}\right) \\ &= \tfrac{j}{8}e^{j9\pi t} + \tfrac{-3j}{8}e^{j3\pi t} + \tfrac{3j}{8}e^{-j3\pi t} + \tfrac{-j}{8}e^{-j9\pi t} \end{aligned} \qquad (3.28)$$

We see that (3.28) contains four frequencies: $\omega = \pm 3\pi$ and $\omega = \pm 9\pi$ rad/s. Since $\gcd(3\pi, 9\pi) = 3\pi$, the fundamental frequency is $\omega_0 = 3\pi$ rad/sec. The Fourier series coefficients are indexed in terms of the fundamental frequency, so

$$a_k = \begin{cases} 0 & \text{for } k = 0 \\ \mp j\frac{3}{8} & \text{for } k = \pm 1 \\ 0 & \text{for } k = \pm 2 \\ \pm j\frac{1}{8} & \text{for } k = \pm 3 \\ 0 & \text{for } k = \pm 4, \pm 5, \pm 6, \ldots \end{cases} \qquad (3.29)$$

This example shows that it is not always necessary to evaluate an integral to obtain the $\{a_k\}$ coefficients. ■

Now we can draw the spectrum (Fig. 3-13) because we know that we have four nonzero a_k components located at the four frequencies: $\omega = \{-9\pi,\ -3\pi,\ 3\pi,\ 9\pi\}$ rad/sec. We prefer to plot the spectrum versus frequency in hertz in this case, so the spectrum lines are at $f = -4.5,\ -1.5,\ 1.5$, and 4.5 Hz. The second harmonic is missing and the third harmonic is at 4.5 Hz.

EXERCISE 3.4: Use the Fourier integral to determine all the Fourier series coefficients of the "sine-cubed" signal. In other words, evaluate the integral

$$a_k = \frac{1}{T_0}\int_0^{T_0} \sin^3(3\pi t)e^{-j(2\pi/T_0)kt}\,dt$$

for all k.

Hints: Find the period first, so that the integration interval is known. In addition, you might find it easier to convert the $\sin^3(\cdot)$ function to exponential form (via the inverse Euler formula for $\sin(\cdot)$) before doing the Fourier integral on each of four different terms. If you then invoke the orthogonality property on each integral, you should get exactly the same answer as (3.29).

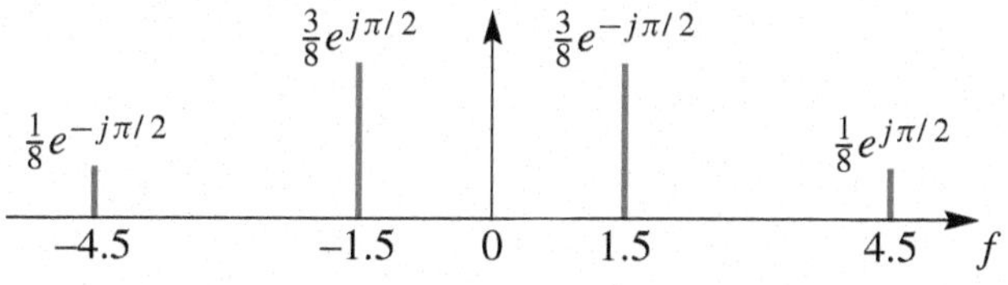

Figure 3-13: Spectrum of the "sine-cubed" signal derived from its Fourier series coefficients. Only the range from −5 to +5 Hz is shown. The complex amplitude of each spectrum line is equal to the Fourier series coefficient a_k for that frequency, kf_0.

EXERCISE 3.5: Make a sketch of the spectrum of the signal defined by:

$$x(t) = \sum_{k=-3}^{3} \frac{1}{1+jk}\, e^{jkt}$$

3-6 Fourier Analysis of Periodic Signals

We can synthesize *any periodic signal* by using a sum of sinusoids (3.27), as long as we constrain the frequencies to be harmonically related. To demonstrate Fourier synthesis of waveshapes that do not look at all sinusoidal, we will work out the details for a square wave and a triangle wave in this section. The resulting formulas for their Fourier coefficients a_k are relatively compact. Figure 3-14 shows the relationship between Fourier analysis and Fourier synthesis using representative plots for the square wave case. If you have access to MATLAB, it is straightforward to write a *Fourier Synthesis Program* that takes a list of frequencies and a list of complex amplitudes and then produces a signal as the sum of several complex exponentials according to the *finite* Fourier synthesis summation formula

$$x_N(t) = \sum_{k=-N}^{N} a_k e^{j(2\pi/T_0)kt} \tag{3.30}$$

This MATLAB programming exercise is described in more detail in the music synthesis lab.

LAB: *#4 Synthesis of Sinusoidal Signals*

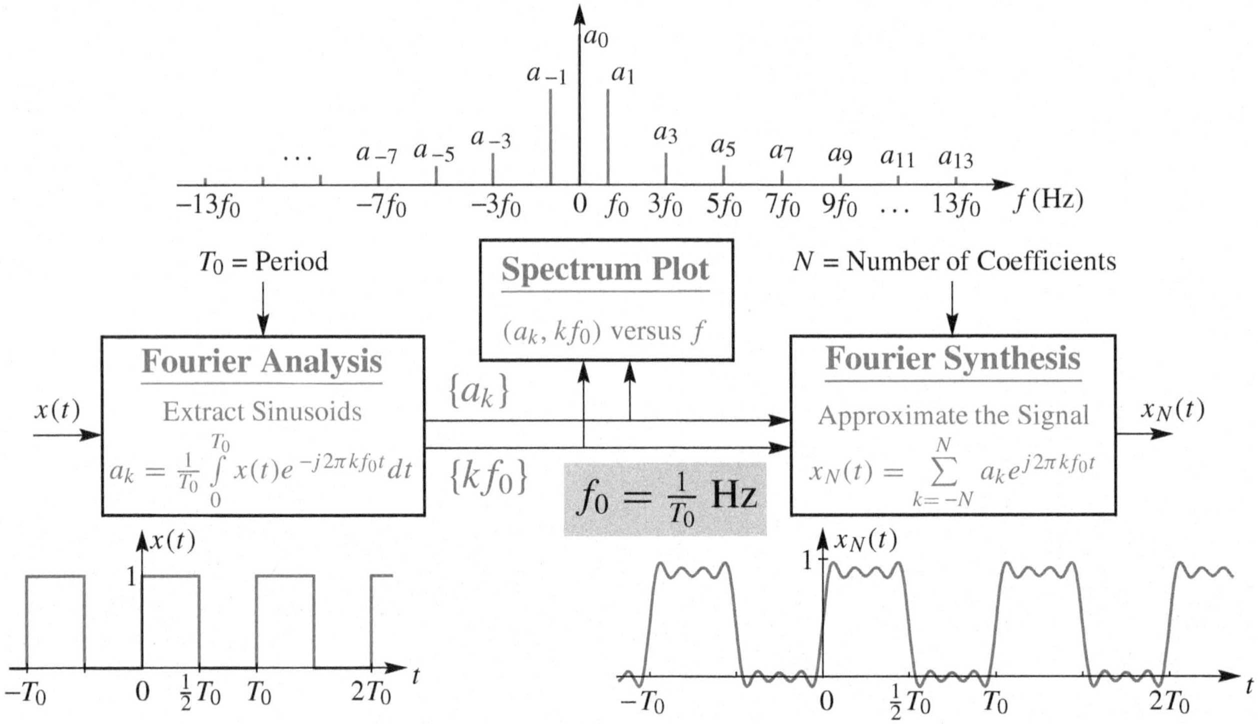

Figure 3-14: Major components of Fourier analysis and synthesis showing the relationship between the original periodic signal $x(t)$, its spectrum, and the synthesized signal $x_N(t)$ that approximates the original.

3-6.1 The Square Wave

The simplest example to consider is the periodic square wave, which is defined for one cycle by

$$s(t) = \begin{cases} 1 & \text{for } 0 \le t < \frac{1}{2}T_0 \\ 0 & \text{for } \frac{1}{2}T_0 \le t \le T_0 \end{cases} \quad (3.31)$$

Figure 3-15 shows a plot of this signal which is called a 50% duty cycle square wave because it is off (equal to zero) during half of its period.

We will derive a formula that depends on k for the complex amplitudes a_k. First of all, we substitute the definition of $x(t)$ into the integral (3.26) and obtain

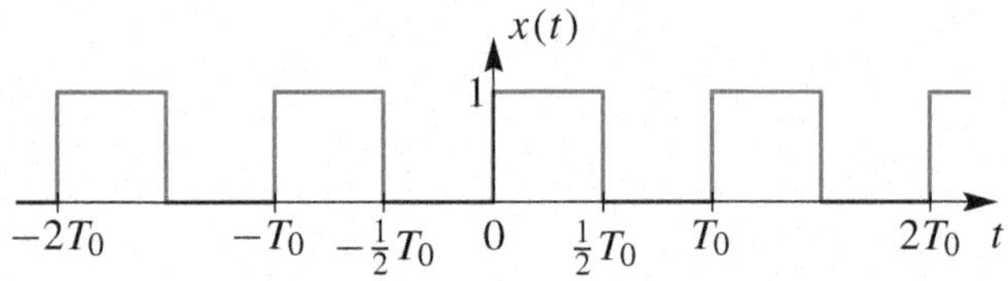

Figure 3-15: Plot of the square-wave signal whose "duty cycle" is 50%.

$$a_k = \frac{1}{T_0}\int_0^{\frac{1}{2}T_0} (1)e^{-j(2\pi/T_0)kt}\,dt$$

The upper limit becomes $\frac{1}{2}T_0$, because the signal $x(t)$ is zero for $\frac{1}{2}T_0 \le t \le T_0$. Next, we perform the integration and simplify:

$$a_k = \left(\frac{1}{T_0}\right) \left. \frac{e^{-j(2\pi/T_0)kt}}{-j(2\pi/T_0)k} \right|_0^{\frac{1}{2}T_0}$$

$$= \left(\frac{1}{T_0}\right) \frac{e^{-j(2\pi/T_0)k(\frac{1}{2}T_0)} - e^{-j(2\pi/T_0)k(0)}}{-j(2\pi/T_0)k}$$

$$= \frac{e^{-j\pi k} - 1}{-j2\pi k}$$

Since $e^{-j\pi} = -1$, we can write the following general formula for the Fourier series coefficients of the square wave.

$$a_k = \frac{1-(-1)^k}{j2\pi k} \qquad \text{for } k \neq 0$$

There is one shortcoming with this formula for a_k; it is not valid when $k = 0$ because k appears in the denominator. Therefore, we must evaluate a_0 separately using (3.22)

$$a_0 = \frac{1}{T_0} \int_0^{\frac{1}{2}T_0} (1)e^{-j0t}\,dt$$

$$= \frac{1}{T_0} \int_0^{\frac{1}{2}T_0} (1)\,dt = \frac{1}{T_0}\left(\tfrac{1}{2}T_0\right) = \tfrac{1}{2}$$

The formula for a_k when $k \neq 0$ has a numerator that is either 0 (for k even) or 2 (for k odd), because $(-1)^k$ alternates between $+1$ and -1. Therefore, the final answer for the Fourier series coefficients of the square wave has three cases:

$$a_k = \begin{cases} \dfrac{1}{j\pi k} & k = \pm1, \pm3, \pm5, \ldots \\ 0 & k = \pm2, \pm4, \pm6, \ldots \\ \frac{1}{2} & k = 0 \end{cases} \tag{3.32}$$

Notice that the formula for a_k does not depend on the period, T_0, but this is not usually the case. Also, the magnitude of these coefficients decreases as $k \to \infty$, so the high frequency terms contribute less when synthesizing the waveform via (3.30).

3-6.1.1 DC Value of a Square Wave

The Fourier series coefficient for $k = 0$ has a special interpretation as the *average value* of the signal $x(t)$. If we repeat the analysis integral (3.26) for the case where $k = 0$, then

$$a_0 = \frac{1}{T_0} \int_0^{T_0} x(t)dt \tag{3.33}$$

The integral is the area under the function $x(t)$ for one period. If we think of the area as a sum and realize that dividing by T_0 is akin to dividing by the number of elements in the sum, we can interpret (3.33) as the average value of the signal. In the specific case of the 50% duty-cycle square wave, the average value is $\frac{1}{2}$ because the signal is equal to $+1$ for half the period and then 0 for the other half. This checks with the earlier calculation that $a_0 = \frac{1}{2}$.

In the synthesis formula (3.27), the a_0 coefficient is an additive constant, so a change in its value will move the plot of the signal up or down vertically. The terminology "DC" comes from electric circuits, where a constant value of current is called direct current, or DC. It is common to call a_0 the DC coefficient, or DC term, in a Fourier expansion. Finally, one should note that the frequency of DC is $f = 0$.

3-6.2 Spectrum for a Square Wave

Figure 3-16 shows the spectrum for the 50% duty cycle square wave analyzed in (3.32) when the fundamental frequency is 25 Hz. Since $a_k = 0$ for k nonzero and even,

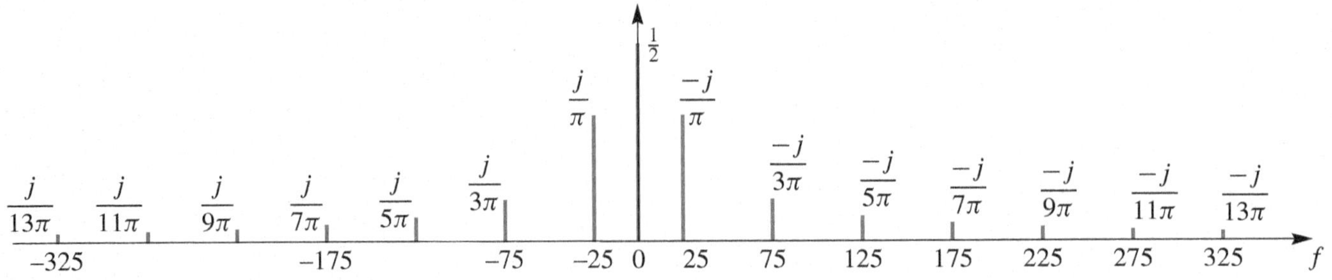

Figure 3-16: Spectrum of a square wave derived from its Fourier series coefficients. Only the range from $k = -13$ to $+13$ is shown. With a fundamental frequency of 25 Hz, this corresponds to frequencies ranging from -325 Hz to $+325$ Hz.

the only frequencies present in the spectrum are the odd harmonics at ± 25, ± 75, ± 125, and so on. The complex amplitudes of the odd harmonics are the Fourier series coefficients, $a_k = -j/(\pi k)$, and these are used as the labels on the spectrum lines in Fig. 3-16. Also the figure shows that the magnitude of these coefficients drops off as $1/k$.

DEMO: *Spectrograms: Simple Sounds*

3-6.3 Synthesis of a Square Wave

Using a simple MATLAB M-file, a synthesis was done via (3.30) with a fundamental frequency of $f_0 = 25$ Hz, and $f_k = kf_0$. The fundamental period is $T_0 = 1/25 = 0.04$ secs. In Fig. 3-17, the plots are shown for three different cases where the number of terms in the sum is $N = 3$, 7, and 17. Notice how the period of the synthesized waveform is always the same, because it is determined by the fundamental frequency.

The synthesis formula (3.30) usually can be simplified to a cosine form. For the particular case of the square wave coefficients (3.32), when we take the DC plus first and third harmonic terms, we get the sum of two cosines plus the constant (DC) level given in equation (3.34). As more harmonic terms are added, a square-wave signal waveshape would be approximated better with this sum of cosines. However, notice what happens in Fig. 3-17 as

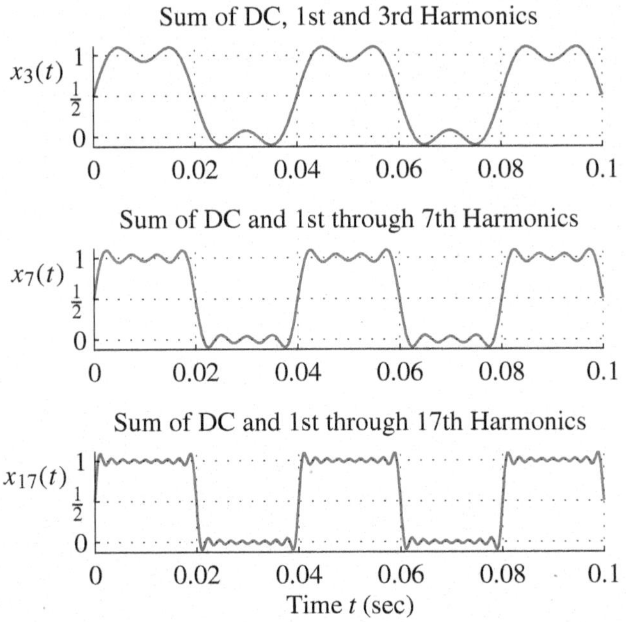

Figure 3-17: Summing harmonic components via (3.30): $N = 3$ (top panel); $N = 7$ (middle) and $N = 17$ (bottom). The DC level of $\frac{1}{2}$ is included in each synthesis.

$$
\begin{aligned}
x_3(t) &= a_{-3}e^{-j3\omega_0 t} + a_{-1}e^{-j\omega_0 t} + a_0 + a_1 e^{j\omega_0 t} + a_3 e^{j3\omega_0 t} \\
&= \frac{1}{-j3\pi}e^{-j3\omega_0 t} + \frac{1}{-j\pi}e^{-j\omega_0 t} + a_0 + \frac{1}{j\pi}e^{j\omega_0 t} + \frac{1}{j3\pi}e^{j3\omega_0 t} \\
&= a_0 + \frac{1}{\pi}(e^{-j\pi/2}e^{j\omega_0 t} + e^{j\pi/2}e^{-j\omega_0 t}) + \frac{1}{3\pi}(e^{-j\pi/2}e^{j3\omega_0 t} + e^{j\pi/2}e^{-j3\omega_0 t}) \\
&= \frac{1}{2} + \frac{2}{\pi}\cos(\omega_0 t - \pi/2) + \frac{2}{3\pi}\cos(3\omega_0 t - \pi/2)
\end{aligned}
\tag{3.34}
$$

N increases; the sum of cosines appears to converge to the constant values +1 and 0, but the convergence is not uniformly good—the "ears" at the discontinuous steps never go away completely. This behavior, which occurs at any discontinuity of a waveform, is called the *Gibbs phenomenon*, and it is one of the interesting subtleties of Fourier theory that is extensively studied in advanced treatments.

LAB: *#4 Synthesis of Sinusoidal Signals*

DEMO: *Fourier Series*

3-6.4 Triangle Wave

Another interesting case that is still relatively simple is that of a triangle wave shown in Fig. 3-18. The mathematical formula for the triangle wave consists of two segments. We have to give the definition of the waveform over exactly one period, so we do that for the time interval $0 \le t \le T_0$:

$$
x(t) = \begin{cases} 2t/T_0 & \text{for } 0 \le t < \frac{1}{2}T_0 \\ 2(T_0 - t)/T_0 & \text{for } \frac{1}{2}T_0 \le t < T_0 \end{cases} \tag{3.35}
$$

where $T_0 = 0.04$ sec in Fig. 3-18. Unlike the square wave, the triangle wave is a continuous signal.

Now we attack the Fourier integral for this case to derive a formula for the $\{a_k\}$ coefficients of the triangle wave. We might suspect from our earlier experience that the DC coefficient has to be found separately, so we do that first. Plugging into the definition with $k = 0$, we obtain

$$
a_0 = \frac{1}{T_0}\int_0^{T_0} x(t)\,dt
$$

If we recognize that the integral over one period is, in fact, the area under the triangle, we get

$$
a_0 = \frac{1}{T_0}(\text{area}) = \frac{1}{T_0}(T_0)(\tfrac{1}{2}) = \tfrac{1}{2} \tag{3.36}
$$

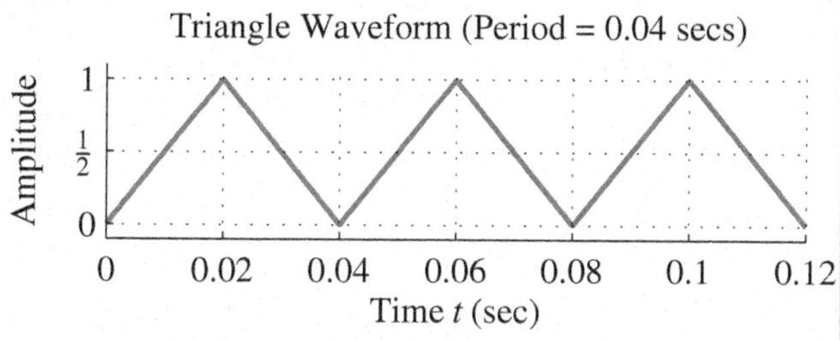

Figure 3-18: Periodic triangle wave.

For the general case where $k \neq 0$, we must break the Fourier series analysis integral into two sections because the signal consists of two pieces:

$$a_k = \frac{1}{T_0} \int_0^{\frac{1}{2}T_0} (2t/T_0) e^{-j(2\pi/T_0)kt}\, dt + \frac{1}{T_0} \int_{\frac{1}{2}T_0}^{T_0} (2(T_0 - t)/T_0) e^{-j(2\pi/T_0)kt}\, dt \tag{3.37}$$

After integration by parts and many tedious algebraic steps, the integral for a_k can be written as

$$a_k = \frac{e^{-jk\pi} - 1}{\pi^2 k^2} \tag{3.38}$$

Since $e^{-jk\pi} = (-1)^k$, the numerator in (3.38) equals either 0 or -2, and we can write the following cases for a_k:

$$a_k = \begin{cases} \dfrac{-2}{\pi^2 k^2} & k = \pm 1, \pm 3, \pm 5, \ldots \\ 0 & k = \pm 2, \pm 4, \pm 6, \ldots \\ \frac{1}{2} & k = 0 \end{cases} \tag{3.39}$$

Once again, this particular formula is independent of T_0, the fundamental period of the triangle wave.

EXERCISE 3.6: Starting from (3.37), derive the formula (3.39) for the Fourier series coefficients of the triangle wave. Use integration by parts to manipulate the integrands which contain terms of the form $te^{-j(2\pi/T_0)kt}$.

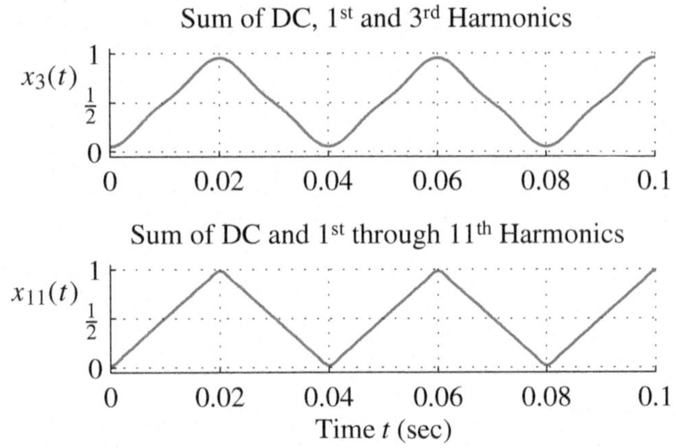

Figure 3-19: Summing harmonic components for the triangle wave via (3.30): first and third harmonics (top panel); up to and including the eleventh harmonic (bottom).

EXERCISE 3.7: Make a plot of the spectrum for the triangle wave (similar to Fig. 3-16 for the square wave). Use the complex amplitudes from (3.39) and assume that $f_0 = 25$ Hz.

3-6.5 Synthesis of a Triangle Wave

The ideal triangle wave in Fig. 3-18 is a continuous signal, unlike the square wave which is discontinuous. Therefore, it is easier to approximate the triangle wave with a finite Fourier sum (3.30). Two cases are shown in Fig. 3-19, for $N = 3$ and 11. The fundamental frequency is equal to $f_0 = 25$ Hz. In the $N = 11$ case the approximation is nearly indistinguishable from the triangularly-shaped waveform. Adding harmonics for $N > 11$ will not improve the synthesis very much. Even the $N = 3$ case is reasonably good, despite using only

DC and two sinuosoidal terms. We can see the reason for this by plotting the spectrum (as in Exercise 3.7), which will show that the high frequency components decrease in size much faster those of the square wave.

EXERCISE 3.8: For the $N = 3$ approximation of the triangle wave, derive the mathematical formula for the sinusoids; similar to what was done in (3.34) for the square wave.

3-6.6 Convergence of Fourier Synthesis

We can think of the finite Fourier sum (3.30) as making an approximation to the true signal; i.e.,

$$x(t) \approx x_N(t) = \sum_{k=-N}^{N} a_k e^{j(2\pi/T_0)kt}$$

In fact, we might hope that with enough complex exponentials we could make the approximation perfect. This leads us to define an error signal, $e_N(t)$, as the difference between the true signal and the synthesis with N terms, i.e., $e_N(t) = x(t) - x_N(t)$. We can quantify the error by measuring a feature of the error, such as its maximum magnitude. This would be called the *worst-case error.*

$$E_{\text{worst}} = \max_{t \in [0, T_0]} |x(t) - x_N(t)| \tag{3.40}$$

Now we can compare the $N = 3$ and $N = 11$ approximations of the triangle wave by comparing their worst-case errors. If Fig. 3-19 is zoomed, these errors can be measured and the result is 0.0497 for $N = 3$ and 0.0168 for $N = 11$. Because the triangle wave is continuous, the maximum error decreases to zero as $N \to \infty$ (i.e., there is no Gibbs' phenomenon). This is not the case for the

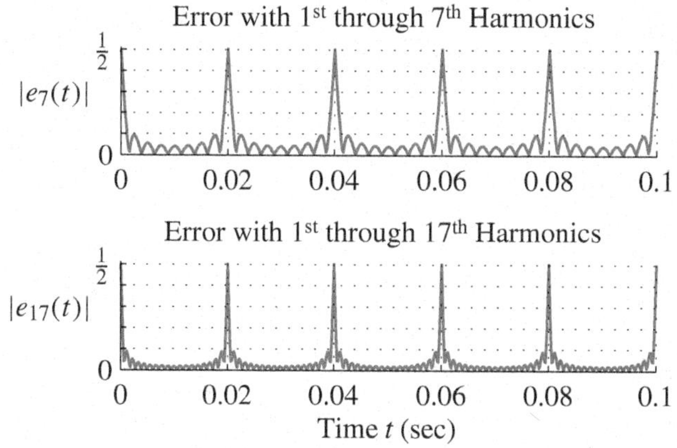

Figure 3-20: Error magnitude when approximating a square wave with a sum of harmonic components via (3.30): $N = 7$ (top panel) and $N = 17$ (bottom).

discontinuous square wave where the maximum error is always half the size of the jump in the waveform right at the discontinuity point with an *overshoot* of about 9% of the size of the discontinuity on either side. This is illustrated in Fig. 3-20 for $N = 7$ and $N = 17$. Fig. 3-20 was generated by using the Fourier coefficients from (3.32) in a MATLAB script to generate a plot of the worst-case error.

3-7 Time–Frequency Spectrum

We have seen that a wide range of interesting waveforms can be synthesized by the equation

$$x(t) = A_0 + \sum_{k=1}^{N} A_k \cos(2\pi f_k t + \phi_k) \tag{3.41}$$

These waveforms range from constants, to cosine signals, to general periodic signals, to complicated-looking signals that are not periodic. One assumption we have made so far is that the amplitudes, phases, and

Figure 3-21: Sheet music notation is a time–frequency diagram.

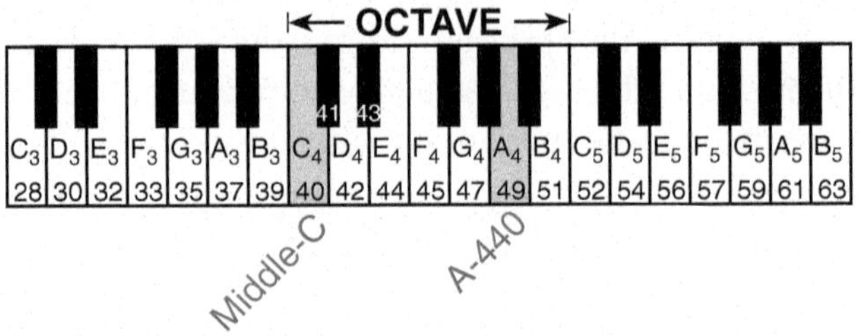

Figure 3-22: Piano keys can be numbered from 1 to 88. Three octaves are shown. Middle C is key 40. A-440 is key 49.

frequencies in (3.41) do not change with time. However, most real-world signals exhibit frequency changes over time. Music is the best example. For very short time intervals, the music may have a "constant" spectrum, but over the long term, the frequency content of the music changes dramatically. Indeed, the changing frequency spectrum is the very essence of music. Human speech is another good example. Vowel sounds, if held for a long time, exhibit a "constant" nature because the vocal tract resonates with its characteristic frequency components. However, as we speak different words, the frequency content is continually changing. In any event, most interesting signals can be modelled as a sum of sinusoids if we let the frequencies, amplitudes, and phases vary with time. Therefore, we need a way to describe such time–frequency variations. This leads us to the concept of a time–frequency spectrum or ***spectrogram***.

The mathematical concept of a time–frequency spectrum is a sophisticated idea, but the intuitive notion of such a spectrum is supported by common, everyday examples. The best example to cite is musical notation (Fig. 3-21). A musical score specifies how a piece is to be played by giving the notes to be played, the time duration of each note, and the starting time of each. The notation itself is not completely obvious, but the horizontal "axis" in Fig. 3-21 is time, while the vertical axis is frequency. The time duration for each note varies depending on whether it is a whole note, half note, quarter note, eighth, sixteenth, etc. In Fig. 3-21 most of the notes are sixteenth notes, indicating that the piece should be played briskly. If we assign a time duration to a sixteenth note, then all sixteenth notes should have the same duration. An eighth note would have twice the duration of a sixteenth note, and a quarter note would be four times longer than a sixteenth note, and so on.

The vertical axis has a much more complicated notation to define frequency. If you look carefully at Fig. 3-21, you will see that the black dots that mark the notes lie either on one of the horizontal lines or in the space between two lines. Each of these denotes a white key on the piano keyboard depicted in Fig 3-22, and each key produces a different frequency tone. The black keys on the piano are denoted by "sharps" (♯) or flats" (♭). Figure 3-21 has a few notes sharped. The musical score is divided into a treble section (the top five lines) and a bass section (the bottom five lines). The vertical reference point for the notes is "middle C," which lies on an invisible horizontal line between the treble and bass sections (key number 40 in Fig. 3-22). Thus the bottom horizontal line in the treble section represents the white key (E) that is two above middle C; that is, key number 44 in Fig. 3-22.

Once the mapping from the musical score to the piano

keys has been made, we can write a mathematical formula for the frequency of each note. A piano, which has 88 keys, is divided into octaves containing twelve keys each. The meaning of the word ***octave*** is a doubling of the frequency. Within an octave, the neighboring keys maintain a constant frequency ratio. Since there are twelve keys per octave, the ratio (r) is

$$r^{12} = 2 \quad \Longrightarrow \quad r = 2^{1/12} = 1.0595$$

With this ratio, we can compute the frequencies of all keys if we have one reference. The convention is that the A key above middle C, called A-440, has frequency 440 Hz. Since A-440 is key number 49 and middle C is key number 40, the frequency of middle C is

$$f_{\text{middle C}} = 440 \times 2^{(40-49)/12} \approx 261.6 \text{ Hz}$$

It is not our objective to explain how to read sheet music, although two of the lab projects on the CD-ROM investigate methods for synthesizing waveforms to create songs and musical sounds. What is interesting about musical notation is that it uses a two-dimensional display to indicate frequency content that changes with time. If we adopt a similar notation, we can specify how to synthesize sinusoids with time-varying frequency content. Our notation is illustrated in Fig. 3-23.

LAB: *#4 Synthesis of Sinusoidal Signals*

LAB: *#5 FM Synthesis for Musical Instruments*

3-7.1 Stepped Frequency

The simplest example of time-varying frequency content is to make a waveform whose frequency stays constant for a short duration and then steps to a higher (or lower) frequency. An example from music would be to play a ***scale*** that would be a succession of notes progressing over one octave. For example, the C-major scale consists of playing the notes {C, D, E, F, G, A, B, C} one after another,

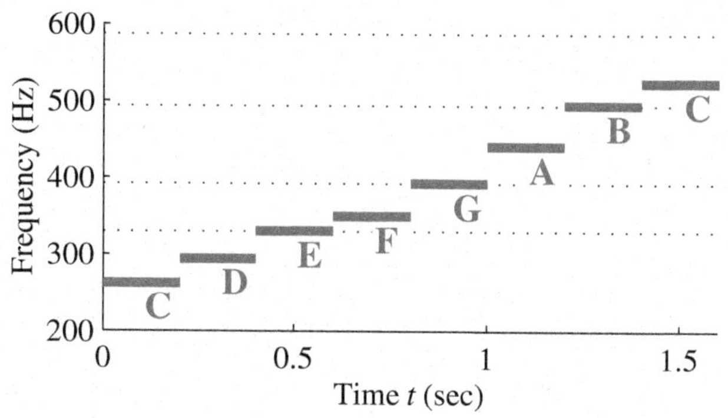

Figure 3-23: Ideal time-frequency diagram for playing the C-major scale. The horizontal dotted lines correspond to the five lines in the treble staff of sheet music (Fig. 3-21).

starting at middle C. This scale is played completely on the white keys. The frequencies of these notes are:

Middle C	D	E	F	G	A	B	C
262 Hz	294	330	349	392	440	494	523

A graphical presentation of the C-major scale is shown in Fig. 3-23. It should be interpreted as follows: Synthesize the frequency 262 Hz for 200 msec, then the frequency 294 Hz during the next 200 msec, and so on. The total waveform duration will be 1.6 sec. In music notation, the notes would be written as in Fig. 3-24(top), where each note is a quarter note.

3-7.2 Spectrogram Analysis

The frequency content of a signal can be considered from two points of view: analysis or synthesis. For example, the ideal time–frequency diagram in Figure 3-23 specifies a rule for synthesizing the C-major scale. Analysis is a more challenging problem, as we saw in Section 3-4, where the Fourier series analysis integral (3.26) was given.

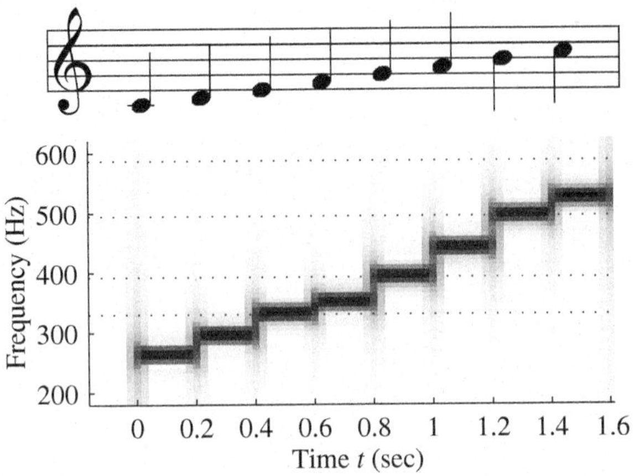

Figure 3-24: Musical notation for the C-major scale, and the corresponding spectrogram computed using MATLAB's `specgram` function.

Analysis for time-varying frequencies is usually considered a subject reserved for advanced graduate courses. One reason is that we cannot write a simple mathematical formula like the Fourier series integral to do the analysis. On the other hand, excellent numerical routines are now available for time–frequency analysis. Specifically, we can compute a *spectrogram*, which is a two-dimensional function of time and frequency that displays the time variation of the spectral content of a signal.

DEMO: *Spectrograms: Real Sounds: Piano*

In MATLAB, the function `specgram` will compute the spectrogram, and its default values work well for most signals.[6] Therefore, it is reasonable to see what sort of output can be produced by the `specgram` function. Figure 3-24 shows the results of applying `specgram` to the stepped-frequency sinusoids that make up the C-major scale. The calculation is performed by doing a frequency analysis on short segments of the signal and plotting the results at the specific time at which the analysis is done. By repeating this process with slight displacements in time, a two-dimensional array is created whose magnitude can be displayed as a grayscale image whose horizontal axis is time and whose vertical axis is frequency. The time axis must be interpreted as the "time of analysis" because the frequency calculation is not instantaneous; rather, it is based on a finite segment of the signal—in this case, 25.6 msec.

It is quite easy to identify the frequency content due to each note, but there are also some interfering artifacts that make the spectrogram in Fig. 3-24 less than ideal. In Chapter 13, we will undertake a discussion of frequency *analysis* in order to explain how the spectrogram is calculated and how one should choose the analysis parameters to get a good result. Even though the spectrogram is a highly advanced idea in signal analysis, its application is relatively easy and intuitive, especially for music signals which are described symbolically by a notation that is very much like a spectrogram.

3-8 Frequency Modulation: Chirp Signals

Section 3-7 revealed the possibility that interesting sounds can be created when the frequency varies as a function of time. In this section, we use a different mathematical formula to create signals whose frequency is time-varying. We will also pursue this idea in Lab #5.

LAB: *#5 FM Synthesis for Musical Instruments*

3-8.1 Chirp or Linearly Swept Frequency

A "chirp" signal is a swept-frequency signal whose frequency changes linearly from some low value to a high one. For example, in the audible region, we might begin at 220 Hz and go up to 2320 Hz. One method

[6] The *SP First* toolbox contains an equivalent function called `spectgr`.

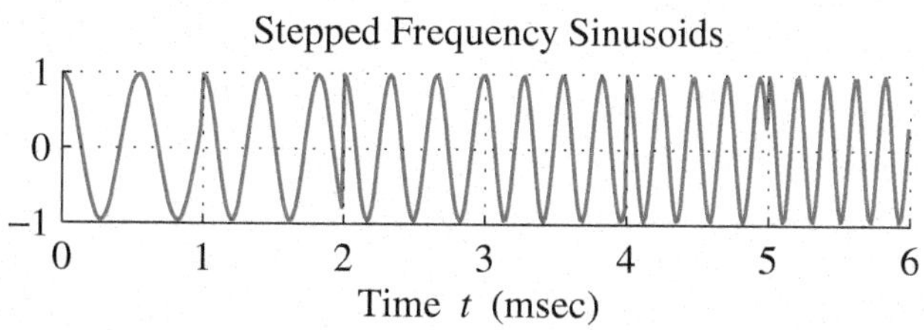

Figure 3-25: Stepped-frequency sinusoid, with six frequencies changing from 30 Hz to 80 Hz in 10 Hz increments. The frequency changes once per millisecond.

for producing such a signal would be to concatenate a large number of short constant-frequency sinusoids, whose frequencies step from low to high. This approach has one notable disadvantage: The boundary between the short sinusoids will be discontinuous unless we are careful to adjust the initial phase of each small sinusoid. Figure 3-25 shows a time waveform where the frequency is being stepped. Notice the jumps at $t = 1, 2, 3, 4, 5$ secs, which, in this case, are caused by using $\phi = 0$ for each small sinusoidal segment.

A better approach is to modify the formula for the sinusoid so that we get a time-varying frequency. Such a formula can be derived from the complex-exponential point of view. If we regard a constant-frequency sinusoid as the real part of a complex (rotating) phasor

$$x(t) = \Re e\{Ae^{j(\omega_0 t+\phi)}\} = A\cos(\omega_0 t + \phi) \qquad (3.42)$$

then the ***angle function***[7] of this signal is the exponent $(\omega_0 t + \phi)$ which obviously changes *linearly* with time. The time derivative of the angle function is ω_0, which equals the constant frequency.

Therefore, we adopt the following general notation for the class of signals with time-varying angle function:

$$x(t) = \Re e\{Ae^{j\psi(t)}\} = A\cos(\psi(t)) \qquad (3.43)$$

where $\psi(t)$ denotes the angle function versus time. For example, we can create a signal with quadratic angle function by defining

$$\psi(t) = 2\pi\mu t^2 + 2\pi f_0 t + \phi \qquad (3.44)$$

Now we can define the ***instantaneous frequency*** for these signals as the slope of the angle function (i.e., its derivative)

$$\omega_i(t) = \frac{d}{dt}\psi(t) \qquad \text{(rad/sec)} \qquad (3.45)$$

where the units of $\omega_i(t)$ are rad/sec, or, if we divide by 2π

$$f_i(t) = \frac{1}{2\pi}\frac{d}{dt}\psi(t) \qquad \text{(Hz)} \qquad (3.46)$$

we obtain Hz. If the angle function of $x(t)$ is quadratic, then its frequency changes *linearly* with time; that is,

$$f_i(t) = 2\mu t + f_0$$

The frequency variation produced by the time-varying angle function is called frequency modulation, and signals of this class are called ***FM signals***. Finally, since the linear variation of the frequency can produce an audible sound similar to a siren or a chirp, the linear FM signals are also called ***chirp*** signals, or simply *chirps*.

DEMO: *Spectrograms: Chirp Sounds*

The process can be reversed because (3.45) states that the instantaneous frequency is the derivative of the angle function $\psi(t)$. Thus, if a certain linear frequency sweep is desired, the actual angle function needed in (3.43) is obtained from the integral of $\omega_i(t)$.

[7]Here we use the term *angle function* to mean the *angle* of the cosine wave. Recall that the constant ϕ is the *phase-shift*.

Example 3-8: Synthesize a Chirp Formula

Suppose we want to synthesize a frequency sweep from $f_1 = 220$ Hz to $f_2 = 2320$ Hz over a 3-second time interval, i.e., the beginning and ending times are $t = 0$ and $t = T_2 = 3$ sec. First of all, it is necessary to create a formula for the instantaneous frequency

$$f_i(t) = \frac{f_2 - f_1}{T_2} t + f_1 = \frac{2320 - 220}{3} t + 220$$

Then we must integrate $2\pi f_i(t)$ to get the angle function:

$$\begin{aligned} \psi(t) &= \int_0^t \omega_i(u)\, du \\ &= \int_0^t 2\pi \left(\frac{2320 - 220}{3} u + 220 \right) du \\ &= 700\pi t^2 + 440\pi t + \phi \end{aligned}$$

where the phase shift, ϕ, is an arbitrary constant. The chirp signal is $x(t) = \cos(\psi(t))$. ■

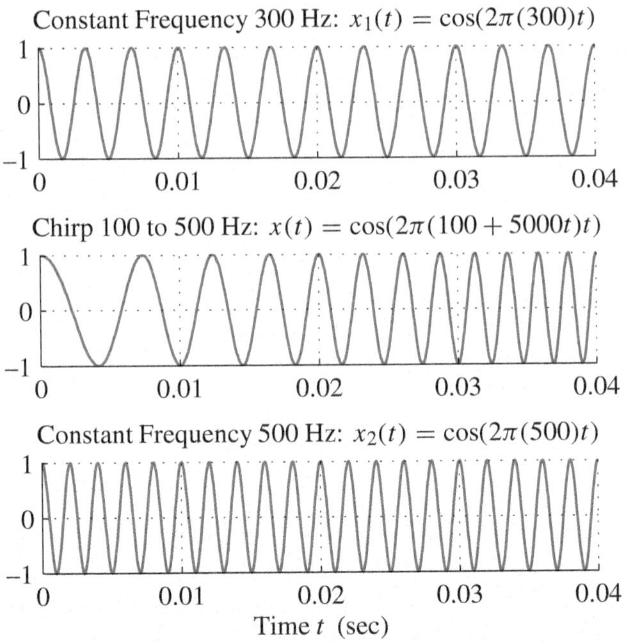

Figure 3-26: Comparing a chirp signal (middle) to constant-frequency sinusoids (top and bottom). Notice where the local frequency of the chirp is equal to one of the sinusoids.

3-8.2 A Closer Look at Instantaneous Frequency

It may be difficult to see why the derivative of the angle function would be the instantaneous frequency. The following experiment provides a clue.

(a) Define a "chirp" signal as in Example 3-8 with the following parameters:

$$f_1 = 100 \text{ Hz}$$
$$f_2 = 500 \text{ Hz}$$
$$T_2 = 0.04 \text{ sec}$$

In other words, determine μ and f_0 in (3.44) to define $x(t)$ so that it sweeps the specified frequency range.

(b) Now make a plot of the signal synthesized in (a). In Fig. 3-26, this plot is the middle panel.

(c) It is difficult to verify whether or not this chirp signal will have the correct frequency content. However, the rest of this experiment will demonstrate that the derivative of the angle function is the "correct" definition of instantaneous frequency. First of all, plot a 300-Hz sinusoid, $x_1(t)$ which is shown in the upper panel of Fig. 3-26.

(d) Finally, generate and plot a 500-Hz sinusoid, $x_1(t)$ as in the bottom panel of Fig. 3-26.

(e) Now compare the three signals in Fig. 3-26 with respect to the frequency content of the chirp. Concentrate on the frequency of the chirp in the time range $0.019 \le t \le 0.021$ sec. Notice that the 300-Hz sinusoid matches the chirp in this time region. Evaluate the theoretical $f_i(t)$ in this region.

(f) It is possible to find another region (near $t = 0.04$s) where the chirp frequency is equal (locally) to 500 Hz.

We have seen that for signals of the form $x(t) = A\cos(\psi(t))$, the instantaneous frequency of the signal is the derivative of the angle function $\psi(t)$. If $\psi(t)$ is constant, the frequency is zero. If $\psi(t)$ is linear, $x(t)$ is a sinusoid at some fixed frequency. If $\psi(t)$ is quadratic, $x(t)$ is a chirp signal whose frequency changes linearly versus time. More complicated variations of $\psi(t)$ can produce a wide variety of signals. One application of FM signals is in music synthesis. This application is illustrated with demos and a lab on the CD-ROM.

DEMO: *Spectrograms: Chirp Sounds*

DEMO: *FM Synthesis*

3-9 Summary and Links

This chapter introduced the concept of the ***spectrum***, in which we represent a signal by its sinusoidal components. The spectrum is a graphical presentation of the complex amplitude for each frequency component in the signal. We showed how complicated signals can be formed from relatively simple spectra, and we presented the essential concepts of the Fourier series so that we could form the spectrum of arbitrary periodic signals. Finally, we ended with a discussion of how the spectrum can vary with time.

At this point, so many different demonstrations and projects can be done that we must limit our list somewhat. Among the laboratory projects on the CD-ROM are three devoted to different aspects of the spectrum. The first (Lab #3) contains exercises on the Fourier series representation of the square wave and a sawtooth wave. The second (Lab #4) requires students to develop a music synthesis program to play a piece such as Bach's "Jesu, Joy of Man's Desiring." This synthesis must be done with sinusoids, but can be refined with extras such as a tapered amplitude envelope. Finally, Lab #5 deals with beat notes, chirp signals, and spectrograms. The second part of this lab involves a music synthesis method based on frequency modulation. The FM synthesis algorithm can produce realistic sounds for instruments such as a clarinet or a drum. In addition, there is one more lab that involves some practical systems that work with sinusoidal signals, such as a Touch-Tone phone. This lab, however, requires some knowledge of filtering so it is reserved for Chapter 7. Write-ups of the labs can be found on the CD-ROM.

DEMO: *Links to Many Demos*

The CD-ROM also contains many demonstrations of sounds and their spectrograms:

(a) Spectrograms of simple sounds such as sine waves, square waves, and other harmonics.

(b) Spectrograms of realistic sounds, including a piano recording, a synthetic scale, and a synthesized music passage done by one of the students who took an early version of this course.

(c) Spectrograms of chirp signals that show how the rate of change of the frequency affects the sound you hear.

(d) An explanation of the FM synthesis method for emulating musical instruments. Several example sounds are included for listening.

(e) Rotating Phasors

(f) Vowel Synthesis

NOTE: *Hundreds of Solved Problems*

Finally, the reader is reminded of the large number of solved homework problems that are available for review and practice on the CD-ROM.

3-10 Problems

P-3.1 A signal composed of sinusoids is given by the equation

$$x(t) = 10\cos(800\pi t + \pi/4) + 7\cos(1200\pi t - \pi/3) - 3\cos(1600\pi t)$$

(a) Sketch the spectrum of this signal, indicating the complex size of each frequency component. Make separate plots for real/imaginary or magnitude/phase of the complex amplitudes at each frequency.

(b) Is $x(t)$ periodic? If so, what is the period?

(c) Now consider a new signal defined as $y(t) = x(t) + 5\cos(1000\pi t + \pi/2)$. How is the spectrum changed? Is $y(t)$ periodic? If so, what is the period?

P-3.2 A signal $x(t)$ has the two-sided spectrum representation shown in Fig. P-3.2.

(a) Write an equation for $x(t)$ as a sum of cosines.

(b) Is $x(t)$ a periodic signal? If so, determine its fundamental period and its fundamental frequency.

(c) Explain why "negative" frequencies are needed in the spectrum.

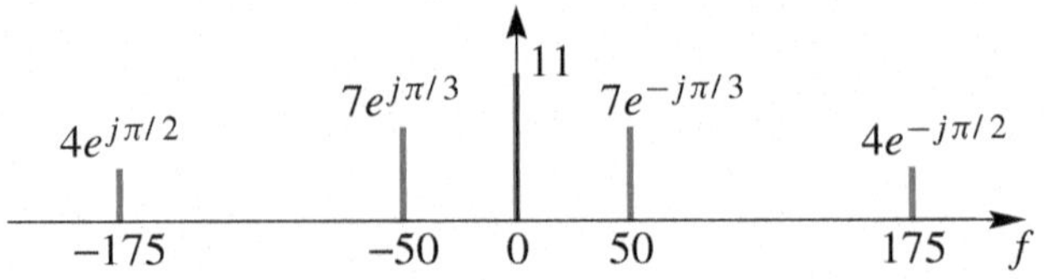

Figure P-3.2

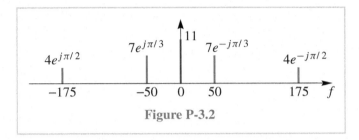

P-3.3 Let $x(t) = \sin^3(27\pi t)$.

(a) Determine a formula for $x(t)$ as the real part of a sum of complex exponentials.

(b) What is the fundamental period for $x(t)$?

(c) Plot the *spectrum* for $x(t)$.

P-3.4 In Section 3-2.2, we discussed a simple example of the "beating" of one cosine wave against another. In this problem, you will consider a more general case. Let

$$x(t) = A\cos[2\pi(f_c - f_\Delta)t] + B\cos[2\pi(f_c + f_\Delta)t]$$

For the case discussed in Section 3-2.2, the amplitudes are equal, $A = B = 1$.

(a) Use phasors to obtain a complex signal $z(t)$ such that

$$x(t) = \Re e\{z(t)\}$$

(b) By manipulating the expression for $z(t)$ and then taking the real part, show that in the more general case above, $x(t)$ can be expressed in the form

$$x(t) = C\cos(2\pi f_\Delta t)\cos(2\pi f_c t) + D\sin(2\pi f_\Delta t)\sin(2\pi f_c t)$$

and find expressions for C and D in terms of A and B. Check your answer by substituting $A = B = 1$.

(c) Find values for A and B so that

$$x(t) = 2\sin(2\pi f_\Delta t)\sin(2\pi f_c t)$$

Plot the spectrum of this signal.

P-3.5 Consider the signal

$$x(t) = 10 + 20\cos(2\pi(100)t + \tfrac{1}{4}\pi) + 10\cos(2\pi(250)t)$$

(a) Using Euler's relation, the signal $x(t)$ defined above can be expressed as a sum of complex exponential signals using the finite Fourier synthesis summation (3.30)

Determine values for f_0, N and all the complex amplitudes, a_k. *It is not necessary to evaluate any integrals to obtain* a_k.

(b) Is the signal $x(t)$ periodic? If so, what is the fundamental period?

(c) Plot the spectrum of this signal versus f in Hz.

P-3.6 An amplitude-modulated (AM) cosine wave is represented by the formula

$$x(t) = [12 + 7\sin(\pi t - \tfrac{1}{3}\pi)]\cos(13\pi t)$$

(a) Use phasors to show that $x(t)$ can be expressed in the form

$$\begin{aligned} x(t) &= A_1\cos(\omega_1 t + \phi_1) \\ &\quad + A_2\cos(\omega_2 t + \phi_2) + A_3\cos(\omega_3 t + \phi_3) \end{aligned}$$

where $\omega_1 < \omega_2 < \omega_3$; i.e., find values of the parameters A_1, A_2, A_3, ϕ_1, ϕ_2, ϕ_3, ω_1, ω_2, ω_3.

(b) Sketch the two-sided spectrum of this signal on a frequency axis. Be sure to label important features of the plot. Label your plot in terms of the numerical values of A_i, ϕ_i, and ω_i.

P-3.7 Consider a signal $x(t)$ such that

$$\begin{aligned} x(t) &= 2\cos(\omega_1 t)\cos(\omega_2 t) \\ &= \cos[(\omega_2 + \omega_1)t] + \cos[(\omega_2 - \omega_1)t] \end{aligned}$$

where $0 < \omega_1 < \omega_2$.

(a) What is the general condition that must be satisfied by $\omega_2 - \omega_1$ and $\omega_2 + \omega_1$ so that $x(t) = x(t + T_0)$, i.e., so that $x(t)$ is periodic with period T_0?

(b) What does the result of (a) imply about ω_1 and ω_2? For example, is ω_2 an integer multiple of ω_1?

P-3.8 A periodic signal is given by the equation

$$\begin{aligned} x(t) &= 2 + 4\cos(40\pi t - \tfrac{1}{5}\pi) \\ &\quad + 3\sin(60\pi t) + 4\cos(120\pi t - \tfrac{1}{3}\pi) \end{aligned}$$

(a) Determine the fundamental frequency ω_0, the fundamental period T_0, the number of terms N, and the coefficients a_k in the finite Fourier representation (3.30) for the signal $x(t)$ above. *It is possible to do this **without** evaluating any integrals.*

(b) Sketch the spectrum of this signal indicating the complex amplitude of each frequency component. Indicate the complex amplitude values at the appropriate frequencies. *There is no need to make separate plots for real/imaginary parts or magnitude/phase.*

(c) Now consider a new signal defined by adding one more sinusoid $y(t) = x(t) + 10\cos(50\pi t - \pi/6)$. How is the spectrum changed? Is $y(t)$ still periodic? If so, what is the fundamental period?

P-3.9 A periodic signal $x(t) = x(t + T_0)$ is described over one period $-T_0/2 \le t \le T_0/2$ by the equation

$$x(t) = \begin{cases} 1 & \text{for } |t| < t_c \\ 0 & \text{for } t_c < |t| \le T_0/2 \end{cases}$$

where $t_c < T_0/2$.

(a) Sketch the periodic function $x(t)$ over the time interval $-2T_0 < t < 2T_0$ for the case $t_c = T_0/4$.

(b) Determine the DC coefficient a_0.

(c) Determine a formula for the Fourier series coefficients a_k in the finite Fourier representation (3.30). Your final result should depend on t_c and T_0.

(d) Sketch the spectrum of $x(t)$ for the case $\omega_0 = 2\pi(100)$ and $t_c = T_0/4$. Use a frequency range from $-10\omega_0$ to $+10\omega_0$.

(e) Sketch the spectrum of $x(t)$ for the case $\omega_0 = 2\pi(100)$ and $t_c = T_0/10$. Use a frequency range from $-10\omega_0$ to $+10\omega_0$.

(f) From your results in (d) and (e), what do you conclude about the relationship between t_c and the relative size of the high-frequency components of $x(t)$? When t_c decreases, do the high-frequency components get bigger or smaller?

P-3.10 The periodic waveform in Fig. P-3.10 has the property of *half-wave symmetry*; i.e., the last half of the period is the negative of the first half. More precisely, signals with half-wave symmetry have the property that

$$x(t + T_0/2) = -x(t) \qquad -\infty < t < \infty \qquad (3.47)$$

In this problem we will show that this condition has an interesting effect on the Fourier series coefficients for the signal.

(a) Suppose that $x(t)$ is a periodic signal with half-wave symmetry and is defined over half a period by

$$x(t) = t \qquad \text{for } 0 \le t < T_0/2$$

where T_0 is the period of the signal. Plot this periodic signal for $-T_0 \le t \le T_0$.

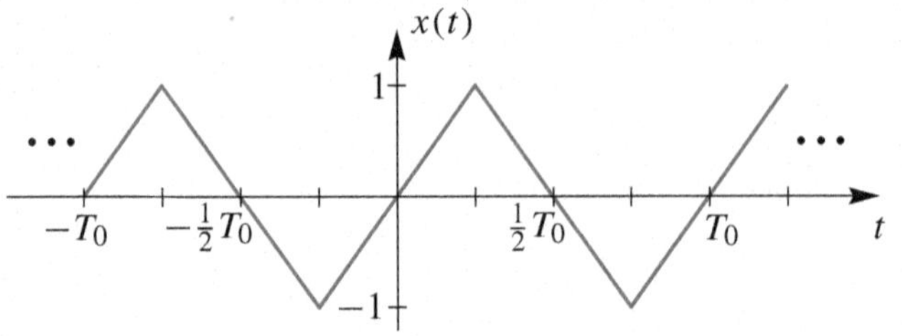

Figure P-3.10

(b) Prove that the DC coefficient a_0 is zero for any periodic signal having half-wave symmetry. *Hint*: Split the integral for a_0 into two parts.

(c) Prove that all the even indexed Fourier series coefficients are zero for a signal with half-wave symmetry; i.e., $a_k = 0$ if k is an even integer.

P-3.11 A signal $x(t)$ is periodic with period $T_0 = 8$. Therefore, it can be represented as a Fourier series of the form

$$x(t) = \sum_{k=-\infty}^{\infty} a_k e^{j(2\pi/8)kt}$$

It is known that the Fourier series coefficients for this representation of a particular signal $x(t)$ are given by the integral

$$a_k = \frac{1}{8}\int_{-4}^{0} (4+t)e^{-j(2\pi/8)kt}\,dt$$

(a) In the integral expression for a_k above, the integrand and the limits define the signal $x(t)$. Determine an equation for $x(t)$ that is valid over one period.

(b) Using the result from part (a), draw a plot of $x(t)$ over the range $-8 \leq t \leq 8$ seconds. Label your plot carefully.

(c) Determine the DC value of $x(t)$.

P-3.12 A periodic signal $x(t)$ with a period $T_0 = 10$ is described *over one period,* $0 \leq t \leq 10$, by the equation

$$x(t) = \begin{cases} 0 & 0 \leq t \leq 5 \\ 2 & 5 < t \leq 10 \end{cases}$$

This signal can be represented by the Fourier series (3.19) which is valid for all time $-\infty < t < \infty$.

(a) Sketch the periodic function $x(t)$ over the time interval $-10 \leq t \leq 20$.

(b) Determine the DC coefficient of the Fourier series, a_0.

(c) Use the Fourier analysis integral (3.21) to find a_1, the *first* Fourier series coefficient (i.e., for $k = 1$).

(d) If we add a constant value of one to $x(t)$, we obtain the signal $y(t) = 1 + x(t)$ with $y(t)$ given over one period by

$$y(t) = \begin{cases} 1 & 0 \leq t \leq 5 \\ 3 & 5 < t \leq 10 \end{cases}$$

This signal can also be represented by a Fourier series, but with different coefficients:

$$y(t) = \sum_{k=-\infty}^{\infty} b_k e^{jk\omega_0 t}$$

Explain how b_0 and b_1 are related to a_0 and a_1. *Note*: You should not have to evaluate any new integrals explicitly to answer this question.

P-3.13 Make a sketch of the spectrum of the signal defined by

$$x(t) = \sin(10t) \left(\sum_{k=-3}^{3} \frac{1}{1+jk} e^{jkt} \right)$$

Hint: use Euler's formula on the $\sin(10t)$ term, and then multiply into the summation in order to get the Fourier series coefficients of $x(t)$.

P-3.14 We have seen that a periodic signal $x(t)$ can be represented by its Fourier series (3.19). It turns out that we can transform many operations on the signal into corresponding operations on the Fourier coefficients a_k. For example, suppose that we want to consider a new periodic signal $y(t) = \frac{dx(t)}{dt}$. What would the Fourier coefficients be for $y(t)$? The answer comes from differentiating the Fourier series representation

$$\begin{aligned} y(t) = \frac{dx(t)}{dt} &= \frac{d}{dt}\left[\sum_{k=-\infty}^{\infty} a_k e^{jk\omega_0 t} \right] \\ &= \sum_{k=-\infty}^{\infty} a_k \frac{d}{dt}\left[e^{jk\omega_0 t}\right] \\ &= \sum_{k=-\infty}^{\infty} a_k \left[(jk\omega_0) e^{jk\omega_0 t}\right] \end{aligned}$$

Thus, we see that $y(t)$ is also in the Fourier series form

$$y(t) = \sum_{k=-\infty}^{\infty} b_k e^{jk\omega_0 t}, \quad \text{where } b_k = (jk\omega_0) a_k$$

so the Fourier series coefficients of $y(t)$ are related to the Fourier series coefficients of $x(t)$ by $b_k = (jk\omega_0)a_k$. This is a nice result, because it allows us to find the Fourier coefficients *without* actually differentiating $x(t)$, and *without* doing any tedious evaluation of integrals to obtain the Fourier coefficients b_k. It is a *general* result that holds for every periodic signal and its derivative.

We can use this style of manipulation to obtain some other useful results for Fourier series. In each case below, use (3.19) as the starting point and express the the given definition for $y(t)$ as a Fourier series and then manipulate the equation so that you can pick off an expression for the Fourier coefficients b_k as a function of the original coefficients a_k.

(a) Suppose that $y(t) = Ax(t)$ where A is a real number; i.e., $y(t)$ is just a scaled version of $x(t)$. Show that the Fourier coefficients for $y(t)$ are $b_k = Aa_k$.

(b) Suppose that $y(t) = x(t - t_d)$ where t_d is a real number; i.e., $y(t)$ is just a delayed version of $x(t)$. Show that the Fourier coefficients for $y(t)$ in this case are $b_k = a_k e^{-jk\omega_0 t_d}$.

P-3.15 Consider the periodic function $x(t)$ plotted in Fig. P-3.15.

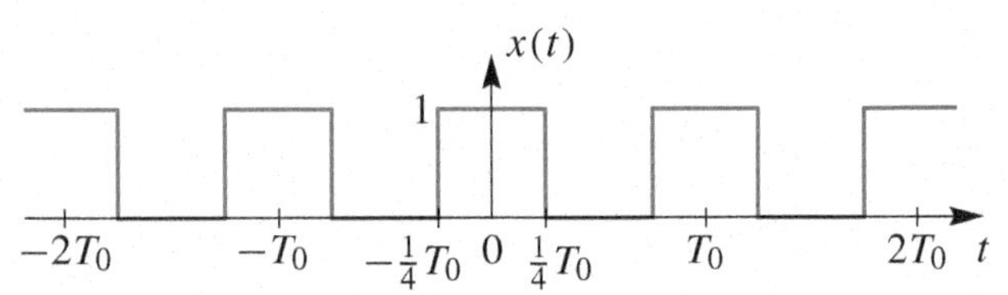

Figure P-3.15

(a) Find the "DC" value a_0 and the other Fourier coefficients a_k for $k \neq 0$ in the Fourier series representation of $x(t)$.

(b) Define a new signal as $y(t) = 2x(t - T_0/2)$, and then sketch the waveform of $y(t)$. Use the results of Problem P-3.14 to write down the Fourier series coefficients b_0 and b_k for $k \neq 0$ for the periodic signal $y(t)$ without evaluating any integrals.

P-3.16 We have seen that musical tones can be modelled mathematically by sinusoidal signals. If you read music or play the piano, you know that the piano keyboard is divided into octaves, with the tones in each octave being twice the frequency of the corresponding tones in the next lower octave. To calibrate the frequency scale, the reference tone is the A above middle C, which is usually called A-440, since its frequency is 440 Hz. Each octave contains 12 tones, and the ratio between the frequencies of successive tones is constant. Thus, the ratio must be $2^{1/12}$. Since middle C is nine tones below A–440, its frequency is approximately $(440)2^{-9/12} \approx 261.6$ Hz. The names of the tones (notes) of the octave starting with middle C and ending with high C are:

Note name	C	$C^\#$	D	$E^\flat$	E	F	$F^\#$
Note number	40	41	42	43	44	45	46
Frequency							
Note name	$F^\#$	G	$G^\#$	A	$B^\flat$	B	C
Note number	46	47	48	49	50	51	52
Frequency				440			

(a) Make a table of the frequencies of the tones of the octave beginning with middle C, assuming that the A above middle C is tuned to 440 Hz.

(b) The above notes on a piano are numbered 40 through 52. If n denotes the note number and f denotes the frequency of the corresponding tone, give a formula for the frequency of the tone as a function of the note number.

(c) A *chord* is a combination of musical notes sounded simultaneously. A *triad* is a three-note chord. The

D-major chord is composed of the tones of D, $F^{\#}$, and A sounded simultaneously. From the set of corresponding frequencies determined in (a), make a sketch of the essential features of the spectrum of the D-major chord assuming that each note is realized by a pure sinusoidal tone. *Do not specify the complex phasors precisely.*

P-3.17 A chirp signal is one that sweeps in frequency from $\omega_1 = 2\pi f_1$ to $\omega_2 = 2\pi f_2$ as time goes from $t = 0$ to $t = T_2$. The general formula for a chirp is

$$x(t) = A\cos(\alpha t^2 + \beta t + \phi) = \cos(\psi(t)) \qquad (3.48)$$

where

$$\psi(t) = \alpha t^2 + \beta t + \phi$$

The derivative of $\psi(t)$ is the ***instantaneous frequency*** which is also the frequency heard if the frequencies are in the audible range.

$$\omega_i(t) = \frac{d}{dt}\psi(t) \qquad \text{radians/sec} \qquad (3.49)$$

(a) For the chirp in (3.48), determine formulas for the beginning frequency (ω_1) and the ending frequency (ω_2) in terms of α, β, and T_2.

(b) For the chirp signal

$$x(t) = \Re e\{e^{j(40t^2+27t+13)}\}$$

derive a formula for the ***instantaneous*** frequency versus time.

(c) Make a plot of the ***instantaneous*** frequency (in Hz) versus time over the range $0 \le t \le 1$ sec.

P-3.18 To see why the derivative of the angle function would be the instantaneous frequency, repeat the experiment of Section 3-8.2.

(a) Use the following parameters to define a chirp signal:

$$f_1 = 1 \text{ Hz}$$

$$f_2 = 9 \text{ Hz}$$

$$T_2 = 2 \text{ sec}$$

In other words, determine α and β in (3.48) to define $x(t)$ so that it sweeps the specified frequency range.

(b) The rest of this problem is devoted to a MATLAB experiment that will demonstrate why the derivative of the angle function is the "correct" definition of instantaneous frequency. First, make a plot of the instantaneous frequency $f_i(t)$ (in Hz) versus time.

(c) Now make a plot of the signal synthesized in (a). Pick a time-sampling interval that is small enough so that the plot is very smooth. Put this plot in the middle panel of a 3×1 subplot, i.e., `subplot(3,1,2)`.

(d) Now generate and plot a 4-Hz sinusoid. Put this plot in the upper panel of a 3×1 subplot, i.e., `subplot(3,1,1)`.

(e) Finally, generate and plot an 8-Hz sinusoid. Put this plot in the lower panel of a 3×1 subplot, i.e., `subplot(3,1,3)`.

(f) Compare the three signals and comment on the frequency content of the chirp. Concentrate on the frequency of the chirp in the time range $1.6 \le t \le 2$ sec. Which sinusoid matches the chirp in this time region? Compare the expected $f_i(t)$ in this region to 4 Hz and 8 Hz.

P-3.19 The plots in Fig. P-3.19 show waveforms on the left and spectra on the right. Match the waveform letter with its corresponding spectrum number. In each case, write the formula for the signal as a sum of sinusoids.

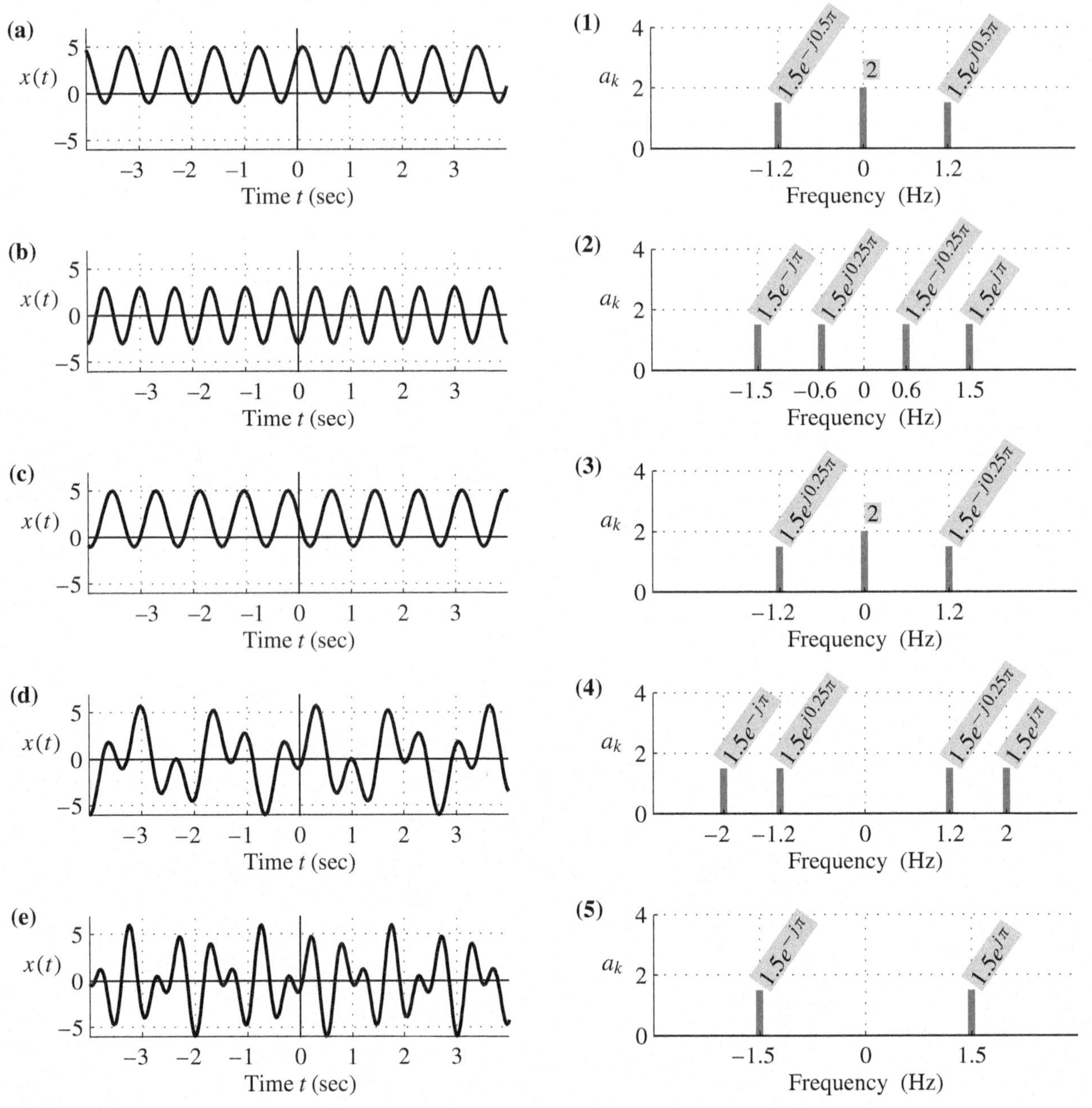
(a)
(b)
(c)
(d)
(e)
x(t)
5
0
−5
−3 −2 −1 0 1 2 3
Time t (sec)
(1)
(2)
(3)
(4)
(5)
a_k
4
2
0
Frequency (Hz)
1.5e^{−j0.5π}
2
1.5e^{j0.5π}
−1.2 0 1.2
1.5e^{−jπ}
1.5e^{j0.25π}
1.5e^{−j0.25π}
1.5e^{jπ}
−1.5 −0.6 0 0.6 1.5
−2 −1.2 0 1.2 2
−1.5 0 1.5

Figure P-3.19

CHAPTER 4

Sampling and Aliasing

This chapter is concerned with the conversion of signals between the analog (continuous-time) and digital (discrete-time) domains. The primary objective of our presentation is an understanding of the ***sampling theorem***, which states that when the sampling rate is greater than twice the highest frequency contained in the spectrum of the analog signal, the original signal can be reconstructed exactly from the samples.

The process of converting from digital back to analog is called ***reconstruction***. A common example is given by audio CDs. Music on a CD is sampled at 44,100 times per second and stored in a digital form from which a CD player reconstructs the continuous (analog) waveform that we listen to. The reconstruction process is basically one of interpolation because we must "fill in" the missing signal values between the sample times t_n by constructing a smooth curve through the discrete-time sample values $x(t_n)$. Although this process may be studied as time-domain interpolation, we will see that a spectrum or frequency-domain view is very helpful in understanding the important issues in sampling.

4-1 Sampling

Sinusoidal waveforms of the form

$$x(t) = A\cos(\omega t + \phi) \tag{4.1}$$

are examples of ***continuous-time*** signals. It is also common to refer to such signals as ***analog*** signals because both the signal amplitude and the time variable are assumed to be real (not discrete) numbers. Continuous-time signals are represented mathematically by functions of time, $x(t)$, where t is a continuous variable. In earlier chapters, we have "plotted" analog waveforms using MATLAB, but actually we did not plot the continuous-time waveform. Instead, we really plotted the waveform only at isolated (discrete) points in time and then connected those points with straight lines. Indeed, digital computers cannot deal with continuous-time signals directly; instead, they must represent and manipulate them numerically (as with MATLAB), or sometimes symbolically (as with *Mathematica* or *Maple*). The key point is that any computer representation is *discrete*. (Recall the discussion in Section 2-4 on p. 15.)

A ***discrete-time signal*** is represented mathematically by an indexed sequence of numbers. When stored in a digital computer, the signal values are held in memory locations, so they would be indexed by memory address. We denote the values of the discrete-time signal as $x[n]$, where n is the integer index indicating the order of the values in the sequence. The square brackets [ ] enclosing the argument n allow us to differentiate between the continuous-time signal $x(t)$ and a corresponding discrete-time signal $x[n]$.[1]

We can obtain discrete-time signals in either of the following ways:

(a) We can ***sample*** a continuous-time signal at equally spaced time instants, $t_n = nT_s$; that is,

$$x[n] = x(nT_s) \qquad -\infty < n < \infty, \qquad (4.2)$$

where $x(t)$ represents any continuously varying signal, e.g., speech or audio. The individual values of $x[n]$ are called *samples* of the continuous-time signal. The fixed time interval between samples, T_s, can also be expressed as a fixed ***sampling rate***, f_s, in samples per second:

$$f_s = \frac{1}{T_s} \quad \text{samples/sec.}$$

[1]The terminology for discrete-time signals is not universal, so we may occasionally use the word *sequence* in place of signal, or the adjective *digital* in place of discrete-time, to refer to $x[n]$.

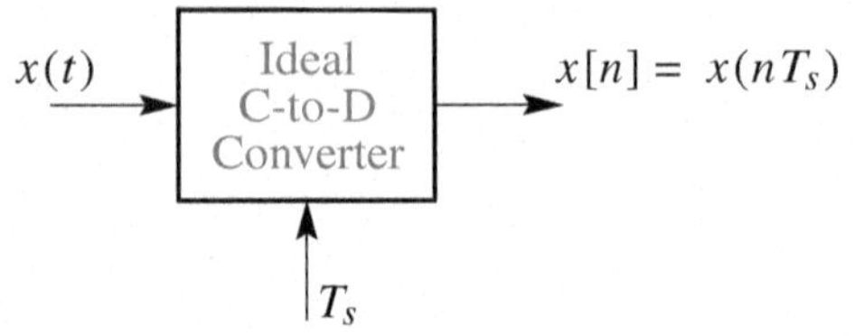

Figure 4-1: Block diagram representation of the ideal continuous-to-discrete (C-to-D) converter. The parameter T_s specifies uniform sampling of the input signal every T_s seconds.

Therefore, an alternative way to write the sequence in (4.2) is $x[n] = x(n/f_s)$.

Sampling can be viewed as a transformation or operation that acts on a continuous-time signal $x(t)$ to produce an output which is a corresponding discrete-time signal $x[n]$. In engineering, it is common to call such a transformation a *system*, and represent it graphically with a block diagram that shows the input and output signals along with a name that describes the system operation. The sampling operation is an example of a system whose input is a continuous-time signal and whose output is a discrete-time signal as shown in Fig. 4-1. The system block diagram of Fig. 4-1 represents the mathematical equation in (4.2) and is called an ***ideal continuous-to-discrete (C-to-D) converter***. Its idealized mathematical form is useful for analysis, but an actual hardware system for doing sampling is an analog-to-digital (A-to-D) converter, which approximates the perfect sampling of the C-to-D converter. A-to-D converters differ from ideal C-to-D converters because of real-world problems such as amplitude quantization to 12 or 16 bits, jitter in the sampling times, and other factors that are difficult to analyze. Since these factors can be made negligible with careful design, we can safely confine our presentation to the ideal C-to-D system.

(b) We can also *compute* the values of a discrete-time signal directly from a formula. A simple example is

$$w[n] = n^2 - 5n + 3$$

which determines the sequence of values $\{3, -1, -3, -3, -1, 3, 9, \ldots\}$ corresponding to the indices $n = 0, 1, 2, 3, 4, 5, 6, \ldots$. Although there might be no explicit underlying continuous-time signal that is being sampled, we will nevertheless often refer to the individual values of the sequence as *samples*. Discrete-time signals described by formulas will be very common in our study of discrete-time signals and systems.

When we plot discrete-time signals, we will use the format shown in Fig. 4-2, which shows eight values (samples) of the sequence $w[n]$. Such plots show clearly that the signal has values only for integer indices; in between, the discrete-time signal is undefined.

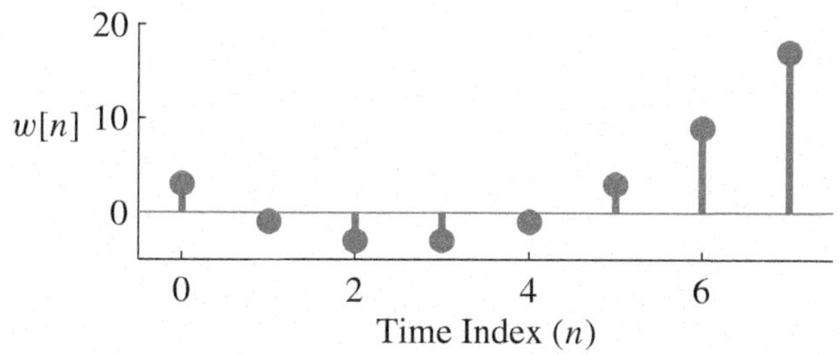

Figure 4-2: Plotting format for discrete-time signals; sometimes called a "lollypop" or "tinker-toy" plot. In MATLAB, the function `stem` will produce this plot, so many students also refer to this as a "stem plot."

4-1.1 Sampling Sinusoidal Signals

Sinusoidal signals are examples of continuous-time signals that exist in the "real world" outside the computer, and for which we can also write a simple mathematical formula. Because more general continuous-time signals can be represented as sums of sinusoids, and because the effects of sampling are easily understood for sinusoids, we will use them as the basis for our study of sampling.

If we sample a signal of the form of (4.1), we obtain

$$\begin{aligned} x[n] &= x(nT_s) \\ &= A\cos(\omega n T_s + \phi) \\ &= A\cos(\hat{\omega} n + \phi), \end{aligned} \tag{4.3}$$

where we have defined $\hat{\omega}$ to be

> ***Normalized Radian Frequency***
>
> $$\hat{\omega} \stackrel{\text{def}}{=} \omega T_s = \frac{\omega}{f_s} \tag{4.4}$$

The signal $x[n]$ in (4.3) is a ***discrete-time cosine signal***, and $\hat{\omega}$ is its ***discrete-time frequency***. We use a "hat" over ω to denote that this is a new frequency variable. It is a normalized version of the continuous-time radian frequency with respect to the sampling frequency. Since ω has units of rad/sec, the units of $\hat{\omega} = \omega T_s$ are *radians*; i.e., $\hat{\omega}$ is a dimensionless quantity. This is entirely consistent with the fact that the index n in $x[n]$ is dimensionless. Once the samples are taken from $x(t)$, the time scale information has been lost. The discrete-time signal is just a sequence of numbers, and these numbers carry no information about the sampling period, T_s, used in obtaining them. An immediate implication of this observation is that an infinite number of continuous-time sinusoidal signals can be transformed into the same discrete-time sinusoid by sampling. All we need to do is change the sampling period inversely proportional to the input frequency of the continuous-time sinusoid. For example, if $\omega = 200\pi$ rad/sec and $T_s = 1/2000$ sec, then $\hat{\omega} = 0.1\pi$ rad. On the other hand, if $\omega = 1000\pi$ rad/sec and $T_s = 1/10000$ sec, $\hat{\omega}$ is still equal to 0.1π rad.

The top panel of Fig. 4-3 shows $x(t) = \cos(200\pi t)$, a continuous-time sinusoid with frequency $f_0 = 100$ Hz. The middle panel of Fig. 4-3 shows the samples taken with sampling period $T_s = 0.5$ msec. The sequence is given by the formula $x[n] = x(nT_s) = \cos(0.1\pi n)$, so the discrete-time radian frequency is $\hat{\omega}_0 = 0.1\pi$. (Since $\hat{\omega}_0$ is dimensionless, it is redundant to specify its units as rad.) The sampling rate in this example is $f_s = 1/T_s = 2000$ samples/sec. The sample values are plotted as discrete points as in the middle panel of Fig. 4-3. The points are not connected by a continuous curve because we do not have any *direct* information about the value of the function between the sample values. In this case, there are 20 sample values per period of the signal, because the sampling frequency (2000 samples/sec) is 20 times higher than the frequency of the continuous signal (100 Hz). From this discrete-time plot, it appears that the sample values alone are sufficient to visually reconstruct a continuous-time cosine wave, but without knowledge of the sampling rate, we cannot tell what the frequency ω should be.

Another example of sampling is shown in the bottom panel of Fig. 4-3. In this case, the 100 Hz sinusoid is sampled at a lower rate ($f_s = 500$ samples/sec) resulting in the sequence of samples $x[n] = \cos(0.4\pi n)$. In this case, the discrete-time radian frequency is $\hat{\omega} = 0.4\pi$. The time between samples is $T_s = 1/f_s = 2$ msec, so there are only five samples per period of the continuous-time signal. We see that without the original waveform superimposed, it would be difficult to discern the precise waveshape of the original continuous-time sinusoid.

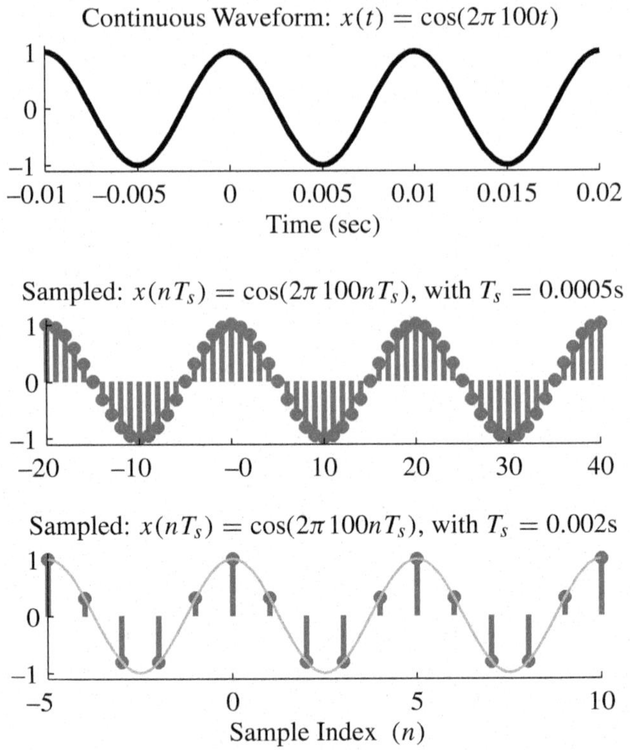

Figure 4-3: A continuous-time 100-Hz sinusoid (top) and two discrete-time sinusoids formed by sampling at $f_s =$ 2000 samples/sec (middle) and at $f_s = 500$ samples/sec (bottom).

EXERCISE 4.1: If the sampling rate is $f_s = 1000$ samples/sec. and the continuous-time signal is $x(t) = \cos(\omega t)$, what value of ω will give a sequence of samples identical to the discrete-time signal shown in the bottom panel of Fig. 4-3?

At first glance, Exercise 4.1 appears to ask a rather simple question because the formula in (4.4) can be solved for ω in terms of f_s and $\hat{\omega}$. Then it is easy to obtain the answer for Exercise 4.1 as $\omega = (0.4\pi)f_s = 400\pi$

rad/sec. But is this the only possible answer? In order to understand this question and find its answer, we need to look more closely at the sampling operation and a concept that we will call ***aliasing***.

4-1.2 The Concept of Aliasing

A simple definition of the word aliasing would involve something like "two names for the same person, or thing." Let us now turn to the question of how aliasing arises in a mathematical treatment of discrete-time signals, specifically for discrete-time sinusoids. The sinusoid $x_1[n] = \cos(0.4\pi n)$ is the mathematical formula for the plot shown in the bottom panel of Fig. 4-3, so it is one "name" that identifies that signal. Now consider another sinusoid, $x_2[n] = \cos(2.4\pi n)$, apparently with a different frequency. To see how a plot of $x_2[n]$ would look, we invoke the simple trigonometric identity $\cos(\theta + 2\pi) = \cos(\theta)$ to obtain

$$\begin{aligned} x_2[n] &= \cos(2.4\pi n) \\ &= \cos(0.4\pi n + 2\pi n) \\ &= \cos(0.4\pi n) \end{aligned}$$

because $2\pi n$ is an integer number of periods of the cosine function. In other words, this phenomenon that we are calling aliasing is solely due to the fact that trigonometric functions are periodic with period 2π.

DEMO: *Sampling Theory Tutorial*

Figure 4-4 shows that the stem plots of these two signals, $x_1[n] = \cos(0.4\pi n)$ and $x_2[n] = \cos(2.4\pi n)$, are identical. Figure 4-4 also shows continuous plots of the signals $\cos(0.4\pi t)$ and $\cos(2.4\pi t)$, which when sampled with $T_s = 1$ would give $x_1[n]$ and $x_2[n]$. It is clear from Fig. 4-4 that the values of these continuous cosine signals are equal at integer values, n. Since $x_2[n] = x_1[n]$ for all integers n, we see that $x_2[n] = \cos(2.4\pi n)$ is another name for the same plot, the same discrete-time signal. It is an *alias*.

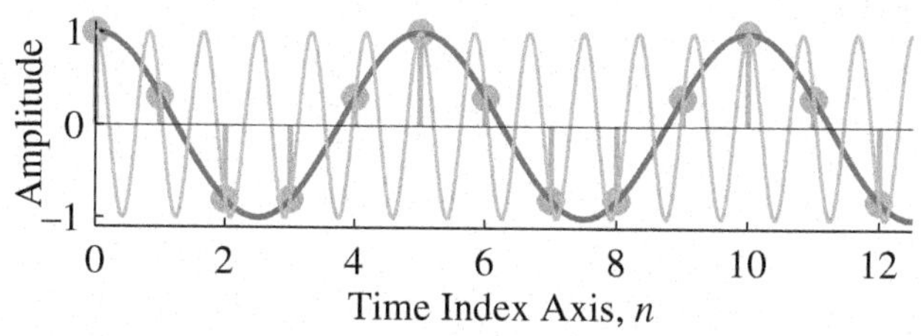

Figure 4-4: Illustration of aliasing: two continuous-time signals drawn through the same samples. The samples belong to two different cosine signals with different frequencies, but the cosine functions have the same values at $n = 0, 1, 2, 3, \ldots$.

The frequency of $x_2[n]$ is $\hat{\omega} = 2.4\pi$, while the frequency of $x_1[n]$ is $\hat{\omega} = 0.4\pi$. When speaking about the frequencies, we say that 2.4π is an *alias* of 0.4π. There are many more frequency aliases as the following problem suggests.

EXERCISE 4.2: Show that $7\cos(8.4\pi n - 0.2\pi)$ is an alias of $7\cos(0.4\pi n - 0.2\pi)$. In addition, find two more frequencies that are aliases of 0.4π rad.

In the previous exercise, it should be easy to see that adding any integer multiple of 2π to 0.4π will give an alias, so the following general formula holds for the frequency aliases:[2]

$$\hat{\omega}_\ell = 0.4\pi + 2\pi\ell \qquad \ell = 0, 1, 2, 3, \ldots \tag{4.5}$$

Since $\hat{\omega} = 0.4\pi$ is the smallest of all the aliases, it is sometimes called the *principal alias*.

However, we are not finished yet. There are other aliases. Another trigonometric identity states that

[2] The integer ℓ could be negative if we allow negative frequencies, but we prefer to avoid that case for the time being.

$\cos(2\pi - \theta) = \cos(\theta)$, so we can generate another alias for $x_1[n] = \cos(0.4\pi n)$ as follows:

$$\begin{aligned} x_3[n] &= \cos(1.6\pi n) \\ &= \cos(2\pi n - 0.4\pi n) \\ &= \cos(0.4\pi n) \end{aligned}$$

The frequency of $x_3[n]$ is 1.6π. A general form for all the alias frequencies of this type would be

$$\hat{\omega}_\ell = -0.4\pi + 2\pi\ell \qquad \ell = 1, 2, 3, \ldots \tag{4.6}$$

For reasons that will become clear later on, these aliases of a negative frequency are called *folded aliases.*

If we examine aliasing for the general discrete-time sinusoid (4.3), an extra complication arises for the folded case as illustrated by the following analysis:

$$\begin{aligned} A\cos((2\pi - \hat{\omega})n - \phi) &= A\cos(2\pi n - \hat{\omega}n - \phi) \\ &= A\cos(-\hat{\omega}n - \phi) \\ &= A\cos(\hat{\omega}n + \phi) \end{aligned} \tag{4.7}$$

Notice that the algebraic sign of the phase angles of the folded aliases must be opposite to the sign of the phase angle of the principal alias.

EXERCISE 4.3: Show that the signal $7\cos(9.6\pi n + 0.2\pi)$ is an alias of the signal $7\cos(0.4\pi n - 0.2\pi)$. It might be instructive to make MATLAB plots of these two signals to verify that the phase must change sign to have identical plots.

In summary, we can write the following general formulas for all aliases of a sinusoid with frequency $\hat{\omega}_0$:

$$\hat{\omega}_0,\ \hat{\omega}_0 + 2\pi\ell,\ 2\pi\ell - \hat{\omega}_0, \qquad (\ell = \text{integer}) \tag{4.8}$$

because the following signals are equal for all n:

$$\begin{aligned} A\cos(\hat{\omega}_0 n + \phi) &= A\cos((\hat{\omega}_0 + 2\pi\ell)n + \phi) \\ &= A\cos((2\pi\ell - \hat{\omega}_0)n - \phi) \end{aligned} \tag{4.9}$$

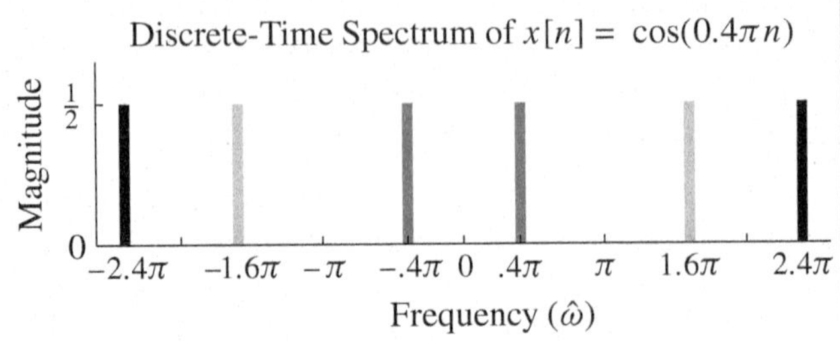

Figure 4-5: Spectrum of a discrete-time sinusoid. The principal alias components are at $\hat{\omega} = \pm 0.4\pi$.

If we were to make a stem plot of all of these signals (with specific numbers for A, $\hat{\omega}$, ϕ and ℓ), we would not be able to tell them apart, as was shown in the example of Fig. 4-4.

4-1.3 Spectrum of a Discrete-Time Signal

We have seen that it is sometimes very useful to represent a continuous-time sinusoidal signal by a spectrum plot. How would we plot the spectrum of a discrete-time signal? Aliasing makes this problematic because a given discrete-time sinusoidal sequence could correspond to an infinite number of different frequencies $\hat{\omega}$. Our approach will be to take this into account by making a plot that explicitly shows that there are many different sinusoids that have the same samples. Figure 4-5 shows that we do this by drawing the spectrum representation of the principal alias along with several more of the other aliases.[3] In Fig. 4-5, the spectrum plot includes a representation of the principal alias $x_1[n] = \cos(0.4\pi n)$, and two of the aliases $x_2[n] = \cos(2.4\pi n)$, and $x_3[n] = \cos(1.6\pi n)$. Recall that from Euler's formula, (3.3) on p. 37, the spectrum of each discrete-time alias signal consists of a positive

[3] We should draw an infinite number of aliases, but we cannot show them all.

frequency component and a corresponding component at negative frequency. Thus, $x_1[n]$ is represented by the components at $\pm 0.4\pi$ and $x_2[n]$ is represented by the black components at $\pm 2.4\pi$, and so forth. Another way of thinking about constructing the plot in Fig. 4-5 is that the spectrum of the principal alias signal consisting of components at $\pm 0.4\pi$ was moved up by 2π and down by 2π, i.e., the spectrum was shifted by integer multiples of 2π.

In the case of the spectrum representation of a continuous-time signal, the assumption was that *all* of the spectrum components were added together to synthesize the continuous-time signal. This is *not* so for the discrete-time case. We show spectrum representations of several of the aliases simply to emphasize the fact that many different frequencies could produce the same time-domain sequence. In the case of a sum of discrete-time sinusoids, such as

$$y_1[n] = 2\cos(0.4\pi n) + \cos(0.6\pi n) \qquad (4.10)$$

we would make a plot like Fig. 4-5 for each of the cosine signals and superimpose them on a single plot. To synthesize the time-domain signal corresponding to a given spectrum representation, we simply need to select one signal from each of the distinct alias sets. For example, (4.10) includes only the principal aliases since 0.4π and 0.6π are both less than π.

EXERCISE 4.4: Make a spectrum plot for the signal of (4.10) similar to Fig. 4-5 showing the principal alias and two other alias frequencies. How would your plot change if the signal was

$$y_2[n] = 2\cos(0.4\pi n) + \cos(2.6\pi n)?$$

4-1.4 The Sampling Theorem

The plots shown in Fig. 4-3 naturally raise the question of how frequently we must sample in order to retain enough information to reconstruct the original continuous-time signal from its samples. The amazingly simple answer is given by the *Shannon sampling theorem*, one of the theoretical pillars of modern digital communications, digital control, and digital signal processing.

> ***Shannon Sampling Theorem***
> A continuous-time signal $x(t)$ with frequencies no higher than f_{max} can be reconstructed exactly from its samples $x[n] = x(nT_s)$, if the samples are taken at a rate $f_s = 1/T_s$ that is greater than $2f_{max}$.

Notice that the sampling theorem involves two issues. First, it talks about reconstructing the signal from its samples, although it never specifies the algorithm for doing the reconstruction. Second, it gives a minimum sampling rate that is dependent on the frequency content of $x(t)$, the continuous-time signal.

The minimum sampling rate of $2f_{max}$ is called the *Nyquist rate*.[4] We can see examples of the sampling theorem in many commercial products. For example, audio CDs use a sampling rate of 44.1 kHz for storing music signals in a digital format. This number is slightly more than two times 20 kHz, which is the generally accepted upper limit for human hearing and perception of musical sounds. In other applications, the Nyquist rate is significant because we are usually motivated to use the lowest possible sampling rate in order to minimize system cost in terms of storage, processing speed per sample, and so on.

The Shannon theorem states that reconstruction of a sinusoid is possible if we have at least two samples

[4] Harry Nyquist and Claude Shannon were researchers at Bell Telephone Laboratories who each made fundamental contributions to the theory of sampling and digital communication during the period from 1920–1950.

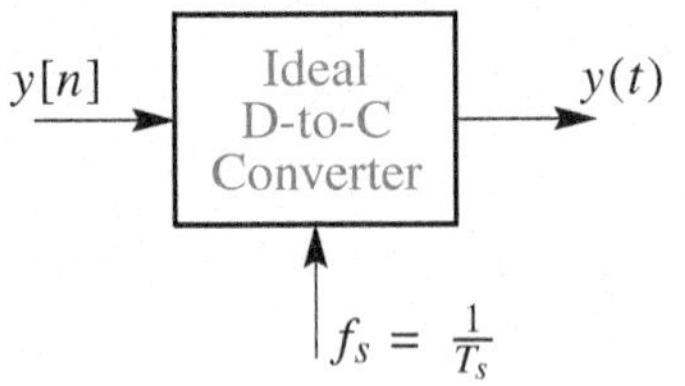

Figure 4-6: Block diagram of the ideal discrete-to-continuous (D-to-C) converter when the sampling rate is $f_s = 1/T_s$. The output signal must satisfy $y(nT_s) = y[n]$.

per period. What happens when we don't sample fast enough? The simple answer is that aliasing occurs. The next sections will delve into this issue by using a spectrum view of the C-to-D conversion process.

4-1.5 Ideal Reconstruction

The sampling theorem suggests that a process exists for reconstructing a continuous-time signal from its samples. This reconstruction process would undo the C-to-D conversion so it is called D-to-C conversion. The ideal discrete-to-continuous (D-to-C) converter is depicted in Fig. 4-6. Since the sampling process of the ideal C-to-D converter is defined by the substitution $t = n/f_s$ as in (4.2), we would expect the same relationship to govern the ideal D-to-C converter; i.e.,

$$y(t) = y[n]\Big|_{n=f_s t} \qquad -\infty < n < \infty \tag{4.11}$$

but this substitution is *only* true when $y(t)$ is a sum of sinusoids. In that special case where $y[n]$ consists of one or more discrete-time sinusoids, such as $y[n] = A\cos(2\pi f_0 n T_s + \phi)$, we can use the substitution in (4.11) to produce $y(t) = A\cos(2\pi f_0 t + \phi)$ at the D-to-C output. This is convenient because we obtain a mathematical formula for $y(t)$, but if we only have a sequence of numbers for $y[n]$ that was obtained by sampling, and we do not know the formula for $y[n]$ or there is no simple formula for the signal, things are not so simple. An actual D-to-A converter involves more than the substitution (4.11), because it must also "fill in" the signal values between the sampling times, $t_n = nT_s$. In Section 4-4 we will see how ***interpolation*** can be used to build an A-to-D converter that *approximates* the behavior of the ideal C-to-D converter. Later on in Chapter 12, we will use Fourier transform theory to show how to build better A-to-D converters by incorporating a lowpass (electronic) filter.

If the ideal C-to-D converter works correctly for a sampled cosine signal, then we can describe its operation as ***frequency scaling***. For $y[n]$ above, the discrete-time frequency is $\hat{\omega} = 2\pi f_0 T_s$, and the continuous-time frequency of $y(t)$ is $\omega = 2\pi f_0$. Thus the relationship appears to be $\omega = \hat{\omega}/T_s = \hat{\omega} f_s$. In fact, we can confirm the same result if we solve the frequency scaling equation (4.4) for ω in terms of $\hat{\omega}$,

$$\omega = \hat{\omega} f_s \tag{4.12}$$

Thus each frequency component in a discrete-time signal could be mapped to a frequency component of the continuous-time output signal. But there is one more issue: the discrete-time signal has aliases—an infinite number of them given by (4.8). Which discrete-time frequency will be used in (4.12)? The selection rule is arbitrary, but the ideal C-to-D converter always selects the lowest possible frequency components (the principal aliases). These frequencies are guaranteed to lie in the range $-\pi < \hat{\omega} \le \pi$, so when converting from $\hat{\omega}$ to analog frequency, the output frequency ***always lies between*** $-\frac{1}{2}f_s$ ***and*** $+\frac{1}{2}f_s$.

DEMO: *Continuous-Discrete Sampling*

In summary, the Shannon sampling theorem guarantees that if $x(t)$ contains no frequencies higher than $f_{\max}$, and if $f_s > 2f_{\max}$, then the output signal $y(t)$ of the ideal

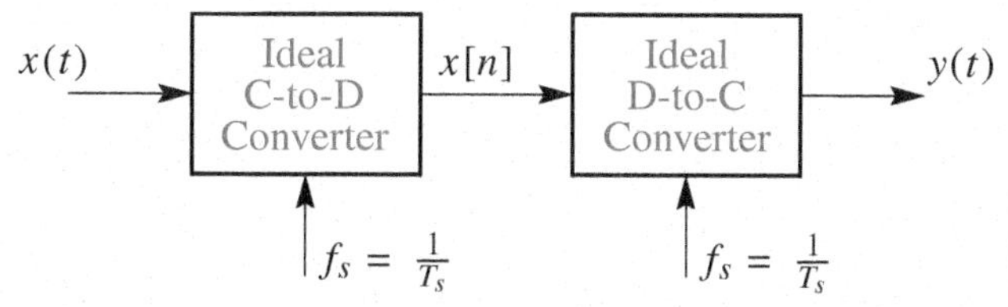

Figure 4-7: Sampling and reconstruction system.

D-to-C converter is *equal* to the signal $x(t)$, the input to the ideal C-to-D converter in Fig. 4-7.

4-2 Spectrum View of Sampling and Reconstruction

We will now present a frequency spectrum explanation of the sampling process. Because the output of the C-to-D converter is a discrete-time signal with an infinite number of aliases, we will use the spectrum diagram that includes *all* of the aliases as discussed in Section 4-1.3. Then the folding and aliasing of frequencies in Fig. 4-7 can be tracked via the movement of spectrum lines in the frequency domain.

DEMO: *Aliasing and Folding*

4-2.1 Spectrum of a Discrete-Time Signal Obtained by Sampling

If a discrete-time sinusoid (4.3) is obtained by sampling, its frequency is given by (4.4). If we include all the aliases as predicted by (4.8), it is necessary to plot an infinite number of spectrum lines to have a thorough representation of the spectrum. Of course, we can only plot a few of them and perhaps indicate that more exist outside our plotting range. Suppose that we start with a continuous-time sinusoid, $x(t) = A\cos(\omega_0 t + \phi)$, whose spectrum consists of *two* spectrum lines at $\pm\omega_0$ with complex amplitudes of $\frac{1}{2}Ae^{\pm j\phi}$.

EXERCISE 4.5: Plot the spectrum of the continuous-time signal $x(t) = \cos(2\pi(100)t + \pi/3)$.

The sampled discrete-time signal

$$\begin{aligned} x[n] = x(n/f_s) &= A\cos((\omega_0/f_s)n + \phi) \\ &= \tfrac{1}{2}Ae^{j\phi}e^{j(\omega_0/f_s)n} + \tfrac{1}{2}Ae^{-j\phi}e^{j(-\omega_0/f_s)n} \end{aligned}$$

also has two spectrum lines at $\hat{\omega} = \pm\omega_0/f_s$, *but it also must contain all the aliases at the following discrete-time frequencies:*

$$\begin{aligned} \hat{\omega} &= \omega_0/f_s + 2\pi\ell \qquad \ell = 0, \pm1, \pm2, \ldots \\ \hat{\omega} &= -\omega_0/f_s + 2\pi\ell \qquad \ell = 0, \pm1, \pm2, \ldots \end{aligned} \tag{4.13}$$

This illustrates the important fact that when a discrete-time sinusoid is derived by sampling, the alias frequencies all are based on the *normalized* value, ω_0/f_s, of the frequency of the continuous-time signal. The alias frequencies are obtained by adding multiples of 2π radians to the normalized value frequency.

The next sections show a set of examples of sampling a continuous-time 100 Hz sinusoid of the form $x(t) = \cos(2\pi(100)t + \pi/3)$. The sampling frequency is varied to show what happens at different sampling rates. The examples in Figs. (4-8–4-11) show the discrete-time spectrum for different cases where f_s is above or below the Nyquist rate. All were constructed with a MATLAB M-file that plots the location of spectrum components involved in the sampling process.

4-2.2 Over-Sampling

In most applications, we try to obey the constraint of the sampling theorem by sampling at a rate higher than twice the highest frequency so that we will avoid the problems of aliasing and folding. This is called *over-sampling*. For example, when we sample the 100 Hz sinusoid, $x(t) = \cos(2\pi(100)t + \pi/3)$ at a sampling rate of $f_s = 500$ samples/sec, we are sampling two and a half times faster than the minimum required by the sampling theorem. The time- and frequency-domain plots are shown in Fig. 4-8. With reference to the C-to-D and D-to-C blocks in Fig. 4-7, the top panel of Fig. 4-8 shows the spectrum of the input signal $x(t)$, while the bottom panel shows the spectrum of $x[n] = \cos(0.4\pi n + \pi/3)$, which is the output of the C-to-D converter. The middle panel shows the time-domain plot of $x(t)$, $x[n]$ and the reconstructed output $y(t)$.

For the continuous-time frequency-domain spectrum representation of the 100 Hz sinusoid (top panel of Fig. 4-8), the frequency axis is measured in hertz. Note that the negative-frequency complex exponential component of the cosine is denoted with a * at the top of its spectrum line. The middle panel shows $x(t)$, $x[n]$, and $y(t)$ together, but keeps the horizontal axis as time in milliseconds. The bottom plot contains the spectrum of the discrete-time signal versus the normalized frequency $\hat{\omega}$. According to the frequency scaling equation (4.4), the input analog frequency of 100 Hz maps to $\hat{\omega} = 2\pi(100)/f_s = 2\pi(100)/500 = 0.4\pi$, so we plot spectrum lines at $\hat{\omega} = \pm 0.4\pi$. Then we also draw all the aliases at

$$\hat{\omega} = 0.4\pi + 2\pi\ell \qquad \ell = 0, \pm 1, \pm 2, \ldots$$
$$\hat{\omega} = -0.4\pi + 2\pi\ell \qquad \ell = 0, \pm 1, \pm 2, \ldots$$

The D-to-C converter transforms the discrete-time spectrum to the continuous-time output spectrum. But there is one complication: the D-to-C converter must select just one pair of spectrum lines from all the possibilities given by (4.13). The selection rule is arbitrary, but in order to be consistent with real D-to-A converter operation, we must assert that the ideal D-to-C converter always selects the lowest possible frequency for each set of aliases. These are what we have called the principal alias frequencies, and in Fig. 4-8 these are the frequency components that fall inside the dashed box shown in the bottom panel; i.e., $|\hat{\omega}| < \pi$. They satisfy the relationship, $-\pi < \hat{\omega} \leq \pi$, and when they are converted from $\hat{\omega}$ to analog frequency, the result is $f = \pm 0.4\pi(f_s/2\pi) = \pm 100$ Hz. Since the frequency components inside the dashed box satisfy $|\hat{\omega}| < \pi$, the spectral lines for the output ***will always lie between*** $-f_s/2$ ***and*** $+f_s/2$.

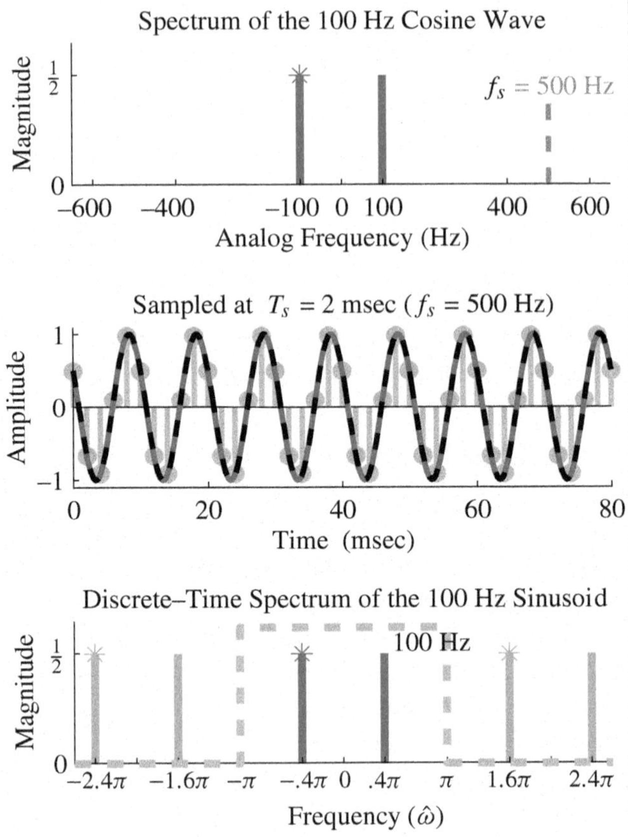

Figure 4-8: Sampling a 100 Hz sinusoid at $f_s = 500$ samples/sec. The spectrum plot (bottom) shows the aliased spectrum components as well as the positive and negative frequency components of the original sinusoid at $\hat{\omega} = \pm 0.4\pi$ rad. The time-domain plot (middle) shows the samples $x[n]$ as gray dots, the original signal $x(t)$ as a continuous orange line, and the reconstructed signal $y(t)$ as a dashed black line.

In summary, for the oversampling case where the original frequency f_0 is less than $f_s/2$, the original waveform will be reconstructed exactly. In the present example, $f_0 = 100$ Hz and $f_s = 500$, so the Nyquist condition of the sampling theorem is satisfied, and the output $y(t)$ equals the input $x(t)$ as shown in the middle panel of Fig. 4-8.

4-2.3 Aliasing Due to Under-Sampling

When $f_s < 2f_0$, the signal is *under-sampled.* For example, if $f_s = 80$ Hz and $f_0 = 100$ Hz, we can show that aliasing distortion occurs. In the top panel of Fig. 4-9, the spectrum of the analog input signal $x(t)$ is shown, along with a dashed line indicating the sampling rate at $f_s = 80$ Hz. The spectrum of the discrete-time signal (in the bottom panel) contains lines at $\hat{\omega} = \pm 2\pi(100)/f_s = \pm 2\pi(100)/80 = \pm 2.5\pi$, as predicted by the frequency scaling equation (4.4). In order to complete the discrete-time spectrum we must also draw all the aliases at

$$\hat{\omega} = 2.5\pi + 2\pi\ell \qquad \ell = 0, \pm 1, \pm 2, \ldots$$
$$\hat{\omega} = -2.5\pi + 2\pi\ell \qquad \ell = 0, \pm 1, \pm 2, \ldots$$

In the middle panel of Fig. 4-9, the 100 Hz sinusoid (solid orange line) is sampled too infrequently to be recognized as the original 100 Hz sinusoid. When we examine the D-to-C process for this case, we use the lowest frequency spectrum lines from the discrete-time spectrum (bottom panel). These are at $\hat{\omega} = \pm 0.5\pi$, so we calculate the output spectrum lines at $f = \pm 0.5\pi(f_s/2\pi) = \pm 80/4 = \pm 20$ Hz. Another way to state this result is to observe that the same samples would have been obtained from a 20 Hz sinusoid. The reconstructed signal is that 20 Hz sinusoid shown as the black dashed line in Fig. 4-9 (middle). Notice that the alias frequency of 20 Hz can be found by subtracting f_s from 100 Hz. Understanding this point is the key to aliasing.

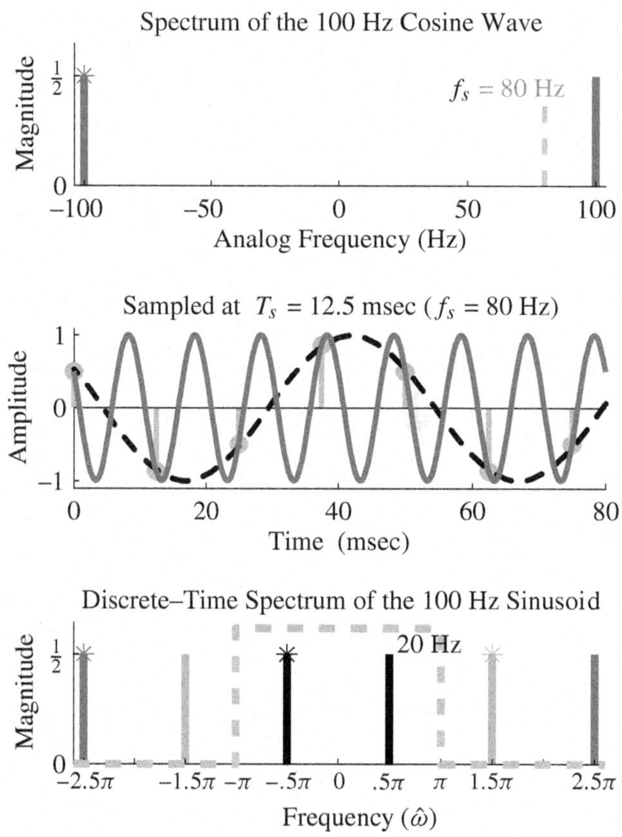

Figure 4-9: Sampling a 100 Hz sinusoid at $f_s = 80$ samples/sec. The time-domain plot (middle) shows the samples $x[n]$ as gray dots, the original sinusoid $x(t)$ as a continuous orange line, and the reconstructed signal $y(t)$ as a dashed black line. In this case, $y(t)$ is a 20 Hz sinusoid that passes through the same sample points.

The aliasing of sinusoidal components can have some dramatic effects. Figure 4-10 shows the case where the sampling rate and the frequency of the sinusoid are the same. Clearly, what happens is that the samples are always taken at the same place on the waveform, so we get the equivalent of sampling a constant (DC), which is the same as a sinusoid with zero frequency. We can

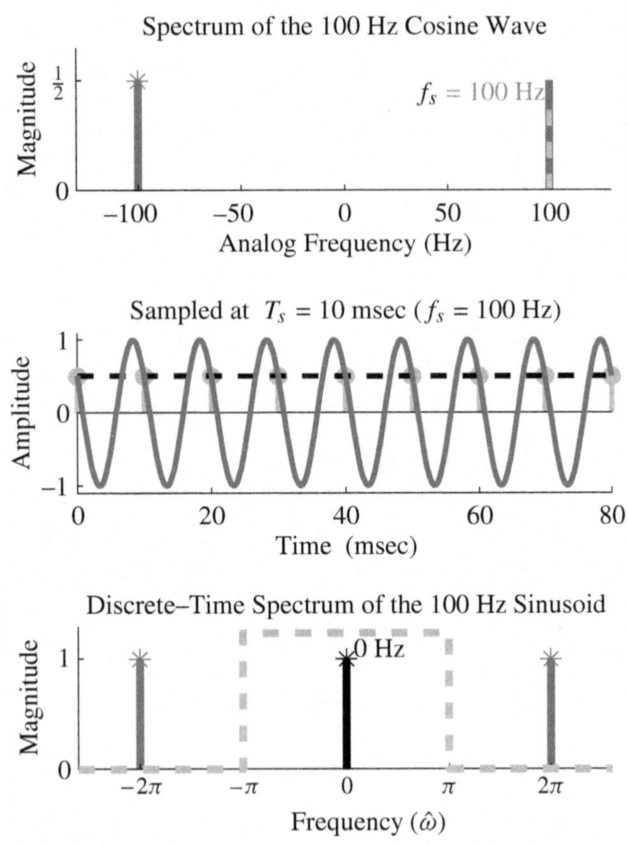

Figure 4-10: Sampling a 100 Hz sinusoid at $f_s = 100$ samples/sec. The time-domain plot (middle) shows the samples $x[n]$ as gray dots, the original sinusoid $x(t)$ as a continuous orange line, and the reconstructed signal $y(t)$ as a dashed black line. The discrete-time spectrum (bottom) contains only one line at $\hat{\omega} = 0$.

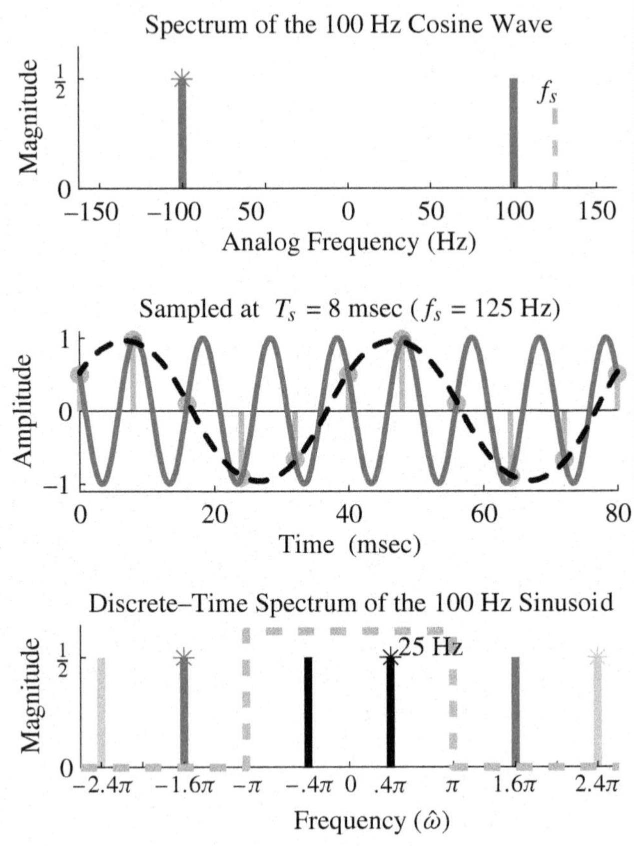

Figure 4-11: Sampling a 100 Hz sinusoid at $f_s = 125$ samples/sec. The time-domain plot (middle) shows the samples $x[n]$ as gray dots, the original sinusoid $x(t)$ as a continuous orange line, and the reconstructed signal $y(t)$ as a dashed black line. In this case, $y(t)$ is a 25 Hz sinusoid that passes through the same sample points.

justify this result in the frequency domain by noting that the discrete-time signal must contain spectrum lines at $\hat{\omega} = \pm 2\pi(100/f_s) = \pm 2\pi$, and also at the aliases separated by $2\pi\ell$. Thus, one of the aliases lands at $\hat{\omega} = 0$, and that is the one reconstructed by the D-to-C converter.

4-2.4 Folding Due to Under-Sampling

Figure 4-11 shows the case where under-sampling leads to folding; here the sampling rate is $f_s = 125$ samples/sec. Once again, the top panel shows the spectrum of the continuous-time signal with spectrum lines at ±100 Hz. The discrete-time spectrum (bottom panel) is constructed by mapping ± 100 Hz to the two spectrum lines at $\hat{\omega} = \pm 2\pi(100/f_s) = \pm 1.6\pi$, and then including all the aliases to get lines at

$$\hat{\omega} = 1.6\pi + 2\pi\ell \qquad \ell = 0, \pm 1, \pm 2, \ldots$$
$$\hat{\omega} = -1.6\pi + 2\pi\ell \qquad \ell = 0, \pm 1, \pm 2, \ldots$$

In this case, an interesting thing happens. The two frequency components between $\pm\pi$ are $\hat{\omega} = \pm 0.4\pi$, but the one at $\hat{\omega} = +0.4\pi$ is an alias of -1.6π. This is an example of folding. The analog frequency of the reconstructed output will be $f = 0.4\pi(f_s/2\pi) = f_s/5 = 25$ Hz. An additional fact about folding is that the sign of the phase of the signal will be changed. If the original 100-Hz sinusoid had a phase of $\phi = +\pi/3$, then the phase of the component at $\hat{\omega} = -1.6\pi$ would be $-\pi/3$, and it follows that the phase of the aliased component at $\hat{\omega} = +0.4\pi$ would also be $-\pi/3$. After reconstruction, the phase of $y(t)$ would be $-\pi/3$. It is possible to observe this "phase-reversal" in the middle panel of Fig. 4-11. The input signal (solid orange line) is going down at $t = 0$, while the reconstructed output (black dashed line) is going up. This means that when we sample a 100 Hz sinusoid at a sampling rate of 125 samples/sec, we get the same samples that we would have gotten by sampling a 25 Hz sinusoid, but with opposite phase.

4-2.5 Maximum Reconstructed Frequency

The preceding examples have one thing in common: the output frequency is always less than $\frac{1}{2}f_s$. For a sampled sinusoid, the ideal D-to-C converter picks the alias frequency closest to $\hat{\omega} = 0$ and maps it to the output analog frequency via $f = \hat{\omega}(f_s/2\pi)$. Since the principal alias is guaranteed to lie between $-\pi$ and $+\pi$, the output frequency will always lie between $-\frac{1}{2}f_s$ and $+\frac{1}{2}f_s$.

LAB: *#3 Chirp Synthesis from Chapter 3*

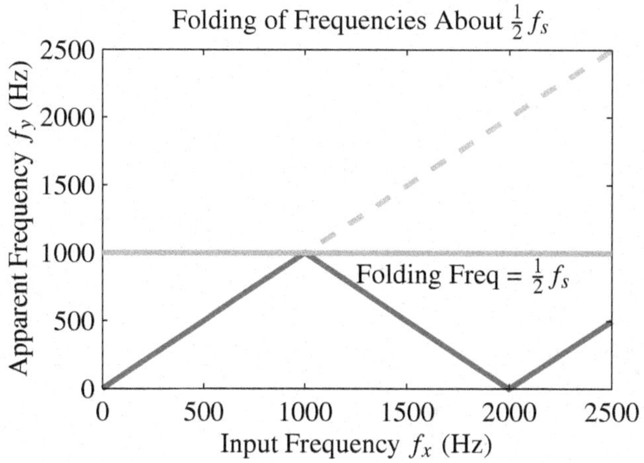

Figure 4-12: Folding of a sinusoid sampled at $f_s = 2000$ samples/sec. The apparent frequency is the lowest frequency of a sinusoid that has exactly the same samples as the input sinusoid.

One striking demonstration of this fact can be made by implementing the system in Fig. 4-7 and using a linear FM chirp signal as the input, and then listening to the reconstructed output signal. Suppose that the instantaneous frequency of the input chirp increases according to the formula $f_i(t) = f_s t$ Hz; i.e., $x(t) = \cos(\pi f_s t^2)$. After sampling, we obtain

$$x[n] = \cos(\pi n^2/f_s)$$

Once $y(t)$ is reconstructed from $x[n]$, what would you hear? It turns out that the output frequency goes up and down as shown in Fig. 4-12.

This happens because we know that the output cannot have a frequency higher than $\frac{1}{2}f_s$, even though the input frequency is continually increasing. If we concentrate on a specific input frequency, we can map it to its normalized discrete frequency, determine all of the aliases, and then figure out the reconstructed frequency from the alias closest to $\hat{\omega} = 0$. There is no easy formula, so we just

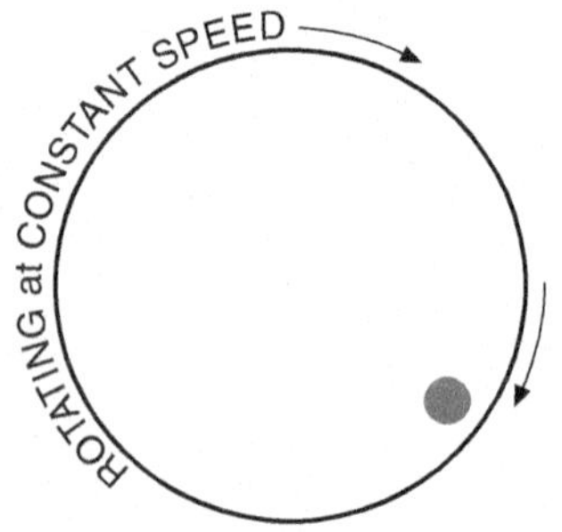

Figure 4-13: The disk attached to the shaft of a motor rotates clockwise at a constant speed.

consider a few separate cases. First of all, when the input frequency goes from 0 to $\frac{1}{2}f_s$, $\hat{\omega}$ will increase from 0 to π and the aliases will not need to be considered. The output frequency will be equal to the input frequency. Now consider the input frequency increasing from $\frac{1}{2}f_s$ to f_s. The corresponding frequency for $x[n]$ increases from $\hat{\omega} = \pi$ to $\hat{\omega} = 2\pi$, and its negative frequency companion goes from $\hat{\omega} = -\pi$ to $\hat{\omega} = -2\pi$. The principal alias of the negative frequency component goes from $\hat{\omega} = +\pi$ to $\hat{\omega} = 0$. The reconstructed output signal will, therefore, have a frequency going from $f = \frac{1}{2}f_s$ to $f = 0$.

Figure 4-12 shows a plot of the output frequency f_y versus the input frequency f_x. As f_x goes from $\frac{1}{2}f_s$ to f_s, the output frequency f_y decreases from $\frac{1}{2}f_s$ to 0. The terminology *folded* frequency comes from the fact that the input and output frequencies are mirror images with respect to $\frac{1}{2}f_s$, and would lie on top of one another if the graph were folded about the $\frac{1}{2}f_s$ line in Fig. 4-12.

4-3 Strobe Demonstration

One effective means for demonstrating aliasing is to use a strobe light to illuminate a spinning object. In fact, this process is used routinely to set the timing of automobile engines and gives a practical example where aliasing is a desirable effect. In our case, we use a disk attached to the shaft of an electric motor that rotates at a constant angular speed (see Fig. 4-13). On this white disk is painted a spot that is easily seen whenever the strobe light flashes. In our particular case, the rotation speed of the motor is approximately 750 rpm, and the flash rate of the strobe light is variable over a wide range.

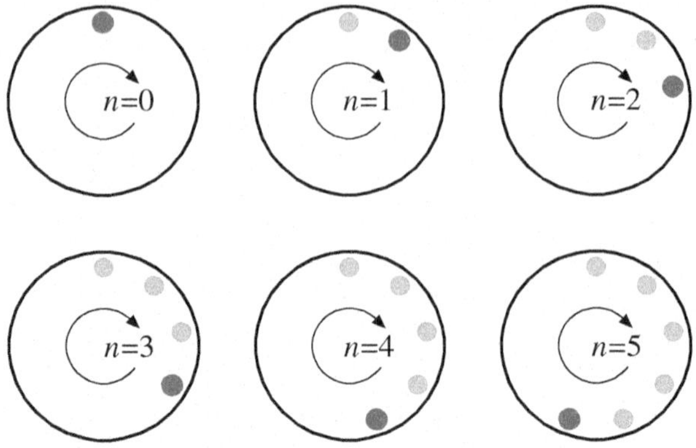

Figure 4-14: Six successive positions of the spot for a very high flashing rate. Light spots indicate previous locations of the dark spot. The disk is rotating clockwise, and the spot appears to move in the same direction. Angular change is $-40°$ per flash.

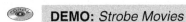

DEMO: *Strobe Movies*

Suppose that the flashing rate is very high, say nine times the rotation rate, i.e., $9 \times 750 = 6750$ rpm. In this case, the disk will not rotate very far between flashes. In fact, for the $9\times$ case, it will move only $360°/9 = 40°$ per flash, as in Fig. 4-14. Since the movement will be clockwise, the angular change from one flash to the next is $-40°$.

If the flash rate of the strobe is set equal to 750 flashes/min, i.e., the rotation rate of the disk, then the spot will appear to stand still. This is aliasing because the spot makes exactly one revolution between flashes and therefore is always at the same position when illuminated by the strobe. It is exactly the same situation illustrated in Fig. 4-10 where the frequency of the sinusoid aliased to zero. This is not the only flash rate for which the spot will appear to stand still. In fact, a slower flash rate that permits two or three or any integer number of complete revolutions between flashes will create the same effect. In our case of a 750-rpm motor, flash rates of 375, 250, $187\frac{1}{2}$, 150, and 125 flashes/min will also work.

DEMO: *Synthetic Strobe Movies*

By using flashing rates that are close to these numbers, we can make the spot move slowly, and we can also control its direction of motion (clockwise or counterclockwise). For example, if we set the strobe for a flashing rate that is just slightly higher than the rotation rate, we will observe another aliasing effect. Suppose that the flashing rate is 806 flashes/min; then the disk rotates slightly less than one full revolution between flashes. We can calculate the movement if we recognize that one complete revolution takes $1/750$ min. Therefore,

$$\begin{aligned}\Delta\theta &= -360^\circ \times \frac{1/806}{1/750} \\ &= -360^\circ \times \frac{750}{806} = -335^\circ = +25^\circ\end{aligned}$$

Once again, the minus sign indicates rotation in a clockwise direction, but since the angular change is almost -360°, we would observe a small positive angular change instead, and the spot would appear to move in the counterclockwise direction, as shown in Fig. 4-15. The fact that one can distinguish between clockwise rotation and counterclockwise rotation is equivalent to saying that positive and negative frequencies have separate physical meanings.

In order to analyze the strobe experiment mathematically, we need a notation for the motion of the spot as a function of time. The spot moves in an x–y coordinate system, so a succinct notation is given by a complex

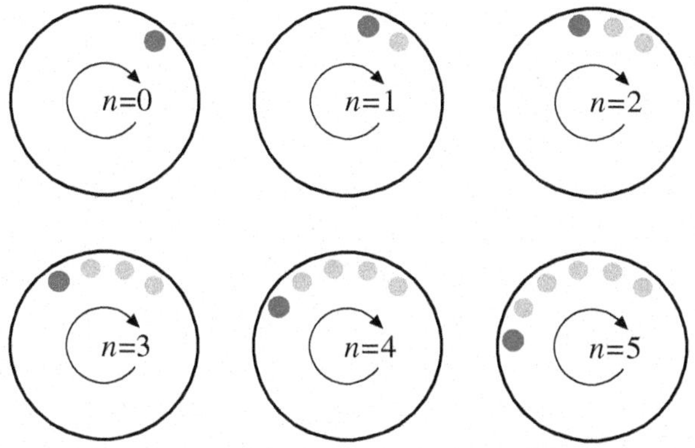

Figure 4-15: Six successive positions of the spot for a flashing rate that aliases the spot motion. Light spots indicate previous locations of the dark spot. The disk is rotating clockwise, but the spot appears to move counterclockwise. Angular change is $+25^\circ$ per flash.

number whose real part is x and whose imaginary part is y. The position of the spot is

$$p(t) = x(t) + jy(t)$$

Furthermore, since the motion of the spot is on a circle of radius r, the correct formula for $p(t)$ is a complex exponential with constant frequency.

$$p(t) = re^{-j(2\pi f_m t - \phi)} \tag{4.14}$$

The minus sign in the exponent indicates clockwise rotation, and the initial phase ϕ specifies the location of the spot at $t = 0$ (see Fig. 4-16). The frequency of the motor rotation f_m is constant, as is the radius r. It will be convenient to set $r = 1$ in what follows so that the formulas will be less cluttered.

The effect of the strobe light is to sample $p(t)$ at a fixed rate given by the flashing rate f_s. Thus the position of the spot at the nth flash can be expressed as the discrete-time signal

$$p[n] = p(t)\Big|_{t=nT_s} = p(nT_s) = p(n/f_s)$$

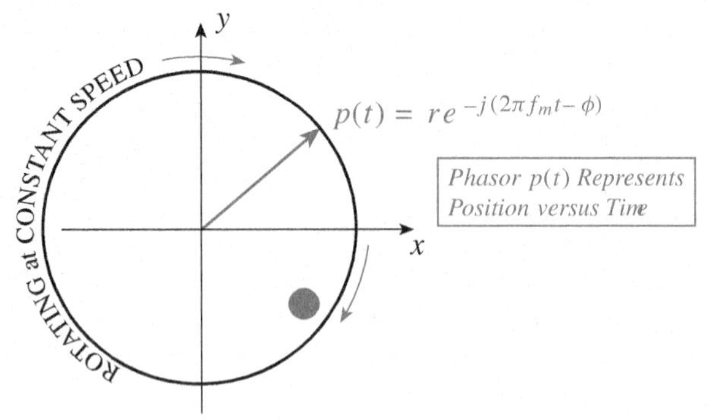

Figure 4-16: The position of the spot on the disk can be expressed as a rotating phasor $p(t) = x(t) + jy(t)$ versus time. The frequency f_m is the rotation rate of the motor in rpm.

Substituting into the complex exponential (4.14) we get

$$p[n] = re^{-j(2\pi(f_m/f_s)n-\phi)} \tag{4.15}$$

If the constraint of the sampling theorem is met (i.e., $f_s > 2|f_m|$), then there will be no aliasing in the experiment. In fact, the angular change from one sample time to the next will be between $-180°$ and $0°$, so the spot will appear to rotate clockwise.

The more interesting cases occur when the flashing rate drops below $2|f_m|$. Then the disk may make one or more revolutions between flashes, which introduces the aliasing phenomenon. Using the sampled position formula (4.15), we can solve the following type of problem: Find *all* possible flashing rates so that the spot will move counterclockwise at a rate of 25° per flash, which is equivalent to one revolution every 14.4 flashes. Assume a constant motor speed of f_m rpm. One twist to this problem is that the two rotation rates are specified in different units.

A systematic approach is possible if we use the property $e^{j2\pi\ell} = 1$, whenever ℓ is an integer. The desired rotation of the spot can be expressed as

$$d[n] = re^{+j(2\pi(25/360)n+\psi)}$$

where the factor $2\pi(25/360)$ equals 25° converted to radians. The initial phase ψ is set equal to ϕ in $p[n]$, so that $d[0] = p[0]$. Then we can equate $p[n]$ and $d[n]$, but we throw in the factor $e^{j2\pi\ell n}$, which is just multiplying by one

$$p[n] = d[n]e^{j2\pi\ell n}$$

Now with the following analysis we can generate an equation that can be solved for the flashing rates:

$$re^{-j(2\pi(f_m/f_s)n-\phi)} = re^{+j(2\pi(25/360)n+\phi)}\, e^{j2\pi\ell n}$$

$$\Rightarrow \quad -\left(2\pi\frac{f_m}{f_s}n - \phi\right) = +\left(2\pi\frac{25}{360}n + \phi\right) + 2\pi\ell n$$

$$-\frac{f_m}{f_s} = \frac{25}{360} + \ell$$

$$\Rightarrow \quad -f_m = f_s\left(\frac{25}{360} + \ell\right)$$

So, finally we can solve for the flashing rate

$$f_s = \frac{-f_m}{(5/72) + \ell} \tag{4.16}$$

where ℓ is any integer. Since we want positive values for the flashing rate, and since there is a minus sign associated with the clockwise rotation rate of the motor $(-f_m)$, we choose $\ell = -1, -2, -3, \ldots$ to generate the different solutions. For example, when the motor rpm is 750, the following flashing rates (in flashes/min) will give the desired spot movement:

ℓ	-1	-2	-3	-4
f_s	805.97	388.49	255.92	190.81

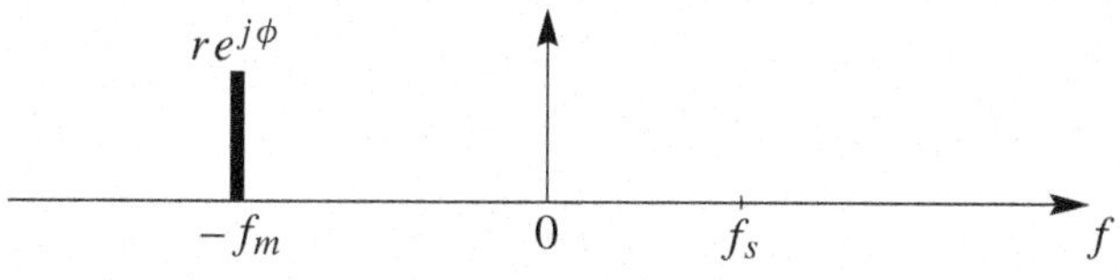

Figure 4-17: Analog spectrum representing the disk spinning clockwise (negative frequency) at f_m rpm. The units for f are cycles/min.

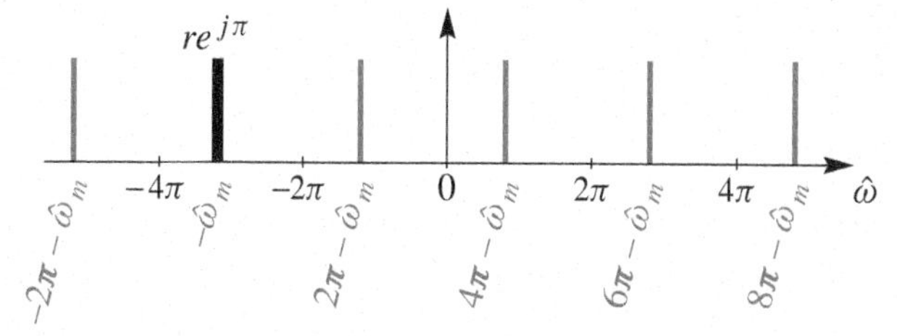

Figure 4-18: Digital spectrum representing the strobed disk spinning clockwise at f_m rpm, but sampled at f_s flashes per minute. The horizontal axis is normalized frequency: $\hat{\omega} = 2\pi f/f_s$. The normalized motor frequency $\hat{\omega}_m = 2\pi(f_m/f_s)$ also appears at the aliases $\hat{\omega}_\ell = 2\pi\ell - \hat{\omega}_m$, where ℓ is an integer.

EXERCISE 4.6: To test your knowledge of this concept, try solving for *all* possible flashing rates so that the spot will move clockwise at a rate of one revolution every 9 flashes. Assume the same motor rotation speed of 750 rpm clockwise. What is the maximum flashing rate for this case?

4-3.1 Spectrum Interpretation

The strobe demo involves the sampling of complex exponentials, so we can present the results in the form of a spectrum diagram rather than using equations, as in the foregoing section. The rotating disk has an analog spectrum given by a single frequency component at $f = -f_m$ cycles/min (see Fig. 4-17).

When the strobe light is applied at a rate of f_s flashes per minute, the resulting spectrum of the discrete-time signal $p[n]$ will contain an infinite number of frequency lines at the frequencies

$$\hat{\omega}_\ell = 2\pi\frac{-f_m}{f_s} + 2\pi\ell = -\hat{\omega}_m + 2\pi\ell \tag{4.17}$$

for $\ell = 0, \pm1, \pm2, \pm3, \ldots$. Equation (4.17) tells us that there are two steps needed to create the spectrum (Fig. 4-18) of the discrete-time signal representing the sampled motion:

(a) Compute a *normalized motor frequency* $\hat{\omega}_m$ by dividing by f_s. Then plot a spectrum line at $\hat{\omega} = -\hat{\omega}_m$. The minus sign represents the clockwise rotation.

(b) Repeat the normalized spectral line for the motor by shifting it by all integer multiples of 2π. This accounts for the aliases of the discrete-time exponential, $p[n]$.

How do we use Fig. 4-18 to explain what we see in the strobe-light experiment? Fig. 4-18 contains an infinite number of spectrum lines, so it offers an infinite number of possibilities for what we will observe. Our visual system, however, must pick one and it is reasonable that we would pick the slowest one. In effect, our human visual system is acting as a D-to-C converter. Thus, in the discrete-time spectrum, only the frequency component closest to $\hat{\omega} = 0$ counts in D-to-C reconstruction, so the strobed signal $p[n]$ appears to be rotating at that lowest normalized frequency. However, one last conversion must be made to give the perceived analog rotation rate in rpm. The discrete-time frequency ($\hat{\omega}$) must be converted back to analog frequency (f). In Fig. 4-18, the alias for $\ell = 2$ is the one closest to zero frequency, so the corresponding analog frequency is

$$f_{\text{spot}} = \frac{1}{2\pi}(\hat{\omega}_2)f_s = \frac{1}{2\pi}(4\pi - \hat{\omega}_m)f_s = 2f_s - f_m$$

This may seem like a roundabout way to say that the observed rotation rate of the spot differs from $-f_m$ by an integer multiple of the sampling rate, but it does provide a systematic method to draw the graphical picture of the relative frequency locations. Finally, the spectrum picture makes it easy to identify the lowest discrete-time frequency as the one that is "reconstructed." We will say more about why this is true in Section 4-4.

The case where the sampling rate is variable and f_m is fixed is a bit harder to solve, but this is the actual situation for our strobed disk experiment. Nonetheless, a graphical approach is still possible because the desired spot frequency defines a line in the discrete-time spectrum, say $\hat{\omega}_d$. This line is the one closest to the origin, so we must add an integer multiple $2\pi\ell$ to $\hat{\omega}_d$ to match the normalized motor rotation frequency.

$$\hat{\omega}_d + 2\pi\ell = \hat{\omega}_m = \frac{-2\pi f_m}{f_s}$$

This equation can be solved for the flashing rate f_s, but the final answer depends on the integer ℓ, which predicts, by the way, that there are many answers for f_s, as we already saw in (4.16). The result is

$$f_s = \frac{-2\pi f_m}{\hat{\omega}_d + 2\pi\ell}$$

4-4 Discrete-to-Continuous Conversion

The purpose of the ideal discrete-to-continuous (D-to-C) converter is to interpolate a smooth continuous-time function through the input samples $y[n]$. Thus, in the special case when $y[n] = A\cos(2\pi f_0 n T_s + \phi)$, and if $f_0 < \frac{1}{2}f_s$, then according to the sampling theorem, the converter should produce

$$y(t) = A\cos(2\pi f_0 t + \phi). \tag{4.18}$$

For sampled sinusoidal signals *only*, the ideal D-to-C converter in effect replaces n by $f_s t$. On the other hand, if $f_0 > \frac{1}{2}f_s$, then we know that aliasing or folding distortion has occurred, and the ideal D-to-C converter will reconstruct a cosine wave with frequency equal to the alias frequency that is less than $\frac{1}{2}f_s$.

4-4.1 Interpolation with Pulses

How does the D-to-C converter work? In this section, we explain how the D-to-C converter does interpolation, and then describe a practical system that is nearly the same as the ideal D-to-C converter. These actual hardware systems, called digital-to-analog (D-to-A) converters, approximate the behavior of the ideal D-to-C system.

A general formula that describes a broad class of D-to-C converters is given by the equation

$$y(t) = \sum_{n=-\infty}^{\infty} y[n]p(t - nT_s), \tag{4.19}$$

where $p(t)$ is the characteristic pulse shape of the converter. Equation (4.19) states that the output signal is produced by adding together many pulses, each shifted in time. In other words, at each sample time $t_n = nT_s$, a pulse $p(t - nT_s)$ is emitted with an amplitude proportional to the sample value $y[n]$ corresponding to that time instant.[5] Note that all the pulses have a common waveshape specified by $p(t)$. If the pulse has duration greater than or equal to T_s, then the gaps between samples will be filled by adding overlapped pulses.

[5] A *pulse* is a continuous-time waveform that is concentrated in time. Typically, a pulse will be nonzero only over a finite interval of time.

Obviously, the important issue is the choice of the pulse waveform $p(t)$. Unfortunately, we do not yet have the mathematical tools required to derive the optimal pulse shape required for exact reconstruction of a waveform $y(t)$ from its samples $y[n] = y(nT_s)$ as is predicted to be possible in the Shannon sampling theorem.

DEMO: *Reconstruction Movies*

This optimal pulse shape can be constructed during a mathematical proof of the sampling theorem. Nonetheless, we will demonstrate the plausibility of (4.19) by considering some simple (suboptimal) examples. Figure 4-19 shows four possible pulse waveforms for D-to-C conversion when $f_s = 200$ Hz.

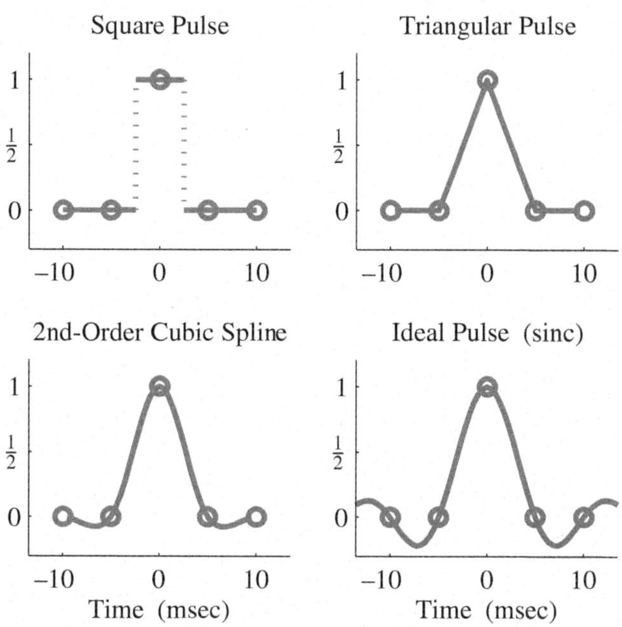

Figure 4-19: Four different pulses for D-to-C conversion. The sampling period is $T_s = 0.005$, i.e., $f_s = 200$ Hz. Note that the duration of the pulses is one, two, or four times T_s, or infinite for the ideal pulse.

4-4.2 Zero-Order Hold Interpolation

The simplest pulse shape that we might propose is a symmetric square pulse of the form

$$p(t) = \begin{cases} 1 & -\frac{1}{2}T_s < t \le \frac{1}{2}T_s \\ 0 & \text{otherwise} \end{cases} \tag{4.20}$$

This pulse is plotted in the upper left panel of Fig. 4-19.

From Fig. 4-19, we see that the total width of the square pulse is $T_s = 5$ ms and that its amplitude is 1. Therefore, each term $y[n]p(t - nT_s)$ in the sum (4.19) will create a flat region of amplitude $y[n]$ centered on $t = nT_s$. This is shown in the top part of Fig. 4-20, which shows the original 83 Hz cosine wave (solid gray line), its samples taken at a sampling rate of 200 samples/sec, and the individual shifted and scaled pulses, $y[n]p(t - nT_s)$ (dashed line). The sum of all the shifted and scaled pulses will be a "stairstep" waveform, as shown in the bottom panel of Fig. 4-20, where the gray curve is the original cosine wave $x(t)$, and the heavy orange curve shows the reconstructed waveform using the square pulse.[6]

DEMO: *Reconstruction Movies*

The space between samples has indeed been filled with a continuous-time waveform; however, it is clear from the lower part of Fig. 4-20 that the reconstructed waveform for the square pulse is a poor approximation of the original cosine wave. Thus, using a square pulse in (4.19) is D-to-C conversion, but not *ideal D-to-C conversion*. Even so, this is a useful model, since many physically realizable digital-to-analog (D-to-A) converters produce outputs that look exactly like this!

[6]Since a constant is a polynomial of zero order, and since the effect of the flat pulse is to "hold" or replicate each sample for T_s sec, the use of a flat pulse is called a ***zero-order hold reconstruction***.

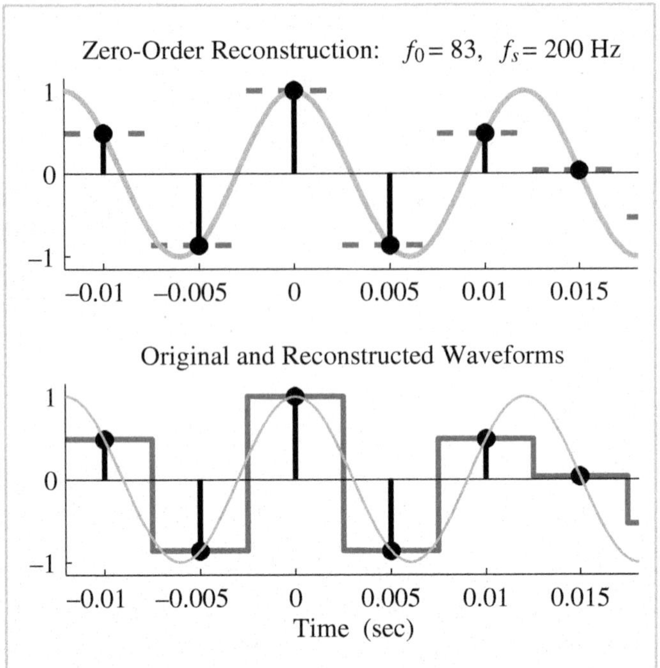

Figure 4-20: D-to-C conversion using a square pulse. Flat regions are characteristic of the zero-order hold.

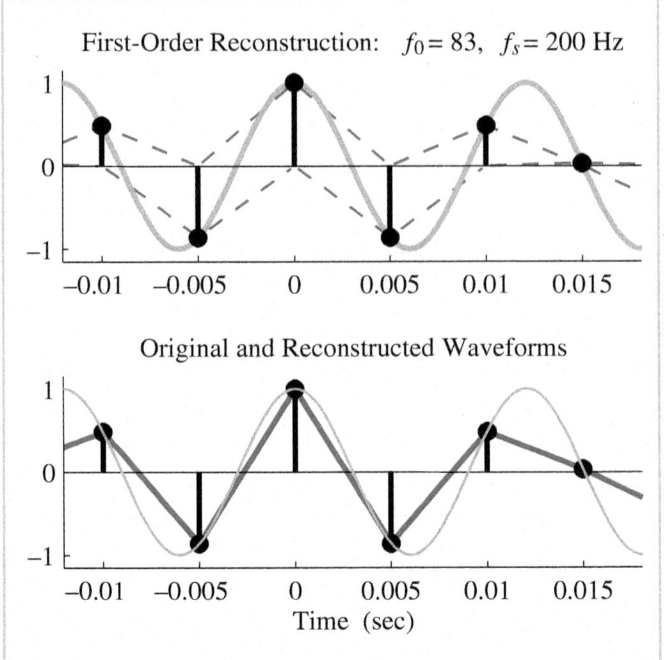

Figure 4-21: D-to-C conversion using a triangular pulse, or linear interpolation

4-4.3 Linear Interpolation

The triangular pulse plotted in the upper right panel of Fig. 4-19 is defined as a pulse consisting of the first-order polynomial (straight line) segments

$$p(t) = \begin{cases} 1 - |t|/T_s & -T_s \le t \le T_s \\ 0 & \text{otherwise} \end{cases} \tag{4.21}$$

Figure 4-21 (top) again shows the original 83-Hz cosine wave (solid gray curve) and its samples, together with the scaled pulses $y[n]p(t - nT_s)$ (dashed orange) for the triangular pulse shape. In this case, the output $y(t)$ of the D-to-C converter at any time t is the sum of the scaled pulses that overlap at that time instant. Since the duration of the pulse is $2T_s$, and they are shifted by multiples of T_s, no more than two pulses can overlap at any given time. The resulting output is shown as the solid orange curve in the bottom panel of Fig. 4-21. Note that the result is that the samples are connected by straight lines. Also note that the values at $t = nT_s$ are exactly correct. This is because the triangular pulses are zero at $\pm T_s$, and only one scaled pulse (with value $y[n]$ at $t = nT_s$) contributes to the value at $t = nT_s$. In this case, we say that the continuous-time waveform has been obtained by *linear interpolation* between the samples. This is a smoother and better approximation of the original waveform (thin gray curve), but there is still significant reconstruction error for this signal.

4-4.4 Cubic Spline Interpolation

A third pulse shape is shown in the lower left panel of Fig. 4-19. This pulse consists of four cubic spline

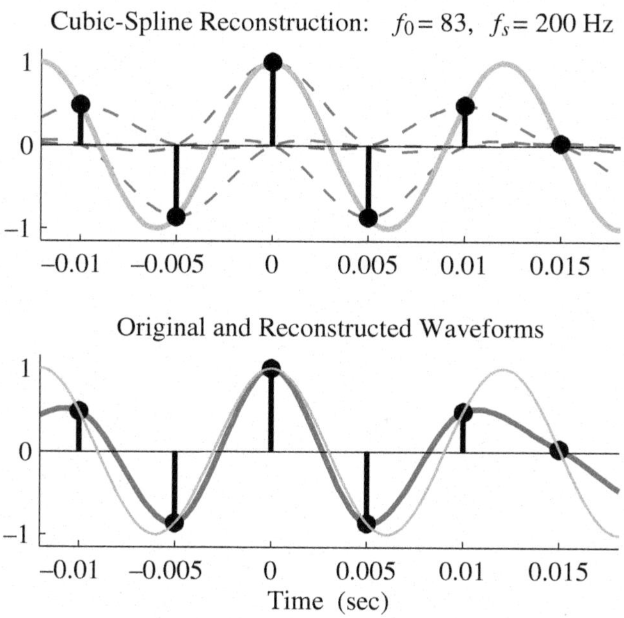

Figure 4-22: D-to-C conversion using a cubic-spline pulse.

(third-order polynomial) segments. Note that it has a duration that is twice that of the triangular pulse and four times that of the square pulse. Also, note that this pulse has zeros at the key locations:

$$p(t) = 0 \qquad \text{for } t = \pm T_s,\ \pm 2T_s$$

and its derivative is smooth at the sample locations. The reconstruction with the cubic-spline pulses is shown in Fig. 4-22. The top panel of Fig. 4-22 shows the original 83-Hz waveform (solid gray), its samples ($f_s = 200$ samples/sec), and the shifted and scaled pulses $y[n]p(t - nT_s)$ (dashed orange). Now note that, for values of t between two adjacent sample times, four pulses overlap and must be added together in the sum (4.19). This means that the reconstructed signal at a particular time instant, which is the sum of these overlapping pulses, depends on the two samples preceding and the two samples following that time instant. The lower panel of Fig. 4-22 shows the original waveform (light gray), the samples, and the output of the D-to-C converter with "cubic-spline pulse" interpolation (solid orange curve). Now we see that the approximation is getting smoother and better, but it is still far from perfect. We will see that this is because the sampling is only 2.4 times the highest frequency.

4-4.5 Over-Sampling Aids Interpolation

From the previous three examples, it seems that one way to make a smooth reconstruction of a waveform such as a sinusoid is to use a pulse $p(t)$ that is smooth and has a long duration. Then the interpolation formula will involve several neighboring samples in the computation of each output value $y(t)$. However, the sampling rate is another important factor. If the original waveform does not vary much over the duration of $p(t)$, then we will also obtain a good reconstruction. One way to ensure this is by over-sampling, i.e., using a sampling rate that is much greater than the frequency of the cosine wave. This is illustrated in Figs. 4-23, 4-24, and 4-25, where the frequency of the cosine wave is still $f_0 = 83$ Hz, but the sampling frequency has been raised to $f_s = 500$ samples/sec. Now the reconstruction pulses are the same shape as in Fig. 4-19, but they are much shorter, since $T_s = 2$ msec instead of 5 msec. The signal changes much less over the duration of a single pulse, so the waveform appears "smoother" and is much easier to reconstruct accurately using only a few samples. Note that even for the case of the square pulse (Fig. 4-23) the reconstruction is better, but still discontinuous; the triangular pulse (Fig. 4-24) gives an excellent approximation; and the cubic-spline pulse gives a reconstruction that is indistinguishable from the original signal on the plotting scale of Fig. 4-25.

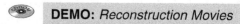

DEMO: *Reconstruction Movies*

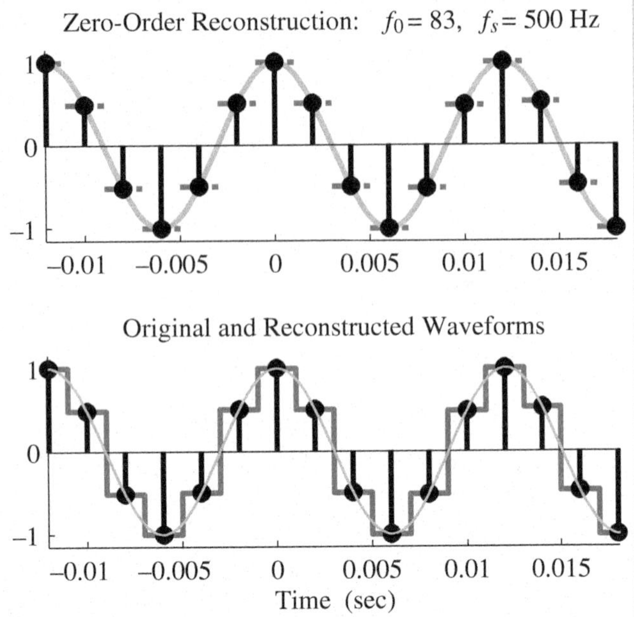

Figure 4-23: D-to-C conversion using a square pulse. The original 83-Hz sinusoid was over-sampled at $f_s = 500$ samples/sec.

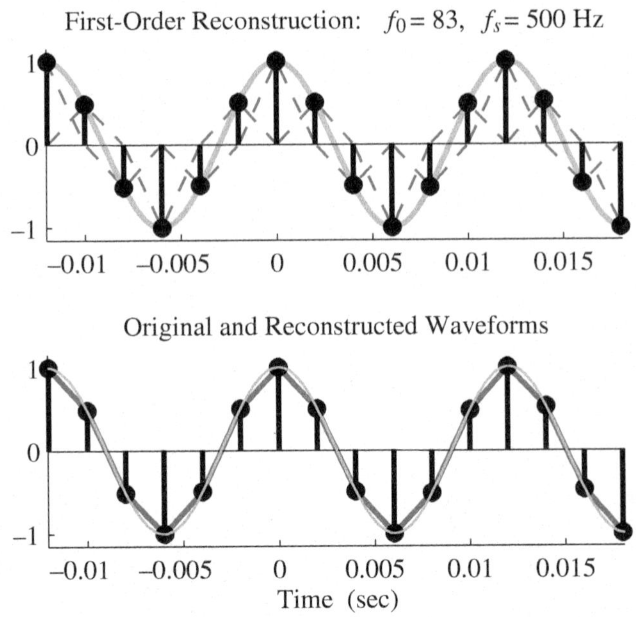

Figure 4-24: D-to-C conversion using a triangular pulse at $f_s = 500$ samples/sec.

Figures 4-23, 4-24, and 4-25 show that over-sampling can make it easier to reconstruct a waveform from its samples. Indeed, this is why audio CD players have over-sampled D-to-A converters! In the CD case, 4× or 2× over-sampling is used to increase the sampling rate before sending the samples to the D-to-A converter. This makes it possible to use a simpler (and therefore less expensive) D-to-A converter to reconstruct an accurate output from a CD player.

4-4.6 Ideal Bandlimited Interpolation

So far in this section, we have demonstrated the basic principles of discrete-to-continuous conversion. We have shown that this process can be approximated very well by a sum of pulses (4.19). One question remains: What is the pulse shape that gives "ideal D-to-C conversion"? In Chapter 12, we will show that it is given by the following equation:

$$p(t) = \frac{\sin \frac{\pi}{T_s} t}{\frac{\pi}{T_s} t} \qquad \text{for } -\infty < t < \infty \tag{4.22}$$

The infinite length of this "pulse" implies that to reconstruct a signal at time t exactly from its samples requires *all* the samples, not just those around that time. The lower right part of Fig. 4-19 shows this pulse over the interval $-2.6T_s < t < 2.6T_s$. It decays outside this interval, but never does reach and stay at zero. Since the pulse has zeros at multiples of T_s, this type of reconstruction is still an interpolation process, called *bandlimited interpolation*. Using this pulse to

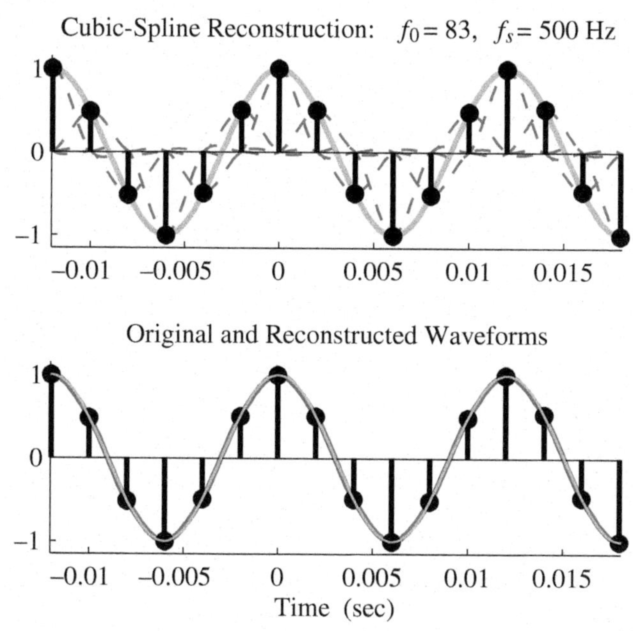

Figure 4-25: D-to-C conversion using a cubic-spline pulse at $f_s = 500$ samples/sec.

reconstruct from samples of a cosine wave will always produce a cosine wave exactly. If the sampling rate satisfies the conditions of the sampling theorem, the reconstructed cosine wave will be identical to the original signal that was sampled. If aliasing occurred in sampling, the ideal D-to-C converter reconstructs a cosine wave with the alias frequency that is less than $f_s/2$.

4-5 The Sampling Theorem

This chapter has discussed the issues that arise in sampling continuous-time signals. Using the example of the continuous-time cosine signal, we have illustrated the phenomenon of aliasing, and we have shown how the original continuous-time cosine signal can be reconstructed by interpolation. All of the discussion of this chapter has been aimed at the goal of establishing confidence in the Shannon sampling theorem, which, because of its central importance to our study of digital signal processing, is repeated here.

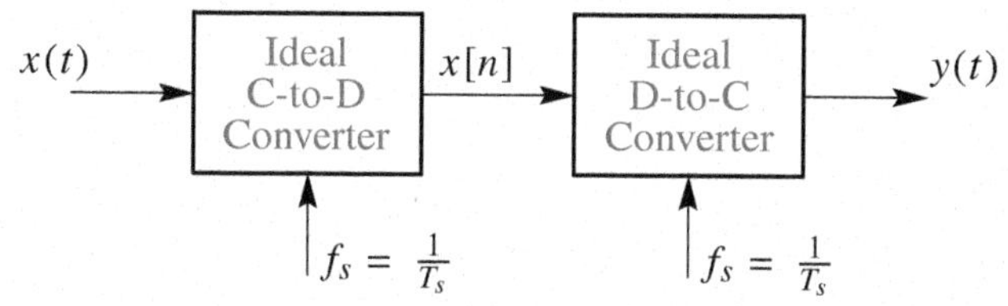

Figure 4-26: Sampling and reconstruction system.

> ***Shannon Sampling Theorem***
> A continuous-time signal $x(t)$ with frequencies no higher than $f_{\max}$ can be reconstructed exactly from its samples $x[n] = x(nT_s)$, if the samples are taken at a rate $f_s = 1/T_s$ that is greater than $2f_{\max}$.

A block diagram representation of the sampling theorem is shown in Fig. 4-26 in terms of the ideal C-to-D and D-to-C converters that we have defined in this chapter. The sampling theorem states that for the sampling and reconstruction system shown in Fig. 4-26, if the input is composed of sinusoidal signals limited to the set of frequencies in the range $0 \le f \le f_{\max}$, then the reconstructed signal is equal to the original signal that was sampled; i.e., $y(t) = x(t)$.

Signals composed of sinusoids such that all frequencies are limited to a "band of frequencies" of the form $0 \le f \le f_{\max}$ are called *bandlimited signals*.[7] Such signals could be represented as

$$x(t) = \sum_{k=0}^{N} x_k(t) \tag{4.23}$$

[7] The corresponding complex exponential signals would be limited to the band $-f_{\max} \le f \le f_{\max}$.

where each of the individual signals is of the form

$$x_k(t) = A_k \cos(2\pi f_k t + \phi_k) \tag{4.24}$$

and it is assumed that the frequencies are ordered so that $f_0 \geq 0$ and $f_N \leq f_{\max}$. As we have seen in Chapter 3, such an additive combination of cosine signals can produce an infinite variety of both periodic and nonperiodic signal waveforms. If we sample the signal represented by (4.23) and (4.24), we obtain

$$x[n] = x(nT_s) = \sum_{k=0}^{N} x_k(nT_s) = \sum_{k=0}^{N} x_k[n] \tag{4.25}$$

where $x_k[n] = A_k \cos(\hat{\omega}_k n + \phi_k)$ with $\hat{\omega}_k = 2\pi f_k / f_s$. That is, if we sample a sum of continuous-time cosines, we obtain a sum of sampled cosines each of which would be subject to aliasing if the sampling rate is not high enough.

The final step in the process of sampling followed by reconstruction in Fig. 4-26 is discrete-to-continuous conversion by interpolation with

$$y(t) = \sum_{n=-\infty}^{\infty} x[n] p(t - nT_s) \tag{4.26}$$

where for perfect reconstruction, $p(t)$ would be given by (4.22). This expression for the reconstructed output is a linear operation on the samples $x[n]$. This means that the total output $y(t)$ will consist of the sum of the outputs due to each of the different components $x_k[n]$. We can see this by substituting (4.25) into (4.26) as in

$$\begin{aligned} y(t) &= \sum_{n=-\infty}^{\infty} \left(\sum_{k=0}^{N} x_k[n] \right) p(t - nT_s) \\ &= \sum_{k=0}^{N} \left(\sum_{n=-\infty}^{\infty} x_k[n] p(t - nT_s) \right) \end{aligned} \tag{4.27}$$

Now since each individual sinusoid is assumed to satisfy the conditions of the sampling theorem, it follows that the D-to-C converter will reconstruct each component perfectly, and therefore we conclude that

$$y(t) = \sum_{k=0}^{N} x_k(t) = x(t)$$

Thus, we have shown that the Shannon sampling theorem applies to *any* signal that can be represented as a bandlimited sum of sinusoids, and since it can be shown that most real-world signals can be represented as bandlimited signals, it follows that the sampling theorem is a very general guide to sampling all kinds of signals. In Chapter 12, we will demonstrate this more convincingly using the Fourier transform.

4-6 Summary and Links

This chapter introduced the concept of sampling and the companion operation of reconstruction. With sampling, the possibility of aliasing always exists, and we have created the strobe demo to illustrate that concept in a direct and intuitive way.

DEMO: *Links to Many Demos*

Lab #6, Digital Images: A/D and D/A, shows how sampling applies to digital images. Some aspects of sampling have already been used in the music synthesis labs that are associated with Chapter 3, because the sounds must be produced by making their samples in a computer before playing them out through a D-to-A converter.

The CD-ROM also contains many demonstrations of sampling and aliasing:

(a) Strobe movie filmed using the natural strobing of a video camera at 30 frames per second.

(b) Synthetic strobe demos produced as MATLAB movies.

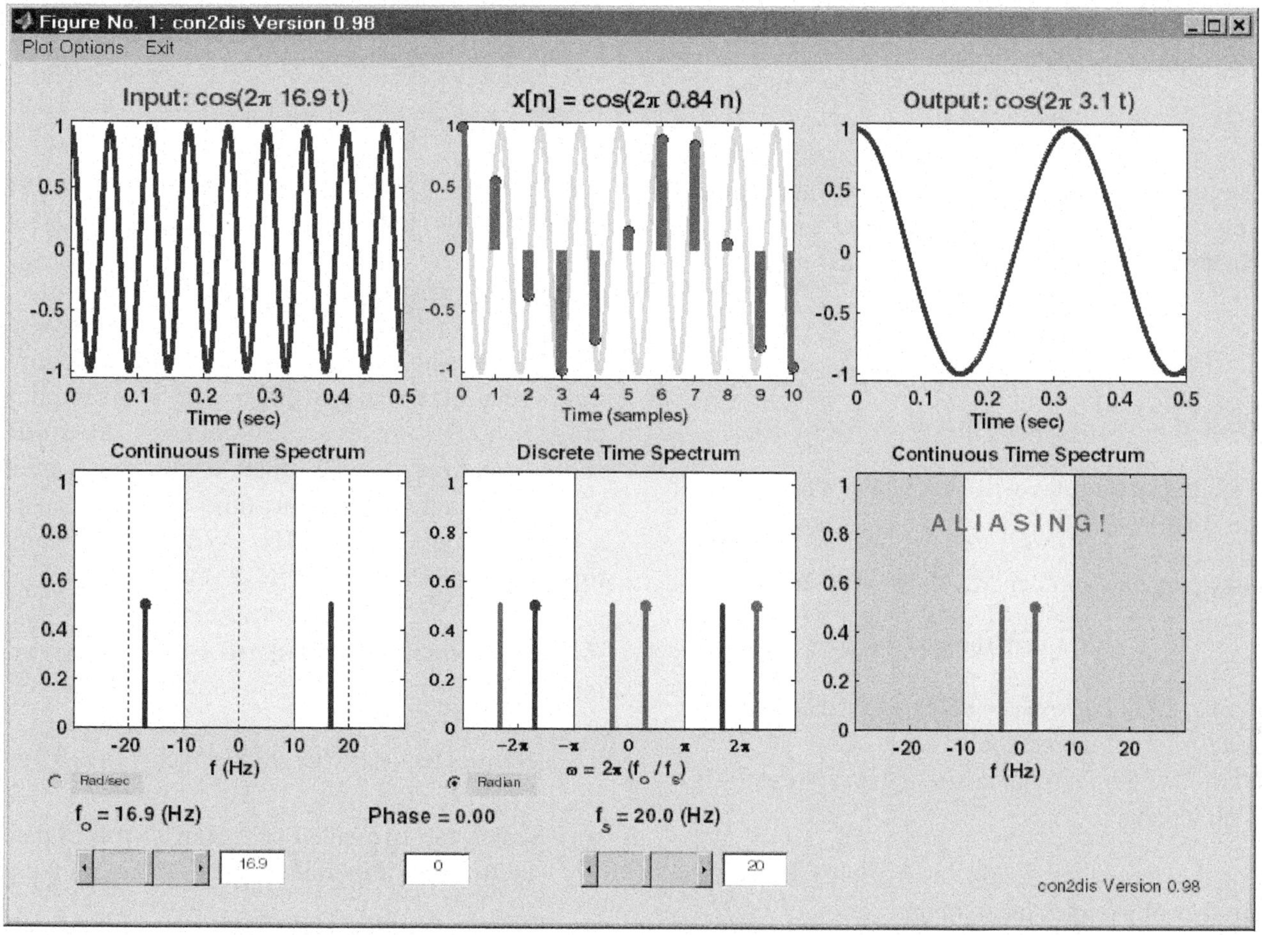

Figure 4-27: Graphical User Interface (GUI) for `con2dis` which illustrates aliasing with simultaneous diagrams in both the time and frequency domains. The user can change the frequencies moving the spectrum lines with a mouse or by entering values in the edit boxes.

(c) Reconstruction movies that show the interpolation process for different pulse shapes and different sampling rates.

(d) A sampling and aliasing MATLAB GUI called `con2dis`, shown in Fig. 4-27.

(e) Aliasing and folding movies that show how the spectrum changes for different sampling rates.

Again, the CD-ROM contains a large number of solved homework problems available for review and practice.

NOTE: *Hundreds of Solved Problems*

4-7 Problems

P-4.1 Consider the cosine wave

$$x(t) = 10\cos(880\pi t + \phi)$$

Suppose that we obtain a sequence of numbers by sampling the waveform at equally spaced time instants nT_s. In this case, the resulting sequence would have values

$$x[n] = x(nT_s) = 10\cos(880\pi nT_s + \phi)$$

for $-\infty < n < \infty$. Suppose that $T_s = 0.0001$ sec.

(a) How many samples will be taken in one period of the cosine wave?

(b) Now consider another waveform $y(t)$ such that

$$y(t) = 10\cos(\omega_0 t + \phi)$$

Find a frequency $\omega_0 > 880\pi$ such that $y(nT_s) = x(nT_s)$ for all integers n.
Hint: Use the fact that $\cos(\theta + 2\pi n) = \cos(\theta)$ if n is an integer.

(c) For the frequency found in (b), what is the total number of samples taken in one period of $x(t)$?

P-4.2 Let $x(t) = 7\sin(11\pi t)$. In each of the following parts, the discrete-time signal $x[n]$ is obtained by sampling $x(t)$ at a rate f_s, and the resultant $x[n]$ can be written as

$$x[n] = A\cos(\hat{\omega}_0 n + \phi)$$

For each part below, determine the values of A, ϕ, and $\hat{\omega}_0$. In addition, state whether or not the signal has been over-sampled or under-sampled.

(a) Sampling frequency is $f_s = 10$ samples/sec.

(b) Sampling frequency is $f_s = 5$ samples/sec.

(c) Sampling frequency is $f_s = 15$ samples/sec.

P-4.3 Suppose that a discrete-time signal $x[n]$ is given by the formula

$$x[n] = 2.2\cos(0.3\pi n - \pi/3)$$

and that it was obtained by sampling a continuous-time signal $x(t) = A\cos(2\pi f_0 t + \phi)$ at a sampling rate of $f_s = 6000$ samples/sec. Determine three different continuous-time signals that could have produced $x[n]$. All these continuous-time signals should have a frequency less than 8 kHz. Write the mathematical formula for all three.

P-4.4 An amplitude-modulated (AM) cosine wave is represented by the formula

$$x(t) = [10 + \cos(2\pi(2000)t)]\cos(2\pi(10^4)t)$$

(a) Sketch the two-sided spectrum of this signal. Be sure to label important features of the plot.

(b) Is this waveform periodic? If so, what is the period?

(c) What relation must the sampling rate f_s satisfy so that $y(t) = x(t)$ in Fig. 4-26?

P-4.5 Suppose that a discrete-time signal $x[n]$ is given by the formula

$$x[n] = 10\cos(0.2\pi n - \pi/7)$$

and that it was obtained by sampling a continuous-time signal at a sampling rate of $f_s = 1000$ samples/sec.

(a) Determine two *different* continuous-time signals $x_1(t)$ and $x_2(t)$ whose samples are equal to $x[n]$; i.e., find $x_1(t)$ and $x_2(t)$ such that $x[n] = x_1(nT_s) = x_2(nT_s)$ if $T_s = 0.001$. Both of these signals should have a frequency less than 1000 Hz. Give a formula for each signal.

(b) If $x[n]$ is given by the equation above, what signal will be reconstructed by an ideal D-to-C converter operating at sampling rate of 2000 samples/sec? That is, what is the output $y(t)$ in Fig. 4-26 if $x[n]$ is as given above?

P-4.6 A nonideal D-to-C converter takes a sequence $y[n]$ as input and produces a continuous-time output $y(t)$ according to the relation

$$y(t) = \sum_{n=-\infty}^{\infty} y[n]p(t - nT_s)$$

where $T_s = 0.1$ second. The input sequence is given by the formula

$$y[n] = \begin{cases} (0.8)^n & 0 \le n \le 5 \\ 0 & \text{otherwise} \end{cases}$$

(a) For the pulse shape

$$p(t) = \begin{cases} 1 & -0.05 \le t \le 0.05 \\ 0 & \text{otherwise} \end{cases}$$

carefully sketch the output waveform $y(t)$.

(b) For the pulse shape

$$p(t) = \begin{cases} 1 - 10|t| & -0.1 \le t \le 0.1 \\ 0 & \text{otherwise} \end{cases}$$

carefully sketch the output waveform $y(t)$.

P-4.7 Suppose that MATLAB is used to plot a sinusoidal signal. The following MATLAB code generates a signal $x[n]$ and plots it. Unfortunately, the time axis of the plot is not labeled properly.

```
Ts = 0.01;
Duration = 0.3;
tt = 0 : Ts : Duration;
Fo = 394;
xx = 9*cos( 2*pi*Fo*tt  + pi/2 );
%
stem( xx )        %<--! NO time axis
```

(a) Make the `stem` plot of the signal with n as the horizontal axis. Either sketch it or plot it using MATLAB.

(b) For the plot above, determine the correct formula for the discrete-time signal in the form

$$x[n] = A\cos(\hat{\omega}_0 n + \phi)$$

(c) Explain how aliasing affects the plot that you see.

P-4.8 The spectrum diagram gives the frequency content of a signal.

(a) Draw a sketch of the spectrum of

$$x(t) = \cos(50\pi t)\,\sin(700\pi t)$$

Label the frequencies and complex amplitudes of each component.

(b) Determine the minimum sampling rate that can be used to sample $x(t)$ without aliasing for any of the components.

P-4.9 The spectrum diagram gives the frequency content of a signal.

(a) Draw a sketch of the spectrum of

$$x(t) = \sin^3(400\pi t)$$

Label the frequencies and complex amplitudes of each component.

(b) Determine the minimum sampling rate that can be used to sample $x(t)$ without aliasing for any of its components.

P-4.10 The intention of the following MATLAB program is to plot 13 cycles of a 13-Hz sinusoid, but it has a bug.

```
Fo = 13;
To = 1/Fo;      %-- Period
Ts = 0.07;
tt = 0 : Ts : (13*To);
xx = real(exp(j*(2*pi*Fo*tt - pi/2)));
%
plot( tt, xx )
xlabel('TIME   (sec)'),  grid on
```

(a) Draw a sketch of the plot that will be produced by MATLAB. Explain how aliasing or folding affects the plot that you see. In particular, what is the *period in the plot,* and how many periods do you observe?

(b) Determine an acceptable value of `Ts` to get a very smooth plot of the desired 13 Hz signal.

P-4.11 An amplitude-modulated (AM) cosine wave is represented by the formula

$$x(t) = [3 + \sin(\pi t)]\cos(13\pi t + \pi/2)$$

(a) Use *phasors* to show that $x(t)$ can be expressed in the form

$$x(t) = A_1\cos(\omega_1 t + \phi_1) + A_2\cos(\omega_2 t + \phi_2) + A_3\cos(\omega_3 t + \phi_3)$$

where $\omega_1 < \omega_2 < \omega_3$; i.e., find values for the parameters A_1, A_2, A_3, ϕ_1, ϕ_2, ϕ_3, ω_1, ω_2, ω_3.

(b) Sketch the two-sided spectrum of this signal on a frequency axis. Be sure to label important features of the plot. Label your plot in terms of the numerical values of A_i, ϕ_i, and ω_i.

(c) Determine the minimum sampling rate that can be used to sample $x(t)$ without aliasing of any of the components.

P-4.12 Refer to Fig. 4-26 for the system with ideal C-to-D and D-to-C converters.

(a) Suppose that the discrete-time signal $x[n]$ is

$$x[n] = 10\cos(0.13\pi n + \pi/13)$$

If the sampling rate is $f_s = 1000$ samples/sec, determine two *different* continuous-time signals $x(t) = x_1(t)$ and $x(t) = x_2(t)$ that could have been inputs to the above system; i.e., find $x_1(t)$ and $x_2(t)$ such that $x[n] = x_1(nT_s) = x_2(nT_s)$ if $T_s = 0.001$.

(b) If the input $x(t)$ is given by the two-sided spectrum representation in Fig. P-4.12, determine a simple formula for $y(t)$ when $f_s = 700$ samples/sec (for both the C-to-D and D-to-C converters).

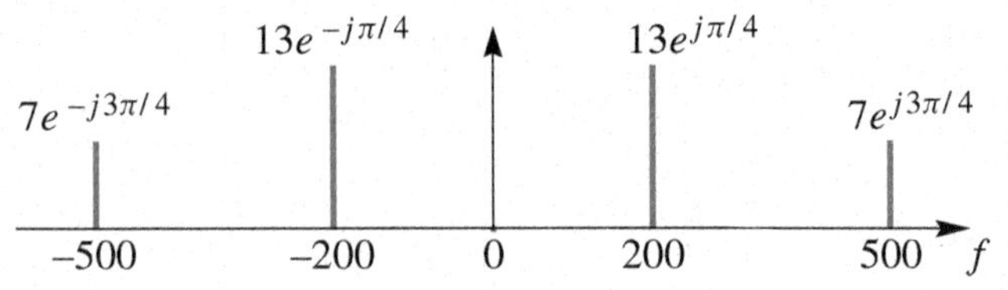

Figure P-4.12

P-4.13 Refer to Fig. 4-26 for the sampling and reconstruction system. Assume that the sampling rates of the C-to-D and D-to-C converters are equal, and the input to the Ideal C-to-D converter is

$$x(t) = 2\cos(2\pi(50)t + \pi/2) + \cos(2\pi(150)t)$$

(a) If the output of the ideal D-to-C converter is equal to the input $x(t)$, i.e.,

$$y(t) = 2\cos(2\pi(50)t + \pi/2) + \cos(2\pi(150)t)$$

what general statement can you make about the sampling frequency f_s in this case?

(b) If the sampling rate is $f_s = 250$ samples/sec., determine the discrete-time signal $x[n]$, and give an expression for $x[n]$ as a sum of cosines. *Make sure that all frequencies in your answer are positive and less than π radians.*

(c) Plot the spectrum of the signal in part (b) over the range of frequencies $-\pi \leq \hat{\omega} \leq \pi$. Make a plot for your answer, labeling the frequency, amplitude, and phase of each spectral component.

(d) If the output of the ideal D-to-C converter is

$$y(t) = 2\cos(2\pi(50)t + \pi/2) + 1$$

determine the value of the sampling frequency f_s. (Remember that the input $x(t)$ is as defined above.)

P-4.14 In the rotating disk and strobe demo, we observed that different flashing rates of the strobe light would make the spot on the disk stand still or move in different directions.

(a) Assume that the disk is rotating *clockwise* at a constant speed of 13 rev/sec. If the flashing rate is 15 times per second, express the movement of the spot on the disk as a complex phasor, $p[n]$, that gives the position of the spot at the nth flash. Assume that the spot is at the top when $n = 0$ (the first flash).

(b) For the conditions in (a), determine the apparent speed (in revolutions per second) and direction of movement of the "strobed" spot.

(c) *Now assume that the rotation speed of the disk is unknown.* If the flashing rate is 13 times per second, and the spot on the disk moves *counterclockwise* by 15 degrees with each flash, determine the rotation speed of the disk (in rev/sec). If the answer is not unique, give *all* possible rotation speeds.

P-4.15 In the rotating disk and strobe demo, we observed that different flashing rates of the strobe light would make the spot on the disk stand still.

(a) Assume that the disk is rotating in the counter-clockwise direction at a constant speed of 720 rpm. Express the movement of the spot on the disk as a rotating complex phasor.

(b) If the strobe light can be flashed at a rate of n flashes per second where n is an integer greater than zero, determine all possible flashing rates such that the disk can be made to stand still.

Note: The only possible flashing rates are 1 per second, 2 per second, 3 per second, etc.

(c) If the flashing rate is 13 times per second, explain how the spot will move, and write a complex phasor that gives the position of the spot at each flash.

P-4.16 The following complex-valued signal is a phasor:

$$z[n] = e^{j\theta[n]}$$

where $\theta[n]$ is the phase.

(a) When the phase changes by a constant amount versus n, the phasor rotates at a constant speed. For the following phasor:

$$z[n] = e^{j(0.08\pi n - 0.25\pi)}$$

make a plot of the phasor locations for $n = 0, 1, 2, 7, 10, 17, 20, 33, 50$, and 99.

(b) What is the period of $z[n]$?

(c) Repeat for the complex phasor that corresponds to the following chirp signal:

$$c[n] = e^{j0.1\pi n^2}$$

In this case, plot the phasor locations for $n = 0, 1, 2, 3, 4$, and 7.

P-4.17 A digital chirp signal is synthesized according to the following formula:

$$x[n] = \Re e\{e^{j\theta[n]}\} = \cos(\pi(0.7 \times 10^{-3})n^2)$$

for $n = 0, 1, 2, \ldots, 200$.

(a) Make a plot of the rotating phasor $e^{j\theta[n]}$ for $n = 10, 50$, and 100.

(b) If this signal is played out through a D-to-A converter whose sampling rate is 8 kHz, make a plot of the instantaneous analog frequency (in Hz) versus time for the analog signal.

(c) If the *constant frequency* digital signal $v[n] = \cos(0.7\pi n)$ is played out through a D-to-A converter whose sampling rate is 8 kHz, what (analog) frequency will be heard?

P-4.18 A discrete-time signal $x[n]$ is known to be a sinusoid

$$x[n] = A\cos(\hat{\omega}_0 n + \phi)$$

The values of $x[n]$ are tabulated for $n = 1, 2, 3, 4$, and 5.

n	1	2	3	4	5
$x[n]$	2.50	0.5226	−1.5451	−3.3457	−4.5677

(a) Plot $x[n]$ versus n.

(b) Prove (with phasors, not trigonometry) the following identity for the cosine signal:

$$\beta = \frac{\cos(n+1)\hat{\omega}_0 + \cos(n-1)\hat{\omega}_0}{\cos n\hat{\omega}_0} \qquad \text{for all } n$$

Determine the value of the constant β.
Note: β does not depend on n, but it might be a function of $\hat{\omega}_0$.

(c) Now determine the numerical values of A, ϕ, and $\hat{\omega}_0$. (This is an easy calculation if you find $\hat{\omega}_0$ first.)

P-4.19 A discrete-time signal $x[n]$ is known to be a sinusoid

$$x[n] = A\cos(\hat{\omega}_0 n + \phi)$$

The values of $x[n]$ are tabulated for $0 \le n \le 4$.

n	0	1	2	3	4
$x[n]$	2.4271	2.9002	2.9816	2.6603	1.9798

Plot $x[n]$, then determine the numerical values of A, ϕ, and $\hat{\omega}_0$.

C H A P T E R 5

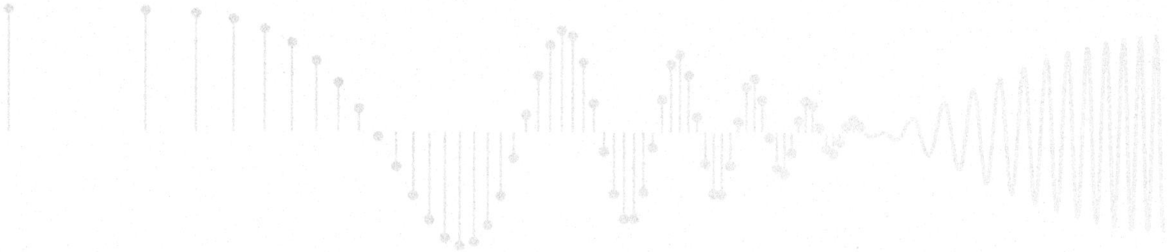

FIR Filters

Up to this point, we have focused our attention on signals and their mathematical representations. In this chapter, we begin to emphasize *systems* or ***filters***. Strictly speaking, a filter is a system that is designed to remove some component or modify some characteristic of a signal, but often the two terms are used interchangeably. In this chapter, we introduce the class of ***FIR (finite impulse response)*** systems, or, as we will often refer to them, *FIR filters*. These filters are systems for which each output sample is the sum of a finite number of weighted samples of the input sequence. We will define the basic input–output structure of the FIR filter as a time-domain computation based on a ***feed-forward*** difference equation. The unit impulse response of the filter will be defined and shown to characterize the filter. The general concepts of linearity and time invariance will also be presented. These properties characterize a wide class of filters that are exceedingly important in both the continuous-time and the discrete-time cases.

Our purpose in this chapter is to introduce the basic ideas of discrete-time systems and to provide a starting point for further study. The analysis of both discrete-time and continuous-time systems is a rich subject of study that is necessarily based on mathematical representations and manipulations.[1] The systems that we introduce in this chapter are the simplest to analyze. The remainder of the text is concerned with extending the ideas of this chapter to other classes of systems and with developing tools for the analysis of other systems.

[1] Systems can be analyzed effectively by the mathematical methods of Fourier analysis, which are introduced in Chapters 9–12 and are covered extensively in more advanced signals and systems texts.

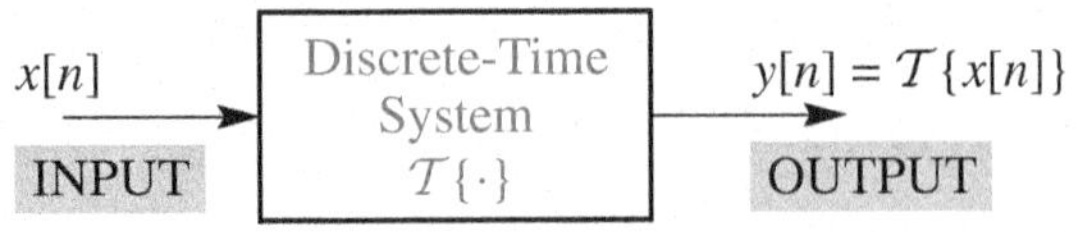

Figure 5-1: Block-diagram representation of a discrete-time system.

5-1 Discrete-Time Systems

A discrete-time system is a computational process for transforming one sequence, called the *input signal*, into another sequence called the *output signal*. As we have already mentioned, systems are often depicted by block diagrams such as the one in Fig. 5-1. In Chapter 4, we used similar block diagrams to represent the operations of sampling and reconstruction. In the case of sampling, the input signal is a continuous-time signal and the output is a discrete-time signal, while for reconstruction the opposite is true. Now we want to begin to study discrete-time systems where the input and output are discrete-time signals. Such systems are very interesting because they can be implemented with digital computation and because they can be designed so as to modify signals in many useful ways.

In general, we represent the operation of a system by the notation

$$y[n] = \mathcal{T}\{x[n]\}$$

which suggests that the output sequence is related to the input sequence by a process that can be described mathematically by an operator $\mathcal{T}$. Since a discrete-time signal is a sequence of numbers, such operators can be described by giving a formula for computing the values of the output sequence from the values of the input sequence. For example, the relation

$$y[n] = (x[n])^2$$

defines a system for which the output sequence values are the square of the corresponding input sequence values. A more complicated example would be the following system definition:

$$y[n] = \max\{x[n], x[n-1], x[n-2]\}$$

In this case, the output depends on three consecutive input values. Since infinite possibilities exist for defining discrete-time systems, it is necessary to limit the range of possibilities by placing some restrictions on the properties of the systems that we study. Therefore, we will begin our study of discrete-time systems in this chapter by introducing a very important class of discrete-time systems called ***FIR filters***. Specifically, we will discuss the representation, implementation, and analysis of discrete-time FIR systems, and illustrate how such systems can be used to modify signals.

5-2 The Running-Average Filter

A simple but useful transformation of a discrete-time signal is to compute a ***moving average*** or ***running average*** of two or more consecutive numbers of the sequence, thereby forming a new sequence of the average values. The FIR filter is a generalization of the idea of a running average. Averaging is commonly used whenever data fluctuate and must be smoothed prior to interpretation. For example, stock-market prices fluctuate noticeably from day to day, or hour to hour. Therefore, one might take an average of the stock price over several days before looking for any trend. Another everyday example concerns credit-card balances where interest is charged on the *average* daily balance.

In order to motivate the general definition of the class of FIR systems, let us consider the simple running average as an example of a system that processes an input sequence to produce an output sequence. To be specific, consider a 3-point averaging method; i.e.,

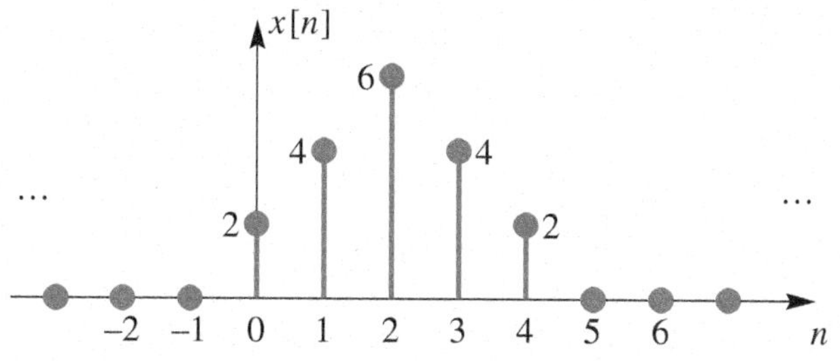

Figure 5-2: Finite-length input signal, $x[n]$.

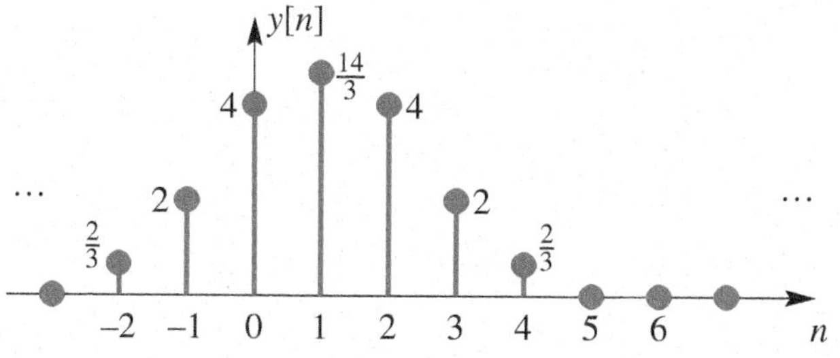

Figure 5-3: Output of running-average filter, $y[n]$.

each value of the output sequence is the sum of three consecutive input sequence values divided by three. If we apply this algorithm to the triangularly shaped sequence shown in Fig. 5-2, we can compute a new sequence called $y[n]$, which is the output of the averaging operator. The sequence in Fig. 5-2 is an example of a ***finite-length*** signal. The ***support*** of such a sequence is the set of values over which the sequence is nonzero; in this case, the support of the sequence is the finite interval $0 \leq n \leq 4$. A 3-point average of the values $\{x[0], x[1], x[2]\} = \{2, 4, 6\}$ gives the answer $\frac{1}{3}(2+4+6) = 4$. This result defines one of the output values. The next output value is obtained by averaging $\{x[1], x[2], x[3]\} = \{4, 6, 4\}$ which, yields a value of 14/3. Before going any further, we should decide on the output indexing. For example, the values 4 and 14/3 could be assigned to $y[0]$ and $y[1]$, *but this is only one of many possibilities.* With this indexing, the equations for computing the output from the input are

$$y[0] = \tfrac{1}{3}(x[0] + x[1] + x[2])$$

$$y[1] = \tfrac{1}{3}(x[1] + x[2] + x[3])$$

which generalizes to the following input–output equation:

$$y[n] = \tfrac{1}{3}(x[n] + x[n+1] + x[n+2]) \qquad (5.1)$$

The equation given in (5.1) is called a ***difference equation***. It is a complete description of the FIR system because we can use (5.1) to compute the entire output signal for all index values $-\infty < n < \infty$. For the input of Fig. 5-2, the result is the signal $y[n]$ tabulated as follows:

n	$n<-2$	-2	-1	0	1	2	3	4	5	$n>5$
$x[n]$	0	0	0	2	4	6	4	2	0	0
$y[n]$	0	$\frac{2}{3}$	2	4	$\frac{14}{3}$	4	2	$\frac{2}{3}$	0	0

Note that the values in orange type in the $x[n]$ row are the numbers involved in the computation of $y[2]$. Also note that $y[n] = 0$ outside of the finite interval $-2 \leq n \leq 4$; i.e., the output also has finite support. The output sequence is also plotted in Fig. 5-3. Observe that the output sequence is longer (has more nonzero values) than the input sequence, and that the output appears to be a somewhat rounded-off version of the input; i.e. it is *smoother* than the input sequence. This behavior is characteristic of the running-average FIR filter.

The choice of the output indexing is arbitrary, but it does matter when speaking about properties of the filter. For example, the filter defined in (5.1) has the property that its output starts (becomes nonzero) before the input

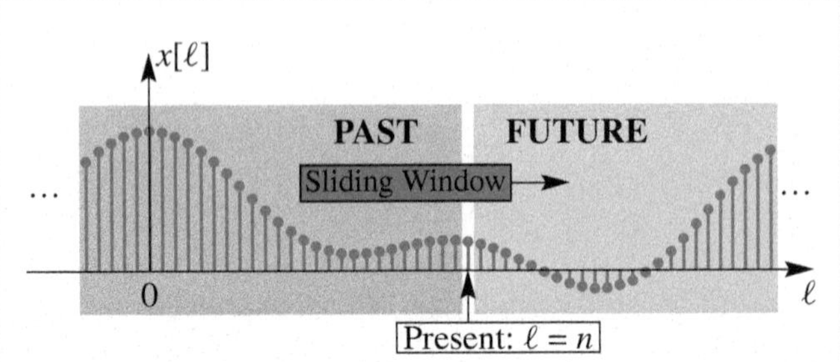

Figure 5-4: The running-average filter calculation at the *present* time ($\ell = n$) uses values within a sliding window. Gray shading indicates the past ($\ell < n$); orange shading, the future ($\ell > n$). Here, the sliding window encompasses values from both the future and the past.

starts. This would certainly be undesirable if the input signal values came directly from an A-to-D converter, as is common in audio signal-processing applications. In this case, n would stand for time, and we can interpret $y[n]$ in (5.1) as the computation of the ***present*** value of the output based on three input values. Since these inputs are indexed as n, $n+1$, and $n+2$, two of them are "in the future." In general, values from either the ***past*** or the ***future*** or both may be used in the computation, as shown in Fig. 5-4. In all cases of a 3-point running average, a ***sliding window*** of three samples determines which three samples are used in the computation of $y[n]$.

A filter that uses only the present and past values of the input is called a ***causal*** filter, implying that the cause does not precede the corresponding effect. Therefore, a filter that uses future values of the input is called ***noncausal***. Noncausal systems cannot be implemented in a real-time application because the input is not yet available when the output has to be computed. In other cases, where stored data blocks are manipulated inside a computer, the issue of causality is not crucial.

An alternative output indexing scheme can produce a 3-point averaging filter that is causal. In this case, the output value $y[n]$ is the average of inputs at n (the present), $n-1$ (one sample previous), and $n-2$ (two samples previous). The difference equation for this filter is

$$y[n] = \tfrac{1}{3}(x[n] + x[n-1] + x[n-2]) \qquad (5.2)$$

The form given in (5.2) is a ***causal running averager***, or it may well be called a *backward average*. Using the difference equation (5.2), we can make a table of all output values over the range $-\infty < n < \infty$. (Notice that now the orange-colored values of $x[n]$ are used in this case to compute $y[4]$ instead of $y[2]$.) The resulting signal $y[n]$ has the same values as before, but its support is now the index interval $0 \le n \le 6$. Observe that the output of the causal filter is simply a shifted version of the output of the previous noncausal filter. This filter is causal because the output depends on only the present and two previous (i.e., past) values of the input. Therefore, the output does not change from zero before the input changes from zero.

n	$n < -2$	-2	-1	0	1	2	3	4	5	6	7	$n > 7$
$x[n]$	0	0	0	2	4	6	4	2	0	0	0	0
$y[n]$	0	0	0	$\frac{2}{3}$	2	4	$\frac{14}{3}$	4	2	$\frac{2}{3}$	0	0

EXERCISE 5.1: Determine the output of a *centralized averager*

$$y[n] = \tfrac{1}{3}(x[n+1] + x[n] + x[n-1])$$

for the input in Fig. 5-2. Is this filter causal or noncausal? What is the support of the output for this input? How would the plot of the output compare to Fig. 5-3?

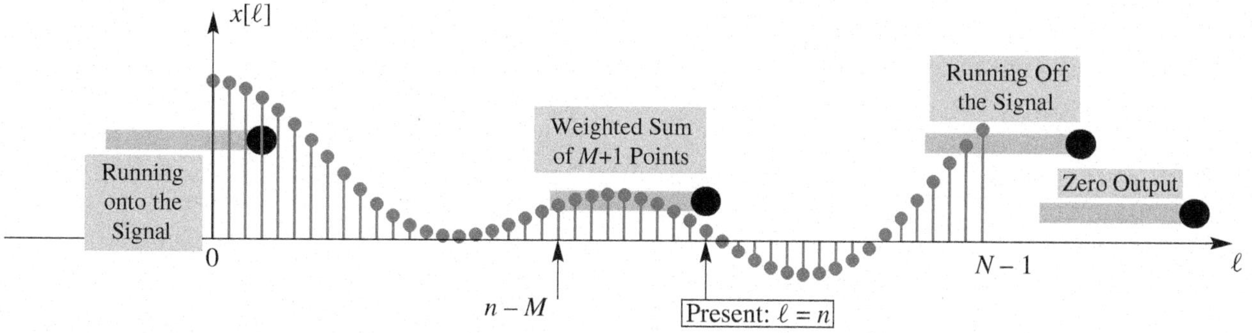

Figure 5-5: Operation of the Mth order causal FIR filter showing various positions of the sliding window of $M+1$ points under which the weighted average is calculated. When the input signal $x[\ell]$ is also finite length (N points), the sliding window will run onto and off of the input data, so the resulting output signal will also have finite length.

5-3 The General FIR Filter

Note that (5.2) is a special case of the general difference equation

$$y[n] = \sum_{k=0}^{M} b_k\, x[n-k] \tag{5.3}$$

That is, when $M = 2$ and $b_k = 1/3$ for $k = 0, 1, 2$, (5.3) reduces to the causal running average of (5.2). If the coefficients b_k are not all the same, then we might say that (5.3) defines a ***weighted running average*** of $M+1$ samples. It is clear from (5.3) that the computation of $y[n]$ involves the samples $x[\ell]$ for $\ell = n, n-1, n-2, \ldots, n-M$; i.e., $x[n], x[n-1], x[n-2]$, etc. Since the filter in (5.3) does not involve future values of the input, the system is causal, and, therefore, the output cannot start before the input becomes nonzero.[2] Figure 5-5 shows that the causal FIR filter uses $x[n]$ and the past M points to compute the output. Figure 5-5 also shows that if the input has finite support ($0 \le \ell \le N-1$), there will be an interval of M samples at the beginning, where the computation will involve fewer than $M+1$ nonzero samples as the sliding window of the filter engages with the input, and an interval of M samples at the end where the sliding window of the filter disengages from the input sequence. It also can be seen from Fig. 5-5 that the output sequence can be as much as M samples longer than the input sequence.

[2]Note that a noncausal system can be represented by (5.3) by allowing negative values of the summation index k.

Example 5-1: FIR Filter Coefficients

The FIR filter is completely defined once the set of filter coefficients $\{b_k\}$ is known. For example, if the $\{b_k\}$ of a causal filter are

$$\{b_k\} = \{3,\ -1,\ 2,\ 1\}$$

then we have a length-4 filter with $M = 3$, and (5.3) expands into a 4-point difference equation:

$$\begin{aligned} y[n] &= \sum_{k=0}^{3} b_k x[n-k] \\ &= 3x[n] - x[n-1] + 2x[n-2] + x[n-3] \end{aligned}$$

■

The parameter M is the ***order*** of the FIR filter. The number of filter coefficients is also called the filter ***length*** (L). The length is one greater than the order; i.e., $L = M+1$. This terminology will make more sense after we have introduced the z-transform in Chapter 7.

EXERCISE 5.2: Compute the output $y[n]$ for the length-4 filter whose coefficients are $\{b_k\} = \{3, -1, 2, 1\}$. Use the input signal given in Fig. 5-2. Verify that the answers tabulated here are correct, then fill in the missing values.

n	$n<0$	0	1	2	3	4	5	6	7	8	$n>8$
$x[n]$	0	2	4	6	4	2	0	0	0	0	0
$y[n]$	0	6	10	18	?	?	?	8	2	0	0

5-3.1 An Illustration of FIR Filtering

To illustrate some of the things that we have learned so far, and to show how FIR filters can modify sequences, consider a signal

$$x[n] = \begin{cases} (1.02)^n + \frac{1}{2}\cos(2\pi n/8 + \pi/4) & 0 \le n \le 40 \\ 0 & \text{otherwise} \end{cases}$$

This signal is shown as the upper plot in Fig. 5-6. We often have real signals of this form; i.e., a component that is the signal of interest (in this case, it may be the slowly varying exponential component $(1.02)^n$) plus another component that is not of interest. Indeed, the second component is often considered to be *noise* that interferes with observation of the desired signal. In this case, we will consider the sinusoidal component $\frac{1}{2}\cos(2\pi n/8 + \pi/4)$ to be noise that we wish to remove. The solid, exponentially growing curve shown in each of the plots in Fig. 5-6 simply connects the sample values of

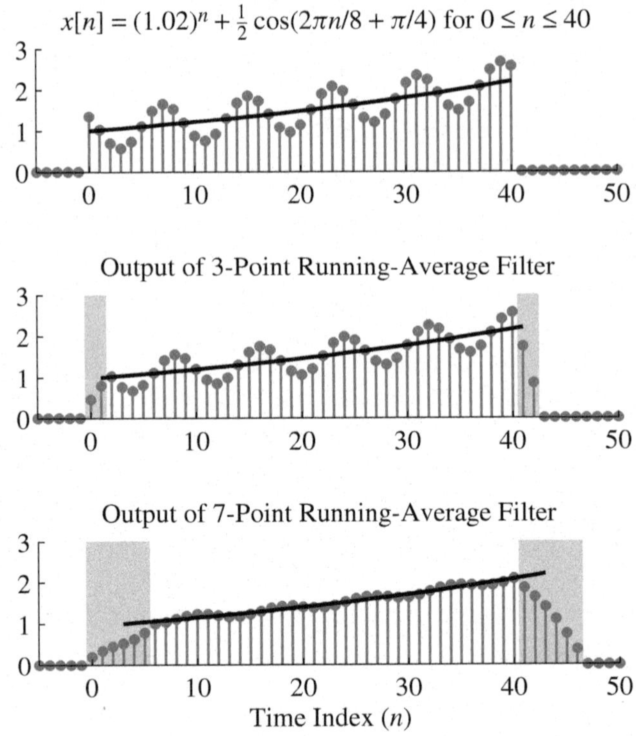

Figure 5-6: Illustration of running-average filtering.

the desired signal $(1.02)^n$ by straight lines for reference in the other two plots.

Now suppose that $x[n]$ is the input to a causal 3-point running averager; i.e.,

$$y_3[n] = \tfrac{1}{3}\left(\sum_{k=0}^{2} x[n-k]\right) \tag{5.4}$$

In this case, $M = 2$ and all the coefficients are equal to 1/3. The output of this filter is shown in the middle plot in Fig. 5-6. We can notice several things about these plots.

(a) Observe that the input sequence $x[n]$ is zero prior to $n = 0$, and from (5.4) it follows that the output must be zero for $n < 0$.

(b) The output becomes nonzero at $n = 0$, and the shaded interval of length $M = 2$ samples at the beginning of the nonzero part of the output sequence is the interval where the 3-point averager "runs onto" the input sequence. For $2 \le n \le 40$, the input samples within the 3-point averaging window are all nonzero.

(c) There is another shaded interval of length $M = 2$ samples at the end (after sample 40), where the filter window "runs off of" the input sequence.

(d) Observe that the size of the sinusoidal component has been reduced, but that the component is not eliminated by the filter. The solid line showing the values of the exponential component has been shifted to the right by $M/2 = 1$ sample to account for the shift introduced by the causal filter.

Clearly, the 3-point running averager has removed some of the fluctuations in the input signal, but we have not recovered the desired component. Intuitively, we might think that averaging over a longer interval might produce better results. The lower plot in Fig. 5-6 shows the output of a 7-point running averager as defined by

$$y_7[n] = \frac{1}{7}\left(\sum_{k=0}^{6} x[n-k]\right) \tag{5.5}$$

In this case, since $M = 6$ and all the coefficients are equal to 1/7, we observe the following:

(a) The shaded regions at the beginning and end of the output are now $M = 6$ samples long.

(b) Now the size of the sinusoidal component is greatly reduced relative to the input sinusoid, and the exponential component is very close to the exponential component of the input (after a shift of $M/2 = 3$ samples).

What can we conclude from this example? First, it appears that FIR filtering can modify signals in ways that may be useful. Second, the length of the averaging interval seems to have a big effect on the resulting output. Third, the running-average filters appear to introduce a shift equal to $M/2$ samples. All of these observations can be shown to apply to more general FIR filters defined by (5.3). However, before we can fully appreciate the details of this example, we must explore the properties of FIR filters in greater detail. We will gain full appreciation of this example only upon the completion of Chapter 6.

5-3.2 The Unit Impulse Response

In this section, we will introduce three new ideas: the unit impulse sequence, the unit impulse response, and the convolution sum. We will show that the impulse response provides a complete characterization of the filter, because the convolution sum gives a formula for computing the output from the input when the unit impulse response is known.

5-3.2.1 Unit Impulse Sequence

The ***unit impulse*** is perhaps the simplest sequence because it has only one nonzero value, which occurs at $n = 0$. The mathematical notation is that of the Kronecker ***delta function***

$$\delta[n] = \begin{cases} 1 & n = 0 \\ 0 & n \neq 0 \end{cases} \tag{5.6}$$

It is tabulated in the second row of this table:

n	...	−2	−1	0	1	2	3	4	5	6	...
$\delta[n]$	0	0	0	1	0	0	0	0	0	0	0
$\delta[n-2]$	0	0	0	0	0	1	0	0	0	0	0

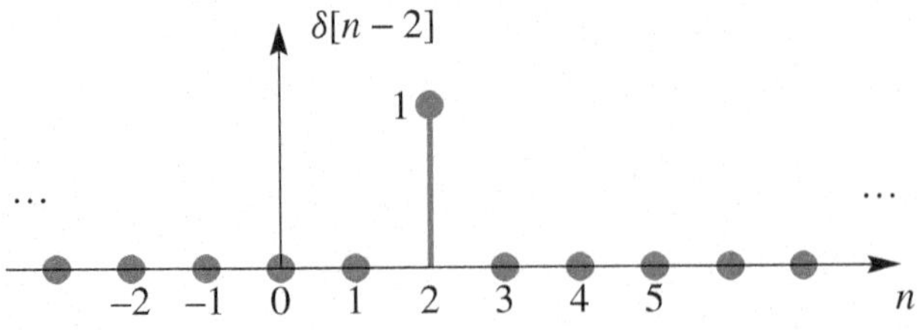

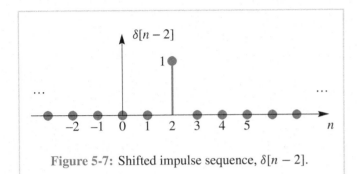

Figure 5-7: Shifted impulse sequence, $\delta[n-2]$.

A shifted impulse such as $\delta[n-2]$ is nonzero when its argument is zero, i.e., $n-2=0$, or equivalently $n=2$. The third row of the table gives the values of the shifted impulse $\delta[n-2]$, and Fig. 5-7 shows a plot of that sequence.

The shifted impulse is a concept that is very useful in representing signals and systems. Consider, for example, the signal

$$\begin{aligned} x[n] &= 2\delta[n] + 4\delta[n-1] + 6\delta[n-2] \\ &\quad + 4\delta[n-3] + 2\delta[n-4] \end{aligned} \tag{5.7}$$

To interpret (5.7), we must observe that the appropriate definition of multiplying a sequence by a number is to multiply each value of the sequence by that number; likewise, adding two or more sequences is defined as adding the sequence values at corresponding positions (times). The following table shows the individual sequences in (5.7) and their sum:

n	...	−2	−1	0	1	2	3	4	5	6	...
$2\delta[n]$	0	0	0	2	0	0	0	0	0	0	0
$4\delta[n-1]$	0	0	0	0	4	0	0	0	0	0	0
$6\delta[n-2]$	0	0	0	0	0	6	0	0	0	0	0
$4\delta[n-3]$	0	0	0	0	0	0	4	0	0	0	0
$2\delta[n-4]$	0	0	0	0	0	0	0	2	0	0	0
$x[n]$	0	0	0	2	4	6	4	2	0	0	0

Figure 5-8: Block diagram showing definition of impulse response.

Clearly, (5.7) is a compact representation of the signal in Fig. 5-2. Indeed, any sequence can be represented in this way. The equation

$$\begin{aligned} x[n] &= \sum_k x[k]\delta[n-k] \\ &= \cdots + x[-1]\delta[n+1] + x[0]\delta[n] \\ &\quad + x[1]\delta[n-1] + x[2]\delta[n-2] + \ldots \end{aligned} \tag{5.8}$$

is true if k ranges over all the nonzero values of the sequence $x[n]$. Equation (5.8) states the obvious: The sequence is formed by using scaled shifted impulses to place samples of the right size at the right positions.

5-3.2.2 Unit Impulse Response Sequence

When the input to the FIR filter (5.3) is a unit impulse sequence, $x[n]=\delta[n]$, the output is, by definition, the ***unit impulse response***, which we will denote by $h[n]$.[3] This is depicted in the block diagram of Fig. 5-8. Substituting $x[n]=\delta[n]$ in (5.3) gives the output $y[n]=h[n]$:

$$h[n] = \sum_{k=0}^{M} b_k\,\delta[n-k] = \begin{cases} b_n & n = 0, 1, 2, \ldots M \\ 0 & \text{otherwise} \end{cases}$$

As we have observed, the sum evaluates to a single term for each value of n because each $\delta[n-k]$ is nonzero

[3] We will usually shorten this to ***impulse response***, with *unit* being understood.

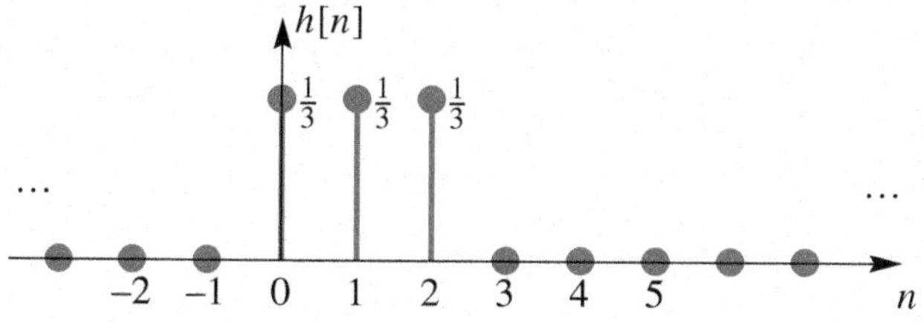

Figure 5-9: Impulse response of 3-point running average filter, $h[n]$.

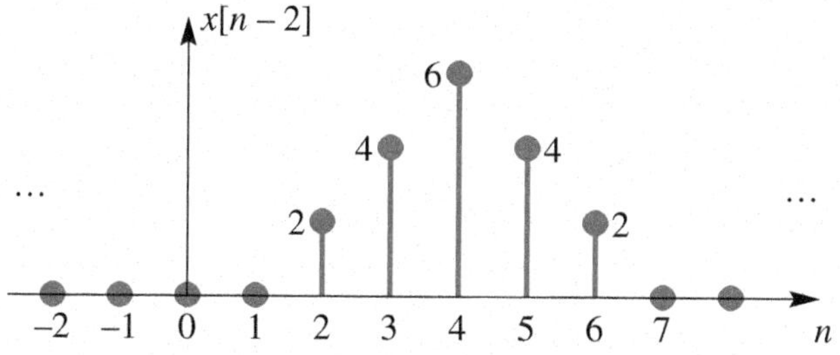

Figure 5-10: Delayed finite-length input signal, $y[n] = x[n-2]$ for input of Fig. 5-2.

only when $n - k = 0$ or $n = k$. In the tabulated form, the impulse response is

n	$n<0$	0	1	2	3	...	M	$M+1$	$n>M$
$x[n] = \delta[n]$	0	1	0	0	0	0	0	0	0
$y[n] = h[n]$	0	b_0	b_1	b_2	b_3	...	b_M	0	0

In other words, the impulse response $h[n]$ of the FIR filter is simply the sequence of difference equation coefficients. Since $h[n] = 0$ for $n < 0$ and for $n > M$, the length of the impulse response sequence $h[n]$ is finite. This is why the system (5.3) is called a ***finite impulse response (FIR)*** system. Figure 5-9 illustrates a plot of the impulse response for the case of the causal 3-point running-average filter.

EXERCISE 5.3: Determine and plot the impulse response of the FIR system

$$y[n] = \sum_{k=0}^{10} k x[n-k]$$

5-3.2.3 The Unit-Delay System

One important system is the operator that performs a delay or shift by an amount n_0

$$y[n] = x[n - n_0] \tag{5.9}$$

When $n_0 = 1$, the system is called a ***unit delay***. The output of the unit delay is particularly easy to visualize. In a plot, the values of $x[n]$ are moved to the right by one time interval. For example, $y[4]$ takes on the value of $x[3]$, $y[5]$ takes on the value of $x[4]$, $y[6]$ is $x[5]$, and so on.

The delay system is actually the simplest of FIR filters; it has only one nonzero coefficient. For example, a system that produces a *delay of 2* has filter coefficients $\{b_k\} = \{0,\ 0,\ 1\}$. The order of this FIR filter is $M = 2$, and its difference equation is

$$\begin{aligned} y[n] &= b_0 \cdot x[n] + b_1 \cdot x[n-1] + b_2 \cdot x[n-2] \\ &= 0 \cdot x[n] + 0 \cdot x[n-1] + 1 \cdot x[n-2] \\ &= x[n-2] \end{aligned}$$

Figure 5-10 shows the output of the delay system with a delay of 2 for the input of Fig. 5-2 on p. 103. The impulse response of the delay system is obtained by substituting $\delta[n]$ for $x[n]$ in (5.9). For the delay-by-2 case,

$$h[n] = \delta[n - n_0] = \delta[n-2] = \begin{cases} 1 & n = 2 \\ 0 & n \neq 2 \end{cases}$$

This impulse response is the signal plotted previously in Fig. 5-7.

5-3.3 Convolution and FIR Filters

A general expression for the FIR filter's output (5.3) can be derived in terms of the impulse response. Since the filter coefficients in (5.3) are identical to the impulse response values, we can replace b_k in (5.3) by $h[k]$ to obtain

$$y[n] = \sum_{k=0}^{M} h[k]\,x[n-k] \tag{5.10}$$

When the relation between the input and the output of the FIR filter is expressed in terms of the input and the impulse response, as in (5.10), it is called a finite ***convolution sum***, and we say that the output is obtained by ***convolving*** the sequences $x[n]$ and $h[n]$.

5-3.3.1 Computing the Output of a Convolution

The method of tabulating values for the output of an FIR filter works for short signals, but lacks the generality needed in more complicated problems. However, there is a simple interpretation of (5.10) that leads to a better algorithm for doing convolution. This algorithm can be implemented using the tableau in Fig. 5-11 that tracks the relative position of the signal values. The example in Fig. 5-11 shows how to convolve $x[n] = \{2,\ 4,\ 6,\ 4,\ 2\}$ with $h[n] = \{3,\ -1,\ 2,\ 1\}$. First of all, we write out the signals $x[n]$ and $h[n]$ on separate rows. Then we use a method similar to "synthetic polynomial multiplication" to form the output as the sum of shifted rows. Each shifted row is produced by multiplying the $x[n]$ row by one of the $h[k]$ values and shifting the result to the right so that it lines up with the $h[k]$ position. The final answer is obtained by summing down the columns.

The justification of this algorithm for evaluating the convolution sum comes from writing out the sum in (5.10) as

$$y[n] = h[0]x[n] + h[1]x[n-1] + h[2]x[n-2] + \ldots$$

n	$n<0$	0	1	2	3	4	5	6	7	$n>7$
$x[n]$	0	2	4	6	4	2	0	0	0	0
$h[n]$	0	3	−1	2	1					
$h[0]x[n]$	0	6	12	18	12	6	0	0	0	0
$h[1]x[n-1]$	0	0	−2	−4	−6	−4	−2	0	0	0
$h[2]x[n-2]$	0	0	0	4	8	12	8	4	0	0
$h[3]x[n-3]$	0	0	0	0	2	4	6	4	2	0
$y[n]$	0	6	10	18	16	18	12	8	2	0

Figure 5-11: Numerical convolution of finite-length signals via synthetic polynomial multiplication

A term such as $x[n-2]$ is the $x[n]$ signal with its values shifted two places to the right. The multiplier $h[2]$ scales the shifted signal $x[n-2]$ to produce the contribution $h[2]x[n-2]$, which is the orange row in the table.

EXERCISE 5.4: Use the "synthetic multiplication" convolution algorithm to compute the output $y[n]$ for the length-4 filter whose coefficients are $\{b_k\} = \{1,\ -2,\ 2,\ -1\}$. Use the input signal given in Fig. 5-2 on p. 103.

Later in this chapter, we will prove that convolution is the fundamental input–output algorithm for a large class of very useful filters that includes FIR filters as a special case. We will show that a general form of convolution that also applies to infinite-length signals is

$$y[n] = \sum_{k=-\infty}^{\infty} h[k]\,x[n-k] \tag{5.11}$$

This convolution sum (5.11) has infinite limits, but reduces to (5.10) when $h[n] = 0$ for $n < 0$ and $n > M$.

5-3.3.2 Convolution in MATLAB

In MATLAB, FIR systems can be implemented by using the `conv( )` function. For example, the following MATLAB statements

```
xx = sin(0.07*pi*(0:50));
hh = ones(11,1)/11;
yy = conv(hh, xx);
```

will evaluate the convolution of the 11-point sequence hh with the 51-point sinusoidal sequence xx. The particular choice for the MATLAB vector hh is actually the impulse response of an 11-point running average system:

$$h[n] = \begin{cases} 1/11 & n = 0, 1, 2, \ldots, 10 \\ 0 & \text{otherwise} \end{cases}$$

That is, all 11 filter coefficients are the same and equal to 1/11.[4]

EXERCISE 5.5: In MATLAB, we can compute only the convolution of finite-length signals. Determine the length of the output sequence computed by the MATLAB convolution above.

We have already hinted that this operation called ***convolution*** is equivalent to polynomial multiplication. In Section 7-5, we will prove that this correspondence is true. At this point, we note that in MATLAB there is no function for multiplying polynomials. Instead, we must know that convolution is equivalent to polynomial multiplication. Then we can represent the polynomials by sequences of their coefficients and use the `conv` function to convolve them, thereby doing polynomial multiplication.

DEMO: *Discrete Convolution*

EXERCISE 5.6: Use MATLAB to compute the following product of polynomials:

$$P(x) = (1 + 2x + 3x^2 + 5x^4)(1 - 3x - x^2 + x^3 + 3x^4)$$

5-4 Implementation of FIR Filters

Recall that the general definition of an FIR filter is

$$y[n] = \sum_{k=0}^{M} b_k\, x[n-k] \tag{5.12}$$

We can see that to use (5.12) to compute the output of the FIR filter, we need the following: (1) a means for multiplying delayed-input signal values by the filter coefficients; (2) a means for adding the scaled sequence values; and (3) a means for obtaining delayed versions of the input sequence. We will find it useful to represent the operations of (5.12) as a block diagram. Such representations will lead to new insights about the properties of the system and about alternative ways to implement the system.

LAB: *#6 Digital Images: A/D and D/A*

5-4.1 Building Blocks

The basic building-block systems we need are the multiplier, the adder, and the unit-delay operator as depicted in Fig. 5-12.

[4]Note that MATLAB indexes all data vectors starting at 1, while we index sequences such as $x[n]$ starting at $n = 0$. This difference must be accounted for in MATLAB programs.

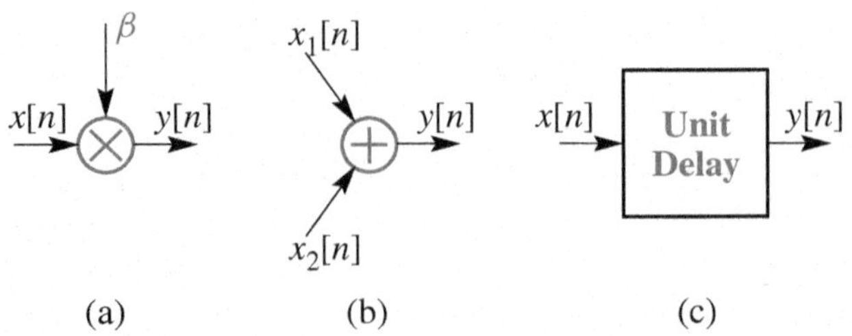

Figure 5-12: Building blocks for making any LTI discrete-time system: (a) multiplier, $y[n] = \beta x[n]$; (b) adder, $y[n] = x_1[n] + x_2[n]$; and (c) unit-delay, $y[n] = x[n-1]$.

5-4.1.1 Multiplier

The first elementary system performs multiplication of a signal by a constant (see Fig. 5-12(a)). The output signal $y[n]$ is given by the rule

$$y[n] = \beta\, x[n]$$

where the coefficient β is a constant. This system can be the hardware multiplier unit in a computer. For a DSP microprocessor, the speed of this multiplier is one of the fundamental limits on the throughput of the digital filtering process. In applications such as image convolution, billions of multiplications per second may have to be performed to implement a good filter, so quite a bit of engineering work has been directed at designing fast multipliers for DSP applications. Furthermore, since many filters require the same sequence of multiplications over and over, pipelining the multiplier also results in a dramatic speed-up of the filtering process.

Notice, by the way, that the simple multiplier is also an FIR filter, with $M = 0$, and $b_0 = \beta$ in (5.3). The impulse response of the multiplier system is simply $h[n] = \beta\delta[n]$.

5-4.1.2 Adder

The second elementary system in Fig. 5-12(b) performs the addition of two signals. This is a different sort of system because it has two inputs and one output. In hardware, the adder is simply the hardware adder unit in the computer. Since many DSP operations require a multiplication followed immediately by an addition, it is common in DSP microprocessors to build a special multiply-accumulate unit, often called a "MADD" or "MAC" unit.[5]

Notice that the adder is a pointwise combination of the values of the two input sequences. It is not an FIR filter, because it has more than one input; however, it is a crucial building block of FIR filters. With many inputs, the adder could be drawn as a multi-input adder, but, in digital hardware, the additions are typically done two inputs at a time.

5-4.1.3 Unit Delay

The third elementary system performs a delay by one unit of time. It is represented by a block diagram, as in Fig. 5-12(c). In the case of discrete-time filters, the time dimension is indexed by integers, so this delay is by one "count" of the system clock. The hardware implementation of the unit delay is actually performed by acquiring a sample value, storing it in memory for one clock cycle, and then releasing it to the output. The delays by more than one time unit that are needed to implement (5.12) can be implemented (from the block-diagram point of view) by cascading several unit delays in a row. Therefore, an M-unit delay requires M memory cells configured as a shift register, which can be implemented as a circular buffer in computer memory.

---

[5]For example, the Texas Instruments C64 DSP can perform up to 2.4 *billion* 16-bit multiply-adds per second (ca. 2001).

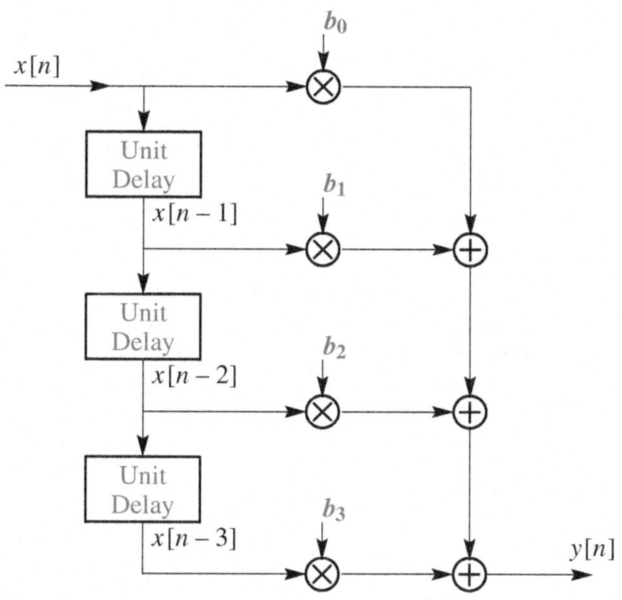

Figure 5-13: Block-diagram structure for the third-order FIR filter.

5-4.2 Block Diagrams

In order to create a graphical representation that is useful for hardware structures, we use ***block-diagram notation***, which defines the interconnection of the three basic building blocks to make more complex structures. In such a directed graph, the nodes (i.e., junction points) are either summing nodes, splitting nodes, or input–output nodes. The connections between nodes are either delay branches or multiplier branches. Figure 5-13 shows the general block diagram for a third-order FIR digital filter ($M = 3$). This structure shows why the FIR filter is also called a ***feed-forward difference equation***, since all paths lead forward from the input to the output. There are no closed-loop paths in the block diagram. In Chapter 8 we will discuss filters with *feedback*, where both input and past output values are involved in the computation of the output.

Strictly speaking, the structure of Fig. 5-13 is a block-diagram representation of the equation

$$y[n] = (((b_0x[n] + b_1x[n-1]) + b_2x[n-2]) + b_3x[n-3])$$

which clearly expands into an equation that can be represented as in (5.12). The input signal is delayed by the cascaded unit delays, each delayed signal is multiplied by a filter coefficient, and the products are accumulated to form the sum. Thus, it is easy to see that there is a one-to-one correspondence between the block diagram and the difference equation (5.12) of the FIR filter, because both are defined by the filter coefficients $\{b_k\}$. A useful skill is to start with one representation and then produce the other. The structure in Fig. 5-13 displays a regularity that makes it simple to define longer filters; the number of cascaded delay elements is increased to M, and then the filter coefficients $\{b_k\}$ are substituted into the diagram. This standard structure is called the ***direct form***. Going from the block diagram back to the difference equation is just as easy, as long as we stick to direct form. Here is a simple exercise to make the correspondence.

EXERCISE 5.7: Determine the difference equation for the block diagram of Fig. 5-14.

5-4.2.1 Other Block Diagrams

Many block diagrams will *implement* the same FIR filter, in the sense that the external behavior from input to output will be the same. Direct form is just one possibility. Other block diagrams would represent a different internal computation, or a different order of computation. In some cases, the internal multipliers might use different

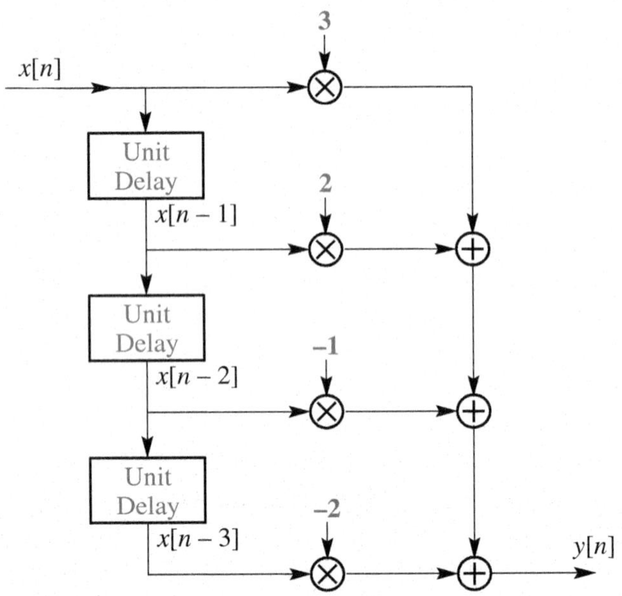

Figure 5-14: Third-order FIR filter in direct form.

coefficients. After we have studied the z-transform in Chapters 7 and 8, we will have the tools to produce many different implementations.

When faced with an arbitrary block diagram, the following four-step procedure may be used to derive the difference equation from the block diagram.

(a) Give a unique signal name to the input of each unit-delay block.

(b) Notice that the output of a unit delay can be written in terms of its input.

(c) At each summing node of the structure, write a signal equation. Use the signal names introduced in steps 1 and 2.

(d) At this point, you will have several equations involving $x[n]$, $y[n]$, and the internal signal names. These can be reduced to one equation involving only $x[n]$ and $y[n]$ by eliminating variables, as is done with simultaneous equations.

Let us try this procedure on the simple but useful example shown in Fig. 5-15. First of all, we observe that the internal signal variables $\{v_1[n], v_2[n], v_3[n]\}$ have been defined in Fig. 5-15 as the inputs to the three delays. Then the delay outputs are, from top to bottom, $v_3[n-1]$, $v_2[n-1]$, and $v_1[n-1]$. We also notice that $v_3[n]$ is a scaled version of the input $x[n]$. In addition, we write the three equations at the output of the three summing nodes:

$$y[n] = b_0x[n] + v_1[n-1]$$
$$v_1[n] = b_1x[n] + v_2[n-1]$$
$$v_2[n] = b_2x[n] + v_3[n-1]$$
$$v_3[n] = b_3x[n]$$

Now we have four equations in five "unknowns." To show that this set of equations is equivalent to the direct form, we can eliminate the $v_i[n]$ by combining the equations in a pairwise fashion.

$$\begin{aligned} v_2[n] &= b_2x[n] + b_3x[n-1] \\ y[n] &= b_0x[n] + b_1x[n-1] + v_2[n-2] \\ &= b_0x[n] + b_1x[n-1] + b_2x[n-2] + b_3x[n-3] \end{aligned}$$

Thus, we have derived the same difference equation as before, so this new structure must be a different way of computing the same thing. In fact, it is widely used and is called the ***transposed form*** for the FIR filter.

EXERCISE 5.8: Explain why Fig. 5-15 is called the *transposed form* of the direct form shown in Fig. 5-13?

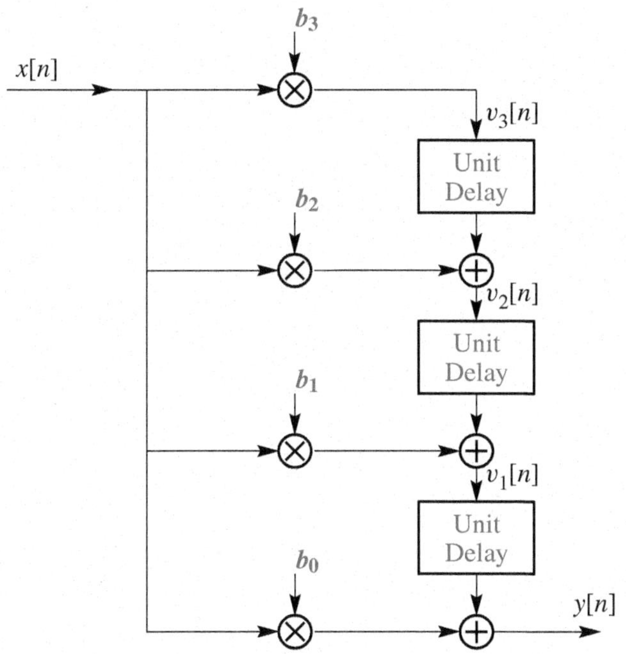

Figure 5-15: Transposed form block diagram structure for the third-order FIR filter.

5-4.2.2 Internal Hardware Details

A block diagram shows dependencies among the different signal variables. Therefore, different block diagrams that implement the same input–output operation may have dramatically different characteristics as far as their internal behavior goes. Several issues come to mind:

(a) The order of computation is specified by the block diagram. In high-speed applications where parallelism or pipelining must be exploited, dependencies in the block diagram represent constraints on the computation.

(b) Partitioning a filter for a VLSI chip would be done (at the highest level) in terms of the block diagram. Likewise, algorithms that must be mapped onto special DSP architectures can be managed by using the block diagram with special compilers that translate the block diagram into optimized code for the DSP chip.

(c) Finite word-length effects are important if the filter is constructed using fixed-point arithmetic. In this case, round-off error and overflow are important real-world problems that depend on the internal order of computation.

Now that we know something about convolution and the implementation of FIR filters, it is time to consider discrete-time systems in a more general way. In the next section, we will show that FIR filters are a special case of the general class of linear time-invariant systems. Much of what we have learned about FIR filters will apply to this more general class of systems.

5-5 Linear Time-Invariant (LTI) Systems

In this section, we discuss two general properties of systems. These properties, ***linearity*** and ***time invariance***, lead to simplifications of mathematical analysis and greater insight and understanding of system behavior. To facilitate the discussion of these properties, it is useful to recall the block-diagram representation of a general discrete-time system shown in Fig. 5-1 on p. 102. This block diagram depicts a transformation of the input signal $x[n]$ into an output signal $y[n]$. It will also be useful to introduce the notation

$$x[n] \longmapsto y[n] \tag{5.13}$$

to represent this transformation. In a specific case, the system is defined by giving a formula or algorithm for computing all values of the output sequence from the values of the input sequence. A specific example is the *square-law* system defined by the rule:

$$y[n] = (x[n])^2 \tag{5.14}$$

Another example is (5.9), which defines the general delay system; still another is (5.12), which defines the general FIR filter and includes the delay system as a special case. We will see that these FIR filters are both linear and time-invariant, while the square-law system is not linear.

5-5.1 Time Invariance

A discrete-time system is said to be *time-invariant* if, when an input is delayed (shifted) by n_0, the output is delayed by the same amount. Using the notation introduced above, we can express this condition as

$$x[n-n_0] \longmapsto y[n-n_0] \tag{5.15}$$

where $x[n] \longmapsto y[n]$. The condition must be true for any choice of n_0, the integer that determines the amount of shift.

A block-diagram view of the time-invariance property is given in Fig. 5-16. In the upper branch, the input is shifted prior to the system; in the lower branch, the output is shifted. Thus, a system can be tested for time-invariance by checking whether or not $w[n] = y[n-n_0]$ in Fig. 5-16.

Consider the example of the square-law system defined by (5.14). If we use the delayed input as the input to the square-law system, we obtain

$$w[n] = (x[n-n_0])^2$$

If $x[n]$ is the input to the square-law system, then $y[n] = (x[n])^2$ and

$$y[n-n_0] = (x[n-n_0])^2 = w[n]$$

so the square-law system is time-invariant.

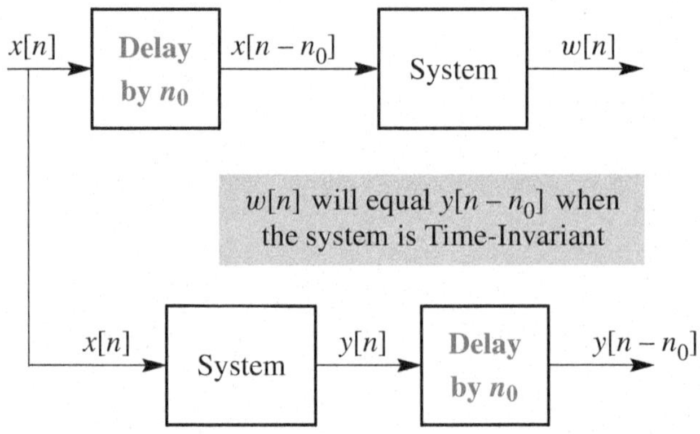

Figure 5-16: Testing time-invariance property by checking the interchange of operations.

A second simple example is the time-flip system, defined by the equation

$$y[n] = x[-n]$$

This system simply reverses the order of "flips" the input sequence about the origin. If we delay the input and then reverse the order about the origin, we obtain

$$w[n] = x[(-n) - n_0] = x[-n-n_0]$$

However, if we first flip the input sequence and then delay it, we obtain a different sequence from $w[n]$; i.e., since $y[n] = x[-n]$

$$y[n-n_0] = x[-(n-n_0)] = x[-n+n_0]$$

Thus, the time-flip system is *not* time-invariant.

EXERCISE 5.9: Test the system defined by the equation

$$y[n] = nx[n]$$

to determine whether it is a time-invariant system.

5-5.2 Linearity

Linear systems have the property that if $x_1[n] \longmapsto y_1[n]$ and $x_2[n] \longmapsto y_2[n]$, then

$$\begin{aligned} x[n] &= \alpha x_1[n] + \beta x_2[n] \\ &\longmapsto \quad y[n] = \alpha y_1[n] + \beta y_2[n] \end{aligned} \tag{5.16}$$

This mathematical condition must be true for any choice of the constants α and β. Equation (5.16) states that if the input consists of a sum of scaled sequences, then the corresponding output is a sum of scaled outputs corresponding to the individual input sequences. A block-diagram view of the linearity property is given in Fig. 5-17, which shows that a system can be tested for the linearity property by checking whether or not $w[n] = y[n]$.

The linearity condition in (5.16) is equivalent to the ***principle of superposition***: If the input is the sum (superposition) of two or more scaled sequences, we can find the output due to each sequence acting alone and then add (superimpose) the separate scaled outputs. Sometimes it is useful to separate (5.16) into two conditions. Setting $\alpha = \beta = 1$ we get the condition

$$x[n] = x_1[n] + x_2[n] \quad \longmapsto \quad y[n] = y_1[n] + y_2[n] \tag{5.17}$$

and using only one scaled input gives

$$x[n] = \alpha x_1[n] \quad \longmapsto \quad y[n] = \alpha y_1[n] \tag{5.18}$$

Both (5.17) and (5.18) must be true in order for (5.16) to be true.

Reconsider the example of the square-law system defined by (5.14). The output $w[n]$ in Fig. 5-17 for this system is

$$w[n] = \alpha(x_1[n])^2 + \beta(x_2[n])^2$$

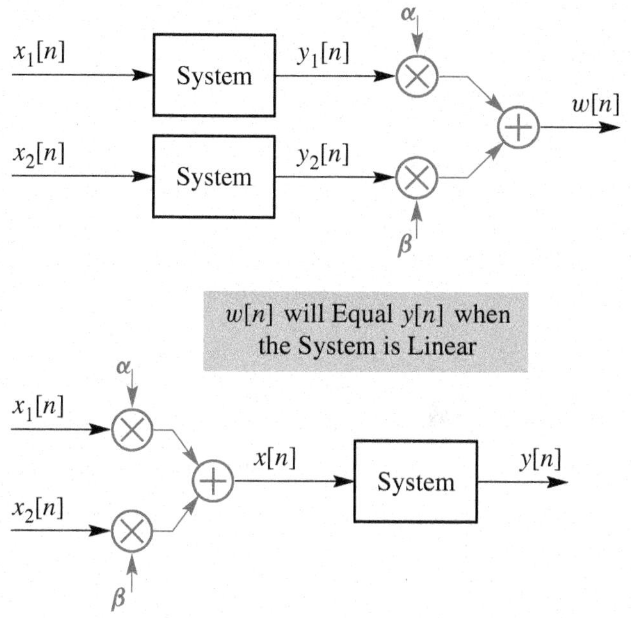

Figure 5-17: Testing linearity by checking the interchange of operations.

while

$$\begin{aligned} y[n] &= (\alpha x_1[n] + \beta x_2[n])^2 \\ &= \alpha^2(x_1[n])^2 + 2\alpha\beta x_1[n]x_2[n] + \beta^2(x_2[n])^2 \end{aligned}$$

Thus, $w[n] \neq y[n]$, and the square-law system has been shown not to be linear. Systems that are not linear are called *nonlinear* systems.

EXERCISE 5.10: Show that the time-flip system $y[n] = x[-n]$ is a linear system.

5-5.3 The FIR Case

FIR Systems described by (5.3) satisfy both the linearity and time invariance conditions. A mathematical proof of time-invariance can be constructed using the procedure depicted in Fig. 5-16. If we define the signal $v[n]$ to be $x[n-n_0]$, then the difference equation relating $v[n]$ to $w[n]$ in the upper branch of Fig. 5-16 is

$$\begin{aligned} w[n] &= \sum_{k=0}^{M} b_k v[n-k] \\ &= \sum_{k=0}^{M} b_k x[(n-k)-n_0] \\ &= \sum_{k=0}^{M} b_k x[(n-n_0)-k] \end{aligned}$$

For comparison, we construct $y[n-n_0]$ in the lower branch

$$y[n] = \sum_{k=0}^{M} b_k x[n-k]$$

$$\Rightarrow \quad y[n-n_0] = \sum_{k=0}^{M} b_k x[n-n_0-k]$$

These two expressions are identical, so the output $w[n]$ is equal to $y[n-n_0]$, and we have proven that the FIR filter is time-invariant.

The linearity condition is simpler to prove. Just substitute into the difference equation (5.3) and collect terms:

$$\begin{aligned} y[n] &= \sum_{k=0}^{M} b_k x[n-k] \\ &= \sum_{k=0}^{M} b_k (\alpha x_1[n-k] + \beta x_2[n-k]) \\ &= \alpha \sum_{k=0}^{M} b_k x_1[n-k] + \beta \sum_{k=0}^{M} b_k x_2[n-k] \\ &= \alpha y_1[n] + \beta y_2[n] \end{aligned}$$

Thus, the FIR filter obeys the principle of superposition; therefore, it is a linear system.

A system that satisfies both properties is called a ***linear time-invariant*** system, or simply ***LTI***. It should be emphasized that the LTI condition is a general condition. The FIR filter is an example of an LTI system. Not all LTI systems are described by (5.3), but all systems described by (5.3) are LTI systems.

5-6 Convolution and LTI Systems

Consider an LTI discrete-time system as depicted in Fig. 5-8. The impulse response $h[n]$ of the LTI system is simply the output when the input is the unit impulse sequence $\delta[n]$. In this section, we will show that the impulse response is a complete characterization for any LTI system, and that convolution is the general formula that allows us to compute the output from the input *for any LTI system.* In our initial discussion of convolution, we considered only finite-length input sequences and FIR filters. Now, we will give a completely general presentation.

5-6.1 Derivation of the Convolution Sum

We begin by recalling from the discussion in Section 5-3.2 that *any* signal $x[n]$ can be represented as a sum of scaled and shifted impulse signals. Each nonzero sample of the signal $x[n]$ multiplies an impulse signal that is shifted to the index of that sample. Specifically, we can write $x[n]$ as follows:

$$\begin{aligned} x[n] &= \sum_{\ell} x[\ell]\delta[n-\ell] \qquad (5.19) \\ &= \cdots + x[-2]\delta[n+2] + x[-1]\delta[n+1] + \\ &\quad x[0]\delta[n] + x[1]\delta[n-1] + x[2]\delta[n-2] + \ldots \end{aligned}$$

In the most general case, the range of summation in (5.19) could be from $-\infty$ to $+\infty$. In (5.19) we have left the range indefinite, realizing that the sum would include all nonzero samples of the input sequence.

A sum of scaled sequences such as (5.19) is commonly referred to as a "linear combination" or superposition of scaled sequences. Thus, (5.19) is a representation of the sequence $x[n]$ as a linear combination of scaled, shifted impulses. Since LTI systems respond in simple and predictable ways to sums of signals and to shifted signals, this representation is particularly useful for our purpose of deriving a general formula for the output of an LTI system.

Figure 5-8 reminds us that the response to the input $\delta[n]$ is, by definition, the impulse response $h[n]$. Time invariance gives us additional information; the response due to $\delta[n-1]$ is $h[n-1]$. In fact, we can write a whole family of input–output pairs as follows:

$$\begin{aligned}
\delta[n] &\longmapsto h[n] \\
\Rightarrow \delta[n-1] &\longmapsto h[n-1] \\
\Rightarrow \delta[n-2] &\longmapsto h[n-2] \\
\Rightarrow \delta[n-(-1)] &\longmapsto h[n-(-1)] = h[n+1] \\
\Rightarrow \delta[n-\ell] &\longmapsto h[n-\ell] \quad \text{for any integer } \ell
\end{aligned}$$

Now we are in a position to use linearity, because (5.19) expresses a general input signal as a linear combination of shifted impulse signals. We can write out a few of the cases:

$$\begin{aligned}
x[0]\delta[n] &\longmapsto x[0]h[n] \\
x[1]\delta[n-1] &\longmapsto x[1]h[n-1] \\
x[2]\delta[n-2] &\longmapsto x[2]h[n-2] \\
x[\ell]\delta[n-\ell] &\longmapsto x[\ell]h[n-\ell] \quad \text{for any integer } \ell
\end{aligned}$$

Then we use superposition to put it all together as

$$x[n] = \sum_{\ell} x[\ell]\delta[n-\ell] \longmapsto y[n] = \sum_{\ell} x[\ell]h[n-\ell] \tag{5.20}$$

The derivation of (5.20) did not assume that either $x[n]$ or $h[n]$ was of finite duration, so, in general, we may need infinite limits on the sum. With this modification we obtain

The Convolution Sum Formula

$$y[n] = \sum_{\ell=-\infty}^{\infty} x[\ell]h[n-\ell] \tag{5.21}$$

This expression represents the convolution operation in the most general sense, so we have proved that ***all LTI systems can be represented by a convolution sum.*** The infinite limits take care of all possibilities, including the special cases where either or both of the sequences are finite in length.

Example 5-2: FIR Convolution

For example, if $h[n]$ is nonzero only in the interval $0 \le n \le M$, then (5.21) reduces to

$$y[n] = \sum_{\ell=n-M}^{n} x[\ell]h[n-\ell] \tag{5.22}$$

because the argument $n-\ell$ must lie in the range $0 \le n-\ell \le M$, so the range for ℓ in (5.21) is restricted to $(n-M) \le \ell \le n$. ■

Fig. 5-18 shows a MATLAB GUI that is available to generate examples of convolution with a variety of simple signals.

DEMO: *Discrete Convolution*

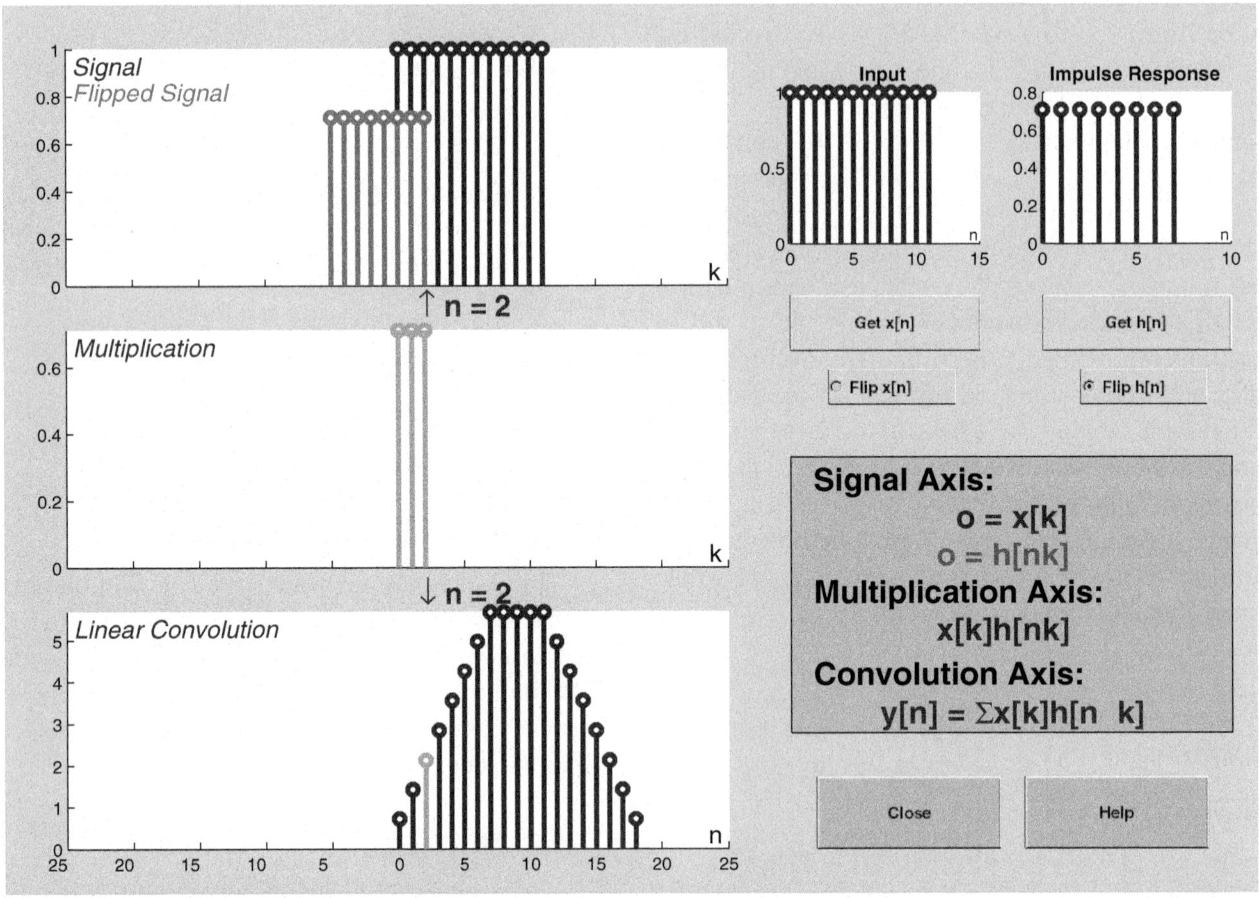

Figure 5-18: Graphical User Interface (GUI) for `dconvdemo` which illustrates the *sliding window* nature of FIR filtering (i.e., convolution). The user can pick the input signal and the impulse response, and then slide the filter over the input signal to observe how the output is constructed.

EXERCISE 5.11: By making the substitution $k = n - \ell$ in (5.22), show that $y[n]$ can also be expressed in the same form as (5.10), $y[n] = \sum_{k=0}^{M} h[k]x[n-k]$

5-6.2 Some Properties of LTI Systems

The properties of convolution are the properties of LTI systems. Thus, it is of interest to explore these properties and relate them to properties of LTI systems.

5-6.2.1 Convolution as an Operator

An interesting aspect of convolution is its algebraic character as an operation between two signals. The convolution of $x[n]$ and $h[n]$ is an operation that will be denoted by $*$, i.e.,

$$y[n] = x[n] * h[n] = \sum_{\ell=-\infty}^{\infty} x[\ell]h[n-\ell] \tag{5.23}$$

We say that the sequence $x[n]$ is *convolved* with the sequence $h[n]$ to produce the output $y[n]$.

The notation $x[n] * h[n]$ is useful because it allows us to think about convolution problems in an operational way. As a simple example, recall that the ideal delay system has impulse response $h[n] = \delta[n - n_0]$. We know that the output of the ideal delay system is $y[n] = x[n - n_0]$. Therefore, it follows that

Convolution with an Impulse

$$x[n] * \delta[n - n_0] = x[n - n_0] \tag{5.24}$$

This is a very important and useful result because (5.24) states that "to convolve any sequence $x[n]$ with an impulse located at $n = n_0$, all we need to do is translate the origin of $x[n]$ to n_0."

5-6.2.2 Commutative Property of Convolution

It is relatively easy to prove that convolution is a commutative operation between two sequences.

$$x[n] * h[n] = h[n] * x[n] \tag{5.25}$$

In fact, the truth of the commutative property is clear from the computational algorithm in Fig. 5-11 on p. 110, where either $x[n]$ or $h[n]$ could be written on the first row, so $x[n]$ and $h[n]$ are interchangeable. In the following algebraic manipulation, we make the change of variables $k = n - \ell$ and sum on the new dummy variable k:

$$\begin{aligned} y[n] &= x[n] * h[n] \\ &= \sum_{\ell=-\infty}^{\infty} x[\ell]h[n-\ell] \\ &= \sum_{k=+\infty}^{-\infty} x[n-k]h[k] \qquad (\textit{Note: } k = n - \ell) \\ &= \sum_{k=-\infty}^{\infty} h[k]x[n-k] = h[n] * x[n] \end{aligned}$$

On the third line in this set of equations, the limits on the sum can be swapped without consequence, because a set of numbers can be added in any order. Thus, we have proved that convolution is a commutative operation, i.e.,

$$y[n] = x[n] * h[n] = h[n] * x[n]$$

5-6.2.3 Associative Property of Convolution

The associative property, which is perhaps less obvious than the commutative property, states that, when we are convolving three signals, we can convolve two of them and then convolve that result with the third signal.

$$(x_1[n] * x_2[n]) * x_3[n] = x_1[n] * (x_2[n] * x_3[n]) \tag{5.26}$$

The algebraic manipulation needed to prove the associative property is tedious, but we give it here to illustrate how convolution expressions can be manipulated into different forms. We assume that $x_1[n]$, $x_2[n]$, and $x_3[n]$ are three arbitrary sequences to be convolved. Then,

$$
\begin{aligned}
x_1[n] * (x_2[n] * x_3[n])& \\
&= \sum_{\ell=-\infty}^{\infty} x_1[\ell] \left(\sum_{k=-\infty}^{\infty} x_2[k] x_3[(n-\ell)-k] \right) \\
&= \sum_{\ell=-\infty}^{\infty} x_1[\ell] \sum_{q=-\infty}^{\infty} x_2[q-\ell] x_3[n-q] \\
&= \sum_{q=-\infty}^{\infty} \sum_{\ell=-\infty}^{\infty} x_1[\ell] x_2[q-\ell] x_3[n-q] \\
&= \sum_{q=-\infty}^{\infty} \left(\sum_{\ell=-\infty}^{\infty} x_1[\ell] x_2[q-\ell] \right) x_3[n-q] \\
&= (x_1[n] * x_2[n]) * x_3[n]
\end{aligned}
$$

In the third line, we used the change of variables $q = \ell + k$. This proves that convolution is an associative operation. The implication of this property for cascaded systems is explored in Section 5-7.

5-7 Cascaded LTI Systems

In a cascade connection of two systems, the output of the first system is the input to the second system, and the overall output of the cascade system is taken to be the output of the second system. Figure 5-19 shows two LTI systems (LTI 1 and LTI 2) in cascade. LTI systems have the remarkable property that two LTI systems in cascade can be implemented in either order. This property is a direct consequence of the commutative and associative properties of convolution, as demonstrated by the following three equivalent expressions that can be

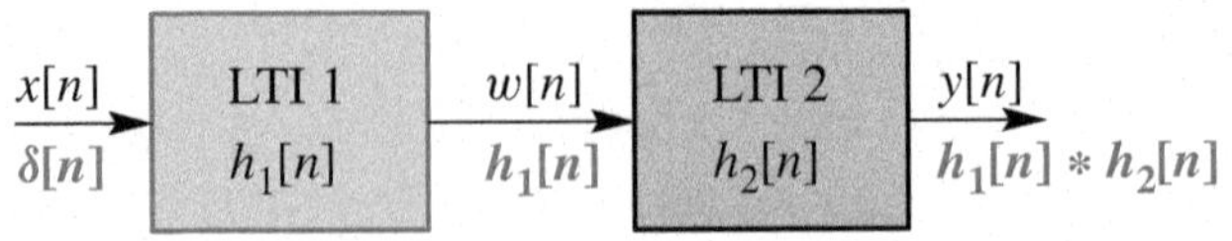

Figure 5-19: A cascade of two LTI systems. The overall impulse response is the convolution of the two individual impulse responses.

obtained by applying the commutative and associative properties of convolution:

$$
\begin{aligned}
y[n] &= (x[n] * h_1[n]) * h_2[n] && (5.27)\\
&= x[n] * (h_1[n] * h_2[n]) && (5.28)\\
&= x[n] * (h_2[n] * h_1[n]) && (5.29)\\
&= (x[n] * h_2[n]) * h_1[n] && (5.30)
\end{aligned}
$$

Equation (5.27) is a mathematical statement of the fact that the second system processes the output of the first, which is $w[n] = x[n] * h_1[n]$. Equation (5.28) shows that the output $y[n]$ is the convolution of the input with a new impulse response $h_1[n] * h_2[n]$. This corresponds to Fig. 5-20(b) with $h[n] = h_1[n] * h_2[n]$. Equation (5.29) uses the commutative property of convolution to show that $h[n] = h_1[n] * h_2[n] = h_2[n] * h_1[n]$. Applying the associative property leads to (5.30), which is the cascade connection in Fig. 5-20(a). Notice that reordering the LTI systems in cascade gives the same final output, i.e., it is correct to label the outputs of all three systems in Figs. 5-19 and 5-20 with the same symbol, $y[n]$, even though the intermediate signals, $w[n]$ and $v[n]$, are different.

Another way to show that the order of cascaded LTI systems does not affect the overall system response is to prove that the impulse response of the two cascade systems is the same. In Fig. 5-19, the impulse input and the corresponding outputs are shown below the arrows. When the input to the first system is an impulse, the output

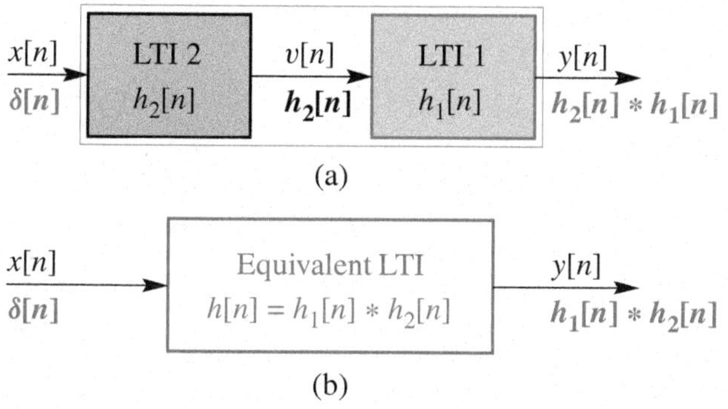

Figure 5-20: Switching the order of cascaded LTI systems. The single LTI system in (b) is equivalent to the cascade connection of the two systems in (a), and also to the cascade in Fig. 5-19.

of LTI 1 is its impulse response, $h_1[n]$, which becomes the input to LTI 2. The output of LTI 2 is, therefore, just the convolution of its input $h_1[n]$ with its impulse response $h_2[n]$. Therefore, the overall impulse response of Fig. 5-19 is $h_1[n] * h_2[n]$. In the same way, we can easily show that the overall impulse response of the other cascade system in Fig. 5-20(a) is $h_2[n] * h_1[n]$. Since convolution is commutative, the two cascade systems have the same impulse response

$$h[n] = h_1[n] * h_2[n] = h_2[n] * h_1[n] \tag{5.31}$$

which is also the impulse response $h[n]$ of the equivalent system. Again, since the overall impulse response is the same for each of the three systems in Figs. 5-19 and 5-20, the output is the same for all three systems for the same input.

Example 5-3: Impulse Response of Cascade

To illustrate the utility of the results that we have obtained for cascaded LTI systems, consider the cascade of two systems defined by

$$h_1[n] = \begin{cases} 1 & 0 \le n \le 3 \\ 0 & \text{otherwise} \end{cases} \qquad h_2[n] = \begin{cases} 1 & 1 \le n \le 3 \\ 0 & \text{otherwise} \end{cases}$$

The results of this section show that the overall cascade system has impulse response

$$h[n] = h_2[n] * h_1[n]$$

Therefore, to find the overall impulse response we must convolve $h_1[n]$ with $h_2[n]$. This can be done by using the polynomial multiplication algorithm of Section 5-3.3.1. In this case, the computation is as follows:

n	$n<0$	0	1	2	3	4	5	6	$n>6$
$h_1[n]$	0	1	1	1	1	0	0	0	0
$h_2[n]$	0	0	1	1	1				
$h_2[0]h_1[n]$	0	0	0	0	0	0	0	0	0
$h_2[1]h_1[n-1]$	0	0	1	1	1	1	0	0	0
$h_2[2]h_1[n-2]$	0	0	0	1	1	1	1	0	0
$h_2[3]h_1[n-3]$	0	0	0	0	1	1	1	1	0
$h[n]$	0	0	1	2	3	3	2	1	0

Therefore, the equivalent impulse response is

$$h[n] = \sum_{k=0}^{6} b_k \delta[n-k]$$

where $\{b_k\}$ is the sequence $\{0, 1, 2, 3, 3, 2, 1\}$.

This result means that a system with this impulse response $h[n]$ can be implemented either by the single difference equation

$$y[n] = \sum_{k=0}^{6} b_k x[n-k] \tag{5.32}$$

where $\{b_k\}$ is the above sequence, or by the pair of difference equations

$$w[n] = \sum_{k=0}^{3} x[n-k] \qquad y[n] = \sum_{k=1}^{3} w[n-k] \tag{5.33}$$

■

Example 5-3 illustrates an important point. There is a significant difference between (5.32) and (5.33). It can be seen that the implementation in (5.33) requires a total of only five additions to compute each value of the output sequence, while (5.32) requires five additions and an additional four multiplications by coefficients that are not equal to unity. On a larger scale (longer filters), such differences in the amount and type of computation can be very important in practical applications of FIR filters. Thus, the existence of alternative equivalent implementations of the same filter is significant.

5-8 Example of FIR Filtering

We conclude this chapter with an example of the use of FIR filtering on a real signal. An example of sampled data that can be viewed as a signal is the Dow-Jones Industrial Average. The DJIA is a sequence of numbers obtained by averaging the closing prices of a selected number of representative stocks. It has been computed since 1897, and is used in many ways by investors and economists. The entire sequence dating back to 1897 makes up a signal that has positive values and is exponentially growing. In order to obtain an example where fine detail is visible, we have selected the segment from 1950 to 1970 of the weekly closing price average to be our signal $x[n]$.[6] This signal is the one showing high variability in Fig. 5-21, where each weekly value is plotted as a distinct point. The smoother curve in Fig. 5-21 is the output of a 51-point causal running averager; i.e.,

$$y[n] = \frac{1}{51}\sum_{k=0}^{50} x[n-k] = x[n] * h[n]$$

[6]The complete set of weekly sampled data for the period 1897–2002 would involve over 5000 samples and would range in value from about 40 to almost 12,000. A plot of the entire sequence would not show the effects that we wish to illustrate.

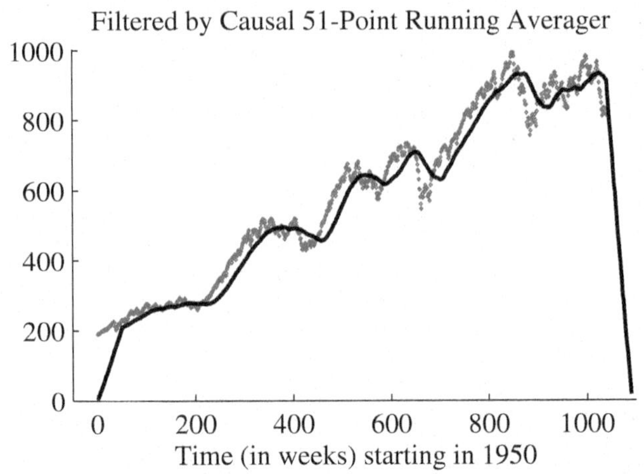

Figure 5-21: DJIA weekly closings filtered by a 51-point running-average FIR filter. Input (orange) and output (black).

where $h[n]$ is the impulse response of the causal 51-point running averager.

Notice that, as we have observed in Section 5-3.1, there is a region at the beginning and end (50 samples, in this case) where the filter is engaging and disengaging the input signal. Also notice that much of the fine-scale variation has been removed by the filter. Finally, notice that the output is shifted relative to the input. In Chapter 6, we will develop techniques that will allow us to show that the shift introduced by this filter is exactly $M/2 = 25$ samples.

In this example, it is important to be able to compare the input and output without the shift. It would be better use the noncausal *centralized* running averager

$$\tilde{y}[n] = \frac{1}{51}\sum_{k=-25}^{25} x[n-k] = x[n] * \tilde{h}[n]$$

where $\tilde{h}[n]$ is the impulse response of the centralized running averager. The output $\tilde{y}[n]$ can be obtained by shifting (advancing) the output of the causal running averager by 25 samples to the left. In terms of our previous discussion of cascaded FIR systems, we could think of the centralized system as a cascade of the causal system with a system whose impulse response is $\delta[n+25]$, i.e.,[7]

$$\begin{aligned}\tilde{y}[n] &= y[n] * \delta[n+25] \\ &= (x[n] * h[n]) * \delta[n+25] \\ &= x[n] * (h[n] * \delta[n+25]) \\ &= x[n] * h[n+25]\end{aligned}$$

Thus, we find that $\tilde{h}[n] = h[n+25]$, i.e., the impulse response of the centralized running averager is a shifted version of the impulse response of the causal running averager. By shifting the impulse response, we remove the delay introduced by the causal system.

EXERCISE 5.12: Determine the impulse response $h[n]$ of the 51-point causal running averager and determine the impulse response $\tilde{h}[n]$ for the 51-point centralized running averager.

The cascade representation of the centralized running averager is depicted in Fig. 5-22. Another way to describe the system in Fig. 5-22 is that the second system *compensates* for the delay of the first, or we might describe the centralized running averager as a *delay-compensated* running-average filter.

Figure 5-23 shows the input $x[n]$ and output $\tilde{y}[n]$ for the delay-compensated running-average filter. It now can be seen that corresponding features of the input and output are well aligned.

[7]Recall from (5.24) that convolution of $y[n]$ with an impulse simply shifts the origin of the sequence to the location of the impulse (in this case, to $n = -25$).

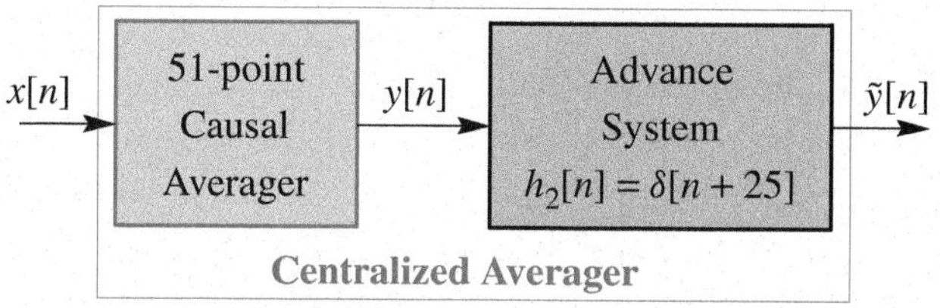

Figure 5-22: Cascade interpretation of centralized running averager in terms of causal running averager with delay compensation.

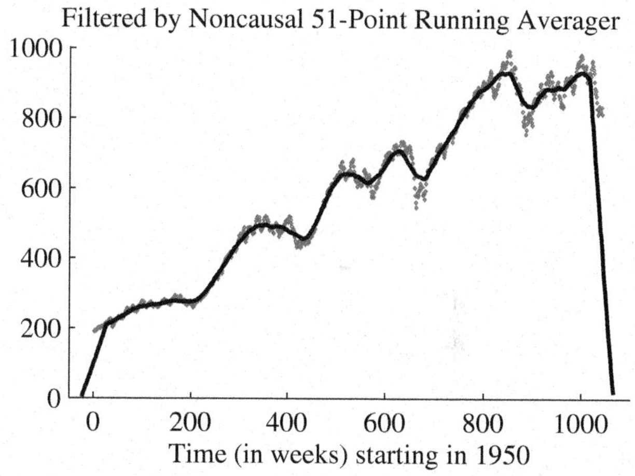

Figure 5-23: Input (orange) and output (black) for delay-compensated 51-point running averager applied to DJIA weekly closings.

As the example in this section illustrates, FIR filters can be used to remove rapid fluctuations in signals. Furthermore, the example shows that it is worthwhile to develop the fundamental mathematical properties of such systems because these properties can be useful in helping us to understand the way such systems work. In Chapter 6, we will further develop our understanding of FIR systems.

5-9 Summary and Links

This chapter introduced the concept of FIR filtering. Among the laboratory projects on the CD-ROM, there is one (Lab #7) that deals with discrete convolution and the effects of sampling. Lab #7 also requires the use of two MATLAB GUIs, one for sampling, CON2DIS, the other for discrete convolution, DCONVDEMO.

LAB: *#7 Sampling, Convolution, and FIR Filtering*

DEMO: *Ch 4, Continuous-Discrete Sampling*

DEMO: *Discrete Convolution*

The CD-ROM also contains individual demonstrations of the properties of linearity and time invariance illustrated by using simple FIR filters to process shifted and scaled sinusoids.

Finally, the reader is once again reminded of the large number of solved homework problems that are available for review and practice on the CD-ROM.

NOTE: *Hundreds of Solved Problems*

5-10 Problems

P-5.1 If the impulse response $h[n]$ of an FIR filter is

$$h[n] = \delta[n-1] - 2\delta[n-4]$$

write the difference equation for the FIR filter.

P-5.2 Evaluate the "running" average

$$y[n] = \frac{1}{L}\sum_{k=0}^{L-1} x[n-k]$$

for the *unit-step* input signal, i.e.,

$$x[n] = u[n] = \begin{cases} 0 & \text{for } n < 0 \\ 1 & \text{for } n \geq 0 \end{cases}$$

(a) Make a plot of $u[n]$ before working out the answer for $y[n]$.

(b) Now compute the numerical values of $y[n]$ over the time interval $-5 \leq n \leq 10$, assuming that $L = 5$.

(c) Make a sketch of the output over the time interval $-5 \leq n \leq 10$, assuming that $L = 5$. Use MATLAB if necessary, but learn to do it by hand also.

(d) Finally, derive a general formula for $y[n]$ that will apply for any length L and for the range $n \geq 0$.

P-5.3 A linear time-invariant system is described by the difference equation

$$y[n] = 2x[n] - 3x[n-1] + 2x[n-2]$$

(a) When the input to this system is

$$x[n] = \begin{cases} 0 & n < 0 \\ n+1 & n = 0, 1, 2 \\ 5-n & n = 3, 4 \\ 1 & n \geq 5 \end{cases}$$

Compute the values of $y[n]$, over the index range $0 \leq n \leq 10$.

(b) For the previous part, plot both $x[n]$ and $y[n]$.

(c) Determine the response of this system to a unit impulse input; i.e., find the output $y[n] = h[n]$ when the input is $x[n] = \delta[n]$. Plot $h[n]$ as a function of n.

P-5.4 A linear time-invariant system is described by the difference equation

$$y[n] = 2x[n] - 3x[n-1] + 2x[n-2]$$

(a) Draw the implementation of this system as a block diagram in direct form.

(b) Give the implementation as a block diagram in transposed direct form.

P-5.5 Consider a system defined by

$$y[n] = \sum_{k=0}^{M} b_k x[n-k]$$

(a) Suppose that the input $x[n]$ is nonzero only for $0 \le n \le N-1$; i.e., it has a support of N samples. Show that $y[n]$ is nonzero at most over a finite interval of the form $0 \le n \le P-1$. Determine P and the support of $y[n]$ in terms of M and N.

(b) Suppose that the input $x[n]$ is nonzero only for $N_1 \le n \le N_2$. What is the support of $x[n]$? Show that $y[n]$ is nonzero at most over a finite interval of the form $N_3 \le n \le N_4$. Determine N_3 and N_4 and the support of $y[n]$ in terms of N_1, N_2, and M.

Hint: Draw a sketch similar to Fig. 5-5 on p. 105.

P-5.6 The unit-step signal "turns on" at $n = 0$, and is usually denoted by $u[n]$.

(a) Make a plot of $u[n]$.

(b) We can use the unit-step sequence to represent other sequences that are zero for $n < 0$. Plot the sequence $x[n] = (0.5)^n u[n]$.

(c) The L-point running average is defined as

$$y[n] = \frac{1}{L}\sum_{k=0}^{L-1} x[n-k]$$

For the input sequence $x[n] = (0.5)^n u[n]$, compute the numerical value of $y[n]$ over the index range $-5 \le n \le 10$, assuming that $L = 4$.

(d) For the input sequence $x[n] = a^n u[n]$, derive a general formula for $y[n]$ that will apply for any value a, for any length L, and for the index range $n \ge 0$. In doing so, you may have use for the formula:

$$\sum_{k=M}^{N} \alpha^k = \frac{\alpha^M - \alpha^{N+1}}{1-\alpha}$$

P-5.7 Answer the following questions about the time-domain response of FIR digital filters:

$$y[n] = \sum_{k=0}^{M} b_k\, x[n-k]$$

When tested with an input signal that is an impulse, $x[n] = \delta[n]$, the observed output from the filter is the signal $h[n]$ shown in Fig. P-5.7. Determine the filter coefficients $\{b_k\}$ of the difference equation for the FIR filter.

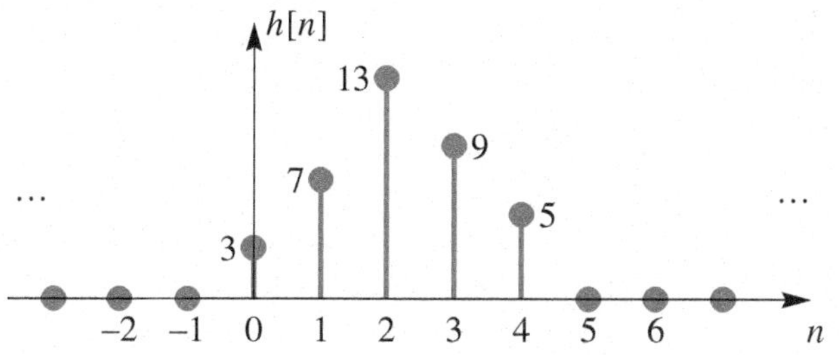

Figure P-5.7

P-5.8 If the filter coefficients of an FIR system are $\{b_k\} = \{13, -13, 13\}$ and the input signal is

$$x[n] = \begin{cases} 0 & \text{for } n \text{ even} \\ 1 & \text{for } n \text{ odd} \end{cases}$$

determine the output signal $y[n]$ for all n. Give your answer as either a plot or a formula.

P-5.9 For each of the following systems, determine whether or not the system is (1) linear, (2) time-invariant, and (3) causal.

(a) $y[n] = x[n]\cos(0.2\pi n)$

(b) $y[n] = x[n] - x[n-1]$

(c) $y[n] = |x[n]|$

(d) $y[n] = Ax[n] + B$, where A and B are constants.

P-5.10 Suppose that $\mathcal{S}$ is a linear, time-invariant system whose exact form is unknown. It is tested by running some inputs into the system, and then observing the output signals. Suppose that the following input–output pairs are the result of the tests:

Input: $x[n]$	**Output:** $y[n]$
$\delta[n] - \delta[n-1]$	$\delta[n] - \delta[n-1] + 2\delta[n-3]$
$\cos(\pi n/2)$	$2\cos(\pi n/2 - \pi/4)$

(a) Make a plot of the signal:

$$y[n] = \delta[n] - \delta[n-1] + 2\delta[n-3].$$

(b) Use linearity and time invariance to find the output of the system when the input is

$$x[n] = 7\delta[n] - 7\delta[n-2]$$

P-5.11 A linear time-invariant system has impulse response

$$h[n] = 3\delta[n] - 2\delta[n-1] + 4\delta[n-2] + \delta[n-4]$$

(a) Draw the implementation of this system as a block diagram in direct form.

(b) Give the implementation as a block diagram in transposed direct form.

P-5.12 For a particular LTI system, when the input is the unit step, $x_1[n] = u[n]$, the corresponding output is

$$y_1[n] = \delta[n] + 2\delta[n-1] - \delta[n-2] = \begin{cases} 0 & n < 0 \\ 1 & n = 0 \\ 2 & n = 1 \\ -1 & n = 2 \\ 0 & n \geq 3 \end{cases}$$

Determine the output when the input to the LTI system is $x_2[n] = 3u[n] - 2u[n-4]$. Give your answer as a formula expressing $y_2[n]$ in terms of known sequences, or give a list of values for $-\infty < n < \infty$.

P-5.13 For the FIR filter $y[n] = x[n] - ax[n-1]$, determine the output signal for two similar cases.

(a) The input signal is $x[n] = a^n u[n]$.

(b) The input signal is $x[n] = a^n(u[n] - u[n-10])$.

P-5.14 One form of the *deconvolution* process starts with the output signal and the filter's impulse response, from which it should be possible to find the input signal.

(a) If the output of an FIR filter with $h[n] = \delta[n-2]$ is

$$y[n] = u[n-3] - u[n-6],$$

determine the input signal, $x[n]$.

(b) If the output of a first-difference FIR filter is

$$y[n] = \delta[n] - \delta[n-4],$$

determine the input signal, $x[n]$.

(c) If the output of a 4-point averager is

$$y[n] = -5\delta[n] - 5\delta[n-2],$$

determine the input signal, $x[n]$.

P-5.15 Another form of the *deconvolution* process starts with the output signal and the input signal, from which it should be possible to find the impulse response.

(a) If the input signal is $x[n] = u[n]$, find the FIR filter that will produce the output $y[n] = u[n-1]$.

(b) If the input signal is $x[n] = u[n]$, find the FIR filter that will produce the output $y[n] = \delta[n]$.

(c) If the input signal is $x[n] = (\frac{1}{2})^n u[n]$, find the FIR filter that will produce the output $y[n] = \delta[n-1]$.

P-5.16 Sometimes it is not possible to solve the *deconvolution* process for a given input–output pair. For example, prove that there is no FIR filter that can process the input $x[n] = \delta[n]+\delta[n-1]$ to give the output $y[n] = \delta[n]$.

P-5.17 Suppose that three systems are connected in cascade. In other words, the output of $\mathcal{S}_1$ is the input to $\mathcal{S}_2$, and the output of $\mathcal{S}_2$ is the input to $\mathcal{S}_3$. The three systems are specified as follows:

$$\mathcal{S}_1: \quad y_1[n] = x_1[n] - x_1[n-1]$$

$$\mathcal{S}_2: \quad y_2[n] = x_2[n] + x_2[n-2]$$

$$\mathcal{S}_3: \quad y_3[n] = x_3[n-1] + x_3[n-2]$$

Thus $x_1[n] = x[n]$, $x_2[n] = y_1[n]$, $x_3[n] = y_2[n]$, and $y[n] = y_3[n]$. We wish to determine the equivalent system that is a single operation from the input $x[n]$ (into $\mathcal{S}_1$) to the output $y[n]$, which is the output of $\mathcal{S}_3$.

(a) Determine the impulse response $h_i[n]$ for each individual subsystem $\mathcal{S}_i$.

(b) Determine the impulse response $h[n]$ of the overall system, i.e., find $h[n]$ so that $y[n] = x[n] * h[n]$.

(c) Write one difference equation that defines the overall system in terms of $x[n]$ and $y[n]$ only.

P-5.18 Consider a system implemented by the following MATLAB program:

```
% xx.mat is a binary file containing the
%    vector of input samples called "xx"
load xx
yy1 = conv(ones(1,4),xx);
yy2 = conv([1, -1, -1, 1],xx);
ww = yy1 + yy2;
yy = conv(ones(1,3),ww);
```

The overall system from input `xx` to output `yy` is an LTI system composed of three LTI systems.

(a) Draw a block diagram of the system that is implemented by the program given above. Be sure to indicate the impulse responses and difference equations of each of the component systems.

(b) The overall system is an LTI system. What is its impulse response, and what is the difference equation that is satisfied by the input $x[n]$ and the output $y[n]$?

CHAPTER 6

Frequency Response of FIR Filters

Chapter 5 introduced the class of FIR discrete-time systems. We showed that the weighted running average of a finite number of input sequence values defines a discrete-time system, and we showed that such systems are linear and time-invariant. We also showed that the impulse response of an FIR system completely defines the system. In this chapter, we introduce the concept of the ***frequency response*** of a linear time-invariant FIR filter and show that the frequency response and impulse response are uniquely related. It is remarkable that for linear time-invariant systems, when the input is a complex sinusoid, the corresponding output signal is another complex sinusoid of exactly the same frequency, but with different magnitude and phase. The frequency-response function over all frequencies summarizes the response of an LTI system by giving the magnitude and phase change experienced by all possible sinusoids. Furthermore, since linear time-invariant systems obey the principle of superposition, the frequency-response function is a complete characterization of the behavior of the system for any input that can be represented as a sum of sinusoids. Since almost any discrete-time signal can be represented by a superposition of sinusoids, the frequency response is sufficient therefore to represent the system for almost any signal.

6-1 Sinusoidal Response of FIR Systems

Linear time-invariant systems behave in a particularly simple way when the input is a discrete-time complex exponential. To see this, consider the FIR system

$$y[n] = \sum_{k=0}^{M} b_k x[n-k] \tag{6.1}$$

and assume that the input is a complex exponential signal with normalized radian frequency $\hat{\omega}$

$$x[n] = Ae^{j\phi}e^{j\hat{\omega}n} \qquad -\infty < n < \infty$$

Recall that this discrete-time signal could have been obtained by sampling the continuous-time signal

$$x(t) = Ae^{j\phi}e^{j\omega t}$$

If $x[n] = x(nT_s)$, then ω and $\hat{\omega}$ are related by $\hat{\omega} = \omega T_s$, where T_s is the sampling period. For such inputs, the corresponding output is

$$\begin{aligned} y[n] &= \sum_{k=0}^{M} b_k\, Ae^{j\phi}e^{j\hat{\omega}(n-k)} \\ &= \left(\sum_{k=0}^{M} b_k e^{-j\hat{\omega}k}\right) Ae^{j\phi}e^{j\hat{\omega}n} \\ &= \mathcal{H}(\hat{\omega})Ae^{j\phi}e^{j\hat{\omega}n} \qquad -\infty < n < \infty \end{aligned} \tag{6.2}$$

where

$$\mathcal{H}(\hat{\omega}) = \sum_{k=0}^{M} b_k e^{-j\hat{\omega}k} \tag{6.3}$$

Because we have represented the frequency of the complex exponential signal as the general symbol $\hat{\omega}$, we have obtained an expression (6.3) that is a function of $\hat{\omega}$. In other words, (6.3) describes the response of the LTI system to a complex exponential signal of *any* frequency $\hat{\omega}$. The quantity $\mathcal{H}(\hat{\omega})$ defined by (6.3) is therefore called the ***frequency-response function*** for the system. (Generally, we shorten this to ***frequency response***.)

However, there is an issue of notation for the frequency response that is dictated by consistency with the z-transform (to be introduced in Chapter 7). In (6.3) the righthand side contains powers of the complex exponential $e^{j\hat{\omega}}$. This is true of many expressions for the frequency response. Therefore, we elect to use the notation $H(e^{j\hat{\omega}})$ instead of $\mathcal{H}(\hat{\omega})$ to emphasize the ubiquity of $e^{j\hat{\omega}}$. The function $H(e^{j\hat{\omega}})$ still depends on the variable $\hat{\omega}$. Furthermore, since the impulse response sequence of an FIR filter is the same as the sequence of filter coefficients, we can express the frequency response in terms of either the filter coefficients b_k or the impulse response $h[k]$; i.e.,

The Frequency Response of an FIR System

$$H(e^{j\hat{\omega}}) = \sum_{k=0}^{M} b_k e^{-j\hat{\omega}k} = \sum_{k=0}^{M} h[k]e^{-j\hat{\omega}k} \tag{6.4}$$

Several important points can be made about (6.2) and (6.4). First of all, the precise interpretation of (6.2) is as follows: When the input is a discrete-time complex exponential signal, the output of an LTI FIR filter is also a discrete-time complex exponential signal with a different complex amplitude, but the same frequency $\hat{\omega}$. The frequency response multiplies the signal, thereby changing the complex amplitude. While it is tempting to express this fact by the mathematical statement $y[n] = H(e^{j\hat{\omega}})x[n]$, it is strongly recommended that this never be done because it is too easy to forget that the mathematical statement is true *only* for complex exponential signals of frequency $\hat{\omega}$. The statement $y[n] = H(e^{j\hat{\omega}})x[n]$ is *meaningless* for any signal other than signals of precisely the form $x[n] = Ae^{j\phi}e^{j\hat{\omega}n}$. *It is very important to understand this point.*

A second important point is that the frequency response $H(e^{j\hat{\omega}})$ is complex-valued so it can be expressed either as $H(e^{j\hat{\omega}}) = |H(e^{j\hat{\omega}})|e^{j\angle H(e^{j\hat{\omega}})}$ or as $H(e^{j\hat{\omega}}) = \Re e\{H(e^{j\hat{\omega}})\} + j\Im m\{H(e^{j\hat{\omega}})\}$. The effect of the LTI system on the magnitude and phase of the input complex exponential signal is determined completely by the frequency response function $H(e^{j\hat{\omega}})$. Specifically, if the input is $x[n] = Ae^{j\phi}e^{j\hat{\omega}n}$, then using the polar form of $H(e^{j\hat{\omega}})$ gives the result

$$\begin{aligned} y[n] &= |H(e^{j\hat{\omega}})|e^{j\angle H(e^{j\hat{\omega}})} \cdot Ae^{j\phi}e^{j\hat{\omega}n} \\ &= (|H(e^{j\hat{\omega}})|A) \cdot e^{j(\angle H(e^{j\hat{\omega}})+\phi)}e^{j\hat{\omega}n} \end{aligned} \tag{6.5}$$

The magnitude and angle form of the frequency response is the most convenient form, since multiplication is most conveniently accomplished in polar form. The angle of the frequency response simply adds to the phase of the input, thereby producing additional phase shift in the complex exponential signal. Since the magnitude of the frequency response multiplies the magnitude of the complex exponential signal, this part of the frequency response controls the size of the output. Thus, $|H(e^{j\hat{\omega}})|$ is also referred to as the *gain* of the system.

Example 6-1: Frequency Response Formula

Consider an LTI system for which the difference equation coefficients are $\{b_k\} = \{1,\ 2,\ 1\}$. Substituting into (6.3) gives

$$H(e^{j\hat{\omega}}) = 1 + 2e^{-j\hat{\omega}} + e^{-j\hat{\omega}2}$$

To obtain formulas for the magnitude and phase of the frequency response of this FIR filter, we can manipulate the equation as follows:

$$\begin{aligned} H(e^{j\hat{\omega}}) &= 1 + 2e^{-j\hat{\omega}} + e^{-j\hat{\omega}2} \\ &= e^{-j\hat{\omega}}\left(e^{j\hat{\omega}} + 2 + e^{-j\hat{\omega}}\right) \\ &= e^{-j\hat{\omega}}\left(2 + 2\cos\hat{\omega}\right) \end{aligned}$$

Since $(2 + 2\cos\hat{\omega}) \geq 0$ for frequencies $-\pi < \hat{\omega} \leq \pi$, the magnitude is $|H(e^{j\hat{\omega}})| = (2+2\cos\hat{\omega})$ and the phase is $\angle H(e^{j\hat{\omega}}) = -\hat{\omega}$. ■

Example 6-2: Complex Exponential Input

Consider the complex input $x[n] = 2e^{j\pi/4}e^{j\pi n/3}$. If this signal is the input to the system of Example 6-1, then $|H(e^{j\pi/3})| = 2+2\cos(\pi/3) = 3$ and $\angle H(e^{j\pi/3}) = -j\pi/3$. Therefore, the output of the system for the given input is

$$\begin{aligned} y[n] &= 3e^{-j\pi/3} \cdot 2e^{j\pi/4}e^{j\pi n/3} \\ &= (3 \cdot 2) \cdot e^{(j\pi/4 - j\pi/3)}e^{j\pi n/3} \\ &= 6e^{-j\pi/12}e^{j\pi n/3} = 6e^{j\pi/4}e^{j\pi(n-1)/3} \end{aligned}$$

Thus, for this system and the given input $x[n]$, the output is equal to the input multiplied by 3, and the phase shift corresponds to a delay of one sample. ■

EXERCISE 6.1: When the sequence of coefficients is symmetrical ($b_0 = b_M$, $b_1 = b_{M-1}$, etc.), the frequency response can be manipulated as in Example 6-1. Following the style of that example, show that the frequency response of an FIR filter with coefficients $\{b_k\} = \{1,\ -2,\ 4\ -2,\ 1\}$ can be expressed as

$$H(e^{j\hat{\omega}}) = [4 - 4\cos(\hat{\omega}) + 2\cos(2\hat{\omega})]e^{-j2\hat{\omega}}$$

6-2 Superposition and the Frequency Response

The principle of superposition makes it very easy to find the output of a linear time-invariant system if the input is a sum of complex exponential signals. This is why the frequency response is so important in the analysis and design of LTI systems.

$$
\begin{aligned}
y[n] &= H(e^{j0})A_0 e^{j0n} + H(e^{j\hat{\omega}_1})\frac{A_1}{2}e^{j\phi_1}e^{j\hat{\omega}_1 n} + H^*(e^{j\hat{\omega}_1})\frac{A_1}{2}e^{-j\phi_1}e^{-j\hat{\omega}_1 n} \\
&= H(e^{j0})A_0 + |H(e^{j\hat{\omega}_1})|e^{j\angle H(e^{j\hat{\omega}_1})}\frac{A_1}{2}e^{j\phi_1}e^{j\hat{\omega}_1 n} + |H(e^{j\hat{\omega}_1})|e^{-j\angle H(e^{j\hat{\omega}_1})}\frac{A_1}{2}e^{-j\phi_1}e^{-j\hat{\omega}_1 n} \\
&= H(e^{j0})A_0 + |H(e^{j\hat{\omega}_1})|\frac{A_1}{2}e^{j(\hat{\omega}_1 n+\phi_1+\angle H(e^{j\hat{\omega}_1}))} + |H(e^{j\hat{\omega}_1})|\frac{A_1}{2}e^{-j(\hat{\omega}_1 n+\phi_1+\angle H(e^{j\hat{\omega}_1}))} \\
&= H(e^{j0})A_0 + |H(e^{j\hat{\omega}_1})|A_1\cos\left(\hat{\omega}_1 n+\phi_1+\angle H(e^{j\hat{\omega}_1})\right)
\end{aligned}
\tag{6.6}
$$

As an example, suppose that the input to an LTI system is a cosine wave with a specific normalized frequency $\hat{\omega}_1$ plus a DC level,

$$x[n] = A_0 + A_1\cos(\hat{\omega}_1 n + \phi_1)$$

If we represent the signal in terms of complex exponentials, the signal is composed of three complex exponential signals,

$$x[n] = A_0 e^{j0n} + \frac{A_1}{2}e^{j\phi_1}e^{j\hat{\omega}_1 n} + \frac{A_1}{2}e^{-j\phi_1}e^{-j\hat{\omega}_1 n}$$

with frequencies $\hat{\omega} = 0$, $\hat{\omega}_1$, and $-\hat{\omega}_1$. By superposition, we can determine the output due to each term separately and then add them to obtain the output $y[n]$ corresponding to $x[n]$. Because the components of the input signal are all complex exponential signals, it is easy to find their respective outputs if we know the frequency response of the system; we just multiply each component by $H(e^{j\hat{\omega}})$ evaluated at the corresponding frequency, i.e.,

$$
\begin{aligned}
y[n] = {} & H(e^{j0})A_0 e^{j0n} \\
& + H(e^{j\hat{\omega}_1})\frac{A_1}{2}e^{j\phi_1}e^{j\hat{\omega}_1 n} + H(e^{-j\hat{\omega}_1})\frac{A_1}{2}e^{-j\phi_1}e^{-j\hat{\omega}_1 n}
\end{aligned}
$$

Note that we have used the fact that a constant signal is a complex exponential with $\hat{\omega} = 0$. If we express $H(e^{j\hat{\omega}_1})$ as $H(e^{j\hat{\omega}_1}) = |H(e^{j\hat{\omega}_1})|e^{j\angle H(e^{j\hat{\omega}_1})}$, then the algebraic steps[1] in (6.6) show that $y[n]$ can finally be expressed as the cosine signal. Notice that the magnitude and phase change of the cosine input signal are taken from the positive frequency part of $H(e^{j\hat{\omega}})$, but also notice that it was crucial to express $x[n]$ as a sum of complex exponentials and then use the frequency response to find the output due to each component separately.

Example 6-3: Cosine Input

For the FIR filter with coefficients $\{b_k\} = \{1, 2, 1\}$, find the output when the input is

$$x[n] = 2\cos\left(\frac{\pi}{3}n - \frac{\pi}{2}\right)$$

The frequency response of the system was determined in Example 6-1 to be

$$H(e^{j\hat{\omega}}) = (2 + 2\cos\hat{\omega})e^{-j\hat{\omega}}$$

---

[1] It is assumed that $H(e^{j\hat{\omega}})$ has the ***conjugate-symmetry*** property $H(e^{-j\hat{\omega}}) = H^*(e^{j\hat{\omega}})$, which is always true when the filter coefficients are real. (See Section 6-4.3.)

$$\begin{aligned} y[n] &= H(e^{j0})X_0 + \sum_{k=1}^{N}\left(H(e^{j\hat{\omega}_k})\frac{X_k}{2}e^{j\hat{\omega}_k n} + H(e^{-j\hat{\omega}_k})\frac{X_k^*}{2}e^{-j\hat{\omega}_k n}\right) \\ &= H(e^{j0})X_0 + \sum_{k=1}^{N}|H(e^{j\hat{\omega}_k})||X_k|\cos\left(\hat{\omega}_k n + \angle X_k + \angle H(e^{j\hat{\omega}_k})\right) \end{aligned} \tag{6.7}$$

Note that $H(e^{-j\hat{\omega}}) = H^*(e^{j\hat{\omega}})$; i.e., $H(e^{j\hat{\omega}})$ has conjugate symmetry. Solution of this problem requires just one evaluation of $H(e^{j\hat{\omega}})$ at the frequency $\hat{\omega} = \pi/3$:

$$\begin{aligned} H(e^{j\pi/3}) &= e^{-j\pi/3}\,(2 + 2\cos(\pi/3)) \\ &= e^{-j\pi/3}\,(2 + 2(\tfrac{1}{2})) = 3e^{-j\pi/3} \end{aligned}$$

Therefore, the magnitude is $|H(e^{j\pi/3})| = 3$ and the phase is $\angle H(e^{j\pi/3}) = -\pi/3$, so the output is

$$\begin{aligned} y[n] &= (3)(2)\cos\left(\frac{\pi}{3}n - \frac{\pi}{3} - \frac{\pi}{2}\right) \\ &= 6\cos\left(\frac{\pi}{3}(n-1) - \frac{\pi}{2}\right) \end{aligned}$$

Notice that the magnitude of the frequency response multiplies the amplitude of the cosine signal, and the phase angle of the frequency response adds to the phase of the cosine signal. This problem could also be solved with the DLTI MATLAB GUI. ■

DEMO: *DLTI Demo*

If the input signal consists of many complex exponential signals, the frequency response can be applied to find the output due to each component separately, and the results added to determine the total output. This is the principle of superposition at work. If we can find a representation for a signal in terms of complex exponentials, the frequency response gives a simple and highly intuitive means for determining what an LTI system does to that input signal. For example, if the input to an LTI system is a real signal and can be represented as

$$\begin{aligned} x[n] &= X_0 + \sum_{k=1}^{N}\left(\frac{X_k}{2}e^{j\hat{\omega}_k n} + \frac{X_k^*}{2}e^{-j\hat{\omega}_k n}\right) \\ &= X_0 + \sum_{k=1}^{N}|X_k|\cos(\hat{\omega}_k n + \angle X_k) \end{aligned}$$

then it follows that if $H(e^{-j\hat{\omega}}) = H^*(e^{j\hat{\omega}})$, the corresponding output is $y[n]$ in (6.7).

That is, each individual complex exponential component is modified by the frequency response evaluated at the frequency of that component.

Example 6-4: Three Sinusoidal Inputs

For the FIR filter with coefficients $\{b_k\} = \{1, 2, 1\}$, find the output when the input is

$$x[n] = 4 + 3\cos\left(\frac{\pi}{3}n - \frac{\pi}{2}\right) + 3\cos\left(\frac{7\pi}{8}n\right) \tag{6.8}$$

The frequency response of the system was determined in Example 6-1, and is the same as the frequency response in Example 6-3. The input in this example differs from that of Example 6-3 by the addition of a constant (DC) term and an additional cosine signal of frequency $7\pi/8$. The solution by superposition therefore requires that we evaluate $H(e^{j\hat{\omega}})$ at frequencies 0, $\pi/3$, and $7\pi/8$, giving

$$\begin{aligned} H(e^{j0}) &= 4 \\ H(e^{j\pi/3}) &= 3e^{-j\pi/3} \\ H(e^{j7\pi/8}) &= 0.1522e^{-j7\pi/8} \end{aligned}$$

Therefore, the output is

$$\begin{aligned} y[n] &= 4 \cdot 4 + 3 \cdot 3\cos\left(\frac{\pi}{3}n - \frac{\pi}{3} - \frac{\pi}{2}\right) \\ &\quad + 0.1522 \cdot 3\cos\left(\frac{7\pi}{8}n - \frac{7\pi}{8}\right) \\ &= 16 + 9\cos\left(\frac{\pi}{3}(n-1) - \frac{\pi}{2}\right) \\ &\quad + 0.4567\cos\left(\frac{7\pi}{8}(n-1)\right) \end{aligned}$$

Notice that, in this case, the DC component is multiplied by 4, the component at frequency $\hat{\omega} = \pi/3$ is multiplied by 3, but the component at frequency $\hat{\omega} = 7\pi/8$ is multiplied by 0.1522. Because the frequency-response magnitude (gain) is so small at frequency $\hat{\omega} = 7\pi/8$, the component at this frequency is essentially *filtered out* of the input signal. ■

The examples of this section illustrate an approach to solving problems that is often called the ***frequency-domain*** approach. As these examples show, we do not need to deal with the ***time-domain*** description (i.e., the difference equation or impulse response) of the system when the input is a complex exponential signal. We can work exclusively with the frequency-domain description (i.e., the frequency–response function), if we think about how the spectrum of the signal is modified by the system rather than considering what happens to the individual samples of the input signal. We will have ample opportunity to visit both the time-domain and the frequency-domain in the remainder of this chapter.

DEMO: *Introduction to FIR Filtering*

6-3 Steady-State and Transient Response

In Section 6-1, we showed that if the input is

$$x[n] = Xe^{j\hat{\omega}n} \qquad -\infty < n < \infty \tag{6.9}$$

where $X = Ae^{j\phi}$, then the corresponding output of an LTI FIR system is

$$y[n] = H(e^{j\hat{\omega}})Xe^{j\hat{\omega}n} \qquad -\infty < n < \infty \tag{6.10}$$

where

$$H(e^{j\hat{\omega}}) = \sum_{k=0}^{M} b_k e^{-j\hat{\omega}k} \tag{6.11}$$

In (6.9), the condition that $x[n]$ be a complex exponential signal existing over $-\infty < n < \infty$ is important. Without this condition, we will not obtain the simple result of (6.10). However, this condition appears to be somewhat impractical. In any practical implementation, we surely would not have actual input signals that exist back to $-\infty$! Fortunately, we can relax the condition that the complex exponential be defined over the doubly infinite interval and still take advantage of the convenience of (6.10). To see this, consider the following "suddenly applied" complex exponential signal that starts at $n = 0$ and is nonzero only for $0 \le n$:

$$x[n] = Xe^{j\hat{\omega}n}u[n] = \begin{cases} Xe^{j\hat{\omega}n} & 0 \le n \\ 0 & n < 0 \end{cases} \tag{6.12}$$

Note that multiplication by the unit-step signal is a convenient way to impose the suddenly applied condition. The output of an LTI FIR system for this input is

$$y[n] = \sum_{k=0}^{M} b_k X e^{j\hat{\omega}(n-k)} u[n-k] \qquad (6.13)$$

By considering different values of n and the fact that $u[n-k] = 0$ for $k > n$, it follows that the sum in (6.13) can be expressed as

$$y[n] = \begin{cases} 0 & n < 0 \\ \left(\displaystyle\sum_{k=0}^{n} b_k e^{-j\hat{\omega}k}\right) X e^{j\hat{\omega}n} & 0 \le n < M \\ \left(\displaystyle\sum_{k=0}^{M} b_k e^{-j\hat{\omega}k}\right) X e^{j\hat{\omega}n} & M \le n \end{cases} \qquad (6.14)$$

That is, when the complex exponential signal is suddenly applied, the output can be considered to be defined over three distinct regions. In the first region, $n < 0$, the input is zero, and therefore the corresponding output is zero, too. The second region is a transition region whose length is M samples (i.e., the *order* of the FIR system). In this region, the complex multiplier of $e^{j\hat{\omega}n}$ depends upon n. This region is often called the ***transient*** part of the output. In the third region, $M \le n$, the output is identical to the output that would be obtained if the input were defined over the doubly infinite interval. That is,

$$y[n] = H(e^{j\hat{\omega}}) X e^{j\hat{\omega}n} \qquad M \le n \qquad (6.15)$$

This part of the output is generally called the ***steady-state*** part. While we have specified that the steady-state part exists for all $n \ge M$, it should be clear that (6.15) holds only as long as the input remains equal to $Xe^{j\hat{\omega}n}$. If, at some time $n > M$, the input changes frequency or goes to zero, another transient region will occur.

Example 6-5: Steady-State Output

A simple example will illustrate the above discussion. Consider the system of Exercise 6.1, whose filter coefficients are the sequence $\{b_k\} = \{1, -2, 4, -2, 1\}$. The frequency response of this system is

$$H(e^{j\hat{\omega}}) = [4 - 4\cos(\hat{\omega}) + 2\cos(2\hat{\omega})]e^{-j2\hat{\omega}}$$

If the input is the suddenly applied cosine signal

$$x[n] = \cos(0.2\pi n - \pi)u[n]$$

we can represent it as the sum of two suddenly applied complex exponential signals. Therefore, the frequency response can be used as discussed in Section 6-2 to determine the corresponding steady-state output. Since $H(e^{j\hat{\omega}})$ at $\hat{\omega} = 0.2\pi$ is

$$H(e^{j\hat{\omega}}) = [4 - 4\cos(0.2\pi) + 2\cos(0.4\pi)]e^{-j2(0.2\pi)},$$

and $M = 4$, the steady-state output is

$$y[n] = 1.382\cos(0.2\pi(n-2) - \pi) \qquad 4 \le n$$

The frequency response has allowed us to find a simple expression for the output everywhere in the steady-state region. If we desire the values of the output in the transient region, we could compute them using the difference equation for the system.

The input and output signals for this example are shown in Fig. 6-1. Since $M = 4$ for this system, the transient region is $0 \le n \le 3$ (indicated by the shaded region), and the steady-state region is $n \ge 4$. Also note that, as predicted by the steady-state analysis above, the signal in the steady-state region is simply a scaled and shifted (by 2 samples) version of the input. ■

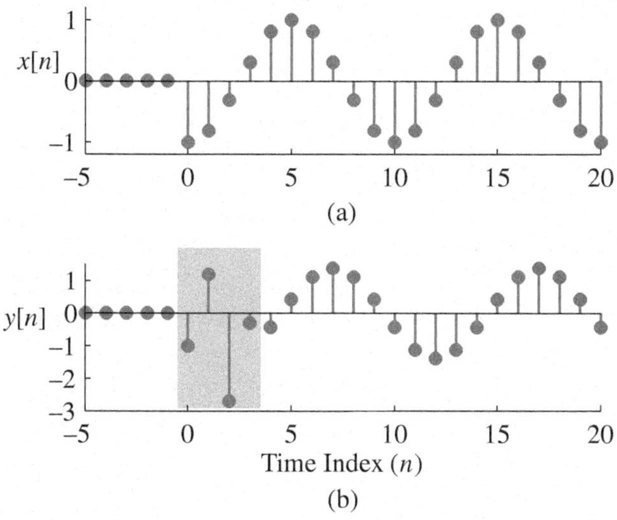

Figure 6-1: (a) Input $x[n] = \cos(0.2\pi n - \pi)u[n]$ and (b) corresponding output $y[n]$ for FIR filter with coefficients $\{1, -2, 4, -2, 1\}$. The transient region is the shaded area in (b). (Note the different vertical scales.)

6-4 Properties of the Frequency Response

The frequency response function $H(e^{j\hat{\omega}})$ is a complex-valued function of the normalized frequency variable $\hat{\omega}$. This function has interesting properties that often can be used to simplify analysis.

6-4.1 Relation to Impulse Response and Difference Equation

$H(e^{j\hat{\omega}})$ can be calculated directly from the filter coefficients $\{b_k\}$. If (6.1) is compared to (6.4), we see that, given the difference equation, it is simple to write down an expression for $H(e^{j\hat{\omega}})$ by noting that each term $b_k x[n-k]$ in (6.1) corresponds to a term $b_k e^{-j\hat{\omega}k}$ or $h[k]e^{-j\hat{\omega}k}$ in (6.4), and vice versa. Likewise, $H(e^{j\hat{\omega}})$ can be determined directly from the impulse response since the impulse response of the FIR system consists of the sequence of filter coefficients; i.e., $h[k] = b_k$ for $k = 0, 1, \ldots, M$. To emphasize this point, we can write the correspondence

Time Domain	$\leftrightarrow$	*Frequency Domain*
$h[n] = \sum_{k=0}^{M} h[k]\delta[n-k]$	$\leftrightarrow$	$H(e^{j\hat{\omega}}) = \sum_{k=0}^{M} h[k]e^{-j\hat{\omega}k}$

The process of going from the difference equation or impulse response to the frequency response is straightforward for the FIR filter. It is also simple to go from the frequency response to the difference equation or to the impulse response if we express $H(e^{j\hat{\omega}})$ in terms of powers of $e^{-j\hat{\omega}}$. These points are illustrated by the following examples.

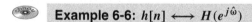

Example 6-6: $h[n] \longleftrightarrow H(e^{j\hat{\omega}})$

Consider the FIR filter defined by the impulse response

$$h[n] = -\delta[n] + 3\delta[n-1] - \delta[n-2]$$

By inspection, the sequence of filter coefficients is $\{b_k\} = \{-1, 3, -1\}$, so the difference equation corresponding to this impulse response is

$$y[n] = -x[n] + 3x[n-1] - x[n-2]$$

and the frequency response of this system is

$$H(e^{j\hat{\omega}}) = -1 + 3e^{-j\hat{\omega}} - e^{-j2\hat{\omega}}$$

■

Example 6-7: Difference Equation from $H(e^{j\hat{\omega}})$

Suppose that the frequency response is given by the equation

$$H(e^{j\hat{\omega}}) = e^{-j\hat{\omega}}\left(3 - 2\cos\hat{\omega}\right)$$

Since $\cos\hat{\omega} = \frac{1}{2}(e^{j\hat{\omega}} + e^{-j\hat{\omega}})$, we can write

$$\begin{aligned} H(e^{j\hat{\omega}}) &= e^{-j\hat{\omega}}\left[3 - 2\left(\frac{e^{j\hat{\omega}} + e^{-j\hat{\omega}}}{2}\right)\right] \\ &= -1 + 3e^{-j\hat{\omega}} - e^{-j\hat{\omega}2} \end{aligned}$$

which corresponds to the following FIR difference equation:

$$y[n] = -x[n] + 3x[n-1] - x[n-2]$$

The impulse response, likewise, is easy to determine directly from $H(e^{j\hat{\omega}})$, when expressed in terms of powers of $e^{-j\hat{\omega}}$. ■

EXERCISE 6.2: Use the inverse Euler formula for sines to find the impulse response and difference equation for $H(e^{j\hat{\omega}}) = 2j\sin(\hat{\omega}/2)e^{-j\hat{\omega}/2}$.

6-4.2 Periodicity of $H(e^{j\hat{\omega}})$

An important property of a discrete-time LTI system is that its frequency response $H(e^{j\hat{\omega}})$ is always a periodic function with period 2π. This can be seen by considering a frequency $\hat{\omega}+2\pi$ where $\hat{\omega}$ is any frequency. Substituting into (6.3) gives

$$\begin{aligned} H(e^{j(\hat{\omega}+2\pi)}) &= \sum_{k=0}^{M} b_k e^{-j(\hat{\omega}+2\pi)k} \\ &= \sum_{k=0}^{M} b_k e^{-j\hat{\omega}k} e^{-j2\pi k} = H(e^{j\hat{\omega}}) \end{aligned}$$

since $e^{-j2\pi k} = 1$ when k is an integer. It is not surprising that $H(e^{j\hat{\omega}})$ should have this property, since, as we have seen in Chapter 4, a change in the input frequency by 2π is not detectable; i.e.,

$$x[n] = Xe^{j(\hat{\omega}+2\pi)n} = Xe^{j\hat{\omega}n}e^{j2\pi n} = Xe^{j\hat{\omega}n}$$

In other words, two complex exponential signals with frequencies differing by 2π cannot be distinguished from their samples alone, so there is no reason to expect a discrete-time system to behave differently for two such frequencies. For this reason, it is always sufficient to specify the frequency response only over an interval of one period, e.g., $-\pi < \hat{\omega} \le \pi$.

6-4.3 Conjugate Symmetry

The frequency response $H(e^{j\hat{\omega}})$ is complex, but usually has a symmetry in its magnitude and phase that allows us to concentrate on just half of the period when plotting. This is the property of ***conjugate symmetry***

$$H(e^{-j\hat{\omega}}) = H^*(e^{j\hat{\omega}}) \tag{6.16}$$

which is true whenever the filter coefficients are real so that $b_k = b_k^*$ (equivalently $h[k] = h^*[k]$). We can prove this property for the FIR case as follows:

$$\begin{aligned} H^*(e^{j\hat{\omega}}) &= \left(\sum_{k=0}^{M} b_k e^{-j\hat{\omega}k}\right)^* \\ &= \sum_{k=0}^{M} b_k^* e^{+j\hat{\omega}k} \\ &= \sum_{k=0}^{M} b_k e^{-j(-\hat{\omega})k} = H(e^{-j\hat{\omega}}) \end{aligned}$$

The conjugate-symmetry property implies that the magnitude function is an even function of $\hat{\omega}$ and the phase is an odd function, i.e.,

$$|H(e^{-j\hat{\omega}})| = |H(e^{j\hat{\omega}})|$$

$$\angle H(e^{-j\hat{\omega}}) = -\angle H(e^{j\hat{\omega}})$$

Similarly, the real part is an even function of $\hat{\omega}$ and the imaginary part is an odd function, i.e.,

$$\Re e\{H(e^{-j\hat{\omega}})\} = \Re e\{H(e^{j\hat{\omega}})\}$$

$$\Im m\{H(e^{-j\hat{\omega}})\} = -\Im m\{H(e^{j\hat{\omega}})\}$$

As a result, plots of the frequency response are often shown only over half a period, $0 \le \hat{\omega} \le \pi$, because the negative frequency region can be constructed by symmetry. These symmetries are illustrated by the plots in Section 6-5.

EXERCISE 6.3: Prove that the magnitude is an even function of $\hat{\omega}$ and the phase is an odd function of $\hat{\omega}$ for a conjugate-symmetric frequency response.

6-5 Graphical Representation of the Frequency Response

Two important points should be emphasized about the frequency response of an LTI system. The first is that for a given system, the frequency response usually varies with frequency, so that sinusoids of different frequencies are treated differently by the system. The second important point is that by appropriate choice of the coefficients, b_k, a wide variety of frequency response shapes can be realized. In order to visualize the variation of the frequency response with frequency, it is useful to plot $H(e^{j\hat{\omega}})$ versus $\hat{\omega}$. We will see that the plot tells us at a glance what the system does to complex exponential signals and sinusoids of different frequencies. Several examples are provided in this section to illustrate the value of plotting the frequency response.

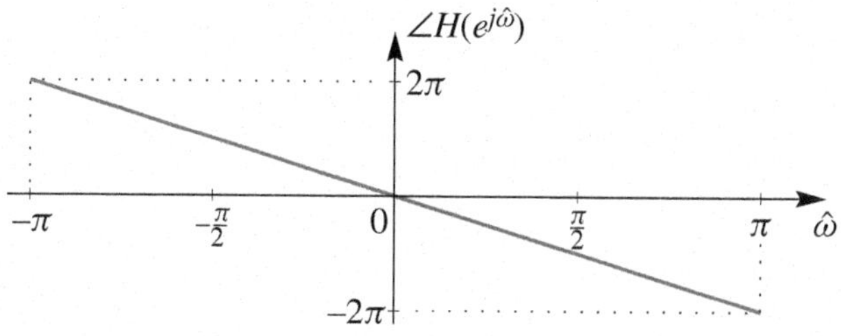

Figure 6-2: Phase response of pure delay ($n_0 = 2$) system, $H(e^{j\hat{\omega}}) = e^{-j2\hat{\omega}}$.

6-5.1 Delay System

The delay system is a simple FIR filter given by the difference equation

$$y[n] = x[n - n_0]$$

It has only one nonzero filter coefficient, $b_{n_0} = 1$, so its frequency response is

$$H(e^{j\hat{\omega}}) = e^{-j\hat{\omega}n_0} \quad (6.17)$$

For this filter, a plot of the frequency response is easy to visualize; the magnitude response is one for all frequencies and the phase is given by the equation of a straight line with a slope equal to $-n_0$, as in Fig. 6-2. As a result, we can associate the property of linear phase with time delay in all filters. Since time delay affects only the time origin of the signal in a predictable way, we often think of linear phase as an ideal phase response.

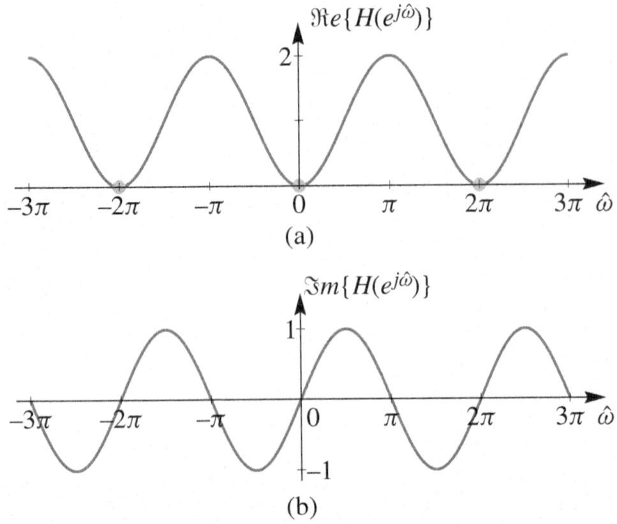

Figure 6-3: (a) Real and (b) imaginary parts for $H(e^{j\hat{\omega}}) = 1 - e^{-j\hat{\omega}}$ over three periods showing periodicity and conjugate symmetry of $H(e^{j\hat{\omega}})$.

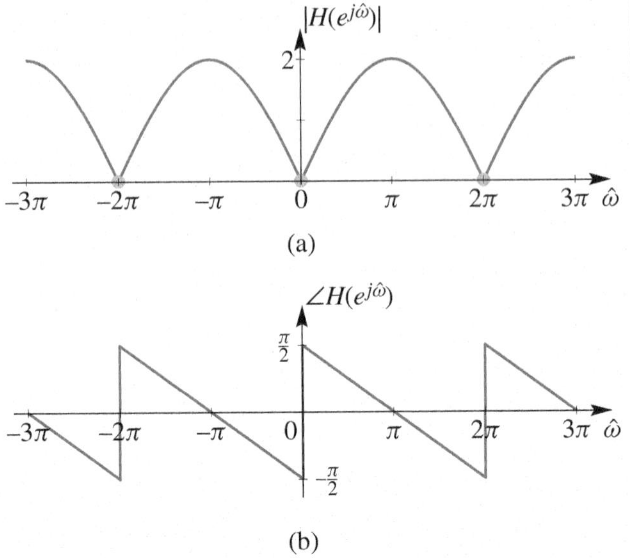

Figure 6-4: (a) Magnitude and (b) phase for $H(e^{j\hat{\omega}}) = 1 - e^{-j\hat{\omega}}$ over three periods showing periodicity and conjugate symmetry of $H(e^{j\hat{\omega}})$.

6-5.2 First-Difference System

As another simple example, consider the first-difference system

$$y[n] = x[n] - x[n-1]$$

The frequency response of this LTI system is

$$H(e^{j\hat{\omega}}) = 1 - e^{-j\hat{\omega}} = 1 - \cos\hat{\omega} + j\sin\hat{\omega}$$

The different parts of the complex representations are

$$\begin{aligned}
\Re e\{H(e^{j\hat{\omega}})\} &= (1 - \cos\hat{\omega}) \\
\Im m\{H(e^{j\hat{\omega}})\} &= \sin\hat{\omega} \\
|H(e^{j\hat{\omega}})| &= [(1-\cos\hat{\omega})^2 + \sin^2\hat{\omega}]^{1/2} \\
&= [2(1-\cos\hat{\omega})]^{1/2} = 2|\sin(\hat{\omega}/2)| \\
\angle H(e^{j\hat{\omega}}) &= \arctan\left(\frac{\sin\hat{\omega}}{1-\cos\hat{\omega}}\right)
\end{aligned}$$

The real and imaginary parts for this example are plotted in Fig. 6-3; the magnitude and phase are plotted in Fig. 6-4. All functions are plotted for $-3\pi < \hat{\omega} < 3\pi$, even though we would normally need to plot the frequency response of a discrete-time system only for $-\pi < \hat{\omega} < \pi$, or (because of conjugate symmetry) $0 \le \hat{\omega} < \pi$. These extended plots verify that $H(e^{j\hat{\omega}})$ is periodic with period 2π, and they verify the conjugate symmetry properties discussed in Section 6-4.3.

The utility of the magnitude and phase plots of $H(e^{j\hat{\omega}})$ can be seen even for this simple example. In Fig. 6-4, $H(e^{j0}) = 0$, so we easily see that the system completely removes components with $\hat{\omega} = 0$ (i.e., DC). Furthermore, we can also see that the system emphasizes the higher frequencies (near $\hat{\omega} = \pi$) relative to the lower frequencies, so it would be called a ***highpass filter***. This is another typical way of thinking about systems in the frequency domain.

The real and imaginary parts and the magnitude and phase can always be determined as demonstrated above by standard manipulations of complex numbers. However, there is a simpler approach for getting the magnitude and phase when the sequence of coefficients is either symmetric or antisymmetric about a central point. The following algebraic manipulation of $H(e^{j\hat{\omega}})$ is possible because the $\{b_k\}$ coefficients satisfy the symmetry condition:

$$b_k = -b_{M-k}$$

The trick, which we have already used in Example 6-1 on p. 132, is to factor out an exponential whose phase is half of the filter order $(M/2)$ times $\hat{\omega}$, and then use the inverse Euler formula to combine corresponding positive- and negative-frequency complex exponentials; i.e.,

$$\begin{aligned} H(e^{j\hat{\omega}}) &= 1 - e^{-j\hat{\omega}} \\ &= e^{-j\hat{\omega}/2}\left(e^{j\hat{\omega}/2} - e^{-j\hat{\omega}/2}\right) \\ &= 2je^{-j\hat{\omega}/2}\sin(\hat{\omega}/2) \\ &= 2\sin(\hat{\omega}/2)e^{j(\pi/2-\hat{\omega}/2)} \end{aligned}$$

The form derived for $H(e^{j\hat{\omega}})$ is almost a valid polar form, but since $\sin(\hat{\omega}/2)$ is negative for $-\pi < \hat{\omega} < 0$, we must write $|H(e^{j\hat{\omega}})| = 2|\sin(\hat{\omega}/2)|$ and absorb the algebraic sign[2] into the phase response for $-\pi < \hat{\omega} < 0$, i.e.,

$$\angle H(e^{j\hat{\omega}}) = \begin{cases} \pi/2 - \hat{\omega}/2 & 0 < \hat{\omega} < \pi \\ -\pi + \pi/2 - \hat{\omega}/2 & -\pi < \hat{\omega} < 0 \end{cases}$$

This formula for the phase is consistent with the phase plot in Fig. 6-4, which exhibits several linear segments.

[2] Remember that $-1 = e^{\pm j\pi}$, so we can add either $+\pi$ or $-\pi$ to the phase for $-\pi < \hat{\omega} < 0$. In this case, we add $-\pi$ so that the resulting phase curve remains between $-\pi$ and $+\pi$ radians for all $\hat{\omega}$.

Notice also that the phase plot has discontinuities at $\hat{\omega} = 0$ and $\hat{\omega} = \pm 2\pi$. The size of these discontinuities is π, since they correspond to a sign change in $H(e^{j\hat{\omega}})$.

Example 6-8: First-Difference Removes DC

Suppose that the input to a first-difference system is $x[n] = 4 + 2\cos(0.3\pi n - \pi/4)$. Since the output is related to the input by the equation $y[n] = x[n] - x[n-1]$, it follows that:

$$\begin{aligned} y[n] &= 4 + 2\cos(0.3\pi n - \pi/4) \\ &\quad - 4 - 2\cos(0.3\pi(n-1) - \pi/4) \\ &= 2\cos(0.3\pi n - \pi/4) - 2\cos(0.3\pi n - 0.55\pi) \end{aligned}$$

From this result, we see that the first-difference system removes the constant value and leaves two cosine signals of the same frequency, which could be combined by phasor addition. However, the solution using the frequency-response function is simpler. Since the first-difference system has frequency response

$$H(e^{j\hat{\omega}}) = 2\sin(\hat{\omega}/2)e^{j(\pi/2-\hat{\omega}/2)}$$

the output of this system for the given input is

$$\begin{aligned} y[n] &= 4H(e^{j0}) \\ &\quad + 2|H(e^{j0.3\pi})|\cos(0.3\pi n - \pi/4 + \angle H(e^{j0.3\pi})) \end{aligned}$$

Therefore, since $H(e^{j0}) = 0$ and

$$\begin{aligned} H(e^{j0.3\pi}) &= 2j\sin(0.3\pi/2)e^{-j0.3\pi/2} \\ &= 0.908e^{j(\pi/2-0.15\pi)} \end{aligned}$$

the output will be

$$\begin{aligned} y[n] &= (0.908)(2)\cos(0.3\pi n - \pi/4 + \pi/2 - 0.3\pi/2) \\ &= 1.816\cos(0.3\pi n + 0.1\pi) \end{aligned}$$

■

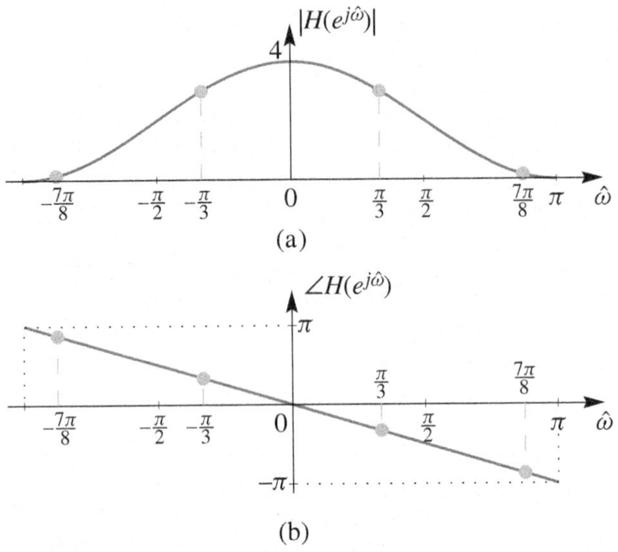

Figure 6-5: Magnitude (a) and phase (b) of system with frequency response $H(e^{j\hat{\omega}}) = (2 + 2\cos\hat{\omega})e^{-j\hat{\omega}}$. Gray dots indicate points where the frequency response is evaluated to calculate the sinusoidal response in Example 6-9.

6-5.3 A Simple Lowpass Filter

In Examples 6-1, 6-3, and 6-4, the system had frequency response

$$H(e^{j\hat{\omega}}) = 1 + 2e^{-j\hat{\omega}} + e^{-j2\hat{\omega}} = (2 + 2\cos\hat{\omega})e^{-j\hat{\omega}}$$

Since the factor $(2+2\cos\hat{\omega}) \geq 0$ for all $\hat{\omega}$, it follows that

$$|H(e^{j\hat{\omega}})| = (2 + 2\cos\hat{\omega})$$

and

$$\angle H(e^{j\hat{\omega}}) = -\hat{\omega}$$

These functions are plotted in Fig. 6-5 for $-\pi < \hat{\omega} < \pi$. Figure 6-5 shows at a glance that the system has a delay of 1 sample and that it tends to favor the low frequencies (close to $\hat{\omega} = 0$) with high gain, while it tends to suppress high frequencies (close to $\hat{\omega} = \pi$). In this case, there is a gradual decrease in gain from $\hat{\omega} = 0$ to $\hat{\omega} = \pi$, so that the midrange frequencies receive more gain than the high frequencies, but less than the low frequencies. Filters with magnitude responses that suppress the high frequencies of the input are called ***lowpass filters***.

Example 6-9: Lowpass Filter

If we repeat Example 6-4, we can show how a plot of $H(e^{j\hat{\omega}})$ makes it easy to find a filter's output for sinusoidal inputs. In Example 6-4, the input was

$$x[n] = 4 + 3\cos\left(\frac{\pi}{3}n - \frac{\pi}{2}\right) + 3\cos\left(\frac{7\pi}{8}n\right) \quad (6.18)$$

and the filter coefficients were $\{b_k\} = \{1,\ 2,\ 1\}$. Fig. 6-5 shows the frequency response of this system, which is a lowpass filter. In order to get the output signal, we must evaluate $H(e^{j\hat{\omega}})$ at frequencies 0, $\pi/3$, and $7\pi/8$ giving

$$\begin{aligned} H(e^{j0}) &= 4 \\ H(e^{j\pi/3}) &= 3e^{-j\pi/3} \\ H(e^{j7\pi/8}) &= 0.1522e^{-j7\pi/8} \end{aligned}$$

These values are the points indicated with gray dots on the graphs of Fig. 6-5. As in Example 6-4, the output is

$$\begin{aligned} y[n] &= 4 \cdot 4 + 3 \cdot 3\cos\left(\frac{\pi}{3}n - \frac{\pi}{3} - \frac{\pi}{2}\right) \\ &\quad + 0.1522 \cdot 3\cos\left(\frac{7\pi}{8}n - \frac{7\pi}{8}\right) \\ &= 16 + 9\cos\left(\frac{\pi}{3}(n-1) - \frac{\pi}{2}\right) \\ &\quad + 0.4567\cos\left(\frac{7\pi}{8}(n-1)\right) \end{aligned}$$

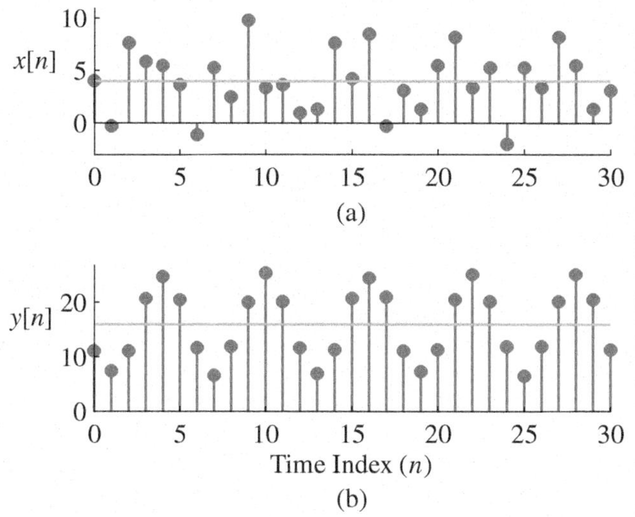

Figure 6-6: Input and output of system with frequency response $H(e^{j\hat{\omega}}) = (2 + 2\cos\hat{\omega})e^{-j\hat{\omega}}$. (a) Segment of the input signal $x[n]$ given by (6.18), and (b) the corresponding segment of the output.

The plot of the frequency response shows that all frequencies around $\hat{\omega} = \pi$ are greatly attenuated by the system. Also, the linear phase plot with slope of -1 indicates that all frequencies experience a time delay of 1 sample.

The output of the simple lowpass filter is the time waveform shown in Fig. 6-6(b). Note that the DC component is indicated in both parts of the figure as a gray horizontal line. The output appears to be the sum of a constant level of 16 plus a cosine that has amplitude 9 and seems to be periodic with period 6. Closer inspection reveals that this is not exactly true because there is a third output component at frequency $\hat{\omega} = 7\pi/8$ which is just barely visible in Fig. 6-6(b). Its size is about 5% of the size of the component with frequency $\hat{\omega} = \pi/3$. ■

6-6 Cascaded LTI Systems

In Section 5-7, we showed that if two LTI systems are connected in cascade (output of the first is input to the second), then the overall impulse response is the convolution of the two individual impulse responses, and therefore the cascade system is equivalent to a single system whose impulse response is the convolution of the two individual impulse responses. In this section, we will show that the frequency response of a cascade connection of two LTI systems is simply the product of the individual frequency responses.

DEMO: *Cascading FIR Filters*

Figure 6-7(a) shows two LTI systems in cascade. To find the frequency response of the overall system (from input $x[n]$ to output $y_2[n]$), we let

$$x[n] = e^{j\hat{\omega}n}$$

Then, the output of the first LTI system is

$$w_1[n] = H_1(e^{j\hat{\omega}})e^{j\hat{\omega}n}$$

and the output of the second system is

$$y_1[n] = H_2(e^{j\hat{\omega}})\left(H_1(e^{j\hat{\omega}})e^{j\hat{\omega}n}\right) = H_2(e^{j\hat{\omega}})H_1(e^{j\hat{\omega}})e^{j\hat{\omega}n}$$

From a similar analysis of Fig. 6-7(b), it follows that

$$y_2[n] = H_1(e^{j\hat{\omega}})\left(H_2(e^{j\hat{\omega}})e^{j\hat{\omega}n}\right) = H_1(e^{j\hat{\omega}})H_2(e^{j\hat{\omega}})e^{j\hat{\omega}n}$$

Since $H_2(e^{j\hat{\omega}})H_1(e^{j\hat{\omega}}) = H_1(e^{j\hat{\omega}})H_2(e^{j\hat{\omega}})$ from the commutative property of complex multiplication, it follows that $y_1[n] = y_2[n]$; i.e., the two cascade systems are equivalent for the same complex exponential input,

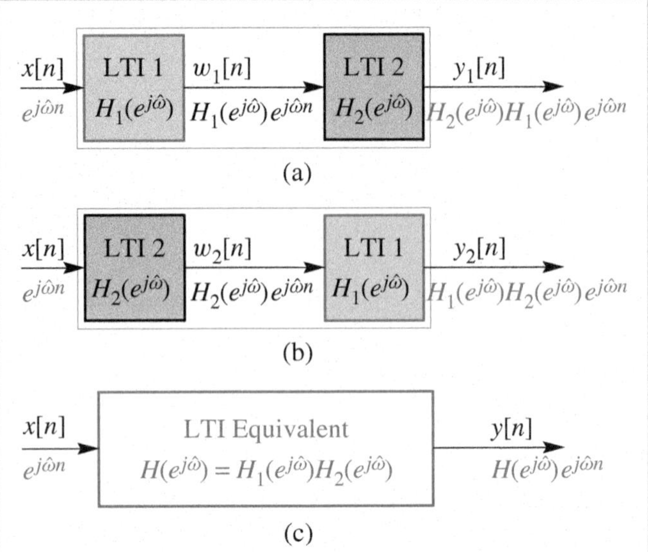

Figure 6-7: Frequency response of three equivalent cascaded LTI systems. All three systems have the same frequency response so that $y[n] = y_1[n] = y_2[n]$ for the same input $x[n]$.

and both of them are equivalent to a single LTI system with frequency response

$$H(e^{j\hat{\omega}}) = H_2(e^{j\hat{\omega}})H_1(e^{j\hat{\omega}}) = H_1(e^{j\hat{\omega}})H_2(e^{j\hat{\omega}}) \quad (6.19)$$

The output $y[n]$ of any system with frequency response $H(e^{j\hat{\omega}})$ given by (6.19) will be the same as either $y_1[n]$ or $y_2[n]$. This is depicted in Fig. 6-7(c).

Recall from Section 5-7 on p. 122 that the overall impulse response is $h_1[n] * h_2[n]$. We can summarize this by the correspondence

Convolution $\longleftrightarrow$ ***Multiplication***
$h_1[n] * h_2[n] \longleftrightarrow H_1(e^{j\hat{\omega}})H_2(e^{j\hat{\omega}})$

(6.20)

That is, convolution of impulse responses corresponds to multiplication of the frequency responses of cascaded systems. The correspondence shown in (6.20) is useful because it provides another way of representing and manipulating LTI systems. This is illustrated by the following example.

Example 6-10: Cascade

Suppose that the first system in a cascade of two systems is defined by the set of coefficients {2, 4, 4, 2} and the second system is defined by the coefficients {1, −2, 1}. The frequency responses of the individual systems are

$$H_1(e^{j\hat{\omega}}) = 2 + 4e^{-j\hat{\omega}} + 4e^{-j\hat{\omega}2} + 2e^{-j\hat{\omega}3}$$

and

$$H_2(e^{j\hat{\omega}}) = 1 - 2e^{-j\hat{\omega}} + e^{-j\hat{\omega}2}$$

The overall frequency response is

$$\begin{aligned} H(e^{j\hat{\omega}}) &= H_1(e^{j\hat{\omega}})H_2(e^{j\hat{\omega}}) \\ &= \left(2+4e^{-j\hat{\omega}}+4e^{-j\hat{\omega}2}+2e^{-j\hat{\omega}3}\right)\left(1-2e^{-j\hat{\omega}}+e^{-j\hat{\omega}2}\right) \\ &= 2 + 0e^{-j\hat{\omega}} - 2e^{-j\hat{\omega}2} - 2e^{-j\hat{\omega}3} + 0e^{-j\hat{\omega}4} + 2e^{-j\hat{\omega}5} \end{aligned}$$

Thus, the overall equivalent impulse response is

$$h[n] = 2\delta[n] - 2\delta[n-2] - 2\delta[n-3] + 2\delta[n-5]$$

■

This example illustrates that convolution of two impulse responses is equivalent to multiplying their corresponding frequency responses. Notice that, for FIR systems, the frequency response is just a polynomial in the variable $e^{-j\hat{\omega}}$. Thus, multiplying two frequency responses requires polynomial multiplication. This result provides the theoretical basis for the "synthetic" polynomial multiplication algorithm discussed in Section 5-3.3.1 on p. 110.

EXERCISE 6.4: Suppose that two systems are cascaded. The first system is defined by the set of coefficients $\{1, 2, 3, 4\}$, and the second system is defined by the coefficients $\{-1, 1, -1\}$. Determine the frequency response and the impulse response of the overall cascade system.

6-7 Running-Average Filtering

A simple linear time-invariant system is defined by the equation

$$y[n] = \frac{1}{L}\sum_{k=0}^{L-1} x[n-k] \tag{6.21}$$

$$= \frac{1}{L}\left(x[n] + x[n-1] + \cdots + x[n-L+1]\right)$$

This system (6.21) is called an *L-point running averager*, because the output at time n is computed as the average of $x[n]$ and the $L-1$ previous samples of the input. The system defined by (6.21) can be implemented in MATLAB for $L = 11$ by the statements:

```
bb = ones(11,1)/11;
yy = conv(bb, xx);
```

where `xx` is a vector containing the samples of the input. The vector `bb` contains the 11 filter coefficients, which are all the same size, in this case.

Using (6.4) on p. 131, frequency response of the L-point running averager is

$$H(e^{j\hat{\omega}}) = \frac{1}{L}\sum_{k=0}^{L-1} e^{-j\hat{\omega}k} \tag{6.22}$$

We can derive a simple formula for the magnitude and phase of the averager by making use of the formula for the sum of the first L terms of a geometric series,

$$\sum_{k=0}^{L-1} \alpha^k = \frac{1-\alpha^L}{1-\alpha} \qquad (\text{for } \alpha \neq 1) \tag{6.23}$$

First of all, we identify $e^{-j\hat{\omega}}$ as α, and then do the following steps:

$$H(e^{j\hat{\omega}}) = \frac{1}{L}\sum_{k=0}^{L-1} e^{-j\hat{\omega}k} \tag{6.24}$$

$$= \frac{1}{L}\left(\frac{1-e^{-j\hat{\omega}L}}{1-e^{-j\hat{\omega}}}\right)$$

$$= \frac{1}{L}\left(\frac{e^{-j\hat{\omega}L/2}(e^{j\hat{\omega}L/2} - e^{-j\hat{\omega}L/2})}{e^{-j\hat{\omega}/2}(e^{j\hat{\omega}/2} - e^{-j\hat{\omega}/2})}\right)$$

$$= \left(\frac{\sin(\hat{\omega}L/2)}{L\sin(\hat{\omega}/2)}\right) e^{-j\hat{\omega}(L-1)/2} \tag{6.25}$$

The numerator and denominator are simplified by using the inverse Euler formula for sines. We will find it convenient to express (6.25) in the form

$$H(e^{j\hat{\omega}}) = D_L(e^{j\hat{\omega}})e^{-j\hat{\omega}(L-1)/2} \tag{6.26}$$

where

$$D_L(e^{j\hat{\omega}}) = \frac{\sin(\hat{\omega}L/2)}{L\sin(\hat{\omega}/2)} \tag{6.27}$$

The function $D_L(e^{j\hat{\omega}})$ is often called the ***Dirichlet*** function, and the subscript L indicates that it comes from an L-point averager. In MATLAB, it can be evaluated with the `diric` function.

6-7.1 Plotting the Frequency Response

The frequency response of an 11-point running-average filter is given by the equation

$$H(e^{j\hat{\omega}}) = D_{11}(e^{j\hat{\omega}})e^{-j\hat{\omega}5} \tag{6.28}$$

where, in this case, $D_{11}(e^{j\hat{\omega}})$ is the Dirichlet function defined by (6.27) with $L = 11$, i.e.,

$$D_{11}(e^{j\hat{\omega}}) = \frac{\sin(11\hat{\omega}/2)}{11\sin(\hat{\omega}/2)} \tag{6.29}$$

As Eq. (6.28) makes clear, the frequency-response function, $H(e^{j\hat{\omega}})$, can be expressed as the product of the real amplitude function $D_{11}(e^{j\hat{\omega}})$ and the complex exponential factor $e^{-j5\hat{\omega}}$. The latter has a magnitude of 1 and a phase angle $-5\hat{\omega}$. Figure 6-8(a) shows a plot of the amplitude function, $D_{11}(e^{j\hat{\omega}})$; the phase function $-5\hat{\omega}$ is in the bottom part of the figure. We use the terminology *amplitude* rather than magnitude, because $D_{11}(e^{j\hat{\omega}})$ can be negative. We can obtain a plot of the magnitude $|H(e^{j\hat{\omega}})|$ by taking the absolute value of $D_{11}(e^{j\hat{\omega}})$. We shall consider the amplitude representation first, because it is simpler to examine the properties of the amplitude and phase functions. Figure 6-8 shows only one period, i.e., $-\pi < \hat{\omega} < \pi$. The frequency response is, of course, periodic with period 2π, so the plots in Fig. 6-8 would simply repeat with that period.

In the case of the 11-point running averager, the phase factor is easy to plot, since it is a straight line with slope of -5. The amplitude factor is somewhat more involved. First note that $D_{11}(e^{-j\hat{\omega}}) = D_{11}(e^{j\hat{\omega}})$; i.e., $D_{11}(e^{j\hat{\omega}})$ is an even function of $\hat{\omega}$ because it is the ratio of two odd functions. Since $D_{11}(e^{j\hat{\omega}})$ is even and periodic with period 2π, we need only to consider its values in the interval $0 \le \hat{\omega} \le \pi$. All others

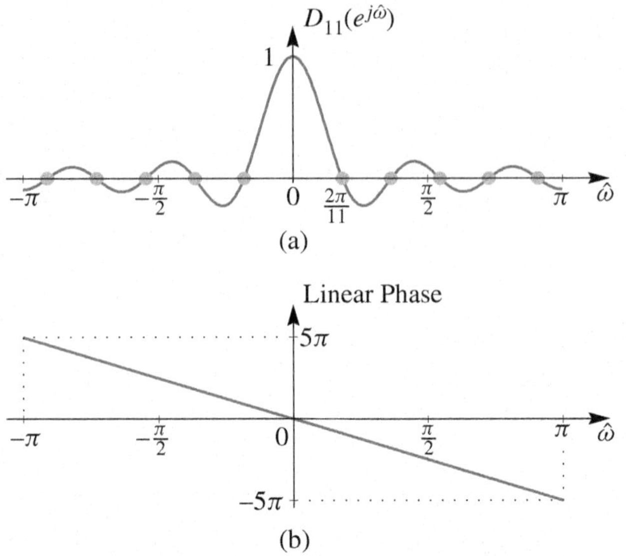

Figure 6-8: (a) *Amplitude* and (b) phase functions for frequency response of 11-point running-average filter. The *amplitude* is a *Dirichlet function,* but it is not the magnitude since it has negative values.

can be inferred from symmetry and periodicity. The numerator is $\sin(11\hat{\omega}/2)$, which, of course, oscillates between $+1$ and -1 and is zero whenever $11\hat{\omega}/2 = \pi k$, where k is an integer; solving for $\hat{\omega}$, $D_{11}(e^{j\hat{\omega}})$ is zero at frequencies $\hat{\omega}_k = 2\pi k/11$ where k is a nonzero integer. The denominator of $D_{11}(e^{j\hat{\omega}})$ is $11\sin(\hat{\omega}/2)$, which is zero at $\hat{\omega} = 0$ and increases to a maximum of eleven at $\hat{\omega} = \pi$. Therefore, $D_{11}(e^{j\hat{\omega}})$ is large around $\hat{\omega} = 0$, where the denominator is small, and it oscillates with decreasing amplitude as $\hat{\omega}$ increases to π. The behavior for $\hat{\omega} = 0$ is of particular interest because, at this frequency, $D_{11}(e^{j\hat{\omega}})$ is indeterminate, i.e.,

$$D_{11}(e^{j0}) = \frac{0}{0}$$

By l'Hôpital's rule, however, it is easily shown that $\lim_{\hat{\omega}\to 0} D_{11}(e^{j\hat{\omega}}) = 1$. Thus, the function $D_{11}(e^{j\hat{\omega}})$ has the following properties:

(a) $D_{11}(e^{j\hat{\omega}})$ is an even function of $\hat{\omega}$ that is periodic with period 2π.

(b) $D_{11}(e^{j\hat{\omega}})$ has a maximum value of one at $\hat{\omega} = 0$.

(c) $D_{11}(e^{j\hat{\omega}})$ decays as $\hat{\omega}$ increases, reaching its smallest nonzero amplitude at $\hat{\omega} = \pm\pi$.

(d) $D_{11}(e^{j\hat{\omega}})$ has zeros at nonzero integer multiples of $2\pi/11$. (In general, the zeros of $D_L(e^{j\hat{\omega}})$ are at nonzero multiples of $2\pi/L$.)

Together, the amplitude and phase plots of Fig. 6-8 completely define the frequency response of the 11-point running-average filter. Normally, however, the frequency response is represented in the form

$$H(e^{j\hat{\omega}}) = |H(e^{j\hat{\omega}})|e^{j\angle H(e^{j\hat{\omega}})}$$

This would require plotting $|H(e^{j\hat{\omega}})|$ and $\angle H(e^{j\hat{\omega}})$ as functions of $\hat{\omega}$. It is easy to see from (6.28) that $|H(e^{j\hat{\omega}})| = |D_{11}(e^{j\hat{\omega}})|$. The top part of Fig. 6-9 shows $|H(e^{j\hat{\omega}})| = |D_{11}(e^{j\hat{\omega}})|$ for the 11-point running-average filter. On the other hand, the phase response, $\angle H(e^{j\hat{\omega}})$, is more complicated to plot than the linear function shown in Fig. 6-8(b). There are two reasons for this:

(a) The algebraic sign of $D_{11}(e^{j\hat{\omega}})$ must be represented in the phase function, since $|H(\hat{\omega})| = |D_{11}(e^{j\hat{\omega}})|$ discards the sign of $D_{11}(e^{j\hat{\omega}})$.

(b) It is generally easiest to plot the *principal value* of the phase function.

The sign of $D_{11}(e^{j\hat{\omega}})$ can be incorporated into the phase by noting that $D_{11}(e^{j\hat{\omega}}) = |D_{11}(e^{j\hat{\omega}})|e^{\pm j\pi}$ whenever $D_{11}(e^{j\hat{\omega}}) < 0$. The principal value of the angle of a complex number is defined to be the angle between $-\pi$ and $+\pi$ radians. Using the result

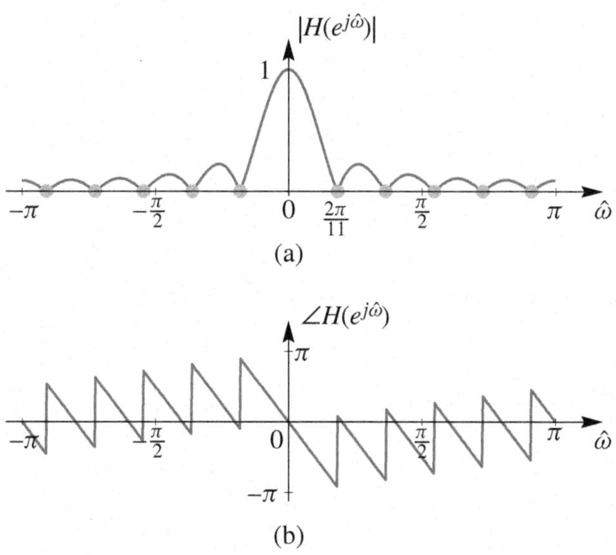

Figure 6-9: Magnitude and phase of frequency response of 11-point running-average filter. Compare to Fig. 6-8.

$$e^{j(\theta\pm 2\pi k)} = e^{j\theta}e^{\pm j2\pi k} = e^{j\theta}$$

where k is any integer, we see that we can add or subtract integer multiples of 2π from the angle of a complex number without changing the value of the complex number. We can always find a multiple of 2π, which, when added to or subtracted from θ, will produce a result in the range $-\pi < \theta \leq +\pi$.

This is called *reducing θ modulo 2π*. The principal value is generally what is computed when an inverse-tangent function is evaluated in MATLAB or other computer languages. In Fig. 6-8(b), we were able to plot an angle whose values were outside the principal value range simply because we had an equation for the angle. A plot like Figure 6-9 could be produced using the following MATLAB code:

```
omega = -pi:(pi/500):pi;
bb = ones(1,11)/11;
HH = freqz(bb,1,omega);
subplot(2,1,1),  plot(omega,abs(HH))
subplot(2,1,2),  plot(omega,angle(HH))
```

The MATLAB function `angle` uses the arctangent to return the principal value of the angle determined by the real and imaginary parts of the elements of the vector HH.

In Fig. 6-9, the phase curve is seen to have discontinuities that occur at the zeros of $D_{11}(e^{j\hat{\omega}})$. These discontinuities are due to the combination of multiples of π radians added to the phase due to the negative sign of $D_{11}(e^{j\hat{\omega}})$ in the intervals $2\pi/11 < |\hat{\omega}| < 4\pi/11$, $6\pi/11 < |\hat{\omega}| < 8\pi/11$, and $10\pi/11 < |\hat{\omega}| < \pi$ and multiples of 2π that are added implicitly in the computation of the principal value. The equation for the phase curve plotted in Fig. 6-9 is as follows for frequencies $0 \le \hat{\omega} \le \pi$:

$$\angle H(e^{j\hat{\omega}}) = \begin{cases} -5\hat{\omega} & 0 \le \hat{\omega} < 2\pi/11 \\ -5\hat{\omega}+\pi & 2\pi/11 < \hat{\omega} < 4\pi/11 \\ -5\hat{\omega}+2\pi & 4\pi/11 < \hat{\omega} < 6\pi/11 \\ -5\hat{\omega}+\pi+2\pi & 6\pi/11 < \hat{\omega} < 8\pi/11 \\ -5\hat{\omega}+4\pi & 8\pi/11 < \hat{\omega} < 10\pi/11 \\ -5\hat{\omega}+\pi+4\pi & 10\pi/11 < \hat{\omega} \le \pi \end{cases}$$

The values for $-\pi < \hat{\omega} < 0$ can be filled in using the fact that $\angle H(e^{-j\hat{\omega}}) = -\angle H(e^{j\hat{\omega}})$.

EXERCISE 6.5: Test yourself to see whether you understand why the principal value of $\angle H(e^{j\hat{\omega}})$ is as shown in Fig. 6-9 for the 11-point moving averager.

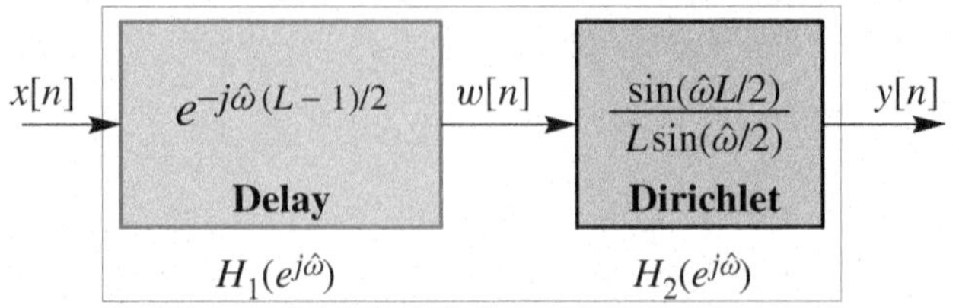

Figure 6-10: Representation of L-point running averager as the cascade of a delay and a real frequency response.

6-7.2 Cascade of Magnitude and Phase

It can be seen from (6.25) that $H(e^{j\hat{\omega}})$ is the product of two functions, i.e., $H(e^{j\hat{\omega}}) = H_2(e^{j\hat{\omega}})H_1(e^{j\hat{\omega}})$ where

$$H_1(e^{j\hat{\omega}}) = e^{-j\hat{\omega}(L-1)/2} \tag{6.30}$$

and

$$H_2(e^{j\hat{\omega}}) = D_L(e^{j\hat{\omega}}) = \frac{\sin(\hat{\omega}L/2)}{L\sin(\hat{\omega}/2)} \tag{6.31}$$

which is the Dirichlet function defined earlier. The component $H_1(e^{j\hat{\omega}})$ contributes only to the phase of $H(e^{j\hat{\omega}})$, and we see that this phase contribution is a linear function of $\hat{\omega}$. Earlier, we saw that a linear phase such as $\angle H_1(e^{j\hat{\omega}}) = -\hat{\omega}(L-1)/2$ corresponds to a time delay of $(L-1)/2$ samples. The linear-phase contribution (with slope of $-(L-1)/2 = -5$) is clearly in evidence in Fig. 6-8(b). The frequency response of the second system $H_2(e^{j\hat{\omega}})$ is real. It contributes to the magnitude of $H(e^{j\hat{\omega}})$, and when it is negative, it also contributes $\pm\pi$ to the phase of $H(e^{j\hat{\omega}})$ causing the discontinuities at multiples of $2\pi/11$.

The product representation suggests the block diagram of Fig. 6-10, which shows that the running averager can be thought of as a cascade combination of a delay followed by a "lowpass filter" that accentuates low frequencies relative to high frequencies. The overall moving average system can only be implemented by the difference equation (6.21) on p. 145. However, the block diagram of Fig. 6-10 is a useful convenience for thinking about the system. The system cannot be implemented by this cascade because $H_2(e^{j\hat{\omega}}) = D_L(e^{j\hat{\omega}})$ can never be implemented by itself. When $(L-1)/2$ is a integer, $w[n] = x[n-(L-1)/2]$ is a delay in Fig. 6-10. The case when $(L-1)/2$ is not an integer requires special interpretation, which will be provided in Section 6-8.2. For the present discussion, we will assume that L is an odd integer, so that $(L-1)/2$ is also an integer.

6-7.3 Experiment: Smoothing an Image

As a simple experiment to show the filtering effect of the running-average system, consider the image at the top of Fig. 6-11. The image is a two-dimensional discrete signal that can be represented as a two-dimensional array of samples $x_i[m,n]$. In an image, each sample value is called a *pixel*, which is shorthand for *picture element.* A single horizontal scan line (at $m = 40$) was extracted from the image yielding the one-dimensional signal $x[n] = x_i[40,n]$, plotted at the bottom of Fig. 6-11. The position in the image from which $x[n]$ was extracted is shown by the colored line in the image. The values in the image signal are all positive integers in the range $0 \le x_i[m,n] \le 255$. These numbers can be represented by 8-bit binary numbers.[3] If you compare the one-dimensional plot to the gray levels in the region around the line, you will see that dark regions in the image have large values (near 255), and bright regions have low values (near zero). This is actually a "negative" image, but that is appropriate since it is a scan of a handwritten homework solution.

An 11-point running averager was applied to $x[n]$, and the input and output were plotted on the same graph

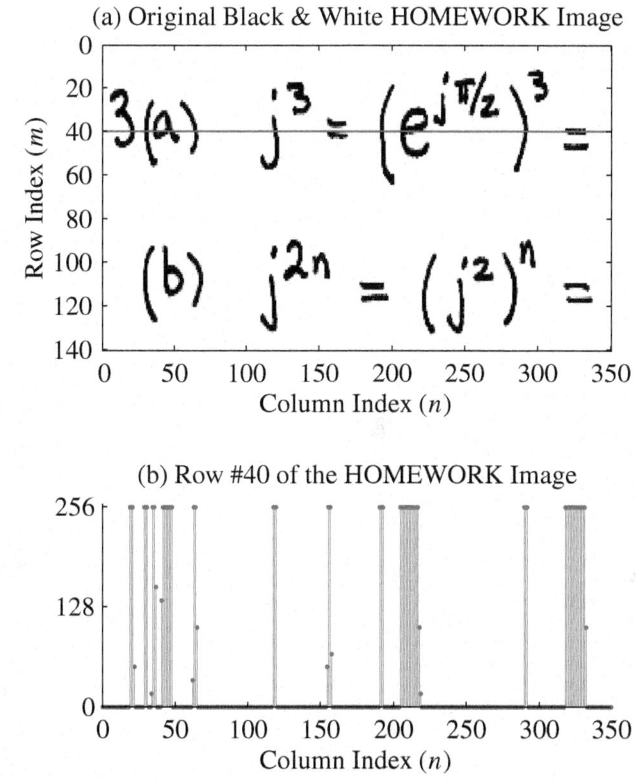

Figure 6-11: "Homework" image and one horizontal scan of the image at row 40.

(Fig. 6-12). Notice that the output $y[n]$ appears to be a smoother version of $x[n]$ with a slight shift to the right. This shift is the 5-sample delay that we expect for an 11-point running averager. The smoothness is a result of the relative attenuation of the higher frequencies in the signal that correspond to the edges of the handwritten characters in the image. To verify the effect of the delay of the system, Fig. 6-13 shows a plot of $w[n] = x[n-5]$ and $y[n]$. Now we see that the output appears to be aligned with the input.

The 11-point averager can be applied first over all the rows and then over all the columns of the image to get

[3]We often say that the total dynamic range of such image values is 8 bits.

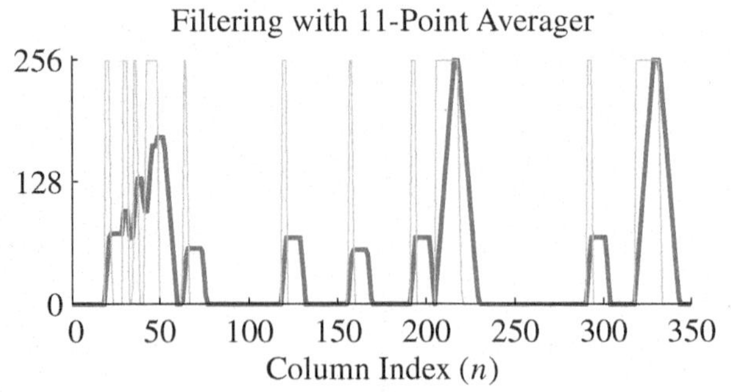

Figure 6-12: Input and output of 11-point running averager. The solid orange line is the output; the thin gray line is the input.

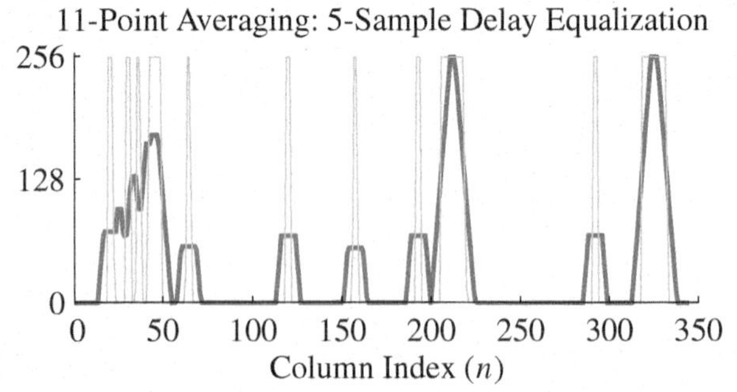

Figure 6-13: Delayed (by 5 samples) input and output of 11-point running averager. The solid orange line is the output; the thin gray line is the delayed input.

a visual assessment of the lowpass filtering operation.[4] Each row is filtered using the one-dimensional averager, then each column of that filtered image is processed. The result is shown in Fig. 6-14, where is it obvious that the lowpass filter has blurred the image. As we have seen, the filter attenuates the high-frequency components of

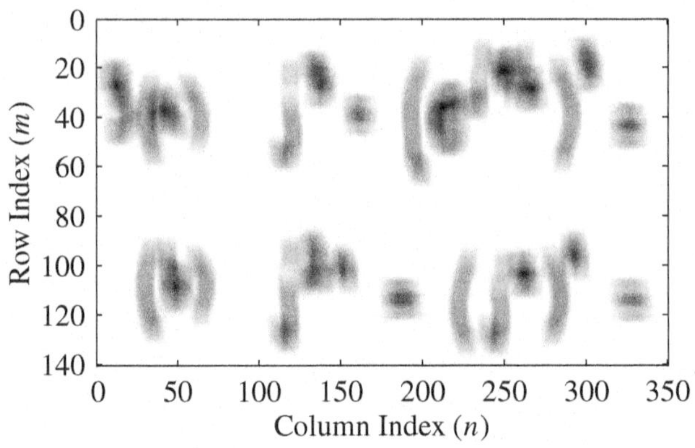

Figure 6-14: Result of filtering both the rows and the columns of the "homework" image with an 11-point running averager. The processed image had to be rescaled so that its values would occupy the entire gray scale range.

the image. Thus, we can conclude that sharp edges in an image must be associated with high frequencies.

As another example of the effect of filtering, the image signal was distorted by adding a cosine signal plus a constant to create a new input:

$$x_1[n] = x[n] + 128\cos(2\pi n/11) + 128$$

This corrupted signal $x_1[n]$ was filtered with the 11-point running averager. The delayed input $x_1[n-5]$ and the corresponding output are shown in Fig. 6-15. By comparing Figs. 6-13 and 6-15, it is clear that the output is the same for $x[n]$ and $x_1[n]$.[5] The reason is clear: $\hat{\omega} = 2\pi/11$ is one of the frequencies that is completely removed by the averaging filter, because $H(e^{j2\pi/11}) = 0$. Since the system is LTI and obeys superposition, the output due to $x_1[n]$ must be the same as the output due to $x[n]$ alone. If the cosine is added to each row of an image, it appears visible as vertical stripes (Fig. 6-16(a)).

[4] In MATLAB, the function `filter2( )` will do this two-dimensional filtering.

[5] A careful comparison reveals that there is a small difference over the transient region $0 \le n \le 9$.

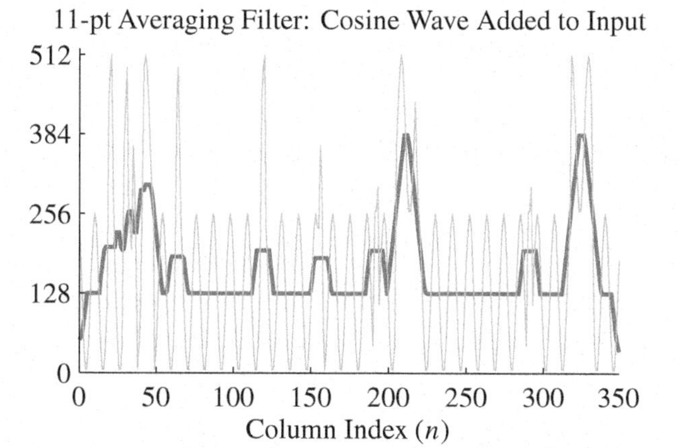

Figure 6-15: Result of 11-point running averager with $\cos(2\pi n/11)$ added to the image scan line. The solid orange line is the output; the thin gray line is the input.

When each row is processed with an 11-point averaging filter, the cosine will be removed, but the image will be blurred horizontally (Fig. 6-16(b)).

DEMO: *Cascading FIR Filters*

6-8 Filtering Sampled Continuous-Time Signals

Discrete-time filters can be used to filter continuous-time signals that have been sampled. In this section, we study the relationship between the frequency response of the discrete-time filter and the effective frequency response applied to the continuous-time signal. When the input to a discrete-time system is a sequence derived by sampling a continuous-time signal, we can use our knowledge of sampling and reconstruction to interpret the effect of the filter on the original continuous-time signal.

Consider the system depicted in Fig. 6-17, and assume that the input is the complex sinusoid

$$x(t) = Xe^{j\omega t}$$

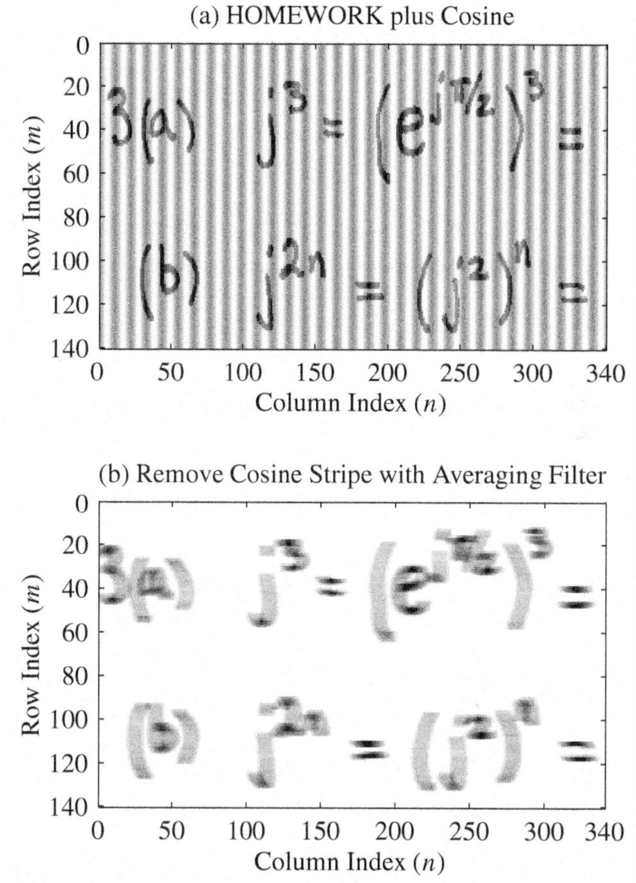

Figure 6-16: (a) "Homework plus cosine" image. The periodic nature of the cosine across each row causes a vertical striping. (b) After filtering the rows of the "homework plus cosine" image with an 11-point running averager, the processed image is blurred, but has no traces of the cosine stripes. (Both input and output were rescaled for 8-bit image display.)

with $X = Ae^{j\phi}$. After sampling, the input sequence to the discrete-time filter is

$$x[n] = x(nT_s) = Xe^{j\omega nT_s} = Xe^{j\hat{\omega}n}$$

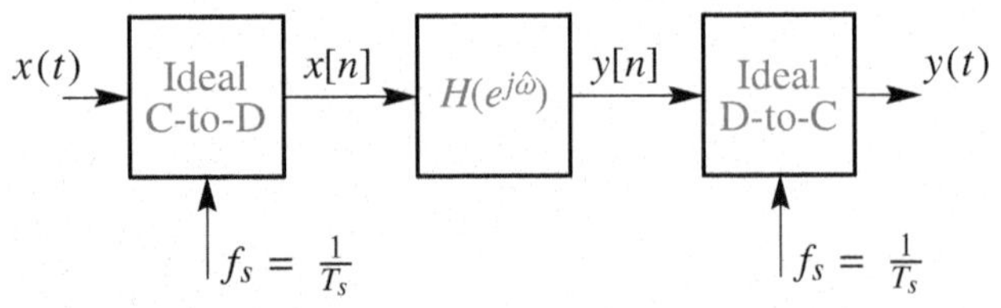

Figure 6-17: System for doing discrete-time filtering of continuous-time signals.

The relationship between the discrete-time frequency $\hat{\omega}$ and continuous-time frequencies ω and f is

$$\hat{\omega} = \omega T_s = \frac{2\pi f}{f_s} \tag{6.32}$$

If the frequency of the continuous-time signal satisfies the condition of the sampling theorem, i.e., $|\omega| < \pi / T_s$, then there will be no aliasing, and the normalized discrete-time frequency is such that $|\hat{\omega}| < \pi$.

The frequency response of the discrete-time system gives us a quick way to calculate the output $y[n]$ in Fig 6-17.

$$y[n] = H(e^{j\hat{\omega}})Xe^{j\hat{\omega}n}$$

If we now make the substitution $\hat{\omega} = \omega T_s$, then we can write $y[n]$ in terms of the analog frequency ω as

$$y[n] = H(e^{j\omega T_s})Xe^{j\omega T_s n}$$

Finally, since no aliasing occurred in the original sampling, the ideal D-to-C converter will reconstruct the original frequency, giving

$$y(t) = H(e^{j\omega T_s})Xe^{j\omega t}$$

Remember that this formula for $y(t)$ is good only for frequencies such that $-\pi/T_s < \omega < \pi/T_s$, and recall that the ideal D-to-C converter reconstructs all digital frequency components in the band $|\hat{\omega}| < \pi$ as analog frequencies in the band $|\omega| < \pi/T_s$. As long as there is no aliasing, the frequency band of the input signal $x(t)$ matches the frequency band of the output $y(t)$. Thus, the overall system of Fig. 6-17 behaves as though it is an LTI continuous-time system whose frequency response is $H(e^{j(\omega T_s)})$.

It is very important to understand this analysis of the system of Fig. 6-17. We have just shown that the system of Fig. 6-17 can be used to implement LTI filtering operations on continuous-time signals. Furthermore, it is clear from this analysis that the block diagram of Fig. 6-17 represents an infinite number of systems. This is true in two ways. First, for any given discrete-time system, we can change the sampling period T_s (avoiding aliasing of the input) and obtain a new system. Alternatively, if we fix the sampling period, we can change the discrete-time system to vary the overall response. In a specific case, all we have to do is select the sampling rate to avoid aliasing, and then design a discrete-time LTI filter whose frequency response $H(e^{j\omega T_s})$ has the desired frequency-selective properties.

EXERCISE 6.6: In general, when we use the system of Fig. 6-17 to filter continuous-time signals, we would want to choose the sampling frequency $f_s = 1/T_s$ to be as low as possible. Why?

6-8.1 Example: Lowpass Averager

As an example, we use the 11-point moving averager

$$y[n] = \frac{1}{11}\sum_{k=0}^{10} x[n-k]$$

as the filter in Fig. 6-17. The frequency response of this discrete-time system is

$$H(e^{j\hat{\omega}}) = \frac{\sin(\hat{\omega}11/2)}{11\sin(\hat{\omega}/2)}e^{-j\hat{\omega}5}$$

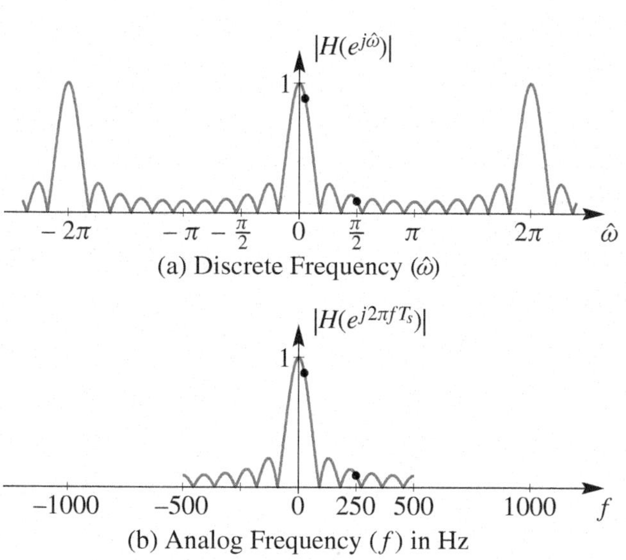

Figure 6-18: Frequency response of 11-point moving averager (a) and equivalent analog-frequency response (b) when used to filter analog signals. The sampling frequency is $f_s = 1000$ Hz, so the maximum analog frequency that can be processed is 500 Hz.

The magnitude of this frequency response is shown in the top part of Fig. 6-18.

When this system is used as the discrete-time system in Fig. 6-17 with sampling frequency $f_s = 1000$, we want to answer two questions: What is the equivalent analog-frequency response, and how would the signal

$$x(t) = \cos(2\pi(25)t) + \sin(2\pi(250)t)$$

be processed by the system?

The frequency-response question is easy. The equivalent analog-frequency response is

$$H(e^{j\omega T_s}) = H(e^{j\omega/1000}) = H(e^{j2\pi f/1000})$$

where f is in Hz. A plot of the equivalent continuous-time frequency response versus f is shown in the bottom part of Fig. 6-18. Note that the frequency response of the overall system stops abruptly at $|f| = f_s/2 = 500$ Hz, since the ideal D-to-C converter does not reconstruct frequencies above $f_s/2$.

The second question is also easy if we track the two frequencies of the input signal through the three systems of Fig. 6-17. The input $x(t)$ contains two frequencies at $\omega = 2\pi(25)$ and $\omega = 2\pi(250)$. Since $f_s = 1000 > 2(250) > 2(25)$, there is no aliasing,[6] so the same frequency components appear in the output signal $y(t)$. The magnitude and phase changes are found by evaluating the equivalent frequency response at 25 and 250 Hz.

$$\begin{aligned} H(e^{j2\pi(25)/1000}) &\\ &= \frac{\sin(\pi(25)(11)/1000)}{11\sin(\pi(25)/1000)} e^{-j2\pi(25)(5)/1000} \\ &= 0.8811 e^{-j\pi/4} \\ H(e^{j2\pi(250)/1000}) &\\ &= \frac{\sin(\pi(250)(11)/1000)}{11\sin(\pi(250)/1000)} e^{-j2\pi(250)(5)/1000} \\ &= 0.0909 e^{-j\pi/2} \end{aligned}$$

These values can be checked against the plots in Fig. 6-18. Thus the final output is

$$\begin{aligned} y(t) = 0.8811&\cos(2\pi(25)t - \pi/4) \\ &+ 0.0909\sin(2\pi(250)t - \pi/2) \end{aligned}$$

The lowpass nature of the filter has greatly attenuated the 250-Hz component, while the 25-Hz component suffered only slight attenuation because it lies in the passband of the filter near 0 Hz.

[6] When there is aliasing, this sort of problem is less straightforward because the output signal $y(t)$ will have different frequency components from the input.

EXERCISE 6.7: Assuming the same input signal and the same discrete-time system, work the example of this section again, but use a sampling rate of $f_s = 500$ Hz.

6-8.2 Interpretation of Delay

We have seen that a frequency response of the form $H(e^{j\hat{\omega}}) = e^{-j\hat{\omega}n_0}$ implies a time delay of n_0 samples. For n_0, an integer, the interpretation of this is straightforward. If the input to the system is $x[n]$, the corresponding output is $y[n] = x[n - n_0]$. However, if n_0 is not an integer, the interpretation is less obvious. An example of where this can occur is the L-point running-average system whose frequency response is

$$H(e^{j\hat{\omega}}) = D_L(e^{j\hat{\omega}})e^{-j\hat{\omega}(L-1)/2}$$

where $D_L(e^{j\hat{\omega}})$ is the real function

$$D_L(e^{j\hat{\omega}}) = \frac{\sin(\hat{\omega}L/2)}{L\sin(\hat{\omega}/2)}$$

Thus, the L-point running averager includes a delay of $\frac{1}{2}(L-1)$ samples. If L is an odd integer, this delay causes the output to be shifted $\frac{1}{2}(L-1)$ samples with respect to the input. However, if L is an even integer, then $\frac{1}{2}(L-1)$ is not an integer. The analyis of this section provides a useful interpretation of this delay factor.

Suppose that the input to the ideal C-to-D converter is

$$x(t) = Xe^{j\omega t}$$

and that there is no aliasing, so that the sampled input to the L-point running averager is

$$x[n] = Xe^{j\omega nT_s} = Xe^{j\hat{\omega}n}$$

where $\hat{\omega} = \omega T_s$. Now, the output of the L-point running-average filter is

$$y[n] = H(e^{j\hat{\omega}})Xe^{j\hat{\omega}n} = D_L(e^{j\hat{\omega}})e^{-j\hat{\omega}(L-1)/2}Xe^{j\hat{\omega}n}$$

Finally, if $\omega < \pi/T_s$ (i.e., no aliasing occurred in the sampling operation), then the ideal D-to-C converter will reconstruct the complex exponential signal

$$\begin{aligned} y(t) &= D_L(e^{j\hat{\omega}})Xe^{-j\hat{\omega}(L-1)/2}e^{j\omega t} \\ &= D_L(e^{j\omega T_s})Xe^{-j\omega T_s(L-1)/2}e^{j\omega t} \\ &= D_L(e^{j\omega T_s})Xe^{j\omega(t-T_s(L-1)/2)} \end{aligned}$$

Thus, regardless of whether or not $\frac{1}{2}(L-1)$ is an integer, the delay factor $e^{-j\hat{\omega}(L-1)/2}$ corresponds to a delay of $\frac{1}{2}T_s(L-1)$ seconds with respect to continuous-time signals sampled with sampling period T_s.

Example 6-11: Time Delay of FIR Filter

To illustrate the effect of noninteger delay with the running-average filter, consider the cosine signal $x[n] = \cos(0.2\pi n)$, which could have resulted in Fig 6-17 from sampling the signal $x(t) = \cos(200\pi t)$ with sampling rate $f_s = 1000$ Hz. Figure 6-19(a) shows $x(t)$ and $x[n]$. If $x[n]$ is the input to a 5-point running-average filter, the steady-state part of the output is

$$\begin{aligned} y_5[n] &= \frac{\sin(0.2\pi(5/2))}{5\sin(0.2\pi/2)}\cos(0.2\pi n - 0.2\pi(2)) \\ &= 0.6472\cos(0.2\pi(n-2)) \end{aligned}$$

For this filter output, the output of the D-to-C converter in Fig. 6-17(b) would be

$$y_5(t) = 0.6472\cos(200\pi(t - 0.002))$$

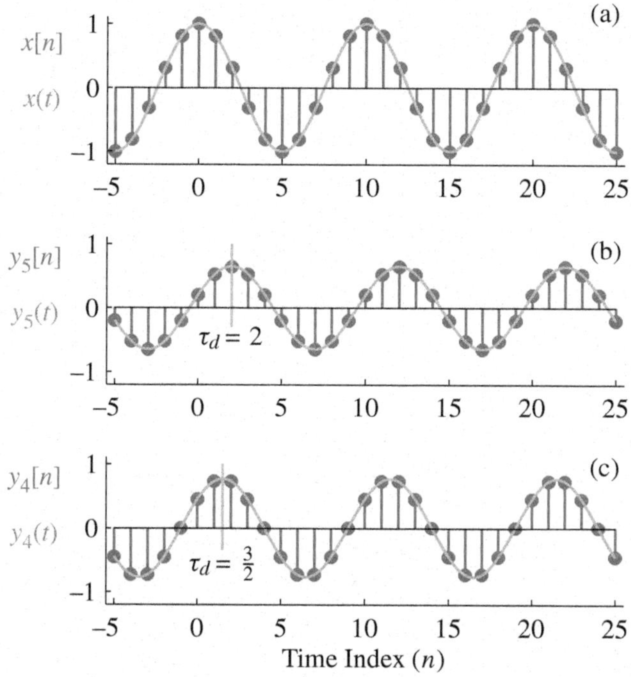

Figure 6-19: Input signal (a), output of 5-point running averager (b), and output of 4-point running averager (c). The solid gray curve is the corresponding continuous-time signal: (a) $x(t)$, (b) $y_5(t)$, and (c) $y_4(t)$.

The delay is 2 samples. On the other hand, if the same signal $x[n]$ is the input to a 4-point running-average system, the steady-state part of the output (Fig. 6-19(c)) is

$$y_4[n] = \frac{\sin(0.2\pi(4/2))}{4\sin(0.2\pi/2)}\cos(0.2\pi n - 0.2\pi(3/2))$$
$$= 0.7694\cos(0.2\pi(n - \tfrac{3}{2}))$$

Now the delay is 3/2 samples, so we cannot write $y_4[n]$ as an integer shift with respect to the input sequence. In this case, the "3/2 samples" delay introduced by the filter can be interpreted in terms of the corresponding output of the D-to-C converter in Fig. 6-17(c), which in this case would be

$$y_4(t) = 0.7694\cos(200\pi(t - 0.0015))$$

Figure 6-19 shows the input and the corresponding outputs $y_5[n]$ and $y_5(t)$ and $y_4[n]$ and $y_4(t)$. In all cases, the solid curve is the continuous-time cosine signal that would be reconstructed by the ideal D-to-C converter for the given discrete-time signal. ■

The following specific points are made by this example:

1. Both outputs are smaller than the input. This is because $D_L(e^{j0.2\pi}) < 1$ for both cases.

2. The gray vertical lines in the lower two panels show the peaks of the output cosine signals that correspond to the peak of the input at $n = 0$. Note that the delay is $(5-1)/2 = 2$ for the 5-point averager and $(4-1)/2 = 3/2$ for the 4-point averager.

3. The effect of the fractional delay is to implement an interpolation of the cosine signal at points halfway between the original samples.

6-9 Summary and Links

This chapter introduced the concept of the frequency response for the class of FIR filters. The frequency response applies to any linear time-invariant system, as we will see in upcoming chapters. The MATLAB GUI shown in Fig. 6-20 embodies the primary result of this chapter. It shows that one evaluation of the frequency response is sufficient to predict how a sinusoid will be processed by an LTI system. When the filter changes, the frequency response changes so sinusoidal frequency components are treated differently.

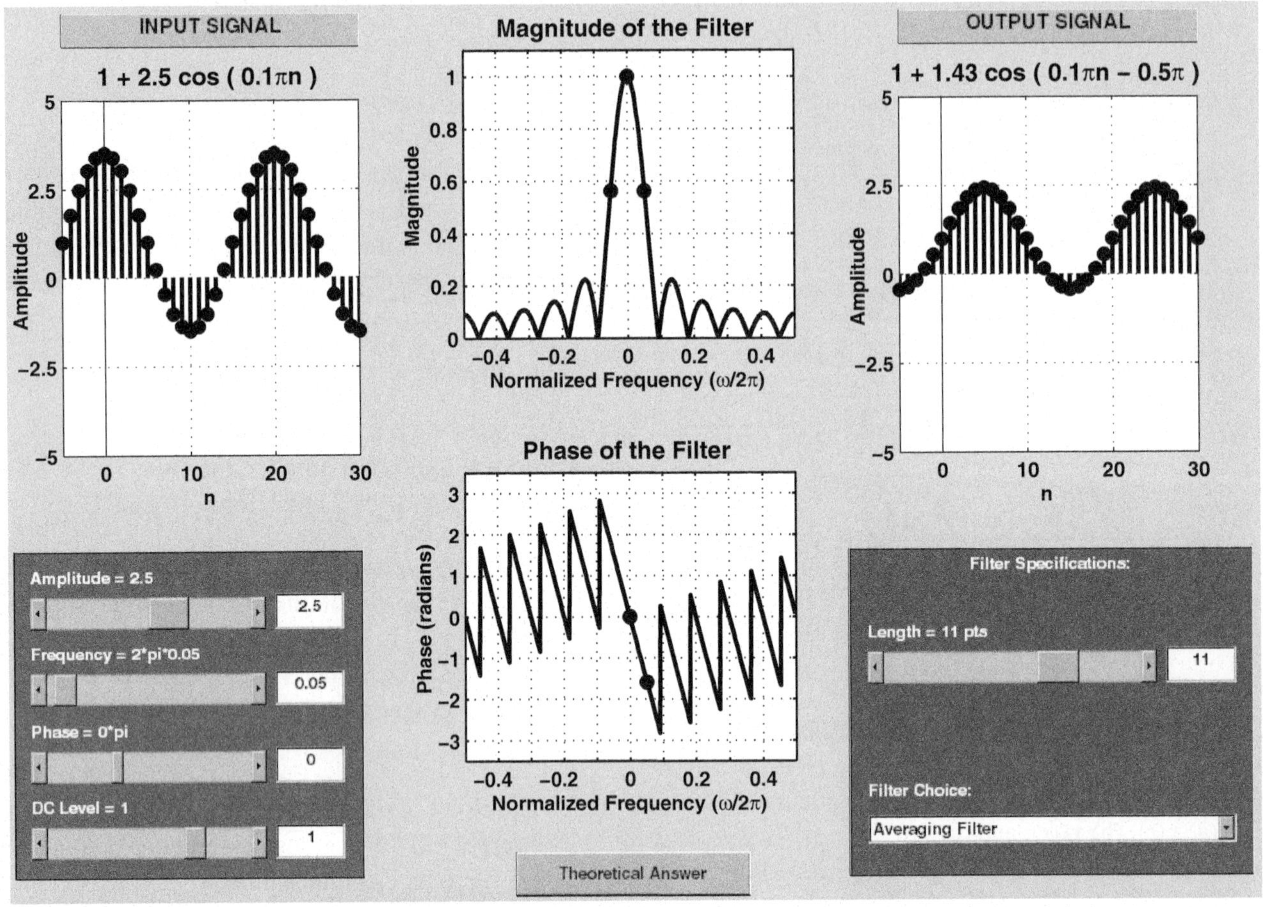

Figure 6-20: Graphical User Interface (GUI) for `DLTIdemo` which illustrates the *sinusoid-in gives sinusoid-out* concept. The user can change the LTI system as well as the input sinusoidal frequency, amplitude and phase.

DEMO: *DLTI: Sinusoid-in gives Sinusoid-out*

This chapter extends the discussion of Chapter 5, which introduced the basics of FIR filtering. The labs, in particular, require the student to be familiar with both chapters. Lab #8 is an experiment with the frequency response of FIR filters.

LAB: *#8 Frequency Response: Band-pass and Nullling Filters*

Two other labs involving FIR filtering are listed under Chapter 7 on the CD-ROM. Lab #9 uses results from Lab #8 to build a system to decode Touch-Tone signals. Finally, Lab #10 uses filters to build a system that determines which note is played on a piano.

LAB: *#9 Encoding and Decoding Touch-Tones*

LAB: *#10 Octave Band Filtering*

The CD-ROM also contains the following demonstrations of lowpass and highpass filtering:

(a) Filtering photographic images to show that lowpass filtering is blurring, and that highpass filtering enhances images.

(b) Cascade processing of images to show that a highpass filtering can undo the blurring effects of a lowpass filter.

DEMO: *Cascading FIR Filters*

(c) Filtering of sound signals to illustrate bass and treble emphasis.

DEMO: *Introduction to FIR Filtering*

Finally, the reader is reminded of the large number of solved homework problems available for review and practice on the CD-ROM.

NOTE: *Hundreds of Solved Problems*

6-10 Problems

P-6.1 Suppose the input signal to an FIR system is

$$x[n] = e^{j(0.4\pi n - 0.5\pi)}$$

If we define a new signal $y[n]$ to be the first difference $y[n] = x[n] - x[n-1]$, it is possible to express $y[n]$ in the form

$$y[n] = Ae^{j(\hat{\omega}_0 n + \phi)}$$

Determine the numerical values of A, ϕ, and $\hat{\omega}_0$.

P-6.2 Suppose that a discrete-time system is described by the input-output relation

$$y[n] = (x[n])^2$$

(a) Determine the output when the input is the complex exponential signal

$$x[n] = Ae^{j\phi}e^{j\hat{\omega}n}$$

(b) Is the output of the form

$$y[n] = H(e^{j\hat{\omega}})Ae^{j\phi}e^{j\hat{\omega}n}$$

If not, why not?

P-6.3 Suppose that a discrete-time system is described by the input-output relation

$$y[n] = x[-n]$$

(a) Determine the output when the input is the complex exponential signal

$$x[n] = Ae^{j\phi}e^{j\hat{\omega}n}$$

(b) Is the output of the form

$$y[n] = H(e^{j\hat{\omega}})Ae^{j\phi}e^{j\hat{\omega}n}$$

If not, why not?

P-6.4 A linear time-invariant system is described by the difference equation

$$y[n] = 2x[n] - 3x[n-1] + 2x[n-2]$$

(a) Find the frequency response $H(e^{j\hat{\omega}})$; then express it as a mathematical formula, in polar form (magnitude and phase).

(b) $H(e^{j\hat{\omega}})$ is a periodic function of $\hat{\omega}$; determine the period.

(c) Plot the magnitude and phase of $H(e^{j\hat{\omega}})$ as a function of $\hat{\omega}$ for $-\pi < \hat{\omega} < 3\pi$. Do this by hand, and then check with the MATLAB function `freqz`.

(d) Find all frequencies $\hat{\omega}$, for which the output response to the input $e^{j\hat{\omega}n}$ is zero.

(e) When the input to the system is $x[n] = \sin(\pi n/13)$, determine the output signal and express it in the form $y[n] = A\cos(\hat{\omega}_0 n + \phi)$.

P-6.5 A linear time-invariant filter is described by the difference equation

$$y[n] = x[n] + 2x[n-1] + x[n-2]$$

(a) Obtain an expression for the frequency response of this system.

(b) Sketch the frequency response (magnitude and phase) as a function of frequency.

(c) Determine the output when the input is $x[n] = 10 + 4\cos(0.5\pi n + \pi/4)$.

(d) Determine the output when the input is the unit-impulse sequence, $\delta[n]$.

(e) Determine the output when the input is the unit-step sequence, $u[n]$.

P-6.6 A linear time-invariant filter is described by the difference equation

$$y[n] = x[n] - x[n-2]$$

(a) Obtain an expression for the frequency response of this system.

(b) Sketch the frequency response (magnitude and phase) as a function of frequency.

(c) Determine the output if the input signal is $x[n] = 4 + \cos(0.25\pi n - \pi/4)$.

(d) The signal $x_1[n] = (4 + \cos(0.25\pi n - \pi/4))\, u[n]$ is zero for $n < 0$, so it "starts" at $n = 0$. If the input signal is $x_1[n]$, for which values of n will the corresponding output be equal to the output obtained in (c)?

P-6.7 For each of the following frequency responses determine the corresponding impulse response

(a) $H(e^{j\hat{\omega}}) = 1 + 2e^{-j3\hat{\omega}}$

(b) $H(e^{j\hat{\omega}}) = 2e^{-j3\hat{\omega}}\cos(\hat{\omega})$

(c) $H(e^{j\hat{\omega}}) = e^{-j4.5\hat{\omega}}\dfrac{\sin(5\hat{\omega})}{\sin(\hat{\omega}/2)}$

P-6.8 The frequency response of a linear time-invariant filter is given by the formula

$$H(e^{j\hat{\omega}}) = (1 + e^{-j\hat{\omega}})(1 - e^{j2\pi/3}e^{-j\hat{\omega}})(1 - e^{-j2\pi/3}e^{-j\hat{\omega}})$$

(a) Write the difference equation that gives the relation between the input $x[n]$ and the output $y[n]$.

(b) What is the output if the input is $x[n] = \delta[n]$?

(c) If the input is of the form $x[n] = Ae^{j\phi}e^{j\hat{\omega}n}$, for what values of $-\pi \le \hat{\omega} \le \pi$ will $y[n] = 0$ for all n?

P-6.9 The frequency response of a linear time-invariant filter is given by the formula

$$H(e^{j\hat{\omega}}) = (1 - e^{-j\hat{\omega}})(1 - \tfrac{1}{2}e^{j\pi/6}e^{-j\hat{\omega}})(1 - \tfrac{1}{2}e^{-j\pi/6}e^{-j\hat{\omega}})$$

(a) Write the difference equation that gives the relation between the input $x[n]$ and the output $y[n]$.

(b) What is the output if the input is $x[n] = \delta[n]$?

(c) If the input is of the form $x[n] = Ae^{j\phi}e^{j\hat{\omega}n}$, for what values of $-\pi < \hat{\omega} \le \pi$ will $y[n] = 0$ for all n?

P-6.10 Suppose that $\mathcal{S}$ is a linear time-invariant system whose exact form is unknown. It has been tested by observing the output signals corresponding to several different test inputs. Suppose that the following input-output pairs were the result of the tests:

Input: $x[n]$	**Output:** $y[n]$
$\delta[n]$	$\delta[n] - \delta[n-3]$
$\cos(2\pi n/3)$	0
$\cos(\pi n/3 + \pi/2)$	$2\cos(\pi n/3 + \pi/2)$

(a) Make a plot of the signal $x[n] = 3\delta[n] - 2\delta[n-2] + \delta[n-3]$.

(b) Determine the output of the system when the input is $x[n] = 3\delta[n] - 2\delta[n-2] + \delta[n-3]$, and make a plot of it.

(c) Determine the output when the input is $x[n] = \cos(\pi(n-3)/3)$.

(d) Is the following statement true or false? Explain.

$$H(e^{j(\pi/2)}) = 0$$

P-6.11 The *Dirichlet* function is defined by

$$D_L(e^{j\hat{\omega}}) = \frac{\sin(L\hat{\omega}/2)}{L\sin(\hat{\omega}/2)}$$

For the case $L = 8$:

(a) Make a plot of $D_8(e^{j\hat{\omega}})$ over the range $-3\pi \le \hat{\omega} \le +3\pi$. Label all the zero crossings.

(b) Determine the period of $D_8(e^{j\hat{\omega}})$.

(c) Find the maximum value of the function.

P-6.12 A linear time-invariant filter is described by the difference equation

$$y[n] = x[n] - 3x[n-1] + 3x[n-2] - x[n-3]$$

(a) Obtain an expression for the frequency response of this system, and, using the fact that $(1-a)^3 = 1 - 3a + 3a^2 - a^3$, show that $H(e^{j\hat{\omega}})$ can be expressed in the form

$$H(e^{j\hat{\omega}}) = 8\sin^3(\hat{\omega}/2)e^{j(-\pi/2 - 3\hat{\omega}/2)}$$

(b) Sketch the frequency response (magnitude and phase) as a function of frequency.

(c) What is the output if the input is $x[n] = 10 + 4\cos(0.5\pi n + \pi/4)$?

(d) What is the output if the input is $x[n] = \delta[n]$?

(e) Use superposition to find the output when $x[n] = 10 + 4\cos(0.5\pi n + \pi/4) + 5\delta[n-3]$.

P-6.13 Suppose that three systems are connected in cascade; i.e., the output of $\mathcal{S}_1$ is the input to $\mathcal{S}_2$, and the output of $\mathcal{S}_2$ is the input to $\mathcal{S}_3$. The three systems are specified as follows:

$$\mathcal{S}_1: \quad y_1[n] = x_1[n] - x_1[n-1]$$
$$\mathcal{S}_2: \quad y_2[n] = x_2[n] + x_2[n-2]$$
$$\mathcal{S}_3: \quad y_3[n] = x_3[n-1] + x_3[n-2]$$

where the output of $\mathcal{S}_i$ is $y_i[n]$ and its input is $x_i[n]$.

(a) Determine the equivalent system that is a single operation from the input $x[n]$ (into $\mathcal{S}_1$) to the output $y[n]$, which is the output of $\mathcal{S}_3$. Thus, $x[n]$ is $x_1[n]$ and $y[n]$ is $y_3[n]$.

(b) Use the frequency response to write one difference equation that defines the overall system in terms of $x[n]$ and $y[n]$ only.

P-6.14 An LTI filter is described by the difference equation

$$y[n] = \frac{1}{4}\Big(x[n] + x[n-1] + x[n-2] + x[n-3]\Big)$$

(a) What is the impulse response $h[n]$ of this system?

(b) Obtain an expression for the frequency response of this system.

(c) Sketch the frequency response (magnitude and phase) as a function of frequency.

(d) Suppose that the input is

$$x[n] = 5 + 4\cos(0.2\pi n) + 3\cos(0.5\pi n)$$

for $-\infty < n < \infty$. Obtain an expression for the output in the form $y[n] = A + B\cos(\hat{\omega}_0 n + \phi_0)$.

(e) Suppose that the input is

$$x_1[n] = [5 + 4\cos(0.2\pi n) + 3\cos(0.5\pi n)]\,u[n]$$

where $u[n]$ is the unit-step sequence. For what values of n will the output $y_1[n]$ be equal to the output $y[n]$ in (d)?

P-6.15 A system for filtering continuous-time signals is shown in Fig. P-6.15.

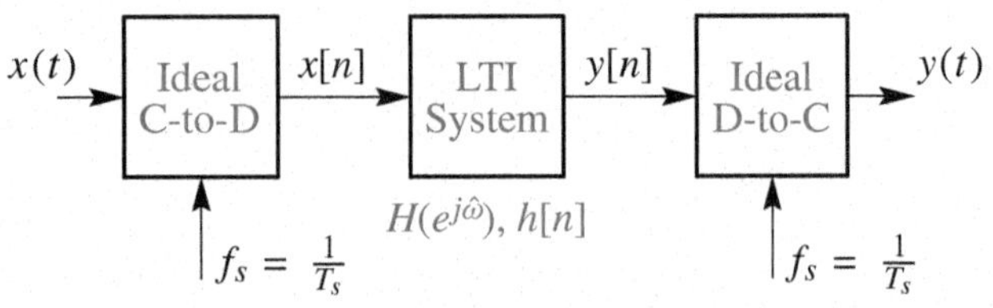

Figure P-6.15

The input to the C-to-D converter in this system is

$$x(t) = 10 + 8\cos(200\pi t) + 6\cos(500\pi t + \pi/4)$$

for $-\infty < t < \infty$. The impulse response of the LTI system is

$$h[n] = \frac{1}{4}\sum_{k=0}^{3}\delta[n-k]$$

If $f_s = 1000$ samples/sec, determine an expression for $y(t)$, the output of the D-to-C converter.

Hint: The results of Problem P-6.14 can be applied here.

P-6.16 The diagram in Fig. P-6.16 depicts a cascade connection of two linear time-invariant systems, where the first system is a 3-point moving averager and the second system is a first difference.

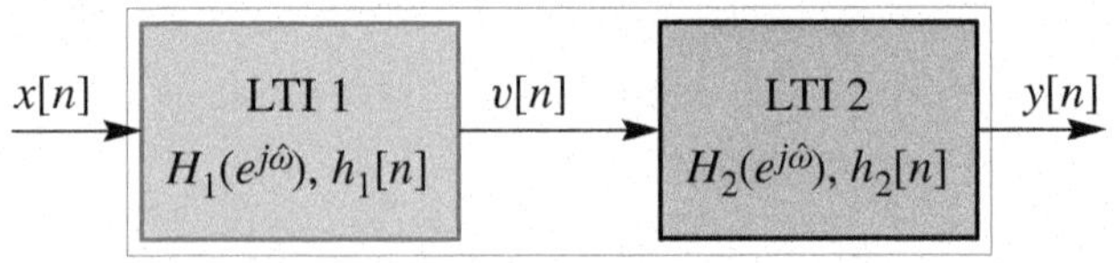

Figure P-6.16

(a) If the input is of the form $x[n] = 10 + x_1[n]$, the output, $y[n]$, of the overall system will be of the form $y[n] = y_1[n]$, where $y_1[n]$ is the output due only to $x_1[n]$. Explain why this is true.

(b) Determine the frequency-response function of the overall cascade system.

(c) Sketch the frequency-response (magnitude and phase) functions of the individual systems and the overall cascade system for $-\pi < \hat{\omega} \leq \pi$.

(d) Obtain a single-difference equation that relates $y[n]$ to $x[n]$ for the overall cascade system.

P-6.17 A linear time-invariant system is described by the difference equation

$$y[n] = -x[n] + 2x[n-2] - x[n-4]$$

(a) Find the impulse response $h[n]$ and plot it.

(b) Determine an equation for the frequency response $H(e^{j\hat{\omega}})$ and express it in the form

$$H(e^{j\hat{\omega}}) = R(e^{j\hat{\omega}})e^{-j\hat{\omega}n_0}$$

where $R(e^{j\hat{\omega}})$ is a real function and n_0 is an integer.

(c) Carefully sketch and label a plot of $|H(e^{j\hat{\omega}})|$ for $-\pi < \hat{\omega} < \pi$.

(d) Carefully sketch and label a plot of the principal value of $\angle H(e^{j\hat{\omega}})$ for $-\pi < \hat{\omega} \leq \pi$.

P-6.18 A linear time-invariant system has frequency response

$$H(e^{j\hat{\omega}}) = (1 - e^{j\pi/2}e^{-j\hat{\omega}})(1 - e^{-j\pi/2}e^{-j\hat{\omega}})(1 + e^{-j\hat{\omega}})$$

The input to the system is

$$x[n] = 5 + 20\cos(0.5\pi n + 0.25\pi) + 10\delta[n-3]$$

Use superposition to determine the corresponding output $y[n]$ of the LTI system for $-\infty < n < \infty$.

P-6.19 For the cascade configuration in Fig. P-6.16, the two systems are defined by

$$H_1(e^{j\hat{\omega}}) = 1 + 2e^{-j\hat{\omega}} + e^{-j\hat{\omega}2} \qquad \text{and}$$
$$h_2[n] = \delta[n] - \delta[n-1] + \delta[n-2] - \delta[n-3]$$

(a) Determine the frequency response $H(e^{j\hat{\omega}})$ for the overall cascade system (i.e., from input $x[n]$ to output $y[n]$). Simplify your answer as much as possible.

(b) Determine and plot the impulse response $h[n]$ of the overall cascade system.

(c) Write down the difference equation that relates $y[n]$ to $x[n]$.

P-6.20 The input to the C-to-D converter in Fig. P-6.15 is

$$x(t) = 10 + 20\cos(\omega_0 t + \pi/3) \qquad -\infty < t < \infty$$

(a) Suppose that the impulse response of the LTI system is $h[n] = \delta[n]$. If $\omega_0 = 2\pi(500)$, for what values of $f_s = 1/T_s$ will it be true that $y(t) = x(t)$?

(b) Now suppose that the impulse response of the LTI system is changed to $h[n] = \delta[n-10]$. Determine the sampling rate $f_s = 1/T_s$ and a range of values for ω_0 so that the output of the overall system is

$$\begin{aligned} y(t) &= x(t - 0.001) \\ &= 10 + 20\cos(\omega_0(t - 0.001) + \pi/3) \end{aligned}$$

for $-\infty < t < \infty$.

(c) Suppose that the LTI system is a 5-point moving averager whose frequency response is

$$H(e^{j\hat{\omega}}) = \frac{\sin(5\hat{\omega}/2)}{5\sin(\hat{\omega}/2)}e^{-j\hat{\omega}2}$$

If the sampling rate is $f_s = 2000$ samples/sec, determine all values of ω_0 such that the output is equal to a constant; i.e., $y(t) = A$ for $-\infty < t < \infty$. Also, determine the constant A in this case.

P-6.21 The frequency response of an LTI system is plotted in Fig. P-6.21

(a) Use these plots to find the output of the system if the input is

$$x[n] = 10 + 10\cos(0.2\pi n) + 10\cos(0.5\pi n)$$

(b) Explain why the phase-response curve is discontinuous at frequencies around $\hat{\omega} = 2\pi(0.17)$ and also around $\hat{\omega} = 0.5\pi$.

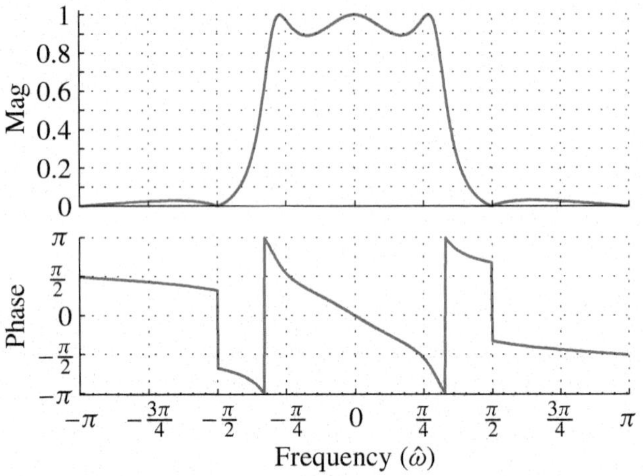

Figure P-6.21

CHAPTER 7

z-Transforms

In this chapter we introduce the z-transform, which brings polynomials and rational functions into the analysis of linear discrete-time systems. We will show that FIR convolution is equivalent to polynomial multiplication and that common algebraic operations, such as multiplying, dividing, and factoring polynomials, can be interpreted as combining or decomposing LTI systems. The most common z-transforms are rational functions, i.e., a numerator polynomial divided by a denominator polynomial. The roots of these polynomials are important, because most properties of digital filters can be restated in terms of the locations of these roots.

The z-transform method is introduced in this chapter for FIR filters and finite-length sequences in general. We will use the FIR case to introduce the important concept of "domains of representation" for discrete-time signals and systems. In this text, we consider three domains of representation of signals and systems: the *n-domain* or *time domain* (the domain of sequences, impulse responses and difference equations), the $\hat{\omega}$-*domain* or *frequency domain* (the domain of frequency responses and spectrum representations), and the z-*domain* (the domain of z-transforms, operators, and poles and zeros).[1] The value of having three different domains of representation is that a difficult analysis in one domain is often much easier in one of the other domains. Therefore, increased understanding will result from developing skills for moving from one representation to

[1]Traditionally, signals and systems texts have identified just two domains: the time domain and the frequency domain. Many authors consider our $\hat{\omega}$-domain and z-domain together as the "frequency domain." This is mainly because, as we will see, the $\hat{\omega}$-domain can be viewed as a special case of the more general z-domain. However, we feel that because of the distinctly different character of the mathematical functions involved in the two domains, there is a distinct advantage in considering the $\hat{\omega}$- and z-domains as separate, but related, points of view.

another. For example, the cascade combination of LTI systems, which in the n-domain seems to require the new (less familiar) technique of convolution, is converted in the z-domain into the more familiar algebraic operation of polynomial multiplication. It is important, however, to note that the "real" or "actual" domain is the n-domain where the signals are generated and processed, and where the implementation of filters takes place. The frequency domain has physical significance when analyzing sound, but is seldom used for implementation. The z-domain exists primarily for its convenience in mathematical analysis and synthesis.

7-1 Definition of the z-Transform

A finite-length signal $x[n]$ can be represented by the relation

$$x[n] = \sum_{k=0}^{N} x[k]\delta[n-k] \tag{7.1}$$

and the z-transform of such a signal is defined by the formula

$$X(z) = \sum_{k=0}^{N} x[k]z^{-k} \tag{7.2}$$

where we will assume that z represents any complex number; i.e., z is the independent (complex) variable of the z-transform $X(z)$. Although (7.2) is the conventional definition of the z-transform,[2] it is instructive to note that $X(z)$ can be written in the form

$$X(z) = \sum_{k=0}^{N} x[k](z^{-1})^k$$

which emphasizes the fact that $X(z)$ is simply a polynomial of degree N in the variable z^{-1}.

[2]Some authors use positive powers of z in the definition of the z-transform, but this convention is not common in signal processing.

When we use (7.2) to determine the z-transform of the signal $x[n]$, we ***transform*** $x[n]$ into a new representation $X(z)$. Indeed, it is often said that we "take the z-transform of $x[n]$." All that we have to do to obtain $X(z)$ is to construct a polynomial whose coefficients are the values of the sequence $x[n]$. Specifically, the kth sequence value is the coefficient of the kth power of z^{-1} in the polynomial $X(z)$. It is just as easy to go from (7.2) back to (7.1). We can recover $x[n]$ from $X(z)$ simply by extracting the coefficient of the kth power of z^{-1} and placing that coefficient in the kth position in the sequence $x[n]$. This operation is sometimes called ***taking the inverse z-transform***. In order to emphasize this unique correspondence between a sequence $x[n]$ and its z-transform, we will use the notation

***n*-Domain**	$\overset{z}{\longleftrightarrow}$	***z*-Domain**
$x[n] = \sum_{k=0}^{N} x[k]\delta[n-k]$	$\overset{z}{\longleftrightarrow}$	$X(z) = \sum_{k=0}^{N} x[k]z^{-k}$

In general, a ***z-transform pair*** is a sequence and its corresponding z-transform, which we will denote as

$$x[n] \overset{z}{\longleftrightarrow} X(z) \tag{7.3}$$

Notice that n is the independent variable of the sequence $x[n]$. Thus, we say that (7.1) represents the signal in the ***n-domain***. Since n is often an index that counts time in a sampled time waveform, we also refer to (7.1) as the ***time-domain*** representation of the signal. Similarly, note that z is the independent variable of the z-transform $X(z)$. Thus, we say that (7.2) represents the signal in the ***z-domain***, and in ***taking the z-transform*** of a signal, we move from the time domain to the z-domain.

As a simple, but very important, example of a z-transform pair, suppose that $x[n] = \delta[n - n_0]$. Then, from the definition, (7.2), it follows that $X(z) = z^{-n_0}$. To emphasize this correspondence we use the notation

$$\begin{array}{ccc} n\text{-Domain} & \stackrel{z}{\longleftrightarrow} & z\text{-Domain} \\ x[n] = \delta[n - n_0] & \stackrel{z}{\longleftrightarrow} & X(z) = z^{-n_0} \end{array} \tag{7.4}$$

When the sequence is defined with numerical values, we can take the z-transform and get a polynomial, as illustrated by the following example.

Example 7-1: z-Transform of a Signal

Consider the sequence $x[n]$ given in the following table:

n	$n < -1$	-1	0	1	2	3	4	5	$n > 5$
$x[n]$	0	0	2	4	6	4	2	0	0

The z-transform of this sequence is

$$X(z) = 2 + 4z^{-1} + 6z^{-2} + 4z^{-3} + 2z^{-4}$$

■

This example shows how to find the z-transform given the sequence. The following example illustrates the inverse z-transform operation (i.e., finding the sequence if we are given its z-transform).

Example 7-2: Inverse z-Transform

Consider the z-transform $X(z)$ given by the equation

$$X(z) = 1 - 2z^{-1} + 3z^{-3} - z^{-5}$$

We can give $x[n]$ in tabular form as in Example 7-1, or we can give an equation for the sequence values as a function of n in the form

$$x[n] = \begin{cases} 0 & n < 0 \\ 1 & n = 0 \\ -2 & n = 1 \\ 0 & n = 2 \\ 3 & n = 3 \\ 0 & n = 4 \\ -1 & n = 5 \\ 0 & n > 5 \end{cases}$$

Alternatively, using the representation (7.1) in terms of impulse sequences, the corresponding sequence $x[n]$ is

$$x[n] = \delta[n] - 2\delta[n - 1] + 3\delta[n - 3] - \delta[n - 5]$$

■

At this point, we have a definition of the z-transform, and we have seen how to find it for a given sequence and how to find the sequence given the z-transform, but why would we want to transform from the n-domain to the z-domain? This is the obvious question to ask at this point, and the remainder of this chapter will attempt to answer it.

7-2 The z-Transform and Linear Systems

The z-transform is indispensable in the design and analysis of LTI systems. The fundamental reason for this has to do with the way that LTI systems respond to the particular input signal z^n for $-\infty < n < \infty$.

7-2.1 The z-Transform of an FIR Filter

Recall that the general difference equation of an FIR filter is

$$y[n] = \sum_{k=0}^{M} b_k\, x[n-k] \tag{7.5}$$

An alternative representation of the input-output relation is the convolution sum

$$y[n] = x[n] * h[n]$$

where $h[n]$ is the impulse response of the FIR filter. Remember that the impulse response $h[n]$ is identical to the sequence of difference equation coefficients b_n, as shown in the following table:

n	$n<0$	0	1	2	...	M	$n>M$
$h[n]$	0	b_0	b_1	b_2	...	b_M	0

which can be represented in the more compact notation,

$$h[n] = \sum_{k=0}^{M} b_k\, \delta[n-k] \tag{7.6}$$

To see why the z-transform is of interest to us for FIR filters, let the input to system in (7.5) be the signal

$$x[n] = z^n \qquad \text{for all } n$$

where z is any complex number. Recall that we have already considered such inputs in Chapter 6, where we used $z = e^{j\hat{\omega}}$. As with our discussion of the frequency response, the qualification "for all n" is an extremely important detail. Because we want to avoid any consideration of what might happen at a starting point such as $n = 0$, we think of this as having the input start at $n = -\infty$, and we assume that for finite values of n, the start-up effects have disappeared, i.e., we are concerned only with the "steady-state" part of the output. For the more general complex exponential input z^n, the corresponding output signal is

$$\begin{aligned} y[n] &= \sum_{k=0}^{M} b_k\, x[n-k] \\ &= \sum_{k=0}^{M} b_k z^{n-k} \\ &= \sum_{k=0}^{M} b_k z^n z^{-k} = \left(\sum_{k=0}^{M} b_k z^{-k}\right) z^n \end{aligned}$$

The term inside the parentheses is a polynomial in z^{-1} whose form depends on the coefficients of the FIR filter. It is called the *system function* of the FIR filter. From our previous definition of the z-transform, this polynomial is observed to be the z-transform of the impulse-response sequence. Using the notation introduced in the previous section, we define the *system function* of an FIR filter to be

$$H(z) = \sum_{k=0}^{M} b_k z^{-k} = \sum_{k=0}^{M} h[k] z^{-k} \tag{7.7}$$

Therefore, we have the following important result:

> ***The system function $H(z)$ is the z-transform of the impulse response.***
>
> $$h[n] = \sum_{k=0}^{M} b_k \delta[n-k] \;\overset{z}{\longleftrightarrow}\; H(z) = \sum_{k=0}^{M} b_k z^{-k}$$

We have just shown that for FIR filters, if the input is z^n for $-\infty < n < \infty$, then the corresponding output is

$$y[n] = h[n] * z^n = H(z) z^n \tag{7.8}$$

That is, the result of convolving the sequence $h[n]$ with the sequence z^n is $H(z)z^n$, where $H(z)$ is the z-transform of $h[n]$. This is a very general statement. In Chapter 8, it will be shown that it applies to any LTI system, not just FIR filters. Thus, the operation of convolution, which really is synonymous with the definition of an LTI system, appears to be closely linked to the z-transform.

Equation (7.7) is general enough to find the ***z*-transform representation** of any FIR filter, because the polynomial coefficients are exactly the same as the filter coefficients $\{b_k\}$ from the FIR filter difference equation (7.5), or, equivalently, the same as the impulse-response sequence from (7.6). Thus, the FIR filter difference equation can be transformed easily into a polynomial in the z-domain simply by replacing each "delay by k" (i.e., $x[n-k]$ in (7.5)) by z^{-k}.

The system function $H(z)$ is a function of the complex variable z. As we have already noted, $H(z)$ in (7.7) is the z-transform of the impulse response, and for the FIR case, it is an Mth-degree polynomial in the variable z^{-1}. Therefore $H(z)$ will have M zeros (i.e., M values z_0 such that $H(z_0)=0$) that (according to the fundamental theorem of algebra) completely define the polynomial to within a multiplicative constant.

Example 7-3: Zeros of System Function

Consider the FIR filter

$$y[n] = 6x[n] - 5x[n-1] + x[n-2]$$

The z-transform system function is

$$\begin{aligned} H(z) &= 6 - 5z^{-1} + z^{-2} \\ &= (3 - z^{-1})(2 - z^{-1}) = 6\,\frac{(z-\frac{1}{3})(z-\frac{1}{2})}{z^2} \end{aligned}$$

Thus, the zeros of $H(z)$ are $\frac{1}{3}$ and $\frac{1}{2}$. Note that the filter

$$w[n] = x[n] - \tfrac{5}{6}x[n-1] + \tfrac{1}{6}x[n-2]$$

has a system function with the same zeros, but the overall constant is 1 rather than 6. This simply means that $w[n] = y[n]/6$. ■

EXERCISE 7.1: Find the system function $H(z)$ of an FIR filter whose impulse response is

$$h[n] = \delta[n] - 7\delta[n-2] - 3\delta[n-3]$$

EXERCISE 7.2: Find the impulse response $h[n]$ of an FIR filter whose system function is

$$H(z) = 4(1 - z^{-1})(1 + z^{-1})(1 + 0.8z^{-1})$$

Hint: Multiply out the factors to get a polynomial and then determine the impulse response by "inverse z-transformation."

7-3 Properties of the z-Transform

In Section 7-1 we gave the general definition of the z-transform, and we showed that for finite-length sequences, it is possible to go uniquely back and forth between the sequence and its z-transform. In this sense, we have demonstrated that the z-transform is a unique representation of *any* finite-length sequence (including the impulse response of an FIR filter). In Section 7-2 we showed that the z-transform arises naturally out of the convolution of the impulse-response sequence with a sequence z^n. In this section, we will explore several properties of the z-transform representation and indicate how the z-transform can be extended to the infinite-length case.

7-3.1 The Superposition Property of the z-Transform

The z-transform is a linear transformation. This is easily seen by considering the sequence $x[n] = ax_1[n] + bx_2[n]$, where both $x_1[n]$ and $x_2[n]$ are assumed to be finite with length less than or equal to N. Using the definition of (7.2), we write

$$
\begin{aligned}
X(z) &= \sum_{n=0}^{N} (ax_1[n] + bx_2[n])\, z^{-n} \\
&= a\sum_{n=0}^{N} x_1[n]z^{-n} + b\sum_{n=0}^{N} x_2[n]z^{-n} \\
&= aX_1(z) + bX_2(z)
\end{aligned}
$$

Thus, we have demonstrated the *superposition property* for the z-transform:

> *The z-transform is linear.*
>
> $$ax_1[n] + bx_2[n] \overset{z}{\longleftrightarrow} aX_1(z) + bX_2(z) \qquad (7.9)$$

This property leads to another way to interpret the z-transform of a finite-length sequence, as illustrated in Example 7-4.

Example 7-4: z-Transform of a Signal

Recall that any finite-length sequence $x[n]$ can be represented as a sum of scaled and shifted impulse sequences as in

$$x[n] = \sum_{k=0}^{N} x[k]\delta[n-k] \qquad (7.10)$$

Furthermore, recall from (7.4) that for a single shifted unit impulse sequence,

$$\delta[n-k] \quad \overset{z}{\longleftrightarrow} \quad z^{-k} \qquad (7.11)$$

Thus, applying (7.11) to each impulse in (7.10) and then adding the individual z-transforms according to (7.9), we obtain as before

$$X(z) = \sum_{k=0}^{N} x[k]z^{-k}$$

■

7-3.2 The Time-Delay Property of the z-Transform

Another important property of the z-transform is that the quantity z^{-1} in the z-domain corresponds to a time shift of 1 in the n-domain. We will illustrate this property with a numerical example. Consider the length-6 signal $x[n]$ defined by the following table of values:

n	$n < 0$	0	1	2	3	4	5	$n > 5$
$x[n]$	0	3	1	4	1	5	9	0

The z-transform of $x[n]$ is the polynomial (in z^{-1})

$$X(z) = 3 + z^{-1} + 4z^{-2} + z^{-3} + 5z^{-4} + 9z^{-5}$$

Recall that the signal values $x[n]$ are the coefficients of the polynomial $X(z)$ and that the exponents correspond to the time locations of the values. For example, the term $4z^{-2}$ indicates that the signal value at $n = 2$ is 4; i.e., $x[2] = 4$.

Now consider the effect of multiplying the polynomial by z^{-1}:

$$
\begin{aligned}
Y(z) &= z^{-1}X(z) \\
&= z^{-1}(3 + z^{-1} + 4z^{-2} + z^{-3} + 5z^{-4} + 9z^{-5}) \\
&= 0z^{0} + 3z^{-1} + z^{-2} + 4z^{-3} + z^{-4} + 5z^{-5} + 9z^{-6}
\end{aligned}
$$

The resulting polynomial $Y(z)$ is the z-transform representation of a signal $y[n]$, which is found by using the polynomial coefficients and exponents in $Y(z)$ to determine the values of $y[n]$ at all time positions. The result is the following table of values for $y[n]$:

n	$n<0$	0	1	2	3	4	5	6	$n>6$
$y[n]$	0	0	3	1	4	1	5	9	0

Each of the signal samples has moved over one position in the table; i.e., $y[n] = x[n-1]$. In general, for any finite-length sequence, multiplication of the z-transform polynomial by z^{-1} simply subtracts one from each exponent in the polynomial, thereby creating a delay of one. Thus, we have the following fundamental relation:

A delay of one sample multiplies the z-transform by z^{-1}.

$$x[n-1] \overset{z}{\longleftrightarrow} z^{-1}X(z) \tag{7.12}$$

which we will refer to as the ***unit-delay property*** of the z-transform .

The unit-delay property can be generalized for the case of shifting by more than one sample by simply applying (7.12) n_0 times. The general result is

Time delay of n_0 samples multiplies the z-transform by z^{-n_0}.

$$x[n-n_0] \overset{z}{\longleftrightarrow} z^{-n_0}X(z) \tag{7.13}$$

7-3.3 A General z-Transform Formula

So far, we have defined the z-transform only for finite-length signals.

$$X(z) = \sum_{n=0}^{N} x[n]z^{-n} \tag{7.14}$$

Our definition assumes that the sequence is nonzero only in the interval $0 \le n \le N$. It is possible to extend the definition to signals of infinite length by simply extending the upper or lower limits to $+\infty$ and $-\infty$, respectively; i.e.,

$$X(z) = \sum_{n=-\infty}^{\infty} x[n]z^{-n} \tag{7.15}$$

However, infinite sums may cause serious mathematical difficulties and require special attention. Summing an infinite number of complex numbers could result in an infinite result. *In mathematical terms, the sum might not converge.* Although we will consider the infinite-length case in Chapter 8, the careful mathematical development of a complete z-transform theory for signal and system analysis is better left to another, more advanced, course.

7-4 The z-Transform as an Operator

The delay property stated in Section 7-3.2 suggests that the quantity z^{-1} is in some sense equivalent to a delay or time shift. This point of view leads to a useful, but potentially confusing, interpretation of the z-transform as an *operator*. To see how this interpretation comes about, we will consider the system function of the unit-delay system.

7-4.1 Unit-Delay Operator

The unit-delay system is one of the basic building blocks for the FIR difference equation. In the time domain, the unit-delay operator $\mathcal{D}$ is defined by

$$y[n] = \mathcal{D}\{x[n]\} = x[n-1] \tag{7.16}$$

It is instructive to find the z-transform representation of this system by letting the input to the unit-delay system be the signal

$$x[n] = z^n \qquad \text{for all } n$$

where z is a complex number. With the z^n input signal, the output of the unit delay is simply

$$\begin{aligned} y[n] &= \mathcal{D}\{x[n]\} \\ &= \mathcal{D}\{z^n\} = z^{n-1} = z^{-1}z^n = z^{-1}x[n] \end{aligned} \tag{7.17}$$

In other words, the input signal is multiplied by z^{-1}, ***in the particular case where*** $x[n] = z^n$.

Strictly speaking, the expression $z^{-1}x[n]$ in (7.17) is misleading, because we must remember that it holds only for $x[n] = z^n$. However, it is common to use the quantity z^{-1} interchangeably with the unit-delay operator symbol $\mathcal{D}$, so that we can say that for *any* input $x[n]$ the action of the unit-delay system is *represented* by the operator z^{-1}; i.e.,

$$y[n] = z^{-1}\{x[n]\} = x[n-1]$$

The brackets enclose the signal operated on by z^{-1} just as in (7.16). Thus, if we are careful in our interpretation, we can use the symbol z^{-1} to stand for the delay operator, and many authors use $\mathcal{D}$ and z^{-1} interchangeably.

We know from the delay property of Section 7-3.2 that if $y[n] = x[n-1]$, then $Y(z) = z^{-1}X(z)$; i.e., for *any* finite-length sequence, z^{-1} multiplies $X(z)$ to produce $Y(z)$. This is the precise way in which z^{-1} represents a unit delay; i.e., it is *not* appropriate to write $z^{-1}x[n]$ without the brackets around $x[n]$, since this mixes the z-domain and the n-domain.

7-4.2 Operator Notation

Consider a system that calculates the ***first difference*** of two successive signal values; i.e.,

$$y[n] = x[n] - x[n-1]$$

The z-transform operator that represents the first-difference system is $(1 - z^{-1})$, because we can write the "operator" equation

$$y[n] = \left(1 - z^{-1}\right)\{x[n]\} = x[n] - x[n-1] \tag{7.18}$$

This equation (7.18) has the following interpretation: The operator "1" leaves $x[n]$ unchanged, and the operator z^{-1} delays $x[n]$ before subtracting it from $x[n]$.

Another simple example would be a system that delays by more than one sample (e.g., by n_d samples):

$$y[n] = x[n - n_d] \qquad n_d \text{ is an integer}$$

In this case, the system function is $H(z) = z^{-n_d}$ and the operator is z^{-n_d}, an obvious generalization of the unit-delay case.

EXERCISE 7.3: Derive the z-transform operator for the first-difference system by working the input $x[n] = z^n$ through the system. Write $y[n]$ as $y[n] = H(z)\{x[n]\}$.

7-4.3 Operator Notation in Block Diagrams

The delay operator concept (7.12) is particularly useful in block diagrams of LTI systems. In a block diagram representation of the FIR filter, the z-transform works as follows: All the unit delays become z^{-1} operators in the transform domain, and, owing to the superposition property of the z-transform, the scalar multipliers and adders are the same as in the time-domain representation. Figure 7-1 shows the z-domain representation of a block diagram for a two-point FIR filter—the z^{-1} operator represents the unit-delay operator.

EXERCISE 7.4: Draw a block diagram similar to Fig. 7-1 for the first difference system: $y[n] = \left(1 - z^{-1}\right)\{x[n]\}$

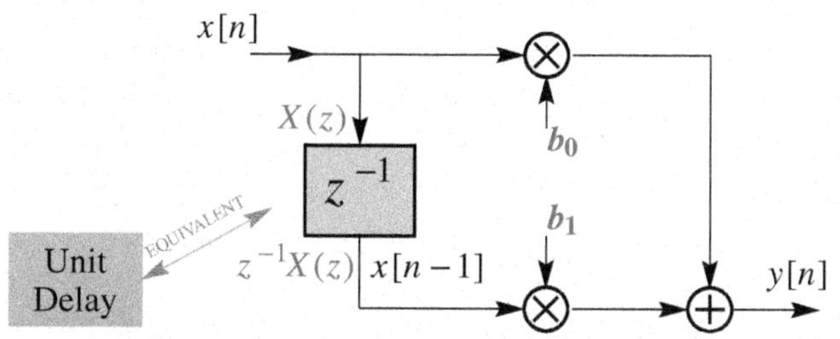

Figure 7-1: Computational structure for a first-order FIR filter whose difference equation is $y[n] = b_0x[n]+b_1x[n-1]$. The block diagram uses z^{-1} to denote the unit delay, because they are equivalent.

7-5 Convolution and the z-Transform

In Section 7-3.2, we observed that a unit delay of a signal in the n-domain is equivalent to multiplication by z^{-1} of the corresponding z-transform in the z-domain. The impulse response of the unit-delay system is

$$h[n] = \delta[n-1]$$

so a delay by one sample is equivalent to the convolution

$$y[n] = x[n] * \delta[n-1] = x[n-1]$$

The system function of the unit-delay system is the z-transform of its impulse response so

$$H(z) = z^{-1}$$

Furthermore, the unit-delay property (7.12) states that delay by one sample multiplies the z-transform by z^{-1}; i.e.,

$$Y(z) = z^{-1}X(z)$$

Therefore, we observe that in the case of the unit delay, the z-transform of the output is equal to the z-transform of the input multiplied by the system function of the LTI system; i.e.,

$$Y(z) = H(z)X(z) \tag{7.19}$$

More importantly, this result (7.19) is true for any LTI system.

To show that convolution is converted into a product of z-transforms (7.19), recall that the discrete convolution of two finite-length sequences $x[n]$ and $h[n]$ is given by the formula:

$$y[n] = x[n] * h[n] = \sum_{k=0}^{M} h[k]x[n-k] \tag{7.20}$$

where M is the order of the FIR filter. To prove the desired result, we can apply the superposition property (7.9) and the general delay property (7.13) to find the z-transform of $y[n]$ as given by (7.20). This leads to

$$\begin{aligned} Y(z) &= \sum_{k=0}^{M} h[k]\left(z^{-k}X(z)\right) \\ &= \left(\sum_{k=0}^{M} h[k]z^{-k}\right) X(z) = H(z)X(z). \end{aligned} \tag{7.21}$$

If $x[n]$ is a finite-length sequence, $X(z)$ is a polynomial, so (7.21) proves that convolution is equivalent to polynomial multiplication. This result is illustrated in the following example.

Example 7-5: Convolution via $H(z)X(z)$

The z-transform method can be used to convolve the following signals:

$$x[n] = \delta[n-1] - \delta[n-2] + \delta[n-3] - \delta[n-4]$$
$$h[n] = \delta[n] + 2\delta[n-1] + 3\delta[n-2] + 4\delta[n-3]$$

The z-transforms of the sequences $x[n]$ and $h[n]$ are:

$$X(z) = 0 + 1z^{-1} - 1z^{-2} + 1z^{-3} - 1z^{-4}$$

$$\text{and} \quad H(z) = 1 + 2z^{-1} + 3z^{-2} + 4z^{-3}$$

Both $X(z)$ and $H(z)$ are polynomials in z^{-1}, so we can compute the z-transform of the convolution by multiplying these two polynomials; i.e.,

$$\begin{aligned} Y(z) &= H(z)X(z) = \\ &(1 + 2z^{-1} + 3z^{-2} + 4z^{-3})(z^{-1} - z^{-2} + z^{-3} - z^{-4}) \\ &= z^{-1} + (-1+2)z^{-2} + (1-2+3)z^{-3} \\ &\quad + (-1+2-3+4)z^{-4} \\ &\quad + (-2+3-4)z^{-5} + (-3+4)z^{-6} + (-4)z^{-7} \\ &= z^{-1} + z^{-2} + 2z^{-3} + 2z^{-4} - 3z^{-5} + z^{-6} - 4z^{-7} \end{aligned}$$

Since the coefficients of any z-polynomial are just the sequence values, with their position in the sequence being indicated by the power of (z^{-1}), we can "inverse transform" $Y(z)$ to obtain

$$\begin{aligned} y[n] &= \delta[n-1] + \delta[n-2] + 2\delta[n-3] + 2\delta[n-4] \\ &\quad - 3\delta[n-5] + \delta[n-6] - 4\delta[n-7] \end{aligned}$$

Now we look at the convolution sum for computing the output. If we write out a few terms, we can detect a pattern that is similar to the z-transform polynomial multiplication.

$$\begin{aligned} y[0] &= h[0]x[0] = 1(0) = 0 \\ y[1] &= h[0]x[1] + h[1]x[0] = 1(1) + 2(0) = 1 \\ y[2] &= h[0]x[2] + h[1]x[1] + h[2]x[0] \\ &= 1(-1) + 2(1) + 3(0) = 1 \\ y[3] &= h[0]x[3] + h[1]x[2] + h[2]x[1] + h[3]x[0] \\ &= 1(1) + 2(-1) + 3(1) = 2 \\ y[4] &= h[0]x[4] + h[1]x[3] + h[2]x[2] + h[3]x[1] \\ &= 1(-1) + 2(1) + 3(-1) + 4(1) = 2 \\ \vdots &= \quad \vdots \end{aligned}$$

Notice how the index of $h[k]$ and the index of $x[n-k]$ sum to the same value (i.e., n) for all products that contribute to $y[n]$. The same thing happens in polynomial multiplication because exponents add.

In Section 5-3.3.1 on p. 110 we demonstrated a synthetic multiplication tableau for evaluating the convolution of $x[n]$ with $h[n]$. Now we see that this is also a process for multiplying the polynomials $X(z)$ and $H(z)$. The procedure is repeated below for the numerical example of this section.

z	z^0	z^{-1}	z^{-2}	z^{-3}	z^{-4}	z^{-5}	z^{-6}	z^{-7}
$x[n],\ X(z)$	0	+1	−1	+1	−1	0	0	0
$h[n],\ H(z)$	1	2	3	4				
$X(z)$	0	+1	−1	+1	−1	0	0	0
$2z^{-1}X(z)$		0	+2	−2	+2	−2	0	0
$3z^{-2}X(z)$			0	+3	−3	+3	−3	0
$4z^{-3}X(z)$				0	+4	−4	+4	−4
$y[n],\ Y(z)$	0	+1	+1	+2	+2	−3	+1	−4]

In the z-transforms $X(z)$, $H(z)$, and $Y(z)$, the power of z^{-1} is implied by the horizontal position of the coefficient in the tableau. Each row is produced by multiplying the $x[n]$ row by one of the $h[n]$ values and shifting the result right by the implied power of z^{-1}. The final answer is obtained by summing down the columns. The final row is the sequence of values of $y[n] = x[n] * h[n]$ or, equivalently, the coefficients of the polynomial $Y(z)$. ■

In this section we have established that convolution and polynomial multiplication are essentially the same thing.[3] Indeed, the most important result of z-transform theory is:

Convolution in the n-domain corresponds to multiplication in the z-domain.

$$y[n] = h[n] * x[n] \quad \stackrel{z}{\longleftrightarrow} \quad Y(z) = H(z)X(z)$$

This result will be seen to have many implications far beyond its use as a basis for understanding and implementing convolution.

EXERCISE 7.5: Use the z-transform of

$$x[n] = \delta[n-1] - \delta[n-2] + \delta[n-3] - \delta[n-4]$$

and the system function $H(z) = 1 - z^{-1}$ to find the output of a first-difference filter when $x[n]$ is the input. Compute your answer by using polynomial multiplication and also by using the difference equation:

$$y[n] = x[n] - x[n-1]$$

What is the degree of the output z-transform polynomial that represents $y[n]$?

7-5.1 Cascading Systems

One of the main applications of the z-transform in system design is its use in creating alternative filters that have exactly the same input–output behavior. An important example is the cascade connection of two or more LTI systems. In block diagram form, the cascade is drawn with the output of the first system connected to the input of the second. The input signal is $x[n]$ and the overall output is $y[n]$. The sequence $w[n]$ is an intermediate signal that can be thought of as temporary storage.

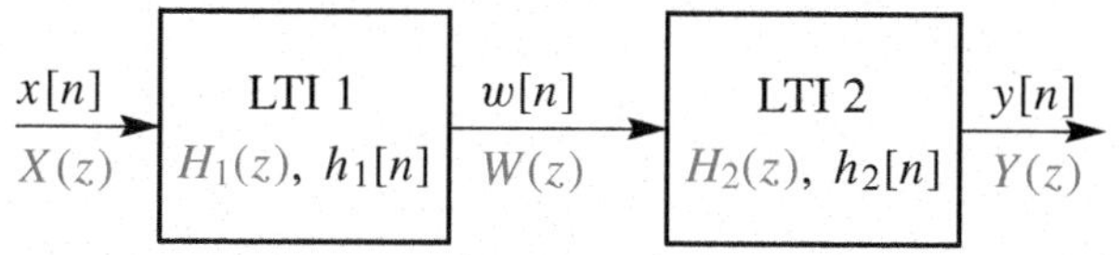

Figure 7-2: Cascade of two LTI systems.

As we have already seen in Section 5-7 on p. 122, if $h_1[n]$ and $h_2[n]$ are the respective impulse responses of the first and second systems, then the overall impulse response from input $x[n]$ to output $y[n]$ in Fig. 7-2 is $h[n] = h_1[n] * h_2[n]$. Therefore, the z-transform of the overall impulse response of the cascade of the two systems is the product of the individual z-transforms of the two impulse responses. That is,

$H(z)$ for a cascade of two LTI systems is the product of the individual system functions.

$$h[n] = h_1[n] * h_2[n] \quad \stackrel{z}{\longleftrightarrow} \quad H(z) = H_1(z)H_2(z)$$

An important consequence of this result follows easily from the fact that multiplication is commutative; i.e., $H(z) = H_1(z)H_2(z) = H_2(z)H_1(z)$. This implies that convolution must also be a commutative operation and that the two systems can be cascaded in either order to obtain the same overall system response.

[3]In MATLAB, there is no special function for multiplying polynomials. Instead, you simply use the convolution function `conv` to multiply polynomials since polynomial multiplication is identical to discrete convolution of the sequences of coefficients.

Example 7-6: $H(z)$ for Cascade

To give a simple example of this idea, consider a system described by the difference equations

$$w[n] = 3x[n] - x[n-1] \qquad (7.22)$$

$$y[n] = 2w[n] - w[n-1] \qquad (7.23)$$

which represent a cascade of two first-order systems as in Fig. 7-2. The output $w[n]$ of the first system is the input to the second system, and the overall output is the output of the second system. The intermediate signal $w[n]$ in (7.22) must be computed prior to being used in (7.23). We can combine the two filters into a single difference equation by substituting $w[n]$ from the first system into the second, which gives

$$\begin{aligned} y[n] &= 2w[n] - w[n-1] \\ &= 2(3x[n] - x[n-1]) - (3x[n-1] - x[n-2]) \\ &= 6x[n] - 5x[n-1] + x[n-2] \qquad (7.24) \end{aligned}$$

Thus we have proved that the cascade of the two first-order systems is equivalent to a single second-order system. It is important to notice that the difference equation (7.24) defines an algorithm for computing $y[n]$ that is different from the algorithm specified by (7.22) and (7.23) together. However, the above analysis shows that with perfectly accurate computation, the outputs of the two different implementations would be exactly the same.

Working out the details of the overall difference equation as we have just done would be extremely tedious if the systems were higher-order. The z-transform simplifies these operations into the multiplication of polynomials. The first-order systems have system functions

$$H_1(z) = 3 - z^{-1} \qquad \text{and} \qquad H_2(z) = 2 - z^{-1}$$

Therefore, the overall system function is

$$H(z) = (3 - z^{-1})(2 - z^{-1}) = 6 - 5z^{-1} + z^{-2}$$

which matches the difference equation in (7.24). Note that, even in this simple example, the z-domain solution is more straightforward than the n-domain solution. ■

EXERCISE 7.6: Use z-transforms to combine the following cascaded systems

$$w[n] = x[n] - x[n-1]$$

$$y[n] = w[n] - w[n-1] + w[n-2]$$

into a single difference equation for $y[n]$ in terms of $x[n]$.

7-5.2 Factoring z-Polynomials

If we can multiply z-transforms to get higher-order systems, we can also factor z-transform polynomials to break down a large system into smaller modules. Since cascading systems is equivalent to multiplying their system functions, the factors of a high-order polynomial $H(z)$ would represent component systems that make up $H(z)$ in a cascade connection.

Example 7-7: Split $H(z)$ into Cascade

Consider the following example

$$H(z) = 1 - 2z^{-1} + 2z^{-2} - z^{-3}$$

One of the roots of $H(z)$ is $z = 1$, so $H_1(z) = (1 - z^{-1})$ is a factor of $H(z)$. The other factor can be obtained by division

$$H_2(z) = \frac{H(z)}{H_1(z)} = \frac{H(z)}{1 - z^{-1}} = 1 - z^{-1} + z^{-2}$$

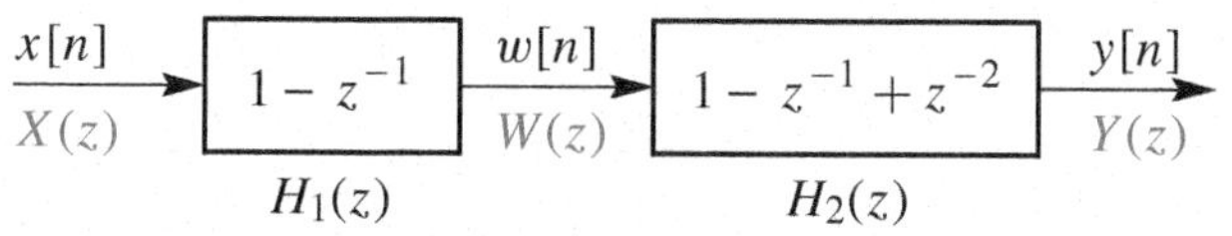

Figure 7-3: Factoring $H(z) = 1 - 2z^{-1} + 2z^{-2} - z^{-3}$ into the product of a first-order system and a second-order system.

The factorization of $H(z)$ as

$$H(z) = (1 - z^{-1})(1 - z^{-1} + z^{-2})$$

gives the cascade shown in the block diagram of Fig. 7-3. The resulting difference equations for the cascade are

$$w[n] = x[n] - x[n-1]$$

$$y[n] = w[n] - w[n-1] + w[n-2]$$

■

7-5.3 Deconvolution

The cascading property leads to an interesting question that has practical application. Can we use the second filter in a cascade to undo the effect of the first filter? What we would like is for the output of the second filter to be equal to the input to the first. Stated more precisely, suppose that we have the cascade of two filters $H_1(z)$ and $H_2(z)$, and $H_1(z)$ is known. Is it possible to find $H_2(z)$ so that the overall system has its output equal to the input? If so, the z-transform analysis tells us that its system function would have to be $H(z) = 1$, so that

$$Y(z) = H_1(z)H_2(z)X(z) = H(z)X(z) = X(z)$$

Since the first system processes the input via convolution, the second filter tries to undo convolution, so the process is called ***deconvolution***. Another term for this is ***inverse filtering***, and if $H_1(z)H_2(z) = 1$, then $H_2(z)$ is said to be the *inverse* of $H_1(z)$ (and vice versa).

Example 7-8: Deconvolution

If we take a specific example, we can generate a solution in terms of z-transforms. Suppose that $H_1(z) = 1 + 0.1z^{-1} - 0.72z^{-2}$. We want $H(z) = 1$, so we require that

$$H_1(z)H_2(z) = 1$$

Since $H_1(z)$ is known, we can solve for $H_2(z)$ to get

$$H_2(z) = \frac{1}{H_1(z)} = \frac{1}{1 + 0.1z^{-1} - 0.72z^{-2}}$$

■

What are we to make of this example? It seems that the deconvolver for an FIR filter must have a system function that is not a polynomial, but a rational function (ratio of two polynomials) instead. This means that the inverse filter for an FIR filter cannot be also an FIR filter, and deconvolution suddenly is not as simple as it appeared. Since we have not yet considered the possibility of anything but polynomial system functions, we cannot give the solution in the form of a difference equation. However, in Chapter 8 we will see that other types of LTI systems exist that do have rational system functions. We will therefore return to the inverse filtering problem in Chapter 8.

7-6 Relationship Between the z-Domain and the $\hat{\omega}$-Domain

The system function $H(z)$ has a functional form that is identical to the form of the frequency response formula $H(e^{j\hat{\omega}})$. This is quite easy to see for the FIR filter if we repeat the formula for the frequency response (6.3) alongside the formula for the system function (7.7):

$\hat{\omega}$-*Domain*	$\overset{z}{\longleftrightarrow}$	z-*Domain*
$H(e^{j\hat{\omega}}) = \sum_{k=0}^{M} b_k e^{-j\hat{\omega}k}$	$\overset{z}{\longleftrightarrow}$	$H(z) = \sum_{k=0}^{M} b_k z^{-k}$

There is a clear correspondence between the z- and $\hat{\omega}$-domains if we make the substitution $z = e^{j\hat{\omega}}$ in $H(z)$.

DEMO: *Three Domains - FIR*

Specifically, it is exceedingly important to note that the connection between $H(e^{j\hat{\omega}})$ and the z-transform $H(z)$ is

$$H(e^{j\hat{\omega}}) = H(z)\big|_{z=e^{j\hat{\omega}}} \tag{7.25}$$

The relationship between the z-domain and the $\hat{\omega}$-domain hinges on the important formula

$$z = e^{j\hat{\omega}} \tag{7.26}$$

To see why this relationship is the key, we need only recall that if the signal z^n is the input to an LTI filter, the resulting output is $y[n] = H(z)z^n$. If the value of z is $z = e^{j\hat{\omega}}$, then

$$y[n] = H(e^{j\hat{\omega}})e^{j\hat{\omega}n}$$

where $H(e^{j\hat{\omega}})$ is obviously the same as what we have called the frequency response.

7-6.1 The z-Plane and the Unit Circle

The notation $H(e^{j\hat{\omega}})$ emphasizes the connection between the $\hat{\omega}$-domain and the z-domain because it indicates explicitly that the frequency response $H(e^{j\hat{\omega}})$ is obtained from the system function $H(z)$ by evaluating $H(z)$ for a specific set of values of z. Recall that since the frequency response is periodic with period 2π, we need only evaluate it over one period, such as $-\pi < \hat{\omega} \leq \pi$. If we substitute these values of $\hat{\omega}$ into (7.26), we see that the corresponding values of z all have unit magnitude and that the angle $\hat{\omega}$ varies from $-\pi$ to $+\pi$. In other words, the values of $z = e^{j\hat{\omega}}$ lie on a circle of radius 1 and range from the point $z = -1$ all the way around the circle and back to the point $z = -1$. Quite naturally, the contour on which all the values of $z = e^{j\hat{\omega}}$ lie is called the ***unit circle***. This is illustrated in Fig. 7-4, which shows the unit circle and a typical point $z = e^{j\hat{\omega}}$, which is at a distance 1 from the origin and at an angle of $\hat{\omega}$ with respect to the real axis of the z-plane.

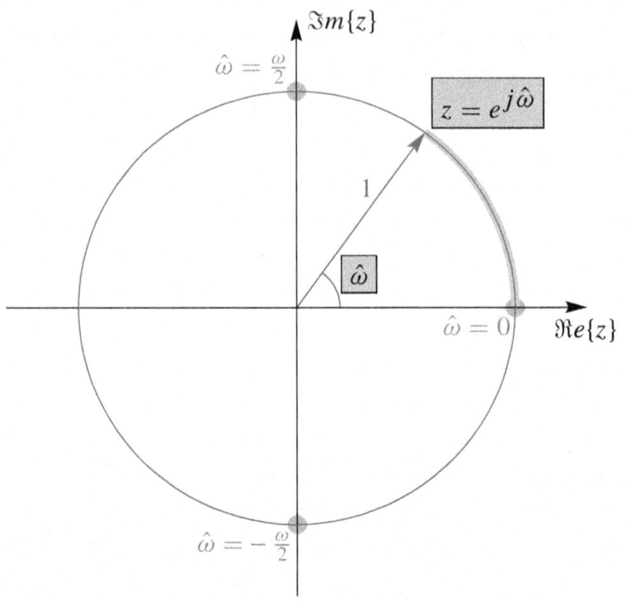

Figure 7-4: The complex z-plane including the unit circle, where $z = e^{j\hat{\omega}}$. The angle around the circle is determined by the frequency $\hat{\omega}$. The orange shading of the unit circle corresponds to a frequency interval from 0 to $\hat{\omega}$. The gray dots on the unit circle are the points $\{-j,\ 1,\ j\}$ which correspond to $\hat{\omega} = \{-\frac{\pi}{2},\ 0,\ \frac{\pi}{2}\}$.

The graphical representation of Fig. 7-4 gives us a convenient way of visualizing the relationship between the $\hat{\omega}$-domain and the z-domain. Because the $\hat{\omega}$-domain lies on a special part of the z-domain — the unit circle — many properties of the frequency response are evident from plots of system function properties in the z-plane. For example, the periodicity of the frequency response is obvious from Fig. 7-4, which shows that evaluating the system function at all points on the unit circle requires moving through an angle of 2π radians. Since frequency $\hat{\omega}$ is equivalent to angle in the z-plane, 2π radians in the z-plane correspond to an interval of 2π radians of frequency. Continuing around the unit circle more times simply cycles through more periods of the frequency response.

7-6.2 The Zeros and Poles of $H(z)$

We have already seen that the system function for an FIR system is essentially determined by its zeros. This is illustrated by the following example.

Example 7-9: Zeros and Poles of $H(z)$

Consider the system function

$$H(z) = 1 - 2z^{-1} + 2z^{-2} - z^{-3},$$

which can be expressed in the following different forms:

$$H(z) = 1 - 2z^{-1} + 2z^{-2} - z^{-3} \quad (7.27)$$

$$= (1 - z^{-1})(1 - e^{j\pi/3}z^{-1})(1 - e^{-j\pi/3}z^{-1}) \quad (7.28)$$

or, if we multiply $H(z)$ by z^3/z^3, we obtain the following two equivalent forms:

$$H(z) = \frac{z^3 - 2z^2 + 2z - 1}{z^3} \quad (7.29)$$

$$= \frac{(z-1)(z-e^{j\pi/3})(z-e^{-j\pi/3})}{z^3} \quad (7.30)$$

Equations (7.27)–(7.30) give four different equivalent forms of $H(z)$. The factored form in (7.30) shows clearly that the zeros of $H(z)$ are at locations $z_1 = 1$, $z_2 = e^{j\pi/3}$, and $z_3 = e^{-j\pi/3} = z_2^*$ in the z-plane. Equation (7.30) also shows that $H(z) \to \infty$ for $z \to 0$. Values of z for which $H(z)$ is undefined (infinite) are called ***poles*** of $H(z)$. In this case, we say that the term z^3 represents three poles at $z = 0$ or that $H(z)$ has a third-order pole at $z = 0$.

We have stated that the poles and zeros determine the system function to within a constant. As an illustration, note that the polynomial $\frac{1}{2}H(z) = 0.5 - z^{-1} + z^{-2} - 0.5z^{-3}$ has exactly the same poles and zeros as $H(z)$ in (7.27). ■

Although it is perhaps less obvious, the locations of both the poles and the zeros are also clear when $H(z)$ is written in the form (7.28), since each factor of the form $(1 - az^{-1})$ always can be expressed as

$$(1 - az^{-1}) = \frac{(z-a)}{z}$$

which shows that each factor of the form $(1 - az^{-1})$ represents a zero at $z = a$ and a pole at $z = 0$. When $H(z)$ contains only negative powers of z, it is usually most convenient to use the representations of the form of (7.27) and (7.28) since the negative powers of z have a direct correspondence to the difference equation and the impulse response.

It is useful to display the zeros and poles of $H(z)$ as points in the complex z-plane. The plot in Fig. 7-5 shows the three zeros and the three poles for Example 7-9. Such a plot is called a ***pole-zero plot***. This plot was generated in MATLAB using the `zplane` function.[4] Each zero location is denoted by a small circle, and

[4] In the *DSP-First toolbox,* there is a function called `zzplane` that will make pole-zero plots.

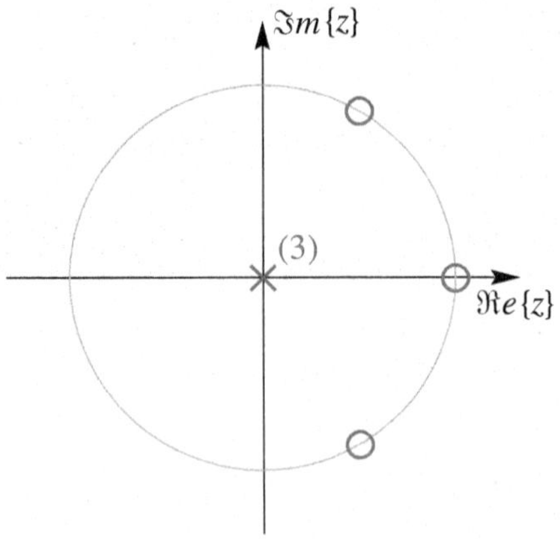

Figure 7-5: Zeros and poles of $H(z)$ marked in a z-plane that includes the unit circle.

the three poles at $z = 0$ are indicated by a single $\times$ with a numeral 3 beside it. In general, when all the poles are not concentrated at $z = 0$, the $\times$ symbol will mark the location of each pole. Since the unit circle is where $H(z)$ is evaluated to obtain the frequency response of the LTI system whose system function is $H(z)$, it is also shown for reference as a gray circle in Fig. 7-5.

7-6.3 Significance of the Zeros of $H(z)$

In Section 7-6.2 we showed that the zeros of a polynomial system function are sufficient to determine $H(z)$ except for a constant multiplier. The system function determines the difference equation of the filter because the polynomial coefficients of $H(z)$ are the coefficients of the difference equation. The difference equation is the direct link between an input $x[n]$ and the corresponding output $y[n]$. However, there are some inputs where knowledge of the zero locations is sufficient to make a precise statement about the output without actually computing it using the difference equation. Such signals are of the form $x[n] = z_0^n$ for all n, where the subscript signifies that z_0 is a particular complex number. In this case, the output is

$$y[n] = H(z_0)z_0^n$$

The quantity $H(z_0)$ is a complex constant, which, through complex multiplication, causes a magnitude and phase change of the input signal z_0^n. In particular, if z_0 is one of the zeros of $H(z)$, then $H(z_0) = 0$ so the output will be zero.

Example 7-10: Nulling Signals with Zeros

For example, when $H(z) = 1 - 2z^{-1} + 2z^{-2} - z^{-3}$, the roots are

$$\begin{aligned} z_1 &= 1 \\ z_2 &= \tfrac{1}{2} + j\tfrac{1}{2}\sqrt{3} = 1e^{j\pi/3} \\ z_3 &= \tfrac{1}{2} - j\tfrac{1}{2}\sqrt{3} = 1e^{-j\pi/3} \end{aligned}$$

As shown in Fig. 7-5, these zeros are all on the unit circle, so complex sinusoids with frequencies $0, \pi/3$, and $-\pi/3$ will be set to zero by the system. That is, the output resulting from each of the following three signals will be zero:

$$\begin{aligned} x_1[n] = (z_1)^n &= 1 \\ x_2[n] = (z_2)^n &= e^{j\pi n/3} \\ x_3[n] = (z_3)^n &= e^{-j\pi n/3} \end{aligned}$$

■

As illustrated by this example, the zeros of the system function that lie on the unit circle correspond to frequencies at which the gain of the system is zero. Thus, complex sinusoids at those frequencies are blocked, or "nulled" by the system.

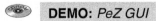

DEMO: *PeZ GUI*

EXERCISE 7.7: Double-check the fact that the inputs $x_1[n]$, $x_2[n]$, and $x_3[n]$ determined in Example 7-10 produce outputs that are zero everywhere by substituting these signals into the difference equation $y[n] = x[n] - 2x[n-1] + 2x[n-2] - x[n-3]$ to show that the complex phasors cancel out for all values of n. Also show that since the filter is linear, it will also null out signals such as $2\cos(\pi n/3)$, which is the sum of $x_2[n]$ and $x_3[n]$.

7-6.4 Nulling Filters

We have just shown that if the zeros of $H(z)$ lie on the unit circle, then certain sinusoidal input signals are removed or nulled by the filter. Therefore, it should be possible to use this result in designing an FIR filter that can null a particular sinusoidal input. Such capability is often needed to eliminate jamming signals in a radar or communications system. Similarly, 60-Hz interference from a power line could be eliminated by placing a null at the correct frequency.

Zeros in the z-plane can remove only signals that have the special form $x[n] = z_0^n$. If we want to eliminate a sinusoidal input signal, then we actually have to remove two signals of the form $z_1^n + z_2^n$; i.e.,

$$x[n] = \cos(\hat{\omega}_0 n) = \tfrac{1}{2}e^{j\hat{\omega}_0 n} + \tfrac{1}{2}e^{-j\hat{\omega}_0 n}$$

Each complex exponential can be removed with a first-order FIR filter, and then the two filters would be cascaded to form the second-order nulling filter that removes the cosine. The second-order FIR filter will have two zeros at $z_1 = e^{j\hat{\omega}_0}$ and $z_2 = e^{-j\hat{\omega}_0}$. The signal z_1^n will be nulled by a filter with system function

$$H_1(z) = 1 - z_1 z^{-1}$$

because $H_1(z_1) = 0$ at $z = z_1$; i.e.,

$$H_1(z_1) = 1 - z_1(z_1)^{-1} = 1 - 1 = 0$$

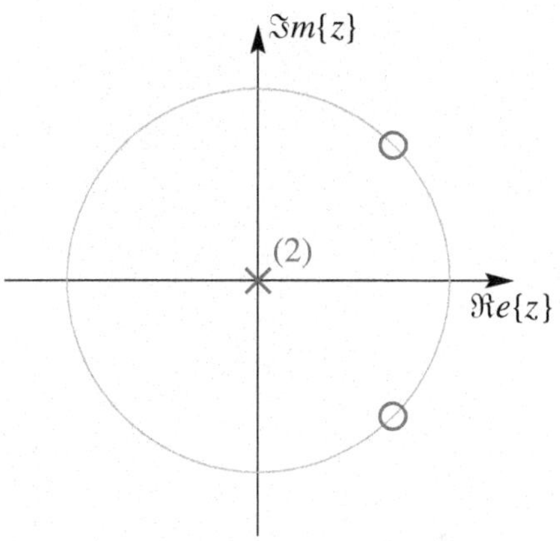

Figure 7-6: Zeros on unit circle for second-order nulling filter to remove sinusoidal components at $\hat{\omega}_0 = \pm\pi/4$. There are two poles at the origin.

Similarly, $H_2(z) = 1 - z_2 z^{-1}$ will remove z_2^n. Thus the second-order nulling filter will be the product

$$\begin{aligned}
H(z) &= H_1(z)H_2(z) \\
&= (1 - z_1 z^{-1})(1 - z_2 z^{-1}) \\
&= 1 - (z_1 + z_2)z^{-1} + (z_1 z_2)z^{-2} \\
&= 1 - (e^{j\hat{\omega}_0} + e^{-j\hat{\omega}_0})z^{-1} + (e^{j\hat{\omega}_0}e^{-j\hat{\omega}_0})z^{-2} \\
&= 1 - 2\cos(\hat{\omega}_0)z^{-1} + z^{-2}
\end{aligned}$$

Figure 7-6 shows the two zeros needed to remove components at $z = e^{\pm j\pi/4}$. For the example depicted in Fig. 7-6, the numerical values for the coefficients of $H(z)$ are

$$H(z) = 1 - 2\cos(\pi/4)z^{-1} + z^{-2} = 1 - \sqrt{2}z^{-1} + z^{-2}$$

Thus the nulling filter that will remove the signal $\cos(0.25\pi n)$ from its input is the FIR filter whose difference equation is

$$y[n] = x[n] - \sqrt{2}\,x[n-1] + x[n-2] \tag{7.31}$$

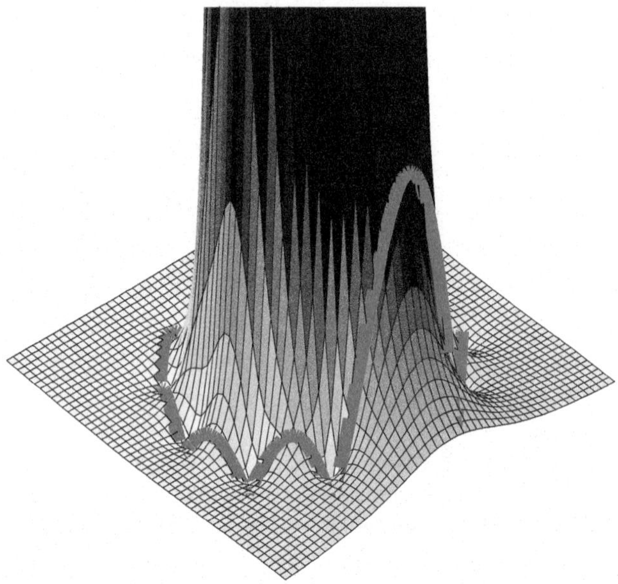

Figure 7-7: z-transform for an FIR filter evaluated over the region $[-1.4 \le \Re e\{z\} \le 1.4] \times [-1.4 \le \Im m\{z\} \le 1.4]$ of the z-plane that includes the unit circle. Values along the unit circle are shown as a colored line where the frequency response (magnitude) is evaluated. The view is from the fourth quadrant, so the point $z = 1$ is in the foreground on the right.

7-6.5 Graphical Relation Between z and $\hat{\omega}$

The equation $z = e^{j\hat{\omega}}$ provides the link between the z-domain and the $\hat{\omega}$-domain. As we have shown in (7.25), the frequency response is obtained by evaluating the system function on the unit circle of the z-plane. This correspondence can be given a useful graphical interpretation. By considering the pole-zero plot of the system function, we can visualize how the frequency response plot of $H(e^{j\hat{\omega}})$ results from evaluating $H(z)$ on the unit circle, and also how it depends on the poles and zeros of $H(z)$. As an example, we show in Fig. 7-7 a plot obtained by evaluating the z-transform magnitude $|H(z)|$ over a region of the z-plane that includes the unit circle as well as values both inside and outside the unit circle. The system in this case is an 11-point running sum; i.e., it is an FIR filter where the coefficients are all equal to one.[5] The system function for this filter is

$$H(z) = \sum_{k=0}^{10} z^{-k} \tag{7.32}$$

In Section 7-7 we will show that the zeros of the system function of this filter are on the unit circle at angles $\hat{\omega} = 2\pi k/11$, for $k = 1, 2, \ldots, 10$. This means that the polynomial in (7.32) can be represented in the form

$$\begin{aligned} H(z) = (1 - e^{j2\pi/11}z^{-1})(1 - e^{j4\pi/11}z^{-1}) \cdots \\ (1 - e^{j18\pi/11}z^{-1})(1 - e^{j20\pi/11}z^{-1}) \end{aligned} \tag{7.33}$$

Recall that each factor of the form $(1 - e^{j2\pi k/11}z^{-1})$ represents a zero at $z = e^{j2\pi k/11}$ and a pole at $z = 0$. Thus, (7.33) displays the 10 zeros of $H(z)$ at $z = e^{j2\pi k/11}$, for $k = 1, 2, \ldots, 10$ and the 10 poles at $z = 0$.

In the magnitude plot of Fig. 7-7, we observe that the zeros pin down the three-dimensional plot around the unit circle. Inside the unit circle, the values of $H(z)$ become very large owing to the poles at $z = 0$. The frequency response $H(e^{j\hat{\omega}})$ is obtained by selecting the values of the z-transform along the unit circle in Fig. 7-7. A plot of $|H(e^{j\hat{\omega}})|$ versus $\hat{\omega}$ is given in Fig. 7-8. The shape of the frequency response can be explained in terms of the zero locations shown in Fig. 7-9 by recognizing that the poles at $z = 0$ push $H(e^{j\hat{\omega}})$ up, but the zeros along the unit circle make $H(e^{j\hat{\omega}}) = 0$ at

[5] This is the same system as the 11-point running-average filter that we discussed in detail in Section 6-7 on p. 145, except that we have omitted the gain constant 1/11.

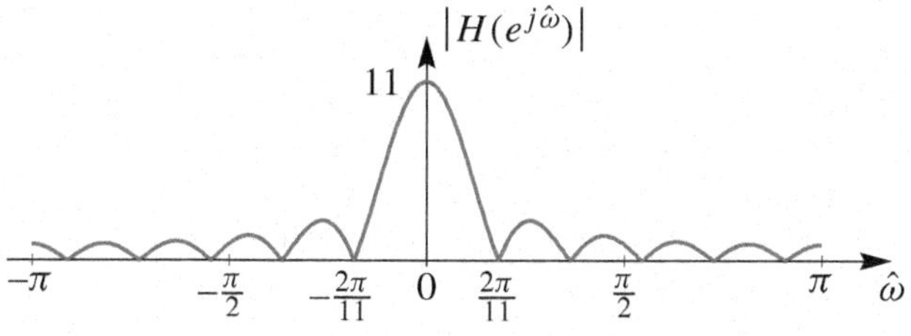

Figure 7-8: Frequency response (magnitude only) for the 11-point running sum. These are the values of $H(z)$ along the unit circle in the z-plane. There are 10 zeros spread out uniformly along the frequency axis.

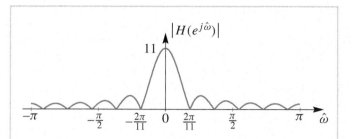

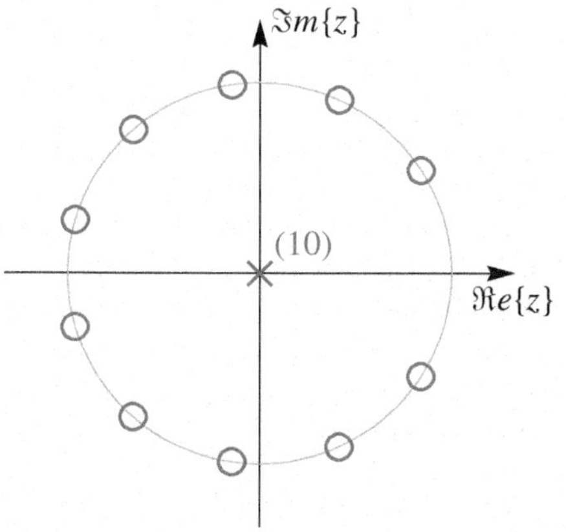

Figure 7-9: Zero and pole distribution for the 11-point running sum. There are 10 zeros spread out uniformly along the unit circle and 10 poles at the origin.

regular intervals except for the region near $\hat{\omega} = 0$ (i.e., around $z = 1$). The unit circle values follow the ups and downs of $H(z)$ as $\hat{\omega}$ goes from $-\pi$ to $+\pi$ with $|z| = 1$.

This example illustrates that an intuitive picture of the frequency response of an LTI system can be visualized from a plot of the poles and zeros of the system function $H(z)$. We simply need to remember that a pole will "push up" the frequency response and a zero will "pull it down." Furthermore, a zero on the unit circle will force the frequency response to be zero at the frequency corresponding to the angular position of the zero.

DEMO: *Three Domains - FIR*

Figure 7-10 shows a frame from a 3-Domain Movie that illustrates the tie between the n, z, and $\hat{\omega}$ domains.

7-7 Useful Filters

Now that we can exploit our new knowledge to design filters with desirable characteristics. In this section, we will look at a special class of bandpass filters (BPFs) that are all close relatives of the running-sum filter.

7-7.1 The L-Point Running-Sum Filter

Generalizing from the previous section, the L-point running-sum filter

$$y[n] = \sum_{k=0}^{L-1} x[n-k]$$

has system function

$$H(z) = \sum_{k=0}^{L-1} z^{-k}$$

Recalling the formula (6.23) from p. 145 for the sum of L terms of a geometric series, $H(z)$ can be represented in the forms

$$H(z) = \sum_{k=0}^{L-1} z^{-k} = \frac{1 - z^{-L}}{1 - z^{-1}} = \frac{z^L - 1}{z^{L-1}(z-1)} \qquad (7.34)$$

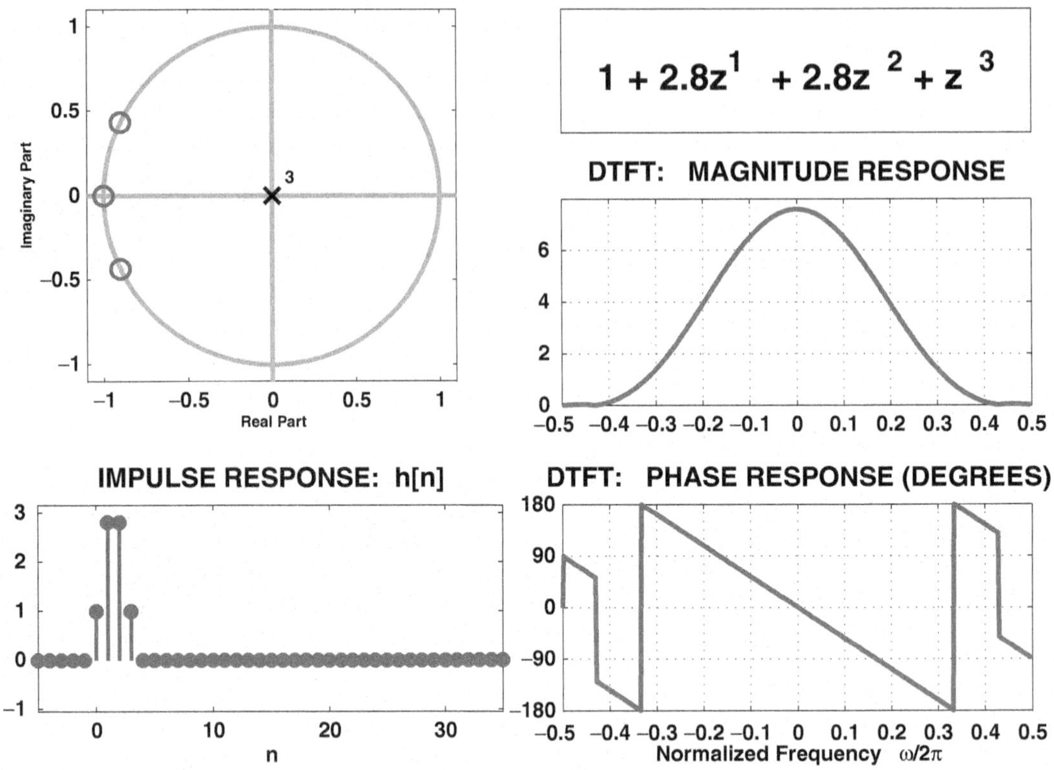

Figure 7-10: One frame of a *3-Domain Movie* showing the different domains: time domain (n), frequency domain ($\hat{\omega}$), the z-transform domain $H(z)$ and a pole-zero plot. In this movie, the zero pair moves around on the unit circle.

The final form for $H(z)$ is a rational function where the numerator polynomial is $z^L - 1$ and the denominator is $z^{L-1}(z-1)$. The zeros of $H(z)$ will be determined by the roots of the numerator polynomial, i.e., the values of z such that

$$z^L - 1 = 0 \qquad \Longrightarrow z^L = 1 \tag{7.35}$$

Since $e^{j2\pi k} = 1$ for k, an integer, it is easy to see by substitution that each of the values

$$z = e^{j2\pi k/L} \quad \text{for } k = 0, 1, 2, \ldots, L-1 \tag{7.36}$$

satisfy (7.35) and therefore these L numbers are the roots of the Lth-order equation in (7.35). Because the values in (7.36) satisfy the equation $z^L = 1$, they are called *the* L^{th} *roots of unity*. The zeros of the denominator in (7.34), which are either $z = 0$ (of order $L - 1$) or $z = 1$, would normally be the poles of $H(z)$. However, since one of the Lth roots of unity is $z = 1$ (i.e., $k = 0$ in (7.36)) that zero of the numerator cancels the corresponding zero of the denominator, so that only the term z^{L-1} really causes a pole of $H(z)$. Therefore, it follows that $H(z)$ can be expressed in the factored form

$$H(z) = \sum_{k=0}^{L-1} z^{-k} = \prod_{k=1}^{L-1}(1 - e^{j2\pi k/L}z^{-1}) \tag{7.37}$$

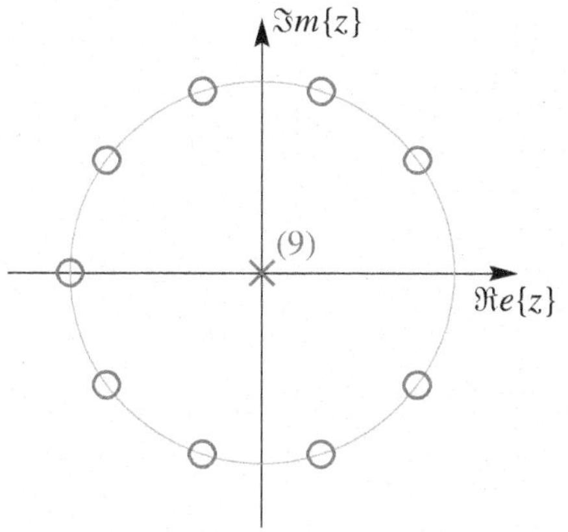

Figure 7-11: Zero and pole distribution for the 10-point running-sum filter. There are nine zeros spread out uniformly along the unit circle, and nine poles at the origin.

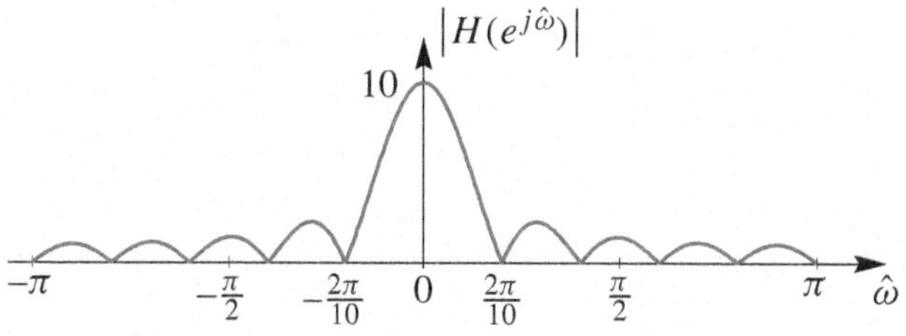

Figure 7-12: Frequency response (magnitude only) for the 10-point running-sum filter. These are the values along the unit circle in the z-plane. There are nine zeros spread out uniformly along the frequency axis.

Example 7-11: $H(e^{j\hat{\omega}})$ from $H(z)$

For a 10-point running-sum filter ($L = 10$), the system function is

$$H(z) = \sum_{k=0}^{9} z^{-k} = \frac{1 - z^{-10}}{1 - z^{-1}} = \frac{z^{10} - 1}{z^9(z - 1)} \quad (7.38)$$

A pole-zero diagram for this case is shown in Fig. 7-11, and the corresponding frequency response for the running-sum filter is shown in Fig. 7-12. The factors of the numerator are the tenth roots of unity, and the zero at $z = 1$ is canceled by the corresponding term in the denominator. This explains why we have only nine zeros around the unit circle with the gap at $z = 1$. The nine zeros around the unit circle in Fig 7-11 show up as zeros along the $\hat{\omega}$ axis in Fig. 7-12 at $\hat{\omega} = 2\pi k/10$, and it is the gap at $z = 1$ that allows the frequency response to be larger at $\hat{\omega} = 0$. The other zeros around the unit circle keep $H(e^{j\hat{\omega}})$ small, thereby creating the "lowpass" filter frequency response shown in Fig. 7-12. ■

7-7.2 A Complex Bandpass Filter

Now we have a new insight that tells us how to control the frequency response of an FIR filter by placing its zeros on the unit circle. This viewpoint makes it easy to create other FIR filters where we control the location of the passband. If we move the passband to a frequency away from $\hat{\omega} = 0$, then we have a filter that passes a small band of frequency components—a bandpass filter, or BPF.

The obvious way to move the passband is to use all but one of the roots of unity as the zeros of an FIR filter. A formula for this new filter is

$$H(z) = \prod_{\substack{k=0 \\ k \neq k_0}}^{L-1} (1 - e^{j2\pi k/L} z^{-1}) \quad (7.39)$$

where the index k_0 denotes the one omitted root at $z = e^{j2\pi k_0/L}$. An example is shown in Fig. 7-13 for $k_0 = 2$ and $L = 10$. In the general case, the passband of the filter whose system function is given by (7.39) should move from the interval around $\hat{\omega} = 0$ to a like interval around

$$\hat{\omega} = \frac{2\pi k_0}{L}$$

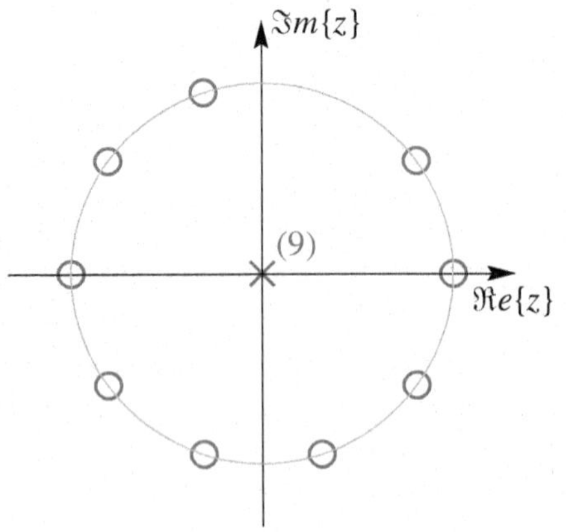

Figure 7-13: Zero and pole distribution for the 10-point complex BPF. The zero at angle $2\pi k_0/L = 2\pi(0.2)$ is the one missing from the tenth roots of unity. The other nine zeros are spread out at equal angles $(2\pi k/10)$ around the unit circle; there are nine poles at the origin.

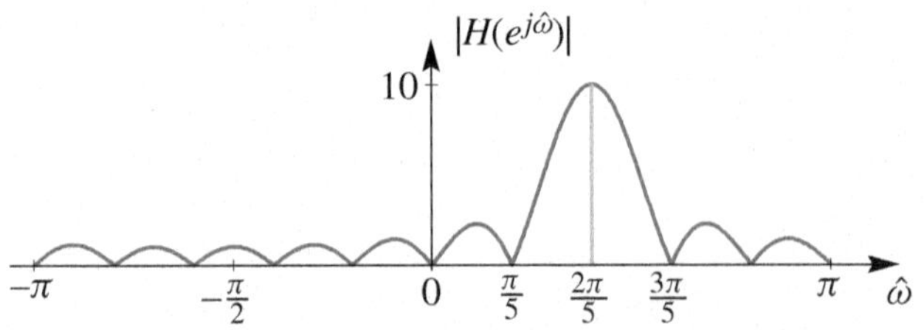

Figure 7-14: Frequency response (magnitude only) for the 10-point complex BPF. These are the values along the unit circle in the z-plane. There are nine zeros spread out uniformly along the frequency axis.

because the zero is missing at that frequency. Figure 7-14 confirms our intuition, because with $k_0 = 2$ the normalized frequency of the peak is $\hat{\omega}/2\pi = k_0/L = 2/10$. This filter is a *bandpass filter*, since frequencies outside the narrow band around $\hat{\omega} = 0.4\pi$ are given much less gain than those in the passband of the filter.

The formula in (7.39) is ideal for seeing how to make the frequency response of a BPF, but it is not very useful for calculating the filter coefficients of the bandpass filter. If one attempts a direct multiplication of the factors in (7.39), nine complex terms must be combined. When all the algebra is finished, the resulting filter coefficients will turn out to be complex-valued. This fact is obvious if we realize that the zeros in Fig. 7-13 *cannot* all be grouped as complex-conjugate pairs.

Another strategy is needed to get the filter coefficients. One idea is to view the zero distribution in Fig. 7-13 as a rotation of the zeros of the running-sum filter in Fig. 7-11. Note that rotation of the z-plane representation will have the corresponding effect of shifting the frequency response along the $\hat{\omega}$-axis by the amount of the rotation. The desired rotation in this case is by the angle $2\pi k_0/L$. So the question is how to move the roots of a polynomial through a rotation. The answer is that we must multiply the kth filter coefficient b_k by $e^{-jk\theta}$ where θ is the desired angle of rotation.

Consider the following general operation on a polynomial $G(z)$:

$$H(z) = G(z/r)$$

Every occurrence of the variable z in the polynomial $G(z)$ is replaced by z/r. The effect of this substitution on the roots of $G(z)$ is to multiply them by r and make these the roots of $H(z)$. For the simple example $G(z) = z^2 - 3z + 2 = (z-2)(z-1)$,

$$\begin{aligned} H(z) = G(z/r) &= (z/r)^2 - 3(z/r) + 2 \\ &= \frac{z^2 - 3rz + 2r^2}{r^2} = \frac{(z-2r)(z-r)}{r^2} \end{aligned}$$

The two roots of $H(z)$ are now $z = 2r$ and $z = r$.

In the case of the complex bandpass filter, $G(z)$ is the running-sum system function

$$G(z) = \sum_{k=0}^{L-1} z^{-k}$$

and the parameter r is a complex exponential $r = e^{j2\pi k_0/L}$. Remember that multiplication by a complex exponential will rotate a complex number through the angle $2\pi k_0/L$. Now it is easy to get the new filter coefficients

$$H(z) = G(z/r) = G(ze^{-j2\pi k_0/L}) = \sum_{k=0}^{L-1} z^{-k} e^{j2\pi k_0 k/L}$$

Thus the filter coefficients of the complex bandpass filter are

$$b_k = e^{j2\pi k_0 k/L} \qquad \text{for } k = 0, 1, 2, \ldots, L-1 \quad (7.40)$$

Another way to determine the frequency response of the complex bandpass filter is to compute it directly, as in

$$\begin{aligned} H(e^{j\hat{\omega}}) &= \sum_{k=0}^{L-1} e^{j2\pi k_0 k/L} e^{-j\hat{\omega}k} \\ &= \sum_{k=0}^{L-1} e^{-j(\hat{\omega}-j2\pi k_0/L)k} \qquad (7.41) \\ &= G(e^{j(\hat{\omega}-j2\pi k_0/L)}) \end{aligned}$$

This equation shows that the frequency response of the system whose filter coefficients are given by (7.40) is a shifted (by $2\pi k_0/L$) version of the frequency response of the L-point running-sum filter.

7-7.3 A Bandpass Filter with Real Coefficients

The obvious disadvantage of the previous strategy is that the resulting filter coefficients (7.41) are complex. We can modify the strategy slightly to get a bandpass filter with real coefficients if we just take the real part of the complex BPF coefficients. Thus the kth filter coefficient is now

$$b_k = \cos(2\pi k_0 k/L) \qquad \text{for } k = 0, 1, \ldots, L-1$$

With these real filter coefficients, the new BPF can be written as the sum of two complex BPFs. By expanding the coefficients of z^{-k} in terms of complex exponentials we obtain

$$\begin{aligned} H(z) &= \sum_{k=0}^{L-1} (\cos(2\pi k_0 k/L))\, z^{-k} \\ &= \sum_{k=0}^{L-1} z^{-k} \left(\tfrac{1}{2}e^{j2\pi k_0 k/L} + \tfrac{1}{2}e^{-j2\pi k_0 k/L}\right) \\ &= \tfrac{1}{2}\sum_{k=0}^{L-1} z^{-k} e^{j2\pi k_0 k/L} + \tfrac{1}{2}\sum_{k=0}^{L-1} z^{-k} e^{-j2\pi k_0 k/L} \\ &= H_1(z) + H_2(z) \end{aligned}$$

where $H_1(z)$ is a complex bandpass filter centered on frequency $2\pi k_0/L$ and $H_2(z)$ is a complex bandpass filter centered on frequency $-2\pi k_0/L$. Figure 7-15 shows the frequency response for $L = 10$ and $k_0 = 2$.

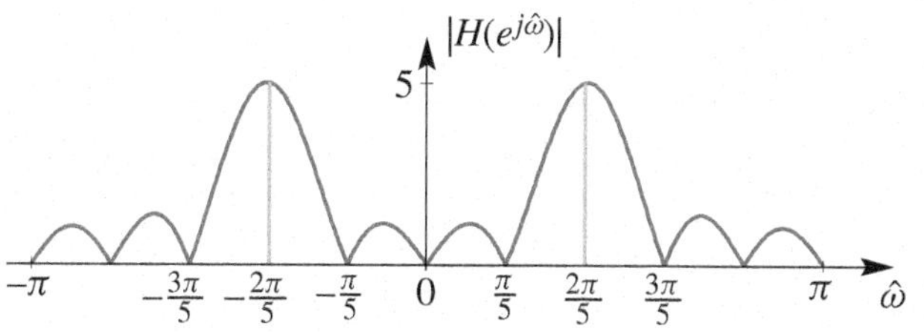

Figure 7-15: Frequency response (magnitude only) for the 10-point real BPF. Notice the two passbands at $\hat{\omega} = \pm 2\pi(2)/10$—one at positive frequency, the other at negative frequency.

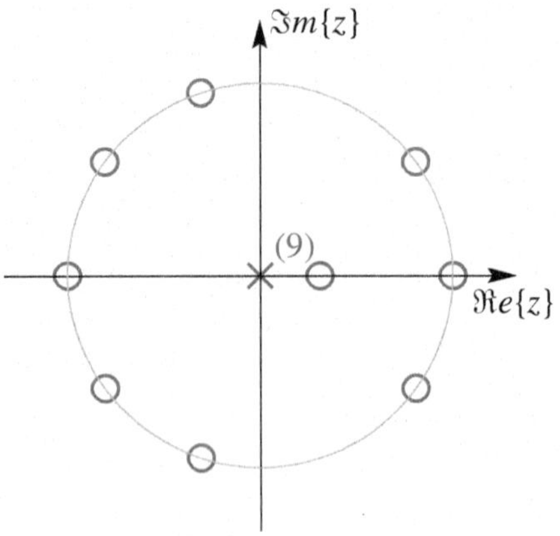

Figure 7-16: Pole-zero distribution for the 10-point real BPF. Of the original 10 roots of unity, two have been dropped off the unit circle at the angles $\pm 4\pi/10$, but a new one appears on the real axis. There are nine poles at the origin.

There are zeros of the frequency response at some of the frequencies $\hat{\omega} = 2\pi k/L$ because both component filters have zeros at all these frequencies except for $\pm 2\pi k_0/L$. As is the case with any real-valued filter, the magnitude of the frequency response exhibits a symmetry about $\hat{\omega} = 0$.

The frequency response in Fig. 7-15 has two peaks at $\hat{\omega} = \pm 4\pi/10 = \pm 2\pi/5$, so there must be two missing zeros on the unit circle at angles $\pm 4\pi/10$. In Fig. 7-16 we see that the two zeros at $z = e^{j4\pi/10} = e^{j2\pi(2)/10}$ and $z = e^{-j4\pi/10} = e^{j2\pi(8)/10}$ have been replaced by a single real zero. Thus, there are eight zeros on the unit circle and one on the real axis for a total of nine zeros, which is the order of the z-transform polynomial. The location of this new zero appears to be at $z = \cos(2\pi k_0/L) = \cos(0.4\pi) = 0.309$, which is the real part of the missing unit-circle zeros.

An algebraic manipulation will uncover the exact location of the new zero. We use the numerator-denominator representation and combine the two terms over a common denominator. To make the notation simpler, let $p = e^{j2\pi k_0/L}$, so that the conjugate is $p^* = e^{-j2\pi k_0/L}$. Then,

$$\begin{aligned} H(z) &= H_1(z) + H_2(z) \\ &= \tfrac{1}{2}\frac{z^L - 1}{z^{L-1}(z-p)} + \tfrac{1}{2}\frac{z^L - 1}{z^{L-1}(z-p^*)} \\ &= \tfrac{1}{2}\frac{(z^L-1)(z-p^*) + (z^L-1)(z-p)}{z^{L-1}(z-p)(z-p^*)} \\ &= \frac{(z^L-1)(z-\tfrac{1}{2}(p+p^*))}{z^{L-1}(z-p)(z-p^*)} \end{aligned}$$

The two factors $(z-p)(z-p^*)$ in the denominator cancel corresponding factors in the numerator polynomial $z^L - 1$, leaving $L-2$ (in this case $L-2 = 8$) of the Lth roots of unity. The term $(z - \tfrac{1}{2}(p+p^*))$ is the new zero at

$$\begin{aligned} z &= \tfrac{1}{2}(p+p^*) \\ &= \tfrac{1}{2}\left(e^{j2\pi k_0/L} + e^{-j2\pi k_0/L}\right) = \cos(2\pi k_0/L) \end{aligned}$$

which is the real part of the canceled zeros.

7-8 Practical Bandpass Filter Design

Although much better filters can be designed by more sophisticated methods, the example of bandpass filter design discussed in Section 7-7 is a useful illustration of the power of the z-transform to simplify the analysis of such problems. Its characterization of a filter by the zeros (and poles) of $H(z)$ is used continually in filter design and implementation. The underlying reason is that the z-transform converts difficult problems involving convolution and frequency response into simple algebraic

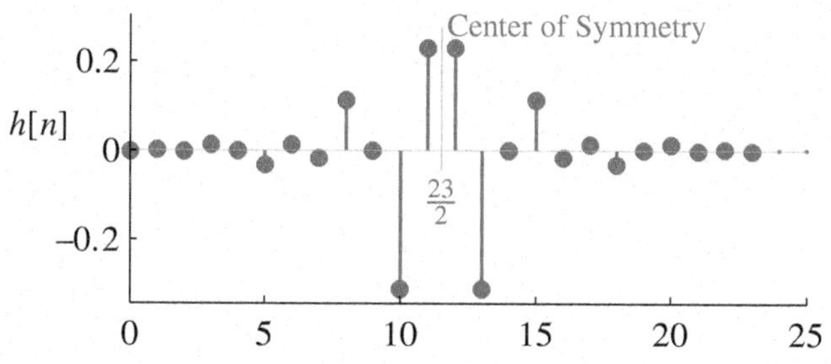

Figure 7-17: Impulse response for the 24-point BPF designed with MATLAB's `fir1` function. These are the FIR filter coefficients $\{b_k\}$, $k = 0, 1, 2, \ldots, 23$, which are needed in the difference equation implementation.

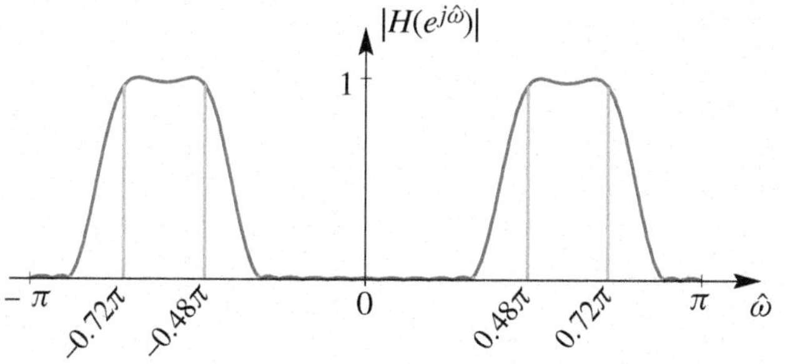

Figure 7-18: Frequency response (magnitude only) for the 24-point BPF. The passbands are the intervals $0.48\pi \leq |\hat{\omega}| \leq 0.72\pi$. The design was performed with `fir1(23,2*[0.2,0.4],kaw)` where `kaw = kaiser(24,2.75)` is a Kaiser window.

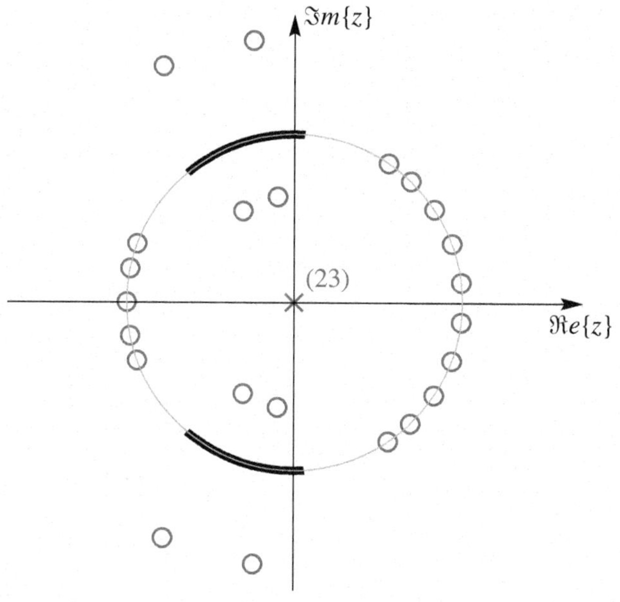

Figure 7-19: Zero and pole distribution for a 24-point BPF designed with MATLAB's `fir1` function. The section of unit circle corresponding to the passband of $H(e^{j\hat{\omega}})$ in Fig. 7-18 is outlined in black.

ideas based on multiplying and factoring polynomials. Thus, skills with basic algebra become essential everyday tools in design.

As a final example of FIR filters, we present a high-order FIR filter that has been designed by a computer-aided filter-design program. Most digital filter design is carried out by software tools that permit much more sophisticated control of the passband and stopband characteristics than we were able to achieve with the simple analysis of Section 7-7. Design programs such as `remez` and `fir1` can be found in the MATLAB software augmented with the Signal Processing Toolbox.

Although it is not our intention to discuss any of these methods or even how they are used, it is interesting to examine the output from the program `fir1` to get a notion of what can be achieved with a sophisticated design method.

The software allows us to specify the frequency range for the passband, and then computes a good approximation to an ideal filter that has unity gain over the passband and zero gain in the stopband. An example is shown in Figs. 7-17, 7-18, and 7-19 for a length–24 FIR bandpass filter.

Table 7-1: Filter coefficients of 24-point FIR bandpass.

$$\begin{aligned}
b_0 &= -0.0108 = b_{23} \\
b_1 &= 0.0037 = b_{22} \\
b_2 &= -0.0052 = b_{21} \\
b_3 &= 0.0300 = b_{20} \\
b_4 &= 0.0000 = b_{19} \\
b_5 &= -0.0526 = b_{18} \\
b_6 &= 0.0164 = b_{17} \\
b_7 &= -0.0219 = b_{16} \\
b_8 &= 0.1275 = b_{15} \\
b_9 &= 0.0000 = b_{14} \\
b_{10} &= -0.3236 = b_{13} \\
b_{11} &= 0.2330 = b_{12}
\end{aligned}$$

The impulse response of the 24-point FIR bandpass filter is given in Fig. 7-17. Notice that the coefficients would be labeled $\{b_k\}$, from $k = 0, 1, 2, \ldots, 23$. The plot shows an obvious symmetry about a midpoint at $n = 23/2$. Table 7-1 lists the values for the impulse response and filter coefficients. These are the coefficients used in the difference equation

$$y[n] = \sum_{k=0}^{23} b_k x[n-k]$$

to implement the filter whose frequency response is shown in Fig. 7-18. Notice the wide passbands in the frequency ranges corresponding to $2\pi(0.24) < |\hat{\omega}| < 2\pi(0.36)$. In these intervals, the gain of the filter deviates only slightly from one, so the amplitudes of sinusoidal signals with these frequencies will not be affected by the filter. We will see in Section 7-9 that this filter has linear phase, so these frequencies will only experience delay.

Also note that in the regions $|\hat{\omega}| < 2\pi(0.16)$ and $2\pi(0.44) < |\hat{\omega}| < 2\pi(0.5)$ the gain is very nearly equal to zero. These regions are the "stopbands" of the filter, since the amplitudes of sinusoids with these frequencies will be multiplied by a very small gain and thus blocked from appearing in the output.

Observe that the frequency response tapers smoothly from the passbands to the stopbands. In these transition regions, sinusoids will be reduced in amplitude according to the gain shown in Fig. 7-18. In many applications, we would want such transition regions to be very narrow. Ideally, we might even want them to have zero width. While this is theoretically achievable, it comes with a high price. It turns out that for an FIR frequency selective (lowpass, bandpass, highpass) filter, the width of the transition region is inversely proportional to M, the order of the system function polynomial. The higher the order, the narrower the transition regions can be, and as $M \to \infty$, the transition regions shrink to zero. Unfortunately, increasing M also increases the amount of computation required to compute each sample of the output, so a trade-off is always required in any practical application.

Figure 7-19 shows the pole-zero plot of the FIR filter. Notice the distinctive pattern of locations of the zeros. In particular, note how the zeros off the unit circle seem to be grouped into groups of four zeros. Indeed, for each zero that is not on the unit circle, there are also zeros at the conjugate location, at the reciprocal location, and at the conjugate reciprocal location. These groups of four zeros are strategically placed by the design process to form the passband of the filter. Similarly, the design process places zeros on (or close to) the unit circle to ensure that the gain of the filter is low in the stopband regions of the frequency axis. Also, note that all complex zeros appear in conjugate pairs, and since the system function is a twenty-third-order polynomial, there are 23 poles at $z = 0$. In Section 7-9, we will show that these properties of the pole-zero distribution are the direct result of the symmetry of the filter coefficients.

7-9 Properties of Linear-Phase Filters

The filter that was discussed in Section 7-8 is an example of a class of systems where the sequence of coefficients (impulse response) has symmetry of the form $b_k = b_{M-k}$, for $k = 0, 1, \ldots, M$. Such systems have a number of interesting properties that are easy to show in the z-domain representation.

7-9.1 The Linear-Phase Condition

FIR systems that have symmetric filter coefficients (and, therefore, symmetric impulse responses) have frequency responses with linear phase. An example that we have already studied is the L-point running sum, whose coefficients are all the same and therefore clearly satisfy the condition $b_k = b_{M-k}$, for $k = 0, 1, \ldots, M$. The example of Section 7-8 also has linear phase because it satisfies the same symmetry condition. To see why linear phase results from this symmetry, let us consider a simple example where the system function is of the form

$$H(z) = b_0 + b_1 z^{-1} + b_2 z^{-2} + b_1 z^{-3} + b_0 z^{-4} \quad (7.42)$$

Thus, $M = 4$ and the length of the sequence is $L = M + 1 = 5$ samples. After working out the frequency response for this special case, the generalization will be obvious. First, observe that we can write $H(z)$ as

$$H(z) = \left(b_0(z^2 + z^{-2}) + b_1(z^1 + z^{-1}) + b_2\right) z^{-2}$$

If M were greater, we would simply have more groups of factors of the form $(z^k + z^{-k})$. Now when we substitute $z = e^{j\hat{\omega}}$, each of these factors will become a cosine term; i.e.,

$$H(e^{j\hat{\omega}}) = \left(2b_0 \cos(2\hat{\omega}) + 2b_1 \cos(\hat{\omega}) + b_2\right) e^{-j\hat{\omega}M/2}$$

In our specific example, we have shown that $H(e^{j\hat{\omega}})$ is of the form

$$H(e^{j\hat{\omega}}) = R(e^{j\hat{\omega}}) e^{-j\hat{\omega}M/2} \quad (7.43)$$

where, in this case, $M = 4$ and $R(e^{j\hat{\omega}})$ is the real function

$$R(e^{j\hat{\omega}}) = \left(2b_0 \cos(2\hat{\omega}) + 2b_1 \cos(\hat{\omega}) + b_2\right)$$

By following this style of analysis for the general case, it is easy to show that (7.43) holds whenever $b_k = b_{M-k}$, for $k = 0, 1, \ldots, M$. In the general case, it can be shown that the result depends on whether the integer M is even or odd

$$R(e^{j\hat{\omega}}) = \begin{cases} b_{\frac{M}{2}} + \sum\limits_{k=0}^{\frac{M-2}{2}} 2b_k \cos[(\frac{M}{2} - k)\hat{\omega}] & M \text{ even} \\ \sum\limits_{k=0}^{\frac{M-1}{2}} 2b_k \cos[(\frac{M}{2} - k)\hat{\omega}] & M \text{ odd} \end{cases} \quad (7.44)$$

Equation (7.43) shows that the frequency response of any symmetric filter has the form of a real amplitude function $R(e^{j\hat{\omega}})$ times a linear-phase factor $e^{-j\hat{\omega}M/2}$. The latter factor, as we have seen in Section 6-5.1 on p. 139, corresponds to a delay of $M/2$ samples. Thus, the analysis presented in Section 6-7 for the running-average filter is typical of what happens in the general symmetric FIR case. The main difference is that by choosing the filter coefficients carefully, as in Section 7-8, we can shape the function $R(e^{j\hat{\omega}})$ to have a much more selective frequency response.

7-9.2 Locations of the Zeros of FIR Linear-Phase Systems

If the filter coefficients satisfy the condition $b_k = b_{M-k}$, for $k = 0, 1, \ldots, M$, it follows that

$$H(1/z) = z^M H(z) \quad (7.45)$$

To demonstrate this "reciprocal property" of linear-phase filters, consider a 4-point system of the form

$$\begin{aligned} H(1/z) &= b_0 + b_1(1/z)^{-1} + b_1(1/z)^{-2} + b_0(1/z)^{-3} \\ &= b_0 + b_1 z^1 + b_1 z^2 + b_0 z^3 \\ &= z^3(b_0 + b_1 z^{-1} + b_1 z^{-2} + b_0 z^{-3}) \\ &= z^3 H(z) \end{aligned}$$

Following the same style for a general M, (7.45) is easily proved for the general case.

The reciprocal property of linear-phase filters is responsible for the distinctive pattern of zeros in the pole-zero plot of Fig. 7-19. The zeros that are not on the unit circle occur as quadruples. Furthermore, these quadruples are responsible for creating the passband of the BPF. Zeros on the unit circle occur in pairs because the complex conjugate partner must be present, and these zeros are mainly responsible for creating the stopband of the filter.

These properties can be shown to be true in general for linear-phase filters. Specifically,

> When $b_k = b_{M-k}$, for $k = 0, 1, \ldots, M$, then if z_0 is a zero of $H(z)$, so are its conjugate, its inverse, and its conjugate inverse; i.e., $\{z_0,\ z_0^*,\ 1/z_0,\ 1/z_0^*\}$ are all zeros of $H(z)$.

The conjugate zeros are included because the filter coefficients, which are also the coefficients of $H(z)$, are real. Therefore, all of the zeros of $H(z)$ must occur in complex-conjugate pairs. The inverse property is true because the filter coefficients are symmetric. Using (7.45), and assuming that z_0 is a zero of $H(z)$, we get

$$H(1/z_0) = z_0^M H(z_0) = 0$$

Since $H(z_0) = 0$, then we must have $H(1/z_0) = 0$ also. Most FIR filters are designed with a symmetry property, so the zero pattern of Fig. 7-19 is typical.

7-10 Summary and Links

The z-transform method was introduced in this chapter for FIR filters and finite-length sequences in general. The z-transform reduces the manipulation of LTI systems into simple operations on polynomials and rational functions. Roots of these z-transform polynomials are quite important because filter properties such as the frequency response can be inferred directly from the root locations.

We also introduced the important concept of "domains of representation" for discrete-time signals and systems. There are three domains: the n-domain or time domain, the $\hat{\omega}$-domain or frequency domain, and the z-domain. With three different domains at our disposal, even the most difficult problems generally can be simplified by switching to one of the other domains.

Among the laboratory projects on the CD-ROM, we have already provided two on the topic of FIR filtering in Chapters 5 and 6. Lab #7 deals with FIR filtering of sinusoidal waveforms, convolution, and deconvolution. Lab #8 deals with the frequency response of common filters such as the first difference and the L-point averager. In Labs #9 and #10, FIR filters will be used in practical systems, such as a touch-tone decoder in Lab #9, and piano note detection in Lab #10. Each of these labs should be easier to understand and simpler to carry out with the newly acquired background on z-transforms.

LAB: *#9 Encoding and Decoding Touch-Tones*

LAB: *#10 Octave Band Filtering*

The CD-ROM also contains some demonstrations of the relationship between the z-plane and the frequency domain and time domain.

(a) A three-domain movie that shows how the frequency response and the impulse response of an FIR filter change as a zero location is moved. Several different filters are demonstrated.

DEMO: *Three Domains - FIR*

(b) A movie that animates the relationship between the z-plane and the unit circle where the frequency response lies.

DEMO: *Z to Freq*

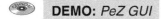

(c) The MATLAB program `PeZ`[6], to facilitate exploring the three domains. The M-files for `PeZ` can be copied and run under MATLAB.

DEMO: *PeZ GUI*

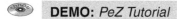

(d) A tutorial on how to run `PeZ`.

DEMO: *PeZ Tutorial*

As in previous chapters, the reader is reminded of the large number of solved homework problems on the CD-ROM that are available for review and practice.

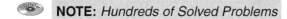

NOTE: *Hundreds of Solved Problems*

7-11 Problems

P-7.1 Use the superposition and time-delay properties to find the z-transforms of the following signals

$$x_1[n] = \delta[n]$$
$$x_2[n] = \delta[n-1]$$
$$x_3[n] = \delta[n-7]$$
$$x_4[n] = 2\delta[n] - 3\delta[n-1] + 4\delta[n-3]$$

P-7.2 Use the superposition and time-delay properties of (7.9) and (7.12) to determine the z-transform $Y(z)$ in terms of $X(z)$ if

$$y[n] = x[n] - x[n-1]$$

and in the process show that for the first difference system, $H(z) = 1 - z^{-1}$.

P-7.3 Suppose that an LTI system has a system function

$$H(z) = 1 + 5z^{-1} - 3z^{-2} + 2.5z^{-3} + 4z^{-8}$$

(a) Determine the difference equation that relates the output $y[n]$ of the system to the input $x[n]$.

(b) Determine and plot the output sequence $y[n]$ when the input is $x[n] = \delta[n]$.

P-7.4 An LTI system is described by the difference equation

$$y[n] = \frac{1}{3}(x[n] + x[n-1] + x[n-2])$$

(a) Determine the system function $H(z)$ for this system.

(b) Plot the poles and zeros of $H(z)$ in the z-plane.

(c) From $H(z)$ obtain an expression for $H(e^{j\hat{\omega}})$, the frequency response of this system.

(d) Sketch the frequency response (magnitude and phase) as a function of frequency for $-\pi \le \hat{\omega} \le \pi$.

(e) What is the output if the input is

$$x[n] = 4 + \cos[0.25\pi(n-1)] - 3\cos[(2\pi/3)n]$$

P-7.5 Consider an LTI system whose system function is the product of five terms

$$H(z) = (1 - z^{-1})(1 - e^{j\pi/2}z^{-1})(1 - e^{-j\pi/2}z^{-1}) \cdots (1 - 0.9e^{j\pi/3}z^{-1})(1 - 0.9e^{-j\pi/3}z^{-1})$$

[6] Originally written by Craig Ulmer, `PeZ` has been updated by Koon Kong.

(a) Write the difference equation that gives the relation between the input $x[n]$ and the output $y[n]$.

Hint: Multiply out the factors of $H(z)$.

(b) Plot the poles and zeros of $H(z)$ in the complex z-plane.

(c) If the input is of the form $x[n] = Ae^{j\phi}e^{j\hat{\omega}n}$, for what values of $-\pi < \hat{\omega} \leq \pi$ will $y[n] = 0$?

P-7.6 The diagram in Fig. P-7.6(a) depicts a cascade connection of two LTI systems; i.e., the output of the first system is the input to the second system and the overall output is the output of the second system.

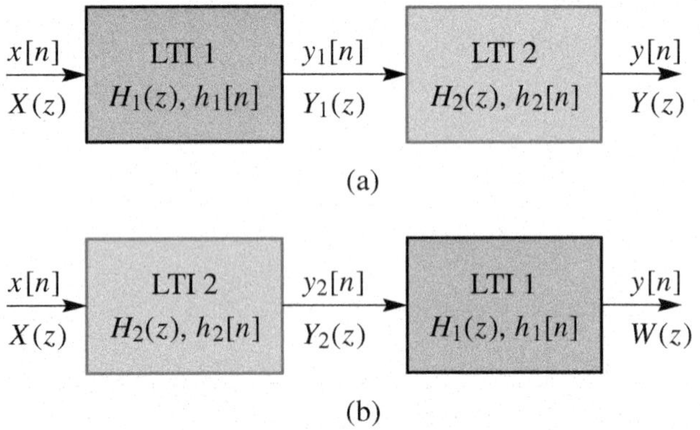

Figure P-7.6

(a) Use z-transforms to show that the system function for the overall system (from $x[n]$ to $y[n]$) is $H(z) = H_2(z)H_1(z)$, where $Y(z) = H(z)X(z)$.

(b) Use the result of (a) to show that the order of the systems is not important; i.e., show that for the same input $x[n]$ into the systems of Figs. P-7.6(a) and P-7.6(b), the overall outputs are the same ($w[n] = y[n]$).

(c) Suppose that System 1 is a 3-point averager described by the difference equation $y_1[n] = \frac{1}{3}(x[n] + x[n-1] + x[n-2])$ and System 2 is described by the system function $H_2(z) = \frac{1}{3}(1 + z^{-1} + z^{-2})$. Determine the system function of the overall cascade system.

(d) Obtain a single difference equation that relates $y[n]$ to $x[n]$ in Fig. P-7.6(a). Is the cascade of two 3-point averagers the same as a 6-point averager? Why would a better term be "weighted averager"?

(e) Plot the poles and zeros of $H(z)$ in the complex z-plane.

(f) From $H(z)$ obtain an expression for the frequency response $H(e^{j\hat{\omega}})$ and sketch the magnitude of the frequency response of the overall cascade system as a function of frequency for $-\pi \leq \hat{\omega} \leq \pi$.

P-7.7 Factor the following polynomial and plot its roots in the complex plane.

$$P(z) = 1 + \tfrac{1}{2}z^{-1} + \tfrac{1}{2}z^{-2} + z^{-3}$$

In MATLAB, see the functions called `roots` and `zplane`.

P-7.8 An LTI system is described by the difference equation

$$y[n] = \frac{1}{4}\{x[n] + x[n-1] + x[n-2] + x[n-3]\}$$
$$= \frac{1}{4}\sum_{k=0}^{3} x[n-k]$$

(a) What is $h[n]$, the impulse response of this system?

(b) Determine the system function $H(z)$ for this system.

(c) Plot the poles and zeros of $H(z)$ in the complex z-plane.

Hint: Remember the Lth roots of unity.

(d) From $H(z)$, obtain an expression for the frequency response $H(e^{j\hat{\omega}})$ of this system.

(e) Sketch the frequency response (magnitude and phase) as a function of frequency (or plot it using `freqz( )` in MATLAB, or `freekz( )` in the *SP-First Toolbox*).

(f) Suppose that the input is

$$x[n] = 5 + 4\cos(0.2\pi n) + 3\cos(0.5\pi n + \pi/4)$$

for $-\infty < n < \infty$. Obtain an expression for the output in the form $y[n] = A + B\cos(\hat{\omega}_0 n + \phi_0)$.

P-7.9 The diagram in Fig. P-7.6(a) depicts a cascade connection of two LTI systems; i.e., the output of the first system is the input to the second system, and the overall output is the output of the second system.

In this problem, assume that both systems in Fig. P-7.6 are 4-point running averagers.

(a) Determine the system function $H(z) = H_1(z)H_2(z)$ for the overall system.

(b) Plot the poles and zeros of $H(z)$ in the z-plane.

Hint: The poles and zeros of $H(z)$ are the combined poles and zeros of $H_1(z)$ and $H_2(z)$.

(c) From $H(z)$, obtain an expression for the frequency response $H(e^{j\hat{\omega}})$ of the overall cascade system.

(d) Sketch the frequency response (magnitude and phase) functions of the overall cascade system for $-\pi < \hat{\omega} \leq \pi$.

(e) Use multiplication of z-transform polynomials to determine the impulse response $h[n]$ of the overall cascade system.

P-7.10 Suppose that an LTI system has system function equal to

$$H(z) = 1 - 3z^{-2} + 2z^{-3} + 4z^{-6}$$

The input to the system is the sequence

$$x[n] = 2\delta[n] + \delta[n-1] - 2\delta[n-2] + 4\delta[n-4]$$

(a) Without actually computing the output, determine from the above information the values of N_1 and N_2 so that the following is true:

$$y[n] = 0 \quad \text{for } n < N_1 \text{ and } n > N_2$$

(b) Use z-transforms and polynomial multiplication to find the sequence $y[n] = x[n] * h[n]$.

P-7.11 The intention of the following MATLAB program is to filter a sinusoid using the `conv` function.

```
omega = pi/6;
nn = [ 0:29 ];
xn = cos(omega*nn - pi/4);
bb = [ 1  0  0  1 ];
yn = conv( bb, xn );
```

(a) Determine $H(z)$ and the zeros of the FIR filter.

(b) Determine a formula for $y[n]$, the signal contained in the vector `yn`. Ignore the first few points, so your formula must be correct only for $n \geq 3$. This formula should give numerical values for the amplitude, phase and frequency of $y[n]$.

(c) Give a value of `omega` such that the output is guaranteed to be zero, for $n \geq 3$.

P-7.12 Suppose that a system is defined by the system function

$$H(z) = (1 - z^{-1})(1 + z^{-2})(1 + z^{-1})$$

(a) Write the time-domain description of this system in the form of a difference equation.

(b) Write a formula for the frequency response of the system.

(c) Derive simple formulas for the magnitude response versus $\hat{\omega}$ and the phase response versus $\hat{\omega}$. These formulas must contain no complex terms and no square roots.

(d) This system can "null" certain input signals. For which input frequencies $\hat{\omega}_0$ is the response to $x[n] = \cos(\hat{\omega}_0 n)$ equal to zero?

(e) When the input to the system is $x[n] = \cos(\pi n/3)$ determine the output signal $y[n]$ in the form:

$$A\cos(\hat{\omega}_0 n + \phi)$$

Give numerical values for the constants A, $\hat{\omega}_0$ and ϕ.

P-7.13 Show that the system defined by the difference equation (7.31) on p. 179 will null *any* sinusoid of the form $A\cos(0.25\pi n + \phi)$ independent of the specific values of A and ϕ.

P-7.14 An LTI system has system function

$$H(z) = (1 + z^{-2})(1 - 4z^{-2}) = 1 - 3z^{-2} - 4z^{-4}$$

The input to this system is

$$x[n] = 20 - 20\delta[n] + 20\cos(0.5\pi n + \pi/4)$$

for $-\infty < n < \infty$. Determine the output of the system $y[n]$ corresponding to the above input $x[n]$. Give an equation for $y[n]$ that is valid for all n.

Note: This is an easy problem if you approach it correctly!

P-7.15 The input to the C-to-D converter in Fig. P-7.15 is

$$x(t) = 4 + \cos(250\pi t - \pi/4) - 3\cos[(2000\pi/3)t]$$

The system function for the LTI system is

$$H(z) = \frac{1}{3}(1 + z^{-1} + z^{-2})$$

If $f_s = 1000$ samples/sec, determine an expression for $y(t)$, the output of the D-to-C converter.

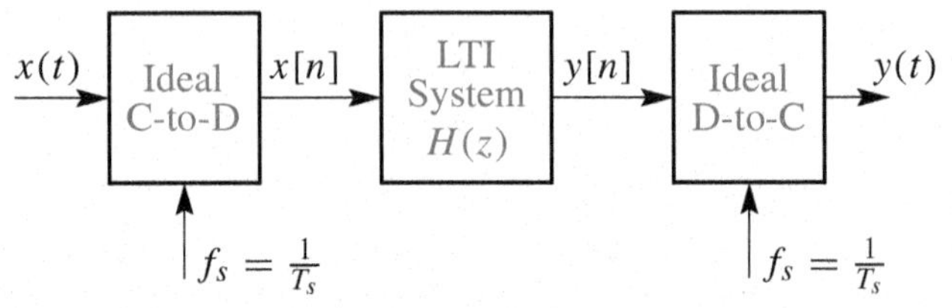

Figure P-7.15

P-7.16 In the cascade system of Fig. P-7.6(a) it is known that the system function $H(z)$ of the overall system is the product of four terms

$$H(z) = (1 - z^{-2})(1 - 0.8e^{j\pi/4}z^{-1})$$
$$(1 - 0.8e^{-j\pi/4}z^{-1})(1 + z^{-2})$$

(a) Determine the poles and zeros of $H(z)$ and plot them in the complex z-plane.

(b) It is possible to determine two system functions $H_1(z)$ and $H_2(z)$ so that: (1) The overall cascade system has the given system function $H(z)$, and (2) the output of the first system is $y_1[n] = x[n] - x[n-4]$. Find $H_1(z)$ and $H_2(z)$.

(c) Determine the difference equation that relates $y[n]$ to $y_1[n]$ for your answer in (b).

P-7.17 In Section 7-9 we showed that symmetric FIR filters have special properties. In this problem, we consider the antisymmetric case where $b_k = -b_{M-k}$, for $k = 0, 1, \ldots, M$. Consider the specific example

$$H(z) = b_0 + b_1 z^{-1} - b_1 z^{-3} - b_0 z^{-4}$$

where in this case, $b_2 = -b_2 = 0$.

(a) Show that, for this example,

$$H(e^{j\hat{\omega}}) = [2b_0 \sin(2\hat{\omega}) + 2b_1 \sin(\hat{\omega})] e^{j(\pi/2 - j\hat{\omega}2)}$$

(b) Show that, for this example,

$$H(1/z) = -z^4 H(z)$$

(c) Generalize these results for any M (i.e., both even and odd).

P-7.18 Refer to Fig. P-7.6(a) which depicts a cascade connection of two LTI systems.

(a) Suppose that System 1 and System 2 both have a "square pulse" impulse response of the form

$$h_1[n] = h_2[n] = \delta[n] + \delta[n-1] + \delta[n-2] + \delta[n-3]$$

Determine the system functions $H_1(z)$ and $H_2(z)$ for the two systems.

(b) Use z-transforms to determine the system function $H(z)$ of the overall system.

(c) Use polynomial multiplication to determine the corresponding impulse response $h[n]$ for the overall system.

(d) Obtain a single difference equation that relates $y[n]$ to $x[n]$ in the cascade system.

(e) Can you see how this result could be used to do linear interpolation of a signal that had been subsampled by four? State a complete procedure based on your results in (a)–(d).

(f) Show that $H_1(z) = H_2(z)$ can be expressed as

$$H_1(z) = H_2(z) = \frac{1 - z^{-4}}{1 - z^{-1}}$$

(g) Find $H(z)$ and plot its poles and zeros in the complex z-plane.

(h) Show that the frequency responses of the two systems are

$$H_1(e^{j\hat{\omega}}) = H_2(e^{j\hat{\omega}}) = \frac{\sin(2\hat{\omega})}{\sin(\hat{\omega}/2)} e^{-j3\hat{\omega}/2}$$

(i) From $H(z)$ found in (g) obtain an expression for the frequency response $H(e^{j\hat{\omega}})$ and sketch the magnitude of the frequency response of the overall cascade system as a function of frequency for $-\pi \le \hat{\omega} \le \pi$.

Hint: Note that $H(e^{j\hat{\omega}}) = H_2(e^{j\hat{\omega}}) H_1(e^{j\hat{\omega}}) = [H_1(e^{j\hat{\omega}})]^2 = [H_2(e^{j\hat{\omega}})]^2$.

C H A P T E R 8

IIR Filters

This chapter introduces a new class of LTI systems that have infinite duration impulse responses. For this reason, systems of this class are often called infinite-impulse-response (IIR) systems or IIR filters. In contrast to FIR filters, IIR digital filters involve previously computed values of the output signal as well as values of the input signal in the computation of the present output. Since the output is "fed back" to be combined with the input, these systems are examples of the general class of ***feedback systems***. From a computational point of view, since output samples are computed in terms of previously computed values of the output, the term ***recursive filter*** is also used for these filters.

The z-transform system functions for IIR filters are rational functions that have both poles and zeros at nonzero locations in the z-plane. Just as for the FIR case, we will show that many insights into the important properties of IIR filters can be obtained directly from the pole-zero representation.

We begin this chapter with the first-order IIR system, which is the simplest case because it involves feedback of only the previous output sample. We show by construction that the impulse response of this system has an infinite duration. Then the frequency response and the z-transform are developed for the first-order filter. After showing the relationship among the three domains of representation for this simple case, we consider second-order filters. These filters are particularly important because they can be used to model ***resonances*** such as would occur in a speech synthesizer, as well as many other natural phenomena that exhibit vibratory behavior. The frequency response for the second-order case can exhibit a narrowband character that leads to the definition of bandwidth and center frequency, both of which can be controlled by appropriate choice of the feedback coefficients of the filter. The analysis and insights developed for the first- and second-order cases are readily generalized to higher-order systems.

8-1 The General IIR Difference Equation

FIR filters are extremely useful and have many nice properties, but that class of filters is not the most general class of LTI systems. This is because the output $y[n]$ is formed solely from a finite segment of the input signal $x[n]$. The most general class of digital filters that can be implemented with a finite amount of computation is obtained when the output is formed not only from the input, but also from previously computed outputs.

DEMO: *IIR Filtering*

The defining equation for this class of digital filters is the difference equation

$$y[n] = \sum_{\ell=1}^{N} a_\ell y[n-\ell] + \sum_{k=0}^{M} b_k x[n-k] \tag{8.1}$$

The filter coefficients consist of two sets: $\{b_k\}$ and $\{a_\ell\}$. For reasons that will become obvious in the following simple example, the coefficients $\{a_\ell\}$ are called the *feedback* coefficients, and the $\{b_k\}$ are called the *feed-forward* coefficients. In all, $N+M+1$ coefficients are needed to define the recursive difference equation (8.1).

Notice that if the coefficients $\{a_\ell\}$ are all zero, the difference equation (8.1) reduces to the difference equation of an FIR system. Indeed, we have asserted that (8.1) defines the most general class of LTI systems that can be implemented with finite computation, so FIR systems must be a special case. When discussing FIR systems we have referred to M as the *order* of the system. In this case, M is the number of delay terms in the difference equation and the degree or order of the polynomial system function. For IIR systems, we have both M and N as measures of the number of delay terms, and we will see that the system function of an IIR system is the ratio of an Mth-order polynomial to an Nth-order polynomial. Thus, there can be some ambiguity as to the order of an IIR system. In general, we will define N, the number of feedback terms, to be the order of an IIR system.

Example 8-1: IIR Block Diagram

Rather than tackle the general form given in (8.1), consider the first-order case where $M = N = 1$; i.e.,

$$y[n] = a_1 y[n-1] + b_0 x[n] + b_1 x[n-1] \tag{8.2}$$

The block diagram representation of this difference equation, which is shown in Fig. 8-1, is constructed by noting that the signal $v[n] = b_0 x[n] + b_1 x[n-1]$ is computed by the left half of the diagram, and we "close the loop" by computing $a_1 y[n-1]$ from the delayed output and adding it to $v[n]$ to produce the output $y[n]$. This diagram clearly shows that the terms feed-forward and feedback describe the direction of signal flow in the block diagram. ■

We will begin by studying a simplified version of the system defined by (8.2) and depicted in Fig. 8-1. This will involve characterizing the filter in each of the three domains: time domain, frequency domain, and z-domain. Since the filter is defined by a time-domain difference equation (8.2), we begin by studying how the difference equation is used to compute the output from the input, and we will illustrate how feedback results in an impulse response of infinite duration.

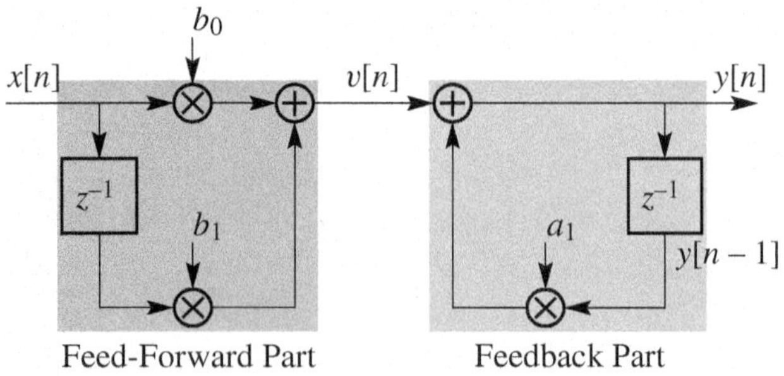

Figure 8-1: First-order IIR system showing one feedback coefficient a_1 and two feed-forward coefficients b_0 and b_1.

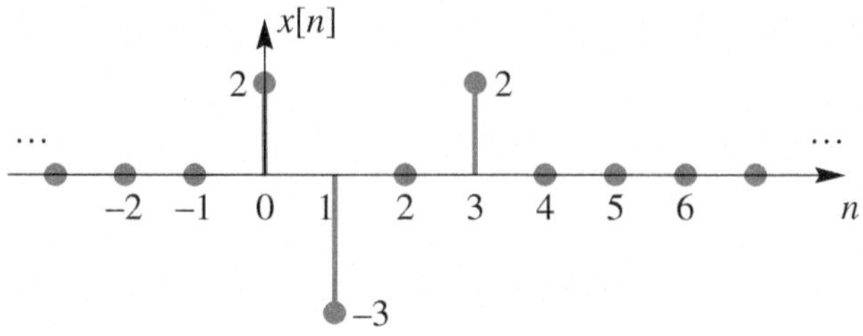

Figure 8-2: Input signal to recursive difference equation.

8-2 Time-Domain Response

To illustrate how the difference equation can be used to implement an IIR system, we will begin with a numerical example. Assume that the filter coefficients in (8.2) are $a_1 = 0.8$, $b_0 = 5$, and $b_1 = 0$, so that

$$y[n] = 0.8y[n-1] + 5x[n] \tag{8.3}$$

and assume that the input signal is

$$x[n] = 2\delta[n] - 3\delta[n-1] + 2\delta[n-3] \tag{8.4}$$

In other words, the total duration of the input is four samples, as shown in Fig. 8-2.

Although it is not a requirement, it is logical that the output sequence values should be computed in normal order (i.e., from left to right in a plot of the sequence). Furthermore, since the input is zero before $n = 0$, it would be natural to assume that $n = 0$ is the starting time of the output. Thus, we will consider computing the output from the difference equation (8.3) in the order $n = 0, 1, 2, 3, \ldots$. For example, the value of $x[0]$ is 2, so we can evaluate (8.3) at $n = 0$ obtaining

$$y[0] = 0.8y[0-1] + 5x[0] = 0.8y[-1] + 5(2) \tag{8.5}$$

Immediately we run into a problem: The value of $y[n]$ at $n = -1$ is not known. This is a serious problem, because no matter where we start computing the output, we will always have the same problem; at any point along the n-axis, we will need to know the output at the previous time $n-1$. If we know the value $y[n-1]$, however, we can use the difference equation to compute the next value of the output signal at time n. Once the process is started, it can be continued indefinitely by iteration of the difference equation. The solution requires the following two assumptions, which together are called the ***initial rest conditions***.

Initial Rest Conditions

1. The input must be assumed to be zero prior to some starting time n_0; i.e., $x[n] = 0$ for $n < n_0$. We say that such inputs are *suddenly applied*.

2. The output is likewise assumed to be zero prior to the starting time of the signal; i.e., $y[n] = 0$ for $n < n_0$. We say that the system is *initially at rest* if its output is zero prior to the application of a suddenly applied input.

These conditions are not particularly restrictive, especially in the case of a real-time system, where a new output must be computed as each new sample of the input is taken. Real-time systems must, of course, be ***causal*** in the sense that the computation of the present output sample must not involve future samples of the input or output. Furthermore, any practical device would have a time at which it first begins to operate. All that is needed is for the memory containing the delayed output samples to be set initially to zero.[1]

With the initial rest assumption, we let $y[n] = 0$ for $n < 0$, so now we can evaluate $y[0]$ as

$$y[0] = 0.8y[-1] + 5(2) = 0.8(0) + 5(2) = 10$$

Once we have started the recursion, the rest of the values follow easily, since the input signal and previous outputs are known.

$$\begin{aligned}
y[1] &= 0.8y[0] + 5x[1] = 0.8(10) + 5(-3) = -7 \\
y[2] &= 0.8y[1] + 5x[2] = 0.8(-7) + 5(0) = -5.6 \\
y[3] &= 0.8y[2] + 5x[3] = 0.8(-5.6) + 5(2) = 5.52 \\
y[4] &= 0.8y[3] + 5x[4] = 0.8(5.52) + 5(0) = 4.416 \\
y[5] &= 0.8y[4] + 5x[5] = 0.8(4.416) + 0 = 3.5328 \\
&\vdots \qquad\qquad \vdots
\end{aligned}$$

This output sequence is plotted in Fig. 8-3 up to $n = 7$.

One key feature to notice in Fig. 8-3 is the structure of the output signal after the input turns off ($n > 3$). For this range of n, the difference equation becomes

$$y[n] = 0.8y[n-1] \qquad \text{for } n > 3$$

Thus the ratio between successive terms is a constant, and the output signal decays exponentially with a rate

[1]In the case of a digital filter applied to sampled data stored in computer memory, the causality condition is not required, but, generally, the output is computed in the same order as the natural order of the input samples. The difference equation could be *recursed* backwards through the sequence, but this would require a different definition of ***initial conditions***.

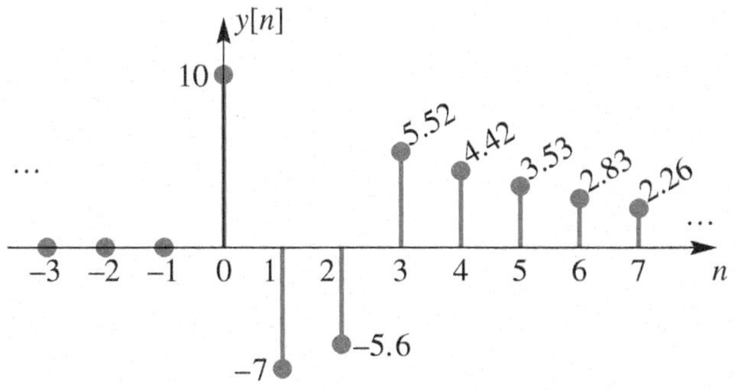

Figure 8-3: Output signal from recursive difference equation of (8.5) for the input of Fig. 8-2. For $n > 3$, the sequence is proportional to $(0.8)^n$, because the input signal ends at $n = 3$.

determined by $a_1 = 0.8$. Therefore, we can write the closed form expression

$$y[n] = y[3](0.8)^{n-3} \qquad \text{for } n \geq 3$$

for the rest of the sequence $y[n]$ once the value for $y[3]$ is known.

8-2.1 Linearity and Time Invariance of IIR Filters

When applied to the general IIR difference equation of (8.1), the condition of initial rest is sufficient to guarantee that the system implemented by iterating the difference equation is both linear and time-invariant. Although the feedback terms make the proof more complicated than the FIR case (see Section 5-5.3 on p. 117), we can show that, for suddenly applied inputs and initial rest conditions, the principle of superposition will hold because the difference equation involves only linear combinations of the input and output samples. Furthermore, since the initial rest condition is always applied just before the beginning of a suddenly applied input, time invariance also holds.

EXERCISE 8.1: Assume that the input to the difference equation (8.3) is $x_1[n] = 10x[n-4]$, where $x[n]$ is given by (8.4) and Fig. 8-2. Use iteration to compute the corresponding output $y_1[n]$ for $n = 0, 1, \ldots, 11$ using the assumption of initial rest. Compare your result to the output plotted in Fig. 8-3, and verify that the system behaves as if it is both linear and time-invariant.

8-2.2 Impulse Response of a First-Order IIR System

In Chapter 5, we showed that the response to a unit impulse sequence characterizes a linear time-invariant system completely. Recall that when $x[n] = \delta[n]$, the resulting output signal, denoted by $h[n]$, is by definition the ***impulse response***. Since all other input signals can be written as a superposition of weighted and delayed impulses, the corresponding output for all other signals can be constructed from weighted and shifted versions of the impulse response, $h[n]$. That is, since the recursive difference equation with initial rest conditions is an LTI system, its output can always be represented as the convolution sum

$$y[n] = \sum_{k=-\infty}^{\infty} x[k]h[n-k] \tag{8.6}$$

Therefore, it is of interest to characterize the recursive difference equation by its impulse response.

To illustrate the nature of the impulse response of an IIR system, consider the first-order recursive difference equation with $b_1 = 0$,

$$y[n] = a_1 y[n-1] + b_0 x[n]. \tag{8.7}$$

By definition, the difference equation

$$h[n] = a_1 h[n-1] + b_0 \delta[n] \tag{8.8}$$

must be satisfied by the impulse response $h[n]$ for all values of n. We can construct a general formula for the impulse response in terms of the parameters a_1 and b_0 by simply constructing a table of a few values and then writing down the general formula by inspection. The following table shows the sequences involved in the computation:

n	$n<0$	0	1	2	3	4	$\cdots$
$\delta[n]$	0	1	0	0	0	0	$\cdots$
$h[n-1]$	0	0	b_0	$b_0(a_1)$	$b_0(a_1)^2$	$b_0(a_1)^3$	$\cdots$
$h[n]$	0	b_0	$b_0(a_1)$	$b_0(a_1)^2$	$b_0(a_1)^3$	$b_0(a_1)^4$	$\cdots$

From this table, we see that the general formula is

$$h[n] = \begin{cases} b_0(a_1)^n & \text{for } n \geq 0 \\ 0 & \text{for } n < 0 \end{cases} \tag{8.9}$$

If we recall the definition of the unit step sequence,

$$u[n] = \begin{cases} 1 & \text{for } n \geq 0 \\ 0 & \text{for } n < 0 \end{cases} \tag{8.10}$$

(8.9) can be expressed in the form

$$h[n] = b_0(a_1)^n u[n] \tag{8.11}$$

where multiplication of $(a_1)^n$ by $u[n]$ provides a compact representation of the conditions $n < 0$ and $n \geq 0$.

Example 8-2: Impulse Response

For the example of (8.3) with $a_1 = 0.8$ and $b_0 = 5$, the impulse response is

$$h[n] = 5(0.8)^n u[n] = \begin{cases} 5(0.8)^n & \text{for } n \geq 0 \\ 0 & \text{for } n < 0 \end{cases} \tag{8.12}$$

This is the impulse response of the system in (8.3). ■

EXERCISE 8.2: Substitute the solution (8.11) into the difference equation (8.8) and verify that the difference equation is satisfied for all values of n.

EXERCISE 8.3: Find the impulse response of the following first-order system:

$$y[n] = 0.5y[n-1] + 5x[n-7]$$

Assume that the system is *at rest* for $n < 0$. Plot the resulting signal $h[n]$ as a function of n. Pay careful attention to where the nonzero portion of the impulse response begins.

A slightly more general problem would be to find the impulse response of the first-order system when a shifted version of the input signal is also included in the difference equation; i.e.,

$$y[n] = a_1 y[n-1] + b_0 x[n] + b_1 x[n-1]$$

Because this system is linear and time-invariant, it follows that its impulse response can be thought of as a sum of two terms as in

$$\begin{aligned} h[n] &= b_0(a_1)^n u[n] + b_1(a_1)^{n-1}u[n-1] \\ &= \begin{cases} 0 & n < 0 \\ b_0 & n = 0 \\ (b_0 + b_1 a_1^{-1})(a_1)^n & n \geq 1 \end{cases} \end{aligned}$$

Notice that the impulse response still decays exponentially with rate dependent only on a_1.

EXERCISE 8.4: Find the impulse response $h[n]$ of the following first-order system:

$$y[n] = -0.5y[n-1] - 4x[n] + 5x[n-1].$$

Plot the resulting impulse response as a function of n.

8-2.3 Response to Finite-Length Inputs

For finite-length inputs, the convolution sum is easy to evaluate for either FIR or IIR systems. Suppose that the finite-length input sequence is

$$x[n] = \sum_{k=N_1}^{N_2} x[k]\delta[n-k]$$

so that $x[n] = 0$ for $n < N_1$ and $n > N_2$. Then it follows from (8.6) that the corresponding output satisfies

$$y[n] = \sum_{k=N_1}^{N_2} x[k]h[n-k]$$

Example 8-3: IIR Response to General Input

As an example, consider again the LTI system defined by the difference equation (8.3), whose impulse response was shown in Example 8-2 to be $h[n] = 5(0.8)^n u[n]$. For the input of (8.4) and Fig. 8-2,

$$x[n] = 2\delta[n] - 3\delta[n-1] + 2\delta[n-3]$$

it is easily seen that

$$\begin{aligned} y[n] &= 2h[n] - 3h[n-1] + 2h[n-3] \\ &= 10(0.8)^n u[n] - 15(0.8)^{n-1}u[n-1] \\ &\quad + 10(0.8)^{n-3}u[n-3] \end{aligned}$$

To evaluate this expression for a specific time index, we need to take into account the different regions over which the individual terms are nonzero. If we do, we obtain

$$y[n] = \begin{cases} 0 & n < 0 \\ 10 & n = 0 \\ 10(0.8) - 15 = -7 & n = 1 \\ 10(0.8)^2 - 15(0.8) = -5.6 & n = 2 \\ 10(0.8)^3 - 15(0.8)^2 + 10 = 5.52 & n = 3 \\ 5.52(0.8)^{n-3} & n > 3 \end{cases}$$

A comparison to the output obtained by iterating the difference equation (see Fig. 8-3 on p. 199) shows that we have obtained the same values of the output sequence by superposition of scaled and shifted impulse responses. ■

Example 8-3 illustrates two important points about IIR systems.

1. The initial rest condition guarantees that the output sequence does not begin until the input sequence begins (or later).

2. Because of the feedback, the impulse response is infinite in extent, and the output due to a finite-length input sequence, being a superposition of scaled and shifted impulse responses, is generally (but not always) infinite in extent. This is in contrast to the FIR case, where a finite-length input always produces a finite-length output sequence.

EXERCISE 8.5: Find the impulse response of the first-order system

$$y[n] = -0.5y[n-1] + 5x[n]$$

and use it to find the output due to the input signal

$$x[n] = \begin{cases} 0 & n < 1 \\ 3 & n = 1 \\ -2 & n = 2 \\ 0 & n = 3 \\ 3 & n = 4 \\ -1 & n = 5 \\ 0 & n > 5 \end{cases}$$

Write a formula that is the sum of four terms, each of which is a shifted impulse response. Assume the initial rest condition. Plot the resulting signal $y[n]$ as a function of n for $0 \le n \le 10$.

8-2.4 Step Response of a First-Order Recursive System

When the input signal is infinitely long, the computation of the output of an IIR system using the difference equation is no different than for an FIR system; we simply continue to iterate the difference equation as long as samples of the output are desired. In the FIR case, the difference equation and the convolution sum are the same thing. This is not true in the IIR case, and computing the output using convolution is practical only in cases where simple formulas exist for both the input and the impulse response. Thus, in general, IIR filters must be implemented by iterating the difference equation. The computation of the response of a first-order IIR system to a unit step input provides a relatively simple illustration of the issues involved.

Again, assume that the system is defined by

$$y[n] = a_1 y[n-1] + b_0 x[n]$$

and assume that the input is the unit step sequence given by

$$u[n] = \begin{cases} 1 & \text{for } n \geq 0 \\ 0 & \text{for } n < 0 \end{cases} \tag{8.13}$$

As before, the difference equation can be iterated to produce the output sequence one sample at a time. The first few values are tabulated here. Work through the table to be sure that you understand the computation.

n	$x[n]$	$y[n]$
$n<0$	0	0
0	1	b_0
1	1	$b_0 + b_0(a_1)$
2	1	$b_0 + b_0(a_1) + b_0(a_1)^2$
3	1	$b_0(1 + a_1 + a_1^2 + a_1^3)$
4	1	$b_0(1 + a_1 + a_1^2 + a_1^3 + a_1^4)$
$\vdots$	1	$\vdots$

From the tabulated values, it can be seen that a general formula for $y[n]$ is

$$\begin{aligned} y[n] &= b_0(1 + a_1 + a_1^2 + \cdots + a_1^n) \\ &= b_0 \sum_{k=0}^{n} a_1^k \end{aligned} \tag{8.14}$$

With a bit of manipulation, we can get a simple closed-form expression for the general term in the sequence $y[n]$. For this we need to recall the formula

$$\sum_{k=0}^{L} r^k = \begin{cases} \dfrac{1 - r^{L+1}}{1 - r} & r \neq 1 \\ L + 1 & r = 1 \end{cases} \tag{8.15}$$

which is the formula for summing the first $L + 1$ terms of a geometric series. Armed with this fact, the formula (8.14) for $y[n]$ (when $a_1 \neq 1$) becomes

$$y[n] = b_0 \frac{1 - a_1^{n+1}}{1 - a_1} \qquad \text{for } n \geq 0, \tag{8.16}$$

Three cases can be identified: $|a_1| > 1$, $|a_1| < 1$, and $|a_1| = 1$. Further investigation of these cases reveals two types of behavior.

1. When $|a_1| > 1$, the term a_1^{n+1} in the numerator will dominate and the values for $y[n]$ will get larger and larger without bound. This is called an ***unstable*** condition and is usually a situation to avoid. We will say more about the issue of stability later in Sections 8-4.2 and 8-8.

2. When $|a_1| < 1$, the term a_1^{n+1} will decay to zero as $n \to \infty$. In this case, the system is ***stable***. Therefore, we can find a limiting value for $y[n]$ as $n \to \infty$

$$\lim_{n\to\infty} y[n] = \lim_{n\to\infty} b_0 \frac{1 - a_1^{n+1}}{1 - a_1} = \frac{b_0}{1 - a_1}$$

3. When $|a_1| = 1$, we might have an unbounded output, but not always. For example, when $a_1 = 1$, (8.14) gives $y[n] = (n + 1)b_0$ for $n \geq 0$, and the output $y[n]$ grows as $n \to \infty$. On the other hand, for $a_1 = -1$, the output alternates; it is $y[n] = b_0$ for n even, but $y[n] = 0$ for n odd.

The MATLAB plot in Fig. 8-4 shows the step response for the filter

$$y[n] = 0.8y[n - 1] + 5x[n]$$

Notice that the limiting value is 25, which can be calculated from the filter coefficients

$$\lim_{n\to\infty} y[n] = \frac{b_0}{1 - a_1} = \frac{5}{1 - 0.8} = 25$$

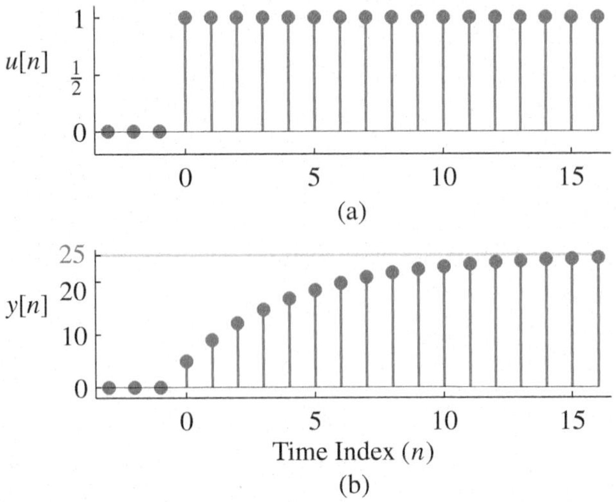

Figure 8-4: (a) Unit-step signal $u[n]$, (b) Step response of a first-order IIR filter.

Now suppose that we try to compute the step response by the convolution sum

$$y[n] = x[n] * h[n] = \sum_{k=-\infty}^{\infty} x[k]h[n-k] \qquad (8.17)$$

Since both the input and the impulse response have infinite durations, we might have difficulty in carrying out the computation. However, the fact that the input and output are given by the formulas $x[n] = u[n]$ and $h[n] = b_0(a_1)^n u[n]$ makes it possible to obtain a result. Substituting these formulas into (8.17) gives

$$y[n] = \sum_{k=-\infty}^{\infty} u[k]b_0(a_1)^{n-k}u[n-k]$$

The $u[k]$ and $u[n-k]$ terms inside the sum will change the limits on the sum because $u[k] = 0$ for $k < 0$ and $u[n-k] = 0$ for $n - k < 0$ (or $n < k$). The final result is

$$y[n] = \begin{cases} 0 & \text{for } n < 0 \\ \sum_{k=0}^{n} b_0(a_1)^{n-k} & \text{for } n \geq 0 \end{cases}$$

Using (8.15) we can write the step response for $n \geq 0$ as

$$\begin{aligned} y[n] &= \sum_{k=0}^{n} b_0(a_1)^{n-k} = b_0(a_1)^n \sum_{k=0}^{n} (a_1)^{-k} \\ &= b_0(a_1)^n \frac{1-(1/a_1)^{n+1}}{1-(1/a_1)} \qquad (8.18) \\ &= b_0 \frac{1-(a_1)^{n+1}}{1-a_1} \end{aligned}$$

which is identical to (8.16), the step response computed by iterating the difference equation. Notice that we were able to arrive at a closed-form expression in this case because of the special nature of the input and impulse response. In general, it would be difficult or impossible to obtain such a closed-form result, but we can always use iteration of the difference equation to compute the output sample-by-sample.

8-3 System Function of an IIR Filter

We saw in Chapter 7 for the FIR case that the system function is the z-transform of the impulse response of the system, and that the system function and the frequency response are intimately related. Furthermore, the following result was shown:

> ***Convolution in the n-domain corresponds to multiplication in the z-domain.***
>
> $y[n] = x[n] * h[n] \quad \stackrel{z}{\longleftrightarrow} \quad Y(z) = X(z)H(z)$

The same is true for the IIR case. The system function for an FIR filter is always a polynomial; however, when the difference equation has feedback, it turns out that the system function $H(z)$ is the ratio of two polynomials. Ratios of polynomials are called ***rational*** functions. In this section we will determine the system function for the example of a first-order IIR system, and show how the system function, impulse response, and difference equation are related.

8-3.1 The General First-Order Case

The general form of the first-order difference equation with feedback is

$$y[n] = a_1 y[n-1] + b_0 x[n] + b_1 x[n-1] \qquad (8.19)$$

Since this equation must be satisfied for all values of n, we can use the property (7.13) on p. 169 to take the z-transform of both sides of the equation to obtain

$$Y(z) = a_1 z^{-1} Y(z) + b_0 X(z) + b_1 z^{-1} X(z)$$

Subtracting the term $a_1 z^{-1} Y(z)$ from both sides of the equation leads to the following manipulations:

$$Y(z) - a_1 z^{-1} Y(z) = b_0 X(z) + b_1 z^{-1} X(z)$$
$$(1 - a_1 z^{-1}) Y(z) = (b_0 + b_1 z^{-1}) X(z)$$

Since the system is an LTI system, it should be true that $Y(z) = H(z)X(z)$, where $H(z)$ is the system function of the LTI system. Solving this equation for $H(z) = Y(z)/X(z)$, we obtain

$$H(z) = \frac{Y(z)}{X(z)} = \frac{b_0 + b_1 z^{-1}}{1 - a_1 z^{-1}} = \frac{B(z)}{A(z)} \qquad (8.20)$$

Thus, we have shown that $H(z)$ for the first-order IIR system is a ratio of two polynomials. The numerator polynomial $B(z)$ is defined by the weighting coefficients $\{b_k\}$ that multiply the input signal $x[n]$ and its delayed versions; the denominator polynomial $A(z)$ is defined by the feedback coefficients $\{a_\ell\}$. That this correspondence is true in general should be clear from the analysis that leads to the formula for $H(z)$. Indeed, the following is true for IIR systems of *any* order:

> *The coefficients of the* ***numerator*** *polynomial of the system function of an IIR system are the coefficients of the* ***feed-forward*** *terms of the difference equation. For the* ***denominator*** *polynomial, the constant term is one, and the remaining coefficients are the negatives of the* ***feedback*** *coefficients.*

In MATLAB, the `filter` function follows this same format. The statement

```
yy = filter(bb,aa,xx)
```

implements an IIR filter, where the vectors `bb` and `aa` hold the filter coefficients for the numerator and denominator polynomials, respectively.

Example 8-4: MATLAB for IIR Filter

The following feedback filter:

$$y[n] = 0.5y[n-1] - 3x[n] + 2x[n-1]$$

would be implemented in MATLAB by

```
yy = filter([-3,2], [1,-0.5], xx)
```

where `xx` and `yy` are the input and output signal vectors, respectively. Notice that the `aa` vector has $-a_1$ for its second element, just like in the polynomial $A(z)$. We can imagine that the filter coefficient multiplying $y[n]$ is 1, so we always have `1` for the first element of `aa`. ■

EXERCISE 8.6: Find the system function (i.e., z-transform) of the following feedback filter:

$$y[n] = 0.5y[n-1] - 3x[n] + 2x[n-1]$$

EXERCISE 8.7: Determine the system function of the system implemented by the following MATLAB statement:

```
yy = filter(5, [1,0.8], xx).
```

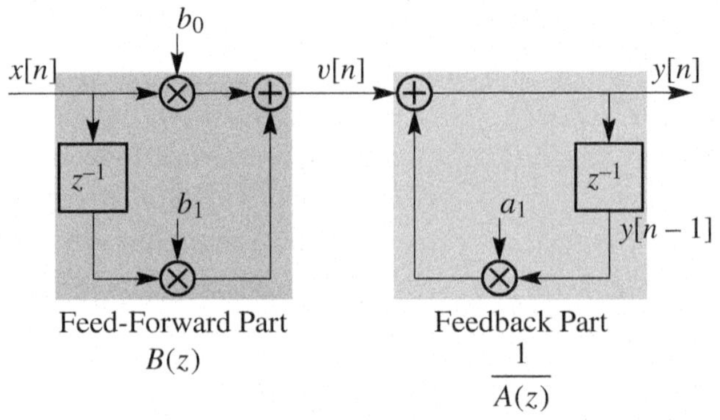

Figure 8-5: First-order IIR system showing one feedback coefficient a_1 and two feed-forward coefficients b_0 and b_1 in Direct Form I structure.

8-3.2 The System Function and Block-Diagram Structures

As we have seen, the system function displays the coefficients of the difference equation in a convenient way that makes it easy to move back and forth between the difference equation and the system function. In this section, we will show that this makes it possible to derive other difference equations, and thus other implementations, by simply manipulating the system function.

8-3.2.1 Direct Form I Structure

To illustrate the connection between the system function and the block diagram, let us return to the block diagram of Fig. 8-1, which is repeated in Fig. 8-5 for convenience. Block diagrams such as Fig. 8-5 are called *implementation structures*, or, more commonly, simply *structures*, because they give a pictorial representation of the difference equations that can be used to implement the system.

Recall that the product of two z-transform system functions corresponds to the cascade of two systems. The system function for the first-order feedback filter can be factored into an FIR piece and an IIR piece, as in

$$\begin{aligned} H(z) &= \frac{b_0 + b_1 z^{-1}}{1 - a_1 z^{-1}} \\ &= \left(\frac{1}{1 - a_1 z^{-1}}\right)\left(b_0 + b_1 z^{-1}\right) = \left(\frac{1}{A(z)}\right) \cdot B(z) \end{aligned}$$

The conclusion to be drawn from this algebraic manipulation is that a valid implementation for $H(z)$ is the pair of difference equations

$$v[n] = b_0 x[n] + b_1 x[n-1] \qquad (8.21)$$

$$y[n] = a_1 y[n-1] + v[n] \qquad (8.22)$$

We see in Fig. 8-5 that the polynomial $B(z)$ is the system function of the feed-forward part of the block diagram, and that $1/A(z)$ is the system function of a feedback part that completes the system. The system implemented in this way is called the ***Direct Form I*** implementation because it is possible to go directly from the system function to this block diagram (or the difference equations that it represents) with no

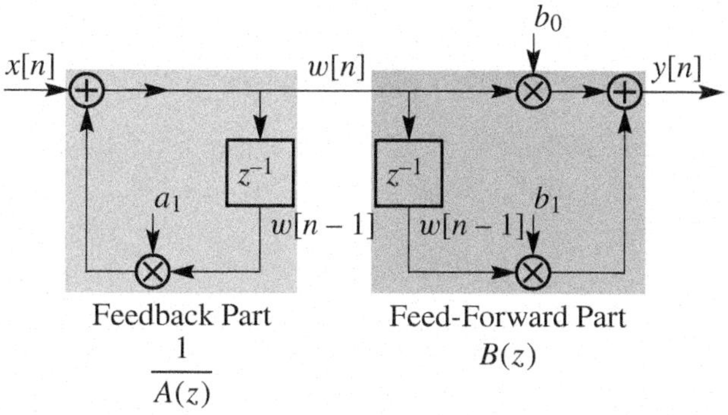

Figure 8-6: First-order IIR system showing one feedback coefficient a_1 and two feed-forward coefficients b_0 and b_1 in a Direct Form II structure.

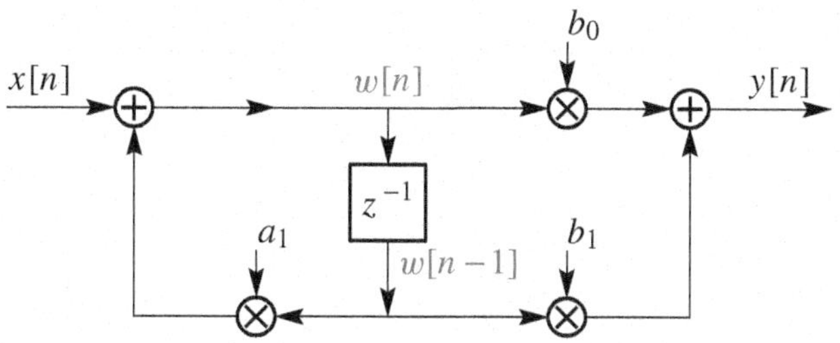

Figure 8-7: First-order IIR system in Direct Form II structure. This is identical to Fig. 8-6, except that the two delays have been merged into one.

other manipulations than to write the numerator and denominator as polynomials in the variable z^{-1}.

8-3.2.2 Direct Form II Structure

We know that for an LTI cascade system, we can change the order of the systems without changing the overall system response. In other words,

$$H(z) = \left(\frac{1}{A(z)}\right) \cdot B(z) = B(z) \cdot \left(\frac{1}{A(z)}\right)$$

Using the correspondences that we have established, leads to the block diagram shown in Fig. 8-6. Note that we have defined a new intermediate variable $w[n]$ as the output of the feedback part and input to the feed-forward part. Thus, the block diagram tells us that an equivalent implementation of the system is

$$w[n] = a_1 w[n-1] + x[n] \qquad (8.23)$$

$$y[n] = b_0 w[n] + b_1 w[n-1] \qquad (8.24)$$

Again, because there is such a direct and simple correspondence between Fig. 8-6 and $H(z)$, this implementation is called the ***Direct Form II*** implementation of the first-order IIR system with system function

$$H(z) = \frac{b_0 + b_1 z^{-1}}{1 - a_1 z^{-1}}$$

The block-diagram representation of Fig. 8-6 leads to a valuable insight. Notice that the input to each of the unit delay operators is the same signal $w[n]$. Thus, there is no need for two delay operations; they can be combined into a single delay, as in Fig. 8-7. Since delay operations are implemented with memory in a computer, the implementation of Fig. 8-7 would require less memory that the implementation of Fig. 8-6. Note, however, that both block diagrams represent the difference equations (8.23) and (8.24).

EXERCISE 8.8: Find the z-transform system function of the following set of cascaded difference equations:

$$w[n] = -0.5w[n-1] + 7x[n]$$

$$y[n] = 2w[n] - 4w[n-1]$$

Draw the block diagrams of this system in both Direct Form I and Direct Form II.

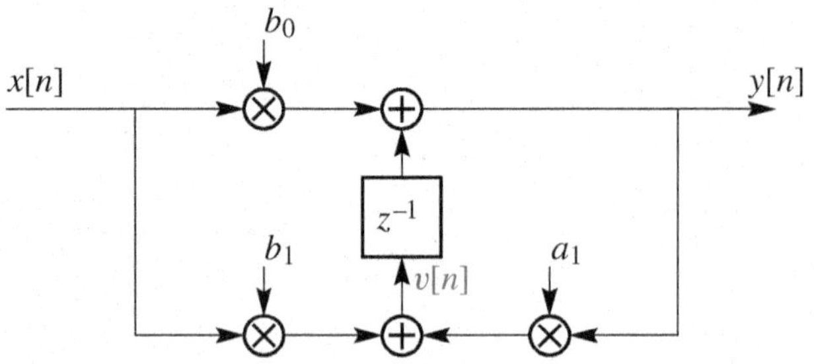

Figure 8-8: Computational structure for a general first-order IIR filter as a transposed Direct Form II structure.

8-3.2.3 The Transposed Form Structure

A somewhat surprising fact about block diagrams like Fig. 8-7 is that if the block diagram undergoes the following transformation:

1. All the arrows are reversed with multipliers being unchanged in value or location.
2. All branch points become summing points, and all summing points become branch points.
3. The input and the output are interchanged.

then the overall system has the same system function as the original system. We will not prove this, but it is true for the kinds of block diagrams that we have just introduced. However, we can use the z-transform to verify that this is true for our simple first-order system.

The feedback structure given in the signal-flow graph of Fig. 8-8 is the ***transposed*** form of the Direct Form II structure shown in Fig. 8-7. In order to derive the actual difference equations, we need to write the equations that are defined by the signal-flow graph. There is an orderly procedure for doing this if we follow two rules:

1. Assign variable names to the inputs of all delay elements. For example, $v[n]$ is used in Fig. 8-8, so the output of the delay is $v[n-1]$.
2. Write equations at all of the summing nodes; there are two in this case.

$$y[n] = b_0 x[n] + v[n-1] \qquad (8.25)$$

$$v[n] = b_1 x[n] + a_1 y[n] \qquad (8.26)$$

The signal-flow graph specifies an actual computation, so (8.25) and (8.26) require three multiplies and two adds at each time step n. Equation (8.25) must be done first, because $y[n]$ is needed in (8.26).

Owing to the feedback, it is impossible to manipulate these equations into one of the other forms by eliminating variables. However, we can recombine these two equations in the z-transform domain to verify that we have the correct system function. First, we take the z-transform of each difference equation obtaining

$$Y(z) = b_0 X(z) + z^{-1} V(z)$$

$$V(z) = b_1 X(z) + a_1 Y(z)$$

Now we eliminate $V(z)$ by substituting the second equation into the first as follows:

$$Y(z) = b_0 X(z) + z^{-1}(b_1 X(z) + a_1 Y(z))$$

$$(1 - a_1 z^{-1}) Y(z) = (b_0 + b_1 z^{-1}) X(z)$$

Therefore, we get

$$H(z) = \frac{Y(z)}{X(z)} = \frac{b_0 + b_1 z^{-1}}{1 - a_1 z^{-1}}$$

Thus, because they have the same system function, (8.25) and (8.26) are equivalent to the Direct Form I (8.21) and (8.22), and to the Direct Form II (8.23) and (8.24).

Why are these different implementations of the same system function of interest to us? They all use the same number of multiplications and additions to compute exactly the same output from a given input. However, this is true only when the arithmetic is perfect. On a computer with finite precision (e.g., 16-bit words), each calculation will involve round-off errors, which means that each implementation will behave slightly differently. In practice, the implementation of high-quality digital filters in hardware demands correct engineering to control round-off errors and overflows.

8-3.3 Relation to the Impulse Response

In the analysis of Section 8-3.1 we implicitly assumed that the system function is the z-transform of the impulse response of an IIR system. While this is true, we have demonstrated only that it is true for the FIR case. In the IIR case, we need to be able to take the z-transform of an infinitely long sequence. As an example of such a sequence, consider $h[n] = a^n u[n]$. Applying the definition of the z-transform from equation (7.15) on p. 169, we would write

$$H(z) = \sum_{n=0}^{\infty} a^n z^{-n} = \sum_{n=0}^{\infty} (az^{-1})^n$$

which is the sum of all the terms in a geometric series where the ratio between successive terms is az^{-1}. Thus, we know that if $|az^{-1}| < 1$, then the sum is finite, and in fact is given by the closed-form expression

$$H(z) = \sum_{n=0}^{\infty} a^n z^{-n} = \frac{1}{1 - az^{-1}}$$

The condition for the infinite sum to equal the closed-form expression can be expressed as $|a| < z$. The values of z in the complex plane satisfying this condition are called the ***region of convergence***. From the preceding analysis, we can state the following exceedingly important z-transform pair:

$$a^n u[n] \quad \overset{z}{\longleftrightarrow} \quad \frac{1}{1 - az^{-1}} \tag{8.27}$$

We will have many occasions to use this result in this chapter.

Example 8-5: $H(z)$ from Impulse Response

As an example of the use of this result, recall that in Section 8-2.2 we showed by iteration that the impulse response of the system

$$y[n] = a_1 y[n-1] + b_0 x[n] + b_1 x[n-1] \tag{8.28}$$

is

$$h[n] = b_0 (a_1)^n u[n] + b_1 (a_1)^{n-1} u[n-1] \tag{8.29}$$

Thus, using the linearity property of the z-transform, the delay property of the z-transform, and the result of (8.27), the system function for this system is

$$\begin{aligned} H(z) &= b_0 \left(\frac{1}{1 - a_1 z^{-1}} \right) + b_1 z^{-1} \left(\frac{1}{1 - a_1 z^{-1}} \right) \\ &= \frac{b_0 + b_1 z^{-1}}{1 - a_1 z^{-1}} \end{aligned} \tag{8.30}$$

which is what we obtained before in Section 8-3.1 on p. 205 by taking the z-transform of the difference equation and solving for $H(z) = Y(z)/X(z)$. ■

8-3.4 Summary of the Method

In this section, we have illustrated some important analysis techniques. We have seen that it is possible to go from the difference equation (8.28) directly to the system function (8.30). We have also seen that, in this simple example, it is possible by ***taking the inverse z-transform*** to go directly from the system function (8.30) to the impulse response (8.29) of the system without the tedious process of iterating the difference equation. We will see that it is possible to do this, in general, by a process of inverse z-transformation based on breaking the z-transform into a sum of terms like the right-hand side of (8.27). Before developing this technique, which will be applicable to higher-order systems, we will continue to focus on the first-order IIR system to illustrate some more important points about the z-transform and its relation to IIR systems.

8-4 Poles and Zeros

An interesting fact about the z-transform system function is that the numerator and denominator polynomials have zeros. These zero locations in the complex z-plane are very important for characterizing the system. Although we like to write the system function in terms of z^{-1} to facilitate the correspondence with the difference equation, it is probably better for finding roots to rewrite the polynomials as functions of z rather than z^{-1}. If we multiply the numerator and denominator by z, we obtain

$$H(z) = \frac{b_0 + b_1 z^{-1}}{1 - a_1 z^{-1}} = \frac{b_0 z + b_1}{z - a_1}$$

In this form, it is easy to find the roots of the numerator and denominator polynomials.

DEMO: *PeZ GUI*

LAB: *PeZ - The z, n, and $\hat{\omega}$ Domains*

DEMO: *PeZ Tutorial*

The numerator has one root at

$$b_0 z + b_1 = 0 \quad \Longrightarrow \quad \text{Root at } z = -\frac{b_1}{b_0}$$

and the denominator has its root at

$$z - a_1 = 0 \quad \Longrightarrow \quad \text{Root at } z = a_1$$

If we consider $H(z)$ as a function of z over the entire complex z-plane, the root of the numerator is a ***zero*** of the function $H(z)$, i.e.,

$$H(z)\big|_{z=-(b_1/b_0)} = 0$$

Recall that the root of the denominator is a location in the z-plane where the function $H(z)$ *blows up*

$$H(z)\big|_{z=a_1} \to \infty$$

so this location ($z = a_1$) is called a ***pole*** of the system function $H(z)$.

EXERCISE 8.9: Find the poles and zeros of the following z-transform system function

$$H(z) = \frac{3 + 4z^{-1}}{1 + 0.5z^{-1}}$$

EXERCISE 8.10: For the z-transform of the following feedback filter

$$y[n] = 0.5y[n-1] - x[n] + 3x[n-1]$$

determine the locations of the pole and zeros.

8-4.1 Poles or Zeros at the Origin or Infinity

When the numerator and denominator polynomials have a different number of coefficients, we can have either zeros or poles at $z = 0$. We saw this in Chapter 7, where FIR systems, whose system functions have only a numerator polynomial, had a number of poles at $z = 0$ equal to the number of zeros of the polynomial. If we count all the poles and zeros at $z = \infty$, as well as $z = 0$, then we can assert that ***the number of poles equals the number of zeros***. Consider the following examples.

Example 8-6: Find Poles and Zeros

The system function of the feedback filter

$$y[n] = 0.5y[n-1] + 2x[n]$$

is found by inspection to be

$$H(z) = \frac{2}{1 - 0.5z^{-1}}$$

When we express $H(z)$ in positive powers of z

$$H(z) = \frac{2z}{z - 0.5}$$

we see that there is one pole at $z = 0.5$ and a zero at $z = 0$. ■

Example 8-7: Zero at $z = \infty$

The system fuction of the feedback filter

$$y[n] = 0.5y[n-1] + 3x[n-1]$$

is found by inspection to be

$$H(z) = \frac{3z^{-1}}{1 - 0.5z^{-1}} = \frac{3}{z - 0.5}$$

The system has one pole at $z = 0.5$, and if we take the limit of $H(z)$ as $z \to \infty$, we get $H(z) \to 0$. Thus, it also has one zero at $z = \infty$. ■

Both of the cases in Examples 8-6 and 8-7 have exactly one pole and one zero, if we count the zero at $z = \infty$.

EXERCISE 8.11: Determine the system function $H(z)$ of the following feedback filter:

$$y[n] = 0.5y[n-1] + 3x[n-2]$$

Show that $H(z)$ has a pole at $z = 0$, as well as $z = 0.5$. In addition, show that $H(z)$ has two zeros at $z = \infty$ by taking the limit as $z \to \infty$.

8-4.2 Pole Locations and Stability

The pole location of a first-order filter determines the shape of the impulse response. In Section 8-3.3 we showed that a system having system function

$$\begin{aligned} H(z) &= b_0\left(\frac{1}{1 - a_1z^{-1}}\right) + b_1z^{-1}\left(\frac{1}{1 - a_1z^{-1}}\right) \\ &= \frac{b_0 + b_1z^{-1}}{1 - a_1z^{-1}} = \frac{b_0(z + b_1/b_0)}{(z - a_1)} \end{aligned}$$

has an impulse response

$$\begin{aligned} h[n] &= b_0(a_1)^n u[n] + b_1(a_1)^{n-1}u[n-1] \\ &= \begin{cases} 0 & \text{for } n < 0 \\ b_0 & \text{for } n = 0 \\ (b_0 + b_1a_1^{-1})a_1^n & \text{for } n \geq 1 \end{cases} \end{aligned}$$

That is, an IIR system with a single pole at $z = a_1$ has an impulse response that is proportional to a_1^n for $n \geq 1$. We see that if $|a_1| < 1$, the impulse response will die out as $n \to \infty$. On the other hand, if $|a_1| \geq 1$, the impulse response will not die out; in fact if $|a_1| > 1$, it will grow without bound. Since the pole of the system function is at $z = a_1$, we see that the location of the pole can tell us whether the impulse response will decay or grow. Clearly, it is desirable for the impulse response to die out, because an exponentially growing impulse response would produce unbounded outputs even if the input samples have finite size. Systems that produce bounded outputs when the input is bounded are called ***stable systems***. If $|a_1| < 1$, the pole of the system function is ***inside*** the unit circle of the z-plane. It turns out that, for the IIR systems we have been discussing, the following is true in general:

> *A causal LTI IIR system with initial rest conditions is stable if all of the poles of its system function lie strictly inside the unit circle of the z-plane.*

Thus, stability of a system can be seen at a glance from a z-plane plot of the poles and zeros of the system function.

Example 8-8: Stability from Pole Location

The system whose system function is

$$H(z) = \frac{1 - 2z^{-1}}{1 - 0.8z^{-1}} = \frac{z - 2}{z - 0.8}$$

has a zero at $z = 2$ and a pole at $z = 0.8$. Therefore, the system is stable. Note that the location of the zero, which is outside the unit circle, has nothing to do with stability of the system. Recall that the zeros correspond to an FIR system that is in cascade with an IIR system defined by the poles. Since FIR systems are always stable, it is not surprising that stability is determined solely by the poles of the system function. ■

EXERCISE 8.12: An LTI IIR system has system function

$$H(z) = \frac{2 + 2z^{-1}}{1 - 1.25z^{-1}}$$

Plot the pole and zero in the z-plane, and state whether or not the system is stable.

8-5 Frequency Response of an IIR Filter

In Chapter 6, we introduced the concept of frequency response $H(e^{j\hat{\omega}})$ as the complex function that determines the amplitude and phase change experienced by a complex exponential input to an LTI system; i.e., if $x[n] = e^{j\hat{\omega}n}$, then

$$y[n] = H(e^{j\hat{\omega}})e^{j\hat{\omega}n} \tag{8.31}$$

In Section 7-6 on p. 175 we showed that the frequency response of an FIR system is related to the system function by

$$H(e^{j\hat{\omega}}) = H(e^{j\hat{\omega}}) = H(z)\big|_{z=e^{j\hat{\omega}}} \tag{8.32}$$

This relation between the system function and the frequency response also holds for IIR systems. However, we must add the provision that the system must be stable in order for the frequency response to exist and to be given by (8.32). This condition of stability is a general condition, but all FIR systems are stable, so up to now we have not needed to be concerned with stability.

Recall that the system function for the general first-order IIR system has the form

$$H(z) = \frac{b_0 + b_1 z^{-1}}{1 - a_1 z^{-1}}$$

where the region of convergence of the system function is $|a_1 z^{-1}| < 1$ or $|a_1| < |z|$. If we wish to evaluate $H(z)$ for $z = e^{j\hat{\omega}}$, then the values of z on the unit circle should be in the region of convergence; i.e., we require $|z| = 1$ to be in the region of convergence of the z-transform. This means that $|a_1| < 1$, which was shown in Section 8-4.2 to be the condition for stability of the first-order system. In Section 8-8 we will give another interpretation of why stability is required for the frequency response to exist. Assuming stability in the first-order case, we get the following formula for the frequency response:

$$H(e^{j\hat{\omega}}) = H(z)\Big|_{z=e^{j\hat{\omega}}} = \frac{b_0 + b_1 e^{-j\hat{\omega}}}{1 - a_1 e^{-j\hat{\omega}}} \tag{8.33}$$

A simple evaluation will verify that (8.33) is a periodic function with a period equal to 2π. This must always be the case for the frequency response of a discrete-time system.

Remember that the frequency response $H(e^{j\hat{\omega}})$ is a complex-valued function of frequency $\hat{\omega}$. Therefore, we can reduce (8.33) to two separate real formulas for the magnitude and the phase as functions of frequency. For the magnitude response, it is expedient to compute the magnitude squared first, and then take a square root if necessary.

The magnitude squared can be formed by multiplying the complex $H(e^{j\hat{\omega}})$ in (8.33) by its conjugate (denoted by H^*). For our first-order example,

$$\begin{aligned}
|H(e^{j\hat{\omega}})|^2 &= H(e^{j\hat{\omega}})H^*(e^{j\hat{\omega}}) \\
&= \frac{b_0 + b_1 e^{-j\hat{\omega}}}{1 - a_1 e^{-j\hat{\omega}}} \cdot \frac{b_0^* + b_1^* e^{+j\hat{\omega}}}{1 - a_1^* e^{+j\hat{\omega}}} \\
&= \frac{|b_0|^2 + |b_1|^2 + b_0 b_1^* e^{+j\hat{\omega}} + b_0^* b_1 e^{-j\hat{\omega}}}{1 + |a_1|^2 - a_1^* e^{+j\hat{\omega}} - a_1 e^{-j\hat{\omega}}} \\
&= \frac{|b_0|^2 + |b_1|^2 + 2\Re e\{b_0^* b_1 e^{-j\hat{\omega}}\}}{1 + |a_1|^2 - 2\Re e\{a_1 e^{-j\hat{\omega}}\}}
\end{aligned}$$

This derivation does not assume that the filter coefficients are real. If the coefficients were real, we would get the further simplification

$$|H(e^{j\hat{\omega}})|^2 = \frac{|b_0|^2 + |b_1|^2 + 2b_0 b_1 \cos(\hat{\omega})}{1 + |a_1|^2 - 2a_1 \cos(\hat{\omega})}$$

This formula is not particularly informative, because it is difficult to use it to visualize the shape of $|H(e^{j\hat{\omega}})|$. However, it could be used to write a program for evaluating and plotting the frequency response. The phase response is even messier. Arctangents would be used to extract the angles of the numerator and denominator, and then the two phases would be subtracted. When the filter coefficients are real, the phase is

$$\phi(\hat{\omega}) = \tan^{-1}\left(\frac{-b_1 \sin\hat{\omega}}{b_0 + b_1 \cos\hat{\omega}}\right) - \tan^{-1}\left(\frac{a_1 \sin\hat{\omega}}{1 - a_1 \cos\hat{\omega}}\right)$$

Again, the formula is so complicated that we cannot gain insight from it directly. In a later section, we will use the poles and zeros of the system function together with the relationship (8.32) to construct an approximate plot of the frequency response without recourse to formulas.

8-5.1 Frequency Response using MATLAB

Frequency responses can be computed and plotted easily by many signal processing software packages. In MATLAB, for example, the function `freqz` is provided for just that purpose.[2] The frequency response is evaluated over an equally spaced grid in the $\hat{\omega}$ domain, and then its magnitude and phase can be plotted. In MATLAB, the functions `abs` and `angle` will extract the magnitude and the angle of each element in a complex vector.

Example 8-9: Plot $H(e^{j\hat{\omega}})$ via MATLAB

Consider the example

$$y[n] = 0.8y[n-1] + 2x[n] + 2x[n-1]$$

In order to define the filter coefficients in MATLAB, we put all the terms with $y[n]$ on one side of the equation, and the terms with $x[n]$ on the other.

$$y[n] - 0.8y[n-1] = 2x[n] + 2x[n-1]$$

[2] The function `freqz` is part of MATLAB's Signal Processing Toolbox. In case a substitute is needed, there is a similar function called `freekz` that is part of the SP-First Toolbox on the CD-ROM.

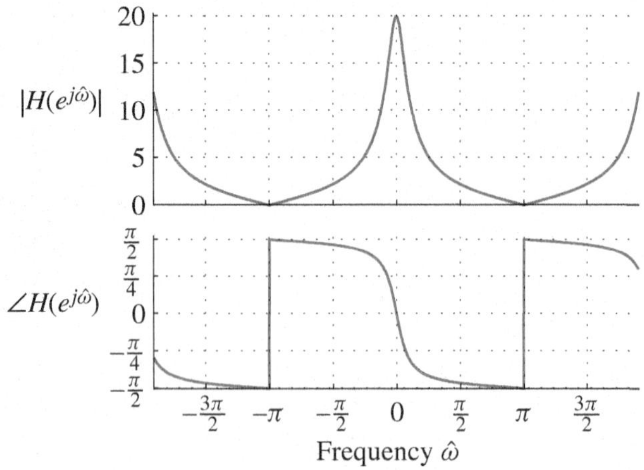

Figure 8-9: Frequency response (magnitude and phase) for a first-order feedback filter. The pole is at $z = 0.8$, and the numerator has a zero at $z = -1$.

Then we read off the filter coefficients and define the vectors `aa` and `bb` as

```
aa = [ 1, -0.8 ]        bb = [ 2, 2 ]
```

Thus, the vectors `aa` and `bb` are in the same form as for the `filter` function. The following call to `freqz` will generate a 401-point vector `HH` containing the values of the frequency response at the vector of frequencies specified by the third argument, `[-6:0.03:6]`.

```
HH = freqz( bb, aa, [-6:0.03:6] );
```

Plots of the resulting magnitude and phase are shown in Fig. 8-9. The frequency interval $-6 \leq \hat{\omega} \leq +6$ is shown so that the 2π-periodicity of $H(e^{j\hat{\omega}})$ will be evident. ■

In this example, we can look for a connection between the poles and zeros and the shape of the frequency response. For this system, we have the system function

$$H(z) = \frac{2 + 2z^{-1}}{1 - 0.8z^{-1}}$$

which has a pole at $z = 0.8$ and a zero at $z = -1$. The point $z = -1$ is the same as $\hat{\omega} = \pi$ because $z = -1 = e^{j\pi} = e^{j\hat{\omega}}|_{\hat{\omega}=\pi}$. Thus, $H(e^{j\hat{\omega}})$ has the value zero at $\hat{\omega} = \pi$, since $H(z)$ is zero at $z = -1$. In a similar manner, the pole at $z = 0.8$ has an effect on the frequency response near $\hat{\omega} = 0$. Since $H(z)$ blows up at $z = 0.8$, the nearby points on the unit circle must have large values. The closest point on the unit circle is at $z = e^{j0} = 1$. In this case, we can evaluate the frequency response directly from the formula to get

$$\begin{aligned} H(e^{j\hat{\omega}})\Big|_{\hat{\omega}=0} &= H(z)\Big|_{z=1} = \frac{2 + 2z^{-1}}{1 - 0.8z^{-1}}\Bigg|_{z=1} \\ &= \frac{2 + 2}{1 - 0.8} = \frac{4}{0.2} = 20 \end{aligned}$$

8-5.2 Three-Dimensional Plot of a System Function

The relationship between $H(e^{j\hat{\omega}})$ and the pole-zero locations of $H(z)$ can be illustrated by making a three-dimensional plot of $H(z)$ and then cutting out the frequency response. The frequency response $H(e^{j\hat{\omega}})$ is obtained by selecting the values of $H(z)$ along the unit circle (i.e., as $\hat{\omega}$ goes from $-\pi$ to $+\pi$, the equation $z = e^{j\hat{\omega}}$ defines the unit circle).

In this section, we use the system function

$$H(z) = \frac{1}{1 - 0.8z^{-1}}$$

to illustrate the relationship between the system function and the frequency response. Figures 8-10 and 8-11 are plots of the magnitude and phase of $H(z)$ over the region $[-1.4, 1.4]\times[-1.4, 1.4]$ of the z-plane. In the magnitude plot of Fig. 8-10, we observe that the pole (at $z = 0.8$) creates a large peak that makes all nearby values very large. At the precise location of the pole, $H(z) \to \infty$, but the grid in Fig. 8-10 does not contain the point ($z = 0.8$),

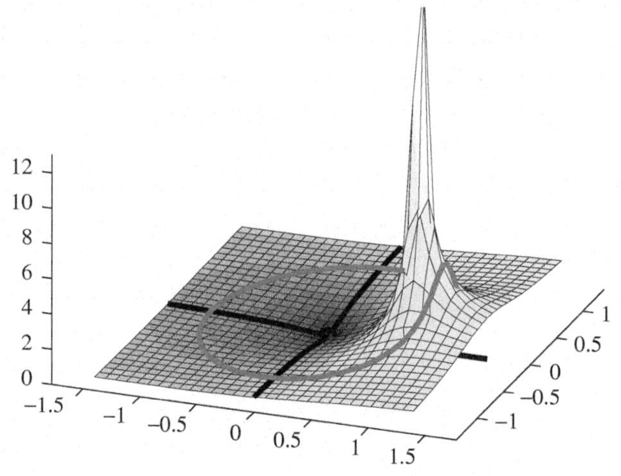

Figure 8-10: z-transform evaluated over a region of the z-plane including the unit circle. Values along the unit circle are shown as a dark line where the frequency response (magnitude) is evaluated. The view is from the fourth quadrant, so the point $z = 1$ is on the right. The first-order filter has a pole at $z = 0.8$ and a zero at $z = 0$.

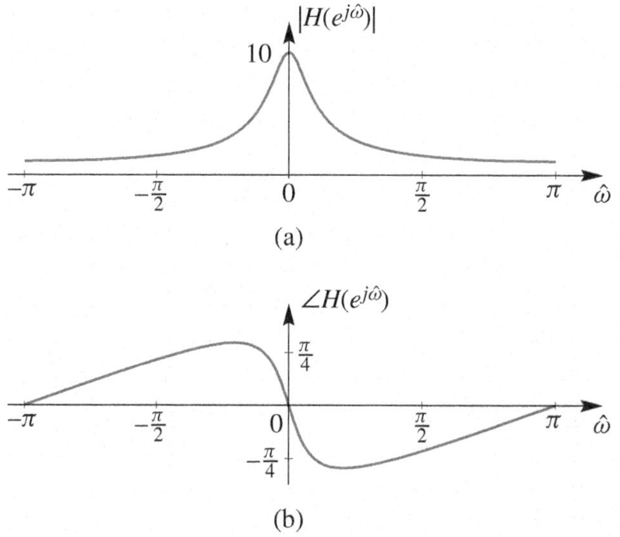

Figure 8-12: Frequency response (magnitude and phase) for a first-order feedback filter. The pole is at $z = 0.8$ and the numerator has a zero at $z = 0$. These are the values of $H(z)$ along the unit circle in the z-plane.

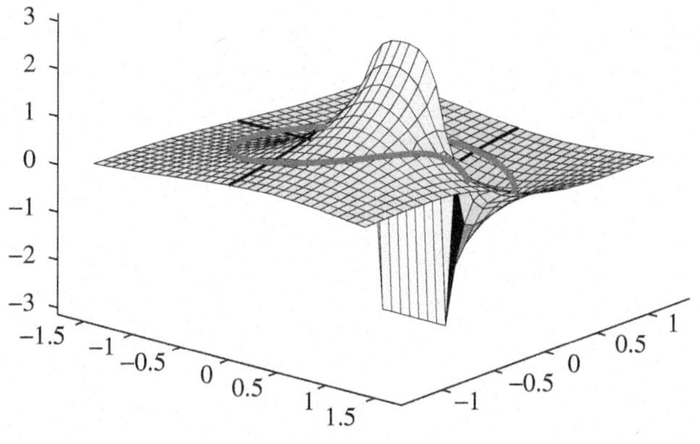

Figure 8-11: Phase of $H(z)$ evaluated over a region of the z-plane that includes the unit circle. View is from the fourth quadrant, so $z = 1$ lies to the right.

so the plot stays within a finite scale. The phase response in Fig. 8-11 also exhibits its most rapid transition at $\hat{\omega} = 0$ which is $z = 1$ (the closest point on the unit circle to the pole at $z = 0.8$).

The frequency response $H(e^{j\hat{\omega}})$ is obtained by selecting the values of the z-transform along the unit circle in Figs. 8-10 and 8-11. Plots of $H(e^{j\hat{\omega}})$ versus $\hat{\omega}$ are given in Fig. 8-12. The shape of the frequency response can be explained in terms of the pole location by recognizing that in Fig. 8-10 the pole at $z = 0.8$ pushes $H(e^{j\hat{\omega}})$ up in the region near $\hat{\omega} = 0$, which is the same as $z = 1$.

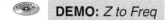

DEMO: *Z to Freq*

The unit circle values follow the ups and downs of $H(z)$ as $\hat{\omega}$ goes from $-\pi$ to $+\pi$.

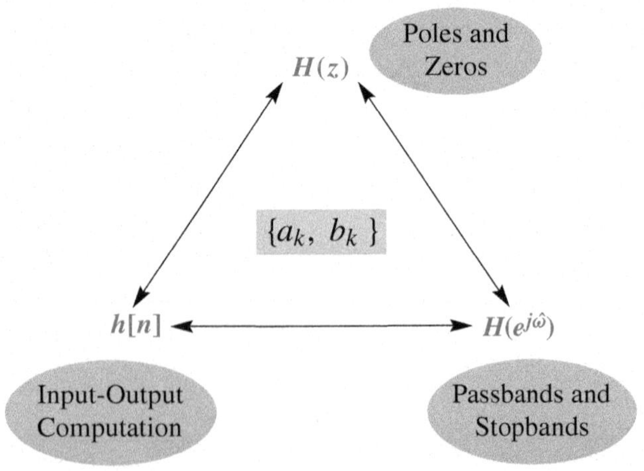

Figure 8-13: Relationship among the n-, z-, and $\hat{\omega}$-domains. The filter coefficients $\{a_k, b_k\}$ play a central role.

8-6 Three Domains

To illustrate the use of the analysis tools that we have developed, we consider the general second-order case.

DEMO: *Three Domains - IIR*

The three domains: n, z and $\hat{\omega}$ are depicted in Fig. 8-13. The defining equation for the IIR digital filter is the feedback difference equation, which, for the second-order case, is

$$\begin{aligned} y[n] = a_1 y[n-1] + a_2 y[n-2] \\ + b_0 x[n] + b_1 x[n-1] + b_2 x[n-2] \end{aligned}$$

This equation provides the algorithm for computing the output signal from the input signal by iteration using the filter coefficients $\{a_1, a_2, b_0, b_1, b_2\}$. It also defines the impulse response $h[n]$.

Following the procedures illustrated for the first-order case, we can also define the z-transform system function directly from the filter coefficients as

$$H(z) = \frac{b_0 + b_1 z^{-1} + b_2 z^{-2}}{1 - a_1 z^{-1} - a_2 z^{-2}}$$

and we can also obtain the frequency response

$$H(e^{j\hat{\omega}}) = \frac{b_0 + b_1 e^{-j\hat{\omega}} + b_2 e^{-j2\hat{\omega}}}{1 - a_1 e^{-j\hat{\omega}} - a_2 e^{-j2\hat{\omega}}}$$

Since the system function is a ratio of polynomials, the poles and zeros of $H(z)$ make up a small set of parameters that completely define the filter.

Finally, the shapes of the passbands and stopbands of the frequency response are highly dependent on the pole and zero locations with respect to the unit circle, and the character of the impulse response can be related to the poles. To make this last point for the general case, we need to develop one more tool—a technique for getting $h[n]$ directly from $H(z)$. This process, which applies to any z-transform and its corresponding sequence, is called the ***inverse z-transform***. The inverse z-transform is developed in the next section.

8-7 The Inverse z-Transform and Some Applications

We have seen how the three domains are connected for the first-order IIR system. Many of the concepts that we have introduced for the first-order system can be extended to higher-order systems in a straightforward manner. However, finding the impulse response from the system function is not an obvious extension of what we have done for the first-order case. We need to develop a process for inverting the z-transform that can be applied to systems with more than one pole. This process is called the ***inverse z-transform***. In this section we will show how to find the inverse for a general rational z-transform.

We will illustrate the process with some examples. The techniques that we develop will then be available for determining the impulse responses of second-order and higher-order systems.

8-7.1 Revisiting the Step Response of a First-Order System

In Section 8-2.4 we computed the step response of a first-order system by both iteration and convolution. Now we will show how the z-transform can be used for the same purpose. Consider a system whose system function is

$$H(z) = \frac{b_0 + b_1 z^{-1}}{1 - a_1 z^{-1}}$$

Recall that the z-transform of the output of this system is $Y(z) = H(z)X(z)$, so one approach to finding the output for a given input $x[n]$ is as follows:

1. Determine the z-transform $X(z)$ of the input signal $x[n]$.
2. Multiply $X(z)$ by $H(z)$ to get $Y(z)$.
3. Determine the inverse z-transform of $Y(z)$ to get the output $y[n]$.

Clearly, this procedure will work and will avoid both iteration and convolution if we can determine $X(z)$ and if we can perform the necessary inverse transformation. Our focus in this section will be on deriving a general procedure for step (3).

In the case of the step response, we see that the input $x[n] = u[n]$ is a special case of the more general sequence $a^n u[n]$; i.e., $a = 1$. Therefore, from (8.27) it follows that the z-transform of $x[n] = u[n]$ is

$$X(z) = \frac{1}{1 - z^{-1}}$$

Table 8-1: Summary of important z-transform properties and pairs.

Short Table Of z-Transforms			
	$x[n]$	$\overset{z}{\longleftrightarrow}$	$X(z)$
1.	$ax_1[n] + bx_2[n]$	$\overset{z}{\longleftrightarrow}$	$aX_1(z) + bX_2(z)$
2.	$x[n - n_0]$	$\overset{z}{\longleftrightarrow}$	$z^{-n_0}X(z)$
3.	$y[n] = x[n] * h[n]$	$\overset{z}{\longleftrightarrow}$	$Y(z) = H(z)X(z)$
4.	$\delta[n]$	$\overset{z}{\longleftrightarrow}$	1
5.	$\delta[n - n_0]$	$\overset{z}{\longleftrightarrow}$	z^{-n_0}
6.	$a^n u[n]$	$\overset{z}{\longleftrightarrow}$	$\dfrac{1}{1 - az^{-1}}$

so $Y(z)$ is

$$\begin{aligned} Y(z) = H(z)X(z) &= \frac{b_0 + b_1 z^{-1}}{(1 - a_1 z^{-1})(1 - z^{-1})} \\ &= \frac{b_0 + b_1 z^{-1}}{1 - (1 + a_1)z^{-1} + a_1 z^{-2}} \end{aligned} \tag{8.34}$$

Now we need to go back to the n-domain by inverse transformation. A standard approach is to use a table of z-transform pairs and simply look up the answer in the table. Our previous discussions in Chapter 7 and earlier in this chapter have developed the basis for a simple version of such a table. A summary of the z-transform knowledge that we have developed so far is given in Table 8-1. Although more extensive tables can be constructed, the results that we have assembled in Table 8-1 are more than adequate for our purposes in this text.

Now let us return to the problem of finding $y[n]$ given $Y(z)$ in (8.34). The technique that we will use is based on the partial fraction expansion[3] of $Y(z)$. This technique is based on the observation that a rational function $Y(z)$ can be expressed as a sum of simpler rational functions; i.e.,

$$\begin{aligned} Y(z) &= H(z)X(z) \\ &= \frac{b_0 + b_1 z^{-1}}{(1 - a_1 z^{-1})(1 - z^{-1})} \qquad (8.35) \\ &= \frac{A}{1 - a_1 z^{-1}} + \frac{B}{1 - z^{-1}} \end{aligned}$$

If the expression on the right is pulled together over a common denominator, it should be possible to find A and B so that the numerator of the resulting rational function will be equal to $b_0 + b_1 z^{-1}$. Equating the two numerators would give two equations in the two unknowns A and B. However, there is a much quicker way. A systematic procedure for finding the desired A and B is based on the observation that, for this example,

$$Y(z)(1 - a_1 z^{-1}) = \frac{b_0 + b_1 z^{-1}}{(1 - z^{-1})} = A + \frac{B(1 - a_1 z^{-1})}{1 - z^{-1}}$$

Then we can evaluate at $z = a_1$ to isolate A; i.e.,

$$\begin{aligned} Y(z)(1 - a_1 z^{-1})\Big|_{z=a_1} &= \frac{b_0 + b_1 z^{-1}}{(1 - z^{-1})}\Bigg|_{z=a_1} \\ &= A + \frac{B(1 - a_1 z^{-1})}{1 - z^{-1}}\Bigg|_{z=a_1} = A \end{aligned}$$

With this result, it follows that

$$A = Y(z)(1 - a_1 z^{-1})\big|_{z=a_1} = \frac{b_0 + b_1 a_1^{-1}}{1 - a_1^{-1}}$$

[3]The ***partial fraction expansion*** is an algebraic decomposition usually presented in calculus for evaluating certain types of integrals.

Similarly, we can find B by

$$B = Y(z)(1 - z^{-1})\big|_{z=1} = \frac{b_0 + b_1}{1 - a_1}$$

Now using the superposition property of the z-transform (entry 1 in Table 8-1), and the exponential z-transform pair (entry 6 in Table 8-1), we can write down the desired answer as

$$y[n] = \left(\frac{b_0 + b_1 a_1^{-1}}{1 - a_1^{-1}}\right) a_1^n u[n] + \left(\frac{b_0 + b_1}{1 - a_1}\right) u[n]$$

which, after some manipulation becomes

$$y[n] = \left(\frac{(b_0 + b_1) - (b_0 a_1 + b_1) a_1^n}{1 - a_1}\right) u[n] \qquad (8.36)$$

If we substitute the value $b_1 = 0$ into (8.36), we get

$$y[n] = b_0 \left(\frac{1 - a_1^{n+1}}{1 - a_1}\right) u[n]$$

which is the same result obtained in Section 8-2.4 both by iteration of the difference equation (8.16) and by convolution (8.18).

With this example, we have established the framework for using the basic properties of z-transforms together with a few basic z-transform pairs to perform inverse z-transformation for any rational z-transform. We summarize this procedure in the following subsection.

8-7.2 A General Procedure for Inverse z-Transformation

Let $X(z)$ be any rational z-transform of degree N in the denominator and M in the numerator. Assuming that

$M < N$, we can find the sequence $x[n]$ that corresponds to $X(z)$ by the following procedure:

Procedure for Inverse z-Transformation ($M < N$)

1. Factor the denominator polynomial of $H(z)$ and express the pole factors in the form $(1 - p_k z^{-1})$ for $k = 1, 2, \ldots, N$.
2. Make a partial fraction expansion of $H(z)$ into a sum of terms of the form

$$H(z) = \sum_{k=1}^{N} \frac{A_k}{1 - p_k z^{-1}}$$

where $A_k = H(z)(1 - p_k z^{-1})\big|_{z=p_k}$

3. Write down the answer as

$$h[n] = \sum_{k=1}^{N} A_k (p_k)^n u[n]$$

This procedure will always work if the poles, p_k, are distinct. Repeated poles complicate the process, but can be handled systematically. We will restrict our attention to the case of non-repeated poles. Furthermore, this procedure can be applied to the inversion of any rational z-transform, not just a system function. We will illustrate the use of this procedure with two examples.

Example 8-10: Inverse z-Transform

Let a z-transform $X(z)$ be

$$X(z) = \frac{1 - 2.1z^{-1}}{1 - 0.3z^{-1} - 0.4z^{-2}} = \frac{1 - 2.1z^{-1}}{(1 + 0.5z^{-1})(1 - 0.8z^{-1})}$$

We wish to write $X(z)$ in the form

$$X(z) = \frac{A}{1 + 0.5z^{-1}} + \frac{B}{1 - 0.8z^{-1}}$$

Continuing the procedure for partial fraction expansion, we obtain

$$A = X(z)(1 + 0.5z^{-1})\big|_{z=-0.5} = \left.\frac{1 - 2.1z^{-1}}{1 - 0.8z^{-1}}\right|_{z=-0.5} = \frac{1 + 4.2}{1 + 1.6} = 2$$

and

$$B = X(z)(1 - 0.8z^{-1})\big|_{z=0.8} = \left.\frac{1 - 2.1z^{-1}}{1 + 0.5z^{-1}}\right|_{z=0.8} = \frac{1 - 2.1/0.8}{1 + 0.5/0.8} = -1$$

Therefore,

$$X(z) = \frac{2}{1 + 0.5z^{-1}} - \frac{1}{1 - 0.8z^{-1}} \tag{8.37}$$

and

$$x[n] = 2(-0.5)^n u[n] - (0.8)^n u[n]$$

Note that the poles at $z = p_1 = -0.5$ and $z = p_2 = 0.8$ give rise to terms in $x[n]$ of the form p_k^n. ■

In Example 8-10, the degree of the numerator is $M = 1$ and the degree of the denominator is $N = 2$. This is important because the partial fraction expansion works only for rational functions such that $M < N$. The next example shows why this is so, and illustrates a method of dealing with this complication.

Example 8-11: Long Division

Let $Y(z)$ be

$$Y(z) = \frac{2 - 2.4z^{-1} - 0.4z^{-2}}{1 - 0.3z^{-1} - 0.4z^{-2}} = \frac{2 - 2.4z^{-1} - 0.4z^{-2}}{(1 + 0.5z^{-1})(1 - 0.8z^{-1})}$$

Now we must add a constant term to the partial fraction expansion, otherwise, we cannot generate the term $-0.4z^{-2}$ in the numerator when we combine the partial fractions over a common denominator. That is, we must assume the following form for $Y(z)$:

$$Y(z) = \frac{A}{1+0.5z^{-1}} + \frac{B}{1-0.8z^{-1}} + C$$

How can we determine the constant C? One way is to perform long division of the denominator polynomial into the numerator polynomial until we get a remainder whose degree is lower than that of the denominator. In this case, the polynomial long division looks as follows:

$$\begin{array}{r|l} & \quad\quad\quad 1 \\ \hline -0.4z^{-2} - 0.3z^{-1} + 1 & -0.4z^{-2} - 2.4z^{-1} + 2 \\ & \underline{-0.4z^{-2} - 0.3z^{-1} + 1} \\ & \quad\quad\;\; -2.1z^{-1} + 1 \end{array}$$

Thus, if we place the remainder $(1 - 2.1z^{-1})$ over the denominator (in factored form), we can write $Y(z)$ as a rational part (fraction) plus the constant 1; i.e.,

$$Y(z) = \frac{1 - 2.1z^{-1}}{(1+0.5z^{-1})(1-0.8z^{-1})} + 1$$

The next step would be to apply the partial fraction expansion technique to the rational part of $Y(z)$. Since the rational part turns out to be identical to $X(z)$ in (8.37) from Example 8-10, the results would be the same as in that example, so we can write $Y(z)$ as

$$Y(z) = \frac{2}{1+0.5z^{-1}} - \frac{1}{1-0.8z^{-1}} + 1$$

Therefore, from Table 8-1,

$$y[n] = 2(-0.5)^n u[n] - (0.8)^n u[n] + \delta[n]$$

Notice again that the time-domain sequence has terms of the form p_k^n. The constant term in the system function generates an impulse, which is nonzero only at $n = 0$ (entry 4 in Table 8-1). ■

8-8 Steady-State Response and Stability

A stable system is one that does not "blow up." This intuitive statement can be formalized by saying that the output of a stable system can always be bounded ($|y[n]| < M_y$) whenever the input is bounded ($|x[n]| < M_x$).[4]

We can use the inverse z-transform method developed in Section 8-7 to demonstrate an important point about stability, the frequency response, and the sinusoidal steady-state response. To illustrate this point, consider the LTI system defined by

$$y[n] = a_1 y[n-1] + b_0 x[n]$$

From our discussion so far, we can state without further analysis that the system function of this system is

$$H(z) = \frac{b_0}{1 - a_1 z^{-1}}$$

and that the impulse response is

$$h[n] = b_0 a_1^n u[n]$$

We can state also that the frequency response is

$$H(e^{j\hat{\omega}}) = H(z)\Big|_{z=e^{j\hat{\omega}}} = \frac{b_0}{1 - a_1 e^{-j\hat{\omega}}}$$

but this is true only if the system is stable ($|a_1| < 1$). The objective of this section is to refine the concept of stability and demonstrate its impact on the response to a sinusoid applied at $n = 0$.

Recall from Section 8-5 and equations (8.31) and (8.33) that the output of this system for a complex exponential input is

$$y[n] = H(e^{j\hat{\omega}_0})e^{j\hat{\omega}_0 n} = \left(\frac{b_0}{1 - a_1 e^{-j\hat{\omega}_0}}\right) e^{j\hat{\omega}_0 n}$$

[4]This definition for stability is called bounded-input, bounded-output stability. The finite constants M_x and M_y can be different.

for $-\infty < n < \infty$. What if the complex exponential input sequence is suddenly applied instead of existing for all n? The z-transform tools that we have developed make it easy to solve this problem. Indeed, the z-transform is ideally suited for situations where the sequences are either finite-length sequences or suddenly applied exponentials. For the suddenly applied complex exponential sequence with frequency $\hat{\omega}_0$

$$x[n] = e^{j\hat{\omega}_0 n}u[n]$$

the z-transform is found from entry 6 of Table 8-1 to be

$$X(z) = \frac{1}{1 - e^{j\hat{\omega}_0}z^{-1}}$$

and the z-transform of the output of the LTI system is

$$\begin{aligned} Y(z) = H(z)X(z) &= \left(\frac{b_0}{1 - a_1 z^{-1}}\right)\left(\frac{1}{1 - e^{j\hat{\omega}_0}z^{-1}}\right) \\ &= \frac{b_0}{(1 - a_1 z^{-1})(1 - e^{j\hat{\omega}_0}z^{-1})} \end{aligned}$$

Using the technique of partial fraction expansion, we can show that

$$Y(z) = \frac{\left(\dfrac{b_0 a_1}{a_1 - e^{j\hat{\omega}_0}}\right)}{1 - a_1 z^{-1}} + \frac{\left(\dfrac{b_0}{1 - a_1 e^{-j\hat{\omega}_0}}\right)}{1 - e^{j\hat{\omega}_0}z^{-1}}$$

Therefore, the output due to the suddenly applied complex exponential sequence is

$$\begin{aligned} y[n] = &\left(\frac{b_0 a_1}{a_1 - e^{j\hat{\omega}_0}}\right)(a_1)^n u[n] \\ &+ \left(\frac{b_0}{1 - a_1 e^{-j\hat{\omega}_0}}\right)e^{j\hat{\omega}_0 n}u[n] \end{aligned} \qquad (8.38)$$

Equation (8.38) shows that the output consists of two terms. One term is proportional to an exponential sequence a_1^n that is solely determined by the pole at $z = a_1$. If $|a_1| < 1$, this term will die out with increasing n, in which case it would be called the ***transient component***. The second term is proportional to the input complex exponential signal, and the constant of proportionality term is $H(e^{j\hat{\omega}_0})$, the frequency response of the system evaluated at the frequency of the suddenly applied complex sinusoid. This complex exponential component is the ***sinusoidal steady-state component*** of the output.

Now we see that the location of the pole of $H(z)$ is crucial if we want the output to reach the sinusoidal steady state. Clearly, if $|a_1| < 1$, then the system is stable and the pole is inside the unit circle. For this condition, the exponential a_1^n dies out and we can state that the limiting value for large n

$$y[n] \rightarrow \left(\frac{b_0}{1 - a_1 e^{-j\hat{\omega}_0}}\right)e^{j\hat{\omega}_0 n} = H(e^{j\hat{\omega}_0})e^{j\hat{\omega}_0 n}$$

Otherwise, if $|a_1| \geq 1$, the term proportional to a_1^n will grow with increasing n and soon dominate the output. The following example gives a specific numerical illustration.

Example 8-12: Transient and Steady-State

If $b_0 = 5$, $a_1 = -0.8$, and $\hat{\omega}_0 = 2\pi/10$, the transient component is

$$\begin{aligned} y_t[n] &= \left(\frac{-4}{-0.8 - e^{j0.2\pi}}\right)(-0.8)^n u[n] \\ &= 2.3351e^{-j0.3502}(-0.8)^n u[n] \\ &= 2.1933(-0.8)^n u[n] - j0.8012(-0.8)^n u[n] \end{aligned}$$

Similarly, the steady-state component is

$$\begin{aligned} y_{ss}[n] &= \left(\frac{5}{1 + 0.8e^{-j0.2\pi}}\right)e^{j0.2\pi n}u[n] \\ &= 2.9188e^{j0.2781}e^{j0.2\pi n}u[n] \\ &= 2.9188\cos(0.2\pi n + 0.2781)\,u[n] \\ &\quad + j\,2.9188\sin(0.2\pi n + 0.2781)\,u[n] \end{aligned}$$

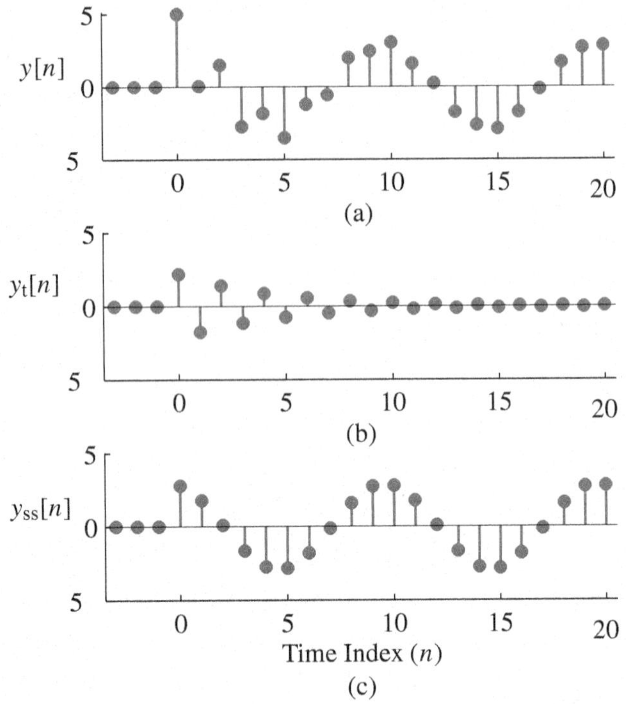

Figure 8-14: Illustration of transient and steady-state responses of an IIR system: (a) Total output $\Re e\{y[n]\}$, (b) decaying transient component $\Re e\{y_t[n]\}$, and (c) steady-state response $\Re e\{y_{ss}[n]\}$.

Figure 8-14 shows the real part of the total output (a), and also the transient component (b), and the steady-state component (c). The signals all start at $n = 0$ when the complex exponential is applied at the input. Note how the transient component oscillates, but dies away, which explains why the steady-state component eventually equals the total output. In Fig. 8-14, $y[n]$ in (a) and $y_{ss}[n]$ in (c) look identical for $n > 15$. ■

On the other hand, if the pole were at $z = 1.1$, the system would be unstable and the output would "blow up"

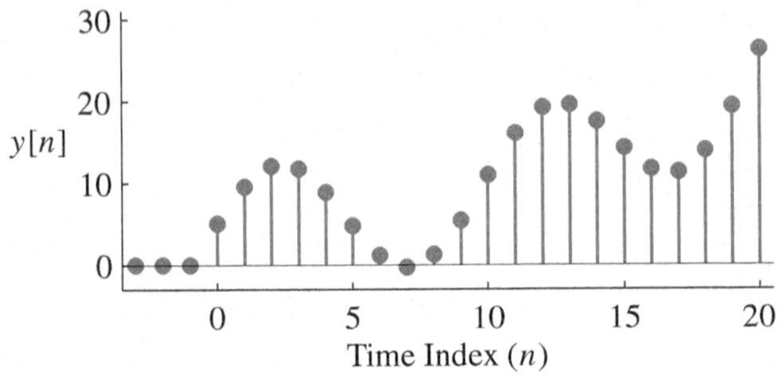

Figure 8-15: Illustration of an unstable IIR system. Pole is at $z = 1.1$.

as shown in Fig. 8-15. In this case, the output contains a term $(1.1)^n$ that eventually dominates and grows without bound.

The result of Example 8-12 can be generalized by observing that the only difference between this example and a system with a higher-order system function is that the total output would include one exponential factor for each pole of the system function as well as the term $H(e^{j\hat{\omega}_0})e^{j\hat{\omega}_0 n}u[n]$. That is, it can be shown that for a suddenly applied exponential input sequence $x[n] = e^{j\hat{\omega}_0 n}u[n]$, the output of an Nth-order IIR $(N > M)$ system will always be of the form

$$y[n] = \sum_{k=1}^{N} A_k(p_k)^n u[n] + H(e^{j\hat{\omega}_0})e^{j\hat{\omega}_0 n}u[n]$$

where the p_ks are the poles of the system function. Therefore, the sinusoidal steady state will exist and dominate in the total response if the poles of the system function all lie strictly inside the unit circle. This makes the concept of frequency response useful in a practical setting where all signals must have a beginning point at some finite time.

8-9 Second-Order Filters

We now turn our attention to filters with two feedback coefficients, a_1 and a_2. The general difference equation (8.1) becomes the second-order difference equation

$$\begin{aligned} y[n] &= a_1 y[n-1] + a_2 y[n-2] \\ &\quad + b_0 x[n] + b_1 x[n-1] + b_2 x[n-2] \end{aligned} \tag{8.39}$$

As before, we can characterize the second-order filter (8.39) in each of the three domains: time domain, frequency domain, and z-domain. We start with the z-transform domain because by now we have demonstrated that the poles and zeros of the system function give a great deal of insight into most aspects of both the time and frequency responses.

DEMO: *PeZ GUI*

8-9.1 z-Transform of Second-Order Filters

Using the approach followed in Section 8-3.1 for the first-order case, we can take the z-transform of the second-order difference equation (8.39) by replacing each delay with z^{-1} (second entry in Table 8-1), and also replacing the input and output signals with their z-transforms:

$$\begin{aligned} Y(z) &= a_1 z^{-1} Y(z) + a_2 z^{-2} Y(z) \\ &\quad + b_0 X(z) + b_1 z^{-1} X(z) + b_2 z^{-2} X(z) \end{aligned}$$

In the z-transform domain, the input-output relationship is $Y(z) = H(z)X(z)$, so we can solve for $H(z)$ by finding $Y(z)/X(z)$. For the second-order filter we get

$$\begin{aligned} Y(z) - a_1 z^{-1} Y(z) - a_2 z^{-2} Y(z) \\ = b_0 X(z) + b_1 z^{-1} X(z) + b_2 z^{-2} X(z) \\ (1 - a_1 z^{-1} - a_2 z^{-2}) Y(z) = (b_0 + b_1 z^{-1} + b_2 z^{-2}) X(z) \end{aligned}$$

which can be solved for $H(z)$ as

$$H(z) = \frac{Y(z)}{X(z)} = \frac{b_0 + b_1 z^{-1} + b_2 z^{-2}}{1 - a_1 z^{-1} - a_2 z^{-2}} \tag{8.40}$$

Thus, the system function $H(z)$ for an IIR filter is a ratio of two second-degree polynomials, where the numerator polynomial depends on the feed-forward coefficients $\{b_k\}$ and the denominator depends on the feedback coefficients $\{a_\ell\}$. It should be possible to work problems such as Exercise 8.13 by simply reading the filter coefficients from the difference equation and then substituting them directly into the z-transform expression for $H(z)$.

EXERCISE 8.13: Find system function $H(z)$ of the following IIR filter:

$$\begin{aligned} y[n] &= 0.5 y[n-1] + 0.3 y[n-2] \\ &\quad - x[n] + 3x[n-1] - 2x[n-2] \end{aligned}$$

Conversely, given the system function $H(z)$, it is a simple matter to write down the difference equation.

EXERCISE 8.14: For the system function

$$H(z) = \frac{1 + 2z^{-1} + z^{-2}}{1 - 0.8z^{-1} + 0.64z^{-2}}$$

write down the difference equation that relates the input $x[n]$ to the output $y[n]$.

Example 8-13: Structure for $H(z)$

The connection between $H(z)$ and the difference equation can be generalized to higher-order filters. If we are given a fourth-order system

$$H(z) = \frac{1 - 3z^{-2}}{1 - 0.8z^{-1} + 0.6z^{-3} + 0.3z^{-4}}$$

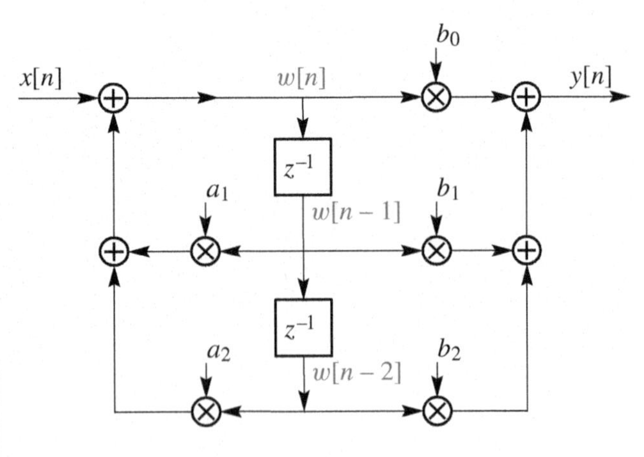

Figure 8-16: Direct Form II (DF-II), an alternative computational structure for the second-order recursive filter.

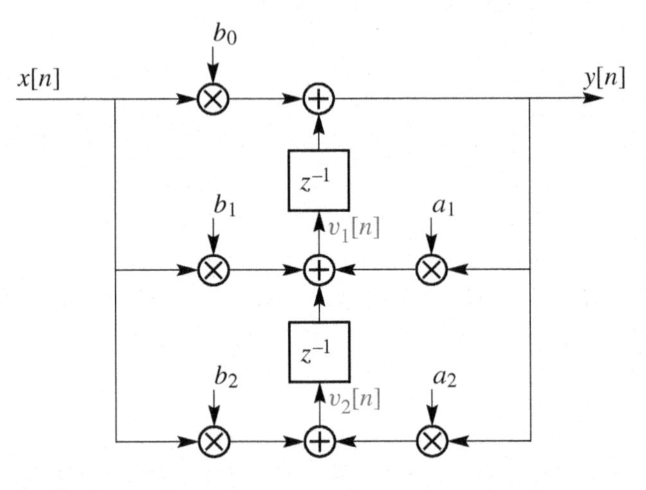

Figure 8-17: Transposed Direct Form II (TDF-II), another computational structure for the second-order recursive filter.

the corresponding difference equation is

$$\begin{aligned} y[n] &= 0.8y[n-1] - 0.6y[n-3] \\ &\quad - 0.3y[n-4] + x[n] - 3x[n-2] \end{aligned}$$

As before, note the sign change in the feedback coefficients, $\{a_k\}$. ■

8-9.2 Structures for Second-Order IIR Systems

The difference equation (8.39) can be interpreted as an algorithm for computing the output sequence from the input. Other computational orderings are possible, and the z-transform has the power to derive alternative structures through polynomial manipulations. Two alternative computational orderings that will implement the system defined by $H(z)$ in (8.40) are given in Figs. 8-16 and 8-17.

In order to verify that the block diagram in Fig. 8-16 has the correct system function, we need to write the equations of the structure at the adders, and then eliminate the internal variable(s). For the case of the Direct Form II in Fig. 8-16, the equations at the output of the summing nodes are

$$\begin{aligned} y[n] &= b_0 w[n] + b_1 w[n-1] + b_2 w[n-2] \\ w[n] &= x[n] + a_1 w[n-1] + a_2 w[n-2] \end{aligned} \tag{8.41}$$

It is impossible to eliminate $w[n]$ in these two equations, unless we work in the z-transform domain. The corresponding z-transform equations are

$$\begin{aligned} Y(z) &= b_0 W(z) + b_1 z^{-1} W(z) + b_2 z^{-2} W(z) \\ W(z) &= X(z) + a_1 z^{-1} W(z) + a_2 z^{-2} W(z) \end{aligned}$$

which can be rearranged into the form

$$\begin{aligned} Y(z) &= (b_0 + b_1 z^{-1} + b_2 z^{-2}) W(z) \\ X(z) &= (1 - a_1 z^{-1} - a_2 z^{-2}) W(z) \end{aligned}$$

Since the system function $H(z)$ is the ratio of $Y(z)$ to $X(z)$, we get

$$H(z) = \frac{Y(z)}{X(z)} = \frac{b_0 + b_1 z^{-1} + b_2 z^{-2}}{1 - a_1 z^{-1} - a_2 z^{-2}}$$

Thus we have proved that the Direct Form II (DF-II) structure in Fig. 8-16 which implements the pair of difference equations (8.41) is identical to the system defined by the single difference equation (8.39).

The Transposed Direct Form II (TDF-II) in Fig. 8-17 can be worked out similarly. The difference equations represented by the block diagram are

$$\begin{aligned} y[n] &= b_0 x[n] + v_1[n-1] \\ v_1[n] &= b_1 x[n] + a_1 y[n] + v_2[n-1] \\ v_2[n] &= b_2 x[n] + a_2 y[n] \end{aligned} \qquad (8.42)$$

Taking the z-transform of each of the three equations gives

$$\begin{aligned} Y(z) &= b_0 X(z) + z^{-1} V_1(z) \\ V_1(z) &= b_1 X(z) + a_1 Y(z) + z^{-1} V_2(z) \\ V_2(z) &= b_2 X(z) + a_2 Y(z) \end{aligned} \qquad (8.43)$$

Using these equations, we can eliminate $V_1(z)$ and $V_2(z)$ as follows:

$$\begin{aligned} Y(z) &= b_0 X(z) + z^{-1}(b_1 X(z) + a_1 Y(z) + z^{-1} V_2(z)) \\ Y(z) &= b_0 X(z) + z^{-1}(b_1 X(z) + a_1 Y(z) \\ &\qquad + z^{-1}(b_2 X(z) + a_2 Y(z))) \end{aligned}$$

Moving all the $X(z)$ terms to the right-hand side and the $Y(z)$ terms to the left-hand side gives

$$(1 - a_1 z^{-1} - a_2 z^{-2}) Y(z) = (b_0 + b_1 z^{-1} + b_2 z^{-2}) X(z)$$

so we get by division

$$H(z) = \frac{Y(z)}{X(z)} = \frac{b_0 + b_1 z^{-1} + b_2 z^{-2}}{1 - a_1 z^{-1} - a_2 z^{-2}}$$

Thus we have shown that the Transposed Direct Form II (TDF-II) is equivalent to the system function for the basic Direct Form-I difference equation in (8.39). Both examples illustrate the power of the z-transform approach in manipulating polynomials that correspond to different structures.

In theory, the system with system function given by (8.40) can be implemented by iterating any of the equations (8.39), (8.41), or (8.42). For example, the MATLAB function `filter` uses the TDF-II structure. However, as mentioned before, the reason for having different block diagram structures is that the order of calculation defined by equations (8.39), (8.41), and (8.42) is different. In a hardware implementation, the different structures will behave differently, especially when using fixed-point arithmetic where rounding error is fed back into the structure. With double-precision floating-point arithmetic as in MATLAB, there is little difference.

EXERCISE 8.15: Draw the block diagram of the Direct Form I difference equation defined by (8.39), and compare it to the other block diagrams in Figs. 8-16 and 8-17.

8-9.3 Poles and Zeros

Finding the poles and zeros of $H(z)$ is less confusing if we rewrite the polynomials as functions of z rather than z^{-1}. Thus, the general second-order rational z-transform would become

$$H(z) = \frac{b_0 + b_1 z^{-1} + b_2 z^{-2}}{1 - a_1 z^{-1} - a_2 z^{-2}} = \frac{b_0 z^2 + b_1 z + b_2}{z^2 - a_1 z - a_2}$$

after multiplying the numerator and denominator by z^2. Recall from algebra the following important property of polynomials:

> ***Property Of Real Polynomials***
>
> *A polynomial of degree N has N roots.*
> *If all the coefficients of the polynomial are real, the roots either must be real or must occur in* ***complex conjugate pairs.***

Therefore, in the second-order case, the numerator and denominator polynomials each have two roots, and there are two possibilities: Either the roots are complex conjugates of each other, or they are both real. We will now concentrate on the roots of the denominator, but exactly the same results hold for the numerator. From the quadratic formula, we get two poles at

$$\frac{a_1 \pm \sqrt{a_1^2 + 4a_2}}{2}$$

When $a_1^2 + 4a_2 \geq 0$, both poles are real; when $a_1^2 + 4a_2 = 0$, they are real and equal. However, when $a_1^2 + 4a_2 < 0$, the square root produces an imaginary result, and we have complex-conjugate poles with values

$$p_1 = \tfrac{1}{2}a_1 + j\tfrac{1}{2}\sqrt{-a_1^2 - 4a_2}$$

$$\text{and} \quad p_2 = \tfrac{1}{2}a_1 - j\tfrac{1}{2}\sqrt{-a_1^2 - 4a_2}$$

In polar form, the complex poles can be expressed as $p_1 = re^{j\theta}$ and $p_2 = re^{-j\theta}$, where the radius r is

$$r = \sqrt{(\tfrac{1}{2}a_1)^2 + \tfrac{1}{4}(-a_1^2 - 4a_2)}$$

$$= \sqrt{\tfrac{1}{4}a_1^2 - \tfrac{1}{4}a_1^2 - a_2} = \sqrt{-a_2}$$

and the angle θ satisfies

$$r\cos\theta = \tfrac{1}{2}a_1 \quad \Longrightarrow \quad \theta = \cos^{-1}\left(\frac{a_1}{2\sqrt{-a_2}}\right)$$

Example 8-14: Complex Poles

The following $H(z)$ has two poles and two zeros.

$$H(z) = \frac{2 + 2z^{-1}}{1 - z^{-1} + z^{-2}} = 2\frac{z^2 + z}{z^2 - z + 1}$$

The poles $\{p_1, p_2\}$ and zeros $\{z_1, z_2\}$ are

$$p_1 = \tfrac{1}{2} + j\tfrac{1}{2}\sqrt{3} = e^{j\pi/3}$$

$$p_2 = \tfrac{1}{2} - j\tfrac{1}{2}\sqrt{3} = e^{-j\pi/3}$$

$$z_1 = 0$$

$$z_2 = -1$$

The system function can be written in factored form as either of the two forms

$$H(z) = \frac{2z(z+1)}{(z - e^{j\pi/3})(z - e^{-j\pi/3})}$$

$$= \frac{2(1 + z^{-1})}{(1 - e^{j\pi/3}z^{-1})(1 - e^{-j\pi/3}z^{-1})}$$

Since the numerator has no z^{-2} term, we have one zero at the origin. As is our custom, we plot these locations in the z-plane and mark the pole locations with **x** and the zeros with **o**. See Fig. 8-18. ■

8-9.4 Impulse Response of a Second-Order IIR System

We have derived the general z-transform system function for the second-order filter

$$H(z) = \frac{B(z)}{A(z)} = \frac{b_0 + b_1 z^{-1} + b_2 z^{-2}}{1 - a_1 z^{-1} - a_2 z^{-2}} \tag{8.44}$$

and we have seen that the denominator polynomial $A(z)$ has two roots that define the poles of the second-order filter. Expressing $H(z)$ in terms of the poles gives

$$H(z) = \frac{b_0 + b_1 z^{-1} + b_2 z^{-2}}{(1 - p_1 z^{-1})(1 - p_2 z^{-1})} \tag{8.45}$$

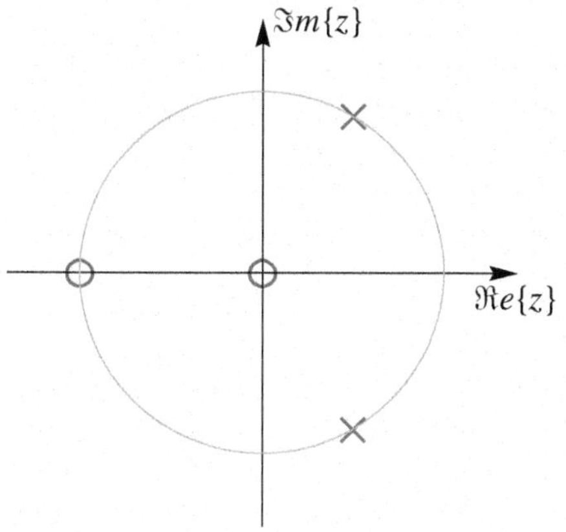

Figure 8-18: Pole-zero plot for system with $H(z) = \dfrac{2+2z^{-1}}{1-z^{-1}+z^{-2}}$. The unit circle is shown for reference.

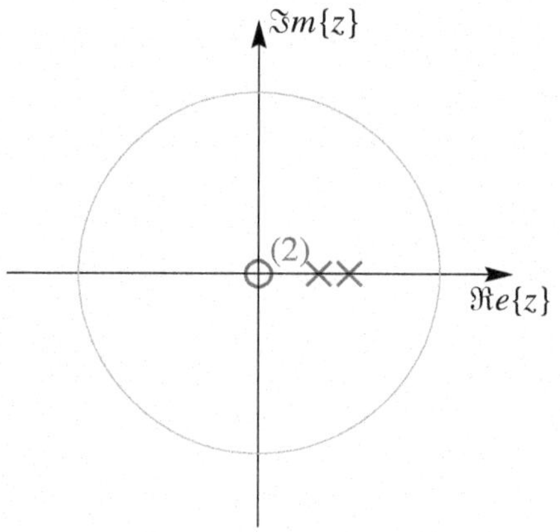

Figure 8-19: Pole-zero plot for system of Example 8-15. The poles are at $z = \frac{1}{2}$ and $z = \frac{1}{3}$; there are two zeros at $z = 0$.

Using the partial fraction expansion technique developed in Section 8-7, we can express the system function (8.44) as

$$H(z) = (-b_2/a_2) + \frac{A_1}{1 - p_1 z^{-1}} + \frac{A_2}{1 - p_2 z^{-1}}$$

where A_1 and A_2 can be evaluated by $A_k = H(z)(1 - p_k z^{-1})|_{z=p_k}$. Therefore, the impulse response will have the form

$$h[n] = (-b_2/a_2)\delta[n] + A_1(p_1)^n u[n] + A_2(p_2)^n u[n]$$

Furthermore, the poles may both be real or they may be a pair of complex conjugates. We will examine these two cases separately.

8-9.4.1 Real Poles

If p_1 and p_2 are real, the impulse response is composed of two real exponentials of the form p_k^n. This case is illustrated by the following example:

Example 8-15: Second-Order: Real Poles

Assume that

$$H(z) = \frac{1}{1 - \frac{5}{6}z^{-1} + \frac{1}{6}z^{-2}} = \frac{1}{(1 - \frac{1}{2}z^{-1})(1 - \frac{1}{3}z^{-1})} \tag{8.46}$$

from which we see that the poles are at $z = \frac{1}{2}$ and $z = \frac{1}{3}$ and that there are two zeros at $z = 0$. The poles and zeros of $H(z)$ are plotted in Fig. 8-19. We can extract the filter coefficients from $H(z)$ and write the following difference equation

$$y[n] = \tfrac{5}{6}y[n-1] - \tfrac{1}{6}y[n-2] + x[n] \tag{8.47}$$

which must be satisfied for any input and its corresponding output. Specifically, the impulse response would satisfy the difference equation

$$h[n] = \tfrac{5}{6}h[n-1] - \tfrac{1}{6}h[n-2] + \delta[n] \tag{8.48}$$

which can be iterated to compute $h[n]$ if we know the values of $h[-1]$ and $h[-2]$, i.e., the values of the impulse response sequence just prior to $n = 0$ where the impulse first becomes nonzero. These values are supplied by the initial rest condition, which means that $h[-1] = 0$ and $h[-2] = 0$. The following table shows the computation of a few values of the impulse response:

n	$n < 0$	0	1	2	3	4	...
$x[n]$	0	1	0	0	0	0	...
$h[n-2]$	0	0	0	1	$\frac{5}{6}$	$\frac{19}{36}$	...
$h[n-1]$	0	0	1	$\frac{5}{6}$	$\frac{19}{36}$	$\frac{65}{216}$	...
$h[n]$	0	1	$\frac{5}{6}$	$\frac{19}{36}$	$\frac{65}{216}$	$\frac{211}{1296}$	...

In contrast to the simpler first-order case, it is very difficult to guess the general n^{th} term for the impulse response sequence. Fortunately, we can rely on the inverse z-transform technique to give us the general formula. Applying the partial fraction expansion to (8.46), we get

$$H(z) = \frac{3}{1 - \frac{1}{2}z^{-1}} - \frac{2}{1 - \frac{1}{3}z^{-1}}$$

which implies that

$$\begin{aligned} h[n] &= 3(\tfrac{1}{2})^n u[n] - 2(\tfrac{1}{3})^n u[n] \\ &= \begin{cases} 3(\frac{1}{2})^n - 2(\frac{1}{3})^n & \text{for } n \geq 0 \\ 0 & \text{for } n < 0 \end{cases} \end{aligned}$$

Since both poles are inside the unit circle, the impulse response dies out for n large; i.e., the system is stable. ■

EXERCISE 8.16: Find the impulse response of the following second-order system:

$$y[n] = \tfrac{1}{4}y[n-2] + 5x[n] - 4x[n-1]$$

Plot the resulting signal $h[n]$ versus n.

8-9.5 Complex Poles

Now let us assume that the coefficients a_1 and a_2 in the second-order difference equation are such that the poles of $H(z)$ are complex. If we express the poles in polar form

$$p_1 = r\,e^{j\theta} \qquad \text{and} \qquad p_2 = r\,e^{-j\theta} = p_1^*$$

it is convenient to rewrite the denominator polynomial in terms of the parameters r and θ. Basic algebra allows us to start from the factored form and derive the polynomial coefficients:

$$\begin{aligned} A(z) &= \left(1 - p_1 z^{-1}\right)\left(1 - p_2 z^{-1}\right) \\ &= \left(1 - r\,e^{j\theta} z^{-1}\right)\left(1 - r\,e^{-j\theta} z^{-1}\right) \\ &= 1 - \left(r\,e^{j\theta} + r\,e^{-j\theta}\right) z^{-1} + r^2 z^{-2} \\ &= 1 - (2r\cos\theta)\,z^{-1} + r^2 z^{-2} \end{aligned} \tag{8.49}$$

The system function is therefore

$$\begin{aligned} H(z) &= \frac{b_0 + b_1 z^{-1} + b_2 z^{-2}}{(1 - re^{j\theta}z^{-1})(1 - re^{-j\theta}z^{-1})} \\ &= \frac{b_0 + b_1 z^{-1} + b_2 z^{-2}}{1 - 2r\cos\theta z^{-1} + r^2 z^{-2}} \end{aligned} \tag{8.50}$$

We can also identify the two feedback filter coefficients as

$$a_1 = 2r\cos\theta \qquad \text{and} \qquad a_2 = -r^2 \tag{8.51}$$

so the corresponding difference equation is

$$\begin{aligned} y[n] &= (2r\cos\theta)y[n-1] - r^2 y[n-2] \\ &\quad + b_0 x[n] + b_1 x[n-1] + b_2 x[n-2] \end{aligned} \tag{8.52}$$

This parameterization is significant because it allows us to see directly how the poles define the feedback terms of the difference equation (8.52). For example, if we want to change the angle of the pole, then we vary the coefficient a_1. Finally, we must remember that (8.51) is valid only for the special case of complex-conjugate poles; when the poles (p_1, p_2) are both real, the filter coefficients are

$$a_1 = p_1 + p_2 \qquad \text{and} \qquad a_2 = p_1 p_2$$

Example 8-16: Invert Complex Poles

Consider the following system

$$y[n] = y[n-1] - y[n-2] + 2x[n] + 2x[n-1]$$

whose system function is

$$\begin{aligned} H(z) &= \frac{2 + 2z^{-1}}{1 - z^{-1} + z^{-2}} \\ &= \frac{2(1 + z^{-1})}{(1 - e^{j\pi/3}z^{-1})(1 - e^{-j\pi/3}z^{-1})} \end{aligned} \tag{8.53}$$

A pole-zero plot for $H(z)$ was already given in Fig. 8-18. Using the partial fraction expansion technique, we can write $H(z)$ in the form

$$\begin{aligned} H(z) &= \frac{\left(\dfrac{2 + 2e^{-j\pi/3}}{1 - e^{-j2\pi/3}}\right)}{1 - e^{j\pi/3}z^{-1}} + \frac{\left(\dfrac{2 + 2e^{j\pi/3}}{1 - e^{j2\pi/3}}\right)}{1 - e^{-j\pi/3}z^{-1}} \\ &= \frac{2e^{-j\pi/3}}{1 - e^{j\pi/3}z^{-1}} + \frac{2e^{j\pi/3}}{1 - e^{-j\pi/3}z^{-1}} \end{aligned}$$

$$\begin{aligned} \text{so } h[n] &= 2e^{-j\pi/3}e^{j(\pi/3)n}u[n] + 2e^{j\pi/3}e^{-j(\pi/3)n}u[n] \\ &= 4\cos\left(2\pi(\tfrac{1}{6})(n-1)\right)u[n] \end{aligned}$$

The two complex exponentials with frequencies $\pm\pi/3$ combine to form the cosine. The impulse response is plotted in Fig. 8-20. ■

An important observation about the system in Example 8-16 is that it produces a pure sinusoid when stimulated by an impulse. Such a system is an example of

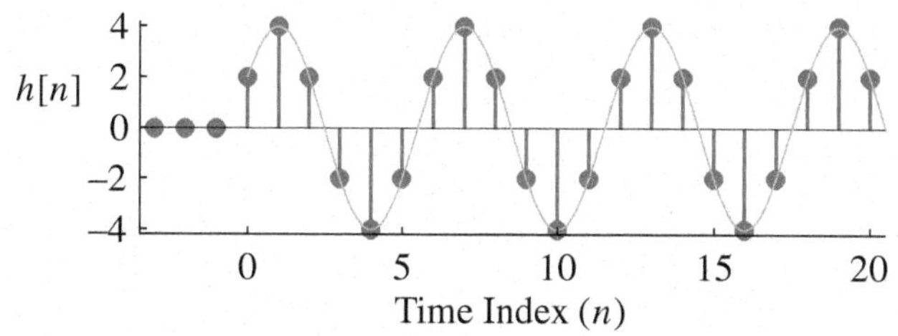

Figure 8-20: Impulse response for system with $A(z) = 1 - z^{-1} + z^{-2}$.

a *sine wave oscillator*. After being stimulated by the single input sample from the impulse, the system continues indefinitely to produce a sinusoid of frequency $\hat{\omega}_0 = 2\pi(\frac{1}{6})$. This frequency is equal to the angle of the poles. A first-order filter (or a filter with all real poles) can only decay (or grow) as $(p)^n$ or oscillate up and down as $(-1)^n$, but a second-order system can oscillate with different periods. This is important when modeling physical signals such as speech, music, or other sounds.

Note that in order to produce the continuing sinusoidal output, the system must have its poles on the unit circle[5] of the z-plane, i.e., $r = 1$. Also note that the angle of the poles is exactly equal to the radian frequency of the sinusoidal output. Thus, we can control the frequency of the sinusoidal oscillator by adjusting the a_1 coefficient of the difference equation (8.52) while leaving a_2 fixed at $a_2 = -1$.

Example 8-17: Poles on Unit Circle

As an example of an oscillator with a different frequency, we can use (8.52) to define a difference equation with prescribed pole locations. If we take

[5]Strictly speaking, a system with poles on the unit circle is unstable, so for some inputs it may blow up, but not for the impulse input.

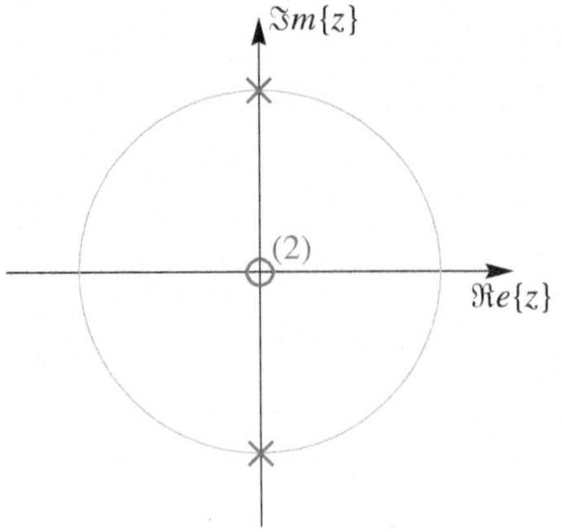

Figure 8-21: Pole-zero plot for system with $H(z) = \dfrac{1}{1+z^{-2}}$. The unit circle is shown for reference.

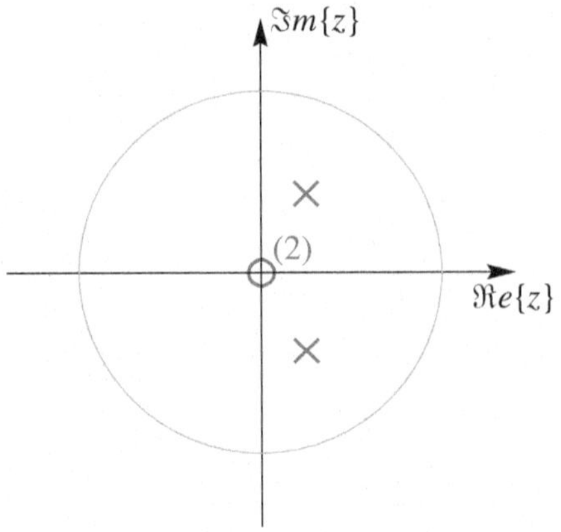

Figure 8-22: Pole-zero plot for system with $H(z) = \dfrac{1}{1-\frac{1}{2}z^{-1}+\frac{1}{4}z^{-2}}$. The unit circle is shown for reference.

$r = 1$ and $\theta = \pi/2$, as shown in Fig. 8-21, we get $a_1 = 2r\cos\theta = 0$ and $a_2 = -r^2 = -1$.

$$y[n] = -y[n-2] + x[n] \tag{8.54}$$

The system function of this system is

$$\begin{aligned} H(z) &= \frac{1}{1+z^{-2}} \\ &= \frac{1}{(1-e^{j\pi/2}z^{-1})(1-e^{-j\pi/2}z^{-1})} \\ &= \frac{\frac{1}{2}}{1-e^{j\pi/2}z^{-1}} + \frac{\frac{1}{2}}{1-e^{-j\pi/2}z^{-1}} \end{aligned}$$

The inverse z-transform gives a general formula for $h[n]$:

$$\begin{aligned} h[n] &= \tfrac{1}{2}e^{j(\pi/2)n}u[n] + \tfrac{1}{2}e^{-j(\pi/2)n}u[n] \\ &= \begin{cases} \cos(2\pi\left(\frac{1}{4}\right)n) & \text{for } n \geq 0 \\ 0 & \text{for } n < 0 \end{cases} \end{aligned} \tag{8.55}$$

Once again, the frequency of the cosine term in the impulse response is equal to the angle of the pole, $\pi/2 = 2\pi(\frac{1}{4})$. ■

If the complex conjugate poles of the second-order system lie on the unit circle, the output oscillates sinusoidally and does not decay to zero. If the poles lie outside the unit circle, the output grows exponentially, whereas if they are inside the unit circle, the output decays exponentially to zero.

Example 8-18: Stable Complex Poles

As an example of a stable system, if we take $r = \frac{1}{2}$ and $\theta = \pi/3$, as shown in Fig. 8-22, we get $a_1 = 2r\cos\theta = 2(\frac{1}{2})(\frac{1}{2}) = \frac{1}{2}$ and $a_2 = -r^2 = -(\frac{1}{2})^2 = -\frac{1}{4}$, and the difference equation (8.52) becomes

$$y[n] = \tfrac{1}{2}y[n-1] - \tfrac{1}{4}y[n-2] + x[n] \tag{8.56}$$

The system function of this system is

$$\begin{aligned} H(z) &= \frac{1}{1-\frac{1}{2}z^{-1}+\frac{1}{4}z^{-2}} \\ &= \frac{\frac{1}{\sqrt{3}}e^{-j\pi/6}}{1-\frac{1}{2}e^{j\pi/3}z^{-1}} + \frac{\frac{1}{\sqrt{3}}e^{j\pi/6}}{1-\frac{1}{2}e^{-j\pi/3}z^{-1}} \end{aligned}$$

and the general formula for $h[n]$ is

$$h[n] = \frac{2}{\sqrt{3}}(\tfrac{1}{2})^n \cos\left(2\pi(\tfrac{1}{6})n - \tfrac{\pi}{6}\right) u[n] \qquad (8.57)$$

In this case, the general formula for $h[n]$ has a decay of $(\frac{1}{2})^n$ multiplying a periodic cosine with period 6. The frequency of the cosine term in the impulse response (8.57) is again the angle of the pole, $\pi/3 = 2\pi/6$; while the decaying term is controlled by the radius of the pole, i.e., $r^n = (\frac{1}{2})^n$. ■

DEMO: *IIR Filtering*

8-10 Frequency Response of Second-Order IIR Filter

Since the frequency response of a stable system is related to the z-transform by

$$H(e^{j\hat{\omega}}) = H(z)\Big|_{z=e^{j\hat{\omega}}}$$

we get the following formula for the frequency response of a second-order system:

$$H(e^{j\hat{\omega}}) = \frac{b_0 + b_1 e^{-j\hat{\omega}} + b_2 e^{-j2\hat{\omega}}}{1 - a_1 e^{-j\hat{\omega}} - a_2 e^{-j2\hat{\omega}}} \qquad (8.58)$$

Since (8.58) contains terms like $e^{-j\hat{\omega}}$ and $e^{-j2\hat{\omega}}$, $H(e^{j\hat{\omega}})$ is guaranteed to be a periodic function with a period of 2π.

The magnitude squared of the frequency response can be formed by multiplying the complex $H(e^{j\hat{\omega}})$ by its conjugate (denoted by $H^*(e^{j\hat{\omega}})$). Rather than work out a general formula, we take a specific numerical example to show the kind of formula that results.

Example 8-19: Frequency Response of Second-Order

Consider the case where the system function is

$$H(z) = \frac{1 - z^{-2}}{1 - 0.9z^{-1} + 0.81z^{-2}}$$

The magnitude squared is derived by multiplying out all the terms in the numerator and denominator of $H(e^{j\hat{\omega}})H^*(e^{j\hat{\omega}})$, and then collecting terms where the inverse Euler formula applies.

$$\begin{aligned}
|H(e^{j\hat{\omega}})|^2 &= H(e^{j\hat{\omega}})H^*(e^{j\hat{\omega}}) \\
&= \frac{1 - e^{-j2\hat{\omega}}}{1 - 0.9e^{-j\hat{\omega}} + 0.81e^{-j2\hat{\omega}}} \cdot \frac{1 - e^{j2\hat{\omega}}}{1 - 0.9e^{j\hat{\omega}} + 0.81e^{j2\hat{\omega}}} \\
&= \frac{2 + 2\cos(2\hat{\omega})}{2.4661 - 3.258\cos\hat{\omega} + 1.62\cos(2\hat{\omega})}
\end{aligned}$$

This formula is useful because it is expressed completely in terms of cosine functions. The procedure is general, so a similar formula could be derived for any IIR filter. Since the cosine is an even function, we can state that any magnitude-squared function $|H(e^{j\hat{\omega}})|^2$ will always be even; i.e.,

$$|H(e^{-j\hat{\omega}})|^2 = |H(e^{j\hat{\omega}})|^2$$

The phase response is a bit messier. If arctangents are used to extract the angle of the numerator and denominator, then the two phases must be subtracted. The filter coefficients in this example are real, so the phase is

$$\begin{aligned}
\phi(\hat{\omega}) = &\tan^{-1}\left(\frac{\sin(2\hat{\omega})}{1 - \cos(2\hat{\omega})}\right) \\
&- \tan^{-1}\left(\frac{0.9\sin\hat{\omega} - 0.81\sin(2\hat{\omega})}{1 - 0.9\cos\hat{\omega} + 0.81\cos(2\hat{\omega})}\right)
\end{aligned}$$

which is an odd function of $\hat{\omega}$. ■

The formulas obtained in this example are too complicated to provide much insight directly. In a later section we will see how to use the poles and zeros of the system function to construct an approximate plot of the frequency response without recourse to such formulas.

8-10.1 Frequency Response via MATLAB

Tedious calculation and plotting of $H(e^{j\hat{\omega}})$ by hand is usually unnecessary if a computer program such as MATLAB is available. The MATLAB function `freqz` is provided for just that purpose. The frequency response can be evaluated over a grid in the $\hat{\omega}$ domain, and then its magnitude and phase can be plotted. In MATLAB, the functions `abs` and `angle` will extract the magnitude and the angle of each element in a complex vector.

Example 8-20: MATLAB for $H(e^{j\hat{\omega}})$

Consider the system introduced in Example 8-19:

$$y[n] = 0.9y[n-1] - 0.81y[n-2] + x[n] - x[n-2]$$

In order to define the filter coefficients in MATLAB, we put all the terms with $y[n]$ on one side of the equation, and the terms with $x[n]$ on the other.

$$y[n] - 0.9y[n-1] + 0.81y[n-2] = x[n] - x[n-2]$$

Then we read off the filter coefficients and define the vectors `aa` and `bb`.

```
aa = [ 1, -0.9, 0.81 ]
bb = [ 1, 0, -1 ]
```

The following call to `freqz` will generate a vector `HH` containing the values of the frequency response at the vector of frequencies specified by the third argument, `[-pi:(pi/100):pi]`.

```
HH = freqz(bb, aa, [-pi:(pi/100):pi])
```

A plot of the resulting magnitude and phase is shown in Fig. 8-23. Since $H(e^{j\hat{\omega}})$ is always periodic with a period of 2π, it is sufficient to make the frequency response plot over the range $-\pi \le \hat{\omega} \le \pi$.

For this example, we can look for a connection between the poles and zeros and the shape of the frequency response. For this $H(z)$ we have

$$H(z) = \frac{1 - z^{-2}}{1 - 0.9z^{-1} + 0.81z^{-2}}$$

which has its poles at $z = 0.9e^{\pm j\pi/3}$ and its zeros at $z = 1$ and $z = -1$. Since $z = -1$ is the same as $z = e^{j\pi}$, we conclude that $H(e^{j\hat{\omega}})$ is zero at $\hat{\omega} = \pi$, because $H(z) = 0$ at $z = -1$; likewise, the zero of $H(z)$ at $z = +1$ is a zero of $H(e^{j\hat{\omega}})$ at $\hat{\omega} = 0$. The poles have angles of $\pm\pi/3$ rad, so the poles have an effect on the frequency response near $\hat{\omega} = \pm\pi/3$. Since $H(z)$ is infinite at $z = 0.9e^{\pm j\pi/3}$, the nearby points on the unit circle (at $z = e^{\pm j\pi/3}$) must have large values. In this case, we can evaluate the frequency response directly from the formula to get

$$H(e^{j\hat{\omega}})\Big|_{\hat{\omega}=\pi/3} = H(z)\Big|_{z=e^{j\pi/3}}$$

$$= \left.\frac{1 - z^{-2}}{1 - 0.9z^{-1} + 0.81z^{-2}}\right|_{z=e^{j\pi/3}}$$

$$= \frac{1 - (-\frac{1}{2} - j\frac{1}{2}\sqrt{3})}{1 - 0.9(\frac{1}{2} - j\frac{1}{2}\sqrt{3}) + 0.81(-\frac{1}{2} - j\frac{1}{2}\sqrt{3})}$$

$$= \frac{|1.5 + j0.5(\sqrt{3})|}{|0.145 + j0.045(\sqrt{3})|} = 10.522$$

This value of the frequency response magnitude is a good approximation to the true maximum value, which actually occurs at $\hat{\omega} = 0.334\pi$. ■

8-10.2 3-dB Bandwidth

The width of the peak of the frequency response in Fig. 8-23 is called the ***bandwidth***. It must be measured at some standard point on the plot of $|H(e^{j\hat{\omega}})|$. The most common practice is to use the 3-dB width,[6] which is calculated as follows:

[6] The term 3-dB means -3 decibels which is $20\log_{10}(1/\sqrt{2})$. It is common practice to plot the $|H(e^{j\hat{\omega}})|$ on a dB scale.

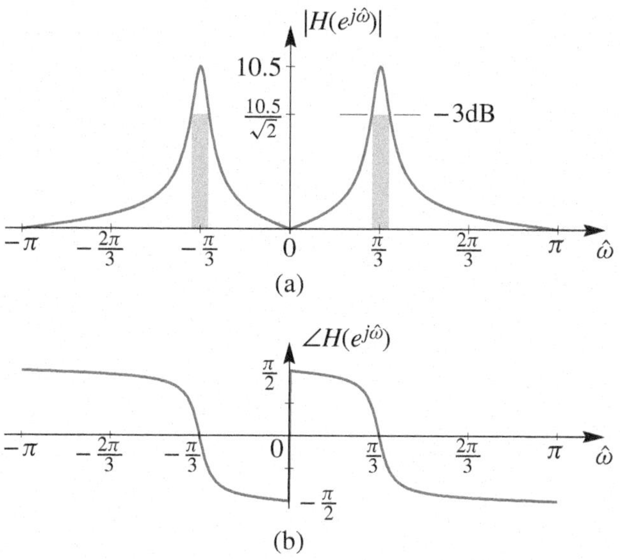

Figure 8-23: Frequency response (magnitude and phase) for a second-order feedback filter. The poles are at $z = 0.9e^{\pm j\pi/3}$ and the numerator has zeros at $z = 1$ and $z = -1$. The shaded region shows the 3-dB bandwidth around the peak at $\hat{\omega} = \pi/3$.

> Determine the peak value for $|H(e^{j\hat{\omega}})|$ and then find the nearest frequency on each side of the peak where the value of the frequency response is $(1/\sqrt{2})H_{\text{peak}}$. The ***3-dB width*** is the difference $\Delta\hat{\omega}$ between these two frequencies.

In Fig. 8-23, the true peak value is 10.526 at $\hat{\omega} = 0.334\pi$, so we look for points where $|H(e^{j\hat{\omega}})| = (1/\sqrt{2})H_{\text{peak}} = (0.707)(10.526) = 7.443$. These occur at $\hat{\omega} = 0.302\pi$ and $\hat{\omega} = 0.369\pi$, so the bandwidth is $\Delta\hat{\omega} = 0.067\pi = 2\pi(0.0335) = 0.2105$ rad.

The 3-dB bandwidth calculation can be carried out efficiently with a computer program, but it is also helpful to have an approximate formula that can give quick "back-of-the-envelope" calculations. An excellent approximation for the second-order case with narrow peaks is given by the formula

$$\Delta\hat{\omega} \approx 2\,\frac{|1-r|}{\sqrt{r}} \tag{8.59}$$

which shows that the distance of the pole from the unit circle $|1-r|$ controls the bandwidth.[7] In Fig. 8-23, the bandwidth (8.59) evaluates to

$$\Delta\hat{\omega} = 2\,\frac{(1-0.9)}{0.95} = \frac{0.2}{0.95} \approx 0.2108 \text{ rad}$$

Thus we see that the approximation is quite good in this case, where the pole is rather close to the unit circle (radius = 0.9).

8-10.3 Three-Dimensional Plot of System Functions

Since the frequency response $H(e^{j\hat{\omega}})$ is the system function evaluated on the unit circle, we can illustrate the connection between the z and $\hat{\omega}$ domains with a three-dimensional plot such as the one shown in Fig. 8-24.

Figure 8-24 shows a plot of the system function $H(z)$ at points inside, outside, and on the unit circle. The peaks located at the poles, $0.85e^{\pm j\pi/2}$, determine the frequency response behavior near $\hat{\omega} = \pm\pi/2$. If the poles were moved closer to the unit circle, the frequency response would have a higher and narrower peak. The zeros at $z = \pm 1$ create valleys that lie on the unit circle at $\hat{\omega} = 0, \pi$.

DEMO: *Z to Freq*

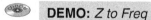

We can estimate any value of $|H(e^{j\hat{\omega}})|$ directly from the poles and zeros. This can be done systematically by writing $H(z)$ in the following form:

$$H(z) = G\,\frac{(z-z_1)(z-z_2)}{(z-p_1)(z-p_2)}$$

[7]This approximate formula for bandwidth is good only when the poles are isolated from one another. The approximation breaks down, for example, when a second-order system has two poles with small angles.

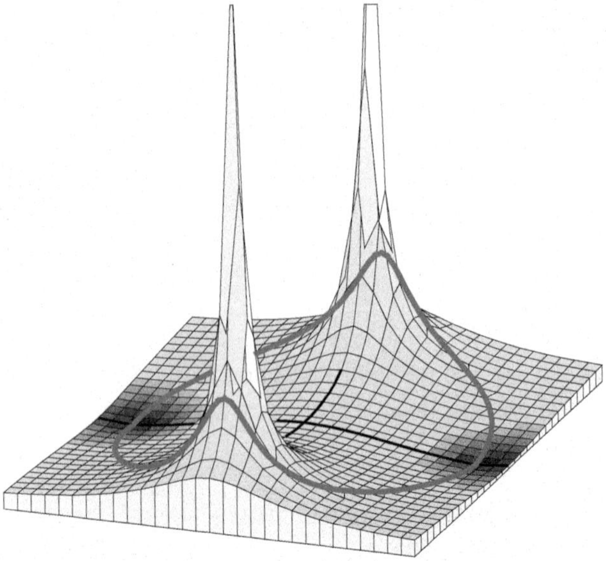

Figure 8-24: z-transform evaluated over a region of the z-plane including the unit circle. The view is from the fourth quadrant, so the $z = 1$ point is at the right. Values along the unit circle are the frequency response (magnitude) for a second-order feedback filter. The poles are at $z = 0.85e^{\pm j\pi/2}$ and the numerator has zeros at $z = \pm 1$.

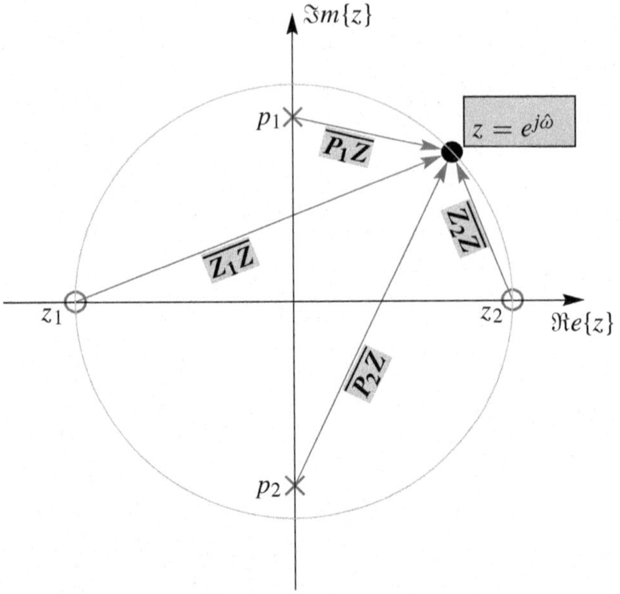

Figure 8-25: z-transform evaluated on the unit circle ($z = e^{j\hat{\omega}}$) by using a product of vector lengths from the poles and zeros. $\overline{P_iZ}$ denotes the vector from the ith pole to ($z = e^{j\hat{\omega}}$), and $\overline{Z_jZ}$ denotes the vector from the jth zero to ($z = e^{j\hat{\omega}}$),

where z_1 and z_2 are the zeros and p_1 and p_2 are the poles of the second-order filter. The parameter G is a gain term that may have to be factored out. Then the magnitude of the frequency response is

$$|H(e^{j\hat{\omega}})| = G\,\frac{|e^{j\hat{\omega}} - z_1|\,|e^{j\hat{\omega}} - z_2|}{|e^{j\hat{\omega}} - p_1|\,|e^{j\hat{\omega}} - p_2|} \tag{8.60}$$

Equation (8.60) has a simple geometric interpretation. Each term $|e^{j\hat{\omega}} - z_i|$ or $|e^{j\hat{\omega}} - p_i|$ is the vector length from the zero z_i or the pole p_i to the unit circle position $e^{j\hat{\omega}}$, shown in Fig. 8-25. The frequency response at a fixed value of $\hat{\omega}$ is the product of G times the product of the lengths of the vectors to the zeros divided by the product of the lengths of the vectors to the poles.

$$|H(e^{j\hat{\omega}})| = G\,\frac{\overline{Z_1Z} \cdot \overline{Z_2Z}}{\overline{P_1Z} \cdot \overline{P_2Z}}$$

As we go around the unit circle, these vector lengths change. When we are on top of a zero, one of the numerator lengths is zero, so $|H(e^{j\hat{\omega}})| = 0$ at that frequency. When we are close to a pole, one of the denominator lengths is very small, so $|H(e^{j\hat{\omega}})|$ will be large at that frequency.

We can apply this geometric reasoning to estimate the magnitude of $H(e^{j\hat{\omega}})$ at $\hat{\omega} = \pi/2$ in Fig. 8-24. We begin by estimating the lengths of the vectors from the zeros and poles to the point $z = e^{j\pi/2}$, which is the same as

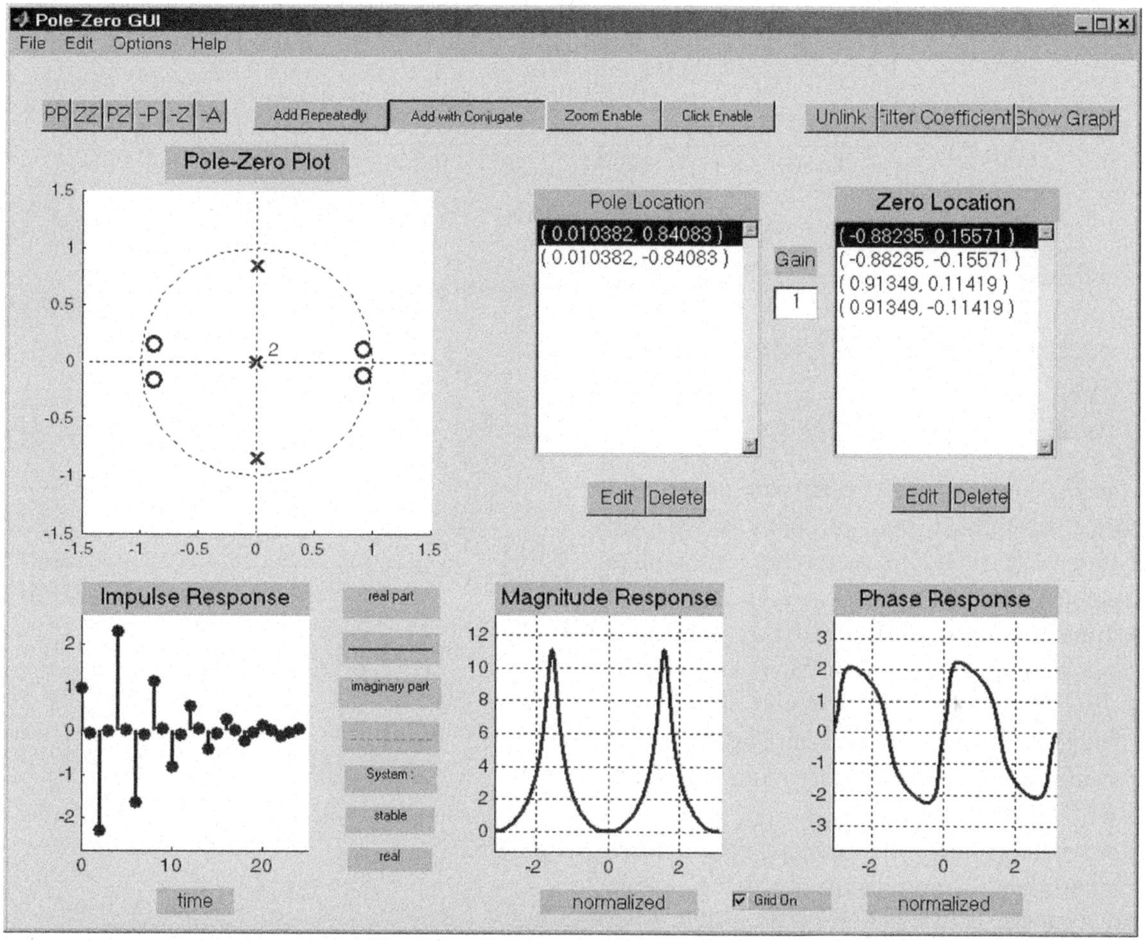

Figure 8-26: Graphical User Interface (GUI) for PeZ which illustrates the *three-domains* concept. The user can enter poles and zeros and see the resulting impulse response and frequency response. In addition, the poles/zero can be moved with a mouse pointer and and the other plots will be updated in real-time.

$z = j$. The lengths of the vectors from the zeros are then divided by the lengths of the vectors from the poles, so we get

$$|H(e^{j\pi/2})| = |H(j)|$$
$$= \frac{|j-1|\,|j+1|}{|j-0.85j|\,|j-(-0.85j)|}$$
$$= \frac{2}{0.15 \times 1.85} = 7.207$$

The gain G was assumed to be 1.

An excellent way to practice with these ideas is to use the MATLAB GUI called PeZ (see Fig. 8-26).

$$H(z) = \frac{0.0798(1+z^{-1}+z^{-2}+z^{-3})}{1-1.556z^{-1}+1.272z^{-2}-0.398z^{-3}} \quad (8.61)$$

$$= \frac{0.0798(1+z^{-1})(1-e^{j\pi/2}z^{-1})(1-e^{-j\pi/2}z^{-1})}{(1-0.556z^{-1})(1-0.846e^{j0.3\pi}z^{-1})(1-0.846e^{-j0.3\pi}z^{-1})} \quad (8.62)$$

$$= -\frac{1}{5} + \frac{0.62}{1-.556z^{-1}} + \frac{0.17e^{j0.96\pi}}{1-.846e^{j0.3\pi}z^{-1}} + \frac{0.17e^{-j0.96\pi}}{1-.846e^{-j0.3\pi}z^{-1}} \quad (8.63)$$

8-11 Example of an IIR Lowpass Filter

First-order and second-order IIR filters are useful and provide simple examples, but, in many cases, we use higher-order IIR filters because they can realize frequency responses with flatter passbands and stopbands and sharper transition regions. The `butter`, `cheby1`, `cheby2`, and `ellip` functions in MATLAB's *Signal Processing Toolbox* can be used to design filters with prescribed frequency-selective characteristics. As an example, consider the system with system function $H(z)$ given by (8.61).

This system is an example of a lowpass ***elliptic filter*** whose numerator and denominator coefficients were obtained using the MATLAB function `ellip`. The exact call was `ellip(3,1,30,0.3)`. Each of the three different forms above is useful: (8.61) for identifying the filter coefficients, (8.62) for sketching the pole-zero plot and the frequency response, and (8.63) for finding the impulse response.[8] Figure 8-27 shows the poles and zeros of this filter. Note that all the zeros are on the unit circle and that the poles are strictly inside the unit circle, as they must be for a stable system.

[8] Factoring polynomials and obtaining the partial fraction expansion was done in MATLAB using the functions `roots` and `residuez`, respectively.

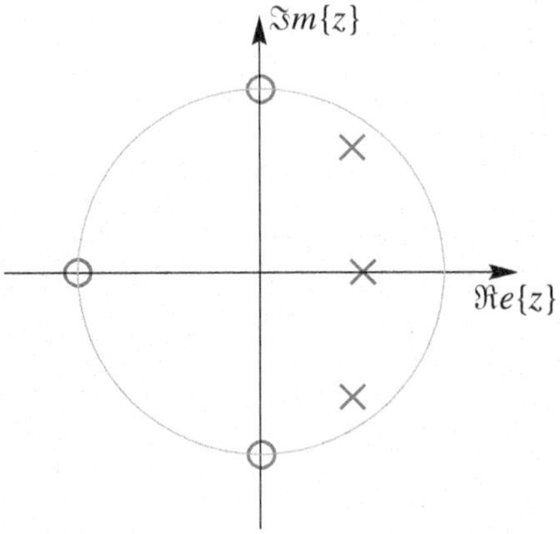

Figure 8-27: Pole-zero plot for a third-order IIR filter (8.62).

EXERCISE 8.17: From (8.61) determine the difference equation (Direct Form I) for implementing this system.

The system function was evaluated on the unit circle using the MATLAB function `freqz`. A plot of this result is shown in Fig. 8-28. Note that the frequency response is large in the vicinity of the poles and small around the zeros. In particular, the passband of the frequency response is $|\hat{\omega}| \leq 2\pi(0.15)$, which corresponds to the poles with angles at $\pm 0.3\pi$. Also, the frequency response

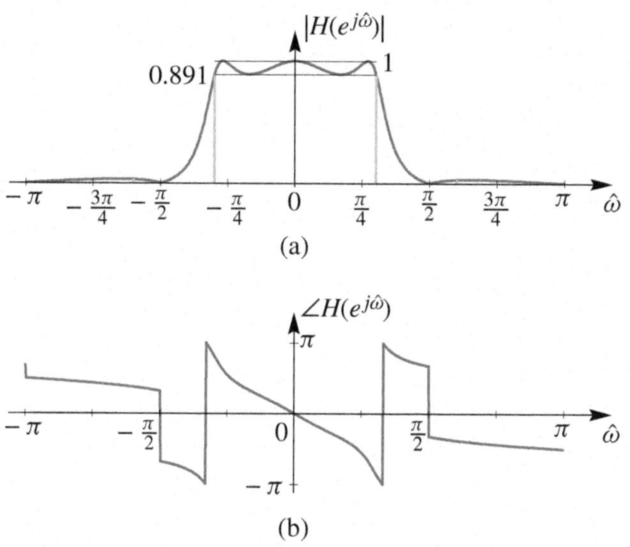

Figure 8-28: Frequency response (magnitude and phase) for a third-order IIR filter.

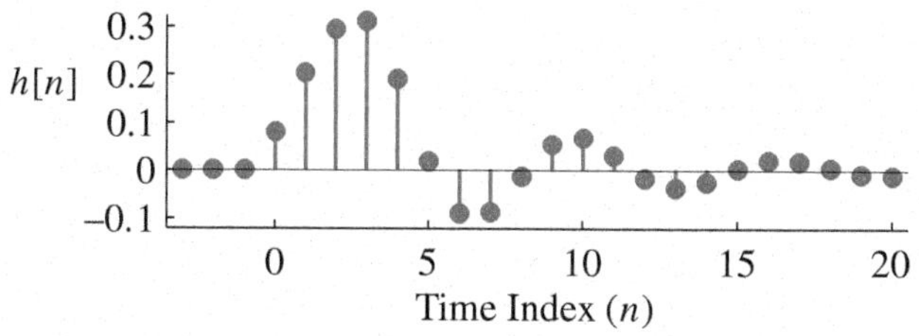

Figure 8-29: Impulse response for a third-order IIR filter.

is exactly zero at $\hat{\omega} = \pm 0.5\pi$ and $\hat{\omega} = \pi$ since the zeros of $H(z)$ are at these angles and lie on the unit circle.

EXERCISE 8.18: From (8.61) or (8.62), determine the value of the frequency response at $\hat{\omega} = 0$.

Finally, Fig. 8-29 shows the impulse response of the system. Note that it oscillates and dies out with increasing n because of the two complex conjugate poles at angles $\pm 0.3\pi$ and radius 0.846. The decaying envelope is $(0.846)^n$.

EXERCISE 8.19: Use the partial fraction form (8.63) to determine an equation for the impulse response of the filter.

Hint: Apply the inverse z-transform.

The elliptic filter example described in this section is a simple example of a practical IIR lowpass filter. Higher-order filters can exhibit much better frequency-selective filter characteristics.

8-12 Summary and Links

The class of IIR filters was introduced in this chapter, along with the z-transform method for filters with poles. The z-transform changes problems about impulse responses, frequency responses, and system structures into algebraic manipulations of polynomials and rational functions. Poles of the system function $H(z)$ turn out to be the most important elements for IIR filters because properties such as the shape of the frequency response or the form of the impulse response can be inferred quickly from the pole locations.

We also continued to stress the important concept of "domains of representation." The n-domain or time domain, the $\hat{\omega}$-domain or frequency domain, and the z-domain give us three domains for thinking about the characteristics of a system. We completed the ties between domains by introducing the inverse z-transform for constructing a signal from its z-transform. As a result, even difficult problems such as convolution can be simplified by working in the most convenient domain (z) and then transforming back to the original domain (n).

Lab #11 is devoted to IIR filters. This lab uses a MATLAB user interface tool called `PeZ` that supports an interactive exploration of the three domains.

LAB: *#11 PeZ - the z, n, and $\hat{\omega}$ Domains*

The `PeZ` tool is useful for studying IIR systems because it presents the user with multiple views of an LTI system: pole-zero domain, frequency response and impulse response. (See Fig. 8-26.) Similar capabilities are now being incorporated into many commercial software packages (e.g., `sptool` in MATLAB).

The CD-ROM also contains the following demonstrations of the relationship between the z-plane and the frequency domain and time domain:

(a) A set of "three-domain" movies that show how the frequency response and impulse response of an IIR filter change as a pole location is varied. Several different filters are demonstrated.

DEMO: *Three Domains - IIR*

(b) A movie that animates the relationship between the z-plane and the unit circle where the frequency response lies.

DEMO: *Z to Freq*

(c) The `PeZ` GUI can be used to construct different IIR and FIR filters.

DEMO: *PeZ GUI*

(d) A tutorial on how to use `PeZ`.

DEMO: *PeZ Tutorial*

(e) A demo that gives more examples of IIR filters.

DEMO: *IIR Filtering*

The reader is again reminded of the large number of solved homework problems on the CD-ROM that are available for review and practice.

NOTE: *Hundreds of Solved Problems*

8-13 Problems

P-8.1 Find the impulse response of the second-order system

$$y[n] = \sqrt{2}y[n-1] - y[n-2] + x[n]$$

Express your answer as separate formulas for the cases where $n < 0$ and $n \geq 0$, thus covering the entire range of n.

P-8.2 Determine a general formula for the Fibonacci sequence by finding the impulse response of the following second-order system:

$$y[n] = y[n-1] + y[n-2] + x[n]$$

P-8.3 Let the input be $x[n] = \delta[n] - \alpha\delta[n-5]$ and the impulse response, $h[n] = 5(0.8)^n u[n]$. If $y[n] = x[n] * h[n]$, determine α so that $y[n] = 0$ for $n \geq 5$.

P-8.4 For the following feedback filters, determine the locations of the poles and zeros and plot the positions in the z-plane.

$$\begin{aligned} y[n] = \tfrac{1}{2}y[n-1] + \tfrac{1}{3}y[n-2] \\ - x[n] + 3x[n-1] - 2x[n-2] \end{aligned}$$

$$\begin{aligned} y[n] = \tfrac{1}{2}y[n-1] - \tfrac{1}{3}y[n-2] \\ - x[n] + 3x[n-1] + 2x[n-2] \end{aligned}$$

In the second case, only the signs on $y[n-2]$ and $x[n-2]$ were changed. Compare the two results.

P-8.5 In this problem, the degrees of the numerator and denominator polynomials are different, so there should be zeros (or poles) at $z = 0$ or $z = \infty$. Determine the poles and zeros of the following filters:

$$y[n] = \tfrac{1}{2}y[n-1] - \tfrac{1}{3}y[n-2] - x[n]$$
$$y[n] = \tfrac{1}{2}y[n-1] - \tfrac{1}{3}y[n-2] - x[n-2]$$
$$y[n] = \tfrac{1}{2}y[n-1] - \tfrac{1}{3}y[n-2] - x[n-4]$$

Plot the positions of the poles and zeros in the z-plane. In all cases, make sure that the number of poles and zeros is the same, by allowing zeros (or poles) at $z = \infty$ or $z = 0$.

P-8.6 Given an IIR filter defined by the difference equation

$$y[n] = -\tfrac{1}{2}y[n-1] + x[n]$$

(a) Determine the system function $H(z)$. What are its poles and zeros?

(b) When the input to the system is three successive impulses

$$x[n] = \begin{cases} +1 & \text{for } n = 0, 1, 2 \\ 0 & \text{for } n < 0 \text{ and } n \geq 3 \end{cases}$$

determine the functional form for the output signal $y[n]$. Assume that the output signal $y[n]$ is zero for $n < 0$.
Hint: Use linearity to find the output as the sum of three terms, each related to the impulse response of the system. Recall that the impulse response of a first-order IIR filter has the form $b_0 a^n$ for $n \geq 0$.

P-8.7 A linear time-invariant filter is described by the difference equation

$$y[n] = -0.8y[n-1] + 0.8x[n] + x[n-1]$$

(a) Determine the system function $H(z)$ for this system. Express $H(z)$ as a ratio of polynomials in z^{-1} (negative powers of z) and also as a ratio of polynomials in positive powers of z.

(b) Plot the poles and zeros of $H(z)$ in the z-plane.

(c) From $H(z)$, obtain an expression for $H(e^{j\hat{\omega}})$, the frequency response of this system.

(d) Show that $|H(e^{j\hat{\omega}})|^2 = 1$ for all $\hat{\omega}$.

P-8.8 Given an IIR filter defined by the difference equation

$$y[n] = -y[n-5] + x[n] \tag{8.64}$$

(a) Determine the system function $H(z)$.

(b) How many poles does the system have? Compute and plot the pole locations.

(c) When the input to the system is the two-point pulse signal:

$$x[n] = \begin{cases} +1 & \text{when } n = 0, 1 \\ 0 & \text{when } n \neq 0, 1 \end{cases}$$

determine the output signal $y[n]$, so that you can make a plot of its general form. Assume that the output signal is zero for $n < 0$.

(d) The output signal is periodic for $n > 0$. Determine the period.

P-8.9 Given an IIR filter defined by the difference equation

$$y[n] = -0.9\, y[n-6] + x[n]$$

(a) Find the z-transform system function for the system.

(b) Find the poles of the system and plot their location in the z-plane.

P-8.10 Given an IIR filter defined by the difference equation

$$y[n] = -\tfrac{1}{2}y[n-1] + x[n]$$

(a) Determine the system function $H(z)$. What are its poles and zeros?

(b) When the input to the system is

$$x[n] = \delta[n] + \delta[n-1] + \delta[n-2]$$

determine the output signal $y[n]$. Assume that $y[n]$ is zero for $n < 0$.

P-8.11 Determine the inverse z-transform of the following:

(a) $H_a(z) = \dfrac{1 - z^{-1}}{1 + 0.77z^{-1}}$

(b) $H_b(z) = \dfrac{1 + 0.8z^{-1}}{1 - 0.9z^{-1}}$

(c) $H_c(z) = \dfrac{z^{-2}}{1 - 0.9z^{-1}}$

(d) $H_d(z) = 1 - z^{-1} + 2z^{-3} - 3z^{-4}$

P-8.12 Determine the inverse z-transform of the following:

(a) $X_a(z) = \dfrac{1 - z^{-1}}{1 - \frac{1}{6}z^{-1} - \frac{1}{6}z^{-2}}$

(b) $X_b(z) = \dfrac{1 + z^{-2}}{1 + 0.9z^{-1} + 0.81z^{-2}}$

(c) $X_c(z) = \dfrac{1 + z^{-1}}{1 - 0.1z^{-1} - 0.72z^{-2}}$

P-8.13 For each of the pole-zero plots in Fig. P-8.13, determine which of the following systems (specified by either an $H(z)$ or a difference equation) matches the pole-zero plot.

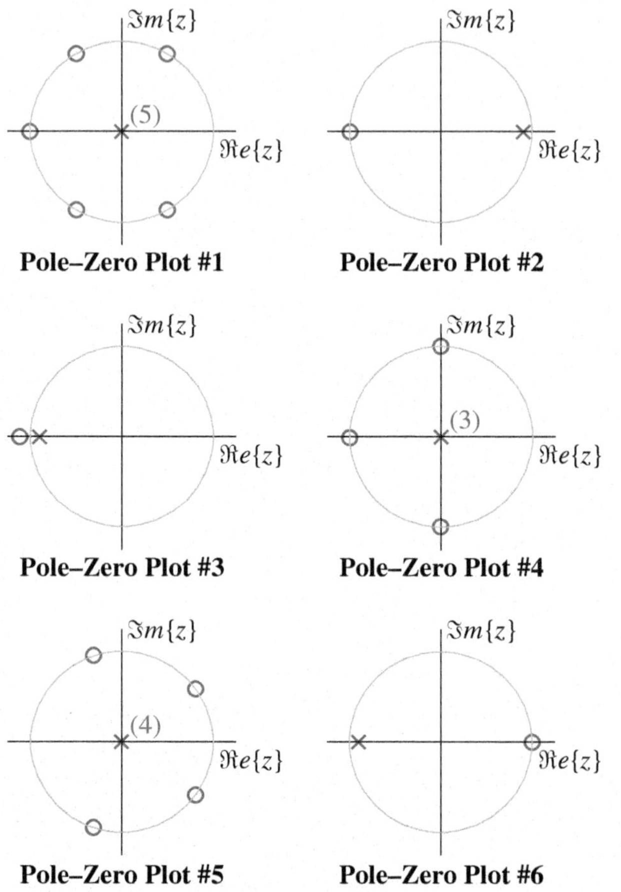

Figure P-8.13

$S_1: \quad y[n] = 0.9y[n-1] + \tfrac{1}{2}x[n] + \tfrac{1}{2}x[n-1]$

$S_2: \quad y[n] = -0.9y[n-1] + 9x[n] + 10x[n-1]$

$S_3: \quad H(z) = \dfrac{\frac{1}{2}(1 - z^{-1})}{1 + 0.9z^{-1}}$

$\mathcal{S}_4: \quad y[n] = \frac{1}{4}x[n] + x[n-1] + \frac{3}{2}x[n-2] + x[n-3] + \frac{1}{4}x[n-4]$

$\mathcal{S}_5: \quad H(z) = 1 - z^{-1} + z^{-2} - z^{-3} + z^{-4}$

$\mathcal{S}_6: \quad y[n] = \sum_{k=0}^{3} x[n-k]$

$\mathcal{S}_7: \quad y[n] = x[n] + x[n-1] + x[n-2] + x[n-3] + x[n-4] + x[n-5]$

P-8.14 For each of the frequency-response plots (A–F) in Fig. P-8.14, determine which of the following systems (specified by either an $H(z)$ or a difference equation) matches the frequency response.

Note: The frequency axis for each plot extends over the range $-\pi \leq \hat{\omega} \leq \pi$.

$\mathcal{S}_1: \quad y[n] = 0.9y[n-1] + \frac{1}{2}x[n] + \frac{1}{2}x[n-1]$

$\mathcal{S}_2: \quad y[n] = -0.9y[n-1] + 9x[n] + 10x[n-1]$

$\mathcal{S}_3: \quad H(z) = \dfrac{\frac{1}{2}(1 - z^{-1})}{1 + 0.9z^{-1}}$

$\mathcal{S}_4: \quad y[n] = \frac{1}{4}x[n] + x[n-1] + \frac{3}{2}x[n-2] + x[n-3] + \frac{1}{4}x[n-4]$

$\mathcal{S}_5: \quad H(z) = 1 - z^{-1} + z^{-2} - z^{-3} + z^{-4}$

$\mathcal{S}_6: \quad y[n] = \sum_{k=0}^{3} x[n-k]$

$\mathcal{S}_7: \quad y[n] = x[n] + x[n-1] + x[n-2] + x[n-3] + x[n-4] + x[n-5]$

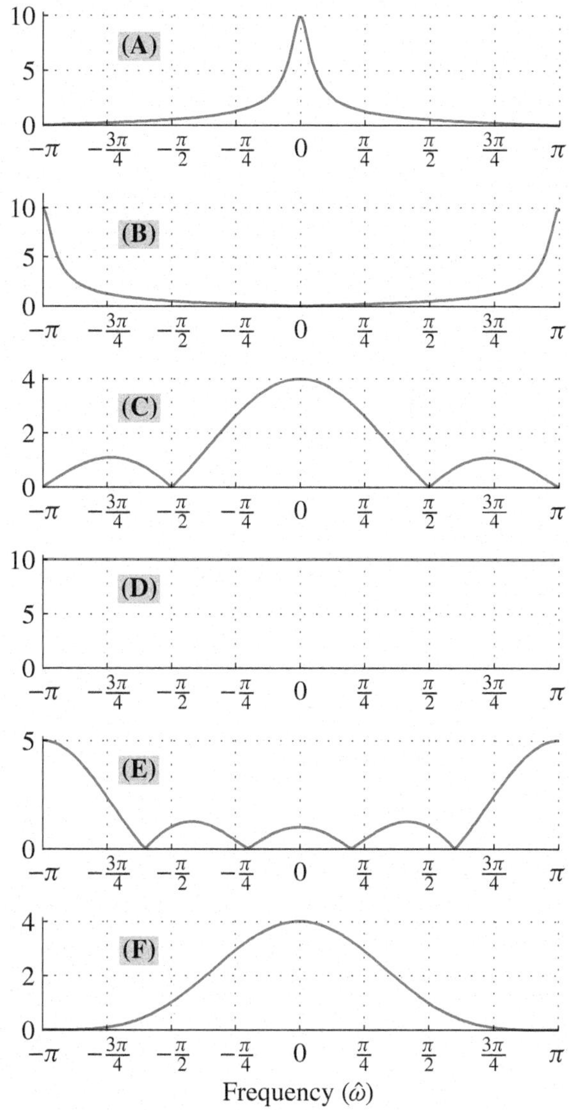

Figure P-8.14

P-8.15 Given an IIR filter defined by the difference equation

$$y[n] = \frac{1}{2}y[n-1] + x[n]$$

(a) When the input to the system is a unit-step sequence, $u[n]$, determine the functional form for the output signal $y[n]$. Use the inverse z-transform method. Assume that the output signal $y[n]$ is zero for $n < 0$.

(b) Find the output when $x[n]$ is a complex exponential that starts at $n = 0$:

$$x[n] = e^{j(\pi/4)n}u[n]$$

(c) From (b), identify the steady-state component of the response, and compare its magnitude and phase to the frequency response at $\hat{\omega} = \pi/4$.

P-8.16 In Fig. P-8.16, match each pole-zero plot (PZ 1–4) with the correct one of five possible frequency responses (A–E). In the pole-zero plots, poles are denoted by **x** and zeros by **o**.

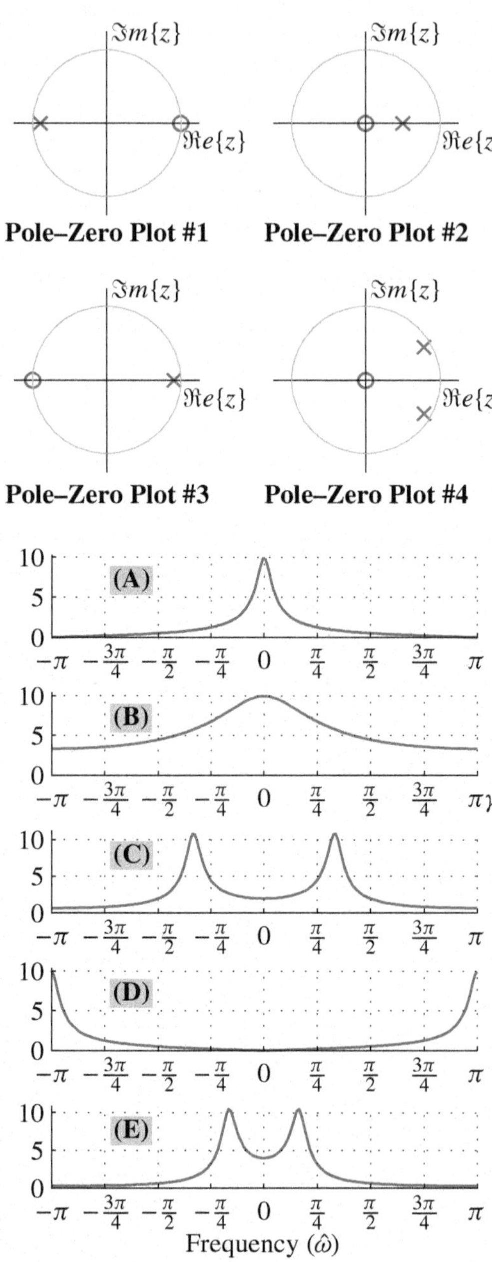

Figure P-8.16

P-8.17

In Fig. P-8.17, match each pole-zero plot (PZ 1–4) with the correct one of five possible impulse responses (A–E). In the pole-zero plots, poles are denoted by **x** and zeros by **o**.

P-8.18 A linear time-invariant filter is described by the difference equation

$$y[n] = 0.8y[n-1] - 0.8x[n] + x[n-1]$$

(a) Determine the system function $H(z)$ for this system. Express $H(z)$ as a ratio of polynomials in z^{-1} and as a ratio of polynomials in z.

(b) Plot the poles and zeros of $H(z)$ in the z-plane.

(c) From $H(z)$, obtain an expression for $H(e^{j\hat{\omega}})$, the frequency response of this system.

(d) Show that $|H(e^{j\hat{\omega}})|^2 = 1$ for all $\hat{\omega}$.

(e) If the input to the system is

$$x[n] = 4 + \cos[(\pi/4)n] - 3\cos[(2\pi/3)n]$$

what can you say, without further calculation, about the form of the output $y[n]$?

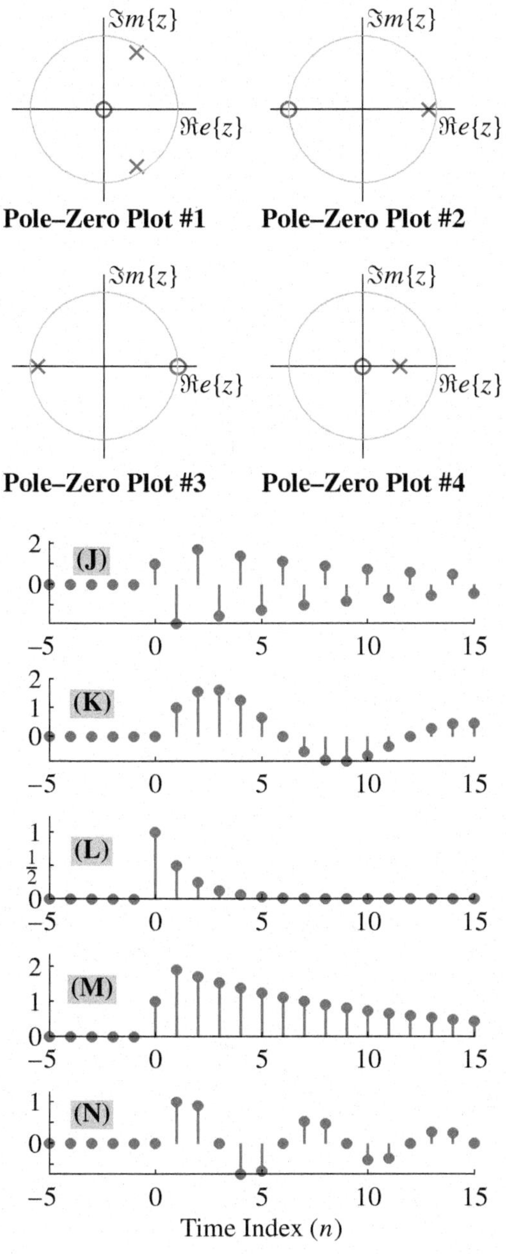

Figure P-8.17

P-8.19 The system function of a linear time-invariant system is given by the formula

$$H(z) = \frac{(1 - z^{-1})(1 - e^{j\pi/2}z^{-1})(1 - e^{-j\pi/2}z^{-1})}{(1 - 0.9e^{j2\pi/3}z^{-1})(1 - 0.9e^{-j2\pi/3}z^{-1})}$$

(a) Write the difference equation that gives the relation between the input $x[n]$ and the output $y[n]$.
Hint: Multiply out the factors of $H(z)$.

(b) Plot the poles and zeros of $H(z)$ in the complex z-plane.
Hint: Express $H(z)$ as a ratio of factored polynomials in z instead of z^{-1}.

(c) If the input is of the form $x[n] = Ae^{j\phi}e^{j\hat{\omega}_0 n}$, for what values of $-\pi \le \hat{\omega}_0 \le \pi$ will $y[n] = 0$?

P-8.20 The input to the C-to-D converter in Fig. P-8.20 is

$$x(t) = 4 + \cos(500\pi t) - 3\cos[(2000\pi/3)t]$$

The system function for the LTI system is

$$H(z) = \frac{(1 - z^{-1})(1 - e^{j\pi/2}z^{-1})(1 - e^{-j\pi/2}z^{-1})}{(1 - 0.9e^{j2\pi/3}z^{-1})(1 - 0.9e^{-j2\pi/3}z^{-1})}$$

If $f_s = 1000$ samples/sec, determine an expression for $y(t)$, the output of the D-to-C converter.

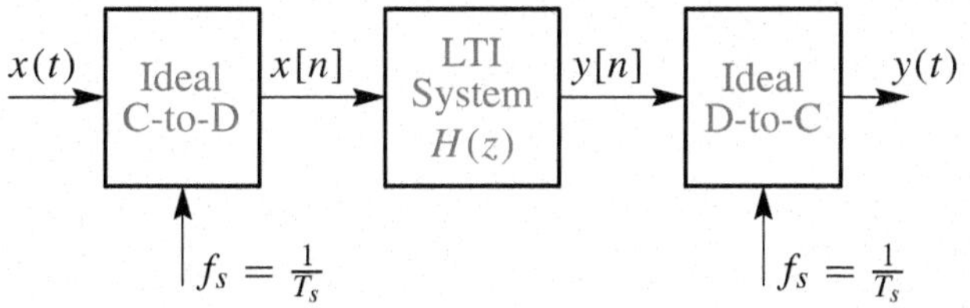

Figure P-8.20

P-8.21 Answer the following questions about the system whose z-transform system function is

$$H(z) = \frac{1 + 0.8z^{-1}}{1 - 0.9z^{-1}}$$

(a) Determine the poles and zeros of $H(z)$.

(b) Determine the difference equation relating the input and output of this filter.

(c) Derive a simple expression (purely real) for the magnitude squared of the frequency response $|H(e^{j\hat{\omega}})|^2$.

(d) Is this filter a lowpass filter or a highpass filter? Explain your answer in terms of the poles and zeros of $H(z)$.

P-8.22 The diagram in Fig. 2 depicts a *cascade connection* of two linear time-invariant systems; i.e., the output of the first system is the input to the second system, and the overall output is the output of the second system.

(a) Use z-transforms to show that the system function for the overall system (from $x[n]$ to $y[n]$) is $H(z) = H_2(z)H_1(z)$, where $Y(z) = H(z)X(z)$.

(b) Suppose that System 1 is an FIR filter described by the difference equation $y_1[n] = x[n] + \frac{5}{6}x[n-1]$, and System 2 is described by the system function $H_2(z) = 1 - 2z^{-1} + z^{-2}$. Determine the system function of the overall cascade system.

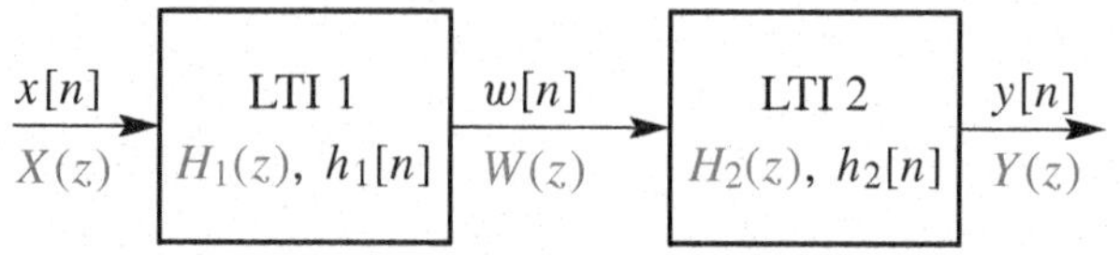

Figure P-8.22

(c) For the systems in (b), obtain a single difference equation that relates $y[n]$ to $x[n]$ in Fig. 2.

(d) For the systems in (b), plot the poles and zeros of $H(z)$ in the complex z-plane.

(e) Derive a condition on $H(z)$ that guarantees that the output signal will always be equal to the input signal.

(f) If System 1 is the difference equation $y_1[n] = x[n] + \frac{5}{6}x[n-1]$, find a system function $H_2(z)$ so that output of the cascaded system will always be equal to its input. In other words, find $H_2(z)$, which will undo the filtering action of $H_1(z)$. This is called *deconvolution* and $H_2(z)$ is the *inverse* of $H_1(z)$.

(g) Suppose that $H_1(z)$ represents a general FIR filter. What conditions on $H_1(z)$ must hold if $H_2(z)$ is to be a stable and causal inverse filter for $H_1(z)$?

P-8.23 Define a discrete-time signal using the formula:

$$y[n] = (0.99)^n \cos(2\pi(0.123)n + \phi)$$

(a) Make a sketch of $y[n]$ versus n, as a "stem" plot. Take the range of n to be $0 \le n \le 20$.

(b) Design an IIR filter that will synthesize $y[n]$. Give your answer in the form of a difference equation with numerical values for the coefficients. Assume that the synthesis will be accomplished by using an impulse input to "start" the difference equation (which is at rest; i.e., has zero initial conditions).

CHAPTER 9

Continuous-Time Signals and LTI Systems

We began this text with an introduction to continuous-time signals and systems, and now we are going to return to that topic. In Chapters 4–8, we focused on discrete-time signals and systems with topics such as sampling, FIR filtering, frequency response, and the z-transform. While discrete-time systems have many virtues, it is generally impossible to avoid some consideration of continuous-time signals and systems since most signals originate in continuous-time form.

For example, musical instruments produce signals in continuous-time, and microphones convert the resulting acoustic pressure wave into a continuously varying electrical signal that can be further processed by amplifiers and filters. When stored on a CD-ROM, a music signal is converted to a discrete-time signal by sampling. However, when we listen to a music signal that is represented on a CD in digital form, it is necessary to convert the samples back to a continuous-time signal that can be processed by the human ear, another continuous-time system. Even when we want to perform analog processing using the combination of A/D converter, digital filter, and D/A converter as in Fig. 6-17 on p. 152, the input is a continuous-time signal $x(t)$ and the output is another continuous-time signal, $y(t)$. Since the overall effect of the system is a transformation from $x(t)$ to $y(t)$, we want to be able to describe and analyze that transformation independent of the fact that we might use a digital filter as part of the implementation.

Another issue is the specification of processing systems in terms of their *ideal* behavior. We will see that communication systems such as radios and modems can be built up from ideal subsystems such as ideal filters and modulators. The advantage of this abstract approach is that we can analyze the system behavior prior to specifying the implementation details of the subsystems. One might call this approach ***Analog Signal***

Processing. In some cases, the preferred implementation of the continuous-time system might involve an analog electronic circuit; in others, a digital implementation would be employed.

Therefore, it is useful to develop a good set of tools for analyzing continuous-time signals and systems, especially in the frequency-domain. The *Fourier Transform* provides the general method for describing the frequency content of continuous-time signals, as well as the frequency response of continuous-time systems. It also sheds new light on the relation between continuous-time and discrete-time signals and systems. This is the goal of Chapters 9–12.

In this chapter, we will begin with a review of our earlier discussion of continuous-time sinusoidal signals and then introduce several other basic signals that are important in the study of continuous-time systems. We will review the concepts of linearity and time-invariance and introduce the convolution representation of linear time-invariant systems. As part of our discussion, we will introduce the unit-impulse signal $\delta(t)$ and its mathematical properties. This signal $\delta(t)$ is necessary to define the impulse response for LTI systems, which is one element of continuous-time convolution.

This chapter, along with Chapters 10 and 11, focuses on the basic theory of continuous-time signals and LTI systems. While it may appear that the results are highly mathematical, and it may not be obvious why they are useful, it is necessary to first build a foundation that can represent and solve real problems. Chapters 12 and 13 discuss a variety of applications of the basic concepts, including filtering, modulation, sampling, and spectrum analysis. Patient study of the fundamentals will be worth the effort!

9-1 Continuous-Time Signals

Continuous-time signals can be expressed as functions of a continuous independent variable. We use the notation $x(t)$ to denote a continuous-time signal. The independent variable, t, is generally associated with time, but that need not always be the case. For example, the theory and techniques that we develop are readily applicable to other situations such as image processing where the independent variable is distance.

It is useful to classify continuous-time signals in terms of their duration, or length. For our purposes, we will define the signal length to be the smallest time interval that contains all the nonzero signal values. In mathematical terminology, this interval is called the *support* of the signal.

9-1.1 Two-Sided Infinite-Length Signals

In Chapters 2 and 3, we studied various types of infinite length signals that were also periodic signals. We emphasized sinusoidal signals of the form

$$x(t) = A\cos(\omega_0 t + \phi) \tag{9.1}$$

which can also be represented as the real part of the complex exponential signal

$$\begin{aligned} z(t) &= Xe^{j\omega_0 t} = Ae^{j\phi}e^{j\omega_0 t} \\ &= A\cos(\omega_0 t + \phi) + jA\sin(\omega_0 t + \phi) \end{aligned} \tag{9.2}$$

where the complex amplitude X is equal to $Ae^{j\phi}$. In both (9.1) and (9.2), the signal is represented by an equation that gives a rule for assigning a signal value to each time value in the range $-\infty < t < \infty$. Another common representation of these signals is a graph such as Fig. 9-1(a).

A wider class of infinite-length signals is the class of periodic signals that satisfy the relationship $x(t) = x(t + T_0)$, where T_0 is the repetition period. One common example of a periodic signal is the square wave shown in Fig. 9-1(b) where the period is $T_0 = \frac{1}{3}$ sec. Finally, it is possible to have an infinite length signal that is not periodic. An example is the decaying exponential signal $x(t) = 5e^{-|t|}$ which has nonzero values over all

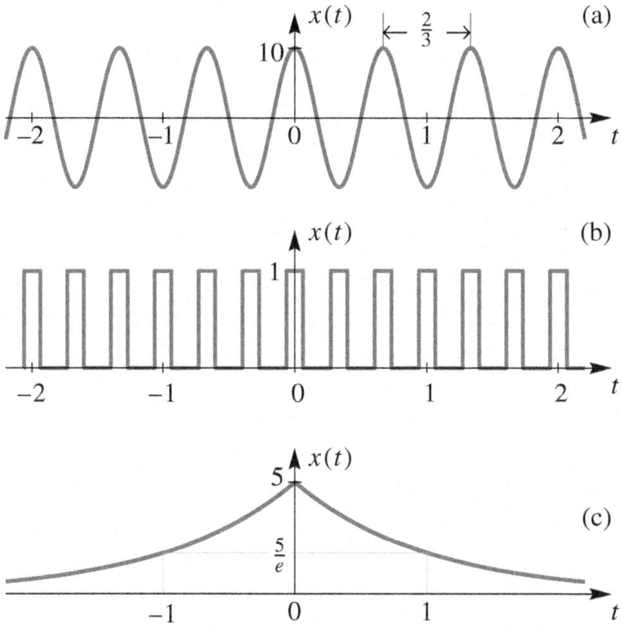

Figure 9-1: Continuous-time signals that are infinitely long: (a) 1.5-Hz Sinusoid with $\omega_0 = 3\pi$ rad/s, $A = 10$ and $\phi = 0$. (b) Square wave with a fundamental frequency of 3 Hz. (c) Two-sided decaying exponential signal.

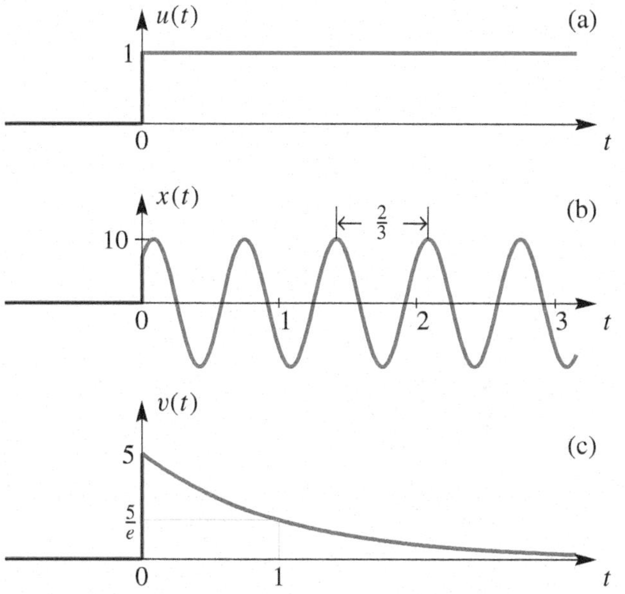

Figure 9-2: One-sided continuous-time signals: (a) Unit step. (b) Cosine wave starting in (9.4) at $t = 0$. (c) Decaying exponential. Each of these signals is zero for $t < 0$.

time, $-\infty < t < \infty$, as shown in Fig. 9-1(c). The signals in Fig. 9-1 are often called ***two-sided signals*** because they are nonzero for both positive and negative time.

9-1.2 One-Sided Signals

One-sided signals have the property that $x(t) = 0$ on a semi-infinite interval. For example, a signal might be zero for $t < t_0$, in which case it is called a ***right-sided signal*** because its graph starts at $t = t_0$ and goes all the way to the right. One-sided signals are necessary because most signals that we generate have a definite starting time—often taken to be $t = 0$. For example, a switch might be used to connect the signal to a system, but prior to closing the switch the system has no input so its input signal is effectively zero. To represent this situation mathematically, it is very useful to define the ***continuous-time unit-step*** signal according to the mathematical equation[1]

$$u(t) = \begin{cases} 1 & t \geq 0 \\ 0 & t < 0 \end{cases} \tag{9.3}$$

which is shown in Fig. 9-2(a). The unit step can then be used to multiply other signals to make a new signal that

[1]The value of the unit step at $t = 0$ is ambiguous. Some authors define $u(0) = \frac{1}{2}$, but we have chosen $u(0) = 1$. In either case, there will be situations where the value of $u(0)$ appears to be ambiguous. This is OK, because mathematics tells us that the value of a function at one isolated point is inconsequential.

starts at $t = 0$. For example in Fig. 9-2(b), a sinusoidal signal that is "turned on" at $t = 0$ would be represented as

$$x(t) = 10\cos(3\pi t - \pi/4)u(t) \tag{9.4}$$

and similarly, the one-sided exponential signal

$$v(t) = 5e^{-at}u(t) = \begin{cases} 5e^{-at} & t \geq 0 \\ 0 & t < 0 \end{cases} \tag{9.5}$$

would appear as in Fig. 9-2(c) for $a = 1$.

Example 9-1: Shifting the Unit Step

In order to get a different starting time, we can define a shifted unit step, e.g., $x(t) = u(t-7)$. Expanding the definition of the unit step, we have

$$x(t) = u(t-7) = \begin{cases} 1 & (t-7) \geq 0 \\ 0 & (t-7) < 0 \end{cases}$$

so a plot of $x(t)$ is zero for $t < 7$, and it makes its transition from zero to one at $t = 7$. ■

9-1.3 Finite-Length Signals

The class of finite-length signals contains signals such as short pulses that can be used to carry binary information. Figure 9-3(a) shows a two-second pulse starting at $t = 2$ and ending at $t = 4$. As in the case of $u(t)$, a finite-length pulse can be used to multiply another signal to form a short burst of that signal, e.g., the finite-length sinusoid in Fig. 9-3(b).

There is a simple relationship between the unit-step signal $u(t)$ and finite-length pulses: shifted versions of the unit-step signal can be subtracted to form a finite-length signal. Thus, the pulse signal $p(t) = u(t-2) - u(t-4)$ has values that are given by

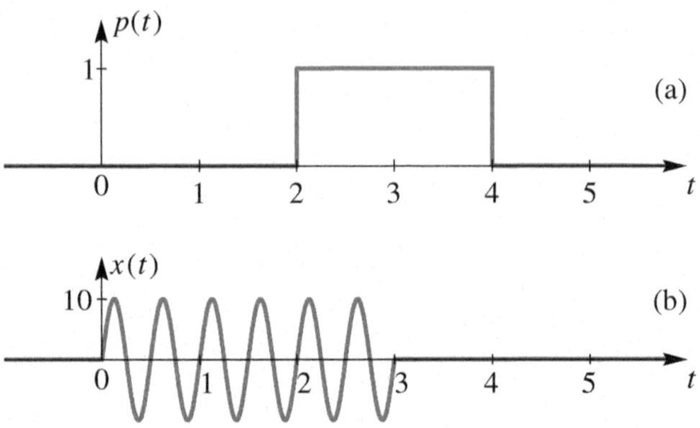

Figure 9-3: Finite-length continuous-time signals: (a) Finite-length pulse with duration = 2 sec. (b) Sinusoidal signal multiplied by a 3-sec. pulse.

$$p(t) = u(t-2) - u(t-4) = \begin{cases} 1 & 2 \leq t < 4 \\ 0 & \text{otherwise} \end{cases} \tag{9.6}$$

This signal is plotted in Fig. 9-3(a). A second way to write the finite-length pulse in terms of $u(t)$ is to use a product notation (see Prob. P-9.1).

EXERCISE 9.1: Write a formula for the finite-length sinusoidal signal of Fig. 9-3(b) using a unit-step notation similar to (9.6) to indicate its finite duration.

9-2 The Unit Impulse

In this section we will introduce a useful concept—a pulse whose duration appears to be so short that it starts and ends at the same instant of time. We will call this signal the ***unit impulse*** signal.

Recall that in the discrete-time case, we defined the *unit impulse sequence* to be

$$\delta[n] = \begin{cases} 1 & n = 0 \\ 0 & n \neq 0 \end{cases}$$

i.e., it is the discrete-time sequence with only one nonzero value. Perhaps the most important use for $\delta[n]$ is its role as a test input to produce the impulse response $h[n]$ of FIR and IIR systems. The impulse response $h[n]$ completely characterizes an LTI discrete-time system through the discrete convolution sum expression (5.11) on p. 110.

Now, for continuous-time systems, we would like to have a similar test signal that would momentarily stimulate the system and give us an "impulse response" that can be used to characterize the continuous-time system if it is LTI. Like $\delta[n]$, the continuous-time impulse must be a very short signal, concentrated at zero. However, the following definition *does not work*

$$\delta(t) = \begin{cases} 1 & t = 0 \\ 0 & t \neq 0 \end{cases} \qquad \text{NO!} \qquad (9.7)$$

because this signal has no strength—its integral is zero. Instead, we will describe $\delta(t)$ as being zero everywhere except for $t = 0$, but also satisfying

$$\int_{-\infty}^{\infty} \delta(t)dt = 1 \qquad (9.8)$$

This does not pin down the value of $\delta(t)$ at $t = 0$, but does define a necessary property that $\delta(t)$ must satisfy.[2]

Since it is problematic to define a function of a continuous variable to be nonzero at only one isolated point, we must take an indirect approach and define the impulse to be the limit of a sequence of continuous functions that become increasingly concentrated at one point. An example is

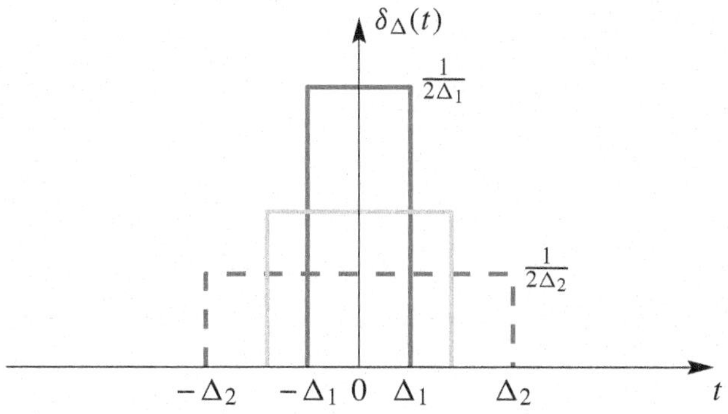

Figure 9-4: Approximating the unit impulse with narrowing pulses.

$$\delta_\Delta(t) = \begin{cases} \frac{1}{2\Delta} & -\Delta < t < \Delta \\ 0 & \text{otherwise} \end{cases} \qquad (9.9)$$

which is a pulse whose width is 2Δ and whose height is $1/(2\Delta)$. As the parameter Δ approaches zero, the pulse gets narrower, but higher. Figure 9-4 shows examples of $\delta_\Delta(t)$ for three different values of Δ. Each of these signals has the property that its total area is one, independent of Δ, so we have

$$\int_{-\infty}^{\infty} \delta_\Delta(\tau)d\tau = 1 \qquad (9.10)$$

The limit of the function $\delta_\Delta(t)$ as $\Delta \to 0$ becomes zero everywhere except at $t = 0$ where its value approaches ∞. We use this as an intuitive definition of the *unit impulse signal*

$$\delta(t) = \lim_{\Delta \to 0} \delta_\Delta(t) \qquad (9.11)$$

[2]The unit impulse for continuous-time signals is hard to define in a mathematically sound fashion because *it is not a function in the normal mathematical sense.* That is, it is not a well-behaved function like a sinusoid or an exponential function. Instead, we will describe the desirable properties for an impulse, and use these as its mathematical definition.

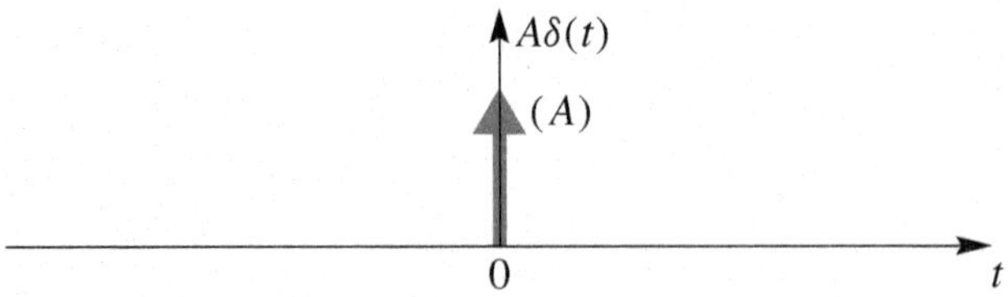

Figure 9-5: Scaled unit-impulse signal is symbolized with an arrow. The area of the impulse is written in parentheses next to the arrowhead.

More important than the limiting value is the fact that the total area of $\delta_\Delta(t)$ remains constant at one regardless of the size of Δ. This unit area property also applies[3] to the limit (9.11), so we conclude that the ***unit-impulse signal*** has the following basic properties:

$$\delta(t) = 0 \qquad \text{for } t \neq 0 \tag{9.12}$$

$$\int_{-\infty}^{\infty} \delta(t)dt = 1 \tag{9.13}$$

In plotting the unit-impulse signal, we will use the style illustrated in Fig. 9-5, which displays $\delta(t)$ as a vertical line with an arrowhead on top. The area (or "size") of the impulse (9.13) is written as the value (A) in parentheses next to the arrowhead. If we write $A\delta(t)$, we say that the area or size of this scaled impulse signal is A because

$$\int_{-\infty}^{\infty} A\delta(t)dt = A\int_{-\infty}^{\infty} \delta(t)dt = A$$

9-2.1 Sampling Property of the Impulse

Even though (9.12) and (9.13) seem adequate to define an impulse, we can develop a better definition by

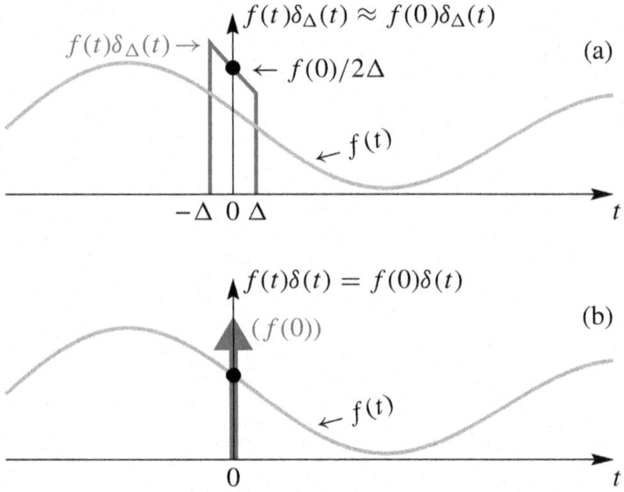

Figure 9-6: (a) Signal $f(t)$ multiplied by a narrow pulse that approximates a unit impulse. (b) Signal $f(t)$ multiplied by a unit impulse.

formulating the *sampling property* of $\delta(t)$. Since the impulse is concentrated at $t = 0$, we consider what happens when we multiply an impulse signal by a continuous signal $f(t)$. As before we use $\delta_\Delta(t)$ and form the product

$$f(t)\delta_\Delta(t) = \begin{cases} f(t)/2\Delta & -\Delta < t < \Delta \\ 0 & \text{otherwise} \end{cases}$$

Figure 9-6(a) shows a continuous signal $f(t)$ and the product $f(t)\delta_\Delta(t)$. This figure shows that in the limit as $\Delta \rightarrow 0$, the only value of $f(t)$ that matters in the product is the value at the single point $t = 0$. From this it follows that

$$\begin{aligned} f(t)\delta(t) &= f(t) \lim_{\Delta\to 0} \delta_\Delta(t) \\ &= \lim_{\Delta\to 0} f(t)\delta_\Delta(t) \\ &= \lim_{\Delta\to 0} f(0)\delta_\Delta(t) = f(0)\delta(t) \end{aligned} \tag{9.14}$$

[3]The unit area property assumes that the following delicate mathematical rearrangement is true $\int \lim_{\Delta\to 0} \delta_\Delta(t)dt = \lim_{\Delta\to 0} \int \delta_\Delta(t)dt$.

In other words, when an impulse multiplies a continuous signal, the value of the continuous signal at $t = 0$ is "sampled" out of the signal, and we are left with only an impulse whose area is scaled by $f(0)$. The notation in (9.14) can be confusing, but we emphasize that the right-hand side must include the impulse and *the area of the impulse* conveys the sample value.

If we want the actual value of $f(t)$ at $t = 0$, then we must integrate (9.14):

$$\int_{-\infty}^{\infty} f(t)\delta(t)dt = \int_{-\infty}^{\infty} f(0)\delta(t)dt$$

$$= f(0)\int_{-\infty}^{\infty} \delta(t)dt = f(0) \qquad (9.15)$$

This is depicted in Fig. 9-6(b) where $(f(0))$ is placed beside the arrowhead to indicate the area of the impulse.

The sampling property can be generalized a little bit if we introduce time shifting. The time-shifted impulse $\delta(t - t_0)$ is called "an impulse at time t_0." Now if we form the product $f(t)\delta(t - t_0)$, the concentration of the impulse is centered at $t = t_0$, so we get

Sampling Property of the Impulse

$$f(t)\delta(t - t_0) = f(t_0)\delta(t - t_0) \qquad (9.16)$$

and also the integral form of the *Sampling Property*

$$\int_{-\infty}^{\infty} f(t)\delta(t - t_0)dt = f(t_0) \qquad (9.17)$$

which is true because the integral of the shifted impulse is also unity. Figure 9-7 shows sampling with two shifted impulses at $t = t_1$ and $t = t_2$.

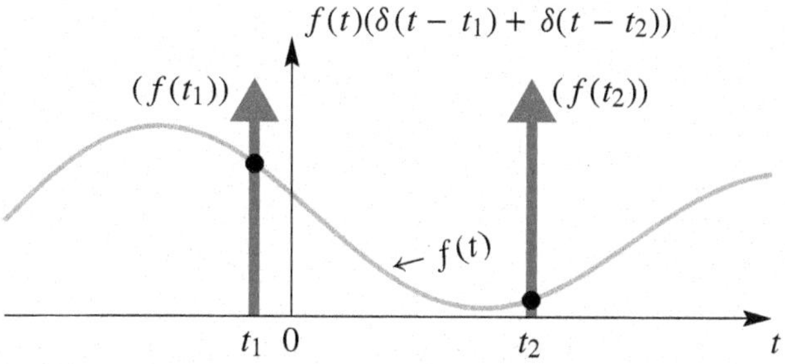

Figure 9-7: Signal $f(t)$ multiplied by two shifted unit impulses. The resulting signal is $f(t)(\delta(t-t_1)+\delta(t-t_2)) = f(t_1)\delta(t - t_1) + f(t_2)\delta(t - t_2)$.

Example 9-2: Sampling with an Impulse

Suppose that we are given the expression $\sin(20\pi t)\delta(t - \frac{1}{80})$ to simplify. The sampling property (9.16) enables the following simplification

$$\begin{aligned}\sin(20\pi t)\delta(t - \tfrac{1}{80}) &= \sin[20\pi(\tfrac{1}{80})]\delta(t - \tfrac{1}{80}) \\ &= \sin(0.25\pi)\delta(t - \tfrac{1}{80}) \\ &= 0.707\delta(t - \tfrac{1}{80})\end{aligned}$$

The important thing to note is that a continuous function (in this case $\sin(20\pi t)$) multiplied by an impulse becomes an impulse with a size dependent only on the value of the continuous function at the time location of the impulse. Using the result of the above manipulation, it follows also that

$$\int_{-\infty}^{\infty} \sin(20\pi t)\delta(t - \tfrac{1}{80})dt = 0.707$$

since the limits of integration include the time at which the impulse occurs. ∎

9-2.2 Mathematical Rigor

The impulse is *not* a function, although some of its properties suggest that it can be treated like one. A rigorous definition of the impulse requires a mathematical theory called "distributions." In this theory the impulse is actually defined as an object that represents the operation of sampling. In our notation, the best definition of the impulse would be the integral form of the sampling property (9.17). In fact, a common warning is that an impulse should only be used within an integral where it can be evaluated to a number. Said another way, it is the area of the impulse $\delta(t - t_0)$ that matters, not its (infinite) value at $t = t_0$.

9-2.3 Engineering Reality

We cannot generate a true impulse signal with electronic equipment, but that does not diminish the utility of the impulse idea. Applications such as flash photography and high-speed data modems need extremely short pulses. Furthermore, we have already seen that the impulse response completely characterizes an LTI system, so it is important to be able to measure the impulse response in the lab with actual narrow-pulse signals. In Section 9-7.4 we will show that a very narrow pulse, which is a normal mathematical function, can produce an output for an LTI system that is virtually identical to the theoretical impulse response of that system.

9-2.4 Derivative of the Unit Step

In calculus we learn that it is not possible to take the derivative of a discontinuous function. However, impulses can be used to expand the concept of differentiation to include discontinuous functions. In this section we will show the close relationship between the impulse $\delta(t)$ and the unit step $u(t)$.

First of all, we must state a generalized "area property" for the impulse

$$\int_a^b \delta(t)dt = \begin{cases} 1 & \text{if } a < 0 \text{ and } b \geq 0 \\ 0 & \text{if } a \geq 0 \text{ or } b < 0 \end{cases} \tag{9.18}$$

where we assume $b > a$.[4] This generalization is reasonable because a definite integral is the area of its integrand between the limits of integration. Since $\delta(t)$ is concentrated at $t = 0$, the area should be one when $\delta(t)$ lies between the limits of integration.

Now we can obtain the relationship between the unit impulse and the unit step. Consider the running integral of $\delta(t)$

$$\int_{-\infty}^{t} \delta(\tau)d\tau \tag{9.19}$$

From the generalized area property (9.18) we get

$$\int_{-\infty}^{t} \delta(\tau)d\tau = \begin{cases} 1 & \text{if } t \geq 0 \\ 0 & \text{if } t < 0 \end{cases} \tag{9.20}$$

But the right-hand side of (9.20) is identical to the definition of the unit step $u(t)$ in (9.3) on p. 247, so

$$u(t) = \int_{-\infty}^{t} \delta(\tau)d\tau \tag{9.21}$$

Next, we recall the Fundamental Theorem of Calculus which states that

$$g(t) = \int_a^t f(\tau)d\tau \quad \Longrightarrow \quad f(t) = \frac{d}{dt}g(t)$$

[4]The situation where one of the limits is zero is indeterminate, so we make an arbitrary distinction in (9.18) in order to be consistent with our definition of the unit step $u(t)$.

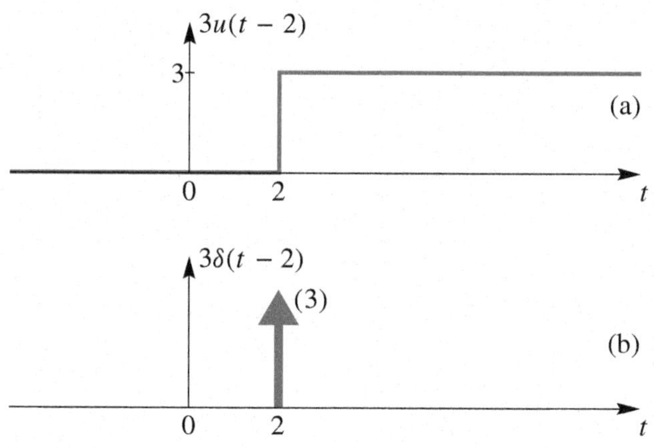

Figure 9-8: (a) Unit step signal $3u(t-2)$ and (b) its derivative, $3\delta(t-2)$.

from which we conclude that the unit impulse acts like the derivative of the unit step

$$\delta(t) = \frac{d}{dt}u(t) \tag{9.22}$$

A quick review of the foregoing derivation would lead to the same properties being true for a shifted impulse

$$\delta(t-t_0) = \frac{d}{dt}u(t-t_0) \tag{9.23}$$

$$u(t-t_0) = \int_{-\infty}^{t} \delta(\tau - t_0)d\tau \tag{9.24}$$

Figure 9-8 shows a shifted scaled unit step and its derivative, a shifted scaled unit impulse.

Armed with a newfound ability to differentiate discontinuous signals, we can use the product rule from calculus to take the derivative of the signal $x(t) = f(t)u(t)$ which is discontinuous at $t = 0$ even when $f(t)$ is differentiable. The signal $x(t)$ will jump from 0 just before $t = 0$ to $f(0+)$ just on the other side of $t = 0$.[5] If we differentiate $x(t)$ using the product rule familiar from calculus and apply the sampling property of the impulse from (9.14), we obtain[6]

$$\begin{aligned} x^{(1)}(t) = \frac{dx(t)}{dt} &= f(t)\frac{du(t)}{dt} + u(t)\frac{df(t)}{dt} \\ &= f(t)\delta(t) + \frac{df(t)}{dt}u(t) \\ &= f(0)\delta(t) + \frac{df(t)}{dt}u(t) \end{aligned}$$

Thus, the derivative of a discontinuous signal contains an impulse whose area is equal to the size of the jump in the signal plus the ordinary derivative of the signal at all other times t where the signal is continuous. If the discontinuity is a negative jump, then the impulse area would be negative.

Example 9-3: Derivative of a Discontinuous Function

Consider the signal $x(t) = e^{-2(t-1)}u(t-1)$ which is discontinuous at $t = 1$. Its derivative is

$$\begin{aligned} \frac{dx(t)}{dt} &= e^{-2(t-1)}\frac{du(t-1)}{dt} + u(t-1)\frac{de^{-2(t-1)}}{dt} \\ &= e^{-2(t-1)}\delta(t-1) - 2e^{-2(t-1)}u(t-1) \\ &= e^{0}\delta(t-1) - 2e^{-2(t-1)}u(t-1) \end{aligned}$$

The graphical representation of this example is shown in Fig. 9-9. Such figures are useful because they show that each discontinuity of the function generates an impulse in the derivative whose size is equal to the size of the discontinuity of the function. ■

[5] The notation $f(0+)$ means $\lim_{t\to 0} f(t)$, but approaching through positive values only.

[6] The compact superscript notation $^{(1)}$ means the first derivative.

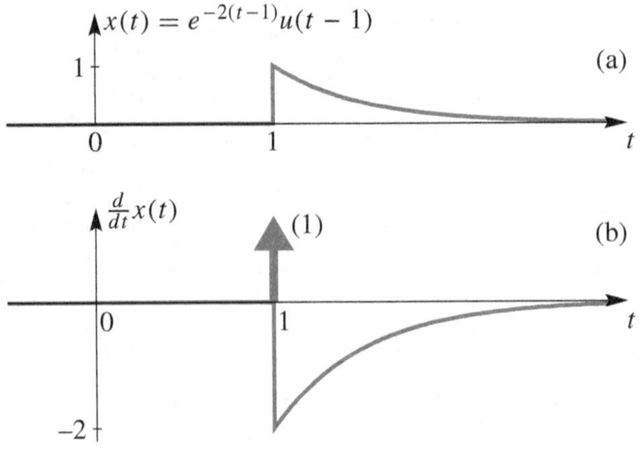

Figure 9-9: (a) Delayed one-sided exponential and (b) its derivative.

9-3 Continuous-Time Systems

Continuous-time systems are natural phenomena or man-made devices that transform one continuous-time signal into another. Mathematically, we say that the system ***operates*** on the input signal $x(t)$ to produce the output signal $y(t)$, and we represent this by a notation such as

$$y(t) = \mathcal{T}\{x(t)\} \tag{9.25}$$

where $\mathcal{T}\{\cdot\}$ is a general symbolic representation of the input/output operator that describes the system. In other words, $\mathcal{T}\{\cdot\}$ is the mathematical rule for assigning an output signal to an input signal. Sometimes it might be possible to state the rule in words, but generally we want a mathematical formula to define how $y(t)$ is produced from $x(t)$. As in the case of discrete-time systems, another compact notation is

$$x(t) \longmapsto y(t) \tag{9.26}$$

which conveys the fact that $y(t)$ is the output of a system when $x(t)$ is the input. A third representation shows the input signal going through a block diagram that contains the symbol for the transformation $\mathcal{T}\{\cdot\}$ as in Fig. 9-10.

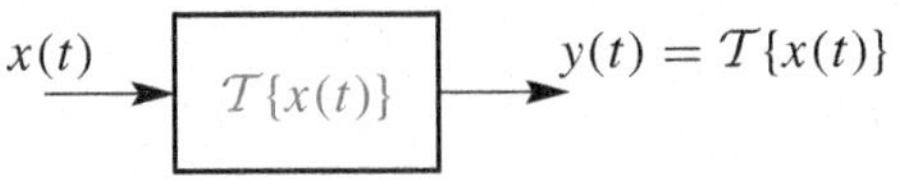

Figure 9-10: Block diagram representation of a continuous-time system.

9-3.1 Some Basic Continuous-Time Systems

A simple example of a continuous-time system is the ***squaring*** system defined by the input/output relation

$$y(t) = [x(t)]^2 \tag{9.27}$$

which states that the output signal value at time t is equal to the square of the input signal value at that same time. Some other systems that have simple mathematical definitions are the ***ideal delay*** system defined by

$$y(t) = x(t - t_d) \tag{9.28}$$

where t_d is the time delay of the system, and the ***ideal differentiator*** system, defined by

$$y(t) = \frac{dx(t)}{dt} \tag{9.29}$$

Finally, we have the ***ideal integrator***, which is defined as the system whose input and output satisfy the *running integral* equation

$$y(t) = \int_{-\infty}^{t} x(\tau)d\tau \tag{9.30}$$

Integrators are useful models for a variety of continuous-time systems such as capacitors and op-amps in electrical circuits.

In each of the four cases given here, the rule for getting the output from the input could have been stated in words. However, it is customary to convert the rule to a mathematical equation in order to facilitate analysis of the operation.

9-3.2 Continuous-Time Outputs

Given the rule for a system's operator, it is often easy to determine the output once the input is known. For example, suppose that the input to the squaring system is $x(t) = -3u(t-2)$. Then the output signal is

$$\begin{aligned} y(t) &= [-3u(t-2)]^2 \\ &= 9[u(t-2)]^2 = 9u(t-2) \end{aligned}$$

Because the unit step is either zero or one, squaring $u(t-2)$ gives zero or one too.

The output of a system is called its ***response***. One particular response is quite important. This is the ***impulse response***, which is the output when the input is $x(t) = \delta(t)$. For example, the impulse response of the ideal integrator can be found by substituting $\delta(t)$ into the system definition (9.30)

$$y(t) = \int_{-\infty}^{t} \delta(\tau)d\tau$$

A previous result (9.21) tells us that the impulse response of the ideal integrator is, in fact, the unit step.

> *The impulse response occurs so often that a specific notation is reserved for that signal. Instead of $y(t)$, the output is denoted by $h(t)$ when the input is an impulse.*

Thus we would say $h(t) = u(t)$ for the integrator.

9-3.3 Analogous Discrete-Time Systems

Recall that in Chapter 5 we have already defined a discrete-time version of the squarer system in (5.14) on p. 116 using a formula similar to (9.27) for relating the input sequence to the output sequence. Likewise, an ideal discrete-time delay system for sequence inputs was defined by (5.9) on p. 109 with the time-shift parameter restricted to be an integer number of samples. On the other hand, while the ideal differentiator is similar to the discrete-time first-difference system, the analogy is not perfect since differentiation is inherently a continuous variable operation involving infinitesimal limits. Likewise, the ideal integrator, defined in (9.30) is similar to a "running sum" discrete-time system defined by the equation

$$y[n] = \sum_{k=-\infty}^{n} x[k] \qquad (9.31)$$

The system defined by (9.31) is also called an *accumulator* system since in computing the output sample $y[n]$ it sums (or accumulates) the values of the input sequence up to and including the value at n. The accumulator simply sums numbers, and this is easily done in hardware with digital arithmetic. The integrator also sums past values of the input signal, but it does the sum in the incremental sense of the Riemann integral.

9-4 Linear Time-Invariant Systems

Just as in the case of discrete-time systems, we can study continuous-time systems by imposing certain properties on the system transformation. In Chapter 5 on *FIR Filters*, Section 5-5, we introduced the properties of linearity and time-invariance for discrete-time systems, and we showed that these two constraints lead to the *convolution sum*

$$y[n] = \sum_{k=-\infty}^{\infty} x[k]h[n-k] \qquad -\infty < n < \infty \qquad (9.32)$$

a general representation that applies to all linear time-invariant (LTI) discrete-time systems. In this section, we will see that linearity and time-invariance also lead to a general relation between the input and output of an LTI continuous-time system. This relation is the *convolution integral.*

9-4.1 Time-Invariance

Suppose that a continuous-time system is represented by

$$x(t) \longmapsto y(t) \tag{9.33}$$

This system is *time-invariant* if, when we delay the input signal by an arbitrary amount t_0, the result is only to delay the output by that same amount. Using the notation of (9.33) the time-invariance condition is

$$x(t-t_0) \longmapsto y(t-t_0) \tag{9.34}$$

This condition must be true for *any* signal $x(t)$ and for *any* real number t_0. Figure 5-16 on p. 116 shows how to test a discrete-time system for time-invariance. An identical method can be used for testing continuous-time systems.

Example 9-4: Squaring is Time-Invariant

Consider the squaring system given by (9.27). When the input is $x(t)$, the output is $y(t) = [x(t)]^2$. Now if the input is $x(t-t_0)$, the corresponding output will be $w(t) = [x(t-t_0)]^2$. Therefore, we see that $w(t) = y(t-t_0)$, so the system is time-invariant. ■

Example 9-5: Integrator is Time-Invariant

The integrator system is also a time-invariant system. To prove this, we replace $x(\tau)$ in (9.30) by $x(\tau - t_0)$ obtaining the output $w(t)$

$$w(t) = \int_{-\infty}^{t} x(\tau - t_0)d\tau \tag{9.35}$$

Now, to prove time-invariance, we must manipulate the integral in (9.35) into a form that is recognizable in terms of the original output $y(t)$. This is done by changing the "dummy variable" of integration to $\sigma = \tau - t_0$. In this substitution, $d\tau$ is replaced by $d\sigma$, the lower limit $\tau = -\infty$ becomes $\sigma = -\infty$, and the upper limit $\tau = t$ becomes $\sigma = t - t_0$. Therefore (9.35) becomes

$$w(t) = \int_{-\infty}^{t-t_0} x(\sigma)d\sigma$$

and it is now clear that $w(t) = y(t-t_0)$, so the integrator system is seen to be time-invariant. ■

Example 9-6: A Time-Varying System

As a third example, consider the *amplitude modulator* system defined by

$$y(t) = x(t)\cos(\omega_c t) \tag{9.36}$$

Such a system is a fundamental component of many radio systems. If we apply the test for time-invariance, we change the input to $x(t-t_0)$ obtaining as output the signal $w(t) = x(t-t_0)\cos(\omega_c t)$, which is *not* equal to $y(t-t_0)$ because to obtain $y(t-t_0)$, both variables t in (9.36) must be replaced by $(t-t_0)$ giving $x(t-t_0)\cos[\omega_c(t-t_0)]$. Therefore, the system is *not time-invariant* or more directly, it is a *time-varying* system. ■

EXERCISE 9.2: The input and output of the time delay system satisfy $y(t) = x(t-t_d)$ where t_d is the delay of the system. Show that the time delay system is a time-invariant system.

9-4.2 Linearity

A continuous-time system is linear if when $x_1(t) \longmapsto y_1(t)$ and $x_2(t) \longmapsto y_2(t)$, then

$$\begin{aligned} x(t) &= \alpha x_1(t) + \beta x_2(t) \\ &\longmapsto y(t) = \alpha y_1(t) + \beta y_2(t) \end{aligned} \tag{9.37}$$

Equation (9.37) expresses the *principle of superposition* for continuous-time linear systems. Figure 5-17 on p. 117 shows how to test a discrete-time system to see if it is linear. We can apply the same procedure to continuous-time systems.

Example 9-7: Squaring is Nonlinear

The squaring system is *not* a linear system, i.e., it is a *nonlinear* system. To see this, note that if $x(t) = \alpha x_1(t) + \beta x_2(t)$, the corresponding output is

$$y(t) = [\alpha x_1(t) + \beta x_2(t)]^2$$

and after some algebra we obtain

$$\begin{aligned} y(t) &= \alpha^2[x_1(t)]^2 + 2\alpha\beta x_1(t)x_2(t) + \beta^2[x_2(t)]^2 \\ &\neq \alpha y_1(t) + \beta y_2(t) \end{aligned}$$

where $y_1(t) = [x_1(t)]^2$ and $y_2(t) = [x_2(t)]^2$. Therefore, the squaring system is nonlinear. ■

Example 9-8: Integrator is Linear

The integrator system is a linear system. To show this, note that if $x(t) = \alpha x_1(t) + \beta x_2(t)$, the corresponding output is

$$\begin{aligned} y(t) &= \int_{-\infty}^{t} x(\tau)d\tau = \int_{-\infty}^{t} [\alpha x_1(\tau) + \beta x_2(\tau)]d\tau \\ &= \alpha \int_{-\infty}^{t} x_1(\tau)d\tau + \beta \int_{-\infty}^{t} x_2(\tau)d\tau \\ &= \alpha y_1(t) + \beta y_2(t) \end{aligned}$$

Thus, the integrator system is a linear system. ■

EXERCISE 9.3: The differentiator system is defined by the input/output relation

$$y(t) = \frac{dx(t)}{dt}$$

Show that the differentiator system is both linear and time-invariant (LTI).

9-4.3 The Convolution Integral

When a continuous-time system is both linear and time-invariant (LTI), the following general statement is true:

> ***Every LTI system can be described by a convolution integral.***
>
> $$y(t) = \int_{-\infty}^{\infty} x(\tau)h(t-\tau)d\tau \qquad (9.38)$$

where $h(t)$ is the impulse response of the system, $x(t)$ is the input and $y(t)$ is the output. Convolution (9.38) is a binary operation between two functions $x(t)$ and $h(t)$ that produces a third function $y(t)$ defined for all time, $-\infty < t < \infty$. This operation is usually written as $y(t) = x(t) * h(t)$, meaning "$x(t)$ is convolved with $h(t)$." We hasten to point out that the operator notation $x(t) * h(t)$ can sometimes be misleading. It disguises the fact that when we write $y(t)$ we really mean "the value of the function y at time t." This value $y(t)$ is obtained by evaluating the integral

$$y(t) = \int_{-\infty}^{\infty} x(\tau)h(t-\tau)d\tau$$

which depends on the parameter t. To obtain all the values of the function y, the integral must be evaluated for each t. Thus, a better notation would be $y = x * h$, emphasizing that the function y is obtained by convolving the two functions x and h. However, the notation $y(t) = x(t) * h(t)$ is generally used to mean the same as (9.38) without much confusion.

First of all, note that the convolution integral is a definite integral, and the limits $-\infty$ to ∞ in the definition (9.38) imply that the integration must be done over all nonzero portions of the integrand $x(\tau)h(t-\tau)$. In specific cases, the limits will be finite, in other cases they might be functions of t. Furthermore, the variables t and τ play distinctly different roles. The variable τ is the "dummy variable" of integration. It disappears when the integral is evaluated. The variable t on the other hand, is the independent variable of the output function $y(t)$ and it will range over all values $-\infty < t < \infty$. In the integral, t is a constant, so, in effect, we must evaluate the integral for each different value of t. We will discuss the evaluation of convolution integrals in more detail in Section 9-7.

Strictly speaking, the statement in (9.38) should read as follows: "A system is LTI if and only if its output can be represented as a convolution." We will not prove the "only if" part of this theorem, but we can verify that the convolution integral does represent a linear and time-invariant system. First assume that $x(t) = \alpha x_1(t) + \beta x_2(t)$. Then the corresponding output is

$$y(t) = \int_{-\infty}^{\infty} x(\tau)h(t-\tau)d\tau$$

$$= \int_{-\infty}^{\infty} [\alpha x_1(\tau) + \beta x_2(\tau)]h(t-\tau)d\tau$$

$$= \alpha \int_{-\infty}^{\infty} x_1(\tau)h(t-\tau)d\tau + \beta \int_{-\infty}^{\infty} x_2(\tau)h(t-\tau)d\tau$$

$$= \alpha y_1(t) + \beta y_2(t)$$

Thus, the operation of convolution of an input $x(t)$ with $h(t)$ is a linear operation.

To verify that convolution is time-invariant, replace the input by $x(\tau - t_0)$ and examine the resulting output

$$w(t) = \int_{-\infty}^{\infty} x(\tau - t_0)h(t-\tau)d\tau$$

By making a substitution $\sigma = \tau - t_0$ for the dummy variable of integration, we transform the integral into

$$w(t) = \int_{-\infty}^{\infty} x(\sigma)h(t-\sigma-t_0)d\sigma$$

$$= \int_{-\infty}^{\infty} x(\sigma)h((t-t_0)-\sigma)d\sigma = y(t-t_0)$$

Therefore, convolution is time-invariant as well as linear, so (9.38) defines an LTI system. Thus, an equivalent statement to (9.38) is the following:

> *For every LTI system, the output $y(t)$ is always equal to $x(t) * h(t)$, the convolution of the input signal $x(t)$ with the system impulse response $h(t)$.*

In other words, once we know the impulse response $h(t)$ of an LTI system, we can use the convolution integral to get the output of the system for ***any*** input.

Example 9-9: Convolve Unit Steps

When the impulse response is a unit step, $h(t) = u(t)$, and the input is also a unit step, $x(t) = u(t)$, the convolution integral becomes

$$\begin{aligned} y(t) &= \int_{-\infty}^{\infty} x(\tau)h(t-\tau)d\tau \\ &= \int_{-\infty}^{\infty} u(\tau)u(t-\tau)d\tau \\ &= \int_{0}^{\infty} 1u(t-\tau)d\tau \\ &= \int_{0}^{t} 1d\tau = \begin{cases} t & t \geq 0 \\ 0 & t < 0 \end{cases} \end{aligned} \tag{9.39}$$

The upper limit becomes t because $u(t-\tau) = 1$ when $t-\tau \geq 0$, or $\tau \leq t$. The final answer (9.39) can be written concisely as

$$u(t) * u(t) = t\,u(t) \tag{9.40}$$

The signal $t u(t)$ is called a ***unit ramp*** because it is linearly increasing with a slope of one.

This example illustrates that one skill in doing a convolution integral is the proper manipulation of the limits of integration prior to performing the actual integration. ■

9-4.4 Properties of Convolution

The operation of convolution, and therefore the class of continuous-time LTI systems, has a number of useful mathematical properties that make it possible to manipulate and simplify complicated LTI systems. Specifically, the operation of convolution is *commutative*, *associative*, and *distributive over addition*. These properties are analogous to similarly named properties of ordinary multiplication and addition, as well as discrete-time convolution.

Commutativity: The commutative property is

$$x(t) * h(t) = h(t) * x(t) \tag{9.41}$$

i.e., we can determine the output of an LTI system by convolving $x(t)$ with $h(t)$ or vice-versa. This property is proved by making the substitution $\sigma = t - \tau$ and $d\sigma = -d\tau$ in the convolution integral and manipulating it as follows:

$$\begin{aligned} y(t) &= \int_{-\infty}^{\infty} x(\tau)h(t-\tau)d\tau \\ &= \int_{\infty}^{-\infty} x(t-\sigma)h(\sigma)(-d\sigma) \\ &= \int_{-\infty}^{\infty} x(t-\sigma)h(\sigma)d\sigma = \int_{-\infty}^{\infty} h(\sigma)x(t-\sigma)d\sigma \end{aligned}$$

Recall that reversing the sign of a definite integral reverses the order of the limits. The last form of the expression for $y(t)$ is by definition $h(t) * x(t)$, so the commutativity of convolution is proved.

Associativity: An operation like convolution is associative if

$$[x(t) * h_1(t)] * h_2(t) = x(t) * [h_1(t) * h_2(t)] \tag{9.42}$$

In plain English, if we have a convolution of three functions, we can do the convolution operations in any order. This is just like multiplication where numbers (or functions) can be multiplied in any order. At this point of our discussion, the proof of this result would involve a tedious manipulation of a double integral. However, in Section 9-8 we will show how this result falls out naturally in the context of manipulating cascaded LTI systems.

Distributivity Over Addition: Convolution also satisfies the relation

$$x(t) * [h_1(t) + h_2(t)] = x(t) * h_1(t) + x(t) * h_2(t) \tag{9.43}$$

The proof of this property is simple if we recall that convolution is a linear operation. Therefore, convolution of $x(t)$ with the sum $h_1(t)+h_2(t)$ must be the sum of the individual convolutions.

Identity Element of Convolution: The question of finding an identity element for convolution is the problem of solving for $x(t)$ in the following equation:

$$x(t) * h(t) = h(t) \tag{9.44}$$

We can show that the answer is the unit-impulse signal by substituting $x(t) = \delta(t)$ in the convolution integral (9.38) and applying the sampling property of the impulse to obtain

$$\int_{-\infty}^{\infty} \delta(\tau)h(t-\tau)d\tau = \int_{-\infty}^{\infty} \delta(\tau)h(t-0)d\tau$$

$$= h(t)\int_{-\infty}^{\infty} \delta(\tau)d\tau = h(t)$$

In making this derivation, it is permissible to move $h(t)$ outside of the integral because the variable t is an independent parameter different from τ, the dummy variable of integration.

We can summarize the identity element in convolution notation, because we have just demonstrated that

$$\delta(t) * h(t) = h(t) \tag{9.45}$$

In other words, the impulse is the identity signal for the operation of convolution, much as 1 is the identity element for ordinary multiplication. Another interpretation of (9.45) is that the output (response) of the LTI system due to a unit impulse input is $h(t)$, but we already knew this fact from the definition of the impulse response!

9-5 Impulse Responses of Basic LTI Systems

If a system is LTI, it is completely characterized by its impulse response $h(t)$. In a practical setting, we might have to resort to measuring a system's impulse response experimentally using a very short pulse that approximates the impulse. In other cases, we have to solve difficult mathematical equations to get $h(t)$. In this section, we will "measure" the impulse responses of several simple LTI systems by simply using $\delta(t)$ as the input to the system to generate the output.

9-5.1 Integrator

The integrator system was defined by the input/output relation[7]

$$y(t) = \int_{-\infty}^{t} x(\tau)d\tau = x^{(-1)}(t)$$

The impulse response of this system is the unit step as shown previously in Equation (9.21). Thus, we can represent the integrator operationally as follows:

> ***A Running Integral is equivalent to convolution with a unit step.***
>
> $$x(t) * u(t) = x^{(-1)}(t)$$

(9.46)

[7]The superscript $^{(-1)}$ means the first anti-derivative, i.e., integral.

9-5.2 Differentiator

The differentiator system defined by

$$y(t) = \frac{dx(t)}{dt} = x^{(1)}(t)$$

is also an LTI system. If we proceed as in the case of the integrator and substitute $\delta(t)$ for $x(t)$, we obtain

$$h(t) = \frac{d\delta(t)}{dt} = \delta^{(1)}(t)$$

The question of what the derivative of an impulse might be can be studied by limiting arguments; however, the concept of a convolution operator gives us a useful interpretation simply by noting that since $\delta^{(1)}(t)$ is the impulse response of the differentiator system, the following operational definition must be true:

Differentiation is convolution with derivative of an impulse.

$$x(t) * \delta^{(1)}(t) = x^{(1)}(t) \tag{9.47}$$

The signal $\delta^{(1)}(t)$ is called the ***doublet***.

Example 9-10: Convolution with Doublet

The convolution of the unit step with the doublet

$$y(t) = u(t) * \delta^{(1)}(t)$$

can be evaluated by using (9.47), and the derivative property of the unit step (9.22); i.e.,

$$y(t) = \frac{d}{dt}u(t) = \delta(t)$$

■

9-5.3 Ideal Delay

The ideal delay system was defined previously as the system whose output is obtained from the input by

$$y(t) = x(t - t_d)$$

where t_d is the amount of time delay in seconds. Substituting $x(t) = \delta(t)$ gives the impulse response of the ideal delay system as

$$h(t) = \delta(t - t_d)$$

Thus, the impulse response of the ideal delay is a delayed impulse, or "an impulse at time $t = t_d$." In our symbolic notation, the input is convolved with $h(t)$ to give the output, so the result is

Time shift is the same as convolution with a shifted impulse.

$$x(t) * \delta(t - t_d) = x(t - t_d). \tag{9.48}$$

This is a very important result that should be treated as one of the fundamental properties of the impulse function.

9-6 Convolution of Impulses

With the time-shift property of (9.48), it is relatively easy to perform the convolution of signals that only contain impulses. In fact, this case is quite similar to discrete-time convolution except that the impulses do not have to be regularly spaced. Here is the basic theorem that we need:

Convolution of Impulses

$$\delta(t - t_1) * \delta(t - t_2) = \delta(t - (t_1 + t_2)) \tag{9.49}$$

In other words, the convolution of two shifted impulses at t_1 and t_2 gives a shifted impulse located at the sum,

$t = t_1 + t_2$. In order to see that this statement is true, we simply need to invoke (9.48); i.e.,

$$\begin{aligned} x(t) * \delta(t - t_2) &= x(t - t_2) \\ \Longrightarrow \delta(t - t_1) * \delta(t - t_2) &= \delta((t - t_2) - t_1) \\ &= \delta(t - (t_1 + t_2)) \end{aligned}$$

where we have substituted $\delta(t - t_1)$ for $x(t)$.

Example 9-11: Convolve Impulses

When combined with the linearity property, we can work problems such as convolving the following two signals:

$$x(t) = \delta(t - 5) + \delta(t + 5)$$

and

$$h(t) = \delta(t - 0.5) - 3\delta(t)$$

$$\begin{aligned} &x(t) * h(t) \\ &\quad = (\delta(t - 5) + \delta(t + 5)) * (\delta(t - 0.5) - 3\delta(t)) \\ &\quad = \delta(t - 5) * \delta(t - 0.5) - 3\delta(t - 5) * \delta(t) \\ &\qquad + \delta(t + 5) * \delta(t - 0.5) - 3\delta(t + 5) * \delta(t) \\ &\quad = \delta(t - 5.5) - 3\delta(t - 5) + \delta(t + 4.5) - 3\delta(t + 5) \end{aligned}$$

So we end up with impulses at $t = 5.5, -4.5$ and ± 5. The area of the impulses at $t = \pm 5$ is -3. ■

Example 9-12: Impulse Convolution Causes Shifting

When one of the signals in a convolution is a regular continuous-time function and the other contains only impulses, a graphical approach often leads to a simple

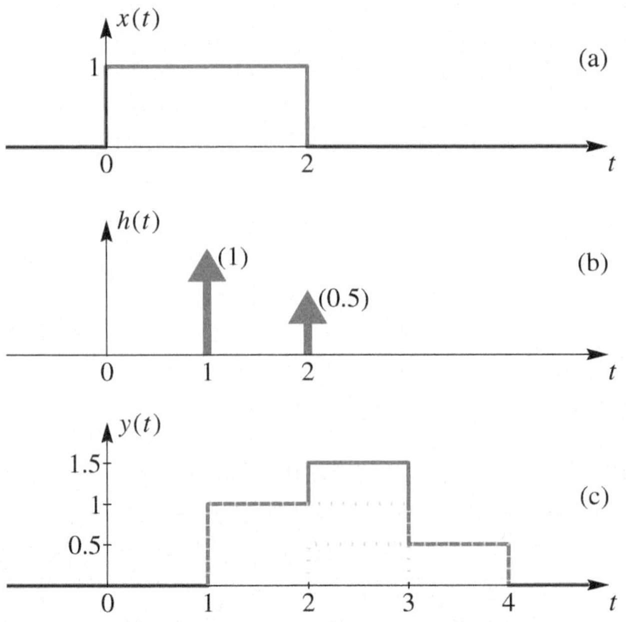

Figure 9-11: Convolution of a pulse with impulses: (a) Pulse input $x(t)$. (b) Impulse response $h(t)$ consisting only of impulses. (c) Resulting output after convolution, $y(t) = x(t) * h(t)$.

solution. Consider the pulse input $x(t)$ and the impulse response $h(t) = \delta(t-1) + 0.5\delta(t-2)$ shown in Fig. 9-11. The equation for the output is simply

$$\begin{aligned} y(t) &= x(t) * [\delta(t - 1) + 0.5\delta(t - 2)] \\ &= x(t - 1) + 0.5x(t - 2) \end{aligned}$$

As Fig. 9-11 shows, such convolutions are easy to do graphically. We simply shift a scaled copy of the continuous signal to the location of each impulse and sum all the shifted and scaled copies. As shown by the dotted lines in Fig. 9-11(c), the two copies overlap in the region $2 \le t < 3$, so the output is $1 + 0.5 = 1.5$ in that interval. ■

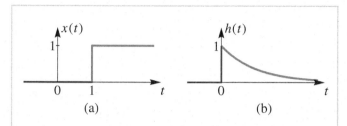

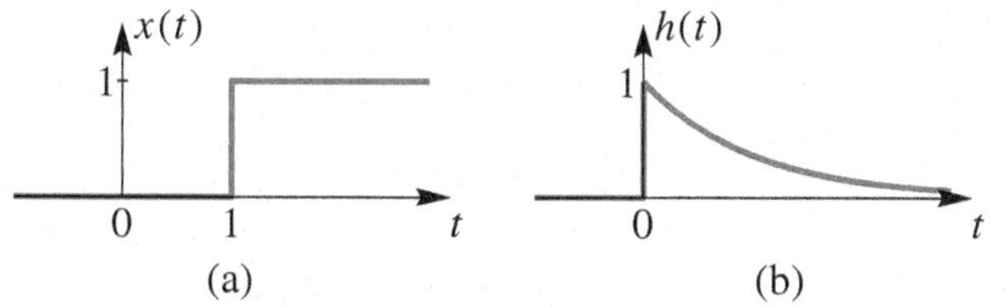

Figure 9-12: Example of convolution: (a) Input is a shifted unit-step signal, (b) Impulse response is a decaying exponential.

9-7 Evaluating Convolution Integrals

In the analysis of continuous-time LTI systems, it is sometimes necessary to evaluate the convolution of two continuous-time signals in detail. We have seen that if one of the two signals is an impulse or a sum of impulses, the evaluation is straightforward; we simply shift the time origin of the other function to the location of the impulse and scale it by the area of the impulse. A more complicated, but nevertheless tractable, situation occurs when both functions are made up of pieces of simple functions like constants or exponentials whose products are readily integrable with a table of integrals. In such cases, care in setting up the limits of integration is crucial to success. We will illustrate this with some examples.

9-7.1 Delayed Unit-Step Input

Suppose that we wish to evaluate a convolution integral

$$y(t) = \int_{-\infty}^{\infty} h(\tau)x(t-\tau)d\tau \qquad (9.50)$$

where the two functions $h(t) = e^{-t}u(t)$ and $x(t) = u(t-1)$ are plotted in Fig. 9-12. Now it is important to point out again that τ is what we call the "dummy variable" of integration in the integral of (9.50) because it disappears when we evaluate at the upper and lower limits. Furthermore, t is the independent variable of $y(t)$. Thus, to compute each value of the function $y(t)$, we must form the product $h(\tau)x(t-\tau)$ and then evaluate the integral in (9.50) for each different value of t.

If we substitute the functions $x(t)$ and $h(t)$ into (9.50), we obtain

$$y(t) = \int_{-\infty}^{\infty} e^{-\tau}u(\tau)u(t-\tau-1)d\tau \qquad (9.51)$$

which is certainly a correct form, but one that still needs a lot of work. In particular, this form does not make it obvious how the limits of integration depend on t. The key to efficient and correct evaluation of a convolution integral of this type is to draw an auxiliary sketch of the two functions whose product is the integrand of the convolution integral. The sketch of $h(\tau)$ is easy; it is just the same graph as $h(t)$ with the axis relabeled. The sketch of $x(t-\tau)$, however, is less obvious, but if we proceed systematically, it is also easily done. To see the steps, consider Fig. 9-13. Figure 9-13(a) shows $x(\tau)$, which is obtained from Fig. 9-12 by simply relabeling the horizontal axis. Figure 9-13(b) shows $g(\tau) = x(-\tau)$, which is the time-reversed version of $x(\tau)$. (We introduce the new symbol $g(\tau)$ simply to emphasize that the time-reversed function is really just a new function of τ.) Finally, in Fig. 9-13(c) we show the function $g(\tau)$ with its time origin shifted by t. Remember that as far as the dummy variable of integration τ is concerned, t is a constant. Now from the definition of $g(\tau)$, we obtain

$$g(\tau - t) = x(-(\tau - t)) = x(t-\tau)$$

i.e., the desired function $x(t-\tau)$ is obtained by first time-reversing $x(\tau)$ and then shifting the result by t. Thus, to

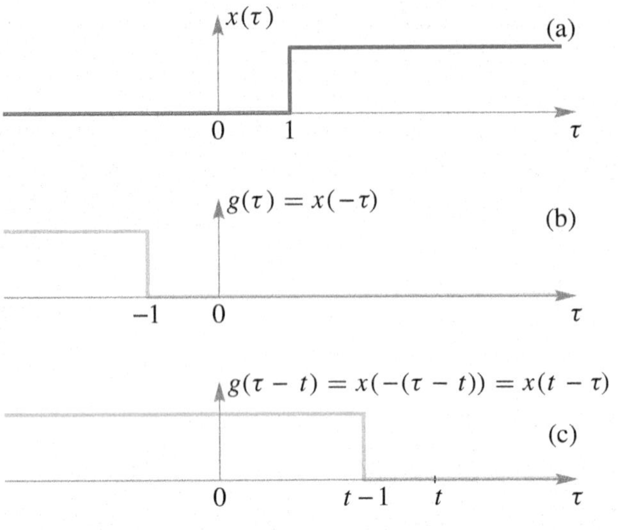

Figure 9-13: Illustration of determining $x(t - \tau)$ in evaluating convolution: (a) Input signal $x(\tau)$. (b) Time reversed input signal $x(-\tau)$. (c) Time reversed input signal shifted by t on the τ axis (shown for $t = 3$). The gray axis indicates a change from t to τ.

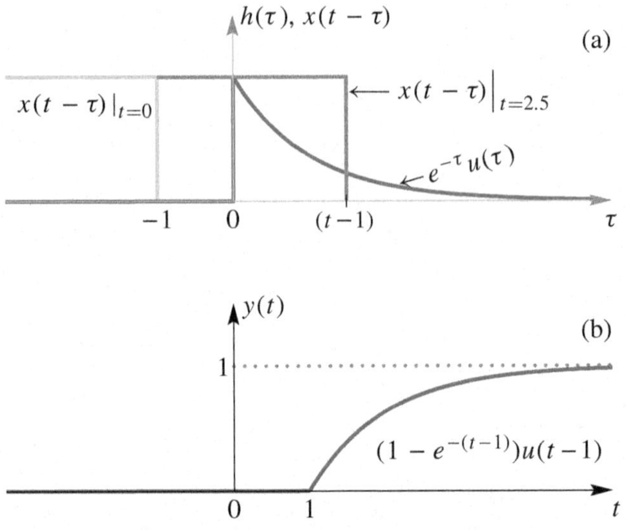

Figure 9-14: Illustration of evaluation of convolution. (a) the signal $h(t)$ and flipped versions of $x(t)$ plotted on the τ axis, (b) the output signal $y(t)$ plotted versus t. The gray axis indicates a change from t to τ.

plot or sketch the graph of $x(t - \tau)$, we must follow the two steps detailed in Fig. 9-13:

1. flip $x(\cdot)$ with respect to the origin

2. then shift the τ-origin to t.

We will generally have to do this for many values of t.

Now we are ready to draw the sketch that shows how $h(\tau)$ and $x(t - \tau)$ interact in evaluating the convolution integral (9.50). Figure 9-14(a) shows a plot of both $h(\tau)$ and $x(t - \tau)$ on the same set of axes for two different choices of t. Note that the graph $h(\tau)$ is identical to the graph $h(t)$ in Fig. 9-12(a) only with its axis relabeled τ instead of t. Also shown in Fig. 9-14 is $x(t - \tau) = u(t - \tau - 1)$ for two different values of t.

Figure 9-14 shows that there are two different regions for evaluating this convolution integral. On the left, we see a "typical" plot of $x(t - \tau)$ that is representative of all values of t such that $t - 1 < 0$. For this case, the nonzero parts of the two functions $h(\tau)$ and $x(t - \tau)$ do not overlap and hence, their product $h(\tau)x(t - \tau)$ is zero for all $t < 1$. Therefore, the convolution integral evaluates to

$$y(t) = 0 \qquad \text{for } t < 1$$

The second typical plot of $x(t - \tau)$ in Fig. 9-14(a) is shown with a gray line. This plot is representative for all values of t such that $t - 1 > 0$, even though it is drawn for the specific value $t = 2.5$. Since the flipped and shifted $x(t - \tau)$ extends infinitely to the left, the nonzero parts of $h(\tau)$ and $x(t - \tau)$ will overlap for all t such that $t - 1 > 0$, or equivalently for $t > 1$. Using the plot of Fig. 9-14, it is straightforward now to evaluate the convolution, because we can see that the product $h(\tau)x(t - \tau)$ equals $e^{-\tau}$ over the range $0 < \tau < t - 1$, meaning that the output for $t > 1$

is given by the integral

$$y(t) = \int_0^{t-1} e^{-\tau} d\tau \qquad \text{for } t > 1$$

The evaluation of this integral involves the following steps:

$$\begin{aligned} y(t) &= \int_0^{t-1} e^{-\tau} d\tau \\ &= -e^{-\tau} \Big]_0^{t-1} \\ &= -e^{-(t-1)} + 1 \\ &= 1 - e^{-(t-1)} \quad \text{for } t > 1 \end{aligned}$$

Finally, we can use the unit-step function to express the zero and nonzero regions of $y(t)$, giving the compact form

$$y(t) = (1 - e^{-(t-1)})u(t-1) \tag{9.52}$$

which is plotted in Fig. 9-14(b).

This example illustrates some important general properties of continuous-time convolution. First, this is an example of the more general case where $x(t) = 0$ for $t < T_0$ and $h(t) = 0$ for $t < T_1$. (In this case $T_0 = 1$ and $T_1 = 0$.) By drawing a sketch similar to Fig. 9-14, it can be shown that if $t < T_0 + T_1$, then the nonzero parts of $x(\tau)$ and $h(t-\tau)$ do not overlap. Therefore, regardless of the detailed nature of the input and impulse response at other values of t, it follows that it is always true that $y(t) = 0$ for $t < T_0 + T_1$. Another important concept that is illustrated by this example is that even if one or both of the signals involved in the convolution is discontinuous, the result of the convolution (output) will be a continuous function of time. This is due to the integration, which has a natural tendency to smooth the signals.

EXERCISE 9.4: Show that the convolution of two exponential signals, $x(t) = e^{-at}u(t)$ and $h(t) = e^{-bt}u(t)$, is

$$y(t) = x(t) * h(t) = \frac{1}{b-a}\left(e^{-at}u(t) - e^{-bt}u(t)\right)$$

if $a \neq b$

EXERCISE 9.5: Convince yourself of the truth of the following statement: If the input contains no impulses, but is discontinuous (like $u(t)$) then the impulse response must contain at least one impulse in order that the output be discontinuous too.

EXERCISE 9.6: Show by drawing pictures like Fig. 9-14(a) that if $x(t) = 0$ for $t < T_1$ and $h(t) = 0$ for $t < T_2$, then the output $y(t) = x(t) * h(t)$ is zero for $t < T_1 + T_2$. On other words, the starting time of $y(t)$ is the sum of the starting times of $x(t)$ and $h(t)$.

EXERCISE 9.7: Repeat the convolution of $x(t) = u(t-1)$ with $h(t) = e^{-t}u(t)$, but this time use the convolution integral form

$$y(t) = \int_{-\infty}^{\infty} x(\tau)h(t-\tau)d\tau$$

Because convolution is commutative, your answer should be the same as (9.52).

In order to provide a tool for studying convolution for continuous-time systems, a MATLAB GUI was created to run different cases where the signals can be selected from a menu of pulses, exponentials, sinusoids and impulses. The screen shot of Fig. 9-15 shows the GUI with a pulse

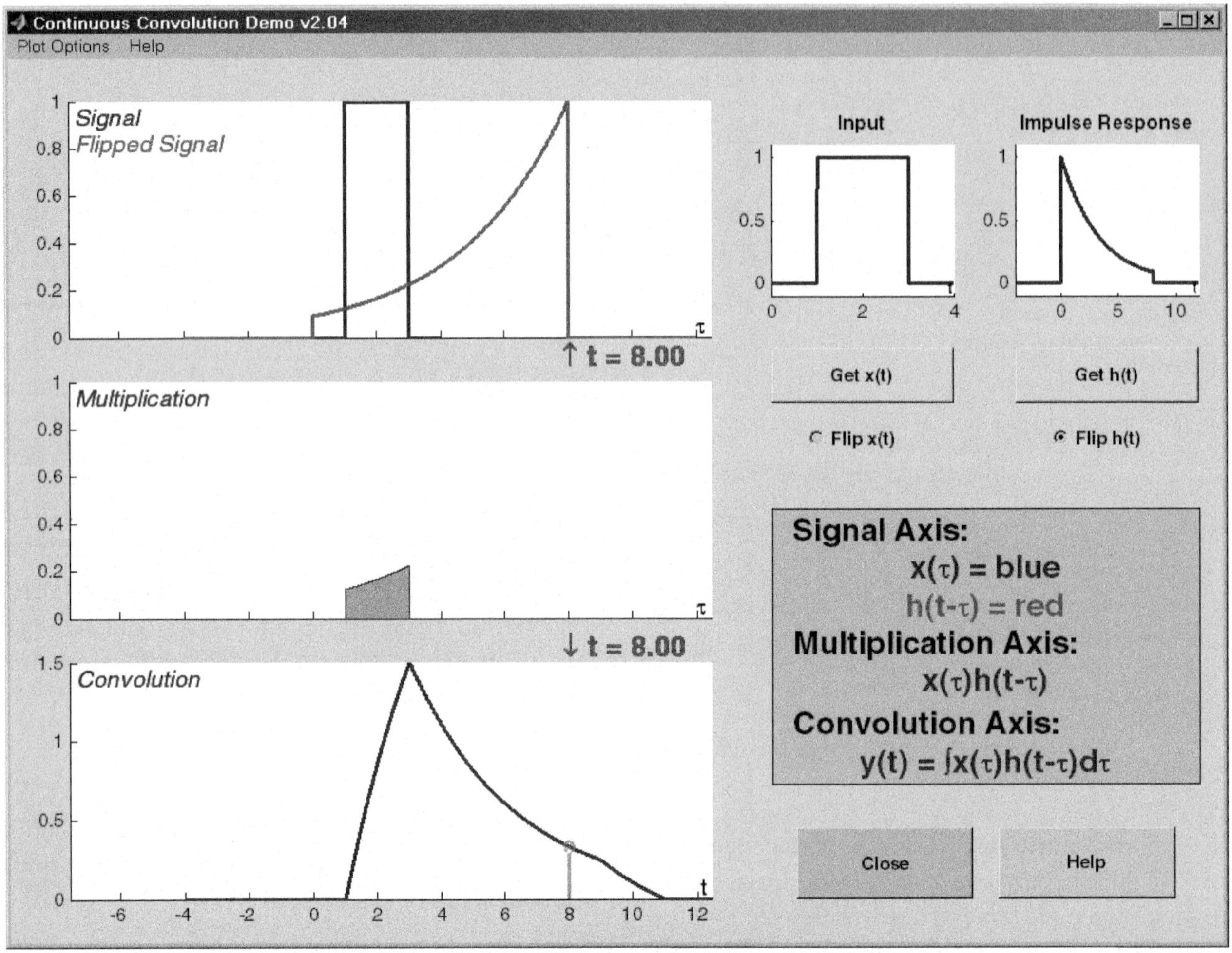

Figure 9-15: Graphical User Interface (GUI) for `cconvdemo` which illustrates the *flip and slide* nature of continuous-time convolution. The user can pick the two signals, and then use the mouse pointer to slide one of the signals (flipped) over the other one. The area of the overlapped signals is shown in the middle panel. The specific example shown here is $x(t) = u(t-1) - u(t-3)$ convolved with $h(t) = e^{-0.3t}[u(t) - u(t-8)]$. For the specific flipped $h(t-\tau)$, the value of the output time parameter is $t = 8$.

convolved with an exponential. The output signal is shown in the bottom panel of the GUI.

DEMO: *Continuous Convolution*

To become skilled at evaluating convolution integrals there is no substitute for practice and studying worked examples. Section 9-7.3 presents a more complicated example, but first, we shall return to the discrete time-

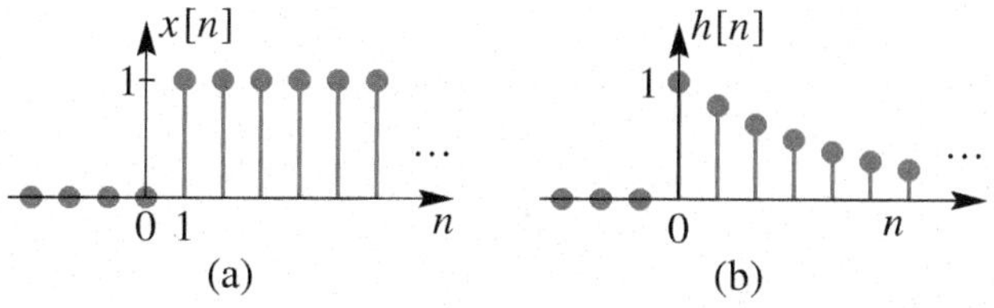

Figure 9-16: Example of convolution: (a) Input is a shifted discrete-time unit-step signal $x[n] = u[n-1]$; (b) Impulse response is a discrete-time decaying exponential $h[n] = a^n u[n]$.

domain and show how what we have just learned can be used in the closed-form evaluation of discrete convolutions.

9-7.2 Evaluation of Discrete Convolution

Recall that the discrete convolution sum is defined as

$$y[n] = \sum_{k=-\infty}^{\infty} h[k]x[n-k] \qquad -\infty < n < \infty \qquad (9.53)$$

If we compare (9.53) to (9.50) we see that the equations look very similar. The integral in (9.50) becomes a discrete sum in (9.53); the dummy variable of integration τ is replaced by the dummy index of summation k; and the independent time variable of the output signal t is replaced by the discrete index n. Although these are significant differences, the basic structure is the same.

We shall continue to explore the analogy by assuming that the input is a shifted discrete-time unit-step sequence $x[n] = u[n-1]$ and the impulse response is a discrete-time exponential sequence $h[n] = a^n u[n]$ as depicted in Fig. 9-16. If we substitute these sequences into (9.53), we obtain

$$y[n] = \sum_{k=-\infty}^{\infty} a^k u[k] u[n-k-1] \qquad (9.54)$$

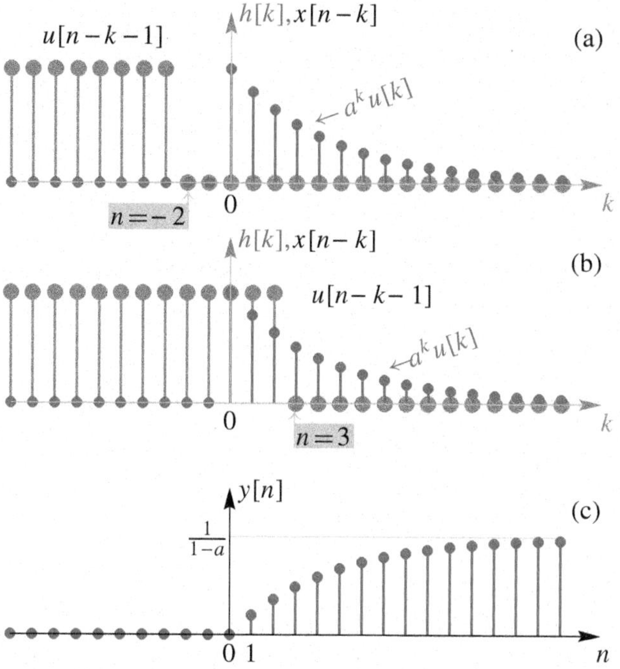

Figure 9-17: Illustration of evaluation of discrete convolution.

which is just as confusing as (9.51) unless we carefully consider how the two sequences interact as we change n. In general, to evaluate (9.53) for a given pair of signals, we need to form the product $h[k]x[n-k]$ for n fixed and sum the values of the resulting product sequence over all k. We must do this for all values $-\infty < n < \infty$. As in the case of continuous-time convolution, it helps if we plot $x[n-k]$ as a function of k for different values of n. We can do this by first time-reversing $x[k]$ to obtain $g[k] = x[-k]$, and then shifting $g[k]$ by n to obtain $g[k-n] = x[-(k-n)] = x[n-k]$. For the given pair of sequences, this is shown in Fig. 9-17(a) where both $h[k] = a^k u[k]$ and $x[n-k] = u[n-k-1]$ are plotted as functions of k. In this particular case $n = -2$. (Note that

$x[n-k]$ is plotted gray, while $h[k]$ is plotted in orange with smaller solid dots.) Similarly, Fig. 9-17(b) again shows both $h[k] = a^k u[k]$ and $x[n-k] = u[n-k-1]$ plotted as functions of k, but this time for $n = 3$. These plots make it easier to evaluate the convolution sum in closed-form. For example, by generalizing from the case depicted in Fig. 9-17(a), we can see that the nonzero parts of the sequences $h[k]$ and $x[n-k]$ do not overlap if $n - 1 < 0$. Therefore, it follows that

$$y[n] = 0 \qquad \text{for } n < 1$$

Furthermore, from Fig. 9-17(b) it follows that for $n - 1 \geq 0$ the output will be

$$y[n] = \sum_{k=0}^{n-1} a^k = \frac{1 - a^n}{1 - a}.$$

The closed form expression for the sum follows from the formula for the sum of n terms of a geometric series. We can combine the results of our evaluation into a single formula for the output sequence obtaining

$$y[n] = \left(\frac{1 - a^n}{1 - a}\right) u[n-1] \qquad (9.55)$$

which is plotted in Fig. 9-17(c).

A comparison of Figs. 9-14(b) and 9-17(c) shows that the output in both cases has the appearance of a "smoothed" version of the input-unit step signal. On the other hand, there are some differences that are mainly due to the fact that in discrete-convolution the signals are sequences rather than functions of a continuous time variable. Notice that in the discrete case the output becomes nonzero immediately when $h[k]$ and $x[n-k]$ first overlap on the k axis. In contrast, for continuous-time signals the first overlap of $h(\tau)$ with $x(t-\tau)$ produces zero area so the output is zero at that time of first overlap. The output signal builds up only as the overlap region achieves a nonzero width. While this difference is significant, it should not obscure the fact that discrete- and continuous-time convolutions are very similar in their effects and in their mathematical structures. Keep this in mind as we consider more examples of continuous-time convolution in the next sections.

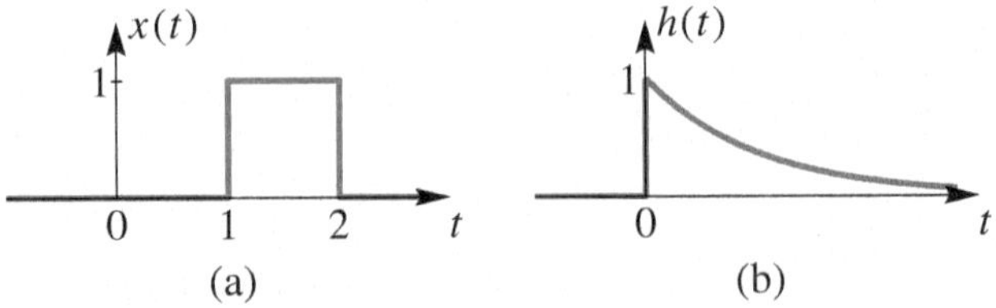

Figure 9-18: Example of convolution: (a) Input. (b) Impulse response.

DEMO: *Discrete Convolution*

9-7.3 Square-Pulse Input

Consider the impulse response $h(t) = e^{-t}u(t)$ and the "pulse" input

$$x(t) = u(t-1) - u(t-2) = \begin{cases} 1 & 1 \leq t < 2 \\ 0 & \text{otherwise} \end{cases}$$

These two functions are shown in Fig. 9-18. We can evaluate the convolution of these two signals without using the "flip and shift" method, if we exploit linearity and time invariance for a previous convolution (9.52) that we have already done. We can summarize that result as

$$e^{-t}u(t) * u(t-1) = (1 - e^{-(t-1)})u(t-1) \qquad (9.56)$$

which is shown in Fig. 9-19(a). Time invariance means that shifting the unit-step function from $u(t-1)$ to $u(t-2)$ will shift the output by one, or

$$e^{-t}u(t) * u(t-2) = (1 - e^{-(t-2)})u(t-2) \qquad (9.57)$$

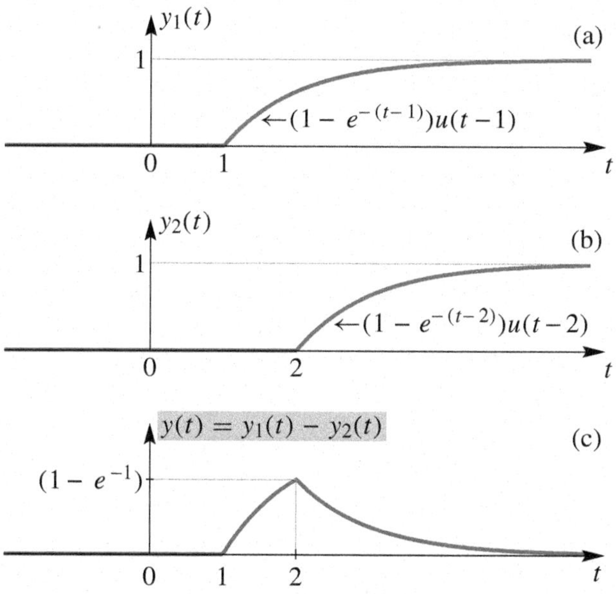

Figure 9-19: Convolution constructed via linearity. (a) Output when the input is $u(t-1)$, (b) output when the input is $u(t-2)$. The output (c) is the difference of the two auxiliary outputs in (a) and (b), because $x(t) = u(t-1) - u(t-2)$.

which is shown in Fig. 9-19(b). Now we use linearity to combine these two results to get the convolution that we want:

$$\begin{aligned} e^{-t}u(t) * [u(t-1) - u(t-2)] \\ = (1-e^{-(t-1)})u(t-1) - (1-e^{-(t-2)})u(t-2) \end{aligned} \quad (9.58)$$

It is hard to visualize the output unless we reduce the output into cases. The complete equation for $y(t)$ is

$$y(t) = \begin{cases} 0 & t < 1 \\ (1-e^{-(t-1)}) & 1 \le t < 2 \\ e^{-(t-2)} - e^{-(t-1)} & 2 \le t \end{cases} \quad (9.59)$$

Alternatively, we can write $y(t) = (1-e^{-1})e^{-(t-2)}$ for the region $(t \ge 2)$ as in Fig. 9-19(c). We can view the output signal in Fig. 9-19(c) as a "smeared out" pulse whose duration is greater than the duration of the input pulse. This is a normal result of convolution that happens when the duration of the impulse response is greater than the duration of the input.

9-7.4 Very Narrow Square Pulse Input

Another example will illustrate how a narrow pulse acts like an impulse. The following pulse signal has a duration of 1 msec, but its area is one:

$$\begin{aligned} x(t) &= 1000[u(t) - u(t-0.001)] \\ &= \begin{cases} 1000 & 0 \le t < 0.001 \\ 0 & \text{otherwise} \end{cases} \end{aligned} \quad (9.60)$$

If we continue to use the same impulse response $h(t) = e^{-t}u(t)$ as in the previous examples, we can use linearity and time-invariance to construct the output in terms of the known convolution (9.52); namely,

$$\begin{aligned} & e^{-t}u(t) * 1000[u(t) - u(t-0.001)] = \\ & 1000[(1-e^{-t})u(t) - (1-e^{-(t-0.001)})u(t-0.001)] \end{aligned} \quad (9.61)$$

Therefore, the complete equation for $y(t)$ is

$$y(t) = \begin{cases} 0 & t < 0 \\ 1000(1-e^{-t}) & 0 \le t < 0.001 \\ 1.0005e^{-t} & t \ge 0.001 \end{cases} \quad (9.62)$$

where 1.0005 is the approximate value of $1000(e^{0.001}-1)$. Figure 9-20 is a plot of $y(t)$ in (9.62), although it is impossible to see any details in the region $(0 \le t < 0.001)$.

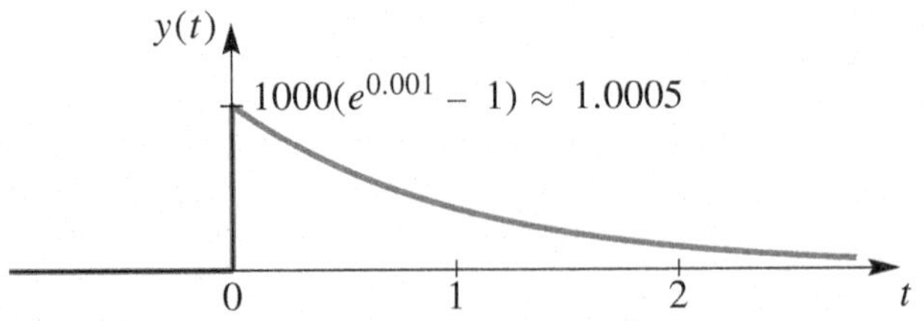

Figure 9-20: Convolution output for a narrow pulse input.

9-7.5 Discussion of Convolution Examples

Two convolution integrals were worked out in the preceding sections based on the convolution done in Section 9-7.1. In all three cases, the impulse response was a decaying exponential which is an appropriate mathematical model for the impulse response of a first-order electrical circuit consisting of a resistor and a capacitor and for other physical systems. The feature that varies among the three examples is the pulse length of the input signal. In Section 9-7.1, the input is an infinitely long signal that turns on at $t = 1$, but lasts forever. In this case, we see that the output of the convolution is also infinitely long, but the transition of the unit-step signal at $t = 1$ has been smoothed out by the convolution with the exponential. The output signal rises gradually from zero to an asymptotic value of 1, but it takes about 5 seconds for the output to be very close to 1 (within 1% of its final value) . Specifically,

$$\begin{aligned} y(6) &= (1 - e^{-(6-1)})u(6-1) \\ &= (1 - e^{-5}) \approx 0.9933 \end{aligned}$$

In Section 9-7.3, the input pulse has a finite duration, but it is not extremely short. In this case, the output is a severely distorted version of the input pulse.

In Section 9-7.4, the input pulse is very short relative to the pulse in the second example, and its area is equal to one. Since this input is nearly an impulse, we might expect that the output should start to act like the impulse response, which it does for $t > 0.001$ secs.

EXERCISE 9.8: Suppose that $x(t) \neq 0$ only for $T_1 \leq t \leq T_2$ and $h(t) \neq 0$ only for $T_3 \leq t \leq T_4$. Draw a sketch of $x(\tau)$ with $h(t-\tau)$ for several values of t as in Fig. 9-14 for "typical" finite-duration functions $x(t)$ and $h(t)$. Use this figure to show that $y(t) = x(t) * h(t)$ also has finite duration given by $y(t) \neq 0$ for $T_5 \leq t \leq T_6$. Find T_5 and T_6 in terms of T_1, T_2, T_3, and T_4.

9-8 Properties of LTI Systems

As we discussed previously for discrete-time systems, linear time-invariant systems have a number of properties that can greatly simplify our work. We will discuss several of these in this section.

9-8.1 Cascade and Parallel Combinations

A cascade of two LTI systems is shown in Fig. 9-21(a). In this configuration, the input $x(t)$ is the input to the first system, whose output is the input to the second system, whose output is the overall output $y_1(t)$.

This figure shows the input to the first system $x(t)$, the output of the first system (and input to the second system) $w_1(t)$, and the overall output (output of the second system) $y_1(t)$ along the top of the arrows connecting the systems. These symbols stand for general inputs and outputs of the system. Below the arrows are the inputs and outputs for an impulse input to the cascade system. For the impulse input, the output of the first system is, by definition, its impulse

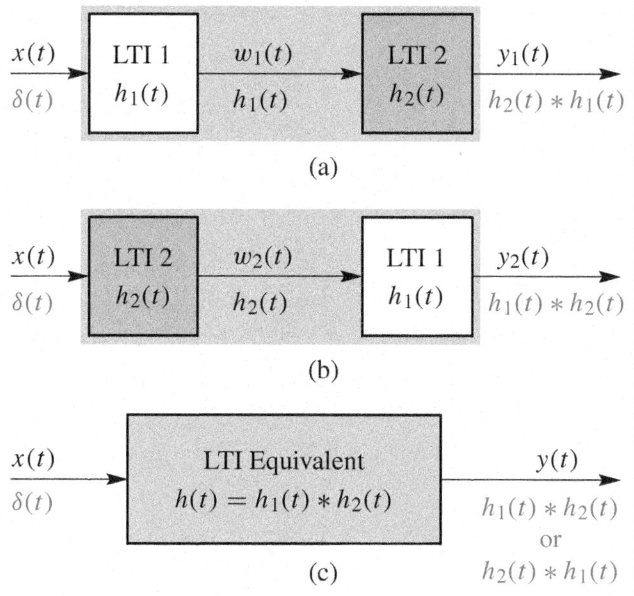

Figure 9-21: Cascade connection of LTI systems: (a) System #2 following System #1. (b) System #1 following System #2. (c) Equivalent system to both (a) and (b).

response $h_1(t)$. Since $h_1(t)$ becomes the input to the second system, the overall output is $h_1(t) * h_2(t)$, and this, by definition, is the impulse response of the overall cascade system. Figure 9-21(b) shows the two systems in reverse order. The same impulse input produces the output $h_2(t) * h_1(t)$ for the overall impulse response of the system. Since, as we have shown, convolution is a commutative operation, it follows that $h_1(t) * h_2(t) = h_2(t) * h_1(t)$, so we conclude that the two cascade systems in Figs. 9-21(a) and (b) have the same overall impulse response from input $x(t)$ to the output $y(t)$. Therefore, $y_1(t) = y_2(t) = y(t)$ whenever the systems have the same input signal, $x(t)$. Therefore, yet another system that has the same overall response $h(t) = h_1(t) * h_2(t) = h_2(t) * h_1(t)$ is the system in Fig. 9-21(c). Thus, the cascade systems

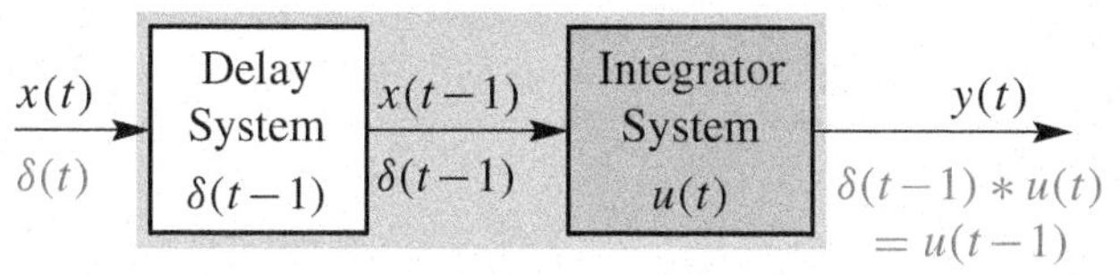

Figure 9-22: Cascade implementation of an integrator with delay of 1.

of Figs. 9-21(a) and (b) can be replaced by the single equivalent system in Fig. 9-21(c). Alternatively, a system like Fig. 9-21(c) whose impulse response can be expressed as $h(t) = h_1(t) * h_2(t)$ can be represented and realized by either of the cascade systems Figs. 9-21(a) or (b).

Example 9-13: Delay-Integrator Cascade

Consider the system whose impulse response is $h(t) = u(t-1)$. Since this impulse response is just a delayed unit-step function, it can be represented by the convolution $h(t) = \delta(t-1) * u(t)$. Thus, the system can be implemented by the cascade of a delay of 1 followed by an integrator as depicted in Fig. 9-22. Also note that the order of the delay and the integrator can be reversed in the cascade configuration without changing the overall impulse response. ■

Figure 9-23 shows a parallel connection of two LTI systems, where the same signal $x(t)$ is the input to both systems, and whose outputs are then added to produce the overall output $y(t)$. As in Fig. 9-21, the upper signal variables denote general inputs and outputs, while below the arrows are the output signals that result when the input is the unit impulse $\delta(t)$.

Note that the overall impulse response is $h_1(t) + h_2(t)$, and the parallel connection of two systems can be replaced by the equivalent system in Fig. 9-23(b).

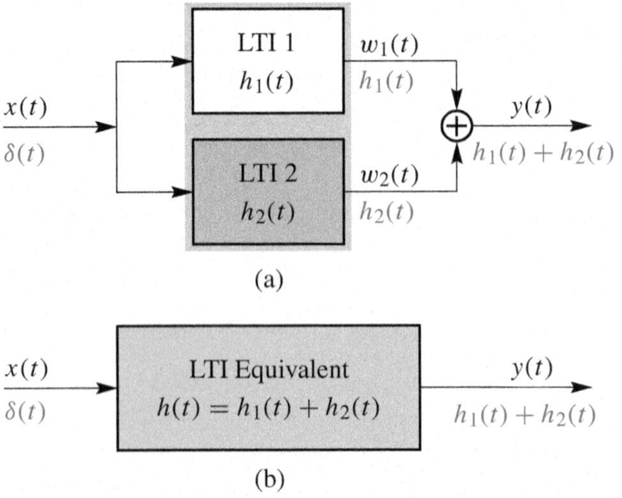

Figure 9-23: Parallel connection of LTI systems: (a) Parallel connection of two systems. (b) Equivalent system to (a).

Example 9-14: Parallel Connection

Suppose that the impulse response of an LTI system is

$$h(t) = u(t-1) - u(t-2) = \begin{cases} 1 & 1 \le t < 2 \\ 0 & \text{otherwise} \end{cases}$$

Using properties of convolution, we can express $h(t)$ as

$$h(t) = u(t) * [\delta(t-1) - \delta(t-2)]$$

so that the system could be implemented by a cascade of an integrator followed by a parallel combination of two delay systems. ■

EXERCISE 9.9: Draw a block diagram representation of the system in Example 9-14 as an integrator followed by a parallel combination of two delay systems.

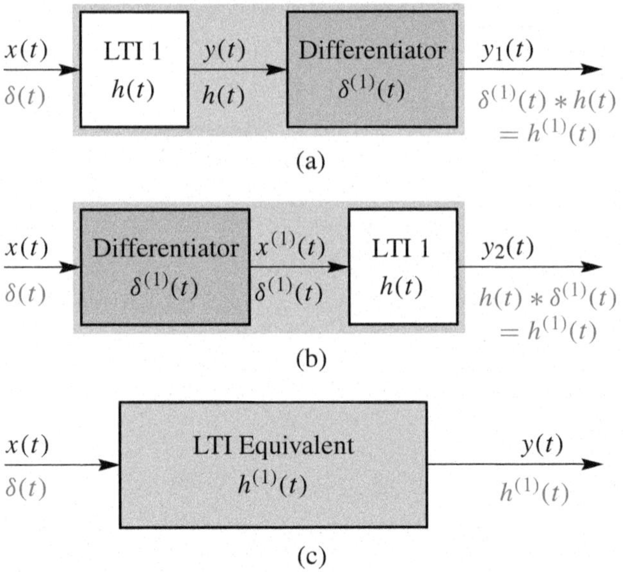

Figure 9-24: Differentiation in convolution: (a) Differentiation of output. (b) Differentiation of input. (c) Differentiation of impulse response (equivalent system to both (a) and (b)).

9-8.2 Differentiation and Integration of Convolution

We can use what we have just learned about the cascade connection of LTI systems to derive a basic property about the derivative of a convolution. First of all, the output in Fig. 9-24(a) is

$$\begin{aligned} y_1(t) = \frac{dy(t)}{dt} &= y(t) * \delta^{(1)}(t) \\ &= [x(t) * h(t)] * \delta^{(1)}(t) \\ &= x(t) * [h(t) * \delta^{(1)}(t)] \\ &= x(t) * h^{(1)}(t) \end{aligned}$$

Likewise, the output in Fig. 9-24(b) is

$$y_2(t) = x^{(1)}(t) * h(t)$$

where the superscript $x^{(1)}$ denotes the first derivative. However, the cascade property implies that $y_2(t) = y_1(t)$, so we obtain the following general result:

> ***Differentiating the input or the impulse response differentiates the result of the convolution.***
>
> $$y^{(1)}(t) = x^{(1)}(t) * h(t) = x(t) * h^{(1)}(t) \quad (9.63)$$

Furthermore, by replacing the differentiators by integrators and doing a similar analysis, we can show that

> ***Integrating the input or the impulse response integrates the result of the convolution.***
>
> $$\int_{-\infty}^{t} y(\tau)d\tau = y^{(-1)}(t) = x(t) * h^{(-1)}(t) \quad (9.64)$$
> $$= x^{(-1)}(t) * h(t)$$

In fact, the results of (9.63) and (9.64) can be generalized to

$$y^{(m+n)}(t) = x^{(m)}(t) * h^{(n)}(t) \quad (9.65)$$

where m and n are positive integers (for derivatives) or negative integers (for integrals). Sometimes (9.65) can be used to simplify or provide alternative approaches for evaluating a convolution. This is illustrated in the following example.

Example 9-15: Derivative-Integral Cascade

Consider the input and impulse response in Section 9-7.3 and plotted in Fig. 9-18 on p. 268. If we differentiate the input and integrate the impulse response before convolution, we will achieve the same result as convolving the two functions directly. In this case, $x(t) = u(t-1) - u(t-2)$ so $x^{(1)}(t) = \delta(t-1) - \delta(t-2)$. To integrate the impulse response, we convolve with the unit-step function and find that $h^{(-1)} = 0$ for $t < 0$ and

$$h^{(-1)}(t) = \int_0^t e^{-\tau} d\tau$$
$$= (1 - e^{-t}) \quad \text{for } t > 0$$

More compactly, $h^{(-1)}(t) = (1 - e^{-t})u(t)$. Now we compute $y(t)$ as follows:

$$\begin{aligned} y(t) &= x^{(1)}(t) * h^{(-1)}(t) \\ &= [\delta(t-1) - \delta(t-2)] * [(1 - e^{-t})u(t)] \\ &= (1 - e^{-(t-1)})u(t-1) - (1 - e^{-(t-2)})u(t-2) \end{aligned}$$

If we look closely at the last expression above, we see that

$$y(t) = \begin{cases} 0 & t < 1 \\ (1 - e^{-(t-1)}) & 1 \le t < 2 \\ e^{-(t-2)} - e^{-(t-1)} & 2 \le t \end{cases}$$

which is identical to (9.59) in Section 9-7.3. ∎

9-8.3 Stability and Causality

The issues of causality and stability are important if we wish to implement any type of system. In this subsection, we will discuss these issues for continuous-time LTI systems and show again that the impulse response completely characterizes and defines the system properties.

Stability For a continuous-time system, where $x(t) \longmapsto y(t)$, we can describe stability in two ways: a general definition based on intuition and also as a precise mathematical condition.

> **General Definition of Stability:** A system is stable if and only if *every* bounded input produces a bounded output.
>
> **Mathematical Condition for Stability:** A system is stable if and only if, for *all bounded* inputs such that $|x(t)| < B < \infty$ for all t, it is also true that $|y(t)| < C < \infty$ for all t, where B and C are constants.

If we specialize to the case of LTI systems, then a necessary and sufficient condition can be based on the impulse response $h(t)$ as follows:

A continuous-time LTI system is stable if and only if:

$$\int_{-\infty}^{\infty} |h(\tau)| d\tau < \infty \tag{9.66}$$

The proof of the sufficiency part of this statement is relatively easy. First, we use the convolution integral to express $|y(t)|$ in terms of the magnitudes of the input $|x(t)|$ and the impulse response $|h(t)|$; i.e.,

$$\begin{aligned} |y(t)| &= \left| \int_{-\infty}^{\infty} h(\tau)x(t-\tau)d\tau \right| \\ &\leq \int_{-\infty}^{\infty} |h(\tau)| \cdot |x(t-\tau)| d\tau \end{aligned} \tag{9.67}$$

Here we have used the fact that the magnitude of the total area under the function $h(\tau)x(t-\tau)$ must be less than or equal to the total area under the magnitude $|h(\tau)x(t-\tau)| = |h(\tau)| \cdot |x(t-\tau)|$ of that function. Now if we assume that the input is bounded, then it must be true that $|x(t-\tau)| < B < \infty$ for all t and τ. It follows that if we replace $|x(t-\tau)|$ by its largest possible value B, we only strengthen the inequality in (9.67). Therefore, we obtain

$$|y(t)| \leq B \int_{-\infty}^{\infty} |h(\tau)| d\tau \tag{9.68}$$

as a bound on the size of $|y(t)|$. From (9.68) it follows that if

$$\int_{-\infty}^{\infty} |h(\tau)| d\tau < \infty$$

then $|y(t)| < C < \infty$, which means that the system is stable. Thus, we have proved that if the system impulse response satisfies (9.66) then it is a stable system. The proof that systems satisfying (9.66) are the only LTI systems that are stable is somewhat tricky and will not be given here.

Example 9-16: Stable System

Consider the LTI system whose impulse response is a square pulse of the form

$$h(t) = u(t+10) - u(t-10) = \begin{cases} 1 & -10 \leq t < 10 \\ 0 & \text{otherwise} \end{cases}$$

To test to see if this system is stable, we simply substitute into (9.66) and check to see if the result is finite. That is

$$\int_{-\infty}^{\infty} |h(\tau)| d\tau = \int_{-10}^{10} d\tau = 20 < \infty$$

Therefore this system is stable. ■

EXERCISE 9.10: The system in Example 9-16 is one example of a system whose impulse response has finite duration. It turns out that all such systems are stable. We define *finite duration* to mean that $h(t) = 0$ for all $t < t_1$ and for all $t > t_2$ where $-\infty < t_1 < t_2 < \infty$. Prove that all such systems are stable if the impulse response is bounded, i.e., $|h(t)| < D < \infty$.

Example 9-17: Unstable System

Suppose a system is defined by the input/output relation

$$y(t) = \frac{1}{x(t)}$$

i.e., the system takes the reciprocal of the input signal. Since division is not a linear operation, this system is *not* LTI. Therefore the stability condition of (9.66) does *not* apply. We must go back to the basic definition of stability, which is that every bounded input must produce a bounded output. To prove that a system is stable, we must test this condition for *every* bounded input. However, if we think that the system might be unstable, we need only find one input that is bounded, but produces an unbounded output, in order to prove that the system is *unstable*. In the case of the reciprocal system, if the input signal is bounded, it could certainly take on the value zero at one or more times, and the output will be infinity at those times. Therefore, the reciprocal system is unstable. ■

Example 9-18: Integrator is Unstable

Now consider the integrator system. Its impulse response is $h(t) = u(t)$, so applying (9.66) leads to

$$\int_{-\infty}^{\infty} |h(\tau)| d\tau = \int_{0}^{\infty} d\tau \rightarrow \infty$$

Therefore the integrator is an unstable system. ■

EXERCISE 9.11: For the integrator system, can you think of a specific input $x(t)$ that is bounded but whose integral grows without bound?

Causality The second property of systems that is important when implementing systems is *causality*. Causality is concerned with precedence in time.

> **General Definition of Causality:** A system is causal if and only if $y(t_0)$, the output at the present time t_0, does *not* depend on the values of the input signal $x(\tau)$ at times in the future, i.e., for $\tau > t_0$.

In other words, a causal system cannot anticipate the future of the input signal. For a continuous-time LTI system with impulse response $h(t)$, the condition for causality simplifies to the following test on the impulse response:

> *A continuous-time LTI system is causal if and only if:*
> $h(\tau) = 0 \quad \text{for } \tau < 0$ (9.69)

To prove that this condition is necessary and sufficient we need two arguments. For necessity, we can show that the system is non-causal if $h(t) \neq 0$ for $t < 0$. Since $h(t)$ is the output when $\delta(t)$ is the input, having $h(t) \neq 0$ for $t < 0$ means that the output *starts before* the input which violates the definition of causality.

For the sufficiency of the condition in (9.69), we need to examine the convolution integral expression for the output of an LTI system at a specific time t_0; i.e.,

$$y(t_0) = \int_{-\infty}^{\infty} x(\tau) h(t_0 - \tau) d\tau \quad (9.70)$$

If (9.69) is true, then $h(t_0 - \tau)$ is zero for $(t_0 - \tau) < 0$, which is the same as

$$h(t_0 - \tau) = 0 \quad \text{for } \tau > t_0$$

Therefore, the integrand in (9.70) is zero for $\tau > t_0$ and we can change the upper limit of the integral to $\tau = t_0$:

$$y(t_0) = \int_{-\infty}^{t_0} x(\tau)h(t_0 - \tau)d\tau \tag{9.71}$$

This integral (9.71) says that $y(t_0)$ depends only on values of the input $x(\tau)$ for $\tau \leq t_0$. Thus we have demonstrated that (9.69) is both necessary and sufficient for causality of an LTI system.

Example 9-19: Non-Causal System

Consider the system of Example 9-16, where $h(t) = 1$ for $-10 \leq t < 10$. Since the system is LTI and the impulse response is nonzero for t between -10 and zero, it is *not* a causal system. ■

Example 9-20: Flip System not Causal

Consider a system defined by the input/output relation

$$y(t) = x(-t)$$

This is a time-reversal system, i.e., it "turns the signal around in time." Equation (9.69) does not apply to this case because the system is only linear, it is *not* time-invariant. Therefore we must resort to the general definition. From the input/output relation, we see that $y(-2) = x(2)$; i.e., we need a "future" value of $x(\tau)$ to compute $y(-2)$. Indeed, at any time $t_0 < 0$, we need a value of the input at a future time $-t_0$ so the system is not causal. ■

EXERCISE 9.12: Is the integrator a causal system?

EXERCISE 9.13: The impulse response of the ideal delay system is $h(t) = \delta(t - t_d)$. For what values of t_d will the delay system be causal?

9-9 Using Convolution to Remove Multipath Distortion

We have studied convolution and its properties in great detail. We will use this concept directly or indirectly at many places in the remaining chapters. Before we proceed to discuss Fourier transforms, let us consider an example of a practical application of the convolution concepts. In many real-life situations it is difficult to record a signal without distortion. This is the case for example when we have "echoes" that can be produced by improperly terminated transmission lines or by recording sound in an environment where sound reflects off walls or other obstructions and multiple copies of the signal return to the microphone after traveling different distances. The simplest way to model this situation would be to define the recorded signal as the sum of the true signal and one reflected component

$$y(t) = x(t) + \alpha x(t - t_d) \tag{9.72}$$

The parameter α, which is usually less than one in magnitude, models the loss of signal strength along the echo path, and t_d would be the amount of additional time for sound to travel the distance of the echo path. When $x(t)$ is an acoustic signal and t_d is 100 msec or so, a human listener would perceive a distinct echo. When t_d is small and there are many reflections the effect would be perceived as a "hollow" sound.

Since convolution with a shifted impulse just shifts and scales the signal, it follows that we can model the introduction of echoes by an LTI system with impulse response

$$h(t) = \delta(t) + \alpha\delta(t - t_d) \quad (9.73)$$

so the recorded signal (9.72) is $y(t) = h(t) * x(t)$. If the transmission environment has more than one additional path, we would simply add more impulses with different scaling parameters and different time delays to define a general LTI echo model in terms of the impulse response

$$h(t) = \sum_{k=0}^{N} \alpha_k \delta(t - t_k) \quad (9.74)$$

Multiple echoes of this sort produce the effect that we call *reverberation.*

When echoes and reverberation occur, it might be necessary to "recover" $x(t)$ from $y(t)$. When we study Fourier analysis later in Chapter 11, we will be able to understand when it is possible to do this in the general case. However, for a single echo it is possible to recover $x(t)$ if one simple condition is satisfied. Indeed, now we will show that it may be possible to recover $x(t)$ by simply passing the available signal $y(t)$ through another LTI system. Such a system would be called an ***inverse system*** for the system whose impulse response is $h(t)$. The operation that we wish to perform is depicted in Fig. 9-25, which shows that the goal of the cascade system is to obtain an overall system whose impulse response is a single impulse. If we write the convolution for the second system in Fig. 9-25, and also substitute the definition of the first system

$$\begin{aligned} x(t) = y(t) * h_i(t) &= [x(t) * h(t)] * h_i(t) \\ &= x(t) * [h(t) * h_i(t)] \end{aligned} \quad (9.75)$$

then we can solve (9.75) for $h_i(t)$ by recalling the identity element of convolution

$$x(t) = x(t) * \delta(t)$$

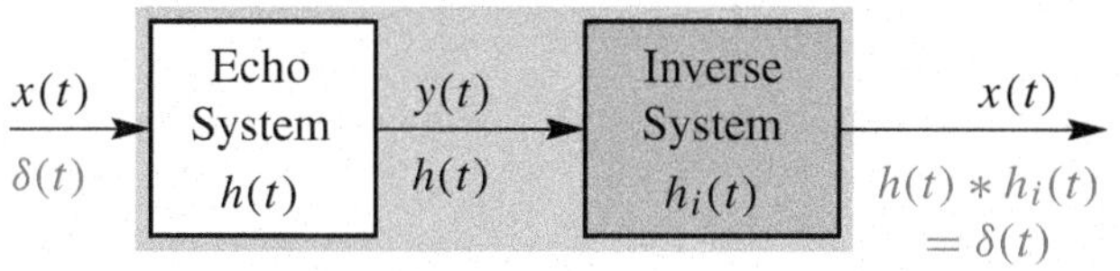

Figure 9-25: Using an LTI system to compensate for an echo.

which leads to

$$\begin{aligned} \delta(t) &= h(t) * h_i(t) \\ &= [\delta(t) + \alpha\delta(t - t_d)] * h_i(t) \end{aligned} \quad (9.76)$$

To find the inverse system, we must solve (9.76) for $h_i(t)$. At this point, we have virtually no tools for solving such convolution equations; however, some intuition and trial-and-error can be used to uncover the solution. First, note that $h_i(t)$ must contain an impulse because we have one on the left-hand side of (9.76). Also, note that we are trying to cancel the echo of size α at delay t_d so let's try including an impulse with size $-\alpha$ at delay t_d; i.e., we propose

$$\tilde{h}_i(t) = \delta(t) - \alpha\delta(t - t_d)$$

for the inverse system.

Substituting this for $h_i(t)$ in (9.76) gives

$$\begin{aligned} h(t) * \tilde{h}_i(t) &= [\delta(t) + \alpha\delta(t - t_d)] * [\delta(t) - \alpha\delta(t - t_d)] \\ &= \delta(t) - \alpha^2\delta(t - 2t_d) \end{aligned}$$

We have gotten rid of the echo of size α at time t_d, but replaced it with an echo of size $-\alpha^2$ at time $2t_d$. This might be an improvement if $|\alpha| < 1$ because the echo size will be smaller, but we still have not completely removed the echo. What would get rid of the echo at $2t_d$? How about adding a second term to our proposed $h_i(t)$? Perhaps we should add another impulse at $2t_d$, as in

$$\tilde{h}_i(t) = \delta(t) - \alpha\delta(t - t_d) + \alpha^2\delta(t - 2t_d)$$

which then produces

$$h(t) * \tilde{h}_i(t) = \delta(t) + \alpha^3\delta(t - 3t_d)$$

As before, if $|\alpha| < 1$ the remaining echo will be smaller. If we keep on doing this, we are led to the solution

$$h_i(t) = \sum_{k=0}^{\infty}(-\alpha)^k\delta(t - kt_d) \qquad (9.77)$$

which is the correct one because it will push the remaining echo off to $t = \infty$ and the echo amplitude will be reduced to zero if $|\alpha| < 1$.

EXERCISE 9.14: Substitute (9.77) into (9.76) and show that all the impulses that result on the right-hand side of the equation cancel except the term $\delta(t)$.

Throughout our discussion, we have repeatedly mentioned the condition $|\alpha| < 1$. In this case, the inverse filter consists of an infinite set of impulses whose sizes decrease exponentially. If $|\alpha| \geq 1$ the sizes of the impulses in the inverse filter impulse response would not decrease. Such a system would be an unstable system and would not be useful in practice. In many situations, however, inverse filters can be derived to at least partially compensate for echoes or other linear distortions. Another notable example is in data communication where adaptive inverse filters are used to compensate for the transmission channel.

9-10 Summary

In this chapter, we have turned our attention to continuous-time signals and systems. We have shown that, just as in the case of discrete-time signals, linear time-invariant continuous-time systems are completely characterized by their response to a unit impulse signal, in the sense that for any given input to the LTI system, the output can be computed as the continuous-time convolution of the input and the impulse response. We showed how to evaluate convolutions of impulses with other signals including other impulse signals, and we gave several examples that illustrate how to evaluate convolution integrals involving piecewise constant or piecewise exponential signals. We also showed that the properties of convolution allow us to manipulate cascade or parallel combinations of LTI systems, and we found that the causality and stability of an LTI system can be tested easily by examining the impulse response. In conclusion, we showed a simple example of how the concept of convolution can be used to represent a practical signal processing problem.

This chapter contains two demos:

(a) The `dconvdemo` demonstrates discrete-time convolution.

DEMO: *Discrete Convolution*

(b) The `cconvdemo` demonstrates continuous-time convolution.

DEMO: *Continuous Convolution*

Both are MATLAB graphical user interfaces (GUIs) that help students visualize the process of convolution. Some features are; 1) Users can choose from a variety of signals; 2) Signals can be dragged around with the mouse and the display is updated in real-time; 3) Tutorial mode allows the student to hide the convolution; 4) Various plot options make both GUIs effective in a lecture/classroom environment. Lab #12 provides a structured framework for students to use both convolution GUIs.

LAB: *#12 Two Convolution GUIs*

The reader is again reminded of the large number of solved homework problems on the CD-ROM that are available for review and practice.

NOTE: *Hundreds of Solved Problems*

9-11 Problems

P-9.1 Equation (9.6) on p. 248 gives one way to write the finite-length pulse $p(t)$ in Fig. 9-3(a) in terms of the unit-step signal $u(t)$. A *second* way is to use a product notation.

(a) Make a plot of the signal $u(t-2)$.

(b) Make a plot of the signal $u(4-t)$. Determine the region where the signal $u(4-t)$ is zero.

(c) Show that the product signal $u(t-2)u(4-t)$ is the same as the signal $p(t)$ in Fig. 9-3(a).

P-9.2 In each of the following cases, state whether or not the continuous-time system is (i) Linear; (ii) Time-invariant; (iii) Stable, and (iv) Causal. In each case, $x(t)$ represents the input and $y(t)$ represents the corresponding output of the system. Provide a brief justification, either in the form of mathematical equations or statements in the form of complete, correct, English sentences. *Remember, in order to show that the system does not have the property, all you have to do is give an example input whose output does not satisfy the condition of the property.*

(a) An *exponentiation system* is defined by the input/output relation

$$y(t) = \exp\{x(t+2)\} = e^{x(t+2)}$$

(b) A *phase modulator* is a system whose input and output satisfy a relation of the form

$$y(t) = \cos[\omega_c t + x(t)]$$

(c) An *amplitude modulator* is a system whose input and output satisfy a relation of the form

$$y(t) = [A + x(t)]\cos(\omega_c t)$$

(d) A system that takes the even part of an input signal is defined by a relation of the form

$$y(t) = \mathcal{E}v\{x(t)\} = \frac{x(t) + x(-t)}{2}$$

P-9.3 Express each of the following in a simpler form:

(a) $\delta(t-10) * [\delta(t+10) + 2e^{-t}u(t) + \cos(100\pi t)]$

(b) $\cos(100\pi t)[\delta(t) + \delta(t-.002)]$

(c) $\dfrac{d}{dt}\left[e^{-2(t-2)}u(t-2)\right]$

(d) $\displaystyle\int_{-\infty}^{t} \cos(100\pi\tau)[\delta(\tau) + \delta(\tau - .002)]d\tau$

Notes: Use properties of the impulse signal $\delta(t)$ and the unit-step signal $u(t)$ to perform the simplifications. Be careful to distinguish between multiplication and convolution. Convolution is denoted by a "star", as in $x(t)*\delta(t-2) = x(t-2)$ and multiplication is usually indicated by juxtaposition as in $x(t)\delta(t-2) = x(2)\delta(t-2)$.

P-9.4 In each of the following cases, simplify the expression as much as possible using the properties of the continuous-time unit-impulse signal. Provide some explanation or intermediate steps for each answer.

(a) $e^{-(t-4)}u(t-4)\delta(t-5)$

(b) $\displaystyle\int_{-\infty}^{t-5} \delta(\tau - 1)d\tau$

(c) $\dfrac{d}{dt}\{e^{-(t-4)}u(t-4)\}$

(d) $\delta(t-1) * \delta(t-2) * \delta(t)$

P-9.5 Find $h(t)$ that satisfies the following equation:

$$[e^{-(t-4)}u(t-4)] * h(t) = 2e^{-t}u(t)$$

P-9.6 Suppose that $x(t) = e^{-4t}u(t)$. Express each of the following in a simpler form:

(a) $x(t)[\delta(t+1) + \delta(t-1)]$

(b) $\displaystyle\int_{-\infty}^{\infty} x(\tau)\delta(\tau - 1)d\tau$

(c) $\displaystyle\int_{-\infty}^{\infty} x(\tau)\delta(t - \tau)d\tau$

(d) $\delta^{(1)}(t) * x(t-1)$

P-9.7 Express each of the following in a simpler form:

(a) $[e^{-3t} + \sin(t)][\delta(t) + \delta(t-1)]$

(b) $\displaystyle\int_{-\infty}^{\infty} \sin(2\pi\tau)\delta(\tau - 1)d\tau$

(c) $\displaystyle\int_{-\infty}^{\infty} \sin(2\pi\tau)\delta(t - \tau)d\tau$

P-9.8 Use the convolution integral to determine the convolution of two exponential signals with the same exponent; namely,

$$y(t) = [e^{-at}u(t)] * [e^{-at}u(t)].$$

Compare this result to Example 9-9 on p. 259 and Exercise 9.4 on p. 265.

P-9.9 Use linearity and time invariance to determine the step response of a system whose impulse response is the following rectangular pulse:

$$h(t) = 5u(t-1) - 5u(t-4)$$

(a) Expand the convolution into two terms and exploit the result from Example 9-9 on p. 259 to get $y(t) = u(t) * h(t)$.

(b) Write the answer for $y(t)$ as separate cases over three different regions of the time axis.

(c) Make a plot of $y(t)$ versus t.

P-9.10 Use linearity and time-invariance to convolve the following two rectangular pulses:

$$x(t) = 2u(t) - 2u(t-2)$$
$$h(t) = 3u(t-5) - 3u(t-1)$$

(a) Expand the convolution into four terms and exploit the result from Example 9-9 to get $y(t) = x(t) * h(t)$.

(b) Write the answer for $y(t)$ as separate cases over five different regions of the time axis.

(c) Make a plot of $y(t)$ versus t.

P-9.11 The impulse response of an LTI continuous-time system is such that $h(t) = 0$ for $t \le T_1$ and for $t \ge T_2$. Show that if $x(t) = 0$ for $t \le T_3$ and for $t \ge T_4$ then $y(t) = x(t) * h(t) = 0$ for $t \le T_5$ and for $t \ge T_6$. In the process of proving this result you should obtain expressions for T_5 and T_6 in terms of T_1, T_2, T_3, and T_4. The best way to work this problem is to draw "flip and slide" pictures for typical input and impulse response signals (e.g., rectangular pulses).

Another way to approach this problem would be to use the `cconvdemo` GUI to generate some examples of convolving finite-duration signals, and then deduce the correct answer from those examples.

P-9.12 The impulse response of a linear time-invariant system is

$$h(t) = \begin{cases} e^{-0.1(t-2)} & 2 \le t < 12 \\ 0 & \text{otherwise} \end{cases}$$

(a) Is the system stable? Justify your answer.

(b) Is the system causal? Justify your answer.

(c) Find the output $y(t)$ when the input is $x(t) = \delta(t-2)$.

P-9.13 The impulse response of a linear time-invariant system is

$$h(t) = \begin{cases} e^{-0.1(t-2)} & 2 \le t < 12 \\ 0 & \text{otherwise} \end{cases}$$

(a) Plot $h(t-\tau)$ as a function of τ for $t = 0, 8$, and 20.

(b) Use convolution to determine the output $y(t)$ when the input is

$$x(t) = \begin{cases} e^{-0.25t} & 0 \le t < 6 \\ 0 & \text{otherwise} \end{cases}$$

Refer to the result of Exercise 9.4 on p. 265, and then exploit linearity and time invariance to expand the convolution into a number of terms that use that result.

(c) Make a plot of the convolution result determined in part (b). Write the output as five separate cases in order to describe $y(t)$ in each part of the plot.

(d) Set this problem up in the `cconvdemo` GUI in order to check your plot of the output $y(t)$.

P-9.14 If the input $x(t)$ and the impulse response $h(t)$ of an LTI system are the following:

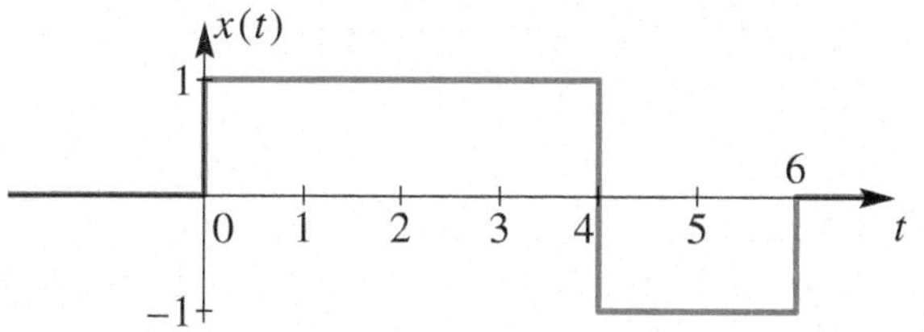

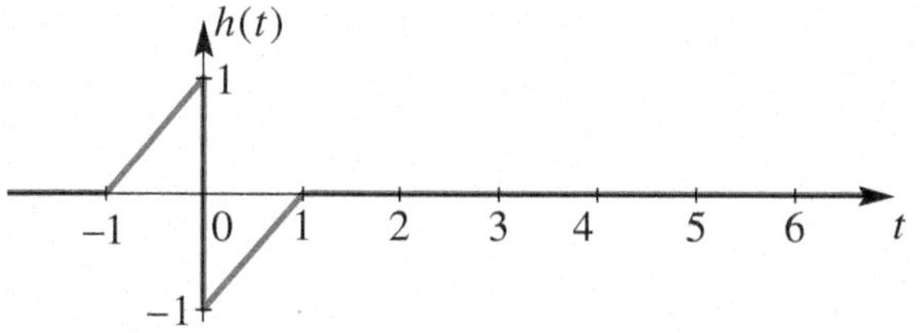

Without evaluating any integrals:

(a) Determine $y(0)$, the value of the output at $t = 0$.

(b) Find all the values of t for which the output is $y(t) = 0$.

P-9.15 A linear time-invariant system has the following impulse response:

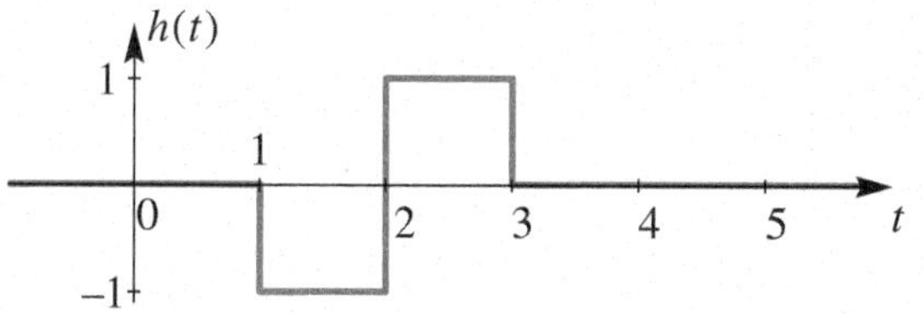

(a) Is the LTI system stable? Give a reason to support your answer.

(b) Plot $h(t-\tau)$ versus τ, for $t = 2$. Label your plot carefully.

(c) If the input is $x(t) = u(t)$, use the convolution integral to find $y(2)$; i.e., $y(t)$ when $t = 2$.

(d) For the input $x(t) = u(t)$ and the given impulse response, the output is $y(t) = 0$ for $t < T_1$ and for $t > T_2$. Find T_1 and T_2. Explain your answers. You may "flip and shift" either $x(t)$ or $h(t)$, whichever leads to the easiest solution.

P-9.16 In the following figure, $x(t)$ is the input and $h(t)$ is the impulse response of an LTI system whose output is $y(t)$:

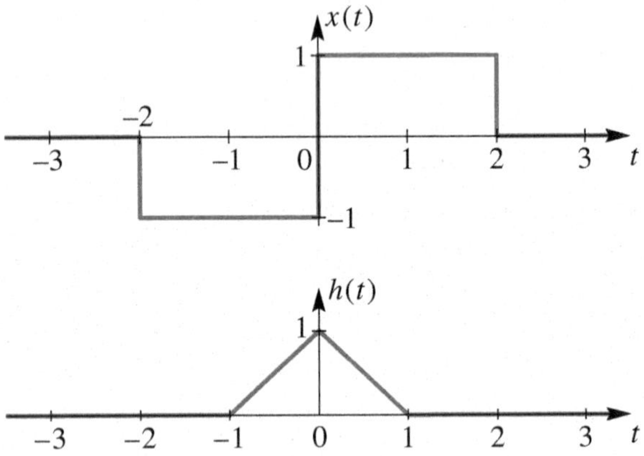

Be sure to find a simple approach to each of the following:

(a) What is the *complete* set of values of t for which $y(t) = 0$?

(b) For what value of t does $y(t)$ have its largest positive value? What is the value at that time?

(c) Carefully sketch the derivative $y^{(1)}(t)$ of the output when the input and impulse response are as given above. You should do this *without* first computing $y(t)$.

P-9.17 A continuous-time system is defined by the input/output relation

$$y(t) = \int_{t-2}^{t+2} x(\tau)d\tau$$

(a) Determine the impulse response, $h(t)$, of this system.

Hint: Substitute $x(\tau) = \delta(\tau)$ and determine values of t for which the integral is nonzero.

(b) Is this a stable system? Explain with a proof (if true) or counter example (if false).

(c) Is it a causal system? Explain with a proof (if true) or counter example (if false).

(d) Use the convolution integral to determine the output of the system when the input is

$$x(t) = u(t + 1)$$

Plot your answer.

P-9.18 A linear time-invariant system has impulse response

$$h(t) = \delta(t) - 3e^{-3t}u(t)$$

Use convolution to find the output $y(t)$ for $-\infty < t < \infty$ when the input is

$$x(t) = u(-t)$$

P-9.19 A linear time-invariant system has impulse response

$$h(t) = \begin{cases} 1 & 2 \le t < 4 \\ 0 & \text{otherwise} \end{cases}$$

(a) Plot $h(-1 - \tau)$ as a function of τ.

(b) Use convolution to find and plot the output $y(t)$ for $-\infty < t < \infty$ when the input is

$$x(t) = u(t) - \delta(t - 2)$$

Hint: Use superposition.

P-9.20 A linear time-invariant system has impulse response: $h(t) = e^{-(t-2)}u(t-2)$

(a) Plot $h(t-\tau)$ versus τ, for $t = 1$. Label your plot.

(b) Is the LTI system causal? Give a reason to support your answer.

(c) Is the system stable? Explain with a proof or counter example.

(d) If the input is $x(t) = u(t)$, then it will be true that the output $y(t)$ is zero for $t \le t_1$. Find t_1.

(e) The rest of the output signal (for $t > t_1$) is nonzero when the input is $x(t) = u(t)$. Use the convolution integral to find the nonzero portion of the output; i.e., find $y(t)$ for $t > t_1$.

P-9.21 Consider the cascade connection of two LTI systems shown in Fig. P-9.21. The first system has impulse response

$$h_1(t) = \begin{cases} e^{-2t} & 0 \le t < 1 \\ 0 & \text{otherwise} \end{cases}$$

and the second system is a differentiator system described by the input/output relation

$$y(t) = \frac{dw(t)}{dt}$$

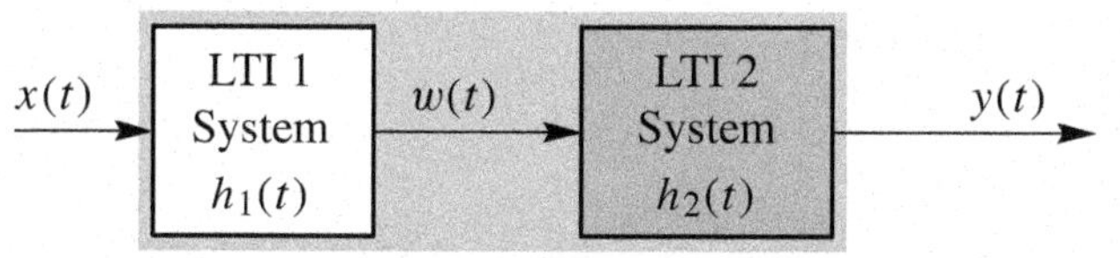

Figure P-9.21

Find the impulse response of the overall system; i.e., find the output $y(t) = h(t)$ when the input is $x(t) = \delta(t)$. Give your answer *both* as an equation and as a carefully labeled sketch.

P-9.22 The block diagram in Fig. P-9.22 depicts two LTI systems in parallel that are cascaded with a third LTI system. The impulse responses of the systems are written within each block of the diagram.

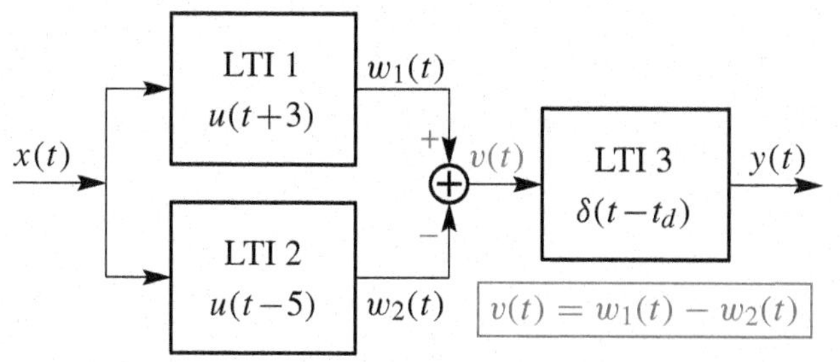

Figure P-9.22

(a) What is the impulse response of the overall LTI system (i.e., from $x(t)$ to $y(t)$)? Give your answer *both* as an equation and as a carefully labeled sketch.

(b) How should the time delay t_d be chosen so that the overall system is causal?

(c) Which systems (#1, #2, #3) are stable? Is the overall system a stable system? Explain.

P-9.23 Consider the cascade of two LTI systems shown in Fig. P-9.21. The first system is described by the input/output relation

$$w(t) = x(t) - x(t-2)$$

and the second system has impulse response

$$h_2(t) = u(t)$$

Find and plot the impulse response of the overall system; i.e., find the output $y(t) = h(t)$ when the input is $x(t) = \delta(t)$.

P-9.24 The system in Fig. P-9.24 is an LTI system.

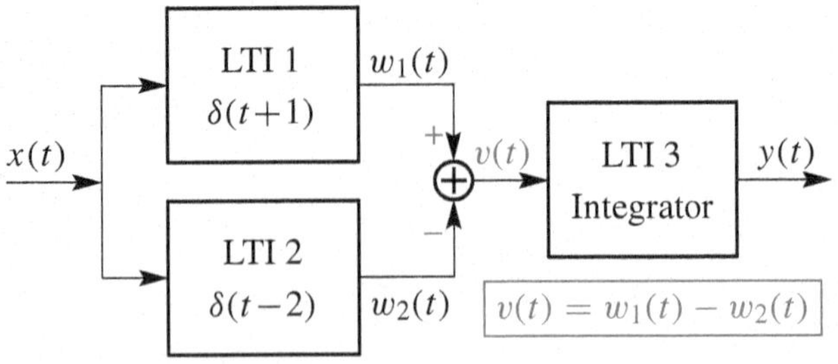

Figure P-9.24

(a) Determine the impulse response of the overall system (i.e., from $x(t)$ to $y(t)$). Give your answer *both* as an equation and as a carefully labeled sketch.

(b) Is the overall cascade system a causal system? Explain.

(c) Is it a stable system? Explain.

P-9.25 Consider the cascade of LTI systems shown in Fig. P-9.25.

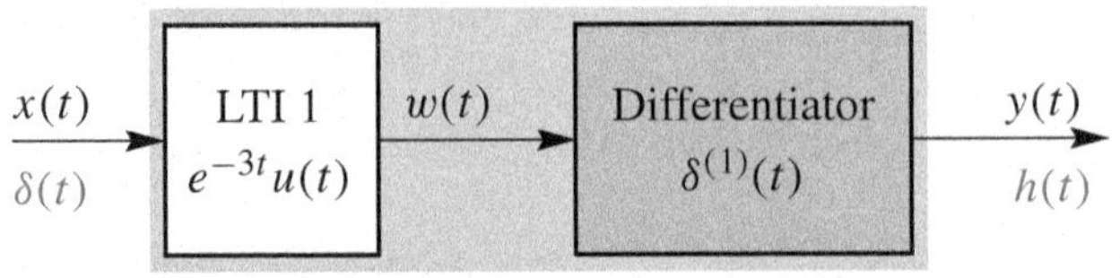

Figure P-9.25

(a) Determine $h(t)$, the impulse response of the overall cascade system.

(b) Determine $y(t)$ when the input is $u(t)$ (i.e., find the unit-step response of the overall system).
Hint: Rearrange the systems to find an easy way.

(c) Determine the output $y(t)$ when the input is $x(t) = u(t) - u(t-10)$. Exploit the LTI properties of the system.

CHAPTER 10

Frequency Response

In Chapter 9 we showed that the operation of continuous-time convolution is synonymous with the definition of a continuous-time LTI system. The convolution integral is easy to apply as an analysis tool when the input signal is an impulse, $\delta(t)$, a step $u(t)$, or combinations of piecewise exponential signals. However, we know from experience with digital filters (as in Chapter 6) that the *frequency response function* is a very efficient way to characterize an LTI system for sinusoidal inputs, so we now set out to do that same characterization for analog filters (i.e., continuous-time systems). We will start with the case where the input signal is a single sinusoid, but then we will show that the same viewpoint and methods apply when the input signal is any periodic waveform.

10-1 The Frequency Response Function for LTI Systems

We want to develop a simple formula for the *sinusoidal response* of an LTI system. The general description of LTI systems tells us that the general output $y(t)$ is obtained from the general input $x(t)$ by the convolution operation defined as

$$y(t) = h(t) * x(t) = \int_{-\infty}^{\infty} h(\tau)x(t-\tau)\,d\tau \qquad (10.1)$$

where the impulse response $h(t)$ is the output when the input is the unit-impulse signal $\delta(t)$.

Rather than tackle the sinusoidal case directly, we take the input signal to be a complex exponential of the form

$$x(t) = Ae^{j\phi}e^{j\omega t} \qquad -\infty < t < \infty \tag{10.2}$$

We will show that the corresponding output signal is another complex exponential *at the same frequency.* We know that this is true for discrete-time LTI systems. To prove that this is also true for continuous-time LTI systems, we can substitute (10.2) into the convolution integral (10.1) and employ the laws of exponents to simplify the expression as follows:

$$\begin{aligned} y(t) &= h(t) * x(t) \\ &= \int_{-\infty}^{\infty} h(\tau)Ae^{j\phi}e^{j\omega(t-\tau)}\,d\tau \\ &= \int_{-\infty}^{\infty} h(\tau)Ae^{j\phi}e^{j\omega t}e^{-j\omega\tau}\,d\tau \\ &= \left(\int_{-\infty}^{\infty} h(\tau)e^{-j\omega\tau}\,d\tau\right)Ae^{j\phi}e^{j\omega t} \\ &= H(j\omega)Ae^{j\phi}e^{j\omega t} \end{aligned} \tag{10.3}$$

The integral within the parentheses reduces to a complex constant when the frequency ω is given a specific value. This complex quantity tells us how the system changes the magnitude and phase of the input signal as it forms the output signal.

Example 10-1: Complex-Exponential Output

Suppose that the input to an LTI system is $x(t) = 10e^{j3\pi t}$, and the value of $H(j3\pi)$ is $2 - j2$. The output can be expressed as $y(t) = Be^{j\theta}e^{j3\pi t}$ if we convert $H(j3\pi)$ to polar form and multiply.

$$y(t) = (2\sqrt{2}e^{-j\pi/4})10e^{j3\pi t} = 20\sqrt{2}e^{j(3\pi t-\pi/4)}$$

Thus the output complex exponential has a magnitude of $20\sqrt{2}$ and a phase of $-\pi/4$. ■

According to (10.3), when the input is a complex exponential signal whose frequency is ω, the output is also a complex-exponential signal having exactly the same frequency. However, the effect of the system on a complex-exponential signal depends on the frequency of the input. This is represented by the fact that we will get different complex values for $H(j\omega)$ as ω changes. If we let the frequency vary, we can define the ***frequency response*** of a continuous-time system as the following function of ω:

Frequency Response of an LTI System

$$H(j\omega) = \int_{-\infty}^{\infty} h(t)e^{-j\omega t}\,dt \tag{10.4}$$

In general, we want the functional form of $H(j\omega)$ so that we can plot (10.4) versus the frequency variable ω, and see the effect of the system on *any* complex-exponential input signal. That is, we need to evaluate the integral in (10.4) to obtain an expression for $H(j\omega)$ in terms of familiar mathematical functions. The following example illustrates this for a real one-sided exponential impulse response.

Example 10-2: Formula for $H(j\omega)$

Suppose that the impulse response is $h(t) = 2e^{-2t}u(t)$. We can determine a formula for the frequency response of the LTI system by evaluating the following integral:

$$H(j\omega) = \int_{-\infty}^{\infty} 2e^{-2t}u(t)e^{-j\omega t}\,dt$$

The steps are

$$H(j\omega) = \int_0^\infty 2e^{-2t-j\omega t}\,dt$$
$$= \frac{2e^{-2t-j\omega t}}{-2-j\omega}\Bigg|_0^\infty = 0 - \frac{2}{-2-j\omega}$$
$$= \frac{2}{2+j\omega} \quad (10.5)$$

The evaluation at the upper limit $t=\infty$ is zero because $\lim_{t\to\infty} |e^{-2t-j\omega t}| = e^{-2t} \to 0$. ■

ω	$\lvert H(j\omega)\rvert$
0	1
2	$2/\sqrt{8} = 0.707$
20	$2/\sqrt{404} \approx 0.0995$
∞	0

Table 10-1: Values of $|H(j\omega)|$ for several values of ω.

10-1.1 Plotting the Frequency Response

The frequency response is a complex-valued function $H(j\omega)$ that depends on ω and is best summarized with two plots: magnitude versus frequency and phase versus frequency. Just as in the case of digital filters, we can talk about passbands and stopbands of analog filters once we are looking at a plot of $H(j\omega)$.

Suppose that we want to characterize the frequency response of the LTI system in Example 10-2, where

$$H(j\omega) = \frac{2}{2+j\omega} \quad (10.6)$$

$H(j\omega)$ is usually a complex-valued function of ω, as is true for Eq. (10.6). In making plots or using $H(j\omega)$ to compute the output due to a complex-exponential input, it is better to express $H(j\omega)$ in polar form. For (10.6) we get the following magnitude and phase functions

$$|H(j\omega)| = \left|\frac{2}{2+j\omega}\right| = \frac{2}{\sqrt{4+\omega^2}} \quad (10.7)$$

$$\angle H(j\omega) = 0 - \angle\{2+j\omega\} = -\tan^{-1}(\omega/2) \quad (10.8)$$

The range of the frequency axis is $-\infty < \omega < \infty$ since continuous-time complex-exponential signals can have any frequency in that range. If we evaluate at a few crucial frequencies such as $\omega = 0, 2, 20, \infty$, we obtain the values in Table 10-1. From these values, we can begin to draw a rough sketch of the magnitude response. Figure 10-1 shows a complete plot of the magnitude and phase of this frequency response function over the range $-8 \le \omega \le 8$.

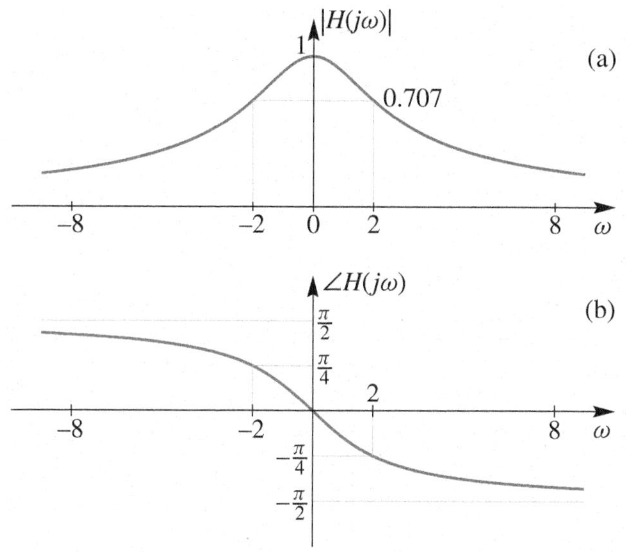

Figure 10-1: Frequency response of a first-order analog filter. (a) Magnitude of frequency response, (b) phase angle of frequency response.

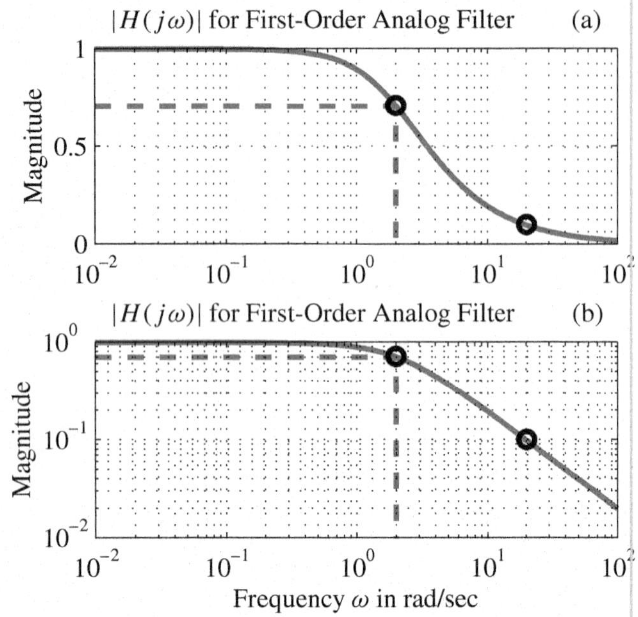

Figure 10-2: (a) MATLAB plot of frequency response of a first-order analog filter, created by the function `[H,W]=freqs(2,[1,2],logspace(-2,2,401))`. The frequency axis is logarithmic. (b) Loglog plot of the frequency response.

10-1.1.1 Logarithmic Plot

In MATLAB we can use `freqs()` to do a complete evaluation over a dense grid and make a plot. Since the frequency axis is infinitely long, it is common to use a logarithmic spacing along the horizontal axis to cover several decades (factors of 10) in frequency. Figure 10-2(a) shows such a plot for the frequency response function of this example. The points $\omega = 2$ and $\omega = 20$ are marked on the graph. Notice that only the positive-frequency part of the magnitude of the frequency response is plotted in Fig. 10-2. This is because the frequency response magnitude for (10.7) is even, i.e., $|H(-j\omega)| = |H(j\omega)|$. In the general case of a system with real impulse response, we will show that $H(-j\omega) = H^*(j\omega)$, where * denotes complex conjugation. Therefore, the magnitude is an even function of ω, $|H(-j\omega)| = |H^*(j\omega)| = |H(j\omega)|$, and we do not need to explicitly plot the entire frequency axis.

Figure 10-2(b) shows the same frequency response, but this time a log scale is used for both magnitude and frequency. Notice that in this case the frequency response appears to fall off linearly with frequency above $\omega = 2$ rad/sec. The curve drops by a multiplicative factor of 0.1 for each factor of 10 increase in frequency. This feature of loglog frequency response plots is very useful for higher-order continuous-time systems where the falloff is linear but faster on a loglog plot. Plots of this type are called Bode plots.

10-1.2 Magnitude and Phase Changes

When the input to an LTI system is a complex-exponential signal with specific frequency ω_0

$$x(t) = Ae^{j\phi}e^{j\omega_0 t} \tag{10.9}$$

then the corresponding output is

$$y(t) = H(j\omega_0)Ae^{j\phi}e^{j\omega_0 t} \tag{10.10}$$

where the frequency response of the LTI system is $H(j\omega)$. Since the frequency response function at $\omega = \omega_0$ evaluates to a complex number that can be expressed in polar form

$$H(j\omega_0) = Me^{j\psi} \tag{10.11}$$

we can rewrite the output (10.10) as

$$y(t) = Me^{j\psi}Ae^{j\phi}e^{j\omega_0 t} = (MA)e^{j(\omega_0 t+\phi+\psi)} \tag{10.12}$$

Notice that we multiply the magnitudes and add the phases.

Example 10-3: Evaluate $H(j\omega)$

Suppose that the frequency response of an LTI system is

$$H(j\omega) = \frac{j\omega}{200\pi + j\omega}$$

To evaluate the frequency response at $\omega = 200\pi$, we just substitute to obtain the complex number

$$\begin{aligned} H(j200\pi) &= \frac{j200\pi}{200\pi + j200\pi} \\ &= \frac{j}{1+j} = \frac{1}{\sqrt{2}} e^{+j\pi/4} \end{aligned}$$

Therefore the magnitude is $M = 1/\sqrt{2}$ and the phase-shift is $\psi = \pi/4$. ■

Example 10-4: Complex-Exponential Input

Now suppose that the input to the system of Example 10-3 is the complex-exponential signal

$$x(t) = 20e^{j0.3\pi} e^{j200\pi t}$$

For the system in Example 10-3, the frequency response integral in (10.4) evaluates to $H(j200\pi) = \frac{1}{\sqrt{2}} e^{+j\pi/4}$ at the input frequency, $\omega_0 = 200\pi$ rad/s. According to (10.10), the output will be

$$\begin{aligned} y(t) &= \frac{1}{\sqrt{2}} e^{+j\pi/4} \left(20e^{j0.3\pi} e^{j200\pi t}\right) \\ &= 10\sqrt{2} e^{j(200\pi t + 0.55\pi)} \end{aligned}$$

■

It is important to note that the numerical values of the magnitude (M) and phase (ψ) of the frequency response will change when the input frequency ω_0 changes, but *the output frequency will be the same as the input frequency.*

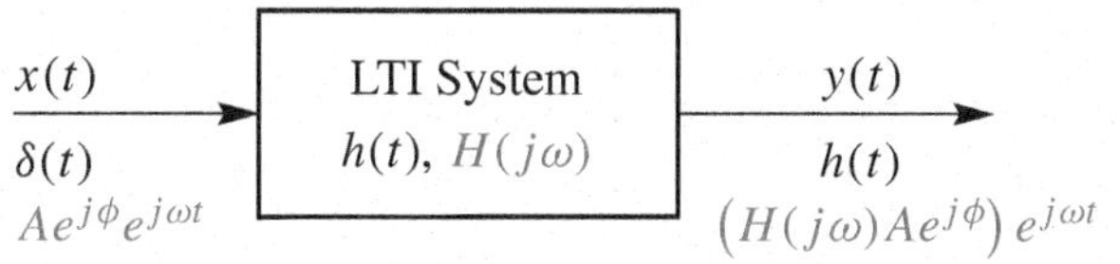

Figure 10-3: Linear time-invariant system with responses to unit impulse and to complex-exponential signals.

The system in Figs. 10-1 and 10-2 is clearly a lowpass filter since it causes complex-exponential signals of low frequencies to be multiplied by values that are approximately one, while high frequencies are multiplied by values close to zero. For the filter in Fig. 10-1, the frequency $\omega = 2$ is taken to be the nominal dividing point between the low frequency passband and the high frequency stopband.

10-2 Response to Real Sinusoidal Signals

The discussion of the previous section can be summarized with Fig. 10-3, which shows the response of an LTI system to a single complex exponential signal. In many cases, we are actually interested in real signals that can be represented as a sum of complex-exponential signals. In this section, we will study the response to such inputs, and show that if the impulse response is real, then the output due to a real signal that can be represented as a sum of complex exponentials can always be expressed as a sum of cosine signals. We will explain why the impulse response must be real-valued in order to apply the magnitude and phase change evaluated from the frequency response. One benefit of this upcoming discussion is that it illustrates the role that linearity plays, and suggests a general approach that underlies the use of frequency responses and Fourier transforms.

10-2.1 Cosine Inputs

Although the derivation of the frequency response assumed the input signal was a complex exponential, $Ae^{j\phi}e^{j\omega t}$, we often are interested in the response to real signals like a cosine signal. We can use our knowledge of the complex-exponential response to derive the sinusoidal response, because the following simple relationship holds when the impulse response of the system is real-valued:

> ***Magnitude and Phase Change of the Sinusoidal Response:***
> If the input to an LTI system is $x(t) = A\cos(\omega t + \phi)$, and if the impulse response is *real-valued,* then the output will be $y(t) = (MA)\cos(\omega t + \phi + \psi)$.

In other words, the sinusoidal signal will keep the same frequency, but it will undergo a magnitude and phase change, just like the complex-exponential case. The numerical values of the magnitude $M = |H(j\omega)|$ and phase-shift $\psi = \angle H(j\omega)$ are determined by the frequency response function evaluated at the input frequency ω.

Strictly speaking, our derivation in the previous section was only valid for complex exponentials, so we must relate the real sinusoid to the complex exponential if we want to take advantage of the magnitude and phase change property. However, we can write a cosine function in terms of complex exponentials by using the inverse Euler formula to obtain

$$A\cos(\omega t + \phi) = \tfrac{1}{2}Ae^{j\phi}e^{j\omega t} + \tfrac{1}{2}Ae^{-j\phi}e^{-j\omega t} \qquad (10.13)$$

One interpretation of (10.13) is that the cosine signal has been decomposed into the sum of two complex-exponential signals with frequencies ω and $-\omega$ and corresponding complex amplitudes $\frac{1}{2}Ae^{j\phi}$ and $\frac{1}{2}Ae^{-j\phi}$. Note that a real cosine signal is obtained because the complex-exponential signals that comprise it are complex conjugates of each other.

Since we are considering only LTI systems, the superposition property applies. Therefore, since the input is the sum of two signals, $x(t) = x_1(t) + x_2(t)$, the output is equal to the sum of the corresponding outputs $y(t) = y_1(t) + y_2(t)$ where $y_1(t)$ is the output due to $x_1(t)$ and $y_2(t)$ is the output due to $x_2(t)$. In our present situation, we need to find the output signals due to

$$x_1(t) = \tfrac{1}{2}Ae^{j\phi}e^{j\omega t} \qquad (10.14)$$

$$x_2(t) = \tfrac{1}{2}Ae^{-j\phi}e^{-j\omega t} \qquad (10.15)$$

and then add the results. Using the frequency response property (10.10), we find that the two outputs are

$$y_1(t) = H(j\omega)\tfrac{1}{2}Ae^{j\phi}e^{j\omega t} \qquad (10.16)$$

$$y_2(t) = H(-j\omega)\tfrac{1}{2}Ae^{-j\phi}e^{-j\omega t} \qquad (10.17)$$

so it follows from the linearity of the LTI system that

$$\begin{aligned} y(t) &= y_1(t) + y_2(t) \\ &= H(j\omega)\tfrac{1}{2}Ae^{j\phi}e^{j\omega t} + H(-j\omega)\tfrac{1}{2}Ae^{-j\phi}e^{-j\omega t} \end{aligned} \qquad (10.18)$$

In other words, the sinusoidal response is the sum of the complex-exponential response at the positive frequency ω and the response at the corresponding negative frequency $-\omega$.

10-2.2 Symmetry of $H(j\omega)$

Since we have declared that there is a simplification when $h(t)$ is real-valued, there must be some simple relationship between $H(j\omega)$ and $H(-j\omega)$. In fact, the relationship is

$$H(-j\omega) = (H(j\omega))^* = H^*(j\omega) \qquad (10.19)$$

That is, the frequency response at $-\omega$ is the complex conjugate of the frequency response at $+\omega$. The proof of this fact is relatively straightforward, because when $h(t)$ is real-valued, it follows that

$$H^*(j\omega) = \left(\int_{-\infty}^{\infty} h(t)e^{-j\omega t}\,dt \right)^*$$
$$= \int_{-\infty}^{\infty} h^*(t)e^{+j\omega t}\,dt$$
$$= \int_{-\infty}^{\infty} h(t)e^{-j(-\omega)t}\,dt = H(-j\omega)$$

In the last line we have used the fact that $h(t)$ is real-valued (so $h^*(t) = h(t)$) to remove the conjugate on $h(t)$, and we have written the exponent with two minus signs to obtain the negative exponent needed for the definition of the frequency response.

The implication of the ***conjugate symmetry property*** (10.19) is that magnitude and phase responses at ω and $-\omega$ are not independent. In fact, if the frequency response at ω is

$$H(j\omega) = Me^{j\psi}$$

then we get

$$H(-j\omega) = H^*(j\omega) = \left(Me^{j\psi}\right)^* = Me^{-j\psi}$$

In other words, the magnitudes are the same, but the phases are opposites.

Now we can return to the problem of simplifying the output signal in (10.18), and write

$$y(t) = y_1(t) + y_2(t) \qquad (10.20)$$
$$= H(j\omega)\tfrac{1}{2}Ae^{j\phi}e^{j\omega t} + H(-j\omega)\tfrac{1}{2}Ae^{-j\phi}e^{-j\omega t}$$
$$= Me^{j\psi}\tfrac{1}{2}Ae^{j\phi}e^{j\omega t} + Me^{-j\psi}\tfrac{1}{2}Ae^{-j\phi}e^{-j\omega t}$$
$$= \tfrac{1}{2}MAe^{j(\omega t+\phi+\psi)} + \tfrac{1}{2}MAe^{-j(\omega t+\phi+\psi)}$$
$$= MA\cos(\omega t + \phi + \psi) \qquad (10.21)$$

where the last line follows from Euler's formula. Summarizing, we have now proven that the sinusoidal signal $x(t) = A\cos(\omega t + \phi)$ will undergo a magnitude and phase change when it is processed by an LTI system whose impulse response is real-valued. Furthermore, the magnitude and phase change can be evaluated numerically from the polar form of the frequency response; i.e.,

$$\boxed{y(t) = |H(j\omega)|A\cos(\omega t + \phi + \angle H(j\omega))} \qquad (10.22)$$

Example 10-5: Sinusoidal Response

Suppose that the frequency response of the LTI system is

$$H(j\omega) = \frac{40\pi}{40\pi + j\omega}$$

Determine the output signal when the input signal is $x(t) = 3\cos(40\pi t - \pi)$.

Solution: We must evaluate $H(j\omega)$ at the frequency $\omega = 40\pi$ rad/sec. It is convenient to work with polar form similar to that derived in Section 10-1.1, i.e.,

$$|H(j40\pi)| = \frac{40\pi}{\sqrt{(40\pi)^2 + \omega^2}}$$
$$= \frac{40\pi}{\sqrt{1600\pi^2 + 1600\pi^2}} = \frac{1}{\sqrt{2}}$$

$$\angle H(j40\pi) = -\tan^{-1}\left(\frac{\omega}{40\pi}\right) = -\tan^{-1}(1) = -\pi/4$$

Now we must use these numerical values to write the formula for the time function, $y(t)$

$$y(t) = 3\left(\frac{1}{\sqrt{2}}\right)\cos(40\pi t - \pi - \pi/4)$$
$$= 2.1213\cos(40\pi t - 5\pi/4)$$

Notice that we multiply the magnitudes and add the phases, but *the frequency remains the same.* ■

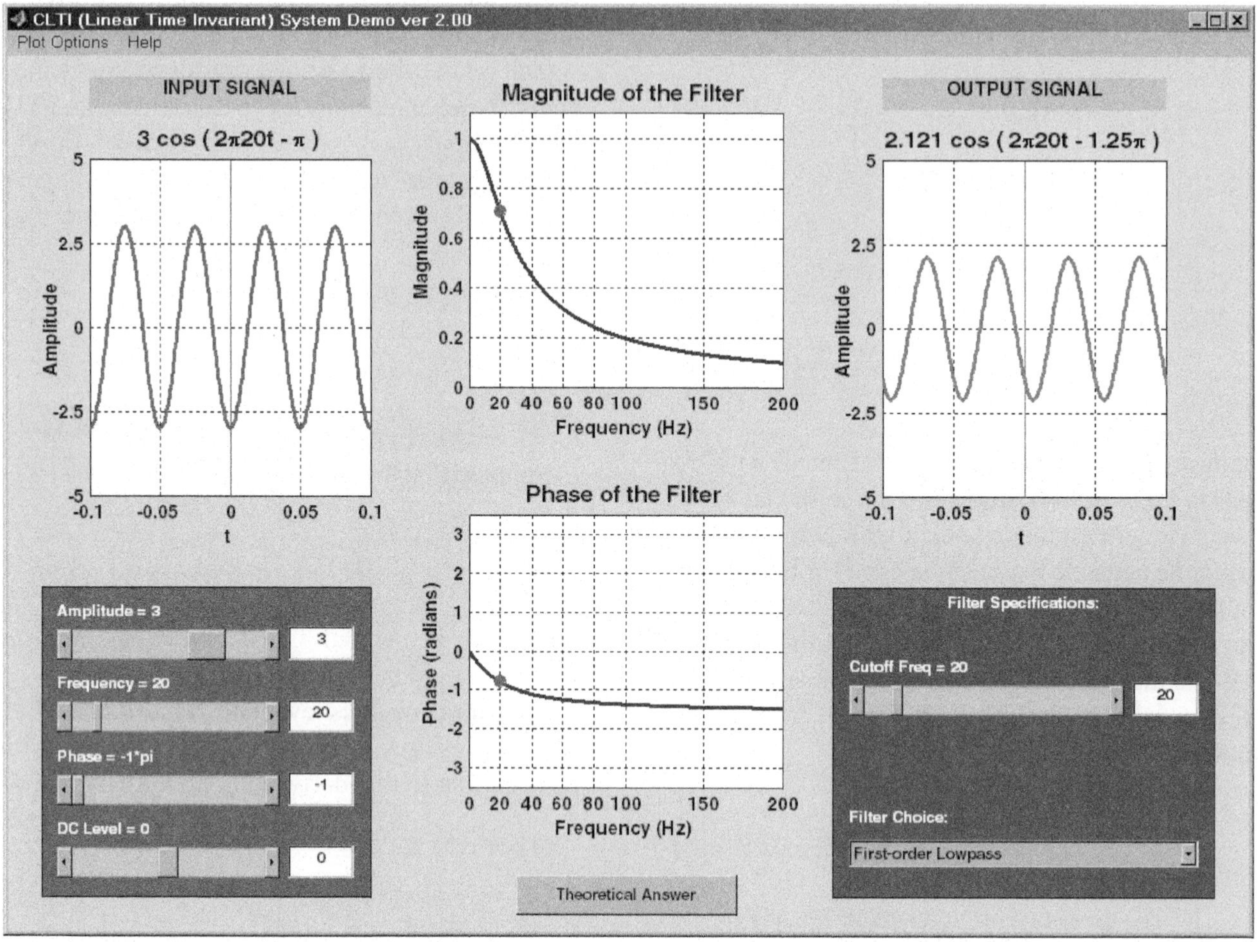

Figure 10-4: Graphical User Interface (GUI) for `CLTIdemo` that illustrates the *sinusoid-in gives sinusoid-out* concept for continuous-time systems. The user can change the LTI system as well as the input sinusoidal frequency, amplitude and phase. All frequencies are given in hertz, e.g., 20 Hz corresponds to 40π rad/sec.

In order to reinforce this important concept of *sinusoid-in gives sinusoid-out* for LTI systems, a MATLAB GUI was created to run different cases where the frequency response can be changed by selecting from a variety of simple filters. Also, the amplitude, phase and frequency of the input sinusoid can be varied. The screen shot of Fig. 10-4 shows the GUI with the filter and input signal chosen to solve the problem in the preceding example. The mathematical formula for the output is shown above the upper right-hand panel in the GUI.

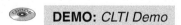

DEMO: *CLTI Demo*

EXERCISE 10.1: Use the frequency response to evaluate the convolution integral

$$y(t) = \int_{-\infty}^{\infty} e^{-\tau}u(\tau)3\cos(t-\tau-\pi)d\tau$$

Hint: determine $h(t)$ from the integral above.

10-2.3 Response to a General Sum of Sinusoids

The use of linearity illustrated in the previous section can be extended to the case where the input is composed of many sinusoids. For a specific example, suppose that the input signal is the weighted sum of N sinusoids

$$x(t) = \sum_{k=1}^{N} A_k \cos(\omega_k t + \phi_k) \quad (10.23)$$

where the frequencies $\{\omega_k\}$ are all different, and each sinusoid has its own amplitude A_k and phase ϕ_k.

If we know the frequency response of the continuous-time system, we can use linearity to construct a formula for the output signal. The strategy is to treat each sinusoidal component separately, and then add all the results together. If we concentrate on the kth input sinusoid, we can compute the corresponding output after evaluating $H(j\omega_k)$; i.e.,

$$y_k(t) = A_k|H(j\omega_k)|\cos(\omega_k t + \phi_k + \angle H(j\omega_k)) \quad (10.24)$$

In this case, we use the magnitude and angle notation in place of M and ψ, but the reader should keep in mind that these are just constants.

Linearity allows us to add all the separate outputs to obtain the final result

$$\begin{aligned} y(t) &= \sum_{k=1}^{N} y_k(t) \\ &= \sum_{k=1}^{N} A_k|H(j\omega_k)|\cos(\omega_k t + \phi_k + \angle H(j\omega_k)) \end{aligned} \quad (10.25)$$

In reality, this formula is not especially useful. Its main value is to convey a very important result; namely, that the individual amplitudes and phases are changed in a predictable manner by the frequency response. Finally, we observe that this result (10.25) does not depend on any special relationship among the frequencies, unlike the results of the next section where all the frequencies are multiples of a fundamental.

Example 10-6: Response for Two Sinusoids

If the frequency response of a continuous-time LTI system is

$$H(j\omega) = \frac{2+j\omega}{1+j\omega}$$

determine the output signal when the input is

$$x(t) = 2\cos t + \cos(3t + 1.57)$$

The answer requires that we evaluate $H(j\omega)$ at $\omega = 1$ and at $\omega = 3$; i.e.,

$$H(j\omega)|_{\omega=1} = \frac{2+j}{1+j} = 1.581e^{-j0.322}$$

$$H(j\omega)|_{\omega=3} = \frac{2+j3}{1+j3} = 1.14e^{-j0.266}$$

Then we can write down the sinusoids that make up the output as

$$y(t) = 3.162\cos(t - 0.322) + 1.14\cos(3t + 1.304)$$

because the magnitude of the first cosine is the product 2(1.581), and the phase of the second cosine is the sum (1.57 − 0.266) = (1.304). ■

EXERCISE 10.2: Suppose that the frequency response of a continuous-time LTI system is

$$H(j\omega) = \frac{(j2 + j\omega)(-j2 + j\omega)}{1 + j\omega} = \frac{4 - \omega^2}{1 + j\omega}$$

and the input is

$$x(t) = 4\cos t + \cos(2t)$$

Determine the output $y(t)$ of the system.

10-2.4 Periodic Input Signals

When the input signal is periodic, it is relatively easy to demonstrate that the output of an LTI system must be periodic. First of all, the mathematical definition of periodicity is

$$x(t) = x(t + T_0) \qquad -\infty < t < \infty \tag{10.26}$$

where T_0 is the period. If T_0 is the *smallest* value that satisfies (10.26), then it is called the *fundamental period* of the signal. If we write the convolution integral for $y(t + T_0)$, we get

$$y(t + T_0) = \int_{-\infty}^{\infty} h(\tau)x(t + T_0 - \tau)\,d\tau \tag{10.27}$$

Now the term $x(t + T_0 - \tau)$ inside the integral is periodic with period T_0 so it can be replaced by

$$x(t + T_0 - \tau) = x(t - \tau) \qquad -\infty < t < \infty \tag{10.28}$$

Thus, the integral (10.27) becomes

$$\begin{aligned} y(t + T_0) &= \int_{-\infty}^{\infty} h(\tau)x(t + T_0 - \tau)\,d\tau \\ &= \int_{-\infty}^{\infty} h(\tau)x(t - \tau)\,d\tau = y(t) \end{aligned} \tag{10.29}$$

and we have demonstrated that the output signal $y(t)$ is periodic with period T_0 if the input to the system is periodic with period T_0. As an aside, we observe that the fundamental period of the output signal might be shorter than T_0 in certain cases.

EXERCISE 10.3: Prove that an LTI system whose impulse response is periodic will always produce output signals that are periodic (assume that the input is not periodic). Note: this type of system is unstable, so the output might not always be bounded. When $y(t)$ is bounded it will be periodic.

The interesting thing about periodic signals is that the theory of Fourier series tells us how to obtain a signal representation as a sum of complex-exponential signals.[1] If we couple this with the frequency response concept that we have introduced in this chapter, we can develop a general approach for determining the output of an LTI system with periodic input that does not involve the evaluation of a convolution integral. Figure 10-5 shows the essentials of this approach. The steps of this approach are summarized as follows:

[1] A review of Chapter 3 might be useful as background for the discussion in the rest of this chapter.

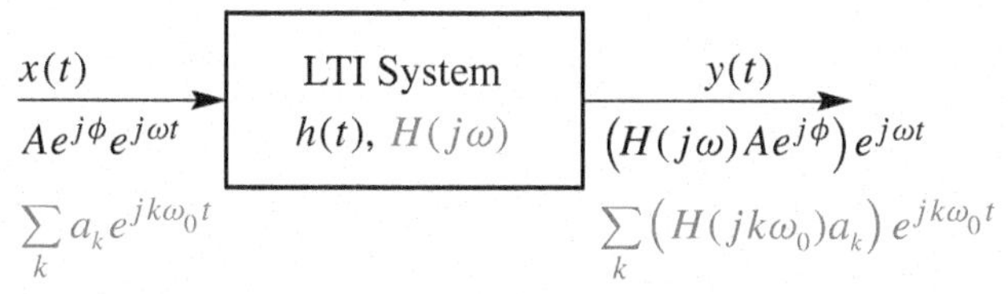

Figure 10-5: Linear time-invariant system with responses to one complex exponential or to a sum of complex exponentials.

1. Express the periodic input signal $x(t)$ as a sum of sinusoids using the method of Fourier series from Chapter 3

$$x(t) = \sum_{k=-\infty}^{\infty} a_k e^{jk\omega_0 t}$$

   where $\omega_0 = 2\pi/T_0$ is the fundamental frequency in radians/second.

2. For the kth complex exponential in the Fourier series sum, use (10.10) on p. 288 to find the corresponding output, which is also a complex exponential at the kth frequency $\omega_k = k\omega_0$

$$y_k(t) = H(jk\omega_0)a_k e^{jk\omega_0 t}$$

3. The complex multiplication of a_k times $H(jk\omega_0)$ produces a new set of Fourier series coefficients

$$b_k = H(jk\omega_0)a_k$$

4. The output signal $y(t)$ is then synthesized from the Fourier series definition

$$y(t) = \sum_{k=-\infty}^{\infty} b_k e^{jk\omega_0 t}$$

The important step is #3, which says that we must *multiply in the frequency-domain*. This is a complex multiplication of a_k times $H(jk\omega_0)$. The most difficult step is #1 which requires the Fourier series analysis integral to extract the coefficients a_k from the periodic signal, $x(t)$.

LAB: *#13 Numerical Evaluation of Fourier Series*

10-3 Ideal Filters

In any practical application of LTI continuous-time systems, the frequency response function $H(j\omega)$ might be derived from an actual analog filter implementation such as an electronic circuit or from a mathematical model of an analog system. However, in the early phases of the design of a system it is often useful to use *ideal filters* that have simple frequency responses that provide ideal frequency-selectivity. We have already encountered ideal filters when considering the frequency response of digital filters in Chapter 6. For analog (continuous-time) filters we must specify $H(j\omega)$ over the entire frequency axis $-\infty < \omega < \infty$.

10-3.1 Ideal Delay System

The concept of time delay is exceedingly important in both discrete- and continuous-time system theory. For continuous-time systems, the ideal delay system is defined by the input/output relation

$$y(t) = x(t - t_d) \qquad (10.30)$$

where t_d is the amount of delay in secs. Thus, an ideal delay system produces an output that is a perfect copy of the input signal, but shifted in time by t_d seconds.

We can find the frequency response of the ideal delay if we first find the impulse response and then use (10.4).

The impulse response, by definition is $h(t) = \delta(t - t_d)$. Therefore, the frequency response is

$$H(j\omega) = \int_{-\infty}^{\infty} \delta(t - t_d)e^{-j\omega t}dt = e^{-j\omega t_d} \qquad (10.31)$$

Another approach that would work in this simple case is to substitute $x(t) = e^{j\omega t}$ into (10.30) and use the definition of the frequency response in (10.3) on p. 286 to determine $H(j\omega)$ from the resulting expression for $y(t)$. This is suggested in Problem P-10.1.

The magnitude of the frequency response of the ideal delay (10.31) is constant (equal to one) independent of frequency and it has a phase of $-\omega t_d$. Since $-\omega t_d$ is the equation of a straight line as a function of ω, we say that the ideal time delay has constant magnitude and ***linear phase***. This means that the ideal delay system does not change the magnitude of an input sinusoid, but it introduces a phase shift that is proportional to the frequency of the sinusoid. This is exactly what is needed to create a perfect delayed copy of the sinusoid or complex exponential of frequency ω.

Example 10-7: Response of Ideal Delay

Suppose that $x(t) = 10e^{j\pi/4}e^{j200\pi t}$ is the input to an ideal delay system with a delay of 0.001 sec. Then the frequency response of the ideal delay is $H(j\omega) = e^{-j\omega 0.001}$, which, when evaluated at the frequency 200π, is

$$H(j200\pi) = e^{-j200\pi(0.001)}$$

The corresponding output is

$$\begin{aligned} y(t) &= H(j200\pi)\left(10e^{j\pi/4}e^{j200\pi t}\right) \\ &= \left(e^{-j200\pi(0.001)}\right)10e^{j\pi/4}e^{j200\pi t} \\ &= \left(e^{-j0.2\pi}\right)10e^{j\pi/4}e^{j200\pi t} \end{aligned}$$

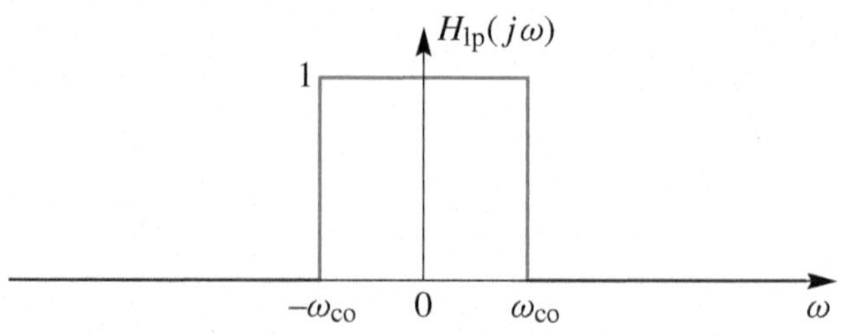

Figure 10-6: Frequency response of an ideal LPF with its cutoff at ω_{co} rad/s.

which we can rewrite ass $y(t) = 10e^{j\pi/4}e^{j200\pi(t-0.001)}$. Thus, as could easily be predicted from the frequency response, the output is equal to the input delayed by 0.001 sec. ■

The frequency response of the ideal time delay system has constant magnitude and linear phase as a function of ω. A ***filter*** is a system that can remove certain frequencies while letting others pass through the system relatively unmodified. We shall now define several ideal ***frequency selective*** filters.

10-3.2 Ideal Lowpass Filter

An ideal lowpass filter (LPF) has a frequency response that consists of two regions: the passband near $\omega = 0$ (DC) where the frequency response is one, and the stopband where it is zero. An ideal lowpass filter is therefore defined as

$$H_{lp}(j\omega) = \begin{cases} 1 & |\omega| \le \omega_{co} \\ 0 & |\omega| > \omega_{co} \end{cases} \qquad (10.32)$$

The frequency ω_{co} is called the *cutoff frequency*, or the band edge of the LPF passband. Figure 10-6 shows a plot of $H(j\omega)$ for the ideal LPF in which we see that $H(j\omega)$ is even symmetric about $\omega = 0$. This property is needed because we know that real-valued impulse responses lead to filters with a conjugate symmetric $H(j\omega)$. Since $H(j\omega)$ is both real and conjugate symmetric, it must be an even function.

Filters like the ideal lowpass filter are impossible to implement. We will see that the impulse response of the ideal lowpass filter is nonzero over the entire time axis. Thus, it is non-causal. However, that does not invalidate the ideal lowpass filter concept; i.e., the idea of selecting the low frequency band and rejecting all other frequencies. Even the filter depicted in Fig. 10-1, which is implementable by a simple electric circuit, might be satisfactory in some applications. In more stringent filtering applications, we simply need to learn how to obtain good approximations to the ideal characteristic. In summary, the process of filter design involves mathematical approximation of the ideal filter with a frequency response that is close enough to the ideal frequency response while corresponding to an implementable filter.

10-3.3 Ideal Highpass Filter

The ideal highpass filter has its stopband centered on low frequencies, and its passband extends from $|\omega| = \omega_{co}$ out to $|\omega| = \infty$.

$$H_{hp}(j\omega) = \begin{cases} 0 & |\omega| \le \omega_{co} \\ 1 & |\omega| > \omega_{co} \end{cases} \tag{10.33}$$

Figure 10-7 shows an ideal HPF with its bandedge at ω_{co} rad/s. In this case the high frequency components of a signal will pass through the filter unchanged while the low frequencies will be completely eliminated. Note that as in the case of the ideal lowpass filter, we have defined the ideal highpass filter so that $H_{hp}(-j\omega) = H^*_{hp}(j\omega)$ so that the corresponding impulse response will be a real function of time.

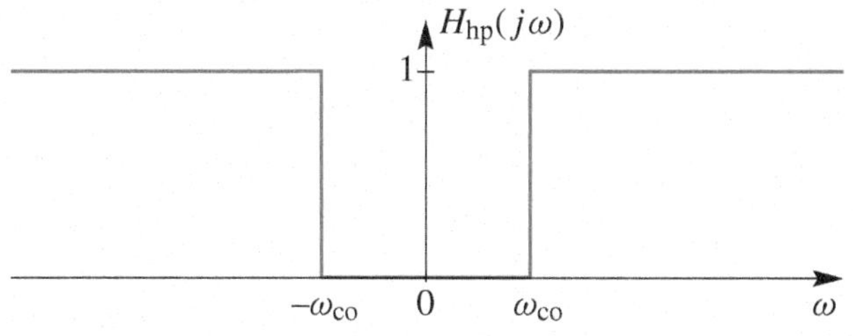

Figure 10-7: Frequency response of an ideal HPF with cutoff at ω_{co} rad/s.

EXERCISE 10.4: If $H_{lp}(j\omega)$ is an ideal LPF with its bandedge at ω_{co} as plotted in Fig. 10-6, show that $G(j\omega)$ defined by $G(j\omega) = 1 - H_{lp}(j\omega)$ is an ideal HPF with the same bandedge.

Hint: Try plotting $G(j\omega)$.

10-3.4 Ideal Bandpass Filter

The ideal bandpass filter has a passband centered away from the low frequency band, so it has two stopbands, one near DC and the other at high frequencies. Two bandedges must be given to specify the ideal BPF, ω_{co1} for the lower bandedge, and ω_{co2} for the upper bandedge.

$$H_{bp}(j\omega) = \begin{cases} 0 & |\omega| < \omega_{co1} \\ 1 & \omega_{co1} \le |\omega| \le \omega_{co2} \\ 0 & |\omega| > \omega_{co2} \end{cases} \tag{10.34}$$

Figure 10-8 shows an ideal BPF with its bandedges at ω_{co1} rad/s and ω_{co2} rad/s. Note again the symmetrical definition of the passbands and stopbands as required to make the corresponding impulse response real. In this case, all frequency components of a signal that lie in the band $\omega_{co1} \le |\omega| \le \omega_{co2}$ are passed unchanged through the filter, while all other frequency components are completely removed.

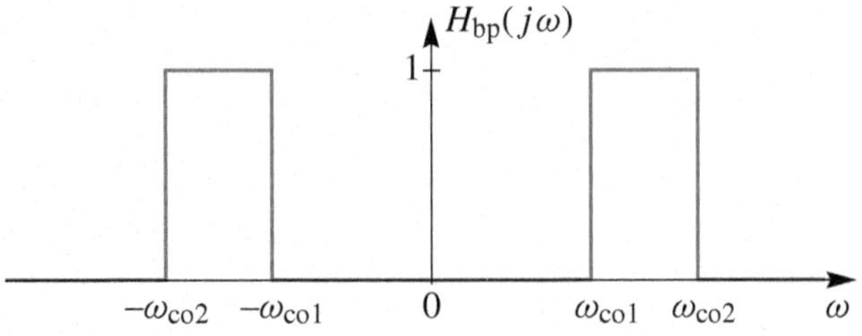

Figure 10-8: Frequency response of an ideal BPF. This filter is bandpass with its passband from $\omega = \omega_{co1}$ to $\omega = \omega_{co2}$ rad/s.

EXERCISE 10.5: If $H_{bp}(j\omega)$ is an ideal BPF with its bandedges at ω_{co1} and ω_{co2}, show by plotting that $G(j\omega)$ defined by $G(j\omega) = 1 - H_{bp}(j\omega)$ could be called an ideal *band-reject* filter. Where is the stopband of the band-reject filter?

10-4 Application of Ideal Filters

Ideal filters are typically used early in the design process to specify the modules in a signal processing system such as a communication receiver. Even though the final system requires implementable filters, the ideal filters simplify the system description. The performance of an actual system can usually be controlled by minimizing the approximation errors between the actual filters and the ideal filters.

As one tangible example of using ideal filters, we consider the task of designing a filter that will convert a square wave into a sinusoid. This might be useful in a practical application because it is relatively easy to generate a square wave with a switch that opens and closes periodically, but it can be difficult to generate a sinusoidal shape directly. The filter needed to solve this problem is a BPF that will extract one of the harmonics of the square wave.

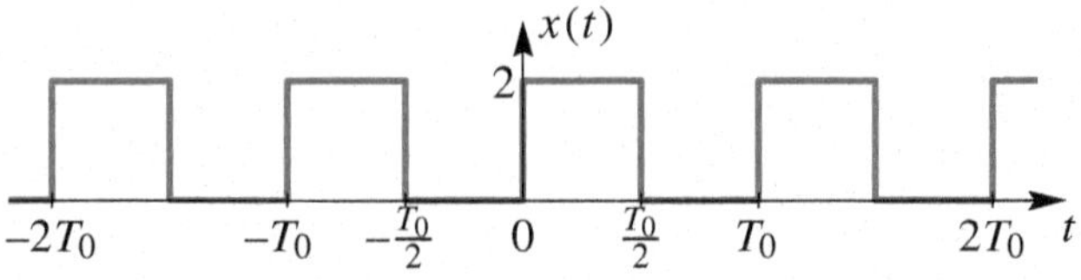

Figure 10-9: Periodic square wave with 50% duty cycle.

Suppose that we can generate a square wave with a fundamental period of $T_0 = 500\ \mu s$ ($f_0 = 2$ kHz or $\omega_0 = 2\pi(2000)$ rad/sec.) Such a signal is shown in Fig. 10-9. This periodic waveform is defined over one period by the equation

$$x(t) = \begin{cases} 2 & 0 \le t < T_0/2 \\ 0 & T_0/2 \le t < T_0 \end{cases} \tag{10.35}$$

In Section 3-4 we showed that periodic waveforms can be represented by a Fourier series of the form

$$x(t) = \sum_{k=-\infty}^{\infty} a_k e^{jk\omega_0 t} \tag{10.36}$$

where the Fourier coefficients are given by the integral

$$a_k = \frac{1}{T_0} \int_{-T_0/2}^{T_0/2} x(t) e^{-jk\omega_0 t} dt \tag{10.37}$$

Substituting the definition of $x(t)$ from (10.35) into (10.37) gives

$$a_k = \frac{1}{T_0} \int_{0}^{T_0/2} 2 e^{-jk\omega_0 t} dt \tag{10.38}$$

If we evaluate the integral in (10.38), we can show that the complex amplitudes in the spectrum are

$$a_k = \begin{cases} \dfrac{2}{j\pi k} & k = \pm1, \pm3, \pm5, \pm7, \ldots \\ 0 & k = \pm2, \pm4, \pm6, \pm8, \ldots \end{cases} \quad (10.39)$$

That is, the Fourier series of the square wave contains only odd harmonics of the fundamental frequency. Thus, the spectrum of the square wave with fundamental frequency of 2 kHz has (radian) frequency components at $\omega = \pm 2\pi(2000)$ rad/sec, $\omega = \pm 2\pi(6000)$ rad/sec, $\omega = \pm 2\pi(10000)$ rad/sec, etc. Since the average value of the square wave (over one period) is not zero, we obtain

$$a_0 = \frac{1}{T_0}\int_0^{T_0/2} 2dt = 1 \quad (10.40)$$

EXERCISE 10.6: Evaluate the integral in (10.38) and show that the result is given by (10.39).

Figure 10-10 shows the spectrum of the square wave for the first six nonzero (positive and negative) frequency components plus a DC component. Note that we have shown the spectrum with vertical lines indicating the relative magnitude of the complex amplitudes. Specific phase information is not shown.

Now we want to apply an LTI filter to the square wave so that the output will be a sinusoid. Two facts can be used to specify an ideal filter that will do this job. First, the desired spectrum of the output signal consists of two frequency components—if we want the output sinusoid at $f = 2$ kHz, or $\omega = 2\pi(2000)$ rad/sec, the required two complex exponential components will be at ± 2 kHz. Second, the action of the filter operating on the input signal can be described as *multiplication of the input spectrum times the frequency response of the filter*. That

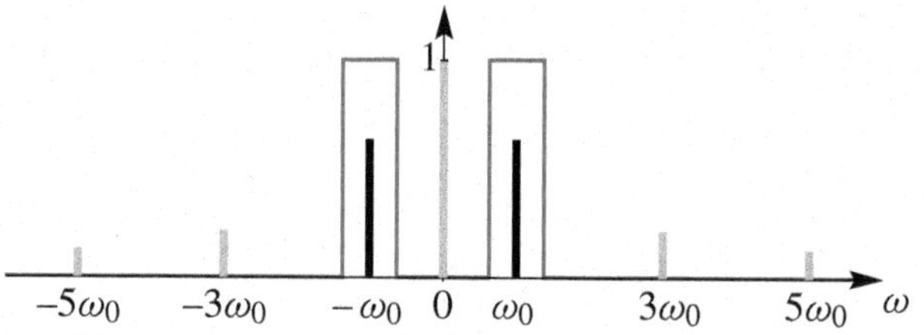

Figure 10-10: Square wave spectrum together with the frequency response of an ideal BPF. The bandpass filter has its passband from $\omega_{co1} = 2\pi(1250)$ rad/sec to $\omega_{co2} = 2\pi(2750)$ rad/sec. The fundamental frequency of the square wave is $\omega_0 = 2\pi(2000)$ rad/sec.

is, if a_k represents the Fourier series coefficients of the input square wave, then the output signal has Fourier coefficients

$$b_k = H(jk\omega_0)a_k \quad (10.41)$$

Perhaps a better view is provided by Fig. 10-10 which shows the frequency response of an ideal BPF superimposed on the spectrum plot of the input square wave. The narrow passband of the BPF is centered around the spectrum lines at $\omega = \pm 2\pi(2000)$ rad/sec (or ± 2 kHz). The ideal passband can have any nonzero width as long as it includes $\omega_0 = 2\pi(2000)$, but does not include any of the other nonzero spectrum components of the input. That is, we require that all other spectrum components lie in the stopbands of the BPF, so that they will be filtered out and the output signal will have only two spectrum lines; i.e., it will be a sinusoid of frequency $2\pi(2000)$. In Fig. 10-10, $\omega_{co1} = 2\pi(1250)$ and $\omega_{co2} = 2\pi(2750)$ rad/sec. It is clear from Fig. 10-10 that for this square-wave input, any ideal BPF will give the same output if $0 < \omega_{co1} < 2\pi(2000)$ and $2\pi(2000) < \omega_{co2} < 2\pi(6000)$.

The complex amplitudes of the two output spectral lines are found by multiplication, and only the $k = \pm 1$ terms of the output Fourier series will be left:

$$b_k = H(j2\pi(2000)k)a_k = \begin{cases} \dfrac{2}{j\pi k} & \text{for } k = \pm 1 \\ 0 & \text{elsewhere} \end{cases}$$

Putting these two together yields the formula for the output sinusoid:

$$\begin{aligned} y(t) &= \frac{2}{j\pi} e^{j2\pi(2000)t} + \frac{2}{j\pi(-1)} e^{-j2\pi(2000)t} \\ &= \frac{4}{\pi}\left(\frac{1}{2j} e^{j2\pi(2000)t} - \frac{1}{2j} e^{-j2\pi(2000)t}\right) \\ &= \frac{4}{\pi}\sin(2\pi(2000)t) = \frac{4}{\pi}\cos(2\pi(2000)t - \pi/2) \end{aligned}$$

The key point of this example is that filtering involves the multiplication of the frequency response times the input spectrum. This provides us with the concept of filters where signal components are either passed through or "filtered out" of an input signal. This principle will be reinforced by our study of Fourier transforms in the following chapters.

EXERCISE 10.7: Plot the frequency response of an ideal BPF such that its output will be

$$y(t) = \sin(2\pi(6000)t)$$

when the input is the square wave of Figs. 10-9 and 10-10.

EXERCISE 10.8: Must the input to the BPF be a square wave, or will any other periodic signal suffice to obtain a 2 kHz sinusoid?

10-5 Time-Domain or Frequency-Domain?

We have shown that a continuous-time LTI system can be represented in the time-domain by its impulse response and in the frequency-domain by its frequency response. Given two such powerful tools, the question naturally arises as to which to use in a given situation. While it is not possible to make iron-clad statements on such issues, it is certainly true that when the input signal can be represented as a sum of sinusoids or complex exponentials that are nonzero for all time $-\infty < t < \infty$, then the frequency response method usually provides the simplest solution. On the other hand, if a signal consists of impulses or step functions or other non-sinusoidal signals, convolution of that signal with the impulse response of the system may be the simplest approach. To illustrate this point consider the following example.

Example 10-8: Superposition

Suppose that an LTI system has impulse response

$$h(t) = \delta(t) - 200\pi e^{-200\pi t} u(t) \tag{10.42}$$

Furthermore, assume that the input is

$$x(t) = 10 + 20\delta(t - 0.1) + 40\cos(200\pi t + 0.3\pi)$$

for $-\infty < t < \infty$. This input consists of three parts: a constant signal, an impulse, and a cosine wave. Because the system is linear, we can take each part separately and use the solution method that is easiest for that part.

For example, we can treat the constant signal as a complex-exponential signal with zero for its frequency.

For this we will need to compute the frequency response. Substituting the impulse response into the definition gives

$$\begin{aligned} H(j\omega) &= \int_{-\infty}^{\infty} [\delta(t) - 200\pi e^{-200\pi t} u(t)] e^{-j\omega t} dt \\ &= \int_{-\infty}^{\infty} \delta(t) e^{-j\omega t} dt - 200\pi \int_{0}^{\infty} e^{-200\pi t} e^{-j\omega t} dt \\ &= 1 - \frac{200\pi}{200\pi + j\omega} \\ &= \frac{j\omega}{200\pi + j\omega} \end{aligned} \tag{10.43}$$

Therefore, the output due to the constant signal is simply

$$10 \longmapsto H(j0)10 = \frac{j0}{200\pi + j0} 10 = 0$$

Similarly, the output due to the cosine signal component can most easily be determined from the frequency response as follows:

$$\begin{aligned} &40\cos(200\pi t + 0.3\pi) \longmapsto \\ &\qquad 40|H(j200\pi)|\cos(200\pi t + 0.3\pi + \angle H(j200\pi)) \end{aligned}$$

If we evaluate $H(j200\pi)$ and compute its magnitude and angle, we find that

$$40\cos(200\pi t + 0.3\pi) \longmapsto \frac{40}{\sqrt{2}}\cos(200\pi t + 0.55\pi)$$

Finally, let us consider the impulse component. At this point we do not know how to represent an impulse as a sum of sinusoids. However, we can still solve the problem of finding the output due to the impulse component because we have been given the *impulse response*. Since convolution with an impulse is very easily evaluated by simply shifting and scaling the impulse response; i.e.,

$$20\delta(t - 0.1) \longmapsto 20h(t - 0.1)$$

Using the given impulse response in (10.42), we obtain

$$\begin{aligned} &20\delta(t - 0.1) \longmapsto \\ &\qquad 20\delta(t - 0.1) - 4000\pi e^{-200\pi(t-0.1)} u(t - 0.1) \end{aligned}$$

The output due to the sum of the input terms is simply the sum of the corresponding outputs because the system is linear and superposition applies. Therefore, the output of the system with impulse response given by (10.42) and frequency response given by (10.43) is

$$\begin{aligned} y(t) &= 0 \\ &\quad + 20\delta(t - 0.1) - 4000\pi e^{-200\pi(t-0.1)} u(t - 0.1) \\ &\quad + \frac{40}{\sqrt{2}}\cos(200\pi t + 0.55\pi) \end{aligned}$$

We have arrived at this solution by applying what we know about LTI systems. While the solution involved several steps, none of the steps involved extensive analysis. Selecting the proper tool was the key. ■

10-6 Summary/Future

In this chapter we have shown how the frequency response and the impulse response of an LTI system are related. We also showed that when the input can be represented as a sum of sinusoidal signals, the frequency response provides a straightforward method of finding the output of an LTI system. In the next chapter, which introduces the concept of the *Fourier transform*, we will show that the frequency response $H(j\omega)$ is the Fourier transform of the impulse response $h(t)$. The theory of Fourier transforms will provide a way to represent a much wider class of signals as a sum of sinusoids thereby opening up new ways that the frequency-domain approach can be used in solving signals and systems problems.

Lab #13 is concerned with numeric computation of Fourier series coefficients using MATLAB's numerical integration capability. The coefficients for several signals are found, and signals are synthesized from their coefficients. Lab #15 uses MATLAB's symbolic capability to find Fourier coefficients.

LAB: *#13 Numerical Evaluation of Fourier Series*

LAB: *#15 Fourier Series (Ch. 12)*

The Fourier series demo automates the work done by hand in Lab #13. In this demo, the student can select from three input waveforms, and select the number of coefficients to compute. The magnitude and phase spectrum are plotted along with the synthesized signal. This is the same demo that was introduced in Chapter 3.

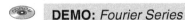

DEMO: *Fourier Series*

`CLTIdemo` is a MATLAB GUI that illustrates the relationship between the input and output of a continuous-time linear time-invariant (LTI) filter when the input is a sinusoidal function. The user is allowed to control the parameters of both the input (amplitude, frequency, phase and DC level) and the filter (cutoff frequency and type of filter).

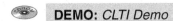

DEMO: *CLTI Demo*

As usual, the reader is reminded of the large number of solved homework problems on the CD-ROM that are available for review and practice.

NOTE: *Hundreds of Solved Problems*

10-7 Problems

P-10.1 Derive the frequency response of an ideal delay system by substituting $x(t) = e^{j\omega t}$ into the system's definition

$$y(t) = x(t - \tfrac{1}{2})$$

In addition, plot the magnitude of the frequency response for this system.

P-10.2 The frequency response of an LTI system is

$$H(j\omega) = \frac{3 - j\omega}{3 + j\omega} e^{-j\omega}$$

(a) Determine the magnitude-squared of the frequency response and reduce it to a simple real-valued form.

(b) Determine the phase of the frequency response $\angle H(j\omega)$.

(c) If the input to the LTI system is $x(t) = 4 + \cos(3t)$ for $-\infty < t < \infty$, what is the corresponding output $y(t)$?

P-10.3 Suppose that the continuous-time LTI system in Fig. P-10.3 is defined by the following input/output relation:

$$y(t) = x(t + 1) + 2x(t) + x(t - 2) \qquad (10.44)$$

(a) Find the impulse response $h(t)$ of the system; i.e., determine the output when the input is an impulse.

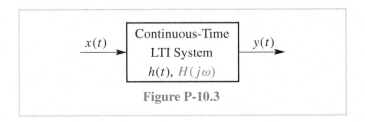

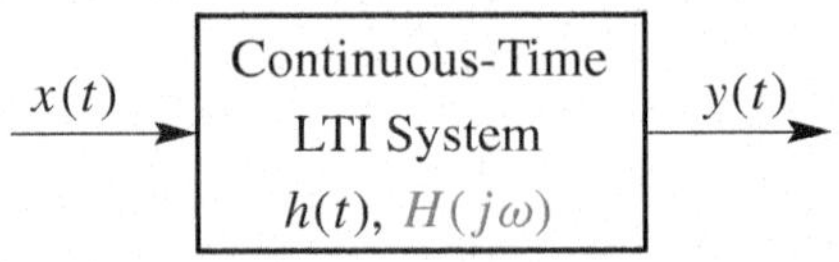

Figure P-10.3

(b) Substitute your answer for $h(t)$ into the integral formula

$$H(j\omega) = \int_{-\infty}^{\infty} h(t)e^{-j\omega t}\,dt$$

to find the frequency response.

(c) Apply the system definition given in Eq. (10.44) directly to the input $x(t) = e^{j\omega t}$ for $-\infty < t < \infty$ and show that $y(t) = H(j\omega)e^{j\omega t}$, where $H(j\omega)$ is as determined in part (b).

P-10.4 The impulse response of a continuous-time linear time-invariant system is

$$h(t) = \delta(t) - 0.1e^{-0.1t}u(t)$$

(a) Find the frequency response $H(j\omega)$ of the system. Express your answer as a single rational function with powers of $(j\omega)$ in the numerator and denominator.

(b) Plot the frequency response magnitude squared

$$|H(j\omega)|^2 = H(j\omega)H^*(j\omega)$$

versus ω. Also plot the phase $\angle H(j\omega)$ as a function of ω.

(c) At what frequency ω does the magnitude squared of the frequency response have its largest value? At what frequency is the magnitude squared of the frequency response equal to one half of its peak value? This is referred to as the *half-power* point of the filter since the magnitude-squared corresponds to power.

(d) Suppose that the input to this system is

$$x(t) = 10 + 20\cos(0.1t) + \delta(t - 0.2)$$

Use superposition to find the output $y(t)$, after finding the response of each term with the easiest method, (impulse response or frequency response).

P-10.5 The input to an LTI system is the periodic pulse wave $x(t)$ depicted in Fig. P-10.5. This input can be represented by the Fourier series

$$x(t) = \sum_{k=-\infty}^{\infty} a_k e^{jk\omega_0 t} \quad \text{where} \quad a_k = \frac{10\sin(\pi k/4)}{\pi k}$$

(a) Determine ω_0 in the Fourier series representation of $x(t)$. Also, write down the integral that must be evaluated to obtain the Fourier coefficients a_k.

(b) Plot the spectrum of the signal $x(t)$; i.e., make a plot showing the a_k's plotted at the frequencies $k\omega_0$ for $-4\omega_0 \le \omega \le 4\omega_0$.

(c) If the frequency response of the system is the ideal *highpass* filter

$$H(j\omega) = \begin{cases} 0 & |\omega| < \pi/8 \\ 1 & |\omega| \ge \pi/8 \end{cases}$$

plot the output of the system, $y(t)$, when the input is $x(t)$ as plotted above.

Hint: First determine which frequency is removed by the filter, and then determine what effect this will have on the waveform.

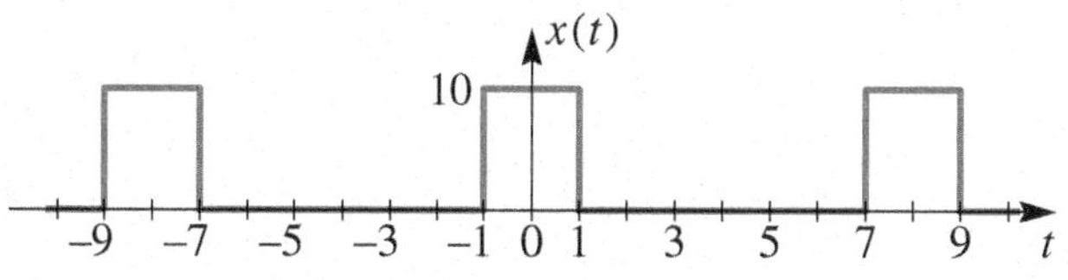

Figure P-10.5

(d) If the frequency response of the system is an ideal lowpass filter

$$H(j\omega) = \begin{cases} 1 & |\omega| < \omega_{co} \\ 0 & |\omega| > \omega_{co} \end{cases}$$

where ω_{co} is the *cutoff frequency*, for what values of ω_{co} will the output of the system have the form

$$y(t) = A + B\cos(\omega_0 t + \phi)$$

where A and B are nonzero?

(e) If the frequency response of the LTI system is $H(j\omega) = 1 - e^{-j2\omega}$, plot the output of the system, $y(t)$, when the input is $x(t)$ as plotted above.
Hint: In this case it will be easiest to determine the impulse response $h(t)$ corresponding to $H(j\omega)$ and from $h(t)$ you can easily find an equation that relates $y(t)$ to $x(t)$. This will allow you to plot $y(t)$.

P-10.6 The input to the LTI continuous-time system in Fig. P-10.6 is defined by the equation

$$x(t) = \sum_{k=-\infty}^{\infty} a_k e^{jk200\pi t}, \quad \text{where } a_k = \begin{cases} \dfrac{1}{\pi k^2} & k \neq 0 \\ 1 & k = 0 \end{cases}$$

(a) Determine the spectrum of $x(t)$ and plot it on a graph that spans the frequency range $-1000\pi \leq \omega \leq 1000\pi$.

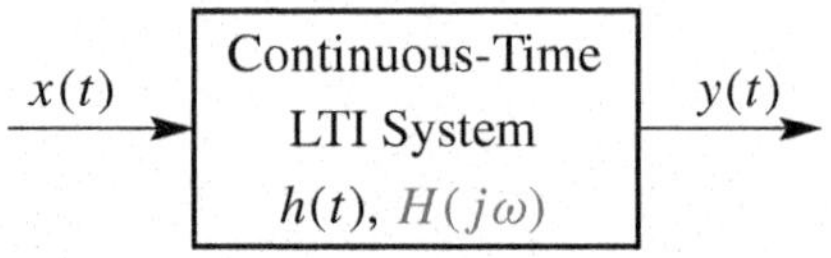

Figure P-10.6

(b) The frequency response of the LTI system is given by the following equation:

$$H(j\omega) = \begin{cases} 10 & |\omega| \leq 300\pi \\ 0 & |\omega| > 300\pi \end{cases}$$

Plot this function on the spectrum plot that you constructed in part (a).

(c) Determine the DC value of the output signal; i.e., determine $b_0 = \dfrac{1}{T_0}\displaystyle\int_0^{T_0} y(t)dt$, where T_0 is the period of the output signal.

(d) Write an equation for $y(t)$. If possible, simplify it to include only cosine functions.

P-10.7 A periodic waveform $x(t)$ with period $T = 4$ is defined over one period by the equation

$$x(t) = e^{-t} \qquad 0 \leq t < 4$$

(a) Carefully sketch $x(t)$.

(b) Determine the fundamental frequency, ω_0, of $x(t)$.

(c) Determine the Fourier series coefficients, a_k, for the above waveform $x(t)$. Give a general formula valid for any integer k.

(d) If $x(t)$ is passed through an ideal LPF whose cutoff frequency is $\omega_{co} = 2\pi/3$ rad/sec, determine the output signal, $y(t)$.

P-10.8 Suppose that the input $x(t)$ to an ideal frequency-selective filter is the square wave depicted in Fig. P-10.8. The Fourier series representation of this waveform is

$$x(t) = \sum_{k=-\infty}^{\infty} a_k e^{jk\omega_0 t} \qquad -\infty < t < \infty$$

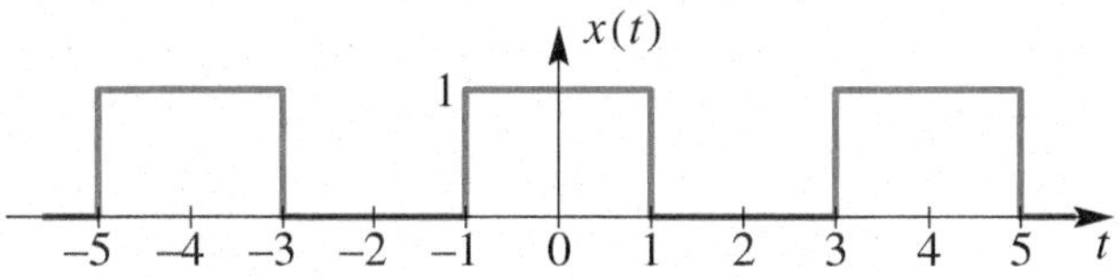

Figure P-10.8

(a) Determine the fundamental frequency ω_0.

(b) Determine the Fourier coefficients a_k for this signal.

(c) State whether or not it is possible that the corresponding output of the LTI system could be the signal

$$y(t) = 2\cos(2\pi t/4) \qquad -\infty < t < \infty$$

If it is impossible, state why. If it is possible, determine the frequency response $H(j\omega)$ of an ideal frequency-selective filter (lowpass, highpass, or bandpass) such that the output is as specified. Draw a carefully labeled sketch of $H(j\omega)$ showing the gain and cutoff frequencies for the filter.

(d) Repeat part (c) when the desired output is

$$y(t) = 2\cos(2\pi t/3)$$

P-10.9 Consider an LTI system whose frequency response $H(j\omega)$ is unknown. The system has a periodic input whose spectrum is shown in Fig. P-10.9.

For each part of this problem, the output of the system is given and the frequency response must be determined by selecting from the list numbered 1–7 below. Choose the frequency response $H(j\omega)$ of the system that could have produced the specific output when the input is the signal with the spectrum in Fig. P-10.9.

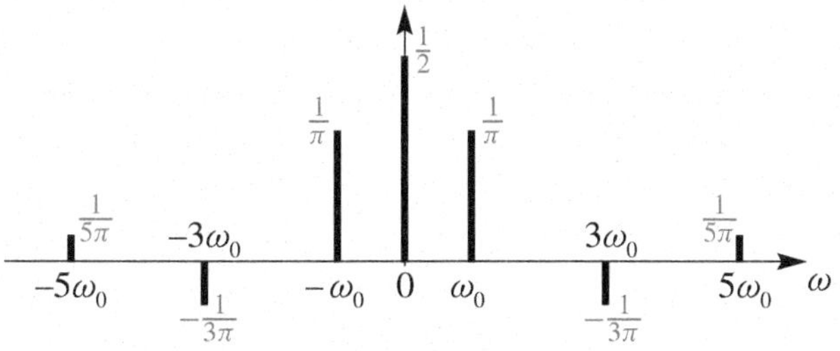

Figure P-10.9

(a) $y(t) = \frac{1}{2}$

(b) $y(t) = \frac{1}{2} + \frac{2}{\pi}\cos[\omega_0(t - \frac{1}{2})]$

(c) $y(t) = \frac{2}{\pi}\cos(\omega_0 t)$

(d) $y(t) = x(t) - \frac{1}{2}$

(e) $y(t) = x(t - \frac{1}{2})$

The possible filters are described by the following equations and graphs (some of these may not be used):

1. $H(j\omega) = \begin{cases} 0 & |\omega| < \frac{1}{2}\omega_0 \\ 1 & |\omega| \geq \frac{1}{2}\omega_0 \end{cases}$

2. $H(j\omega) = e^{-j\omega/2}$

3. $H(j\omega) = \frac{1}{2}[1 + \cos(\omega T_0)]$

4.

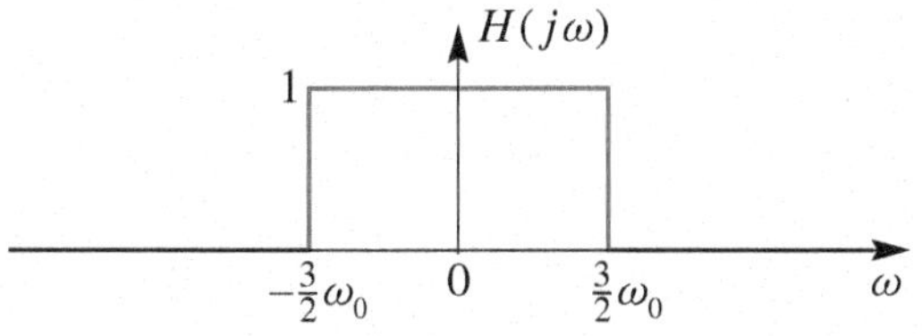

5. $H(j\omega) = \begin{cases} 1 & |\omega| \le \frac{1}{2}\omega_0 \\ 0 & |\omega| > \frac{1}{2}\omega_0 \end{cases}$

6. $H(j\omega) = \begin{cases} e^{-j\omega/2} & |\omega| \le \frac{3}{2}\omega_0 \\ 0 & |\omega| > \frac{3}{2}\omega_0 \end{cases}$

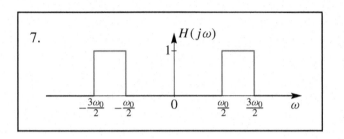

7.

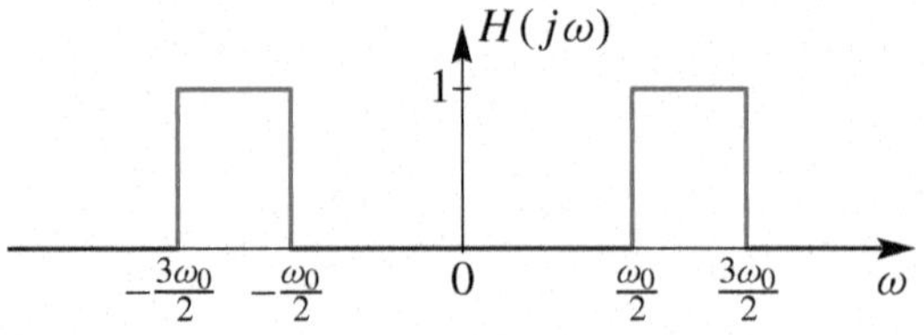

CHAPTER 11

Continuous-Time Fourier Transform

In previous chapters, our presentation of the spectrum was limited to sinusoids and periodic signals. Now we want to develop a ***completely general definition of the frequency spectrum*** that will apply to *any* signal $x(t)$. Armed with this powerful tool, we will be able to: (1) define a precise notion of bandwidth for a signal, (2) explain the inner workings of modern communication systems which are able to transmit many signals simultaneously by sharing the available bandwidth, and (3) define filtering operations that are needed to separate signals in such frequency-shared systems. There are many other applications of the Fourier transform, so it is safe to say that Fourier analysis provides the rigorous language needed to define and design modern engineering systems.

From almost the beginning of this text we have relied on the notion of representing a continuous-time signal by its spectrum, which we have defined up to now to be simply the magnitude, phase, and frequency information that is required to completely specify a sum of cosine signals. In particular, we showed in Chapter 3 that *if a continuous-time signal is periodic* with period T_0, then it can be represented as *a sum of cosines* with all frequencies being integer multiples of the fundamental frequency $\omega_0 = 2\pi/T_0$, and we showed that the Fourier series integral could be used to find the amplitudes and phase angles when we know an equation for the waveform $x(t)$ over one period.

In Chapter 10 we also discussed one of the most important properties of LTI systems, which is that a complex-exponential input of frequency ω produces an output that is also a complex exponential of that same frequency, but modified in amplitude and phase by the frequency response $H(j\omega)$ of the LTI system. This is a profoundly important result that holds for both discrete-time and continuous-time LTI systems. So far, however, we have been limited to input signals that are either individual complex exponentials or sums of complex exponentials of different isolated frequencies. In this chapter, we show how to use the frequency response concept with more general signals, but first we must introduce the concept of the ***Fourier transform***.[1] Therefore, this chapter focuses on the continuous-time Fourier transform and its properties. We will discuss many useful results and show many examples of how the Fourier transform can be used as an aid in thinking about and analyzing continuous-time signals and systems problems.

11-1 Definition of the Fourier Transform

The Fourier transform is defined by the following pair of equations:

Forward
Continuous-Time Fourier Transform

$$X(j\omega) = \int_{-\infty}^{\infty} x(t)e^{-j\omega t}dt \tag{11.1}$$

and

Inverse
Continuous-Time Fourier Transform

$$x(t) = \frac{1}{2\pi}\int_{-\infty}^{\infty} X(j\omega)e^{j\omega t}d\omega \tag{11.2}$$

[1]There is also a Fourier transform for discrete-time signals $x[n]$, called the Discrete-Time Fourier Transform (DTFT). Although we will not discuss the DTFT in this chapter, most of its properties are identical to the ones we will study for the Continuous-Time Fourier Transform (CTFT). The DTFT will be introduced in Chapter 12 and will play a major role in Chapter 13.

The forward transform is an ***analysis*** integral because it extracts spectrum information from the signal, $x(t)$; the inverse transform is a ***synthesis*** integral because it is used to create the time-domain signal from its spectral information. As has been our custom in this book, we will refer to $X(j\omega)$ as the ***frequency-domain*** representation because we can interpret (11.2) as a "sum" of infinitely many complex-exponential signals with $X(j\omega)$ controlling the amplitudes and phases of these signals. Likewise, $x(t)$ is the ***time-domain*** representation of the signal. We indicate this (one-to-one) relationship between the two domains as

Time-Domain		***Frequency-Domain***
$x(t)$	$\stackrel{\mathcal{F}}{\longleftrightarrow}$	$X(j\omega)$

(11.3)

The notation $\stackrel{\mathcal{F}}{\longleftrightarrow}$ signifies that it is possible to go back and forth uniquely between the time-domain and the frequency-domain.

If we are given $x(t)$ as a mathematical function, we can determine the corresponding spectrum function $X(j\omega)$ by evaluating the analysis integral in (11.1). In other words, (11.1) defines a mathematical operation for *transforming* $x(t)$ into a new equivalent representation $X(j\omega)$. Hence the name *Fourier transform* is used for $X(j\omega)$. It is common to say that we ***take the Fourier transform of*** $x(t)$, meaning that we determine $X(j\omega)$ so that we can use the frequency-domain representation of the signal.

Similarly, given $X(j\omega)$ as a mathematical function, we can determine the corresponding time function $x(t)$ using (11.2) by again evaluating an integral—the synthesis integral. Thus, (11.2) defines the *inverse Fourier transform* operation; i.e., the operation that goes from the frequency-domain to the time-domain. Again, common usage is to say that we ***take the inverse Fourier transform*** when we evaluate (11.2).

Example 11-1: Forward Fourier Transform

As an example of determining the Fourier transform, consider the one-sided exponential signal

$$x(t) = e^{-7t}u(t) \tag{11.4}$$

Substituting this function into (11.1) gives

$$\begin{aligned} X(j\omega) &= \int_0^\infty e^{-7t}e^{-j\omega t}dt \\ &= \int_0^\infty e^{-(7+j\omega)t}dt \\ &= \left.\frac{e^{-7t}e^{-j\omega t}}{-(7+j\omega)}\right|_0^\infty \\ &= \frac{1}{7+j\omega} \end{aligned} \tag{11.5}$$

Note that the value of the integral at the upper limit is zero because the magnitude of $e^{-7t}e^{-j\omega t}$ is equal to e^{-7t} which goes to zero as $t \to \infty$.[2] Thus, we have derived the following Fourier transform pair (in the notation of (11.3)):

Time-Domain		***Frequency-Domain***
$e^{-7t}u(t)$	$\overset{\mathcal{F}}{\longleftrightarrow}$	$\dfrac{1}{7+j\omega}$

(11.6)

Notice that in evaluating the integral in (11.5) no restrictions were placed on ω, so we conclude that all frequencies, $-\infty < \omega < \infty$, are required to represent $e^{-7t}u(t)$ by the Fourier transform. ■

It is rare that we actually evaluate the Fourier transform integral (11.1), or the inverse transform integral directly. Instead, the usual route for solving signal processing problems is to use a known Fourier transform pair along with properties of the Fourier transform to get the solution. For example, the Fourier transform is a one-to-one mapping from $x(t)$ to $X(j\omega)$ which means that each transform $X(j\omega)$ corresponds to only one time function $x(t)$, and each time function $x(t)$ corresponds to only one transform $X(j\omega)$. The following example illustrates how uniqueness can be exploited.

Example 11-2: Unique Inverse

Consider the problem of evaluating the following integral:

$$\frac{1}{2\pi}\int_{-\infty}^{\infty}\left(\frac{1}{7+j\omega}\right)e^{-j3\omega}d\omega = ? \tag{11.7}$$

This integral is difficult, if not impossible, to evaluate by ordinary methods of integral calculus. However, the integral is a special case of an inverse transform integral, so the uniqueness of the Fourier transform representation guarantees that we can be confident in writing

$$\frac{1}{2\pi}\int_{-\infty}^{\infty}\left(\frac{1}{7+j\omega}\right)e^{j\omega t}d\omega = e^{-7t}u(t) \tag{11.8}$$

All we have to do is remember the Fourier transform pair in (11.6). Uniqueness guarantees that there is only one time function that goes with a given Fourier transform.

Finally, we can "do the integral" in (11.7) by taking the special case of $t = -3$ in (11.8) which means that we evaluate $e^{-7t}u(t)$ at $t = -3$ to get the answer of zero! It would be very difficult to obtain this answer by the ordinary methods of integral calculus. ■

Example 11-2 suggests that we can do Fourier transforms by table lookup. In other words, we can use previously derived results to simplify operations involving Fourier transforms. In Section 11-4, we will

[2] Recall $|e^{-7t}e^{-j\omega t}| = |e^{-7t}||e^{-j\omega t}|$, and $|e^{-j\omega t}| = 1$.

derive a number of Fourier transform pairs that are later summarized in Table 11-2 of Section 11-9.[3] But first, we will show how the Fourier transform is a natural extension of the Fourier series.

11-2 Fourier Transform and the Spectrum

It is possible to think of an integral as a sum.[4] In our case, we can interpret the synthesis integral (11.2) as a "sum" of complex-exponential signals, each having a different complex amplitude if we write the integrand as

$$\left[\frac{1}{2\pi}X(j\omega)d\omega\right]e^{j\omega t}$$

An integral is a special "sum" because it is the "sum" of infinitesimally small quantities. The Fourier transform function $X(j\omega)$ is the complex amplitude multiplying the complex exponential $e^{j\omega t}$, while the fraction $\frac{1}{2\pi}$ is a constant necessary for proper normalization in the frequency units of rad/sec, and the quantity $d\omega$ gives the complex amplitude its infinitesimal nature. Thus, $X(j\omega)$ carries the amplitude and phase information for all the frequencies required to synthesize $x(t)$ as a "sum" of complex-exponential signals. In the inverse Fourier transform, *all* frequencies $-\infty < \omega < \infty$ are required to represent the signal, so $X(j\omega)$ is a continuous function of ω, not just a countable set like the Fourier series coefficients $\{a_k\}$ used in the periodic case.

[3]Computer software that does symbolic algebra can be programmed to do most Fourier transforms if it can access a large table of "known transform pairs" and "known transform properties" as rules.

[4]Recall the Riemann integral which is derived as the limit of summing rectangular areas that approximate the area under a curve.

11-2.1 Limit of the Fourier Series

Another way to see how the Fourier transform is a sum of infinitesimal complex exponentials is to examine the Fourier series in the limit where the spacing between frequencies is very small.[5] Recall that the Fourier series representation of a periodic signal $x_{T_0}(t)$ is given by the pair of equations

$$a_k = \frac{1}{T_0}\int_{-T_0/2}^{T_0/2} x_{T_0}(t)e^{-jk\omega_0 t}dt \qquad (11.9)$$

and

$$x_{T_0}(t) = \sum_{k=-\infty}^{\infty} a_k e^{jk\omega_0 t} \qquad (11.10)$$

Now consider a finite-duration signal $x(t)$ that is *not periodic*, and therefore does not have a Fourier series representation. Nonetheless, we can use $x(t)$ to define one period of a periodic function as long as the period (T_0) of the periodic function is longer than the duration of $x(t)$. As a tangible example of this process, suppose that we start with a finite-duration rectangular pulse

$$x(t) = \begin{cases} 1 & -\frac{1}{2}T < t < \frac{1}{2}T \\ 0 & \text{otherwise} \end{cases} \qquad (11.11)$$

If $T_0 > T$, then we can form a periodic function with period T_0 by repeating copies of $x(t)$ every T_0 secs. A convenient way to express this "periodic replication" process is to write an infinite sum of time-shifted copies of $x(t)$

$$x_{T_0}(t) = \sum_{n=-\infty}^{\infty} x(t - nT_0)$$

[5]This limiting argument is not a rigorous derivation of the Fourier transform; instead, it is a plausibility argument that suggests the correct form of the transform integral and provides a useful interpretation.

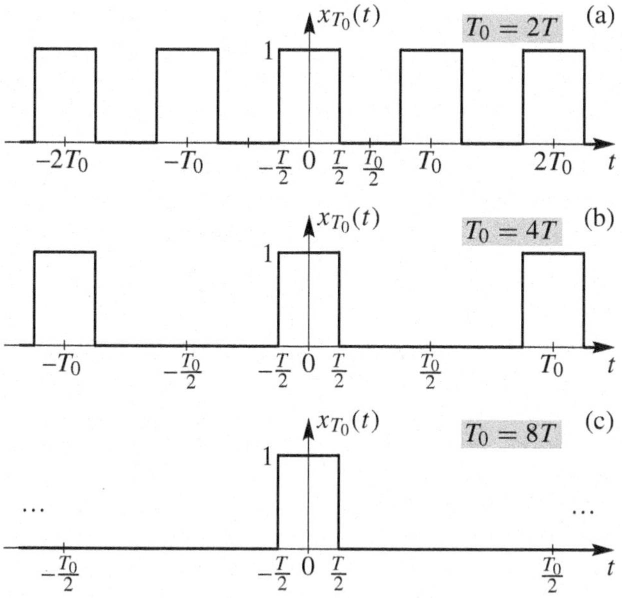

Figure 11-1: Periodic signals formed from finite duration signal $x(t)$: (a) $T_0 = 2T$, (b) $T_0 = 4T$, (c) $T_0 = 8T$. In (c) the repetition is not visible on the plot.

where $x(t - nT_0)$ is the copy centered at $t = nT_0$. Figure 11-1 shows three examples with different values of T_0 for the rectangular pulse signal in (11.11). Note that as T_0 increases relative to T, the periodic signal $x_{T_0}(t)$ becomes equal to $x(t)$ over a longer time interval, so we can claim that

$$\lim_{T_0 \to \infty} x_{T_0}(t) = x(t) \qquad \text{for } -\infty < t < \infty$$

Now we examine the Fourier series for each of these periodic signals and take the limit. The fundamental frequency of the periodic signal $x_{T_0}(t)$ is $\omega_0 = 2\pi/T_0$, and the Fourier series has spectrum lines at integer multiples of ω_0. Hence, ω_0 is the spacing between spectral lines. As $T_0 \to \infty$, we find that the frequencies contained in the spectrum of $x_{T_0}(t)$ become infinitely dense because the spacing between frequencies is ω_0 and

$$\lim_{T_0 \to \infty} \omega_0 = \lim_{T_0 \to \infty} 2\pi/T_0 = 0$$

The effect on the Fourier series representation is shown by rewriting (11.9) and (11.10) as

$$(a_k T_0) = \int_{-T_0/2}^{T_0/2} x_{T_0}(t) e^{-jk\omega_0 t} dt \tag{11.12}$$

and

$$x_{T_0}(t) = \frac{1}{2\pi} \sum_{k=-\infty}^{\infty} (a_k T_0) e^{jk\omega_0 t} \left(\frac{2\pi}{T_0}\right) \tag{11.13}$$

As $T_0 \to \infty$, the fundamental frequency ω_0 gets very small and the set $\{k\omega_0\}$ defines a very dense set of points on the frequency axis that approaches the continuous variable ω. As a result, we can claim that (11.12) approaches (11.1), if we make the identifications $a_k T_0 \to X(j\omega)$ and $x_{T_0}(t) \to x(t)$. Similarly, $\omega_0 = 2\pi/T_0 \to d\omega$ and the sum in (11.13) becomes an integral in the limit, equal to (11.2). This is illustrated for the examples of Fig. 11-1 by the spectra plotted in Fig. 11-2. In this case, the quantities $a_k T_0$ are given by[6]

$$\begin{aligned} a_k T_0 &= \int_{-T/2}^{T/2} e^{-jk\omega_0 t} dt \\ &= -\frac{1}{jk\omega_0} e^{-jk\omega_0 t} \Big|_{-T/2}^{T/2} = -\frac{e^{-jk\omega_0 T/2} - e^{jk\omega_0 T/2}}{jk\omega_0} \\ &= \frac{\sin(k\omega_0 T/2)}{k\omega_0/2} \end{aligned}$$

As Fig. 11-2 shows, the frequencies $k\omega_0$ get closer and closer together as $T_0 \to \infty$, and eventually become

[6]The quantity $a_k T_0$ is bounded, even though $T_0 \to \infty$.

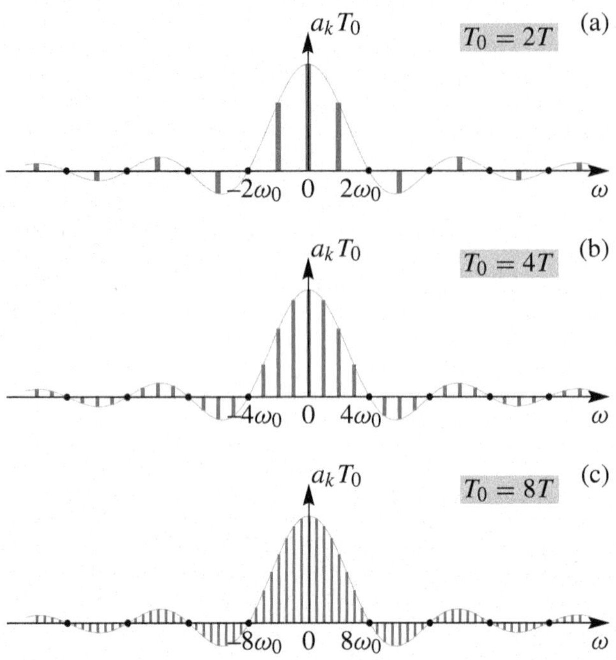

Figure 11-2: Spectra of periodic signals formed from the finite duration signal $x(t)$ for (a) $T_0 = 2T$, (b) $T_0 = 4T$, (c) $T_0 = 8T$.

dense in the interval $-\infty < \omega < \infty$. The quantities $a_k T_0$ approach the continuous envelope function

$$X(j\omega) = \lim_{k\omega_0 \to \omega} \frac{\sin(k\omega_0 T/2)}{k\omega_0/2} = \frac{\sin(\omega T/2)}{\omega/2} \qquad (11.14)$$

which can be seen in the spectrum plots of Fig. 11-2.

11-3 Existence and Convergence of the Fourier Transform

The Fourier transform (11.1) and its inverse (11.2) are, in general, integrals with infinite limits. When an integral has infinite limits, it is said to be an *improper* integral to convey the concern that such an infinite sum of even infinitesimally small quantities might not *converge* to a finite result. When one or both of the limits are infinite, the integral can still be finite if the integrand is well-behaved (e.g., it approaches zero fast enough as $t \to \pm\infty$).

Not every function $x(t)$ has a Fourier transform representation. To aid in our use of the Fourier transform it would be helpful to be able to determine whether the Fourier transform exists or not. As a simple condition for convergence of the integral, we can check the magnitude of $X(j\omega)$. The following manipulations result in a bound on the size of $|X(j\omega)|$ that can be used to obtain a *sufficient condition* for existence of the Fourier transform:

$$|X(j\omega)| = \left| \int_{-\infty}^{\infty} x(t) e^{-j\omega t} dt \right|$$

$$\leq \int_{-\infty}^{\infty} |x(t) e^{-j\omega t}| dt = \int_{-\infty}^{\infty} |x(t)| dt$$

The last step follows from the fact that $|e^{-j\omega t}| = 1$ for all t and ω. Thus, a *sufficient* condition[7] for the existence of the Fourier transform ($|X(j\omega)| < \infty$) is

Sufficient Condition for Existence of $X(j\omega)$

$$\int_{-\infty}^{\infty} |x(t)| dt < \infty \qquad (11.15)$$

[7]The sufficient condition is equivalent to the stability condition when the signal is an impulse response, $h(t)$. Thus we might say that "stable signals" are guaranteed to have a Fourier transform.

The condition of (11.15) states that *if* the function $x(t)$ is *absolutely integrable*, then the Fourier transform can in principle be found by evaluating (11.1). It is therefore a sufficient condition, but it turns out that (11.15) is *not* a necessary condition. We will see many examples of functions that do not satisfy (11.15) for which we can still obtain a useful Fourier transform representation, particularly if we are willing to allow impulse signals (which we know are not well-behaved functions) in the time- and frequency-domain representations of our signals. We will give some examples in the next section.

11-4 Examples of Fourier Transform Pairs

Now that we have shown that the Fourier transform can represent non-periodic signals in much the same way that the Fourier series represents periodic signals, we can begin to develop a library of Fourier transform pairs that will be useful for solving problems.

11-4.1 Right-Sided Real Exponential Signals

The signal $x(t) = e^{-at}u(t)$ is called a *right-sided exponential signal* because it is nonzero only on the right side of its graph as shown in Figure 11-3(a). In Example 11-1 we did a specific case ($a = 7$) of the following Fourier transform pair:

Time-Domain		**Frequency-Domain**	
$e^{-at}u(t)$	$\overset{\mathcal{F}}{\longleftrightarrow}$	$\dfrac{1}{a + j\omega}$	(11.16)

The absolute integrability condition of (11.15) shows that a must be positive in order to have a legitimate transform pair in (11.16). If we substitute the function $x(t) = e^{-at}u(t)$ into (11.15) we obtain

$$\int_0^\infty |e^{-at}|dt = \int_0^\infty e^{-at}dt = -\frac{1}{a}e^{-at}\bigg|_0^\infty$$

This result will be finite only if e^{-at} at the upper limit of ∞ is bounded, which is true only if $a > 0$. Thus, the right-sided exponential signal is guaranteed to have a Fourier transform if it dies out with increasing t, which requires $a > 0$. Otherwise, if $a \leq 0$, the right-sided exponential either does not die out or it increases without bound and, therefore, does not have a Fourier transform when $a \leq 0$.

EXERCISE 11.1: Confirm that (11.16) is a valid Fourier transform pair even when a is a complex number, as long as $\Re e\{a\} > 0$.

The Fourier transform $X(j\omega) = 1/(a + j\omega)$ is a complex function of ω. To plot it we could plot the real and imaginary parts versus ω, or we could plot the magnitude and phase angle as functions of frequency. The real and imaginary parts are found from the following algebraic manipulation:

$$X(j\omega) = \frac{1}{a + j\omega} = \frac{1}{a + j\omega}\left(\frac{a - j\omega}{a - j\omega}\right)$$
$$= \frac{a}{a^2 + \omega^2} + \frac{-j\omega}{a^2 + \omega^2}$$

Therefore, the real and imaginary parts are

$$\Re e\{X(j\omega)\} = \frac{a}{a^2 + \omega^2}$$
$$\Im m\{X(j\omega)\} = -\frac{\omega}{a^2 + \omega^2}$$

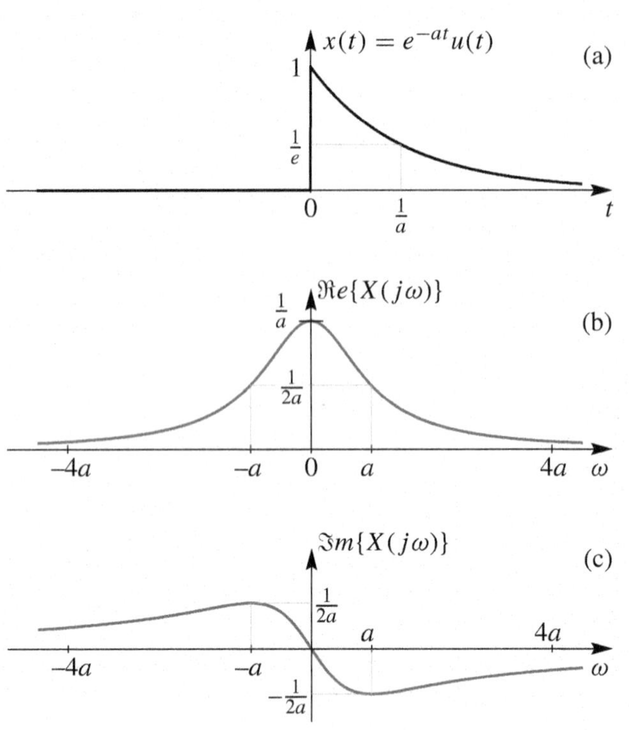

Figure 11-3: Right-sided exponential signal; (a) Time function $x(t) = e^{-at}u(t)$, and (b) Real part of $X(j\omega)$; (c) imaginary part of $X(j\omega)$.

Figures 11-3(b) and 11-3(c) show the real and imaginary parts for this Fourier transform (see Fig. 11-12 on p. 325 for plots of the magnitude and phase).

EXERCISE 11.2: The signal $x(t) = e^{bt}u(-t)$ is an example of a *left-sided real exponential signal.* Sketch this signal for $b > 0$ and then show that the Fourier transform of this signal is

$$X(j\omega) = \frac{1}{b - j\omega}$$

if $b > 0$. Also, using (11.15), show that the Fourier transform does not exist if $b \le 0$.

EXERCISE 11.3: Sketch graphs of the real and imaginary parts of $X(j\omega)$ from Exercise 11.2 as functions of ω and compare them to Figures 11-3(b) and 11-3(c).

11-4.1.1 Bandwidth and Decay Rate

Figure 11-3 illustrates a fundamental property of Fourier transform representations—the inverse relation between time and frequency. First of all, notice that the parameter a controls the rate of decay of the exponential in the time-domain. As a increases, the exponential dies out more quickly, making $x(t)$ more concentrated near $t = 0$. In the frequency-domain, on the other hand, as a increases, the Fourier transform spreads out.

EXERCISE 11.4: Make sketches or MATLAB plots like Fig. 11-3 for $a = 1$ and $a = 2$.

The inverse relationship between the time-domain and frequency-domain occurs consistently. Signals that are short in time duration are spread out in frequency; i.e., they have a wide "bandwidth". Conversely, signals that are spread out in time are concentrated in frequency. A formal statement of this property is the scaling property presented in Section 11-5.1 on p. 322.

11-4.2 Rectangular Pulse Signals

One common mathematical signal model is the ***rectangular pulse*** signal defined by

$$x(t) = \begin{cases} 1 & -\frac{1}{2}T \le t < \frac{1}{2}T \\ 0 & \text{otherwise} \end{cases} \tag{11.17}$$

In this definition, the pulse length is T and the signal is symmetric around $t = 0$. Figure 11-4(a) shows a plot of $x(t)$. We can determine the Fourier transform by substituting this function into (11.1), which gives

$$X(j\omega) = \int_{-T/2}^{T/2} e^{-j\omega t}\,dt$$

$$= \left.\frac{e^{-j\omega t}}{-j\omega}\right|_{-T/2}^{T/2} = \frac{e^{-j\omega T/2} - e^{j\omega T/2}}{-j\omega}$$

$$= \frac{\sin(\omega T/2)}{\omega/2} \qquad (11.18)$$

Not surprisingly, this is identical to the result (11.14) obtained by the limiting argument of Section 11-2. Thus, we have derived the following Fourier transform pair:

Time-Domain		***Frequency-Domain***
$\left[u(t+\tfrac{1}{2}T) - u(t-\tfrac{1}{2}T)\right]$	$\overset{\mathcal{F}}{\longleftrightarrow}$	$\dfrac{\sin(\omega T/2)}{\omega/2}$

(11.19)

where (11.17) is represented more compactly using the unit-step signal.

The Fourier transform of the rectangular pulse signal (shown in Fig. 11-4(b)) is called a ***sinc function***. It is a widely used function in the theory of signals and systems. The formal definition of a "sinc" function is[8]

$$\text{sinc}(\theta) = \frac{\sin(\pi\theta)}{\pi\theta} \qquad (11.20)$$

Even though some authors use the sinc notation defined in (11.20)) to simplify the appearance of some Fourier transform expressions, the sinc(·) function sometimes requires awkward scalings of its argument to make it fit simple situations. For example, in terms of the sinc function notation, the transform of the rectangular pulse is $X(j\omega) = T\,\text{sinc}(\omega T/2\pi)$. We will not use the strict sinc function notation here except to use the name "sinc function" as a generic designation for any function of the general form $\sin(\cdot)/(\cdot)$.

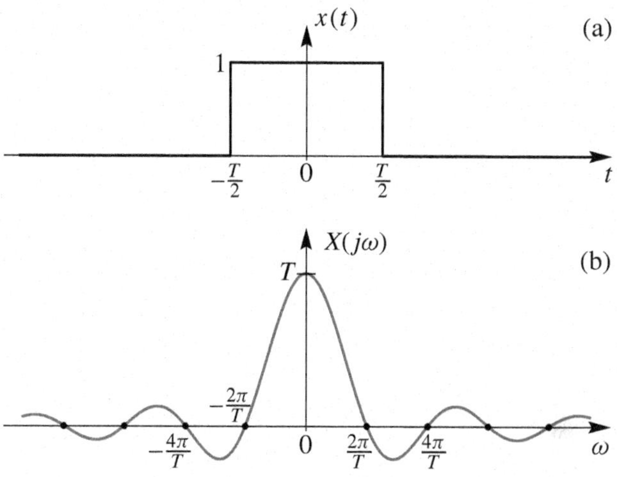

Figure 11-4: Fourier transform of a rectangular pulse. (a) Time function $x(t) = u(t+\frac{1}{2}T) - u(t-\frac{1}{2}T)$, and (b) Corresponding Fourier transform $X(j\omega)$ is a sinc function.

Several important properties of the sinc function are evident in Fig. 11-4(b):

1. The value at $\omega = 0$ is $X(j0) = T$. When we attempt to evaluate the sinc formula (11.18) at $\omega = 0$, we obtain the indeterminate form $\frac{0}{0}$. However, using L'Hôpital's rule from calculus, we obtain

$$X(j0) = \lim_{\omega\to 0}\left(\frac{\sin(\omega T/2)}{\omega/2}\right)$$

$$= \lim_{\omega\to 0}\left(\frac{T/2\cos(\omega T/2)}{1/2}\right) = T$$

[8] MATLAB has a function called `sinc()` in its Signal Processing Toolbox.

Note that we could also use the small angle approximation for the sine function ($\sin\theta \approx \theta$) to obtain the same result since

$$\lim_{\omega\to 0}\left(\frac{\sin(\omega T/2)}{\omega/2}\right) \approx \frac{(\omega T/2)}{\omega/2} = T$$

2. The zeros of the sinc function are at nonzero integer multiples of $\omega = 2\pi/T$, where T is the total duration of the pulse. Notice that the sinc function oscillates between positive and negative values. Specifically, it crosses zero at regular intervals because we have the function $\sin(\omega T/2)$ in the numerator. The zeros of the sinc function (11.18) are the values of ω where the numerator of $X(j\omega)$ becomes zero. These are the values of ω such that $\sin(\omega T/2) = 0$. Since $\sin(\theta) = 0$ for $\theta = n\pi$ where n is an integer, it follows that $X(j\omega) = 0$ for $\omega T/2 = n\pi$ or $\omega = (2\pi/T)n$, as is shown in Figure 11-4.

3. Because of the ω in the denominator of $X(j\omega)$, the function "dies out" with increasing ω, but only as fast as $1/\omega$.

4. $X(j\omega)$ is an even function; i.e., $X(j\omega) = X(-j\omega)$.

$$X(-j\omega) = \frac{\sin(-\omega T/2)}{-\omega/2} = \frac{-\sin(\omega T/2)}{-\omega/2} = X(j\omega)$$

Thus the real even-symmetric rectangular pulse has a real even-symmetric Fourier transform.

These properties of the sinc function will be exploited in a variety of applications of the Fourier transform.

11-4.3 Bandlimited Signals

The notion of bandwidth is central to our understanding and use of Fourier transforms. Intuitively, a frequency function that is concentrated near $\omega = 0$ has a small width

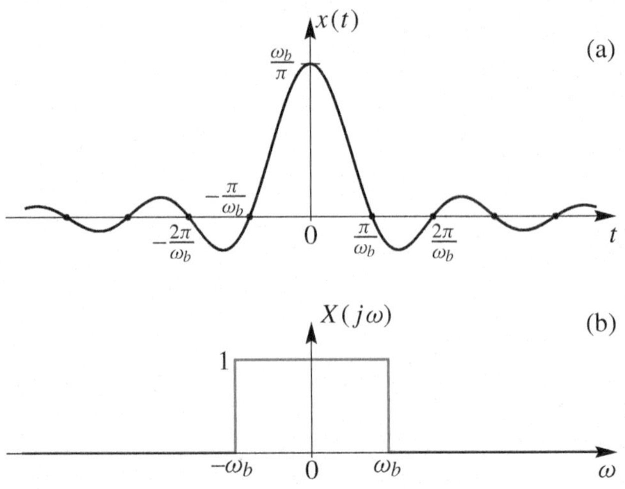

Figure 11-5: Fourier transform of sinc function: (a) Time function $x(t) = \sin(\omega_b t)/(\pi t)$, and (b) corresponding Fourier transform $X(j\omega) = u(\omega + \omega_b) - u(\omega - \omega_b)$.

in ω. We call this width the bandwidth. This leads us to define a *bandlimited signal* as one whose Fourier transform satisfies the condition $X(j\omega) = 0$ for $|\omega| > \omega_b$ with $\omega_b < \infty$. The frequency ω_b is called the *bandwidth* of the bandlimited signal.[9]

One example of an ideally bandlimited Fourier transform is a rectangular shape in the frequency-domain:

$$X(j\omega) = \begin{cases} 1 & -\omega_b \le \omega \le \omega_b \\ 0 & \text{otherwise} \end{cases} \tag{11.21}$$

This Fourier transform is depicted in Fig. 11-5(b). Since the transform is zero for $|\omega| > \omega_b$ it is obviously *bandlimited.*

We want to determine the time-domain signal that has this Fourier transform, i.e., we need to evaluate the

[9]When the Fourier transform in (11.21) is used as a frequency response, ω_b would be called the *cutoff frequency*.

inverse transform integral. In this case, we can find the corresponding time function $x(t)$ by substituting (11.21) into (11.2) obtaining

$$\begin{aligned} x(t) &= \frac{1}{2\pi}\int_{-\omega_b}^{\omega_b} e^{j\omega t}\,d\omega \\ &= \frac{1}{2\pi}\,\frac{1}{jt}e^{j\omega t}\Big|_{-\omega_b}^{\omega_b} = \frac{e^{j\omega_b t}-e^{-j\omega_b t}}{j2\pi t} \\ &= \frac{\sin(\omega_b t)}{\pi t} \end{aligned} \tag{11.22}$$

Figure 11-5(a) shows this time function, which has the form of a sinc function. Therefore, it has the same properties discussed in Section 11-4.2 except that now the sinc function is in the time-domain. In this case, the signal has a peak value of ω_b/π at $t = 0$, and the zero crossings are spaced at nonzero multiples of π/ω_b. Once again, we note the inverse relationship between time width and frequency width. If we increase ω_b, the bandwidth is greater, but the first zero crossing in the time-domain moves closer to $t = 0$ so the time-width is smaller. Now we can add this Fourier transform pair to our list of known transform relations:

Time-Domain		***Frequency-Domain***
$\dfrac{\sin(\omega_b t)}{\pi t}$	$\overset{\mathcal{F}}{\longleftrightarrow}$	$[u(\omega+\omega_b)-u(\omega-\omega_b)]$

(11.23)

11-4.4 Impulse in Time or Frequency

The impulse time-domain signal is the most concentrated time signal that we can have. Therefore, we might expect that its Fourier transform will have a very wide bandwidth, and it does. The Fourier transform of $A\delta(t)$ contains all frequencies in equal amounts. To see this, substitute $x(t) = A\delta(t)$ into (11.1) to obtain

$$\begin{aligned} X(j\omega) &= \int_{-\infty}^{\infty} A\delta(t)e^{-j\omega t}dt \\ &= \int_{-\infty}^{\infty} Ae^{-j\omega 0}\delta(t)dt \\ &= A\int_{-\infty}^{\infty}\delta(t)dt \\ &= A \end{aligned} \tag{11.24}$$

The resulting Fourier transform pair depicted in Fig. 11-6 is

Time-Domain		***Frequency-Domain***
$A\delta(t)$	$\overset{\mathcal{F}}{\longleftrightarrow}$	A

(11.25)

Likewise, we can examine an impulse in frequency, if we define the Fourier transform of a signal to be $X(j\omega) = 2\pi\delta(\omega)$. Then we can show by substitution into (11.2) that $x(t) = 1$ for all t and thereby obtain the Fourier transform pair

Time-Domain		***Frequency-Domain***
1	$\overset{\mathcal{F}}{\longleftrightarrow}$	$2\pi\delta(\omega)$

(11.26)

The constant signal $x(t) = 1$ for all t has only one frequency, namely DC, and we see that its transform is an impulse concentrated at $\omega = 0$.

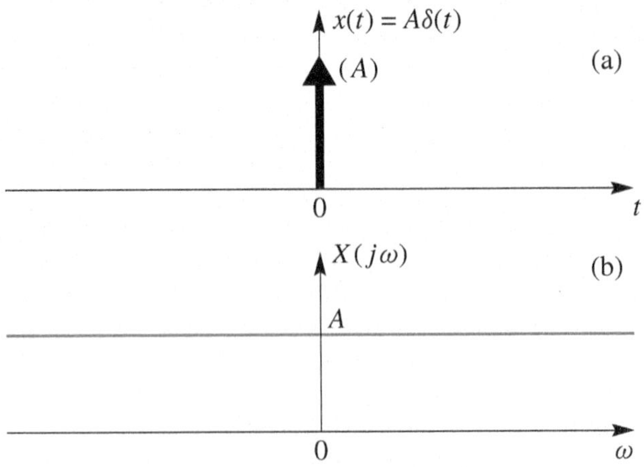

Figure 11-6: Fourier transform of an impulse signal; (a) time function $x(t) = A\delta(t)$, and (b) corresponding Fourier transform $X(j\omega) = A$. Since the impulse is the signal that is most concentrated in time, its Fourier transform is the most spread out in frequency.

EXERCISE 11.5: Make a plot in the style of Fig. 11-6 for the signal $x(t) = 1$ and its corresponding Fourier transform $X(j\omega) = 2\pi\delta(\omega)$.

11-4.5 Sinusoids

In this section and the next, we will show how to determine the Fourier transform of a periodic signal. This may at first seem contradictory since we know that periodic functions can be represented by a Fourier series, which is an ordinary sum of complex exponentials rather than an integral. However, there are distinct advantages for bringing this class of signals under the general Fourier transform umbrella.

The example of the constant signal just discussed in the previous section provides a clue as to how to do this. Suppose that the Fourier transform of a signal is an impulse at $\omega = \omega_0$, $X(j\omega) = 2\pi\delta(\omega - \omega_0)$. By substituting into the inverse transform integral (11.2), we obtain

$$\begin{aligned} x(t) &= \frac{1}{2\pi}\int_{-\infty}^{\infty} 2\pi\delta(\omega - \omega_0)e^{j\omega t}d\omega \\ &= e^{j\omega_0 t}\int_{-\infty}^{\infty} \delta(\omega - \omega_0)d\omega \\ &= e^{j\omega_0 t} \end{aligned} \tag{11.27}$$

from which we write the following transform pair:

Time-Domain		*Frequency-Domain*
$e^{j\omega_0 t}$	$\overset{\mathcal{F}}{\longleftrightarrow}$	$2\pi\delta(\omega - \omega_0)$

(11.28)

This result is not unexpected. It says that a complex-exponential signal of frequency ω_0 has a Fourier transform that is nonzero only at the frequency ω_0. The result in (11.28) is the basis for including all periodic functions in our Fourier transform framework. To go one step further, consider the signal

$$x(t) = A\cos(\omega_0 t + \phi) \qquad -\infty < t < \infty$$

Using Euler's relation we can rewrite this signal as

$$x(t) = \tfrac{1}{2}Ae^{j\phi}e^{j\omega_0 t} + \tfrac{1}{2}Ae^{-j\phi}e^{-j\omega_0 t} \tag{11.29}$$

Since integration is linear, it follows that the Fourier transform of a sum of two or more signals is the sum of their corresponding Fourier transforms. That is, if $x_1(t)$ and $x_2(t)$ have Fourier transforms $X_1(j\omega)$ and $X_2(j\omega)$, respectively, and a and b are constants, we can state the following *linearity property:*

Time-Domain		*Frequency-Domain*
$ax_1(t) + bx_2(t)$	$\overset{\mathcal{F}}{\longleftrightarrow}$	$aX_1(j\omega) + bX_2(j\omega)$

(11.30)

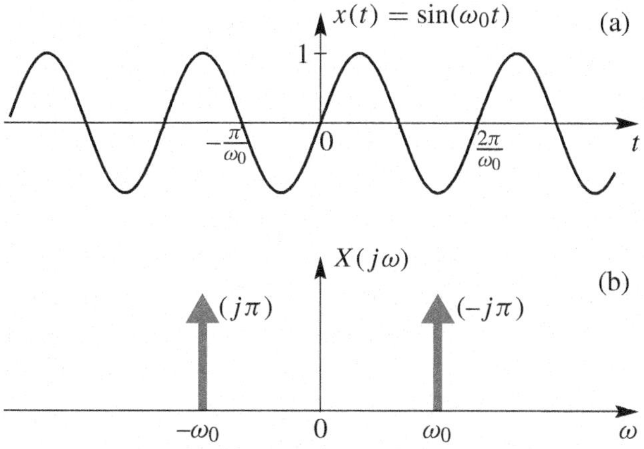

Figure 11-7: Fourier transform of a sinusoid: (a) Time function $x(t) = \sin(\omega_0 t) = \cos(\omega_0 t - \pi/2)$, and (b) corresponding Fourier transform $X(j\omega) = -j\pi\delta(\omega - \omega_0) + j\pi\delta(\omega + \omega_0)$.

Thus, using the transform pair in (11.28), the Fourier transform of the real sinusoid $x(t)$ in (11.29) is

$$X(j\omega) = 2\pi\delta(\omega - \omega_0)(\tfrac{1}{2})Ae^{j\phi} + 2\pi\delta(\omega + \omega_0)(\tfrac{1}{2})Ae^{-j\phi} \quad (11.31)$$

So we have another Fourier transform pair

Time-Domain	***Frequency-Domain***
$A\cos(\omega_0 t + \phi) \overset{\mathcal{F}}{\longleftrightarrow}$	$\pi A e^{j\phi}\delta(\omega - \omega_0) + \pi A e^{-j\phi}\delta(\omega + \omega_0)$

(11.32)

Figure 11-7 shows the case of a sine function and its Fourier transform, which is the special case $A = 1$ and $\phi = -\pi/2$. Note that the size (area) of the impulse at negative frequency $-\omega_0$ is the complex conjugate of the size of the impulse at the positive frequency. We will see that complex conjugate symmetry between the positive and negative frequency components is a general property of the Fourier transform of a *real* signal.

At this point, we comment that a sine wave is one example of a signal that has a Fourier transform, even though the existence condition (11.15) is violated. As stated before, (11.15) is a sufficient condition, but not a necessary condition, for having a Fourier transform.

11-4.6 Periodic Signals

Now we are ready to obtain a general formula for the Fourier transform of any periodic function for which a Fourier series exists. Recall that if $x(t) = x(t + T_0)$ for all t, then the signal is periodic with period T_0 and it can be represented by the sum of complex exponentials

$$x(t) = \sum_{k=-\infty}^{\infty} a_k e^{jk\omega_0 t} \quad (11.33)$$

where $\omega_0 = 2\pi/T_0$ and

$$a_k = \frac{1}{T_0}\int_{-T_0/2}^{T_0/2} x(t)e^{-jk\omega_0 t}dt \quad (11.34)$$

Since the Fourier transform of a sum is the sum of corresponding Fourier transforms, and we know the Fourier transforms of the exponential signals from (11.28), it follows that the Fourier transform of $x(t)$ in (11.33) is

$$X(j\omega) = \sum_{k=-\infty}^{\infty} 2\pi a_k \delta(\omega - k\omega_0) \quad (11.35)$$

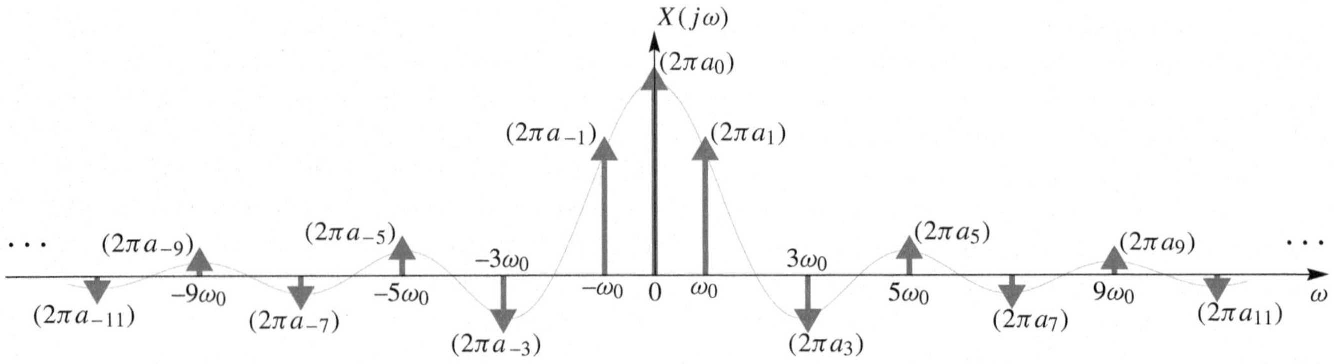

Figure 11-8: Fourier transform of the square wave shown in Fig. 11-9. The transform $X(j\omega)$ has regularly spaced impulses at $\omega = 2\pi k/T_0$.

Thus, any periodic signal with fundamental frequency ω_0 is represented by the following Fourier transform pair as illustrated in Fig. 11-8 for the square wave in Fig.11-9.

Time-Domain		***Frequency-Domain***
$\sum_{k=-\infty}^{\infty} a_k e^{jk\omega_0 t}$	$\overset{\mathcal{F}}{\longleftrightarrow}$	$\sum_{k=-\infty}^{\infty} 2\pi a_k \delta(\omega - k\omega_0)$

(11.36)

EXERCISE 11.6: Substitute (11.35) into the inverse Fourier transform integral in (11.2) and show that when evaluated using the properties of the impulse function, the result is identical to (11.33).

By showing that we can represent any periodic function by a Fourier transform, we have formalized the spectrum concept that we have been using since Chapter 3. The key ingredient is the impulse function which allows us to define Fourier transforms that are zero at all but a discrete set of frequencies. The following example illustrates how we take the Fourier transform of the square wave signal.

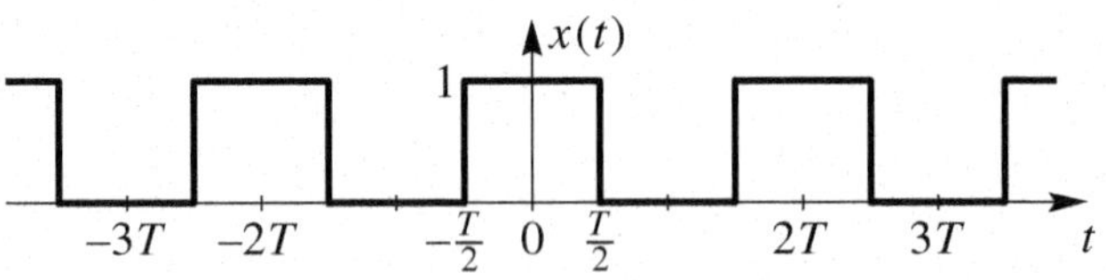

Figure 11-9: Signal $x(t)$ is a 50% duty cycle square wave whose period is $T_0 = 2T$. Its transform is shown in Fig. 11-8.

Example 11-3: Square Wave Transform

Figure 11-9 shows a periodic square wave where $T_0 = 2T$. If we substitute this function into (11.34) we obtain

$$a_k = \frac{1}{T_0}\int_{-T/2}^{T/2} e^{-jk\omega_0 t}dt$$

$$= \frac{1}{T_0}\frac{e^{-jk\omega_0 t}}{(-jk\omega_0)}\bigg|_{-T/2}^{T/2} = \frac{\sin(k\omega_0 T/2)}{k\omega_0 T_0/2} \qquad (11.37)$$

where $\omega_0 = 2\pi/T_0$. We also obtain the DC coefficient by evaluating the integral

$$a_0 = \frac{1}{T_0}\int_{-T/2}^{T/2} dt = \frac{T}{T_0} = \frac{1}{2} \tag{11.38}$$

After substituting $(\omega_0 T) = (2\pi/T_0)(T_0/2) = \pi$ into (11.37), we obtain

$$a_k = \begin{cases} \dfrac{\sin(\pi k/2)}{\pi k} & k \neq 0 \\ \frac{1}{2} & k = 0 \end{cases} \tag{11.39}$$

If we substitute (11.39) into (11.35) we obtain the following equation for the Fourier transform of a periodic square wave:

$$X(j\omega) = \pi\delta(\omega) + \sum_{\substack{k=-\infty \\ k\neq 0}}^{\infty} \left(\frac{2\sin(\pi k/2)}{k}\right)\delta(\omega - k\omega_0) \tag{11.40}$$

Figure 11-8 shows the Fourier transform of the square wave for the case $T_0 = 2T$. The Fourier coefficients are zero for even multiples of ω_0, so there are no impulses at those frequencies. Any periodic signal with fundamental frequency ω_0 will have a transform with a similar appearance—impulses at integer multiples of ω_0, but with different sizes dictated by the a_k coefficients. ■

Example 11-4: Transform of Impulse Train

As another example of finding the Fourier transform of a periodic signal, let us consider the periodic impulse train

$$p(t) = \sum_{n=-\infty}^{\infty} \delta(t - nT_s) \tag{11.41}$$

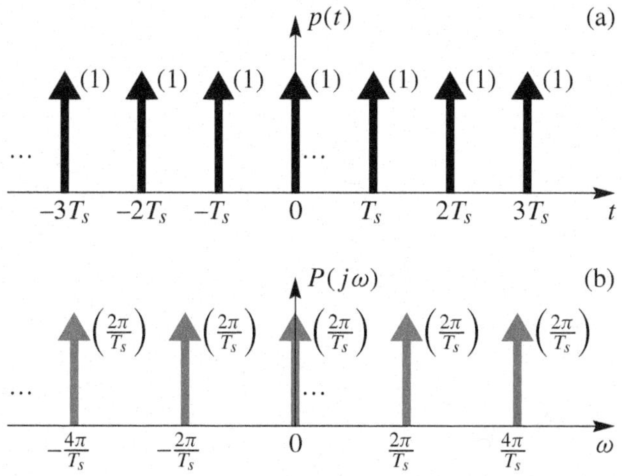

Figure 11-10: Periodic impulse train: (a) Time-domain signal $p(t)$; and (b) Fourier transform $P(j\omega)$. Regular spacing in the frequency-domain is $\omega_s = 2\pi/T_s$

where the period is denoted by T_s. This signal, which will be useful in Chapter 12 in deriving the sampling theorem, is plotted in Fig. 11-10(a). Because $x(t)$ is periodic with period T_s, we can also express (11.41) as a Fourier series

$$p(t) = \sum_{k=-\infty}^{\infty} a_k e^{jk\omega_s t} \tag{11.42}$$

where $\omega_s = 2\pi/T_s$. To determine the Fourier coefficients $\{a_k\}$, we must evaluate the Fourier series integral over one convenient period; i.e.,

$$\begin{aligned} a_k &= \frac{1}{T_s}\int_{-T_s/2}^{T_s/2} \delta(t)e^{-jk\omega_s t}dt \\ &= \frac{1}{T_s}\int_{-T_s/2}^{T_s/2} \delta(t)dt = \frac{1}{T_s} \end{aligned} \tag{11.43}$$

The Fourier coefficients for the periodic impulse train are all the same size. Now in general, the Fourier transform of a periodic signal represented by a Fourier series as in (11.42) is of the form

$$P(j\omega) = \sum_{k=-\infty}^{\infty} 2\pi a_k \delta(\omega - k\omega_s)$$

Substituting (11.43) into the general expression for $P(j\omega)$, we obtain

$$P(j\omega) = \sum_{k=-\infty}^{\infty} \left(\frac{2\pi}{T_s}\right) \delta(\omega - k\omega_s) \qquad (11.44)$$

Therefore, the Fourier transform of a periodic impulse train is also a periodic impulse train. ■

11-5 Properties of Fourier Transform Pairs

The Fourier transform has a set of mathematical properties that help to simplify problem solving. In this section and in Sections 11-6, 11-7, and 11-8, we will derive several of the most useful properties and illustrate their use with some examples. The properties of the Fourier transform are summarized in Table 11-3 on p. 339.

11-5.1 The Scaling Property

We first encountered the inverse relationship between bandwidth and time duration in Section 11-4.1.1. Another example of this relationship is the rectangular pulse function and its Fourier transform, the sinc function. Figure 11-11 shows two cases where the pulse length differs by a factor of two. Notice that the longer pulse in Fig. 11-11(a) has a narrower transform shown in Fig. 11-11(b). This example suggests the following general property of the Fourier transform:

Stretching a time signal will compress its Fourier transform.

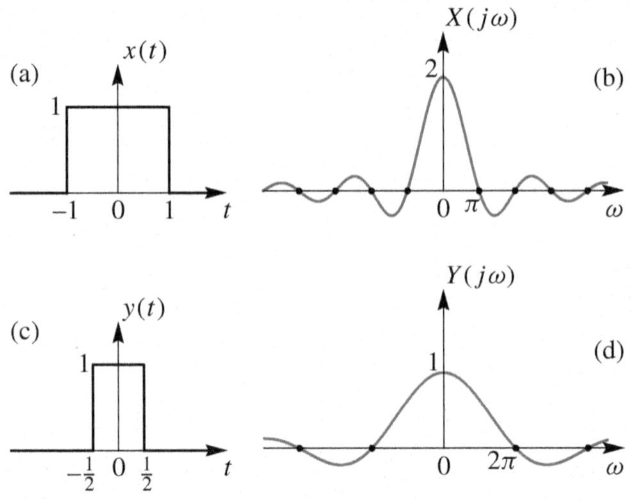

Figure 11-11: Two rectangular pulses (a) and (c) and their corresponding transforms (b) and (d). These cases exhibit the inverse relationship between the time-domain and frequency-domain as dictated by the scaling property of the Fourier transform.

and the converse:

Compressing a time signal will stretch its Fourier transform.

To prove these general statements, consider the signal $y(t) = x(at)$ where a is a real number. If $a > 1$, then a plot of $y(t)$ will have a shorter duration than $x(t)$. For example, the signal $y(t)$ in Fig. 11-11(c) can be defined as

$$y(t) = x(2t)$$

in terms of $x(t)$ in Fig. 11-11(a). Now we can determine $Y(j\omega)$ in terms of $X(j\omega)$. To do this, note that by definition,

$$Y(j\omega) = \int_{-\infty}^{\infty} x(at) e^{-j\omega t} dt \qquad (11.45)$$

If we make the change of variables $\lambda = at$, then the integral becomes

$$Y(j\omega) = (1/a)\int_{-\infty}^{\infty} x(\lambda)e^{-j\omega\lambda/a}d\lambda \tag{11.46}$$

if $a > 0$. However, if $a < 0$, then the limits of integration must be swapped and we generate a minus sign as follows:

$$Y(j\omega) = \int_{\infty}^{-\infty} x(\lambda)e^{-j\omega\lambda/a}d\lambda/a$$

$$= -(1/a)\int_{-\infty}^{\infty} x(\lambda)e^{-j\omega\lambda/a}d\lambda \tag{11.47}$$

We can write the two cases as one because the factor $-(1/a)$ is always positive when $a < 0$; i.e.,

$$Y(j\omega) = (1/|a|)\int_{-\infty}^{\infty} x(\lambda)e^{-j(\omega/a)\lambda}d\lambda \tag{11.48}$$

Thus, we have derived the scaling property of Fourier transforms

Time-Domain		***Frequency-Domain***
$y(t) = x(at)$	$\overset{\mathcal{F}}{\longleftrightarrow}$	$Y(j\omega) = \dfrac{1}{\lvert a\rvert}X(j\omega/a)$

(11.49)

The scaling property is very helpful when describing the bandwidth of a signal. It leads to the general conclusion that short duration signals must be wideband, as shown in Fig. 11-11. This fact can be formalized via the "uncertainty principle" which states that the time duration of a signal and its bandwidth cannot both be made arbitrarily small. Indeed, we have seen that the impulse time signal, which has zero duration, has a Fourier transform with infinite bandwidth.

Example 11-5: Flip Property

As an example of the use of (11.49), suppose that we flip $x(t)$ in time, i.e., $y(t) = x(-t)$. We can show that the resulting Fourier transform is also flipped in frequency. Since $y(t) = x(-t)$, we use $a = -1$ in (11.49) to get

$$Y(j\omega) = \frac{1}{|-1|}X(j(\omega/(-1))) = X(-j\omega)$$

■

Thus we have derived the flip property of the Fourier transform:

Time-Domain		***Frequency-Domain***
$x(-t)$	$\overset{\mathcal{F}}{\longleftrightarrow}$	$X(-j\omega)$

(11.50)

Example 11-6: Left-Sided Signal

Consider the right-sided exponential signal $x(t) = e^{-bt}u(t)$, where $b > 0$. We know that its Fourier transform is

$$X(j\omega) = \frac{1}{b + j\omega}$$

Now define the left-sided exponential signal $w(t) = x(-t) = e^{bt}u(-t)$. From (11.50) and Example 11-5 it follows that

$$W(j\omega) = \frac{1}{b - j\omega}$$

which is what we would obtain by evaluating the Fourier transform integral directly for $w(t)$. ■

EXERCISE 11.7: The Fourier transform integral for $w(t)$ in Example 11-6 is

$$W(j\omega) = \int_{-\infty}^{0} e^{bt} e^{-j\omega t} dt$$

Evaluate this integral and show that the result is the same as was obtained in Example 11-6.

11-5.2 Symmetry Properties of Fourier Transform Pairs

Often the Fourier transform will exhibit a symmetry that leads to various simplifications. For example, with an even-symmetric transform, $X(j\omega) = X(-j\omega)$, the values for negative ω are equal to those for positive ω, so a plot of $X(j\omega)$ would only have to be done for $\omega \geq 0$ to convey all the information in $X(j\omega)$. Along similar lines, when computing values of a symmetric $X(j\omega)$, often the computation can be halved.

In order to exhibit the various symmetries of the Fourier transform and their correspondence to symmetries of the time-domain signal, we examine the impact of the flip and conjugate operators on the real and imaginary parts of the Fourier transform of a real signal. If we take the complex conjugate of $X(-j\omega)$, we obtain

$$X^*(-j\omega) = \left(\int_{-\infty}^{\infty} x(t) e^{-(-j\omega)t} dt \right)^* = \int_{-\infty}^{\infty} x^*(t) e^{-j\omega t} dt$$

That is, $X^*(-j\omega)$ is the Fourier transform of $x^*(t)$. Therefore the following property is true for any $x(t)$:

Time-Domain		*Frequency-Domain*
$x^*(t)$	$\stackrel{\mathcal{F}}{\longleftrightarrow}$	$X^*(-j\omega)$

(11.51)

We can use (11.51) to derive many interesting symmetries for real signals. If $x(t)$ is real, then $x(t) = x^*(t)$ and, therefore, after taking the Fourier transform it must be true that $X(j\omega) = X^*(-j\omega)$ or equivalently, for the real and imaginary parts we have:

$$\Re e\{X(j\omega)\} = \Re e\{X^*(-j\omega)\} \quad (11.52)$$

$$\Im m\{X(j\omega)\} = \Im m\{X^*(-j\omega)\} \quad (11.53)$$

Finally, we get the properties that apply when $x(t)$ is real,

$$\Re e\{X(j\omega)\} = \Re e\{X(-j\omega)\} \quad (11.54)$$

$$\Im m\{X(j\omega)\} = -\Im m\{X(-j\omega)\} \quad (11.55)$$

In other words, when $x(t)$ is real, the real part of its Fourier transform is an *even* function of ω and the imaginary part is an *odd* function of ω. These symmetries are evident in Figs. 11-3(b) and 11-3(c) on p. 314. The condition $X^*(-j\omega) = X(j\omega)$ also implies that the polar form representation $X(j\omega) = |X(j\omega)|e^{j\angle X(j\omega)}$ has the following symmetries (see Problem P-11.16):

$$|X(j\omega)| = |X(-j\omega)| \quad (11.56)$$

$$\angle X(j\omega) = -\angle X(-j\omega) \quad (11.57)$$

In other words, the magnitude has even symmetry and the phase angle has odd symmetry if $x(t)$ is real.

Example 11-7: Magnitude/Phase Symmetries

As another illustration of symmetry, recall that the Fourier transform of $x(t) = e^{-at}u(t)$ is

$$X(j\omega) = \frac{1}{a + j\omega}$$

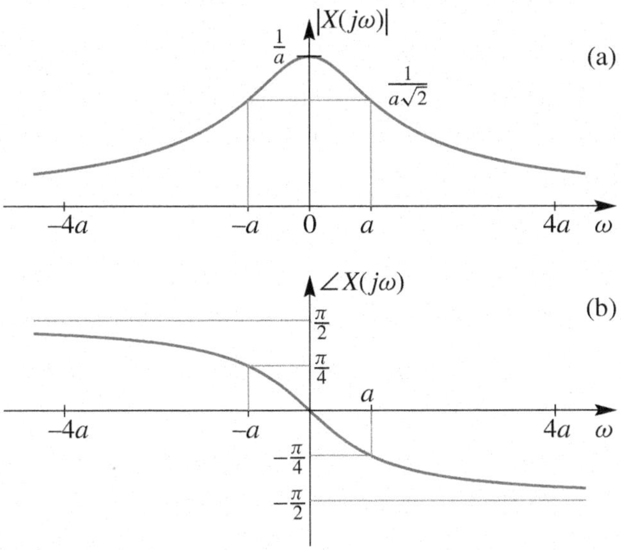

Figure 11-12: Magnitude and phase angle of Fourier transform of right-sided exponential signal: (a) Magnitude of $X(j\omega)$ is even; (b) phase angle of $X(j\omega)$ is odd.

The magnitude is the square root of the magnitude-squared and the phase is extracted via the arctangent

$$\begin{aligned}|X(j\omega)| &= [X(j\omega)X^*(j\omega)]^{1/2} \\ &= \frac{1}{(a^2+\omega^2)^{1/2}} = |X(-j\omega)|\end{aligned}$$

$$\angle X(j\omega) = -\arctan\left(\frac{\omega}{a}\right) = -\angle X(-j\omega)$$

The even and odd symmetries of the magnitude and phase given in the above equations are illustrated in the plots in Fig. 11-12. In this case it would have been sufficient to plot the magnitude and phase for $\omega \geq 0$. ■

If $x(t)$ has multiple symmetries, then $X(j\omega)$ is more constrained. Table 11-1 lists four types of combined symmetries. Suppose that the signal has even symmetry

Table 11-1: Symmetries of the Fourier transform.

$x(t)$	$X(j\omega)$
Real, Even	Real, Even
Real, Odd	Imaginary, Odd
Imaginary, Even	Imaginary, Even
Imaginary, Odd	Real, Odd

so that $x(t) = x(-t)$. Then the flip property (11.50) implies that $X(j\omega) = X(-j\omega)$. If $x(t)$ is also real, we know that $X^*(j\omega) = X(-j\omega)$. Combining these two, we conclude that when $x(t)$ is real and also has even symmetry, $X(j\omega) = X^*(j\omega)$ which implies that $X(j\omega)$ is real. In addition, we can say that $X(j\omega)$ is a real and even function of ω if $x(t)$ is a real, even function of time. By a similar argument, it can be shown that if $x(t)$ is a real odd function so that $x(t) = -x(-t)$, then $X(j\omega)$ is imaginary and odd, i.e., $X(j\omega) = -X(-j\omega) = -X^*(j\omega)$.

Example 11-8: Two-Sided Exponential

Consider the two-sided exponential signal

$$x(t) = e^{-at}u(t) + e^{at}u(-t) = e^{-a|t|}$$

which is the real even time function plotted in Fig. 11-13(a). Using (11.16) and (11.50), it follows that

$$X(j\omega) = \frac{1}{a+j\omega} + \frac{1}{a-j\omega} = \frac{2a}{a^2+\omega^2}$$

Observe in Fig. 11-13(b) that $X(j\omega)$ is also a real even function of ω. ■

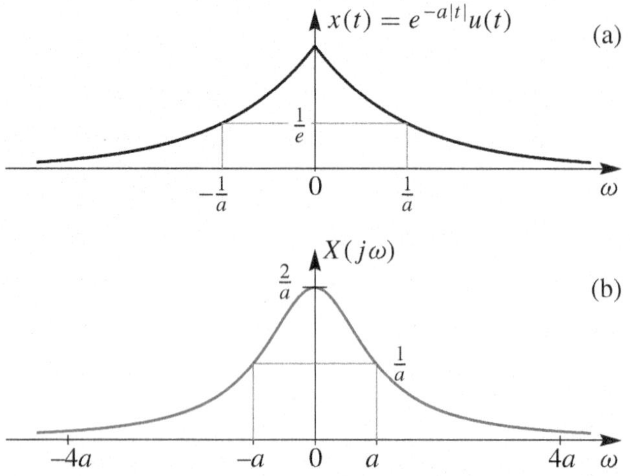

Figure 11-13: Two-sided exponential signal: (a) Time function $x(t) = e^{-a|t|}$; and (b) corresponding Fourier transform $X(j\omega)$.

EXERCISE 11.8: For the real odd function $x(t) = e^{at}u(-t) - e^{-at}u(t)$, show that the Fourier transform is

$$X(j\omega) = \frac{2j\omega}{a^2 + \omega^2}$$

which is an imaginary odd function of ω.

11-6 The Convolution Property

Previously, we derived several Fourier transform pairs. With a table like Table 11-2 in Section 11-9 and a good knowledge of some other Fourier transform relationships that we refer to as *properties of the Fourier transform*, we can solve many signals and systems problems by combining time-domain analysis and frequency-domain analysis. In this section and Section 11-8, we will consider the convolution and multiplication properties, which are perhaps the most important and most powerful of all the Fourier transform relationships. In the course of our discussions we will show how to derive many other useful properties from them. As a starting point for our discussion of the convolution property, we will revisit the concept of frequency response in the new context of the Fourier transform.

11-6.1 Frequency Response

Recall that for continuous-time LTI systems, when the input is $x(t) = e^{j\omega t}$ for all t, then the output is $y(t) = H(j\omega)e^{j\omega t}$. In Chapter 10, we showed in (10.3) on p. 286 that the frequency response of an LTI system is related to the impulse response of the system by

$$H(j\omega) = \int_{-\infty}^{\infty} h(t)e^{-j\omega t}dt \qquad (11.58)$$

Now we recognize that the frequency response is, in fact, the Fourier transform of the impulse response of the system. Therefore, given the impulse response $h(t)$, we can find the frequency response by taking the Fourier transform of $h(t)$. Likewise, given the frequency response, we can find the impulse response by simply evaluating the inverse Fourier transform integral (11.2). If $H(j\omega)$ happens to be a member of our table of Fourier transforms, we can simply look up the answer.

Recall that a sufficient condition for the existence of the Fourier transform of the impulse response was given (11.15) on p. 312 as

$$\int_{-\infty}^{\infty} |h(t)|dt < \infty$$

which is also the condition for stability of the LTI system. Hence, all stable systems have a frequency response function. An equivalent statement is:

> ***Existence of the frequency response and stability of the system are equivalent conditions.***

Example 11-9: Rational Fourier Transform

To illustrate how a known transform pair (from Table 11-2 on p. 338 can be used to find the frequency response given the impulse response, suppose that

$$h(t) = 2e^{-2t}u(t) - e^{-t}u(t) \tag{11.59}$$

We know the transform of the one-sided exponential, so linearity gives the result as two terms

$$H(j\omega) = \frac{2}{2+j\omega} - \frac{1}{1+j\omega} \tag{11.60a}$$

$$= \frac{j\omega}{(2+j\omega)(1+j\omega)} \tag{11.60b}$$

Likewise, if we wanted to start with $H(j\omega)$ and obtain $h(t)$, we could easily do so when $H(j\omega)$ is given in the form of (11.60a) by using the right-sided exponential entry in Table 11-2 on p. 338 to go from the frequency-domain to the time-domain. However, if $H(j\omega)$ is given in the form of (11.60b) we would first need to put it into the form of (11.60a). We show in Section 11-6.3.3 how to do this using the method of partial fraction expansion. ■

The relationship between the impulse response and the frequency response suggests a more general property of the Fourier transform, which we will now discuss.

11-6.2 Fourier Transform of a Convolution

The input and output of an LTI system are related through the convolution integral[10]

$$y(t) = \int_{-\infty}^{\infty} x(\tau)h(t-\tau)d\tau \tag{11.61}$$

We will show that the Fourier transform of a convolution is the product of the Fourier transforms of the two functions that are convolved; i.e.,

Time-Domain	***Frequency-Domain***
$y(t) = x(t) * h(t) \overset{\mathcal{F}}{\longleftrightarrow}$	$Y(j\omega) = X(j\omega)H(j\omega)$

(11.62)

In words, (11.62) states the following:

> ***Convolution in the time-domain corresponds to multiplication in the frequency-domain.***

To prove (11.62), we will take the Fourier transform of both sides of (11.61) obtaining

$$Y(j\omega) = \int_{-\infty}^{\infty} y(t)e^{-j\omega t}dt$$

$$= \int_{-\infty}^{\infty}\left(\int_{-\infty}^{\infty} x(\tau)h(t-\tau)d\tau\right)e^{-j\omega t}dt \tag{11.63}$$

Now assuming that we can interchange the order of integration in (11.63), we obtain

$$Y(j\omega) = \int_{-\infty}^{\infty} x(\tau)\left(\int_{-\infty}^{\infty} h(t-\tau)e^{-j\omega t}dt\right)d\tau \tag{11.64}$$

---

[10] Remember that we use the symbolic notation $y(t) = x(t) * h(t)$ when we want to express (11.61) in a compact form.

The integral in the parenthesis can be manipulated into a recognizable form by making a substitution of variables $\sigma = t - \tau$ as follows:

$$\left(\int_{-\infty}^{\infty} h(t-\tau)e^{-j\omega t}dt\right) = \left(\int_{-\infty}^{\infty} h(\sigma)e^{-j\omega\sigma}d\sigma\right)e^{-j\omega\tau}$$
$$= H(j\omega)e^{-j\omega\tau} \tag{11.65}$$

We then substitute (11.65) back into (11.64) and we obtain finally

$$Y(j\omega) = \int_{-\infty}^{\infty} x(\tau)\left(H(j\omega)e^{-j\omega\tau}\right)d\tau$$
$$= H(j\omega)\int_{-\infty}^{\infty} x(\tau)e^{-j\omega\tau}d\tau$$
$$= H(j\omega)X(j\omega) \tag{11.66}$$

which is the result we have been aiming for. This is the convolution property—one of the most important properties of the Fourier transform.

11-6.3 Examples of the Use of the Convolution Property

To illustrate the power of the convolution property, we will present several examples where a combination of time-domain and frequency-domain analysis can lead to simple solutions to linear systems problems.

11-6.3.1 Convolution of Two Bandlimited Functions

Consider the case where both the input to and the impulse response of an LTI system are sinc functions

$$x(t) = \frac{2\sin(20\pi t)}{\pi t} \quad \text{and} \quad h(t) = \frac{5\sin(10\pi t)}{\pi t}$$

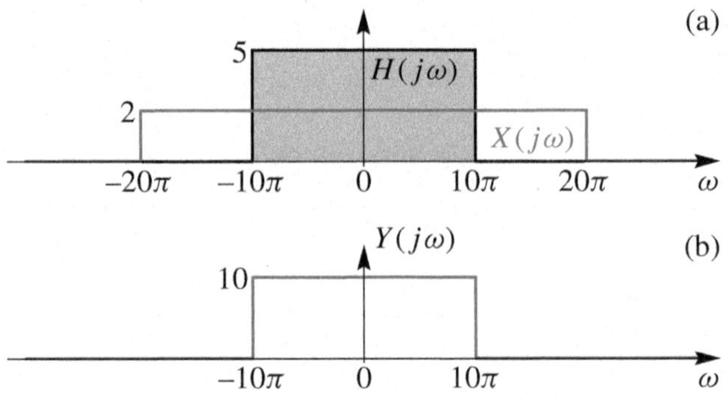

Figure 11-14: (a) Fourier transforms of a bandlimited input signal and the frequency response of an ideal LPF (shaded). (b) Output transform is the product $Y(j\omega) = H(j\omega)X(j\omega)$.

The output of this system for the given input is given by the convolution integral

$$y(t) = x(t) * h(t)$$
$$= \int_{-\infty}^{\infty}\left(\frac{2\sin(20\pi\tau)}{\pi\tau}\right)\left(\frac{5\sin[10\pi(t-\tau)]}{\pi(t-\tau)}\right)d\tau$$

which is impossible to evaluate by the methods of calculus. However, this convolution is easy to perform if we transform to the frequency-domain. In Fig. 11-14(a), we plot the two Fourier transforms $X(j\omega)$ and $H(j\omega)$ together. The two Fourier transforms as determined from Table 11-2 are rectangles, so the product of the two Fourier transforms is equal to another rectangle that has the lower cutoff frequency as in Fig. 11-14(b). In other words, it is easy to see from the graph that

$$Y(j\omega) = H(j\omega)X(j\omega)$$
$$= 5[u(\omega+10\pi) - u(\omega-10\pi)]\cdot$$
$$2[u(\omega+20\pi) - u(\omega-20\pi)]$$
$$= 10[u(\omega+10\pi) - u(\omega-10\pi)]$$

In this case, the LTI system "filters" the Fourier transform of the input by clipping the bandwidth of the input transform to $\pm 10\pi$. When we transform $Y(j\omega)$ back to the time-domain, we can write the somewhat remarkable convolution formula

$$\begin{aligned} y(t) &= x(t) * h(t) \\ &= \int_{-\infty}^{\infty} \left(\frac{2\sin(20\pi\tau)}{\pi\tau} \right) \left(\frac{5\sin[10\pi(t-\tau)]}{\pi(t-\tau)} \right) d\tau \\ &= \frac{10\sin(10\pi t)}{\pi t} \end{aligned}$$

Without the use of Fourier transforms, we would not have been able to obtain this result. While it may look like a somewhat specialized and contrived example, it is not. The concept of a system "passing" all frequencies up to a cutoff frequency and rejecting all frequencies above that cutoff is the very essence of ***filtering***.

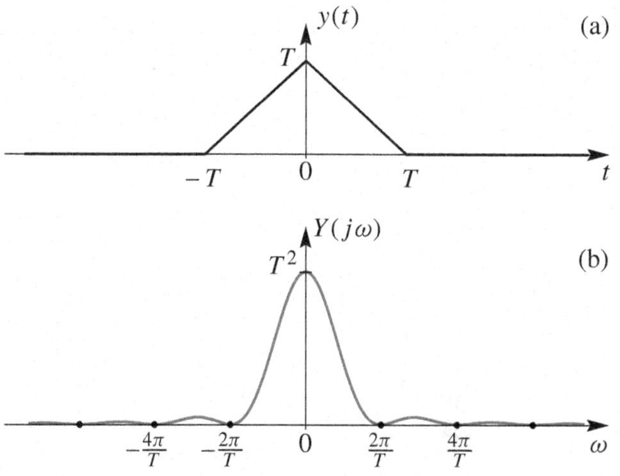

Figure 11-15: (a) Triangle resulting from convolution of two rectangular pulses. (b) Fourier transform is the product of two sinc functions.

11-6.3.2 Product of Two Sinc Functions

Lest the previous example give the false impression that it is always easier to evaluate a convolution by using the Fourier transform, consider the situation when both $x(t)$ and $h(t)$ are the same rectangular pulses centered on $t = 0$ with a total width of T; i.e.,

$$x(t) = h(t) = u(t + \tfrac{1}{2}T) - u(t - \tfrac{1}{2}T) \qquad (11.67)$$

Using (11.19) the corresponding Fourier transforms are

$$X(j\omega) = H(j\omega) = \frac{\sin(\omega T/2)}{(\omega/2)}$$

so it follows that if $y(t) = x(t) * h(t)$, then

$$Y(j\omega) = X(j\omega)H(j\omega) = T^2 \left(\frac{\sin(\omega T/2)}{(\omega T/2)} \right)^2$$

Figure 11-15(b) shows the Fourier transform of the output. Therefore, it follows that the inverse transform is

$$y(t) = x(t) * h(t) = \frac{1}{2\pi} \int_{-\infty}^{\infty} T^2 \left(\frac{\sin(\omega T/2)}{(\omega T/2)} \right)^2 e^{j\omega t} d\omega$$

This appears to be another integral that we cannot evaluate. However, the convolution of two rectangular pulses in the time-domain is rather easy—the answer is the triangularly shaped pulse shown in Fig. 11-15(a).

EXERCISE 11.9: Show by evaluating the convolution of two identical pulses, each one defined by (11.67), that $y(t)$ is the triangle function shown in Fig. 11-15(a).

EXERCISE 11.10: Use convolution to find the time function that corresponds to the Fourier transform

$$Y(j\omega) = \left(\frac{\sin(2\omega)}{(\omega/2)}\right)\left(\frac{\sin(\omega)}{(\omega/2)}\right)$$

11-6.3.3 Partial Fraction Expansions

In many cases, we are interested in the convolution of signals that are one-sided exponentials—such signals are common in electric circuits and other physical systems described by differential equations. For an example, consider the case where

$$x(t) = \delta(t) - e^{-t}u(t)$$

$$\text{and} \quad h(t) = e^{-2t}u(t)$$

From Table 11-2 we obtain the Fourier transforms

$$X(j\omega) = 1 - \frac{1}{1+j\omega} = \frac{j\omega}{1+j\omega}$$

$$\text{and} \quad H(j\omega) = \frac{1}{2+j\omega}$$

In order to evaluate the convolution, $y(t) = x(t) * h(t)$, we work in the transform domain and multiply the transforms

$$\begin{aligned} Y(j\omega) &= X(j\omega)H(j\omega) \\ &= \left(\frac{j\omega}{1+j\omega}\right)\left(\frac{1}{2+j\omega}\right) \\ &= \frac{j\omega}{(1+j\omega)(2+j\omega)} \end{aligned} \qquad (11.68)$$

Now we have the Fourier transform of $y(t)$, but we might not immediately recognize what its inverse would be. This $Y(j\omega)$ is not one of the Fourier transform pairs that we have tabulated, but we should recognized the factors in the denominator as characteristic of one-sided exponential signals. When faced with this situation, it is often possible to do some algebraic manipulation of the mathematical expression so as to express it as a sum (or product) of simpler terms that might be in our table of transforms. Equation (11.68) is already in a product form in this case and the product form would lead us back to doing convolution. If we look carefully at the righthand side of (11.68), we observe that $Y(j\omega)$ is a rational function of the complex variable $(j\omega)$.[11] Therefore, we can decompose $Y(j\omega)$ into a partial fraction expansion to obtain a ***sum of first-order terms*** whose inverse Fourier transforms are recognizable. Specifically, we want to represent $Y(j\omega)$ in the following form:

$$Y(j\omega) = \frac{A}{1+j\omega} + \frac{B}{2+j\omega} \qquad (11.69)$$

where A and B are constants to be determined. Polynomial algebra tells us that if we place the righthand side of (11.69) over a common denominator, we could generate the term $j\omega$ needed in the numerator of (11.68).[12] To find the constant A, we can multiply both sides of (11.69) by $(1 + j\omega)$ and we obtain

$$(1+j\omega)Y(j\omega) = \frac{A(1+j\omega)}{1+j\omega} + \frac{B(1+j\omega)}{2+j\omega}$$

If we now substitute $Y(j\omega)$ from (11.68), we can cancel the factor $(1 + j\omega)$ when it occurs in both numerator and

[11] A rational function has a numerator and a denominator, both of which are polynomials; in this case, polynomials with powers such as $(j\omega)^n$.

[12] However, we could not generate $(j\omega)^2$ or higher powers. The partial fraction expansion only works if the numerator has lower degree than the denominator. If this is not the case, we must use polynomial division to obtain a rational function that is suitable for partial fraction expansion.

denominator and we obtain

$$\frac{j\omega}{2+j\omega} = A + \frac{B(1+j\omega)}{2+j\omega} \tag{11.70}$$

Next we substitute $(j\omega) = -1$ into (11.70) to get A

$$\frac{(-1)}{2+(-1)} = A + \frac{1+(-1)}{2+(-1)} = A = -1$$

The same method can be used to find that $B = 2$, if we multiply by $(2 + j\omega)$. Therefore, we obtain the partial fraction expansion

$$Y(j\omega) = -\frac{1}{1+j\omega} + \frac{2}{2+j\omega} \tag{11.71}$$

From (11.71) we can now do the inverse transform by "table lookup"

$$y(t) = -e^{-t}u(t) + 2e^{-2t}u(t)$$

The partial fraction expansion can be used to convert any rational function of $(j\omega)$ into a sum of terms like (11.71), but there are two stumbling blocks. First, the factors of the denominator must be distinct (see Prob. P-11.10 for inverting $1/(a + j\omega)^2$) and, second, the numerator degree must be lower than the denominator degree. As the next example shows, however, we can use polynomial long division to reduce the numerator degree prior to performing the partial fraction expansion.

Example 11-10: Partial Fraction Expansion

As an example of a case where long division is required, consider the Fourier transform

$$Y(j\omega) = \frac{1-\omega^2+j\omega}{2-\omega^2+3j\omega} \tag{11.72}$$

The first step is to express $Y(j\omega)$ as a ratio of polynomials in $j\omega$. We can do this by noting that $(j\omega)^k = \omega^k j^k$ so if we replace ω^k by $(j\omega)^k/j^k$ we will not change the value of the expression, but we will obtain a rational function in terms of powers of $(j\omega)$. In the example above, we get

$$\begin{aligned} Y(j\omega) &= \frac{1-(j\omega)^2/j^2+j\omega}{2-(j\omega)^2/j^2+3j\omega} \\ &= \frac{1+(j\omega)+(j\omega)^2}{2+3(j\omega)+(j\omega)^2} \end{aligned}$$

Now we are ready to do the polynomial long division as shown below

$$\begin{array}{r|l} & \quad 1 \\ \hline (j\omega)^2+3(j\omega)+2 & (j\omega)^2+\ (j\omega)+1 \\ & \underline{(j\omega)^2+3(j\omega)+2} \\ & \quad -2(j\omega)-1 \end{array}$$

With the quotient and remainder from the long division, we can now express $Y(j\omega)$ as

$$\begin{aligned} Y(j\omega) &= 1 - \frac{1+2(j\omega)}{2+3(j\omega)+(j\omega)^2} & \text{(11.73a)} \\ &= 1 - \frac{1+2j\omega}{(1+j\omega)(2+j\omega)} & \text{(11.73b)} \end{aligned}$$

The second term of $Y(j\omega)$ in (11.73b) can be decomposed into a sum via the partial fraction expansion method

$$-\frac{1+2j\omega}{(1+j\omega)(2+j\omega)} = \frac{A}{1+j\omega} + \frac{B}{2+j\omega} \tag{11.74}$$

In order to find A, we multiply both sides by $(1 + j\omega)$ to obtain

$$-\frac{(1+2j\omega)(1+j\omega)}{(1+j\omega)(2+j\omega)} = \frac{A(1+j\omega)}{1+j\omega} + \frac{B(1+j\omega)}{2+j\omega}$$

Next we cancel the factors $(1 + j\omega)$ wherever possible

$$-\frac{1+2j\omega}{(2+j\omega)} = A + \frac{B(1+j\omega)}{2+j\omega}$$

Now we can substitute $(j\omega) = -1$ on both sides to eliminate the B term

$$-\frac{1-2}{2-1} = A + \frac{0}{2-1}$$

and we get $A = 1$. Similarly, we can find $B = -3$ by multiplying both sides of (11.74) by $(2 + j\omega)$ and evaluating at $j\omega = -2$. The final additive representation for $Y(j\omega)$ in (11.72) is

$$Y(j\omega) = 1 + \frac{1}{1+j\omega} - \frac{3}{2+j\omega}$$

Recalling that the inverse Fourier transform of 1 is $\delta(t)$, it follows that the corresponding time function is

$$y(t) = \delta(t) + e^{-t}u(t) - 3e^{-2t}u(t)$$

■

11-7 Basic LTI Systems

Knowing that convolution is transformed into multiplication by the Fourier transform, it is worthwhile to revisit some of the basic LTI operators that were introduced in Section 9-5. In this section, we will discuss the time delay and differentiation systems, and illustrate how these operations can be combined to form more complex systems.

11-7.1 Time Delay

The ideal time delay system is defined by the relation

$$y(t) = x(t - t_d) \qquad (11.75)$$

Its impulse response is $h(t) = \delta(t - t_d)$, so its frequency response can be obtained by evaluating the Fourier transform integral as in

$$H(j\omega) = \int_{-\infty}^{\infty} \delta(t - t_d)e^{-j\omega t}dt$$

$$= e^{-j\omega t_d} \int_{-\infty}^{\infty} \delta(t - t_d)dt = e^{-j\omega t_d}$$

We can state the result of this analysis in two different ways. First, we have derived the general Fourier transform pair

Time-Domain		***Frequency-Domain***
$\delta(t - t_d)$	$\overset{\mathcal{F}}{\longleftrightarrow}$	$e^{-j\omega t_d}$

(11.76)

Second, by virtue of the convolution property we can state the following general delay property of Fourier transforms:

Time-Domain		***Frequency-Domain***
$x(t - t_d)$	$\overset{\mathcal{F}}{\longleftrightarrow}$	$e^{-j\omega t_d}X(j\omega)$

(11.77)

Example 11-11: Time-Delay Property

As an example of the use of this property, consider the signal

$$x(t) = e^{-2t}u(t - 3)$$

Although we have not seen this signal in exactly this form before, by a slight modification we can obtain a delayed and scaled version of a signal that is in Table 11-2. Note that an equivalent form is

$$x(t) = e^{-6}e^{-2(t-3)}u(t - 3)$$

which shows that we have the signal $e^{-2t}u(t)$ delayed by 3 and multiplied by the constant e^{-6}. Therefore, we can apply the time-delay property and use the transform of the one-sided exponential (the first entry in Table 11-2) to obtain

$$X(j\omega) = \frac{e^{-6}e^{-j\omega 3}}{2 + j\omega} \tag{11.78}$$

as the Fourier transform of the original signal. ■

Time delay is a common operation in systems, and it is exceedingly helpful to know that wherever it appears in a frequency-domain expression, $e^{-j\omega t_d}$ implies delay by t_d in the time-domain.

11-7.2 Differentiation

The ideal k-th derivative system is defined by the relation

$$y(t) = \frac{d^k x(t)}{dt^k} \tag{11.79}$$

For $k = 1$ the system is a differentiator whose impulse response is, by definition, $h(t) = \delta^{(1)}(t)$, the first derivative of the impulse. Its frequency response can be obtained most easily by constructing the output when the input is $x(t) = e^{j\omega t}$. Taking this approach we get

$$y(t) = \frac{d}{dt}\left(e^{j\omega t}\right) = (j\omega)e^{j\omega t}$$

so it follows that

Frequency Response of Ideal Differentiator

$$H(j\omega) = j\omega \tag{11.80}$$

Also, since the frequency response is the Fourier transform of the impulse response, we have, incidentally, derived the following Fourier transform pair:

Time-Domain		***Frequency-Domain***	
$\delta^{(1)}(t)$	$\overset{\mathcal{F}}{\longleftrightarrow}$	$j\omega$.	(11.81)

If we take a second derivative (the cascade of two first derivatives), another factor of $(j\omega)$ is generated giving $(j\omega)^2$, and for the k-th derivative system, the frequency response would be $H(j\omega) = (j\omega)^k$. Since this result holds for any input that has a Fourier transform, we obtain the following general derivative property of Fourier transforms:

Time-Domain		***Frequency-Domain***	
$\dfrac{d^k x(t)}{dt^k}$	$\overset{\mathcal{F}}{\longleftrightarrow}$	$(j\omega)^k X(j\omega)$	(11.82)

Example 11-12: Derivative Property

Suppose that we want to find the Fourier transform of the derivative of the signal in Example 11-11; i.e.,

$$y(t) = \frac{dx(t)}{dt} = \frac{d}{dt}\left(e^{-2t}u(t-3)\right)$$

One approach would be to differentiate the signal and then determine the Fourier transform of the result. Using the product rule, the derivative is

$$\begin{aligned} y(t) &= e^{-2t}\delta(t-3) - 2e^{-2t}u(t-3) \\ &= e^{-6}\delta(t-3) - 2e^{-6}e^{-2(t-3)}u(t-3) \end{aligned}$$

Using results from Example 11-11 on p. 332 and the Fourier transform of a delayed impulse (11.76) on p. 332, we obtain

$$Y(j\omega) = e^{-6}e^{-j\omega 3} - \frac{2e^{-6}e^{-j\omega 3}}{2 + j\omega}$$

Placing the whole expression over a common denominator gives a form that shows we could have just multiplied $X(j\omega)$ (11.78) by $(j\omega)$ since

$$Y(j\omega) = \frac{e^{-6}(j\omega)e^{-j\omega 3}}{2 + j\omega} = j\omega X(j\omega)$$

In other words, we would have obtained the same answer by applying the differentiation property (11.82) directly to the result of Example 11-11. ■

Differentiation is also a basic operation in systems, and it is very useful to know that whenever $(j\omega)$ multiplies a frequency-domain expression, differentiation is implied in the time-domain. In the next section we will see how the differentiation property can help us gain insight into a large class of useful systems.

11-7.3 Systems Described by Differential Equations

Suppose that the frequency response of a system is

$$H(j\omega) = \frac{(j\omega)}{(1+(j\omega))(2+(j\omega))} = \frac{(j\omega)}{2+3(j\omega)+(j\omega)^2}$$

From the convolution property, we know that

$$H(j\omega) = \frac{Y(j\omega)}{X(j\omega)} = \frac{(j\omega)}{2+3(j\omega)+(j\omega)^2}$$

which can be expressed in an alternative form if we cross-multiply both sides by each of the denominators and cancel terms where possible

$$[(j\omega)^2+3(j\omega)+2]Y(j\omega) = (j\omega)X(j\omega) \qquad (11.83)$$

Since (11.83) relates the input transform $X(j\omega)$ to the corresponding output transform $Y(j\omega)$, it follows that we can inverse transform each term (11.83) to obtain the time-domain relationship between $x(t)$ and $y(t)$

$$\frac{d^2y(t)}{dt^2} + 3\frac{dy(t)}{dt} + 2y(t) = \frac{dx(t)}{dt}$$

which must hold for all t. This is a differential equation that is satisfied by the input and output of the system.

Example 11-13: Transform a Differential Equation

Likewise, if we were given a differential equation, we could use the differentiation property to take the Fourier transform of both sides of the differential equation. Starting with

$$\frac{d^2y(t)}{dt^2} + 2\frac{dy(t)}{dt} = \frac{dx(t)}{dt} + 3x(t)$$

we would obtain

$$(j\omega)^2Y(j\omega) + 2(j\omega)Y(j\omega) = (j\omega)X(j\omega) + 3X(j\omega)$$

If we factor out $Y(j\omega)$ on the left-hand side and $X(j\omega)$ on the right-hand side, we obtain

$$[(j\omega)^2 + 2(j\omega)]Y(j\omega) = [(j\omega)+3]X(j\omega)$$

which can be solved for $H(j\omega)$ as

$$H(j\omega) = \frac{Y(j\omega)}{X(j\omega)} = \frac{(j\omega)+3}{(j\omega)^2+2(j\omega)}$$

■

The preceding example can be generalized for higher-order differential equations. Specifically, if the input $x(t)$ and the output $y(t)$ satisfy a differential equation of the form

$$\sum_{k=0}^{N} a_k \frac{d^k y(t)}{dt^k} = \sum_{k=0}^{M} b_k \frac{d^k x(t)}{dt^k} \qquad (11.84)$$

then the frequency response of the system is of the form

$$H(j\omega) = \frac{\displaystyle\sum_{k=0}^{M} b_k (j\omega)^k}{\displaystyle\sum_{k=0}^{N} a_k (j\omega)^k} \qquad (11.85)$$

Equations (11.84) and (11.85) constitute an important transform pair. Given (11.84), we can obtain (11.85) by noting that the terms involving $x(t)$ in the differential equation determine the numerator polynomial, while the terms involving $y(t)$ in the differential equation determine the denominator polynomial. Alternatively, using the derivative property of Fourier transforms to go from the frequency-domain to the time-domain, the numerator polynomial in (11.85) defines the algebraic structure of the terms involving $x(t)$ in the differential equation, and the denominator polynomial in (11.85) defines the algebraic structure of the terms involving $y(t)$ in the differential equation in (11.84).

Differential equations such as (11.84) arise as a description of a large class of physical systems that includes linear lumped parameter electrical circuits and mechanical systems. With (11.84) and (11.85), the Fourier transform becomes a very powerful tool for analyzing linear physical systems.[13]

11-8 The Multiplication Property

One thing that may be obvious at this point is that we often observe a *duality* between the form of a time-domain function and the corresponding Fourier transform. An example is the rectangular pulse signal and its Fourier transform, the sinc function. When the time- and frequency-domains are swapped and the time function is a sinc function, then the Fourier transform has a rectangular pulse shape. In most cases, the detailed scaling of the function may differ, but the basic duality is evident. This is due to the great similarity between (11.1) and (11.2); the direct and inverse Fourier transforms differ only in a minus sign in the exponent and a scale factor.

The implication of duality is that many of the properties of the Fourier transform also swap between domains. Recall that

> ***Convolution of time functions corresponds multiplication of Fourier transforms.***

Therefore, it is not surprising that

> ***Multiplication of time functions corresponds convolution of Fourier transforms.***

In this section, we will prove this result.

11-8.1 The General Signal Multiplication Property

To prove the multiplication property, consider the signal $y(t) = p(t)x(t)$. We want to determine $Y(j\omega)$ in terms of $P(j\omega)$ and $X(j\omega)$. To do this, note that by definition,

$$Y(j\omega) = \int_{-\infty}^{\infty} p(t)x(t)e^{-j\omega t}dt \tag{11.86}$$

We will now substitute the inverse Fourier transform[14]

$$x(t) = \frac{1}{2\pi}\int_{-\infty}^{\infty} X(j\lambda)e^{j\lambda t}d\lambda \tag{11.87}$$

into (11.86) to obtain

$$Y(j\omega) = \int_{-\infty}^{\infty} p(t)\left(\frac{1}{2\pi}\int_{-\infty}^{\infty} X(j\lambda)e^{j\lambda t}d\lambda\right)e^{-j\omega t}dt$$

[13] The Laplace transform is more convenient for these types of problems because the polynomials in $(j\omega)$ become polynomials in a single variable s. However, operations such as partial fraction expansion are identical in form.

[14] We use λ instead of ω for the dummy variable in (11.87) because ω is the independent variable in (11.86).

Interchanging the order of integration yields

$$Y(j\omega) = \frac{1}{2\pi}\int_{-\infty}^{\infty} X(j\lambda)\left(\int_{-\infty}^{\infty} p(t)e^{-j(\omega-\lambda)t}dt\right)d\lambda$$
$$= \frac{1}{2\pi}\int_{-\infty}^{\infty} X(j\lambda)P(j(\omega-\lambda))d\lambda \qquad (11.88)$$
$$= \frac{1}{2\pi}X(j\omega) * P(j\omega) \qquad (11.89)$$

Thus, we have derived the multiplication property of Fourier transforms

Time-Domain		***Frequency-Domain***
$y(t) = p(t)x(t)$	$\overset{\mathcal{F}}{\longleftrightarrow}$	$Y(j\omega) = \frac{1}{2\pi}X(j\omega) * P(j\omega)$

(11.90)

The multiplication property is very useful for describing a class of systems called amplitude modulators. For this reason, we also refer to it as the *modulation property* of Fourier transforms. The basic amplitude modulator system is depicted in Figure 11-16.

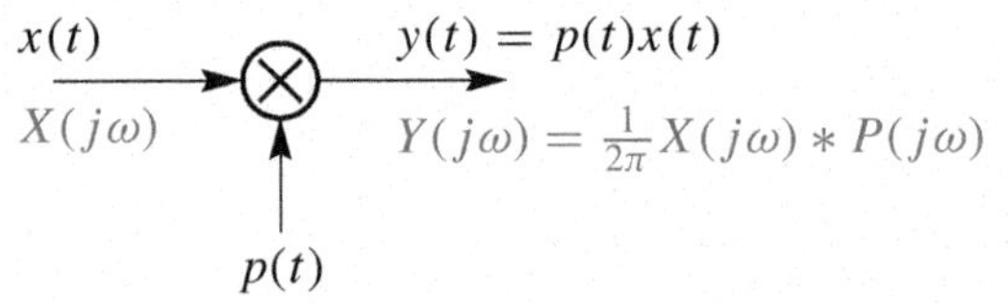

Figure 11-16: Block diagram representation of an amplitude modulator.

EXERCISE 11.11: Prove that the modulator system in Fig. 11-16 is linear, but not time-invariant.

11-8.2 The Frequency Shifting Property

We obtain some particularly useful results when $p(t)$ is periodic. For example, if $p(t) = e^{j\omega_0 t}$, then its Fourier transform is $P(j\omega) = 2\pi\delta(\omega-\omega_0)$. Therefore, the Fourier transform of the output of the modulator, $y(t) = x(t)e^{j\omega_0 t}$, is

$$Y(j\omega) = \frac{1}{2\pi}X(j\omega) * 2\pi\delta(\omega-\omega_0)$$
$$= X(j(\omega-\omega_0)) \qquad (11.91)$$

That is, multiplication of a signal $x(t)$ by a complex exponential of frequency ω_0 simply shifts the Fourier transform of $x(t)$ by ω_0 (to the right on a graph). Since this holds for any signal $x(t)$, we have derived the Fourier transform property that we call the *frequency shifting* property:

Time-Domain		***Frequency-Domain***
$x(t)e^{j\omega_0 t}$	$\overset{\mathcal{F}}{\longleftrightarrow}$	$X(j(\omega-\omega_0))$

(11.92)

Likewise, suppose that $p(t) = \cos(\omega_0 t)$. Then it is easy to prove that the positive and negative frequency components of the cosine lead to the following *cosine modulation* property:

Time-Domain		***Frequency-Domain***
$x(t)\cos(\omega_0 t)$	$\overset{\mathcal{F}}{\longleftrightarrow}$	$\frac{1}{2}X(j(\omega-\omega_0)) + \frac{1}{2}X(j(\omega+\omega_0))$

(11.93)

where $x(t)$ is again any signal that has a Fourier transform.

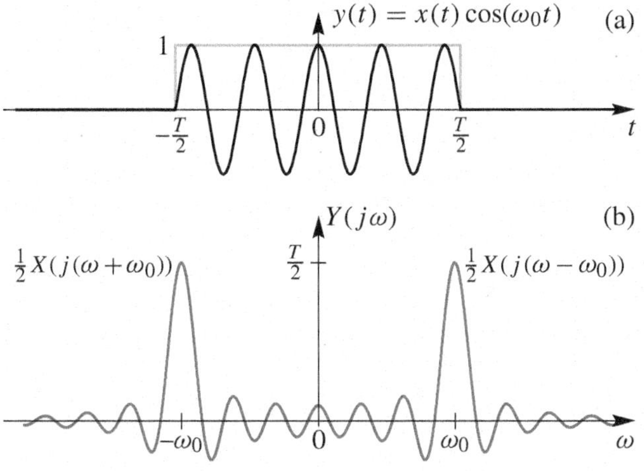

Figure 11-17: Fourier transform of pulse modulated cosine wave: (a) Time function $y(t) = x(t)\cos(\omega_0 t)$. (b) Corresponding Fourier transform $Y(j\omega) = \frac{1}{2}X(j(\omega - \omega_0)) + \frac{1}{2}X(j(\omega + \omega_0))$.

Example 11-14: Finite-Length Sinusoid

As an example of the use of (11.93), suppose that $x(t)$ is the rectangular pulse signal $x(t) = [u(t + T/2) - u(t - T/2)]$, which has Fourier transform

$$X(j\omega) = \frac{\sin(\omega T/2)}{(\omega/2)}$$

Then if we form the signal $y(t) = x(t)\cos(\omega_0 t)$, by (11.93) the corresponding Fourier transform is

$$\begin{aligned} Y(j\omega) &= \tfrac{1}{2}X(j(\omega - \omega_0)) + \tfrac{1}{2}X(j(\omega + \omega_0)) \\ &= \frac{\sin((\omega - \omega_0)T/2)}{(\omega - \omega_0)} + \frac{\sin((\omega + \omega_0)T/2)}{(\omega + \omega_0)} \end{aligned}$$

Figure 11-17(a) shows the product signal $y(t)$ and Fig. 11-17(b) shows the corresponding Fourier transform. Since $X(j\omega)$ is approximately bandlimited to low frequencies, the pulse modulated cosine wave has a Fourier transform that is concentrated near $\pm\omega_0$. The locations of the shifted copies of $X(j\omega)$ are indicated in Fig. 11-17(b). ■

The modulation property of Fourier transforms is extremely useful for analyzing amplitude modulation systems and for deriving the sampling theorem. We will study these and other applications in Chapter 12.

11-9 Table of Fourier Transform Properties and Pairs

In this chapter, we have derived a number of useful transform pairs, and we have also derived several important properties of Fourier transforms. Table 11-2 on p. 338 includes the Fourier transform pairs that we have derived in this section as well as some transform pairs (such as for $u(t)$) that we did not derive.

The basic properties of the Fourier transform are what make it convenient to use in designing and analyzing systems, so they are given in Table 11-3 on p. 339 for easy reference.

11-10 Using the Fourier Transform for Multipath Analysis

In Section 9-9 we presented a time-domain analysis of the problem of multipath transmission of signals. We modeled the multipath with the simple system

$$y(t) = x(t) + \alpha x(t - t_d) \tag{11.94}$$

where α is the amplitude of the echo and t_d its location. While we were able to arrive at the correct inverse filter for removing the echo, our approach was not systematic and relied mostly on trial and error. The Fourier transform provides a much more powerful tool for analyzing such problems.

Table 11-2: Basic Fourier transform pairs.

Table of Fourier Transform Pairs	
Time-Domain: $x(t)$	***Frequency-Domain:*** $X(j\omega)$
$e^{-at}u(t) \quad (a>0)$	$\dfrac{1}{a+j\omega}$
$e^{bt}u(-t) \quad (b>0)$	$\dfrac{1}{b-j\omega}$
$u(t+\frac{1}{2}T)-u(t-\frac{1}{2}T)$	$\dfrac{\sin(\omega T/2)}{\omega/2}$
$\dfrac{\sin(\omega_b t)}{\pi t}$	$[u(\omega+\omega_b)-u(\omega-\omega_b)]$
$\delta(t)$	1
$\delta(t-t_d)$	$e^{-j\omega t_d}$
$u(t)$	$\pi\delta(\omega)+\dfrac{1}{j\omega}$
1	$2\pi\delta(\omega)$
$e^{j\omega_0 t}$	$2\pi\delta(\omega-\omega_0)$
$A\cos(\omega_0 t+\phi)$	$\pi Ae^{j\phi}\delta(\omega-\omega_0)+\pi Ae^{-j\phi}\delta(\omega+\omega_0)$
$\cos(\omega_0 t)$	$\pi\delta(\omega-\omega_0)+\pi\delta(\omega+\omega_0)$
$\sin(\omega_0 t)$	$-j\pi\delta(\omega-\omega_0)+j\pi\delta(\omega+\omega_0)$
$\displaystyle\sum_{k=-\infty}^{\infty} a_k e^{jk\omega_0 t}$	$\displaystyle\sum_{k=-\infty}^{\infty} 2\pi a_k\delta(\omega-k\omega_0)$
$\displaystyle\sum_{n=-\infty}^{\infty} \delta(t-nT)$	$\displaystyle\frac{2\pi}{T}\sum_{k=-\infty}^{\infty}\delta(\omega-\frac{2\pi}{T}k)$

Table 11-3: Basic Fourier transform properties.

Table of Fourier Transform Properties		
Property Name	***Time-Domain:*** $x(t)$	***Frequency-Domain:*** $X(j\omega)$
Linearity	$ax_1(t) + bx_2(t)$	$aX_1(j\omega) + bX_2(j\omega)$
Conjugation	$x^*(t)$	$X^*(-j\omega)$
Time-Reversal	$x(-t)$	$X(-j\omega)$
Scaling	$x(at)$	$\frac{1}{\lvert a\rvert}X(j(\omega/a))$
Delay	$x(t - t_d)$	$e^{-j\omega t_d}X(j\omega)$
Modulation	$x(t)e^{j\omega_0 t}$	$X(j(\omega - \omega_0))$
Modulation	$x(t)\cos(\omega_0 t)$	$\frac{1}{2}X(j(\omega - \omega_0)) + \frac{1}{2}X(j(\omega + \omega_0))$
Differentiation	$\dfrac{d^k x(t)}{dt^k}$	$(j\omega)^k X(j\omega)$
Convolution	$x(t) * h(t)$	$X(j\omega)H(j\omega)$
Multiplication	$x(t)p(t)$	$\frac{1}{2\pi}X(j\omega) * P(j\omega)$

If we apply the Fourier transform to both sides of (11.94) and use the delay property, we obtain

$$Y(j\omega) = X(j\omega) + \alpha e^{-j\omega t_d}X(j\omega)$$

Then we factor out $X(j\omega)$ to obtain

$$Y(j\omega) = \left(1 + \alpha e^{-j\omega t_d}\right)X(j\omega) \tag{11.95}$$

Applying the convolution property of the Fourier transform to (11.95) leads us to the conclusion that the echo system is LTI and that its frequency response is given by

$$H(j\omega) = 1 + \alpha e^{-j\omega t_d} \tag{11.96}$$

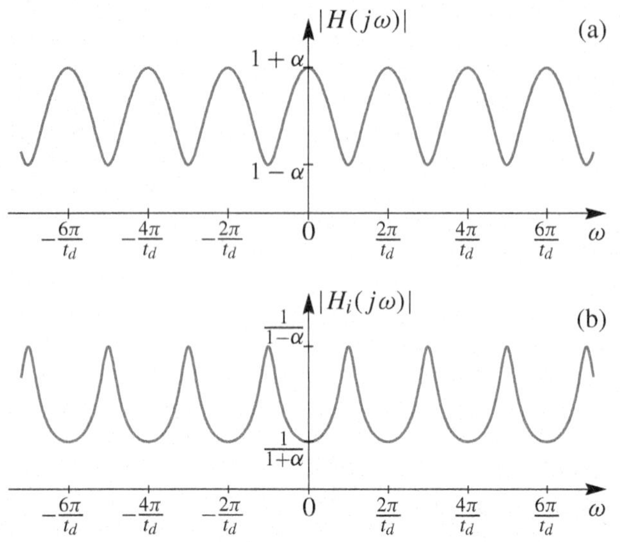

Figure 11-18: Frequency responses of (a) echo system, $|H(j\omega)|$, and (b) its inverse, $|H_i(j\omega)|$.

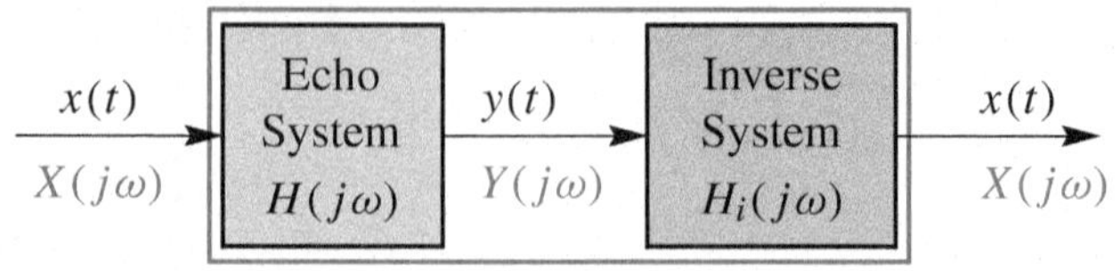

Figure 11-19: Using an LTI system to compensate for an echo so that the final output is equal to the input.

EXERCISE 11.12: Using the delay property and the fact that the Fourier transform of an impulse is the constant 1, show that the impulse response of the echo system is

$$h(t) = \delta(t) + \alpha\delta(t - t_d)$$

Compare your answer to (9.73) on p. 277 in Chapter 9.

EXERCISE 11.13: Show that the frequency response $H(j\omega)$ in (11.96) is periodic in ω with period $2\pi/t_d$.

A plot of the frequency response of the echo system is shown in Fig. 11-18(a). Observe that $|H(j\omega)|$ shown in Fig. 11-18(a) is periodic in ω with period $2\pi/t_d$. Furthermore, note that $|H(j\omega)|$ has dips or nulls at $\omega = \pm(2k-1)\pi/t_d$ where $k = 1, 2, \ldots$. Each dip in the frequency response goes down to $1-\alpha$, so if $\alpha = 1$, the frequency response would be zero and the dip becomes a null. For sound echoes with delay on the order of 5 msec (corresponding to a transmission path distance of a few feet), these nulls will occur at cyclic frequencies on the order of $(k-\frac{1}{2})/t_d = 100, 300, 500, \ldots$ Hz. That is, only a few of these nulls occur in the band of frequencies that is most important for speech intelligibility. Humans perceive the effect of these nulls as a "hollow barrel" effect, which generally does not degrade intelligibility. However, for delays longer than 50 msec., humans hear the delayed signal $\alpha x(t - t_d)$ as a distinct echo. Furthermore, the spacing between nulls becomes less than 20 Hz, so there are a large number of nulls inside the audible speech band. This can degrade speech communication significantly.

As discussed before, if we want to remove the echo, we can attempt to determine an inverse filter or compensator that can be cascaded with the echo system as in Fig. 11-19. In this case, we have indicated the frequency-domain relations as the signal passes through the cascade system. As we will see, the solution is rather straightforward in the frequency-domain. In particular, the final output of the cascade system is desired to be

$$X(j\omega) = H_i(j\omega)Y(j\omega) = H_i(j\omega)H(j\omega)X(j\omega)$$

We must design the inverse filter $H_i(j\omega)$ so that it will remove the effects of $H(j\omega)$, so it must satisfy the equation

$$H_i(j\omega)H(j\omega) = 1 \tag{11.97}$$

or

$$H_i(j\omega) = \frac{1}{H(j\omega)} = \frac{1}{1 + \alpha e^{-j\omega t_d}} \tag{11.98}$$

Equation (11.97) states that $H_i(j\omega)$ must be the multiplicative inverse of $H(j\omega)$.

Figure 11-18(b) shows $|H_i(j\omega)|$ as a function of ω. Since $|H_i(j\omega)|$ must be the multiplicative inverse of $|H(j\omega)|$, it follows that $|H_i(j\omega)|$ must have peaks where $|H(j\omega)|$ has dips and vice versa. Clearly the case $\alpha \approx 1$ is problematic in a practical setting since the gain of the inverse filter would have to be very large. Thus, any noise that was added to the signal after the echo system, would be greatly amplified by the peaks of the inverse filter.

Our frequency-domain analysis has produced several insights into the nature of an inverse filter to remove echoes. In fact, it also provides a relatively easy way to find the impulse response of the inverse filter. If we want to find the impulse response from the frequency response, we must take the inverse Fourier transform of $H_i(j\omega)$. Unfortunately, the transform in (11.98) is not in our table, and it would be difficult to evaluate the inverse transform integral directly. However, we can base the inversion on a well-known formula for the sum of terms of a geometric series, which was also used for z-transforms in Chapter 8.

$$\sum_{k=0}^{\infty} \beta^k = \frac{1}{1-\beta} \tag{11.99}$$

The condition for this sum to converge is $|\beta| < 1$. Comparing the righthand side of (11.98) to (11.99) leads to the equation

$$\begin{aligned} H_i(j\omega) &= \frac{1}{1 + \alpha e^{-j\omega t_d}} \\ &= \sum_{k=0}^{\infty} (-\alpha e^{-j\omega t_d})^k \\ &= \sum_{k=0}^{\infty} (-\alpha)^k e^{-j\omega k t_d} \end{aligned} \tag{11.100}$$

in which the sum converges if $|-\alpha e^{-j\omega t_d}| < 1$, or equivalently, if $|\alpha| < 1$. Now each term in the sum on the right-hand side in (11.100) represents a delayed impulse in the time-domain, so we can write

$$h_i(t) = \sum_{k=0}^{\infty} (-\alpha)^k \delta(t - k t_d) \tag{11.101}$$

which is identical to the result derived in the time-domain in Section 9-9.

11-11 Summary

In this chapter, we introduced the continuous-time Fourier transform and derived a number of its properties. We focused on the convolution and multiplication properties since these are key results for the study of the LTI systems, modulation systems, and sampling systems that will be studied in the next chapter. The `CLTIdemo` is a MATLAB GUI that illustrates the convolution-multiplication relationship between the input and output spectra of a continuous-time linear time-invariant (LTI) filter when the input is a sinusoidal function.

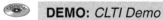

DEMO: *CLTI Demo*

The Fourier series demo, introduced in Chapter 3, can be used to show how frequency spacing of $X(j\omega)$ decreases as the period of $x(t)$ increases.

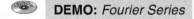

DEMO: *Fourier Series*

The reader is reminded of the large number of solved homework problems on the CD-ROM that are available for review and practice.

NOTE: *Hundreds of Solved Problems*

11-12 Problems

P-11.1 Draw a plot of $y(t) = x(3t)$ when the signal $x(t)$ is the 2-sec pulse

$$x(t) = u(t+1) - u(t-1)$$

In addition, determine the duration of $y(t)$.

P-11.2 The delay property of Fourier transform states that if $X(j\omega)$ is the Fourier transform of $x(t)$, then

$$x(t-t_d) \stackrel{\mathcal{F}}{\longleftrightarrow} e^{-j\omega t_d} X(j\omega)$$

Use this property to find the Fourier transforms of the following signals:

(a) $x(t) = \delta(t+1) + 2\delta(t) + \delta(t-1)$

(b) $x(t) = \dfrac{\sin(100\pi(t-2))}{\pi(t-2)}$

(c) $x(t) = e^{-t}u(t) - e^{-t}u(t-4)$

P-11.3 For each of the following cases, use the table of known Fourier transform pairs to find $x(t)$ when

(a) $X(j\omega) = \dfrac{j\omega}{0.1 + j\omega} e^{-j\omega 0.2}$

(b) $X(j\omega) = 2 + 2\cos(\omega)$

(c) $X(j\omega) = \dfrac{1}{1+j\omega} - \dfrac{1}{2+j\omega}$

(d) $X(j\omega) = j\delta(\omega - 100\pi) - j\delta(\omega + 100\pi)$

P-11.4 In each of the following cases, determine the Fourier transform, or inverse Fourier transform. Give your answer as a simple formula or a plot. Explain each answer by stating which property and transform pair you used.

(a) Find $X(j\omega)$ when $x(t) = \begin{cases} 1 & 0 \le t < 4 \\ 0 & \text{otherwise} \end{cases}$

(b) Find $s(t)$ when

$$S(j\omega) = 4\pi\delta(\omega) + 2\pi\delta(\omega - 10\pi) + 2\pi\delta(\omega + 10\pi)$$

(c) Find $r(t)$ when

$$R(j\omega) = \frac{j\omega}{4 + j2\omega} = \frac{1}{2} - \frac{2}{4 + j2\omega}$$

(d) Find $Y(j\omega)$ when $y(t) = \delta(t+1) + 2\delta(t) + \delta(t-1)$

P-11.5 Determine the Fourier transform of $H(j\omega)$ when

$$h(t) = \frac{d}{dt}\left\{\frac{\sin(4\pi t)}{\pi t}\right\}$$

In addition, plot $|H(j\omega)|$ versus ω.

P-11.6 In each of the following cases, use known Fourier transform pairs together with Fourier transform properties to complete the following Fourier transform pair relationships:

(a) $x(t) = u(t+3)u(3-t) \stackrel{\mathcal{F}}{\longleftrightarrow} X(j\omega) =$

(b) $x(t) = \sin(4\pi t)\sin(50\pi t) \stackrel{\mathcal{F}}{\longleftrightarrow} X(j\omega) =$

(c) $x(t) = \dfrac{\sin 4\pi t}{\pi t} \sin(50\pi t) \stackrel{\mathcal{F}}{\longleftrightarrow} X(j\omega) =$

(d) $x(t) = \qquad \stackrel{\mathcal{F}}{\longleftrightarrow} X(j\omega) = \left(\frac{\sin(200\omega)}{\omega}\right)^2$

(e) $x(t) = \qquad \stackrel{\mathcal{F}}{\longleftrightarrow} X(j\omega) = \cos^2(\omega)$

P-11.7 In the following, the time-domain signal $x(t)$ is given. Use the tables of Fourier transforms and Fourier transform properties to determine the Fourier transform for each of the signals. You may give your answer either as an equation or a carefully labeled plot, whichever is most convenient.

(a) $x(t) = \frac{d}{dt}\left(\frac{10\sin(200\pi t)}{\pi t}\right)$

(b) $x(t) = \frac{2\sin(400\pi t)}{\pi t}\cos(2000\pi t)$

(c) Periodic impulse train

$$x(t) = \sum_{n=-\infty}^{\infty} \delta(t - 10n)$$

P-11.8 In the following, the Fourier transform $X(j\omega)$ is given. Use the tables of Fourier transforms and Fourier transform properties to determine the inverse Fourier transform for each case. You may give your answer either as an equation or a carefully labeled plot, whichever is most convenient.

(a) $X(j\omega) = \frac{e^{-j\omega 3}}{2 + j\omega}$

(b) $X(j\omega) = \frac{j\omega}{2 + j\omega}$

(c) $X(j\omega) = \frac{(j\omega)}{2 + j\omega}e^{-j\omega 3}$

(d) $X(j\omega) = \left(\frac{2\sin(\omega)}{\omega}\right)\sum_{k=-\infty}^{\infty}\left(\frac{\pi}{5}\right)\delta(\omega - 2\pi k/10)$

Hint: Work part (d) as a convolution problem in the time-domain.

P-11.9 Prove that for a real $h(t)$, the magnitude $|H(j\omega)|$ has even symmetry and the phase angle $\angle H(j\omega)$ has odd symmetry, i.e.,

$$|H(j\omega)| = |H(-j\omega)|$$
$$\angle H(j\omega) = -\angle H(-j\omega)$$

If $h(t)$ is purely imaginary, describe how the symmetries of the magnitude and phase change.

P-11.10 Determine the inverse Fourier transform of

$$X(j\omega) = \frac{5}{(3 + j\omega)^2}$$

Although there is no table entry for this case, the convolution property can be applied if $X(j\omega)$ is written as a product.

P-11.11 The impulse response of an LTI system is given by

$$h(t) = \frac{10\sin[4\pi(t-1)]}{\pi(t-1)}$$

(a) Make a detailed and accurately labeled sketch of $h(t)$. Mark the important amplitudes and time locations of peaks and zero crossings.

(b) Determine the Fourier transform $H(j\omega)$ of this impulse response; recall that $H(j\omega)$ is also the frequency response of the system.

(c) Make detailed plots of $|H(j\omega)|$ and $\angle H(j\omega)$ versus ω. Label your plots carefully.

P-11.12 The impulse response of the system in Fig. P-11.12 is

$$h(t) = \frac{4\sin(\omega_{co}t)}{\pi t}$$

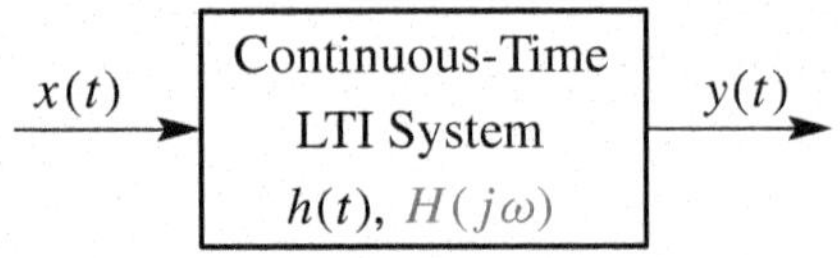

Figure P-11.11

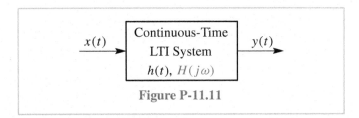

and the input to the system is a periodic signal (with period $T_0 = 1$) given by the equation

$$x(t) = \sum_{n=-\infty}^{\infty} \delta(t-n)$$

(a) Determine the Fourier transform $X(j\omega)$ of the input signal. Plot $X(j\omega)$ over the range $-6\pi < \omega < 6\pi$.

(b) Determine the frequency response $H(j\omega)$ of the system and plot $|H(j\omega)|$ on the same graph as $X(j\omega)$ for the case $\omega_{co} = 5\pi$.

(c) Use your plot in (b) to determine $y(t)$, the output of the LTI system for the given input $x(t)$ when the cutoff frequency is $\omega_{co} = 5\pi$.

(d) How would you choose ω_{co} so that the output is a constant; i.e., $y(t) = C$ for all t. What is the constant C?

P-11.13 The frequency response of an ideal lowpass LTI system is

$$H(j\omega) = \begin{cases} 10\,e^{-j0.0025\omega} & |\omega| < 1000\pi \\ 0 & |\omega| > 1000\pi \end{cases}$$

In each of the following cases, determine the Fourier transform of the input signal and then use frequency-domain methods to determine the corresponding output signal.

(a) Using frequency-domain methods, determine the output of the system when the input is

$$x(t) = \cos(200\pi t) + \frac{2\sin(2000\pi t)}{\pi t}$$

(b) Determine the output if the input is

$$x(t) = \cos(200\pi t) + \frac{2\sin(2000\pi t)}{\pi t} + \cos(3000\pi t)$$

(c) Determine the output if the input is

$$x(t) = \cos(200\pi t) + 2\delta(t)$$

(d) What is it about this system that accounts for the results you obtained in parts (a)–(c) above?

P-11.14 The input to an LTI system such as Fig. P-11.12 is defined by the equation

$$x(t) = \sum_{k=-\infty}^{\infty} a_k e^{jk200\pi t}, \quad \text{where } a_k = \begin{cases} \dfrac{1}{\pi k^2} & k \neq 0 \\ 1 & k = 0 \end{cases}$$

(a) Determine the Fourier transform of $x(t)$ and plot it on a graph that spans the frequency range $-1000\pi \le \omega \le 1000\pi$.

(b) Suppose that the frequency response of the LTI system is given by the following equation:

$$H(j\omega) = \begin{cases} 10 & |\omega| < 300\pi \\ 0 & |\omega| > 300\pi \end{cases}$$

Plot this function on the graph that you constructed in part (a).

(c) Determine the DC value of the output signal; i.e., determine $b_0 = \dfrac{1}{T_0}\displaystyle\int_0^{T_0} y(t)dt$, where T_0 is the period of the output signal.

(d) Write an equation for $y(t)$. If possible, simplify it to include only cosine functions.

P-11.15 Let $x(t)$ be a triangular pulse defined by

$$x(t) = \begin{cases} 1 - |t| & |t| < 1 \\ 0 & \text{elsewhere} \end{cases}$$

(a) By taking the derivative of $x(t)$, use the derivative property to find the Fourier transform of $x(t)$.
Hint: Express the derivative as a sum of two pulses, one with an amplitude of one, the other with an amplitude of minus one. From the tables of Fourier transform pairs and properties, write down the Fourier transform of the derivative of $x(t)$. Then, use the derivative property in the frequency-domain to find $X(j\omega)$.

(b) Find the Fourier transform of $x(t)$ by differentiating $x(t)$ twice and using the second-derivative property. Compare this result to the result of part (a).

P-11.16 The symmetries of the Fourier transform are quite useful. In Section 11-5.2, the conjugate-symmetry property was used to show that for a real $x(t)$ the real part of its transform is even and the imaginary part is odd. For this problem, assume that $x(t)$ is a real signal with Fourier transform, $X(j\omega)$.

(a) Prove that the magnitude $|X(j\omega)|$ is an even function of ω; i.e., $|X(-j\omega)| = |X(j\omega)|$.

(b) Prove that the phase $\angle X(j\omega)$ is an odd function of ω; i.e., $\angle X(-j\omega) = -\angle X(j\omega)$.

P-11.17 Prove that the Fourier transform of the unit-step signal, $u(t)$, is

$$U(j\omega) = \frac{1}{j\omega} + K\delta(\omega)$$

where $K = \pi$. At first glance, it might seem that the impulse term $K\delta(\omega)$ should be zero, but show that $K \neq 0$ because the following signal $s(t)$ is an odd symmetric signal.

$$s(t) = \begin{cases} u(t) - \frac{1}{2} & t \neq 0 \\ 0 & t = 0 \end{cases}$$

Note: Even though $s(t)$ and $u(t) - \frac{1}{2}$ differ at one isolated point, $t = 0$, they still have the same Fourier transform.

CHAPTER 12

Filtering, Modulation, and Sampling

In Chapter 11, we introduced the continuous-time Fourier transform, and demonstrated how to transform back and forth between the time- and frequency-domains. We emphasized the convolution and multiplication properties of the Fourier transform because they are the most general properties. In this chapter, we will show how these two basic properties of the Fourier transform can be used to study several applications, including linear filtering, sinewave modulation, and sampling. By providing a mathematical representation that also has a simple graphical interpretation, the Fourier transform greatly enhances our understanding of these important signal processing functions. Filtering is represented by multiplication by the frequency response function, which leads us to visualize how different parts of the input frequency spectrum are treated by the system. Amplitude modulation and sampling are easily visualized in terms of moving the spectrum from one frequency location to another. This chapter has many practical examples that are readily understood in terms of these two operations.

12-1 Linear Time-Invariant Systems

In Chapter 11 we showed that time-domain convolution is transformed into frequency-domain multiplication by the Fourier transform. Since LTI systems are completely defined by convolution, the convolution property of Fourier transforms is essential for understanding LTI systems, as well as simplifying their analysis.

Figure 12-1 depicts the time-domain and frequency-domain representations of an LTI system. Note that we have indicated the time-domain representation above the arrows and the corresponding frequency-domain representation below. Since the Fourier transform allows us to go back and forth between the time-domain and the frequency-domain with ease, we have greatly expanded

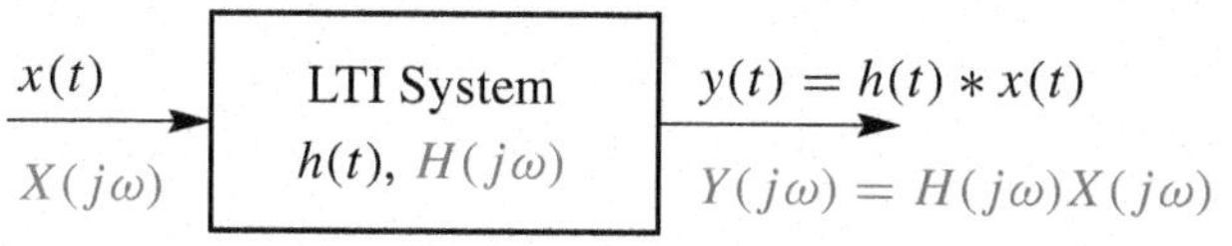

Figure 12-1: Linear time-invariant system depicted in both the time domain and frequency domain.

our options for analyzing LTI systems. In problem solving, a key step will be to choose the representation that simplifies the analysis the most.

12-1.1 Cascade and Parallel Configurations

In Chapter 9 we showed that cascade and parallel configurations of LTI systems can be combined and simplified into a single equivalent system. In this section, we will revisit this topic using a frequency-domain analysis.

Figure 12-2(a) shows a cascade connection of two LTI systems. In the frequency-domain, the output of the first system, $W_1(j\omega) = H_1(j\omega)X(j\omega)$, is the input to the second system. Therefore, the output of the second system, which is also the overall output, is

$$Y(j\omega) = H_2(j\omega)W_1(j\omega) = H_2(j\omega)H_1(j\omega)X(j\omega)$$

If we define $H(j\omega) = H_2(j\omega)H_1(j\omega)$, then $Y(j\omega)$ can be written in the form $Y(j\omega) = H(j\omega)X(j\omega)$.

Equivalent frequency response of a cascade of LTI systems.

$$\begin{aligned} H(j\omega) &= H_2(j\omega)H_1(j\omega) \\ &= H_1(j\omega)H_2(j\omega) \end{aligned} \tag{12.1}$$

The two forms of (12.1) are equivalent because we can multiply complex numbers such as $H_1(j\omega)$ and $H_2(j\omega)$ in either order (commutative property). The implication

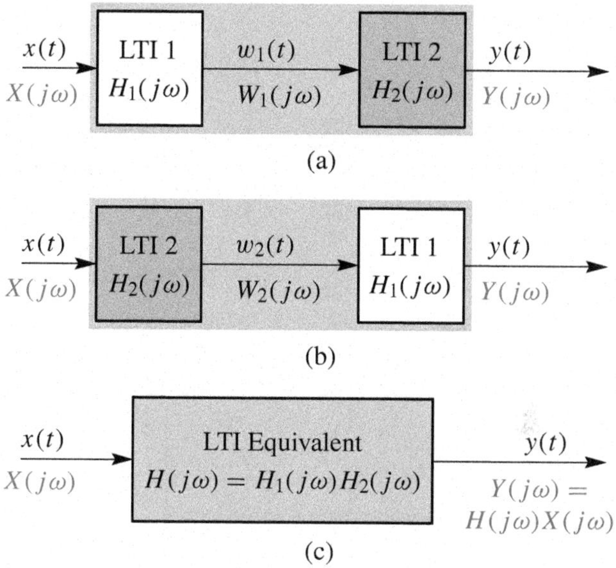

Figure 12-2: Frequency-domain representation of cascade connection of LTI systems: (a) System #2 following System #1. (b) System #1 following System #2. (c) Equivalent system to both (a) and (b).

of the fact that we can multiply the frequency response functions in either order is that we can cascade the two LTI systems in either order and obtain the same overall frequency response. Therefore, Figs. 12-2(a), (b), and (c) all have the same frequency response (12.1) and are equivalent from the point of view of the input $x(t)$ and output $y(t)$. We have, of course, already shown this using the commutative property of convolution in Section 9-8.

In a parallel configuration, the input is fed into both systems simultaneously and their outputs are added to form the overall output as depicted in Fig. 12-3(a). The frequency-domain description of the parallel configuration is

$$\begin{aligned} Y(j\omega) &= H_1(j\omega)X(j\omega) + H_2(j\omega)X(j\omega) \\ &= [H_1(j\omega) + H_2(j\omega)]X(j\omega) \end{aligned}$$

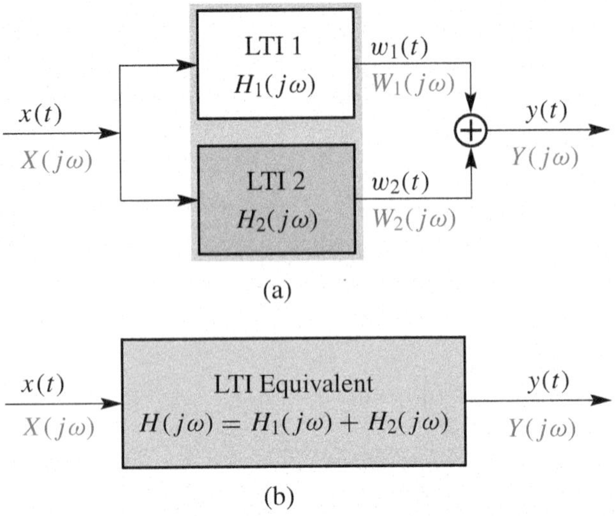

Figure 12-3: Frequency-domain representation of parallel connection of LTI systems: (a) Parallel connection of two systems, and (b) equivalent system to (a).

so $Y(j\omega)$ is again in the form $Y(j\omega) = H(j\omega)X(j\omega)$, leading to the following general result:

Equivalent frequency response of parallel LTI systems.

$$H(j\omega) = H_1(j\omega) + H_2(j\omega) \tag{12.2}$$

That is, Figs. 12-3(a) and (b) have the same overall frequency response (12.2) and are therefore equivalent from the point of view of the input $x(t)$ and output $y(t)$.

12-1.2 Ideal Delay

In Chapter 11 we showed that the frequency response of an ideal delay system is

$$H(j\omega) = e^{-j\omega t_d} \tag{12.3}$$

Whenever the term $e^{-j\omega t_d}$ occurs in a frequency response equation, it can be associated with a time delay of t_d seconds. Delay is a common operation in both discrete-time and continuous-time systems. In continuous-time systems, the ideal delay is often used to model the delays that naturally arise in physical systems such as propagation delays of acoustic and electrical signals.

In many cases, we use a cascade representation of several LTI systems to separate the effects of time-delay from other filtering effects caused primarily by the magnitude of the frequency response. When the phase of $H(j\omega)$ varies linearly with frequency, i.e. $\angle H(j\omega) = -\omega t_d$, the phase effect is particularly easy to understand—it simply corresponds to a time delay of t_d seconds. We will illustrate this decomposition with the following specific example.

Example 12-1: Running Integral

The cascade/parallel system depicted in Fig. 12-4(a) is a cascade of an integrator with a parallel combination of a straight-through connection and a time delay of T seconds. We will determine the frequency response of this system in two ways. First, if we find the impulse response of the overall system, then we can evaluate the Fourier transform integral to determine the frequency response. Since the impulse response of the straight-through connection is just $\delta(t)$, the impulse response of the parallel system is $h_1(t) = \delta(t) - \delta(t-T)$. The impulse response of the integrator is $h_2(t) = u(t)$. Therefore, the impulse response of the overall system is

$$\begin{aligned} h(t) &= [\delta(t) - \delta(t-T)] * u(t) \\ &= u(t) - u(t-T) \\ &= \begin{cases} 1 & 0 \le t < T \\ 0 & \text{otherwise} \end{cases} \end{aligned} \tag{12.4}$$

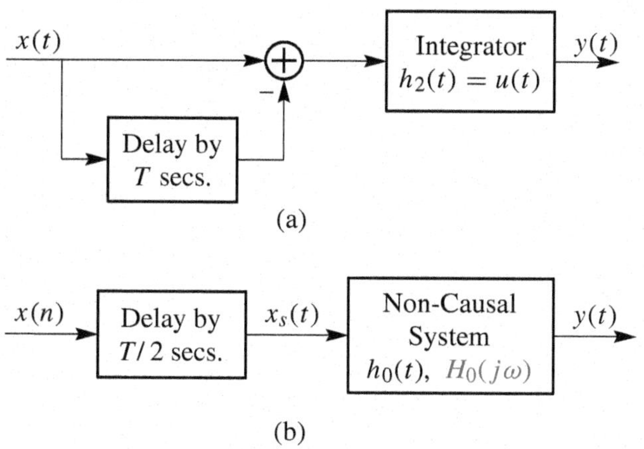

Figure 12-4: (a) Example of a cascade/parallel system involving an integrator and delay; (b) Equivalent cascade of delay and non-causal system.

Since we have the impulse response, we can also write the input/output relation for the overall system as

$$y(t) = x(t) * h(t) = \int_{t-T}^{t} x(\tau)d\tau$$

The limits of this integral vary with time but the integration is always over a time interval of length T. Thus, an appropriate name for the system of Fig. 12-4(a) is the *running integral* system.

To obtain the frequency response of the running integral system, we substitute (12.4) into the Fourier transform integral to obtain

$$\begin{aligned} H(j\omega) &= \int_{-\infty}^{\infty} h(t)e^{-j\omega t}dt = \int_{0}^{T} e^{-j\omega t}dt \\ &= \frac{1-e^{-j\omega T}}{j\omega} = \left(\frac{e^{j\omega T/2}-e^{-j\omega T/2}}{2j(\omega/2)}\right)e^{-j\omega T/2} \\ &= \frac{\sin(\omega T/2)}{(\omega/2)}e^{-j\omega T/2} \end{aligned} \quad (12.5)$$

A second approach to finding the frequency response of the system in Fig. 12-4(a) is to work directly with the frequency responses of the individual systems and then combine them using the rules in (12.1) and (12.2). The frequency response of the parallel part is

$$H_1(j\omega) = 1 - e^{-j\omega T}$$

and the frequency response of the integrator is, from Table 11-2,

$$H_2(j\omega) = \pi\delta(\omega) + \frac{1}{j\omega}$$

Therefore, the frequency response of the overall system in Fig. 12-4(a) is the product

$$\begin{aligned} H(j\omega) &= H_1(j\omega)H_2(j\omega) \\ &= \left(1-e^{-j\omega T}\right)\left(\pi\delta(\omega)+\frac{1}{j\omega}\right) \\ &= \pi(1-e^{-j\omega T})\delta(\omega) + \left(1-e^{-j\omega T}\right)\left(\frac{1}{j\omega}\right) \\ &= 0\delta(\omega) + \frac{1-e^{-j\omega T}}{j\omega} \\ &= \frac{\left(e^{+j\omega T/2}-e^{-j\omega T/2}\right)}{j\omega}e^{-j\omega T/2} \\ &= \frac{\sin(\omega T/2)}{(\omega/2)}e^{-j\omega T/2} \end{aligned}$$

which is identical to the previous result (12.5).

If we define the *amplitude function* as

$$H_0(j\omega) = \frac{\sin(\omega T/2)}{(\omega/2)} \quad (12.6)$$

we can express $H(j\omega)$ in the form

$$H(j\omega) = H_0(j\omega)\, e^{-j\omega T/2} \quad (12.7)$$

The representation of (12.7) is interesting because its product form suggests that the system can be thought of as the cascade of a time delay of $T/2$ seconds with a system whose frequency response $H_0(j\omega)$ is depicted in Fig. 12-4(b). From the third entry in Table 11-2, the system $H_0(j\omega)$ has a non-causal impulse response

$$h_0(t) = u(t + T/2) - u(t - T/2)$$

so the overall impulse response of the cascade system in Fig. 12-4(b) is

$$h(t) = h_0(t) * \delta(t - T/2) = u(t) - u(t - T)$$

which is the same as determined in (12.4). ■

The block diagram of Fig. 12-4(a) is composed of causal systems, and hence, it could be a basis for implementing the system. Certainly, the overall impulse response $h(t)$ corresponds to a causal system. On the other hand, the system in Fig. 12-4(b) contains a system with a non-causal impulse response $h_0(t)$. Even though it could not be the basis for an implementation of the system, the advantage of the representation in Fig. 12-4(b) is that it directs us to think about the system in a new way. We can think about the time delay of $T/2$ somewhat independently of the amplitude effects that are specified by $H_0(j\omega)$. Figure 12-5(a) shows the impulse response of the system, Fig. 12-5(b) shows $H_0(j\omega)$, and Fig. 12-5(c) shows the linear phase component that corresponds to a time delay of $T/2$ seconds.

Note that the plots of Fig. 12-5(b,c) are not standard magnitude and phase frequency response plots because $H_0(j\omega)$ can be both positive and negative, and because the plot of the linear phase component is allowed to increase without the normal "wrap around" that results when the principal value of a phase angle is computed. In Fig. 12-6 we plot the magnitude and the principal value of the phase. The magnitude is determined only by $H_0(j\omega)$ since $|H(j\omega)| = |H_0(j\omega)|$. However, in addition to the linear phase component $-\omega T/2$, $H_0(j\omega)$ also contributes to the phase of $H(j\omega)$. At every frequency where $H_0(j\omega)$ is negative, an odd multiple of π radians must be added to the phase to account for the negative sign. Furthermore, the phase would be computed (e.g., in MATLAB) modulo 2π. This is illustrated in Fig. 12-6(b). Note that the sections of the phase curve between discontinuities are still linear functions of ω. The jumps of 2π radians are due to the "wrap around" of computing the phase modulo 2π, while the jumps of π radians are due to the sign change at the zeros of $H_0(j\omega)$.

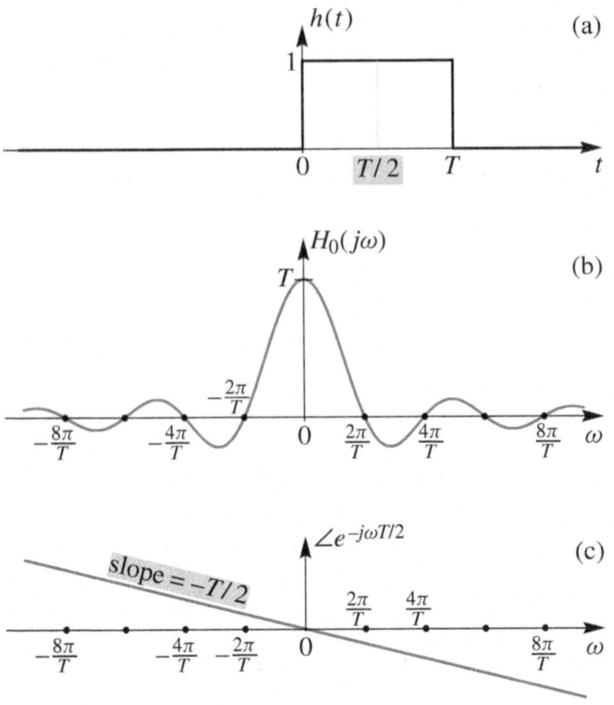

Figure 12-5: Example of cascade/parallel system: (a) Impulse response $h(t)$, (b) amplitude function $H_0(j\omega)$, and (c) linear phase component of $H(j\omega)$ is $-\omega T/2$.

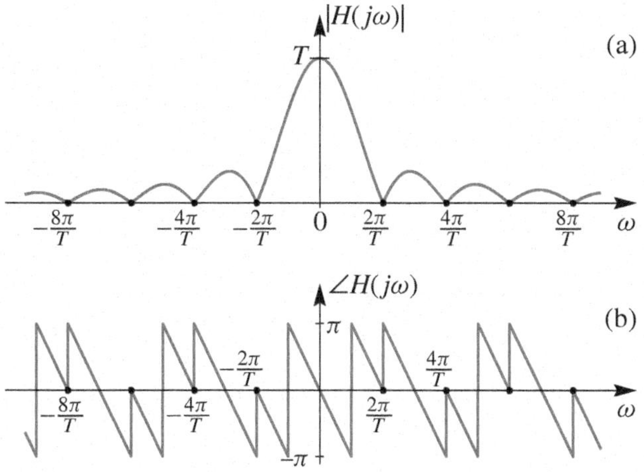

Figure 12-6: Standard plot of frequency response of cascade/parallel system: (a) Magnitude of frequency response $|H(j\omega)| = |H_0(j\omega)|$, and (b) principal value of phase.

12-1.3 Frequency Selective Filters

Example 12-1 illustrates how frequency-domain analysis can be used to manipulate parallel and cascade systems into different forms and how new forms of the system may lend insight into the performance of that system. The system in Example 12-1 is a "lowpass" system or "lowpass filter", abbreviated LPF. This is because the magnitude of the Fourier transform of the output,

$$\begin{aligned}|Y(j\omega)| &= |H_0(j\omega)||e^{-j\omega T/2}||X(j\omega)| \\ &= |H_0(j\omega)||X(j\omega)|\end{aligned}$$

is controlled only by $|H_0(j\omega)|$. In Fig. 12-6(a) we see that frequency components in the bands $\omega \geq 2\pi/T$ and $\omega \leq -2\pi/T$ are attenuated (reduced in size) with respect to components in the low frequency band $-2\pi/T \leq \omega \leq 2\pi/T$.

As we have already discussed in Section 10-3, it is useful to idealize the concept of a lowpass filter and other *frequency selective* filters. In this section, we review and extend the definitions of ideal frequency selective filters.

12-1.3.1 Ideal Lowpass Filter

The ideal lowpass filter is defined by the frequency response function

$$H_{\text{lp}}(j\omega) = \begin{cases} 1 & -\omega_{\text{co}} \leq \omega \leq \omega_{\text{co}} \\ 0 & \text{otherwise} \end{cases} \tag{12.8}$$

The frequency ω_{co}, beyond which the frequency response is identically zero, is called the ***cutoff frequency*** of the ideal LPF.

The ideal lowpass filter is an idealization of a large class of real systems that occur naturally, or that are carefully designed to emphasize their lowpass property. One of the most important conceptual uses is that the output of an ideal lowpass filter is a perfectly bandlimited signal regardless of the nature of the input signal. That is,

$$\begin{aligned}Y(j\omega) &= H_{\text{lp}}(j\omega)X(j\omega) \\ &= \begin{cases} X(j\omega) & -\omega_{\text{co}} \leq \omega \leq \omega_{\text{co}} \\ 0 & \text{otherwise} \end{cases}\end{aligned}$$

Using Table 11-2 in Chapter 11, it follows from the definition of $H_{\text{lp}}(j\omega)$ that the impulse response of the ideal lowpass filter is

$$h_{\text{lp}}(t) = \frac{\sin(\omega_{\text{co}}t)}{\pi t} \tag{12.9}$$

Following the result of Example 12-1, the ideal lowpass filter with delay t_d would have frequency response

$$\begin{aligned}H_{\text{lpd}}(j\omega) &= H_{\text{lp}}(j\omega)e^{-j\omega t_d} \\ &= \begin{cases} e^{-j\omega t_d} & -\omega_{\text{co}} \leq \omega \leq \omega_{\text{co}} \\ 0 & \text{otherwise} \end{cases}\end{aligned} \tag{12.10}$$

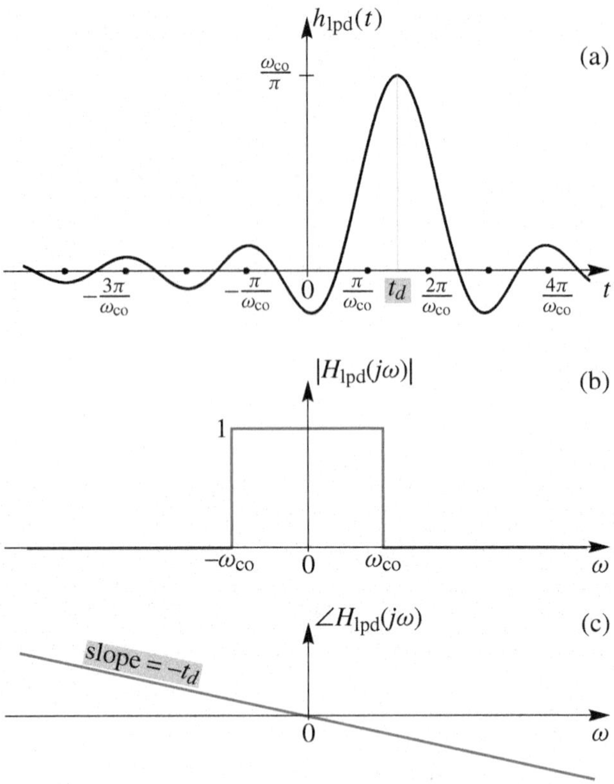

Figure 12-7: Ideal lowpass filter with delay t_d: (a) Impulse response $h_{lpd}(t)$, (b) magnitude of frequency response $|H_{lpd}(j\omega)| = H_{lp}(j\omega)$, and (c) phase of the frequency response (i.e., $\angle H_{lpd}(j\omega)$, is $-\omega t_d$).

and its impulse response would be

$$h_{lpd}(t) = h_{lp}(t - t_d) = \frac{\sin[\omega_{co}(t - t_d)]}{\pi(t - t_d)} \qquad (12.11)$$

The impulse response and frequency response of the ideal lowpass filter with delay are shown in Fig. 12-7. Figure 12-7(a) shows the impulse response (12.11), and Figs. 12-7(b) and (c) show the magnitude and phase respectively of the frequency response as defined in (12.10). Note the similarity between Figs. 12-5 and 12-7. The impulse responses are both symmetrical and pulse-like and both are centered to the right of $t = 0$. The magnitude of the frequency response exhibits a lowpass nature in both cases, and the phase is linear, corresponding to a delay or right shift of the impulse response. In the case of the ideal lowpass filter, the amplitude function in Fig. 12-7(b) is non-negative, so we do not have to be concerned about any additional phase components.

There are other major differences between the running integral LPF in Fig. 12-5 and the ideal LPF in Fig. 12-7. While the magnitude responses are both "lowpass", the running integral frequency response dies out gradually with increasing frequency while the frequency response of the ideal lowpass filter drops abruptly to zero above the cutoff frequency ω_{co}. On the other hand, the running integral filter is causal, but we see from Fig. 12-7(a) that the ideal lowpass filter is non-causal no matter how much delay is inserted. In a very real sense, the fact that the magnitude of the frequency response of the running integral filter deviates from the ideal response is a direct consequence of the fact that the system can be made causal if we allow at least $T/2$ seconds of delay. It turns out that a causal system cannot have a frequency response that is identically zero over a band of frequencies of nonzero width.[1] This means that the ideal lowpass filter cannot be a causal system even if we incorporate time delay. This does not mean, however, that the ideal LPF with delay is a useless concept. Indeed, since we can approximate its characteristics with high precision using realizable systems, it serves as a very useful abstraction that simplifies the analysis of complicated systems.

12-1.3.2 Other Ideal Frequency Selective Filters

Figure 12-8 reminds us of the frequency responses of three other types of ideal frequency selective filters that were introduced in Section 10-3. As in the case

[1] Proof of this general property is beyond the scope of this text.

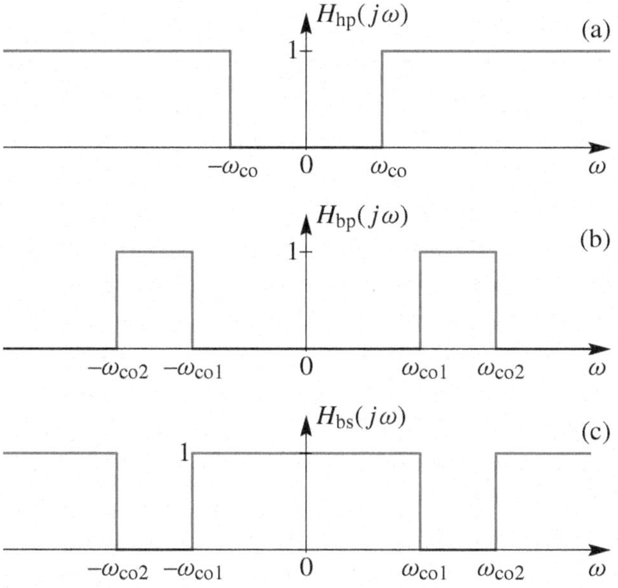

Figure 12-8: Ideal frequency selective filters: (a) Ideal highpass filter, (b) ideal bandpass filter, and (c) ideal bandstop filter.

of the ideal lowpass filter, these filters also are non-causal systems, but they serve as useful abstractions and idealizations of what can be approximated with causal systems.

Figure 12-8(a) is the frequency response of an ideal highpass filter defined by

$$H_{\text{hp}}(j\omega) = \begin{cases} 0 & -\omega_{\text{co}} \leq \omega \leq \omega_{\text{co}} \\ 1 & \text{otherwise} \end{cases} \tag{12.12}$$

In words, the ideal highpass filter passes frequencies above the cutoff frequency ω_{co} and completely rejects frequencies below the cutoff. Another useful representation of the frequency response of the ideal highpass filter with cutoff frequency ω_{co} is

$$H_{\text{hp}}(j\omega) = 1 - H_{\text{lp}}(j\omega) \tag{12.13}$$

From (12.13) it follows that the impulse response of the ideal highpass filter is

$$h_{\text{hp}}(t) = \delta(t) - h_{\text{lp}}(t) = \delta(t) - \frac{\sin(\omega_{\text{co}}t)}{\pi t} \tag{12.14}$$

EXERCISE 12.1: Determine the frequency response $H_{\text{hpd}}(j\omega)$ and impulse response $h_{\text{hpd}}(t)$ for an ideal highpass filter with delay t_d.

The ideal bandpass filter can be represented in terms of ideal lowpass filters in several ways as suggested in some problems at the end of the chapter (e.g., Probs. P-12.1, P-12.2, and P-12.3). Also, each of the ideal filters can be defined with time delay by simply multiplying their frequency responses by $e^{-j\omega t_d}$ or delaying their impulse responses by t_d.

The ideal frequency selective filters that we have defined above are exceedingly useful idealizations of filters that can be implemented in real systems. We will see that lowpass and bandpass filters are essential ingredients in many communications and signal processing systems.

12-1.4 Example of Filtering in the Frequency-Domain

To illustrate how ideal filters can modify a signal, consider the periodic square wave as the input signal. Recall (11.40) in Example 11-3 on p. 320 where we showed that the periodic square wave of Fig. 12-9(b,c) has the Fourier transform

$$X(j\omega) = \pi\delta(\omega) + \sum_{\substack{k=-\infty \\ k\neq 0}}^{\infty} \frac{2\sin(\pi k/2)}{k}\delta(\omega - k\omega_0) \tag{12.15}$$

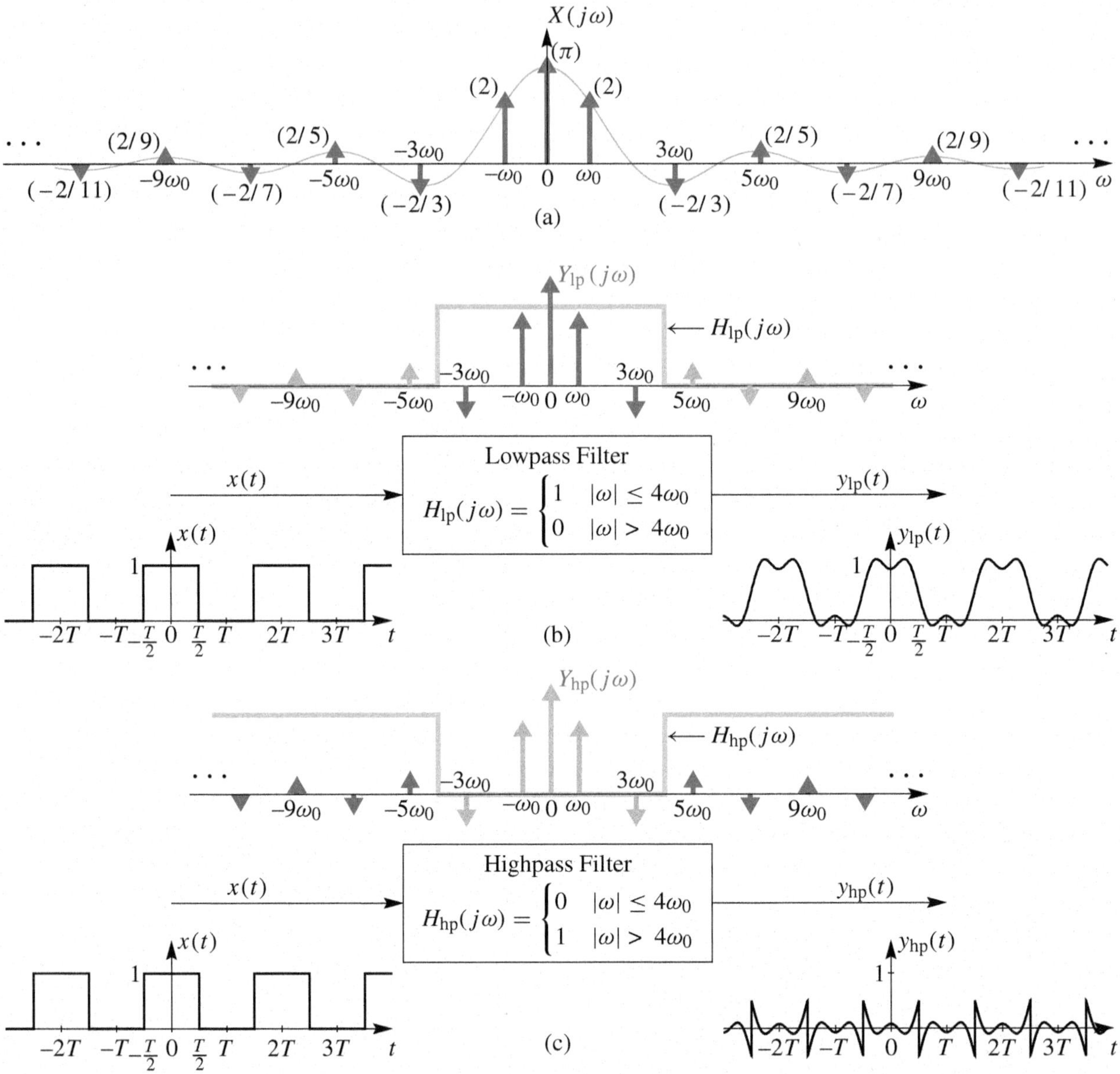

Figure 12-9: Example of ideal lowpass and highpass filtering: (a) Fourier transform $X(j\omega)$ of the input square wave signal $x(t)$ showing impulses at multiples of ω_0. (b) Ideal LPF $H_{\text{lp}}(j\omega)$ processes $x(t)$ to give the output signal $y_{\text{lp}}(t)$. Frequency response of LPF is drawn on top of Fourier transform $X(j\omega)$. Impulses in $X(j\omega)$ removed by LPF are shown in gray. (c) Ideal highpass filtering of the input square wave signal $x(t)$. Frequency response of ideal HPF $H_{\text{hp}}(j\omega)$ is drawn on top of $X(j\omega)$. Impulses in $X(j\omega)$ removed by HPF are shown in gray.

which consists of impulses at integer multiples of ω_0 as depicted in Fig. 12-9(a). If this signal is the input to an ideal lowpass filter with cutoff frequency ω_{co} such that $3\omega_0 < \omega_{co} < 5\omega_0$, as depicted in Fig. 12-9(b), then the Fourier transform of the output contains five spectrum lines (impulses)

$$\begin{aligned}Y(j\omega) &= H_{lp}(j\omega)X(j\omega)\\ &= \pi\delta(\omega) + 2\delta(\omega-\omega_0) + 2\delta(\omega+\omega_0)\\ &\quad - \frac{2}{3}\delta(\omega-3\omega_0) - \frac{2}{3}\delta(\omega+3\omega_0) \qquad (12.16)\end{aligned}$$

The inverse Fourier transform of (12.16) is

$$\begin{aligned}y_{lp}(t) &= \frac{1}{2} + \frac{1}{\pi}e^{j\omega_0 t} + \frac{1}{\pi}e^{-j\omega_0 t} - \frac{1}{3\pi}e^{j3\omega_0 t} - \frac{1}{3\pi}e^{-j3\omega_0 t}\\ &= \frac{1}{2} + \frac{2}{\pi}\cos(\omega_0 t) - \frac{2}{3\pi}\cos(3\omega_0 t) \qquad (12.17)\end{aligned}$$

This output signal is plotted at the right side of Fig. 12-9(b). Notice that the effect of the ideal lowpass filter is to "smooth out" the sharp corners of the square wave. If we decrease ω_{co}, the output would be even smoother. If we increase ω_{co}, the output would eventually approach the input since the entire spectrum would be included. As we have seen in Chapter 3, however, the characteristic "overshoot" and "undershoot" will remain at the discontinuities of the square wave, even though the slope of the output signal at the original discontinuities will continuously increase as ω_{co} increases.

EXERCISE 12.2: For the signal in (12.15), determine a cutoff frequency of an ideal lowpass filter such that its output will be a constant

$$y_{lp}(t) = A \qquad -\infty < t < \infty$$

What is the constant A?

If the square wave is the input to an ideal highpass filter with cutoff frequency ω_{co} such that $3\omega_0 < \omega_{co} < 5\omega_0$, as depicted in Fig. 12-9(c), then the corresponding output signal is given by the waveform plot at the righthand side of Fig. 12-9(c). In this case, we see that the sharp edges of the input are "roughened" by the ideal highpass filter.

EXERCISE 12.3: Show that the output of the ideal highpass filter in Fig. 12-9(c) is equal to the difference between the input square wave and the output of the ideal lowpass filter in Fig. 12-9(b); i.e.,

$$\begin{aligned}y_{hp}(t) &= x(t) - y_{lp}(t)\\ &= x(t) - \left(\frac{1}{2} + \frac{2}{\pi}\cos(\omega_0 t) - \frac{2}{3\pi}\cos(3\omega_0 t)\right)\end{aligned}$$

where $x(t)$ is the input square wave.

EXERCISE 12.4: Sketch a plot of the output of the ideal highpass filter if the input is the square wave in Fig. 12-9(c) and the cutoff frequency of the highpass filter is such that $0 < \omega_{co} < \omega_0$.

LAB: *#14, Design with Fourier Series*

12-1.5 Compensation for the Effect of an LTI Filter

Frequency selective filters are designed to remove certain frequencies while keeping others unmodified. Another type of filter would be designed to *compensate* for effects that can be modeled as LTI systems. Often a signal $x(t)$ is only observable as the output of an LTI system. This is true, for example, when a measurement instrument distorts the signal that it measures. Another example occurs when an audio signal is recorded in a reverberant room. In such a situation, the LTI system would model

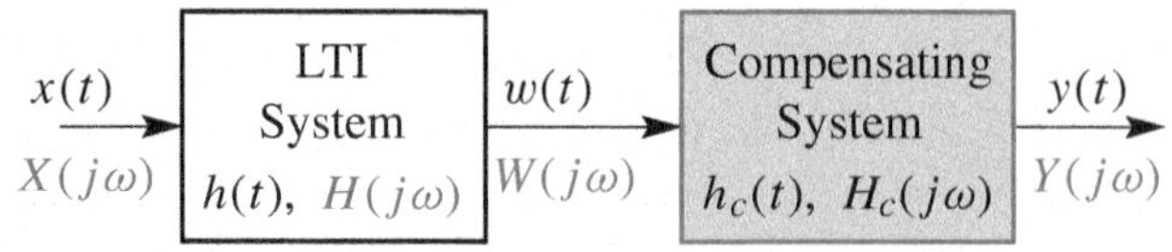

Figure 12-10: Compensation of an LTI system by cascading with another LTI system.

the impulse response of the room cascaded with the response of the microphone, amplifier, and recording system. The room impulse response would consist of shifted impulses that represent the time delays of echoes from the walls, floor and ceiling. In such cases, the LTI system represents a distortion that is imposed on the input signal, and the problem is to undo or compensate for the distortion if possible. We have already discussed a simple example of this type in Sections 9-9 and 11-10. In this section we will consider a somewhat more general formulation of the problem.

To see how we can compensate for the effects of an LTI system in some cases but not others, consider the cascade system shown in Fig. 12-10. The first system represents the distorting system having impulse response $h(t)$ and frequency response $H(j\omega)$. We wish to use the second system to modify the frequency response characteristics of the first system, $H(j\omega)$, so that the overall frequency response is some desired function $H_d(j\omega)$. From (12.1) it follows that the overall frequency response for the cascade system will be

$$H_d(j\omega) = H_c(j\omega)H(j\omega) \tag{12.18}$$

so if $H(j\omega)$ is given, we require that

$$H_c(j\omega) = \frac{H_d(j\omega)}{H(j\omega)} \tag{12.19}$$

If $H_c(j\omega)$ satisfies (12.19), then (12.18) will be true and we can say that the system $H(j\omega)$ has been *compensated perfectly*. Therefore, $H_c(j\omega)$ is the frequency response of the *compensator* system for $H(j\omega)$.

Notice that (12.18) implies that

$$h_d(t) = h(t) * h_c(t) \tag{12.20}$$

In this equation, we are given $h(t)$ and the desired impulse response $h_d(t)$, and we want to ***deconvolve*** to get $h_c(t)$. In general, solving (12.20) in the time-domain would be difficult, if not impossible. However, our frequency-domain solution (12.19) is simple because we have replaced convolution with multiplication. In fact, the frequency-domain solution was deceptively simple; we just divided one Fourier transform by another and assumed that the result was a *valid* Fourier transform. In many cases this is either not true, or it requires a more careful analysis. This point will be illustrated by some examples, but before we consider these examples, it is worthwhile mentioning the special case of ***inverse filtering***. In this case, we want to completely undo the effects of the system $H(j\omega)$, which means that we want the overall frequency response to be $H_d(j\omega) = 1$. Equivalently, we want $h_d(t) = \delta(t)$ so that $Y(j\omega) = X(j\omega)$ in Fig. 12-10, and also $y(t) = x(t)$. If we substitute this desired frequency response into (12.19), we obtain

$$H_c(j\omega) = \frac{1}{H(j\omega)} = H_i(j\omega) \tag{12.21}$$

In this special case, the compensating system is an exact inverse of the given system, and in the frequency-domain, the frequency response of the compensator is the reciprocal of $H(j\omega)$. The existence of an inverse system is not always guaranteed. We need to test $H_i(j\omega)$ to determine if it is a valid Fourier transform of a causal and stable impulse response $h_i(t)$. We have already illustrated inverse filtering for the case of an echo system via time-domain analysis in Sections 9-9 and via frequency-domain analysis in Section 11-10. Some additional examples will illustrate specific points about inverse filtering.

Example 12-2: Find an Inverse Filter

A simple example for which it is straightforward to find the inverse system is when the first system has impulse response $h(t) = e^{-at}u(t)$ with $a > 0$. In this case, the frequency response of the first system is

$$H(j\omega) = \frac{1}{a + j\omega}$$

and

$$H_i(j\omega) = \frac{1}{H(j\omega)} = a + j\omega$$

This is a valid Fourier transform, which corresponds to the impulse response

$$h_i(t) = a\delta(t) + \frac{d\delta(t)}{dt}$$

■

EXERCISE 12.5: Verify by convolution that for the system in Example 12-2,

$$h(t) * h_i(t) = e^{-at}u(t) * \left(a\delta(t) + \frac{d\delta(t)}{dt}\right) = \delta(t)$$

Example 12-3: No Inverse for Ideal LPF

The ideal lowpass filter, whose frequency response is

$$H_{\text{lp}}(j\omega) = \begin{cases} 1 & -\omega_{\text{co}} \le \omega \le \omega_{\text{co}} \\ 0 & \text{otherwise} \end{cases}$$

does not have an inverse since $H_{\text{lp}}(j\omega) = 0$ for $|\omega| > \omega_{\text{co}}$ and

$$H_i(j\omega) = \frac{1}{H(j\omega)}$$

This would require that $H_i(j\omega) = \infty$ for all $|\omega| > \omega_{\text{co}}$. This is clearly not a valid Fourier transform under any circumstances, so the inverse system does not exist for the ideal lowpass filter. This is not a surprising result when we recall that the ideal lowpass filter multiplies the Fourier transform of all input signals by zero for $|\omega| > \omega_{\text{co}}$. Therefore, it would be impossible to uniquely recover the portion of the input Fourier transform for frequencies above the cutoff frequency of the ideal lowpass filter. ■

Example 12-4: Inverse of Time Delay

The ideal time-delay system has frequency response of

$$H(j\omega) = e^{-j\omega t_d}$$

which means that the inverse system has frequency response of

$$H_i(j\omega) = \frac{1}{e^{-j\omega t_d}} = e^{j\omega t_d}$$

This is a valid Fourier transform, which corresponds to the impulse response

$$h_i(t) = \delta(t + t_d)$$

That is, the effect of the delay system can be compensated exactly if we can *advance* the output of the delay system by the amount of the delay. In a practical sense, the problem with this inverse system is that it is not causal and hence could not be implemented. ■

As the above examples show, determining an inverse system is full of difficulties. In cases like the ideal delay, the inverse system is noncausal. In some cases, like ideal frequency selective filters, an inverse system simply does not exist. In other cases, like the running average filter, which has zeros at isolated frequencies rather than over a contiguous band, a causal inverse system can be found, but it will have infinite gain at all the zeros of the original frequency response, and its gain will increase with frequency. In a practical setting, large gain at high frequencies will amplify any high-frequency noise that was added to the signal before it can be sent through the inverse filter. Such noise is almost always present and when magnified by the inverse filter, it makes the overall output useless. Even with all these difficulties in finding inverse filters, the general problem of recovering the input to an LTI system arises in many applications. Thus, much work has been done to develop approximate approaches that are less sensitive to noise and that do not require exact knowledge of the distorting LTI system. Unfortunately, these methods are beyond the scope of this introductory text, but the concept of inverse filtering provides a valuable example of this class of signal processing problems.

12-2 Sinewave Amplitude Modulation

In Section 12-1 we have explained linear filtering by exploiting the convolution property of Fourier transforms which states that convolution in the time-domain corresponds to multiplication in the frequency-domain. Now we will turn our attention to the dual result that multiplication in the time-domain corresponds to convolution in the frequency-domain. In the important case where one of the multiplied signals is periodic, the multiplication property implies that the spectrum of the other signal is shifted in frequency. In Chapter 11, we indicated that the frequency shift property would be useful for describing a class of communication

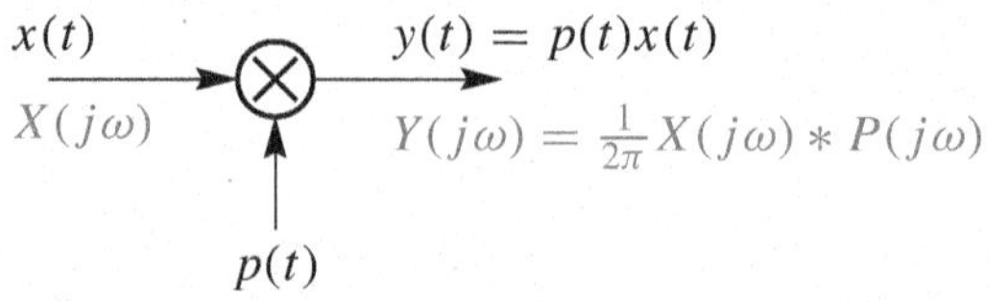

Figure 12-11: Block diagram representation of a general amplitude modulator.

systems called amplitude modulators. The time-domain and frequency-domain representations of the general amplitude modulator system are shown again in Fig. 12-11. In this section, we will consider the important special case when $p(t)$ is a sinusoidal signal.

12-2.1 Double-Sideband Amplitude Modulation

When $p(t) = \cos(\omega_c t)$, the amplitude modulator system is described by the input/output relation

DSBAM Definition

$$y(t) = x(t)\cos(\omega_c t) \tag{12.22}$$

which is depicted in Fig. 12-12. For reasons that will soon be clear, the system of Fig. 12-12 is called a *double-sideband amplitude modulator* (DSBAM). In

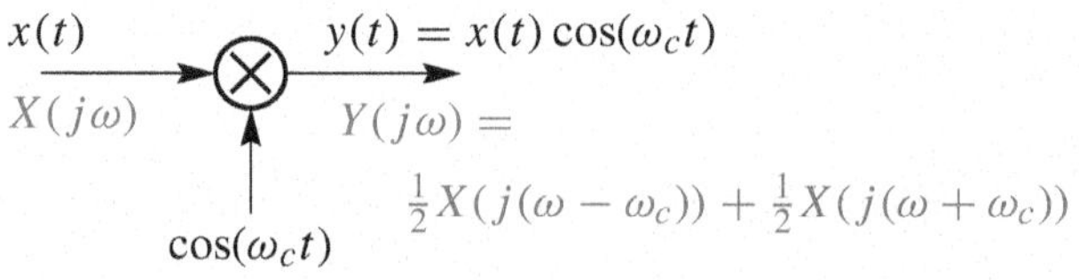

Figure 12-12: Block diagram representation of a sinusoidal amplitude modulator.

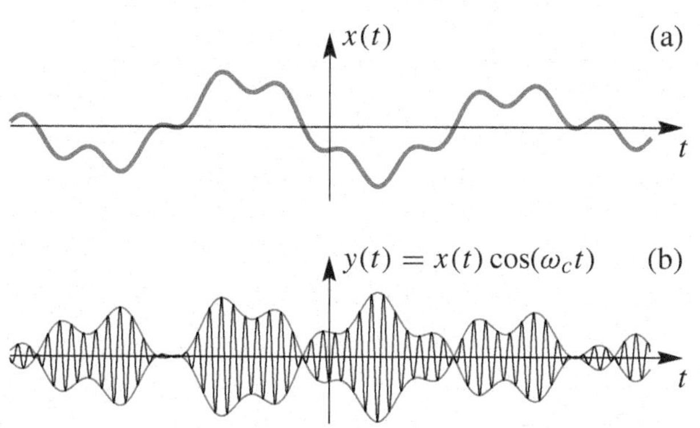

Figure 12-13: Waveform of double-sideband amplitude modulation: (a) slowly varying input waveform $x(t)$, and (b) corresponding DSBAM waveform $y(t) = x(t)\cos(\omega_c t)$.

this context, $\cos(\omega_c t)$ is called the ***carrier***, ω_c the ***carrier frequency***, and $x(t)$ the ***modulating signal***. When $x(t)$ is slowly varying compared to the carrier $\cos(\omega_c t)$, (12.22) has the simple interpretation that $x(t)$ modulates (varies) the amplitude of the carrier $\cos(\omega_c t)$. Figure 12-13 illustrates the nature of the DSBAM time-domain waveform for a slowly varying $x(t)$. In the plot of Fig. 12-13(b), notice that the carrier oscillates much faster than the modulating signal. This allows us to see an ***envelope*** which is a smooth curve through the peaks of $y(t)$. The envelope displays the variation of $|x(t)|$, not $x(t)$, because whenever $x(t)$ is negative, a phase shift of π radians is introduced. This phase change can be seen in Fig. 12-13(b) at the time instants where $x(t)$ changes sign.

EXERCISE 12.6: Show that the amplitude modulator is a linear system. In addition, show that it is *not* a time-invariant system.

The frequency-domain representation of the amplitude modulator system is derived by expressing $y(t)$ as

$$\begin{aligned} y(t) &= x(t)\cos(\omega_c t) \\ &= x(t)\left(\frac{e^{j\omega_c t} + e^{-j\omega_c t}}{2}\right) \\ &= \tfrac{1}{2}x(t)e^{j\omega_c t} + \tfrac{1}{2}x(t)e^{-j\omega_c t} \end{aligned} \qquad (12.23)$$

Applying the frequency shift property to each term of (12.23) leads to

$$Y(j\omega) = \tfrac{1}{2}X(j(\omega - \omega_c)) + \tfrac{1}{2}X(j(\omega + \omega_c)) \qquad (12.24)$$

That is, the Fourier transform of the output of the sinusoidal modulator is composed of two frequency-shifted copies of the Fourier transform of the input.

EXERCISE 12.7: Derive the DSBAM frequency-domain representation in (12.24) using the general signal multiplication property of the Fourier transform. This will require convolution of $X(j\omega)$ with the Fourier transform of $\cos(\omega_c t)$.

The DSBAM frequency-domain description is illustrated Fig. 12-14. Figure 12-14(a) depicts a "typical" bandlimited Fourier transform $X(j\omega)$. It is typical in the sense that $X(j\omega) = 0$ for $|\omega| > \omega_b$. For now, we are not particularly concerned with the exact shape of the function in the band $|\omega| < \omega_b$, so we are going to use a triangle shape since it is easy to draw. The main point is that scaled copies of $X(j\omega)$ are shifted to the locations $\omega = \pm\omega_c$. If $X(j\omega)$ is bandlimited as shown in Fig. 12-14(a), then the copies $\frac{1}{2}X(j(\omega \pm \omega_c))$ do not overlap when they are added. Therefore, they retain all the information in the original signal in a clearly recognizable form in Fig. 12-14(b).[2] The condition

[2]Figures like Fig. 12-14 are very useful in understanding modulation systems, so it is a good idea to become familiar with making sketches in this style; i.e., with $X(j\omega)$ represented by a "typical" bandlimited function such as a triangle.

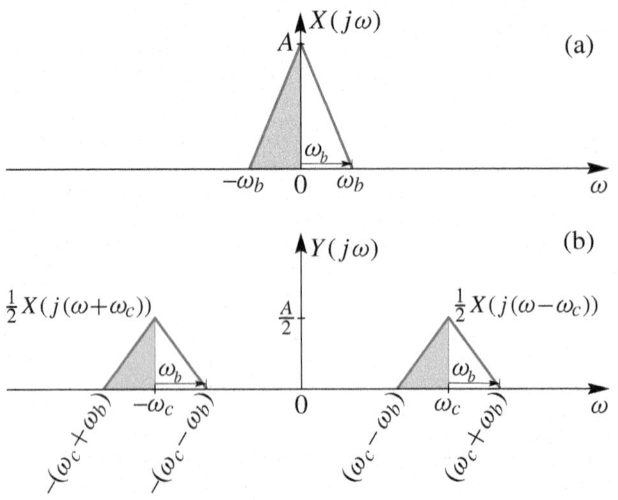

Figure 12-14: Illustration of sinusoidal amplitude modulation: (a) "Typical" Fourier transform of an input, and (b) corresponding Fourier transform of an output.

that guarantees that the two copies do not overlap is $(\omega_c - \omega_b) > 0$ or $\omega_c > \omega_b$. Note that we have shaded the negative frequency part of $X(j\omega)$ so that we can follow where each part of the spectrum has moved. The transform in Fig. 12-14(a) is called a *baseband* spectrum because it is centered at $\omega = 0$.

Figure 12-14(b) can be used to explain why this type of modulation is called double-sideband amplitude modulation. The baseband spectrum has frequency components on the positive side and also the negative side—these are called ***sidebands***. When the amplitude modulation process shifts $X(j\omega)$ to ω_c, both sidebands are moved up to ω_c. The lower sideband corresponds to the negative frequencies of $X(j\omega)$ and upper sideband corresponds to the positive frequencies of $X(j\omega)$. Likewise, both sidebands are shifted down to $-\omega_c$. As a result, the spectrum in Fig. 12-14(b) occupies twice as much frequency width as the original baseband spectrum. This wastes space in the frequency spectrum, but DBSAM is popular because it can be received with a very simple system (Fig. 12-15). On the other hand, it is possible to create a single sideband system (SSBAM) that does not require extra spectral width. See Problem P-12.4 on p. 381.

Before continuing with our discussion of amplitude modulation, we should address the obvious question of why we would be interested in modulating a sinusoid with another signal $x(t)$. The quick answer is that baseband signals do not propagate very far. If $x(t)$ is a voice signal, we know that it can only be heard for a short distance, even if we were to shout. However, with DSBAM, that same voice signal can be converted to a very high carrier frequency and transmitted to a remote location using electromagnetic (radio) waves. For example, we could take a lowpass filtered voice signal, whose spectrum is concentrated at frequencies below 5 kHz, and create a signal whose spectrum is concentrated in a 10 kHz band around a much higher frequency such as 750 kHz which is an assignable AM broadcast frequency. In its high frequency radio form, the signal can sometimes propagate for thousands of miles. Since amplitude modulation moves the frequency spectrum of the signal to a new band of frequencies, and since the original Fourier transform representation appears to be preserved in $Y(j\omega)$, it should be possible to recover the original signal from $y(t)$ by converting the signal back to baseband. A second virtue of the modulated form of the signal is that a single propagation channel can be shared by a multiplicity of bandlimited signals if different carrier frequencies are assigned to different broadcasters. We will discuss this shared frequency system further in Section 12-2.3, but for now we will return to our analysis of DSBAM.

The process of recovering the original signal $x(t)$ from the amplitude modulation signal $y(t)$ is called ***demodulation***. The possibility of signal recovery is what makes radio broadcasting systems feasible. In

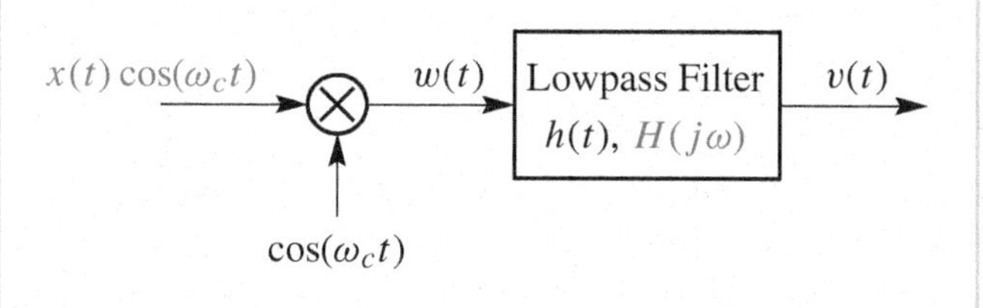

Figure 12-15: Block diagram of sinusoidal demodulator. The input to the system is the transmitted signal $y(t) = x(t)\cos(\omega_c t)$. The multiplier is called a *mixer*.

order to recover $x(t)$, we must re-center the Fourier transform copies on zero frequency. This can be done by multiplying again with $\cos(\omega_c t)$ and then filtering the result, as shown in Fig. 12-15. To see how the system of Fig. 12-15 works, note that since $w(t) = y(t)\cos(\omega_c t)$, it follows that $W(j\omega)$ has two copies of $Y(j\omega)$

$$W(j\omega) = \tfrac{1}{2}Y(j(\omega-\omega_c)) + \tfrac{1}{2}Y(j(\omega+\omega_c))$$

Substituting (12.24) for $Y(j\omega)$ shows that $W(j\omega)$ has *three copies* of $X(j\omega)$

$$\begin{aligned} W(j\omega) &= \\ &\tfrac{1}{2}\left[\tfrac{1}{2}X(j(\omega-\omega_c-\omega_c)) + \tfrac{1}{2}X(j(\omega-\omega_c+\omega_c))\right] \\ &+ \tfrac{1}{2}\left[\tfrac{1}{2}X(j(\omega+\omega_c-\omega_c)) + \tfrac{1}{2}X(j(\omega+\omega_c+\omega_c))\right] \\ &= \tfrac{1}{4}X(j(\omega-2\omega_c)) + \tfrac{1}{2}X(j\omega) + \tfrac{1}{4}X(j(\omega+2\omega_c)) \end{aligned} \quad (12.25)$$

Figure 12-16(a) shows $Y(j\omega)$ and Fig. 12-16(b) shows $W(j\omega)$ for the example illustrated in Fig. 12-14. Note that we have obtained a baseband term, $\frac{1}{2}X(j\omega)$, in $W(j\omega)$. If the condition $(2\omega_c - \omega_b) > \omega_b$ is satisfied, or equivalently $\omega_c > \omega_b$, it will be possible to extract $X(j\omega)$ with an ideal lowpass filter of the form

$$H(j\omega) = \begin{cases} G & -\omega_{co} \le \omega \le \omega_{co} \\ 0 & \text{otherwise} \end{cases} \quad (12.26)$$

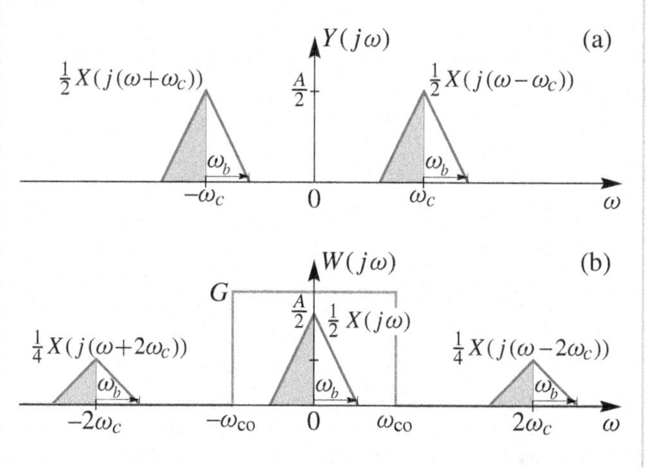

Figure 12-16: Illustration of demodulation of sinusoidal amplitude modulation: (a) "Typical" Fourier transform of DSBAM signal $Y(j\omega)$. (b) Corresponding Fourier transform of output of the mixer, $W(j\omega)$. The gray rectangular box is the frequency response of the ideal lowpass filter required to extract $X(j\omega)$.

where the gain of the LPF should be $G = 2$ and the cutoff frequency should satisfy $\omega_b < \omega_{co} < 2\omega_c - \omega_b$. Notice that this entire explanation has been carried out in the frequency-domain to obtain $V(j\omega) = X(j\omega)$, but it follows that $v(t) = x(t)$.

While we have shown that we can modulate and demodulate a signal $x(t)$ on a sinusoidal carrier, we have overlooked numerous practical considerations. Most important is the fact that it is extremely difficult to synchronize the frequencies and phases of two sinusoidal oscillators that are not in the same physical location. This is illustrated by the modulation/demodulation system in Fig. 12-17, which shows the general case where the carrier frequencies at the modulator (in the transmitter) and the mixer (in the receiver) are different; also there could be a relative phase shift of ϕ between the two carriers. To illustrate the problems inherent in Fig. 12-17, we will state two results in the form of exercises.

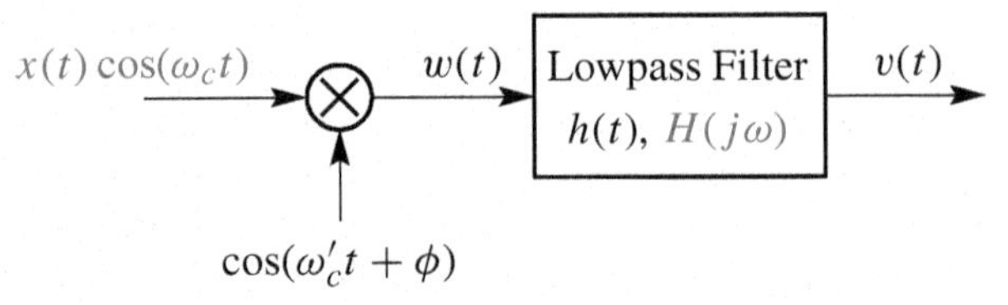

Figure 12-17: Block diagram of a demodulator whose mixer differs in frequency and phase from the transmitter. The input to the mixer is the transmitted signal $y(t) = x(t)\cos(\omega_c t)$.

EXERCISE 12.8: Consider the special case of a bandlimited input where $X(j\omega) = 0$ for $|\omega| > \omega_b$, and $\omega_c > \omega_b$. If we have no frequency difference between the transmitter and receiver, i.e., $\omega_c = \omega_c'$, but do have a phase difference, $\phi \neq 0$, show that $v(t)$ in Fig. 12-17 is

$$v(t) = x(t)\cos\phi \qquad (12.27)$$

The result stated in Exercise 12.8 says that if the carriers in Fig. 12-17 have the same frequency, but are out of phase, the output will be proportional to $x(t)$, but with an amplitude scale factor that depends on ϕ. For example, if $\phi = 0$ the carriers are exactly synchronized and we obtain the best result, i.e., $v(t) = x(t)$. However, if $\phi = \pi/2$ we get no output, i.e., $v(t) = 0$, which is clearly unacceptable. In some situations, the phase difference might vary with time so that the amplitude would fade in and out depending on the momentary value of ϕ.

EXERCISE 12.9: Show that if $\phi = 0$, but $\omega_c - \omega_c' = \Delta$, then for bandlimited inputs such that $\omega_c > \omega_b$ the output of the demodulator is

$$v(t) = x(t)\cos(\Delta t) \qquad (12.28)$$

This result assumes that the cutoff frequency of the lowpass filter in (12.26) satisfies $\omega_{co} > \omega_b + \Delta$.

From (12.28) we see that if the carriers are not exactly matched in frequency, the Fourier transform of $x(t)$ is not shifted all the way back to where it originated. In the time-domain, this effect is easily described if Δ is small. This will be similar to the beat note phenomenon discussed in Chapter 3 in that the factor $\cos(\Delta t)$ will act like a slowly varying amplitude causing the signal $x(t)$ to fade in and out at a rate determined by Δ.

Equations (12.27) and (12.28) show that any attempt to demodulate using synchronized carriers has some serious practical limitations. For this reason, another form of amplitude modulation is used in the everyday AM broadcast system where it must be possible to build receivers very economically.

12-2.2 DSBAM with Transmitted Carrier (DSBAM-TC)

One way to avoid the problems of synchronous demodulation of AM signals is to transmit the carrier signal along with the modulated carrier. The input/output relation for a DSBAM system with transmitted carrier (DSBAM-TC) is

DSBAM-TC Definition

$$y(t) = [C + x(t)]\cos(\omega_c t) \qquad (12.29)$$

The constant C is chosen to satisfy the relation $C \geq \max\{|x(t)|\}$, so that $C + x(t) \geq 0$ for all t. If we define $e(t) = C + x(t)$, then $e(t)$ is the envelope of the peaks of the modulated cosine wave because it is nonnegative. Figure 12-18 illustrates the DSBAM-TC modulation process in the time-domain. Figure 12-18(a) shows a signal $x(t)$, Fig. 12-18(b) shows the corresponding positive envelope function $e(t)$, and Fig. 12-18(c) shows the resulting modulator output $y(t)$. If we expand the expression (12.29) for $y(t)$, we obtain

$$y(t) = C\cos(\omega_c t) + x(t)\cos(\omega_c t) \qquad (12.30)$$

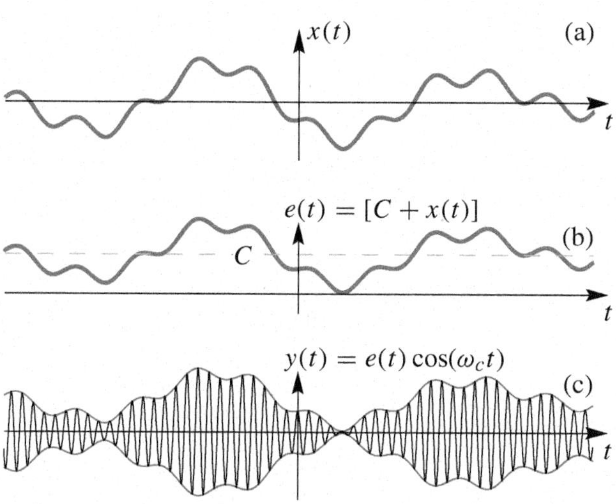

Figure 12-18: Illustrative waveform of double-sideband amplitude modulation with transmitted carrier: (a) Input signal $x(t)$, (b) envelope signal $e(t)$, and (c) DSBAM-TC output signal.

which shows that the DSBAM-TC signal contains a cosine of amplitude C plus a component that is identical to the DSBAM signal. This is why this type of modulation is called ***double-sideband amplitude modulation with transmitted carrier***. Notice that in this case, the signal $y(t)$ is nonzero even when $x(t) = 0$.

EXERCISE 12.10: Show that the DSBAM-TC modulation system defined by the input/output relation (12.29) is *not* a linear system. Also show that it is *not* a time-invariant system.

The Fourier transform of the envelope signal $e(t) = C + x(t)$ is

$$E(j\omega) = 2\pi C\delta(\omega) + X(j\omega) \qquad (12.31)$$

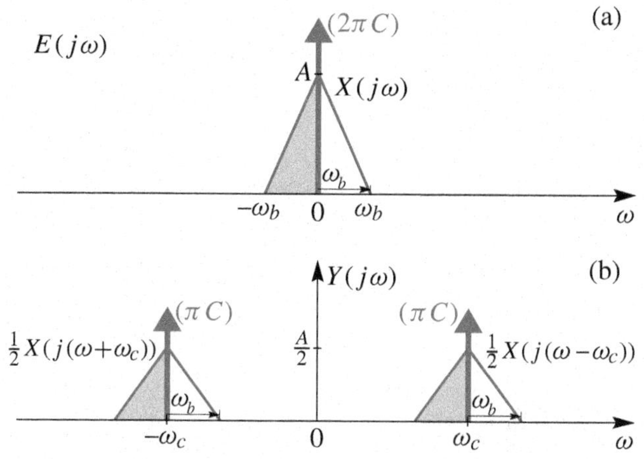

Figure 12-19: Illustration of double-sideband amplitude modulation with transmitted carrier: (a) Fourier transform of envelope $E(j\omega)$, and (b) Fourier transform of DSBAM-TC output signal.

From (12.29), the frequency shift property, and (12.31) it follows that

$$\begin{aligned} Y(j\omega) &= \tfrac{1}{2}E(j(\omega-\omega_c)) + \tfrac{1}{2}E(j(\omega+\omega_c)) \\ &= \pi C\delta(\omega-\omega_c) + \pi C\delta(\omega+\omega_c) \\ &\quad + \tfrac{1}{2}X(j(\omega-\omega_c)) + \tfrac{1}{2}X(j(\omega+\omega_c)) \end{aligned} \qquad (12.32)$$

Figure 12-19 depicts the Fourier transform of a typical transmitted carrier AM signal. Figure 12-19(a) shows $E(j\omega)$ as the sum of a typical bandlimited input transform plus the impulse due to the constant C, while Fig. 12-19(b) shows the frequency shifted spectrum of the DSBAM-TC signal.

EXERCISE 12.11: Show that if $y(t) = [C + x(t)]\cos(\omega_c t)$ is the input to the demodulator in Fig. 12-15, then the output would be $v(t) = C + x(t)$.

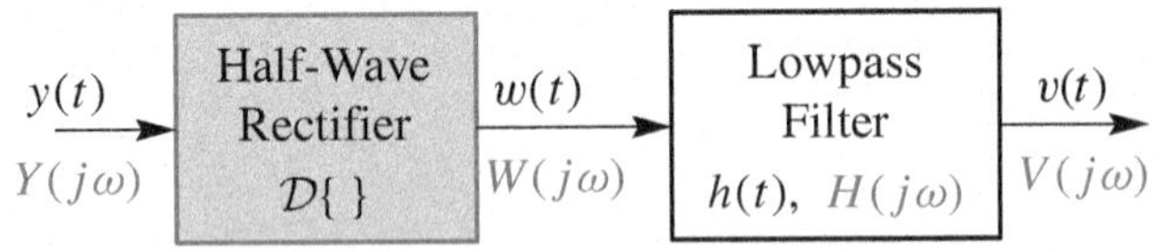

Figure 12-20: DSBAM-TC demodulator. The cascade of the half-wave rectifier and LPF functions as an *envelope detector*.

In Exercise 12.11, you can demonstrate that the envelope signal $e(t)$ in a DSBAM-TC signal could be demodulated by synchronous demodulation; however, this is what we are trying to avoid. By transmitting the carrier signal and ensuring that the envelope $e(t)$ is always non-negative, it becomes possible to demodulate with a much simpler system, such as the ***envelope detector*** in Fig. 12-20 consisting of a half-wave rectifier followed by a LPF. The half-wave rectifier system in Fig. 12-20 is defined by the relation

$$w(t) = \mathcal{D}\{y(t)\} = \begin{cases} y(t) & \text{if } y(t) > 0 \\ 0 & \text{if } y(t) \leq 0 \end{cases} \tag{12.33}$$

EXERCISE 12.12: Show that the half-wave rectifier is a nonlinear system.

The half-wave rectifier "chops off" the negative parts of the input waveform, leaving a positive signal. Since $e(t) \geq 0$ for all t, the envelope of the half-wave rectified signal is also the desired envelope function $e(t)$. This is illustrated in Fig. 12-21(a), which depicts the output of the half-wave rectifier for the DSBAM-TC signal shown in Fig. 12-18(c).

Now we undertake an analysis of the DSBAM-TC envelope detector to show how the Fourier transform can explain the inner workings of this system. The output of the half-wave rectifier for the DSBAM-TC wave is

$$w(t) = \mathcal{D}\{y(t)\} = \mathcal{D}\{e(t)\cos\omega_c t\} \tag{12.34}$$

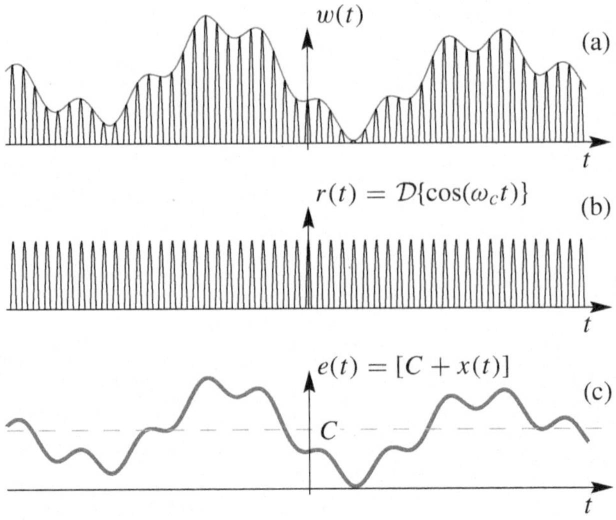

Figure 12-21: Illustrative waveforms in envelope detector: (a) Output of half-wave rectifier, $w(t)$, (b) half-wave rectified cosine $\mathcal{D}\{\cos\omega_c t\}$, and (c) envelope signal $e(t)$. The output of the lowpass filter will be proportional to $e(t)$.

Because $e(t) \geq 0$, $e(t)$ plays no role in determining the sign of $y(t)$, and (12.34) can be written as

$$w(t) = e(t)\mathcal{D}\{\cos\omega_c t\} = e(t)r(t) \tag{12.35}$$

where $r(t)$ is a periodic half-wave rectified cosine wave

$$r(t) = \mathcal{D}\{\cos\omega_c t\}$$

shown in Fig. 12-21(b). Figure 12-22 provides a more detailed plot of the half-wave rectified cosine wave. Since this signal is periodic with period $T_c = 2\pi/\omega_c$, we can write it in the form of a Fourier series

$$r(t) = \sum_{k=-\infty}^{\infty} a_k e^{jk\omega_c t} \tag{12.36}$$

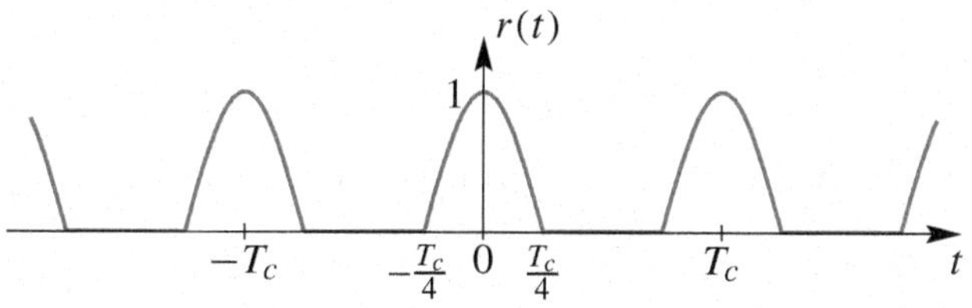

Figure 12-22: Detailed view of half-wave rectified cosine ($T_c = 2\pi/\omega_c$).

EXERCISE 12.13: Show that the Fourier coefficients for the half-wave rectified cosine wave are

$$a_k = \frac{1}{T_c}\int_{-T_c/4}^{T_c/4} \cos(\omega_c t)e^{-jk\omega_c t}dt = \begin{cases} \dfrac{\cos(\pi k/2)}{\pi(1-k^2)} & k \neq \pm 1 \\ \frac{1}{4} & k = \pm 1 \end{cases} \quad (12.37)$$

Now we can find the Fourier transform of the output of the half-wave rectifier. Using the Fourier series (12.37) for $r(t)$, we can write

$$w(t) = e(t)\sum_{k=-\infty}^{\infty} a_k e^{jk\omega_c t} = \sum_{k=-\infty}^{\infty} a_k e(t)e^{jk\omega_c t} \quad (12.38)$$

from which it follows by the frequency shift property that

$$W(j\omega) = \sum_{k=-\infty}^{\infty} a_k E(j(\omega - k\omega_c)) \quad (12.39)$$

Figure 12-23(a) shows a typical Fourier transform of a bandlimited envelope signal $e(t) = C + x(t)$. The impulse is the Fourier transform of the constant C and the triangle represents $X(j\omega)$, the Fourier transform of $x(t)$. Figure 12-23(b) is a plot of the Fourier transform $R(j\omega)$ using the Fourier coefficients from (12.37). Figure 12-23(c) depicts the corresponding Fourier transform $W(j\omega)$ of the output of the half-wave rectifier consisting of frequency-shifted scaled copies of $E(j\omega)$. Observe that if $\omega_c - \omega_b > \omega_b$, a condition that is easily met in the AM broadcast system, then $W(j\omega)$ contains a perfect (scaled) copy of $E(j\omega)$, which can be extracted using a LPF $H(j\omega)$ with gain $G = 1/a_0 = \pi$ and cutoff frequency such that $\omega_b < \omega_{co} < \omega_c - \omega_b$. An example of an appropriate $H(j\omega)$ is shown in Fig. 12-23(c) as a gray rectangular box. Under these

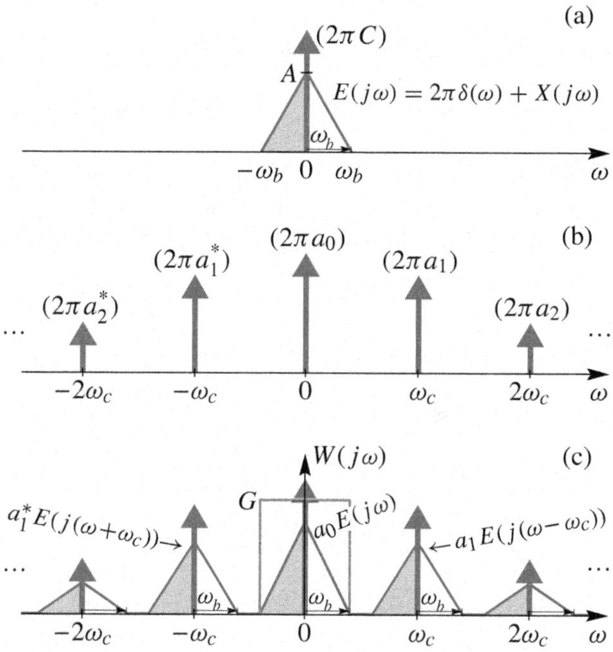

Figure 12-23: Fourier transforms in DSBAM-TC demodulation: (a) Fourier transform $E(j\omega)$ of envelope, (b) Fourier transform $R(j\omega)$ of half-wave rectified cosine, and (c) Fourier transform $W(j\omega)$ of half-wave rectified DSBAM-TC waveform.

conditions, the Fourier transform of the output of the lowpass filter in Fig. 12-20 will be $V(j\omega) = E(j\omega)$ so the output time function will be $v(t) = e(t) = C + x(t)$. Thus the envelope detector can recover the input signal $x(t)$ plus an additive constant.

In a practical setting, the combination of half-wave rectifier and lowpass filter is generally approximated very well by an electronic circuit consisting only of a diode, a resistor, and a capacitor. For this reason, AM receivers can be made very economically, and this is why DSBAM-TC has been used for many years in the AM broadcast system. There is a penalty, however, because DSBAM-TC requires extra power to transmit the carrier signal.

12-2.3 Frequency Division Multiplexing

In Sections 12-2.1 and 12-2.2 we showed that the Fourier spectrum of a bandlimited signal can be shifted by amplitude modulation to a higher frequency band as required for radio transmission of voice, music, or data. A second, and perhaps equally important aspect of this frequency shift is that by choosing several carrier frequencies such that the shifted spectra of two or more bandlimited signals do not overlap, we can transmit a multiplicity of signals over the same channel (e.g., the atmosphere, outer space, or fiber-optic cable). This stacking of signals in the frequency-domain is called ***frequency-division multiplexing*** or ***FDM***.[3] The essence of frequency-division multiplexing is depicted in Fig. 12-24, which shows two bandlimited signals that modulate different sinusoidal carriers of frequencies ω_{c1} and ω_{c2}. The pair of DSBAM signals $y_1(t) = x_1(t)\cos(\omega_{c1}t)$ and $y_2(t) = x_2(t)\cos(\omega_{c2}t)$ are superimposed (added) and so their Fourier transforms are added too.[4] The plot in Fig. 12-24(b) depicts the resulting Fourier transform $Y(j\omega) = Y_1(j\omega) + Y_2(j\omega)$. Note that the copies of the original bandlimited Fourier transforms do not overlap, as long as we obey the condition that $\omega_{c2} > \omega_{c1} + 2\omega_b$.

The reason that we require no overlap of the shifted frequency spectra of the input signals is that we want to be able to recover each of the individual signals $y_1(t)$ and $y_2(t)$ from the sum. The process of sorting out the individual signals is called ***demultiplexing***. If the shifted spectra do not overlap, demultiplexing can be accomplished with bandpass filters as in Fig. 12-24(c). For example, to recover the signal $x_1(t)$, we need the bandpass filter in Fig. 12-25(a), which selects the channel containing the spectral information for $y_1(t)$. Once $y_1(t)$ has been extracted from $y(t)$ by bandpass filtering, we can demodulate it with an appropriate demodulator. Figure 12-24 assumes DSBAM modulation, but if DSBAM-TC is used, then the demodulator would be an envelope detector. In Fig. 12-24(c), the second channel is demultiplexed by the filter in Fig. 12-25(b) followed by the DSBAM demodulator.

The system of Fig. 12-24 can be generalized to more than two channels simply by including more carrier frequencies that are far enough apart. The carrier frequencies of an FDM system for multiplexing two or more signals each with highest frequency ω_b (using amplitude modulation) must satisfy the condition

$$\omega_{c(k+1)} > \omega_{ck} + 2\omega_b$$

In the government regulated broadcast spectrum, portions of the frequency-domain are set aside for specific applications and modulation techniques. For example, the range of carrier frequencies for the AM broadcast system is 530 kHz to 1620 kHz with the channels (AM radio stations) separated by 10 kHz. This means that the input voice and music signals must be bandlimited

[3] In optics, the term wavelength-division multiplexing (WDM) is equivalent to FDM.

[4] In radio broadcasting, the signals are naturally superimposed because each station radiates its signal into the atmosphere. In addition, we are assuming that the transmission medium is linear, so the signals add together.

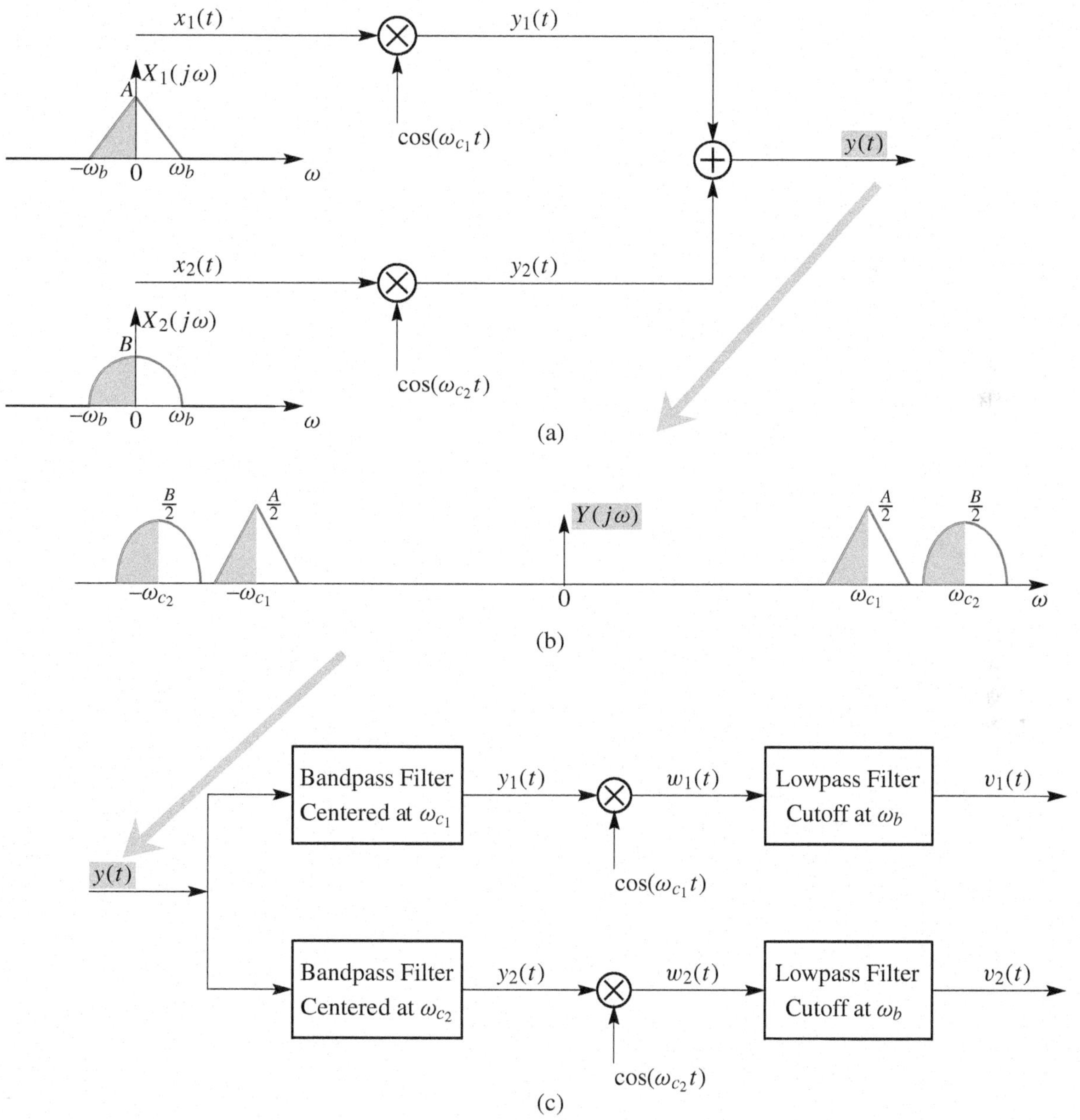

Figure 12-24: Frequency-division multiplexing. The carrier frequencies must be far enough apart so that the frequency shifted spectra in $Y(j\omega)$ do not overlap.

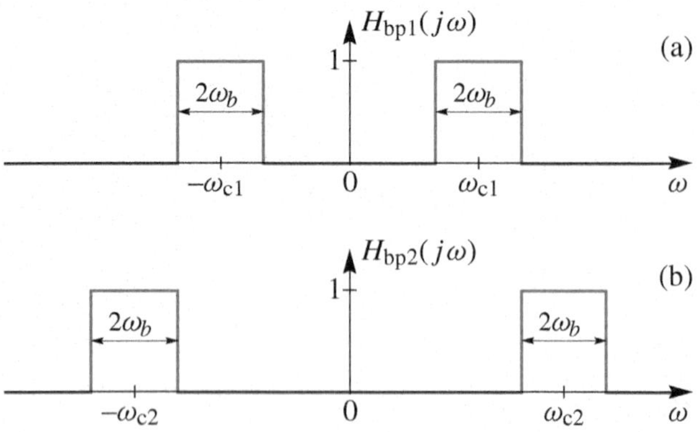

Figure 12-25: Bandpass filters for demultiplexing: (a) Channel centered on ω_{c1}, and (b) channel centered on ω_{c2}.

to nominally 5 kHz before modulation. In the case of FM broadcasting, the carrier frequencies range from 88.1 MHz to 107.9 MHz, and the input signals to an FM modulator are bandlimited to 15 kHz. The FM modulation process expands the bandwidth in addition to shifting the resulting spectrum to be centered on the carrier frequency. Therefore, it is necessary to separate FM channels by 200 kHz so that the broadened spectra do not overlap and they can be demultiplexed by bandpass filtering. In broadcast applications such as these, a listener only needs to "tune in" to one channel at a time. Therefore, the demultiplexer generally consists of a single bandpass filter whose center frequency can be adjusted to the desired carrier frequency.

LAB: *#16 AM Communication System*

12-3 Sampling and Reconstruction

At this point, we have sufficient understanding of the Fourier transform to undertake a thorough discussion of sampling and reconstruction of bandlimited signals. As in the case of sinusoidal modulation, we will rely heavily on the multiplication, frequency shift, and convolution properties of the continuous-time Fourier transform.

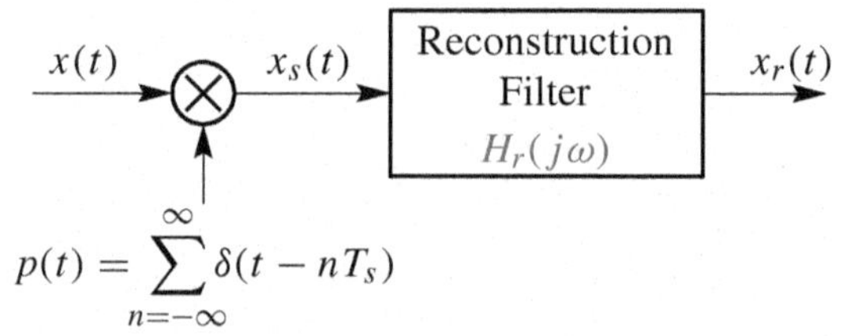

Figure 12-26: Block diagram of periodic impulse train modulation followed by signal reconstruction.

12-3.1 The Sampling Theorem and Aliasing

The Shannon/Nyquist sampling theorem was originally presented in Chapter 4. Now we will derive this famous theorem, but we will take a somewhat indirect approach. First of all, we must show that sampling is equivalent to amplitude modulation of a periodic impulse train. The signal being sampled, $x(t)$ in Fig. 12-26, multiplies the impulse train $p(t)$ to produce a new signal $x_s(t)$.

$$\begin{aligned}
x_s(t) &= x(t)p(t) \\
&= x(t) \sum_{n=-\infty}^{\infty} \delta(t - nT_s) \qquad (12.40) \\
&= \sum_{n=-\infty}^{\infty} x(t)\delta(t - nT_s) \\
&= \sum_{n=-\infty}^{\infty} x(nT_s)\delta(t - nT_s) \qquad (12.41)
\end{aligned}$$

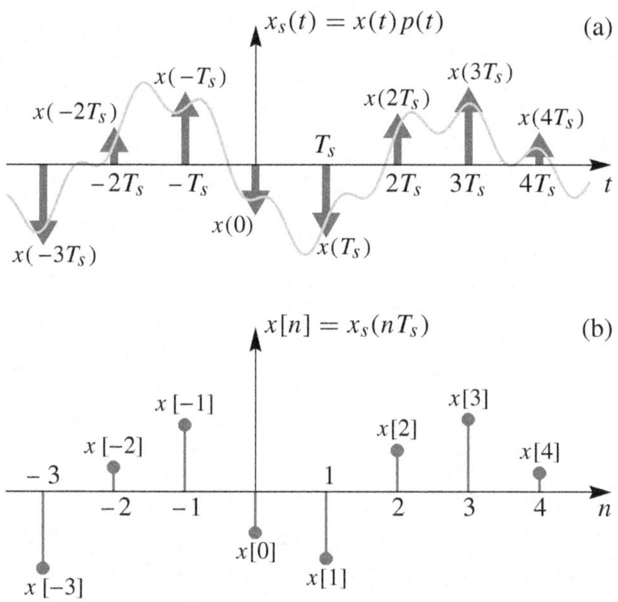

Figure 12-27: Illustration of sampling by modulating a periodic impulse train: (a) Signal $x(t)$ (solid line) and modulated impulse train $x_s(t)$, and (b) sequence of samples $x[n] = x(nT_s)$.

Equation (12.41) states that the periodic impulse train in Fig. 12-26 "samples" the signal $x(t)$ at the times at which the impulses occur. The resulting continuous-time signal $x_s(t)$ is an impulse train where the sizes (areas) of the impulses are the values of $x(t)$ at the times $t_n = nT_s$. This is depicted in Fig. 12-27(a). If we extracted the sample values from the impulses and arranged them in a sequence, we would have the discrete-time sequence of samples $x[n] = x(nT_s)$ depicted in Fig. 12-27(b). Thus, the signal $x_s(t)$ can be said to represent the result of sampling the continuous-time signal $x(t)$.

To obtain a frequency-domain representation of the sampled signal, we can begin by deriving an expression for the Fourier transform of the signal $x_s(t)$. To do this, we represent the periodic impulse train in terms of its Fourier series. The Fourier series coefficients for the periodic impulse train were shown in Example 11-4 on p. 321 to be $a_k = 1/T_s$ for $-\infty < k < \infty$. Therefore, we can write as an equivalent expression to (12.40),

$$x_s(t) = x(t) \sum_{k=-\infty}^{\infty} \frac{1}{T_s} e^{jk\omega_s t} = \frac{1}{T_s} \sum_{k=-\infty}^{\infty} x(t) e^{jk\omega_s t} \tag{12.42}$$

where $\omega_s = 2\pi/T_s$. Now we apply the frequency shift property of the Fourier transform to obtain

$$X_s(j\omega) = \frac{1}{T_s} \sum_{k=-\infty}^{\infty} X(j(\omega - k\omega_s)) \tag{12.43}$$

Example 12-5: Use of Multiplication Property

The result in (12.43) could also be derived by applying the multiplication property of Fourier transforms to (12.40). Since it was shown in Example 11-4 that

$$P(j\omega) = \sum_{k=-\infty}^{\infty} \left(\frac{2\pi}{T_s}\right) \delta(\omega - \omega_s)$$

and since $x_s(t) = x(t)p(t)$, it follows that

$$X_s(j\omega) = \frac{1}{2\pi} X(j\omega) * \sum_{k=-\infty}^{\infty} \left(\frac{2\pi}{T_s}\right) \delta(\omega - k\omega_s)$$

$$= \frac{1}{T_s} \sum_{k=-\infty}^{\infty} X(j(\omega - k\omega_s))$$

which is the same result we obtained in (12.43). ■

Since the sampling rate is $\omega_s = 2\pi/T_s$, we can state the following important result:

> ***The Fourier transform of a periodic impulse train modulated by $x(t)$ is composed of periodically repeated copies of $X(j\omega)$.*** (12.44)
>
> $$X_s(j\omega) = \frac{1}{T_s}\sum_{k=-\infty}^{\infty} X(j(\omega - 2\pi k/T_s))$$

This periodic repetition of the spectrum is illustrated in Fig. 12-28. Figure 12-28(a) shows a typical bandlimited Fourier transform representing $X(j\omega)$. Figure 12-28(b) shows the periodic replication of $X(j\omega)$ according to (12.44). Note that Fig. 12-28(b) was constructed with the assumption that $\omega_s = 2\pi/T_s > 2\omega_b$, so the shifted copies of $X(j\omega)$ do not overlap. It can be seen that for the example shown in Fig. 12-28(b), we retain a perfect copy of $X(j\omega)$ (scaled by $1/T_s$) in the additive combination that constitutes the periodic transform function $X_s(j\omega)$.

Figure 12-28(c), on the other hand, illustrates the case where $\omega_s = 2\pi/T_s < 2\omega_b$. In this case, the negative frequencies of $X(j(\omega - \omega_s))$ overlap with the positive frequencies of $X(j\omega)$, etc. This is, of course, what we have called ***aliasing (folding) distortion***. By generalizing from Figs. 12-28(b,c), it can be seen that aliasing distortion can be avoided if the following two conditions hold:

1. The signal to be sampled must be bandlimited so that $X(j\omega) = 0$ for $|\omega| \geq \omega_b$.

2. The sampling frequency must satisfy the condition $\omega_s = 2\pi/T_s \geq 2\omega_b$.

As we will see in the next section, if these two conditions hold, we can reconstruct the signal uniquely from its samples using an ideal lowpass filter with gain T_s and cutoff frequency $\omega_s/2$. This is, of course, the essence of what we called the ***Shannon/Nyquist sampling theorem*** in Chapter 4.

12-3.2 Bandlimited Signal Reconstruction

To see how a bandlimited signal can be reconstructed from its samples, we return to Fig. 12-26, which shows the signal being reconstructed by an LTI lowpass filter with frequency response $H_r(j\omega)$

$$\begin{aligned} X_r(j\omega) &= H_r(j\omega)X_s(j\omega) \qquad (12.45)\\ &= H_r(j\omega)\frac{1}{T_s}\sum_{k=-\infty}^{\infty} X(j(\omega - k\omega_s)) \end{aligned}$$

The individual Fourier transforms in (12.45) are illustrated in Fig. 12-29, where it is assumed that the sampling rate is high enough to avoid aliasing distortion. If the frequency response of the filter is ideal

$$H_r(j\omega) = \begin{cases} T_s & |\omega| \leq \omega_s/2 = \pi/T_s \\ 0 & |\omega| > \omega_s/2 = \pi/T_s \end{cases} \qquad (12.46)$$

as depicted in Fig. 12-29(b), the filter will extract $X(j\omega)$ from the sum of shifted copies in Fig. 12-29(a) and give the final result $X_r(j\omega) = X(j\omega)$ in Fig. 12-29(c). Therefore, when the input signal is bandlimited and the sampling rate is at least twice the highest frequency of the input signal, then the reconstruction is perfect and we have $x_r(t) = x_s(t) * h_r(t) = x(t)$.

Example 12-6: Spectrum During Sampling

As an example of sampling and reconstruction of a bandlimited signal, consider

$$x(t) = \frac{1}{3\pi} + \frac{1}{3\pi}\cos(\pi t + \pi/2)$$

whose Fourier transform is

$$X(j\omega) = \frac{2}{3}\delta(\omega) + \frac{j}{3}\delta(\omega - \pi) - \frac{j}{3}\delta(\omega + \pi)$$

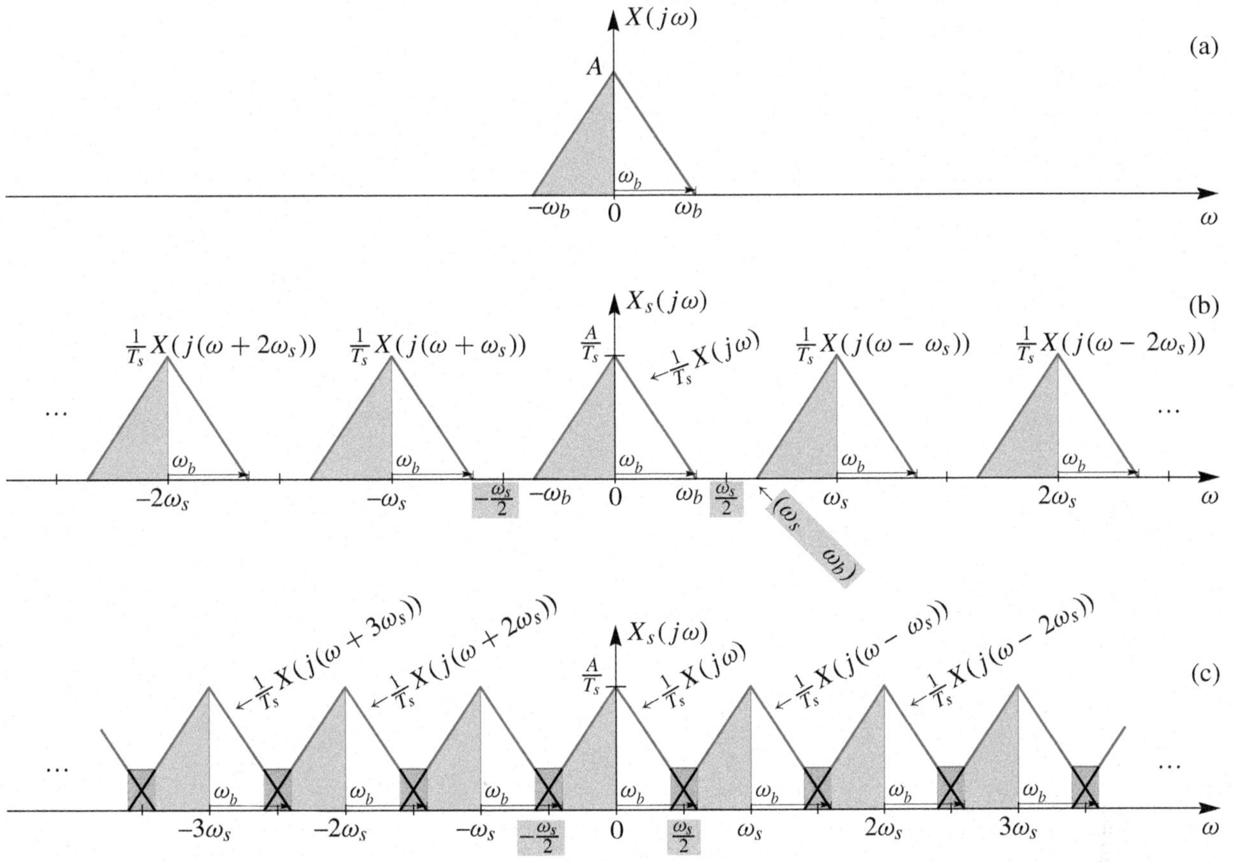

Figure 12-28: Illustration of sampling and aliasing in the frequency-domain. (a) Fourier transform of signal $x(t)$, (b) Fourier transform of sampled signal $x_s(t) = p(t)x(t)$ with $\omega_s > 2\omega_b$, and (c) Fourier transform of sampled signal $x_s(t) = p(t)x(t)$ showing aliasing distortion when $\omega_s < 2\omega_b$.

The highest frequency in this signal is π rad/sec, so any sampling rate such that $\omega_s = 2\pi/T_s > 2\pi$ will guarantee that we can exactly reconstruct the signal $x(t)$ from its samples $x[n] = x(nT_s)$. Figure 12-30(a) shows $X(j\omega)$ and Fig. 12-30(b) shows $X_s(j\omega)$ for $\omega_s = 6\pi$, or $T_s = \frac{1}{3}$ sec. The light colored box in Fig. 12-30(b) is the frequency response of the ideal reconstruction filter $H_r(j\omega)$ for a sampling rate of 6π rad/s; in other words, its cutoff frequency is 3π rad/s. Note that the product of $X_r(j\omega) = H_r(j\omega)X_s(j\omega)$ will be exactly equal to the original Fourier transform $X(j\omega)$ as expected. Therefore, the output of the ideal reconstruction filter in Fig. 12-30 will be

$$x_r(t) = x(t) = \frac{1}{3\pi} + \frac{1}{3\pi}\cos(\pi t + \pi/2)$$

■

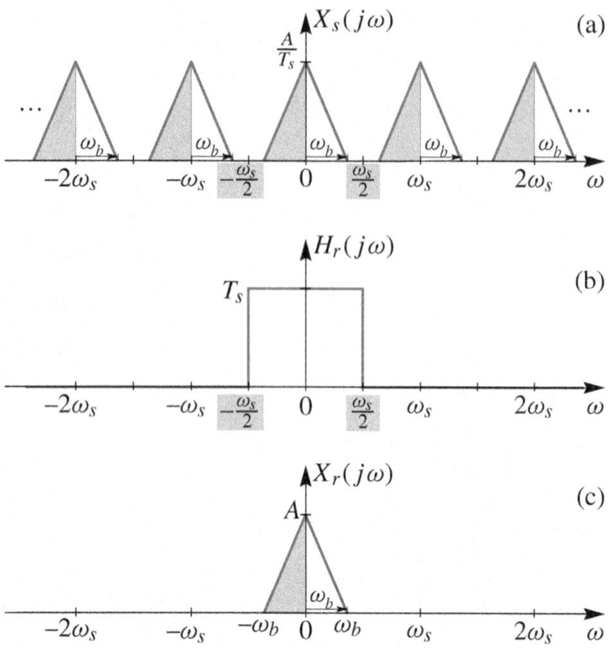

Figure 12-29: Illustration of frequency-domain representation of reconstruction from sampled signal: (a) Fourier transform of sampled signal $x_s(t)$, (b) ideal reconstruction filter, and (c) Fourier transform of output of ideal bandlimited reconstruction filter.

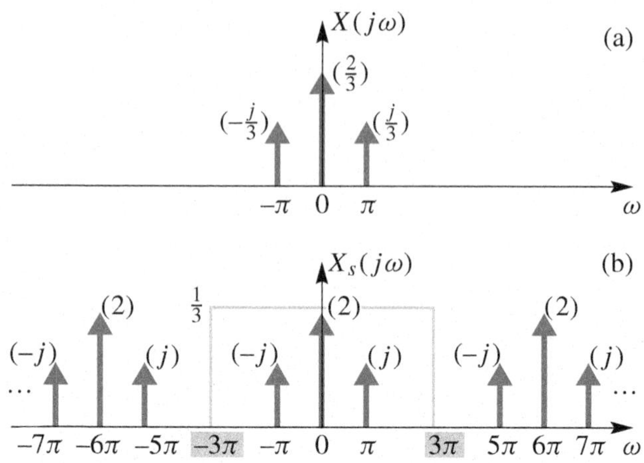

Figure 12-30: Illustration of frequency-domain representation of sampling: (a) Fourier transform of signal $x(t) = \frac{1}{3\pi} + \frac{1}{3\pi}\cos(\pi t + \pi/2)$, and (b) Fourier transform of sampled signal $x_s(t)$. Light colored box is the ideal reconstruction filter with cutoff frequency at $\omega = 3\pi$.

EXERCISE 12.14: Repeat Example 12-6 for the case of $\omega_s = 1.5\pi$. Plot $X_s(j\omega)$ as in Fig. 12-30(b) and give an equation for the signal that is reconstructed by the ideal reconstruction filter for $\omega_s = 1.5\pi$.

12-3.3 Bandlimited Interpolation

Figure 12-31(a) shows the impulse response (12.47) of the ideal reconstruction filter. The corresponding frequency response (12.46) is shown in Fig. 12-31(b).

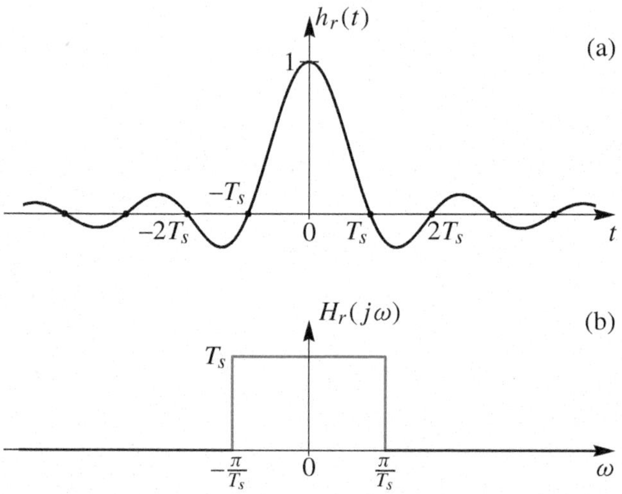

Figure 12-31: Ideal bandlimited reconstruction filter: (a) Impulse response $h_r(t)$, and (b) frequency response $H_r(j\omega)$.

Note that the ideal lowpass reconstruction filter has gain T_s and cutoff frequency $\omega_{co} = \omega_s/2 = \pi/T_s$. The impulse response, which is

$$h_r(t) = \frac{\sin \frac{\pi}{T_s} t}{\frac{\pi}{T_s} t} \tag{12.47}$$

has a maximum value of 1 at $t = 0$ and it is zero at nonzero integer multiples of T_s in Fig. 12-31(a). The time-domain expression for the output of the reconstruction filter is obtained by substituting (12.41) on p. 368 for $x_s(t)$ into

$$x_r(t) = x_s(t) * h_r(t)$$

This substitution leads to

$$\begin{aligned} x_r(t) &= x_s(t) * h_r(t) \\ &= \left(\sum_{n=-\infty}^{\infty} x(nT_s)\delta(t - nT_s) \right) * h_r(t) \\ &= \sum_{n=-\infty}^{\infty} x(nT_s) h_r(t - nT_s) \end{aligned} \tag{12.48}$$

That is, the continuous-time reconstructed signal $x_r(t)$ consists of an infinite sum of scaled and shifted impulse responses of the reconstruction filter, as discussed previously in Section 4-4 in Chapter 4. When (12.47) is substituted into (12.48), we obtain the following result:

> ***A bandlimited signal $x(t)$ such that $X(j\omega) = 0$ for $|\omega| \geq \omega_b$ can be reconstructed exactly from samples taken with sampling rate $\omega_s = 2\pi/T_s \geq 2\omega_b$ via the bandlimited interpolation formula:***
>
> $$x_r(t) = \sum_{n=-\infty}^{\infty} x(nT_s) \frac{\sin\left[\frac{\pi}{T_s}(t - nT_s)\right]}{\frac{\pi}{T_s}(t - nT_s)}$$ (12.49)

The process of bandlimited interpolation is illustrated in Fig. 12-32, which shows the impulse train with impulses $x(nT_s)\delta(t - nT_s)$ modulated by the sample values and the corresponding outputs $x(nT_s)h_r(t - nT_s)$ for each impulse.

Note that since $h_r(t)$ has infinite extent, the contribution from each sample overlaps everywhere with the contributions from all other samples, so that each sample value contributes to the reconstruction of $x(t)$ at all values of t. However, since $h_r(0) = 1$, and $h_r(t_n) = 0$ for $t_n = \pm T_s, \pm 2T_s, \ldots$, it is always true that $x_r(nT_s) = x(nT_s)$ regardless of whether aliasing distortion occurred in the original sampling. If no aliasing distortion has occurred, then our frequency-domain argument shows that (12.49) recreates the original signal perfectly for all values of t.

12-3.4 Ideal C-to-D and D-to-C Converters

The block diagram of Fig. 12-26 has been a convenient basis for an analysis that has demonstrated the truth of the Shannon/Nyquist sampling theorem. Furthermore, it has led us to a method of reconstructing a bandlimited signal from samples of a continuous-time signal. This block diagram is only loosely related to how sampling and signal reconstruction are actually implemented in practice. A representation that is closer to reality, but nevertheless still idealized, is shown in Fig. 12-33. In Fig. 12-33(a) the ***Ideal Continuous-to-Discrete (C-to-D) Converter*** is represented as an impulse train modulator followed by a system that simply lifts the sample values off the impulses and places them in a sequence. That is, the ideal C-to-D converter is defined by the input/output relation

> ***The Ideal C-to-D Converter***
>
> $x[n] = x(nT_s)$ for n an integer. (12.50)

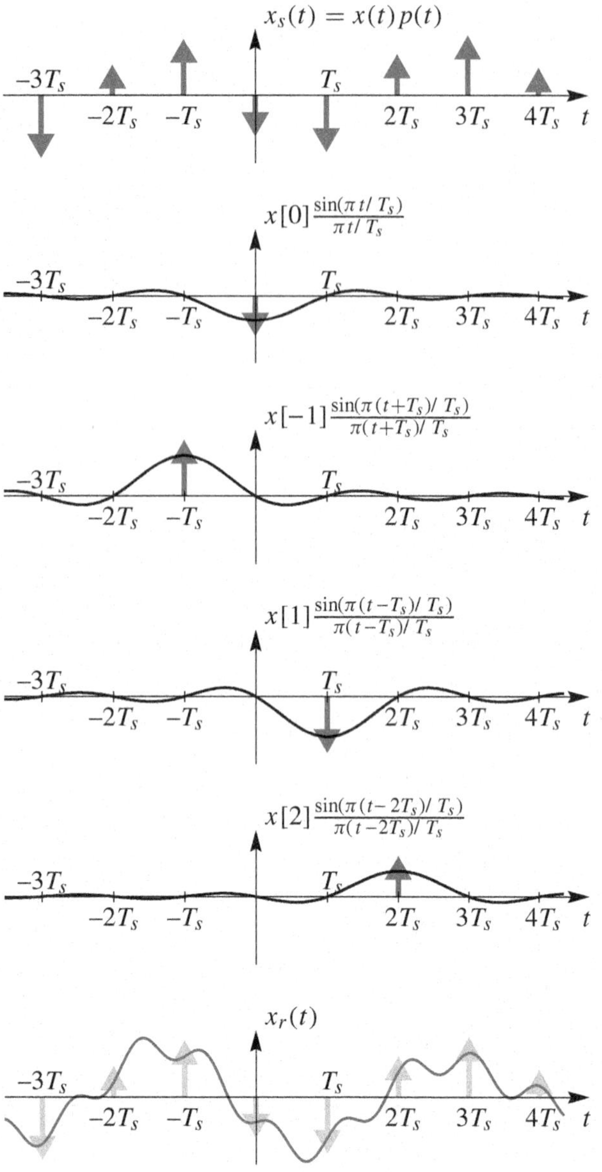

Figure 12-32: Illustration of time-domain reconstructed signal as the sum of the individual weighted and shifted sinc functions.

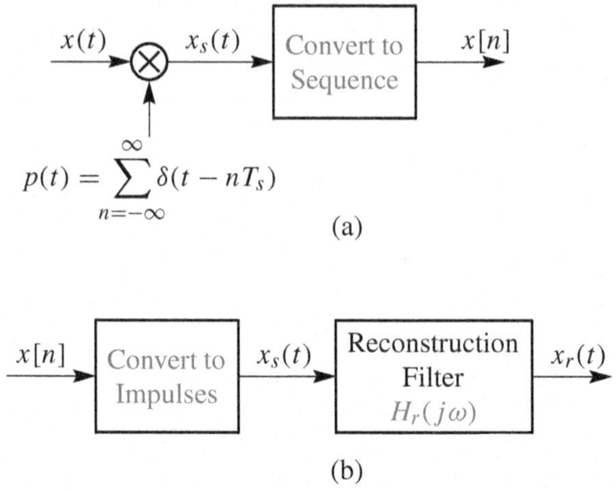

Figure 12-33: Definitions of (a) ideal C-to-D, and (b) D-to-C converters.

The input to the ideal C-to-D converter is a continuous-time signal and the output is a sequence of samples. This was shown in Fig. 12-27(b) on p. 369. The ideal C-to-D converter is an idealization of a practical hardware system that usually consists of a sample-and-hold circuit (which captures the input voltage at one instant of time and holds that value for T_s sec) followed by an analog-to-digital (A-to-D) converter (which obtains a numerical representation of the voltage). The major difference is that an A-to-D converter produces sample values that are "quantized" to a finite number of values. For example, a 16-bit A-to-D converter produces 65536 different possible values for its output samples. If the amplitude quantization is fine enough, this 16-bit limit is often negligible. More detailed analysis of quantization effects can be found in advanced texts on digital signal processing.

In Fig. 12-33(b) the ***Ideal Discrete-to-Continuous (D-to-C) Converter*** is the cascade of two systems. The first system takes an input sequence and makes a modulated impulse train, which is then used as the input to the second system, an ideal bandlimited reconstruction filter. Therefore, the input/output relation for the ideal D-to-C converter is

The Ideal D-to-C Converter

$$x_r(t) = \sum_{n=-\infty}^{\infty} x[n] \frac{\sin\left[\frac{\pi}{T_s}(t - nT_s)\right]}{\frac{\pi}{T_s}(t - nT_s)} \tag{12.51}$$

In a cascade connection of the C-to-D and D-to-C converters of Fig. 12-33, the operation of creating a sequence from the impulse train would be followed immediately by the operation of creating an impulse train from the sample sequence. Thus, a cascade connection of the systems in parts (a) and (b) of Fig. 12-33 is completely equivalent to the system in Fig. 12-26. Therefore, our analysis of Fig. 12-26 has shown that if the C-to-D converter samples a bandlimited signal *at a high enough sampling rate*, the ideal D-to-C converter will reconstruct the original signal exactly.

12-3.5 The Discrete-Time Fourier Transform

The main purpose of defining the ideal C-to-D and D-to-C converter systems is to eliminate the need to explicitly consider the modulated impulse train $x_s(t)$. The signal $x_s(t)$ has served us well as the basis for representing sampling and reconstruction, but usually we do not actually have such a signal available to us—we only have the discrete-time sequence of samples as depicted in Fig. 12-27(b). As a final step towards easing our dependence on the modulated impulse train, let us recall that we have already obtained the following two forms for $x_s(t)$:

$$x_s(t) = x_c(t) \sum_{n=-\infty}^{\infty} \delta(t - nT_s)$$

$$= x_c(t) \sum_{k=-\infty}^{\infty} \frac{1}{T_s} e^{jk\omega_s t} \tag{12.52}$$

$$= \sum_{n=-\infty}^{\infty} x_c(nT_s)\delta(t - nT_s) \tag{12.53}$$

where $x_c(t)$ is the continuous-time signal being sampled (the subscript c is added for emphasis in this section). Equation (12.52) allowed us to apply the frequency shift property to obtain the Fourier transform of $x_s(t)$

$$X_s(j\omega) = \frac{1}{T_s} \sum_{k=-\infty}^{\infty} X_c(j(\omega - k\omega_s)) \tag{12.54}$$

which is a periodic function of ω with period $\omega_s = 2\pi/T_s$. Equation (12.54) gives $X_s(j\omega)$ in terms of shifted copies of $X_c(j\omega)$, the Fourier transform of the continuous-time signal $x_c(t)$. Using (12.53) and the delay property of Fourier transforms, it follows that an equivalent relation for $X_s(j\omega)$ is

$$X_s(j\omega) = \sum_{n=-\infty}^{\infty} x_c(nT_s)e^{-j\omega nT_s} = \sum_{n=-\infty}^{\infty} x[n]e^{-j\omega nT_s} \tag{12.55}$$

which expresses $X_s(j\omega)$ in terms of the *samples* of $x_c(t)$. Note again that this form of $X_s(j\omega)$ is periodic in ω with period $\omega_s = 2\pi/T_s$ because the complex exponential functions are periodic. Now we can eliminate $X_s(j\omega)$ by equating (12.54) and (12.55) to obtain

$$\sum_{n=-\infty}^{\infty} x[n]e^{-j\omega nT_s} = \frac{1}{T_s} \sum_{k=-\infty}^{\infty} X_c(j(\omega - 2\pi k/T_s)) \tag{12.56}$$

The left-hand side of (12.56) is starting to look like the z-transform of the sequence of samples $x[n] = x_c(nT_s)$, which by definition is

$$X(z) = \sum_{n=-\infty}^{\infty} x[n]z^{-n} \tag{12.57}$$

If we make the substitution $z = e^{j\omega T_s}$, we get

$$\sum_{n=-\infty}^{\infty} x[n]e^{-j\omega n T_s} = X(z)\Big|_{z = e^{j\omega T_s}} = X(e^{j\omega T_s}) \tag{12.58}$$

Our objective is to get a function of the discrete-time frequency $\hat{\omega}$ which is defined by the frequency normalization

$$\hat{\omega} = \omega T_s \tag{12.59}$$

Thus we can define a new transform, called the *discrete-time Fourier transform* or *DTFT*, to be

Discrete-Time Fourier Transform

$$X(e^{j\hat{\omega}}) = \sum_{n=-\infty}^{\infty} x[n]e^{-j\hat{\omega}n} \tag{12.60}$$

The DTFT is related to the z-transform via $X(e^{j\hat{\omega}}) = X(z)\big|_{z = e^{j\hat{\omega}}}$.

EXERCISE 12.15: Show that $X(e^{j\hat{\omega}})$ is periodic in $\hat{\omega}$ with period 2π, and show that $X(e^{j\omega T_s})$ is periodic in ω with period $2\pi/T_s$.

Using the definition (12.60), we can state the following fundamental relation between the DTFT of a sequence of samples $x[n] = x_c(nT_s)$ and the continuous-time Fourier transform (CTFT) of $x(t)$:

Relation between the DTFT and the CTFT in sampling.

$$X(e^{j\omega T_s}) = X(e^{j\hat{\omega}})\Big|_{\hat{\omega} = \omega T_s} = \frac{1}{T_s}\sum_{k=-\infty}^{\infty} X_c(j(\omega - 2\pi k/T_s)). \tag{12.61}$$

12-3.6 The Inverse DTFT

In (12.51) of Section 12-3.4, we showed that the output of an ideal D-to-C converter with a sequence of input samples $x[n] = x_c(nT_s)$ is

$$x_r(t) = \sum_{m=-\infty}^{\infty} x[m]\frac{\sin\frac{\pi}{T_s}(t - mT_s)}{\frac{\pi}{T_s}(t - mT_s)} \tag{12.62}$$

which has CTFT

$$X_r(j\omega) = \begin{cases} T_s X(e^{j\omega T_s}) & |\omega| \le \pi/T_s \\ 0 & |\omega| > \pi/T_s \end{cases}$$

This means that another way of writing $x_r(t)$ is

$$x_r(t) = \frac{1}{2\pi}\int_{-\pi/T_s}^{\pi/T_s} T_s X(e^{j\omega T_s})e^{j\omega t}d\omega$$

Now, if we use (12.62) to compute $x_r(nT_s)$, we obtain

$$x_r(nT_s) = \sum_{m=-\infty}^{\infty} x[m]\frac{\sin\pi(n - m)}{\pi(n - m)} = x[n]$$

since all terms are zero except one, i.e.,

$$\frac{\sin\pi(n - m)}{\pi(n - m)} = \begin{cases} 1 & n = m \\ 0 & n \ne m \end{cases}$$

Therefore, even if the original Fourier transform is not bandlimited, we have just shown that $x_r(nT_s) = x[n] = x_c(nT_s)$. Therefore, it follows that

$$x_c(nT_s) = \frac{1}{2\pi} \int_{-\pi/T_s}^{\pi/T_s} T_s X(e^{j\omega T_s}) e^{j\omega T_s n} d\omega \qquad (12.63)$$

If we make the substitution $\hat{\omega} = \omega T_s$, (12.63) can be written as

Inverse DTFT

$$x[n] = \frac{1}{2\pi} \int_{-\pi}^{\pi} X(e^{j\hat{\omega}}) e^{j\hat{\omega}n} d\hat{\omega}. \qquad (12.64)$$

Equation (12.64) gives the inverse DTFT in terms of normalized frequency $\hat{\omega}$, and thus it is the partner to the DTFT equation (12.60).

Example 12-7: DTFT of a sinc Function

Consider the bandlimited signal

$$x_c(t) = \frac{\sin(\omega_b t)}{\pi t}$$

If we sample this signal with sampling period T_s, we obtain the sequence

$$x[n] = x_c(nT_s) = \frac{\sin(\omega_b n T_s)}{\pi n T_s} = \frac{\sin(\hat{\omega}_b n)}{\pi n T_s} \qquad (12.65)$$

where $\hat{\omega}_b = \omega_b T_s$. The Fourier transform of the continuous-time signal $x_c(t)$ is

$$X_c(j\omega) = \begin{cases} 1 & |\omega| \le \omega_b \\ 0 & \text{otherwise} \end{cases}$$

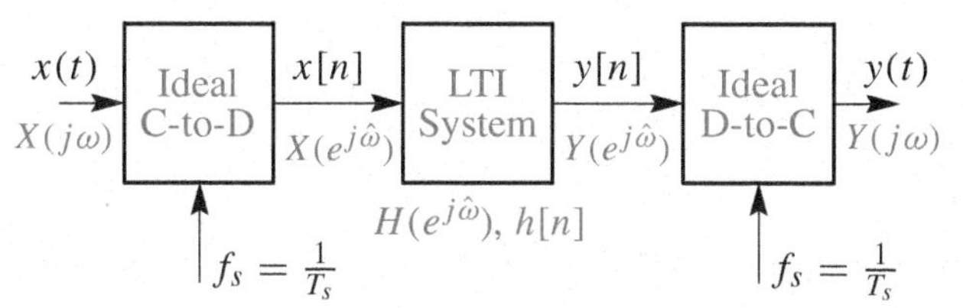

Figure 12-34: Discrete time filtering of a continuous-time signal.

Equation (12.61) gives

$$X(e^{j\omega T_s}) = \frac{1}{T_s} \sum_{n=-\infty}^{\infty} \frac{\sin(\hat{\omega}_b n)}{\pi n} e^{-j\omega n T_s}$$

$$= \frac{1}{T_s} \sum_{k=-\infty}^{\infty} X_c(j(\omega - k\omega_s))$$

and we can match terms to discover that the DTFT of a sinc function is a rectangle:

$$X(e^{j\hat{\omega}}) = \sum_{n=-\infty}^{\infty} \frac{\sin(\hat{\omega}_b n)}{\pi n} e^{-j\hat{\omega}n} = \begin{cases} 1 & |\hat{\omega}| \le \hat{\omega}_b \\ 0 & \text{otherwise} \end{cases}$$ ■

12-3.7 Discrete-Time Filtering of Continuous-Time Signals

One of the most important concepts related to sampling is the idea of filtering a continuous-time signal by sampling the signal, filtering it with a discrete-time LTI system, and then converting back to a continuous-time output signal. This idea is depicted in Fig. 12-34. Our knowledge of sampling and bandlimited reconstruction enables us to understand the operation of this system.

Recall from Chapters 7 and 8 that the discrete-time system can be described by the z-transform relation

$$Y(z) = H(z)X(z) \qquad (12.66)$$

or equivalently, by the DTFT relation

$$Y(e^{j\hat{\omega}}) = H(e^{j\hat{\omega}})X(e^{j\hat{\omega}}) \tag{12.67}$$

The output of the D-to-C converter is given by the time-domain relation

$$y(t) = \sum_{n=-\infty}^{\infty} y[n]h_r(t - nT_s) \tag{12.68}$$

where $h_r(t)$ is given by (12.47) on p. 373. Now the frequency-domain representation of the D-to-C converter is easily derived by taking the *continuous-time* Fourier transform of (12.68), obtaining

$$\begin{aligned} Y(j\omega) &= \sum_{n=-\infty}^{\infty} y[n]H_r(j\omega)e^{-j\omega n T_s} \\ &= H_r(j\omega)\sum_{n=-\infty}^{\infty} y[n]e^{-j\omega T_s n} \\ &= H_r(j\omega)Y(e^{j\omega T_s}) \end{aligned} \tag{12.69}$$

If we substitute (12.67) and (12.61) into (12.69), we obtain

$$\begin{aligned} Y(j\omega) &= H_r(j\omega)H(e^{j\omega T_s})X(e^{j\omega T_s}) \\ &= H_r(j\omega)H(e^{j\omega T_s})\frac{1}{T_s}\sum_{k=-\infty}^{\infty} X_c(j(\omega - 2\pi k/T_s)) \end{aligned} \tag{12.70}$$

While (12.70) looks quite complex, it is not really too difficult to interpret. First of all, the summation is a sum of shifted copies of $X_c(j\omega)$. By construction, that sum is necessarily periodic with period $\omega_s = 2\pi/T_s$ in the variable ω. Furthermore, if $X_c(j\omega) = 0$ for $|\omega| > \omega_b$ and $\omega_s > 2\omega_b$, then the shifted copies do not overlap, and one period of the function is exactly equal to $(1/T_s)X_c(j\omega)$ in the frequency interval $-\pi/T_s < \omega < \pi/T_s$. The middle term, $H(e^{j\omega T_s})$, is also periodic with period $\omega_s = 2\pi/T_s$. The left-hand term, $H_r(j\omega)$ has a gain of T_s and is zero outside the interval $|\omega| \leq \omega_s/2 = \pi/T_s$. Therefore, if no aliasing distortion occurs, the output Fourier transform is

$$Y(j\omega) = \begin{cases} H(e^{j\omega T_s})X_c(j\omega) & |\omega| \leq \pi/T_s \\ 0 & \text{otherwise} \end{cases} \tag{12.71}$$

Equation (12.70) is illustrated in Fig. 12-35. Figure 12-35(a) shows a typical bandlimited Fourier transform representing $X_c(j\omega)$ and 12-35(b) shows the discrete-time Fourier transform $X(e^{j\omega T_s})$ of the sampled signal $x[n] = x_c(nT_s)$, while Fig. 12-35(c) shows $X(e^{j\hat{\omega}})$ as a function of normalized frequency $\hat{\omega}$. The frequency response of the digital filter is shown in Fig. 12-35(d) versus $\hat{\omega}$. The filter shown in Fig. 12-35(d) is a discrete-time ideal lowpass filter with cutoff frequency $0 < \hat{\omega}_{\text{co}} < \pi$. One period of $H(e^{j\hat{\omega}})$ is defined by

$$H(e^{j\hat{\omega}}) = \begin{cases} 1 & |\hat{\omega}| \leq \hat{\omega}_{\text{co}} \\ 0 & \hat{\omega}_{\text{co}} < |\hat{\omega}| < \pi \end{cases} \tag{12.72}$$

where $H(e^{j\hat{\omega}})$ is periodic with period 2π. Now $Y(e^{j\hat{\omega}}) = H(e^{j\hat{\omega}})X(e^{j\hat{\omega}})$ so in this example, the high frequencies are removed by the filter as shown in Fig. 12-35(e). The ideal reconstruction filter is also shown as the gray box in Fig. 12-35(e). Since it removes all frequencies outside the band $|\omega| \leq \pi/T_s$, the final output Fourier transform is the transform shown in Fig. 12-35(f). As illustrated by the example of Fig. 12-35, the net effect of the system of Fig. 12-34 is that the Fourier transform of the output has the form

$$Y(j\omega) = H_{\text{eff}}(j\omega)X_c(j\omega) \tag{12.73}$$

where the overall *effective* frequency response is

$$H_{\text{eff}}(j\omega) = H(e^{j\omega T_s}) \tag{12.74}$$

Equation (12.73) holds *only* if $X_c(j\omega) = 0$ for $|\omega| \geq \pi/T_s$; in other words, only if the input is bandlimited

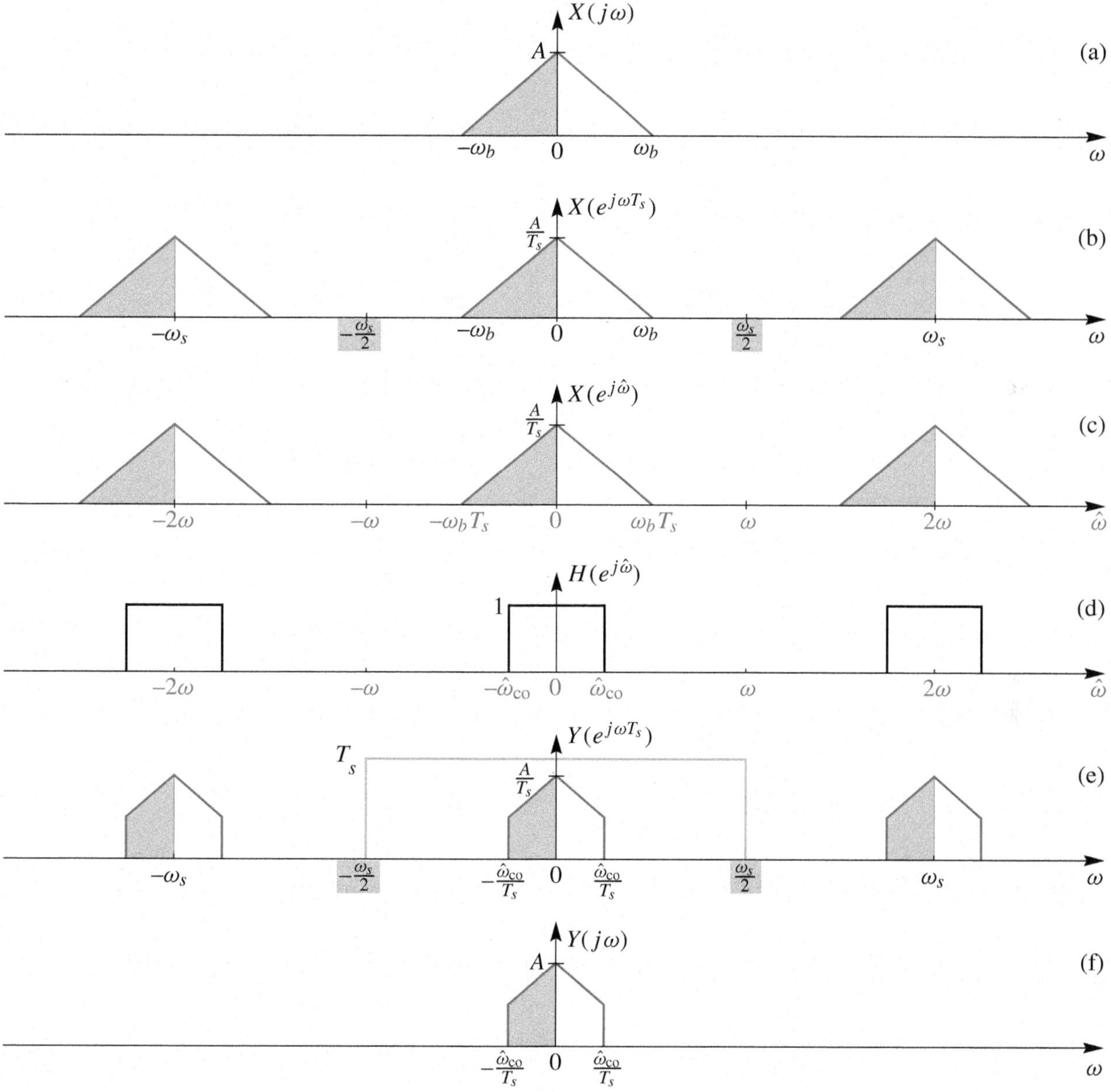

Figure 12-35: Illustration of discrete-time filtering of a sampled bandlimited signal: (a) Fourier transform of continuous-time bandlimited signal $x_c(t)$, (b) discrete-time Fourier transform (DTFT) of output signal from ideal C-to-D converter plotted as a function of ω, (c) DTFT of output signal from ideal C-to-D converter plotted as a function of normalized frequency $\hat{\omega}$, (d) frequency response of ideal lowpass digital filter (plotted vs. $\hat{\omega}$), (e) DTFT of output signal after digital lowpass filter and frequency response of ideal reconstruction filter (plotted vs. ω), and (f) Fourier transform of output of D-to-C converter.

and it is sampled at a high enough sampling rate to avoid aliasing distortion. In the case of the discrete-time ideal LPF in (12.72), the effective frequency response is

$$H_{\text{eff}}(j\omega) = \begin{cases} 1 & |\omega| \le \hat{\omega}_{\text{co}}/T_s \\ 0 & \text{otherwise} \end{cases}$$

This result illustrates an important point implicit in (12.74)—the effective frequency response of the overall system of Fig. 12-34 depends on the sampling period T_s. In the case of the ideal lowpass filter of Fig. 12-35(d), we see that the cutoff frequency of the overall effective system is $\omega_{\text{co}} = \hat{\omega}_{\text{co}}/T_s$. Thus, if the sampling rate increases, the effective cutoff frequency increases proportionally, and vice versa.

EXERCISE 12.16: In the example of Fig. 12-35, the input signal is oversampled, i.e., $\omega_s = 2\pi/T_s > 2\omega_b$. In Fig. 12-35(f), some of the high frequencies are cut off by the ideal lowpass digital filter. However, assuming that $\hat{\omega}_{\text{co}}$ is fixed, it should be possible to choose ω_s so that $y(t) = x_c(t)$ for the example shown. Give a condition that relates T_s, $\hat{\omega}_{\text{co}}$, and ω_b so that $y(t) = x_c(t)$ if $X_c(j\omega)$ is bandlimited as shown in Fig. 12-35(a).

Example 12-8: $H_{\text{eff}}(j\omega)$ for Time Delay

Let the discrete-time system be a 10-sample delay system defined by $y[n] = x[n-10]$. The frequency response of this system is

$$H(e^{j\hat{\omega}}) = e^{-j\hat{\omega}10}$$

Therefore, the overall effective frequency response for bandlimited input signals that are sampled above the Nyquist rate is

$$H_{\text{eff}}(j\omega) = e^{-j\omega 10 T_s}$$

which, of course, is the frequency response of an ideal time delay system with delay $t_d = 10T_s$ sec. ∎

Example 12-9: $H_{\text{eff}}(j\omega)$ for First-Difference

Suppose that the discrete-time system is a first-difference system described by $y[n] = x[n] - x[n-1]$. The corresponding frequency response is

$$H(e^{j\hat{\omega}}) = 1 - e^{-j\hat{\omega}} = 2j\sin(\hat{\omega}/2)e^{-j\hat{\omega}/2}$$

Then if this system is used as the discrete-time system in Fig. 12-34, and no aliasing distortion occurs in sampling the bandlimited input signal, the overall effective frequency response is

$$H_{\text{eff}}(j\omega) = 2j\sin(\omega T_s/2)e^{-j\omega T_s/2}$$ ∎

EXERCISE 12.17: Determine the frequency response of the discrete-time filter that, when used in the system of Fig. 12-34, will give an overall effective frequency response

$$H_{\text{eff}}(j\omega) = \begin{cases} 1 & 1000\pi < |\omega| < 2000\pi \\ 0 & \text{otherwise} \end{cases}$$

when the sampling rate is $\omega_s = 10000\pi$ rad/s.

12-4 Summary

In this chapter, we have discussed three very important applications of Fourier transform concepts—linear filtering, amplitude modulation, and sampling. We have shown that the multiplication and convolution properties make the Fourier transform an ideal tool for analysis of these systems.

Several of the labs in this chapter deal with the Fourier series. Labs #14a and #14b show that Fourier series analysis is a powerful method for predicting the response of an LTI system when the input is a periodic signal. The design in Lab #14a is for a power supply with control over the ripple. In Lab #14b students design a bandpass filter that produces a desired output sinusoid while reducing the amount of distortion. Lab #15 shows how to find Fourier series coefficients using MATLAB's Symbolic Toolbox (based on Maple). Then the symbolic Fourier series is used to analyze the power supply from Lab #14a. In Lab #16, students explore Amplitude Modulation so that they can implement and demonstrate a functional AM system that operates on a voice signal. Finally, in Labs #17 and #18, the design and implementation of a modem based on FSK (Frequency Shift Keying) is explored.

LAB: *#14a, #14b, #15, #16, #17, and #18*

As usual, the reader is reminded of the large number of solved homework problems on the CD-ROM that are available for review and practice.

NOTE: *Hundreds of Solved Problems*

12-5 Problems

P-12.1 An ideal bandpass filter has a passband from ω_{co1} to ω_{co2}, as depicted in Fig. 12-8(b) on p. 353. Show that its impulse response can be represented as a cosine modulated lowpass filter impulse response as in

$$h_{bp}(t) = \frac{\sin(\omega_{co}t)}{\pi t}\cos(\omega_0 t) \tag{12.75}$$

Find ω_0 and ω_{co} in (12.75) in terms of the bandpass cutoff frequencies ω_{co1} and ω_{co2}. Express $H_{bp}(j\omega)$ in terms of $H_{lp}(j\omega)$ to obtain the frequency-domain representation of (12.75).

P-12.2 If an ideal bandpass filter has a passband from ω_{co1} to ω_{co2}, as depicted in Fig. 12-8(b) on p. 353, show that its impulse response can be represented as the difference of two ideal lowpass filter impulse responses

$$h_{bp}(t) = \frac{\sin(\omega_{co2}t)}{\pi t} - \frac{\sin(\omega_{co1}t)}{\pi t} \tag{12.76}$$

P-12.3 If an ideal bandpass filter has a passband from ω_{co1} to ω_{co2}, as depicted in Fig. 12-8(b) on p. 353, show that its impulse response can be represented as the convolution of the impulse responses of an ideal highpass filter and an ideal lowpass filter as in

$$h_{bp}(t) = \left[\delta(t) - \frac{\sin(\omega_{co1}t)}{\pi t}\right] * \frac{\sin(\omega_{co2}t)}{\pi t} \tag{12.77}$$

where ω_{co1} and ω_{co2} are the bandpass cutoff frequencies. What is the frequency-domain representation of (12.77)?

P-12.4 Consider the amplitude modulation system in Fig. P-12.4(a) with a bandlimited input signal $x(t)$ as

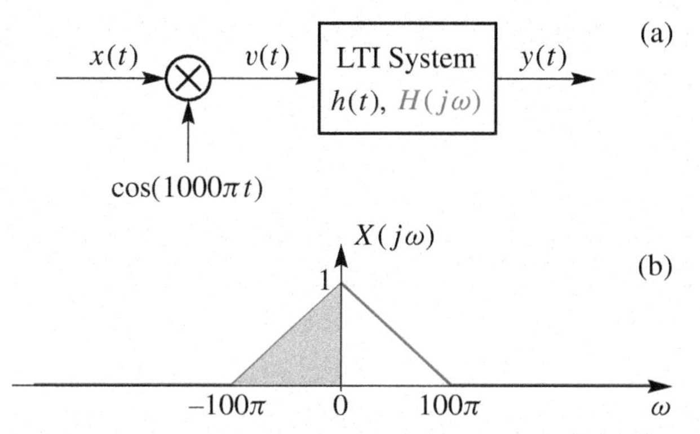

Figure P-12.4

depicted in Fig. P-12.4(b). Assume that the LTI system has the frequency response of an ideal bandpass filter

$$H(j\omega) = \begin{cases} 1 & 900\pi < |\omega| < 1000\pi \\ 0 & \text{otherwise} \end{cases}$$

(a) Plot the Fourier transform $H(j\omega)$ of the ideal BPF specified above. Make the plot for both negative and positive frequencies.

(b) Plot the Fourier transform $V(j\omega)$ of the signal $v(t)$ at the output of the multiplier.

(c) Plot the Fourier transform $Y(j\omega)$ of the output signal $y(t)$ from the filter. *Do not try to find* $y(t)$.

(d) The output signal is called a "single-sideband" signal. Can you justify this terminology?

Note that the negative frequency portion of the Fourier transform $X(j\omega)$ in Fig. P-12.4 (b) is shaded. Mark the corresponding region or regions in your plots of $V(j\omega)$ and $Y(j\omega)$.

P-12.5 Refer to the single-sideband (SSB) system of Fig. P-12.4. Draw a block diagram of a system that will recover the original input signal $x(t)$ from the SSB signal $y(t)$.

P-12.6 Consider the amplitude modulation system in Fig. P-12.6(a) with a bandlimited input signal whose Fourier transform is depicted in Fig. P-12.6(b). Assume that the LTI system has the frequency response of an ideal lowpass filter

$$H(j\omega) = \begin{cases} 2 & |\omega| \le \omega_m \\ 0 & \text{otherwise} \end{cases}$$

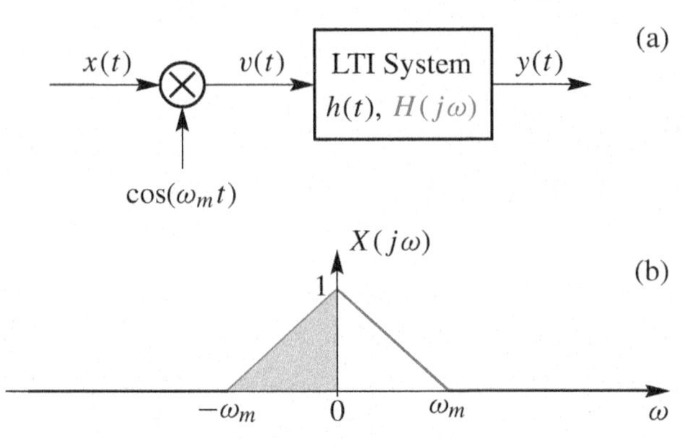

Figure P-12.6

(a) Plot the Fourier transform $V(j\omega)$ of the signal $v(t)$ at the output of the multiplier. Clearly indicate the frequencies in $V(j\omega)$ that correspond to the shaded region in $X(j\omega)$ in Fig. P-12.6 (b).

(b) Plot the Fourier transform $Y(j\omega)$ of the output signal $y(t)$ from the lowpass filter. *Do not try to find* $y(t)$.

(c) Basing your answer on the plot obtained in part (b), give a written statement of what this system does to the spectrum of the input $X(j\omega)$.

(d) How could you recover $x(t)$ from $y(t)$? Draw a block diagram of your solution.

P-12.7 The system in Fig. P-12.7(a) is called a *quadrature modulation system*. It is a method of sending two bandlimited signals over the same channel. The demodulator in Fig. P-12.7(b) will recover one of the two input signals. Assume that both input signals are bandlimited such that the maximum frequency is ω_m; i.e., $X_1(j\omega) = 0$ for $|\omega| \ge \omega_m$ and $X_2(j\omega) = 0$ for $|\omega| \ge \omega_m$, where $\omega_m \ll \omega_c$.

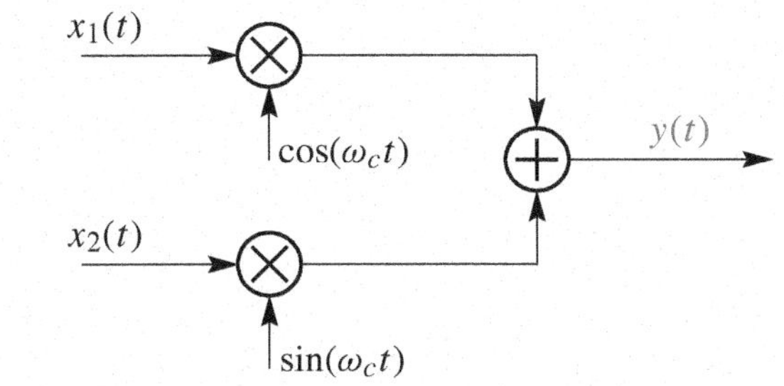

(a)

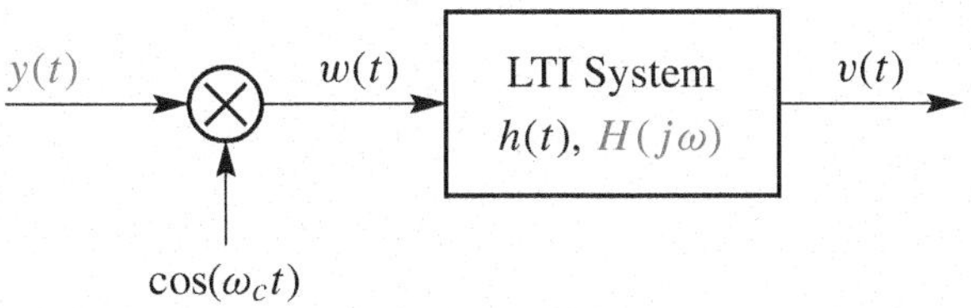

(b)

Figure P-12.7

(a) The output of the quadrature modulator in Fig. P-12.7(a) is $y(t) = x_1(t)\cos(\omega_c t) + x_2(t)\sin(\omega_c t)$. Determine an expression for the Fourier transform $Y(j\omega)$ in terms of $X_1(j\omega)$ and $X_2(j\omega)$. Make a sketch of $Y(j\omega)$. Assume simple (but different) shapes for the bandlimited Fourier transforms $X_1(j\omega)$ and $X_2(j\omega)$, and use them in making your sketch of $Y(j\omega)$.

(b) From the expression found in part (a) and the sketch that you drew, you should see that $Y(j\omega) = 0$ for $|\omega| \le \omega_a$ and for $|\omega| \ge \omega_b$. Determine ω_a and ω_b.

(c) Given the trigonometric identities $2\sin\theta\cos\theta = \sin 2\theta$, $2\sin^2\theta = (1 - \cos 2\theta)$, and $2\cos^2\theta = (1 + \cos 2\theta)$, show that in the demodulator in Fig. P-12.7(b), the output of the mixer is

$$w(t) = \tfrac{1}{2}x_1(t)[1 - \cos(2\omega_c t)] + \tfrac{1}{2}x_2(t)\sin(2\omega_c t)$$

(d) The signal $w(t)$ as determined in part (c) is the input to an LTI system. Determine the frequency response of that system so that its output is $v(t) = x_1(t)$. Give your answer as a carefully labeled plot of $H(j\omega)$.

(e) Draw a block diagram of a demodulator system whose output will be $x_2(t)$ when its input is $y(t)$. This requires that you change the mixer.

P-12.8 Consider the DSBAM demodulation system in Fig. P-12.8 where

$$H(j\omega) = \begin{cases} 2 & |\omega| \le \omega_{co} \\ 0 & |\omega| > \omega_{co} \end{cases}$$

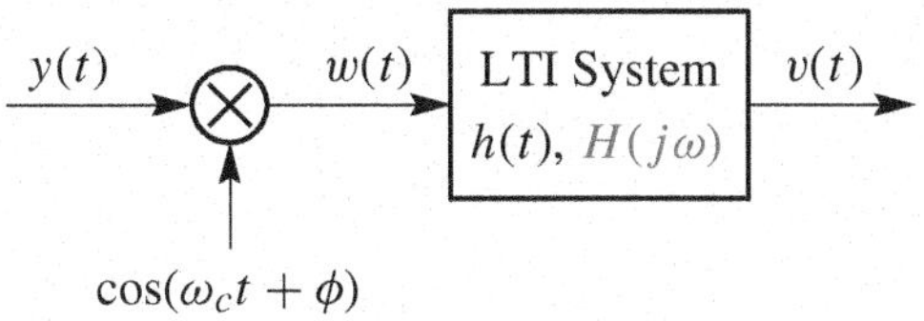

Figure P-12.8

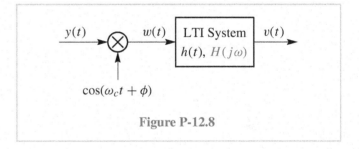

We have shown that if $x(t)$ has a bandlimited Fourier transform such that $X(j\omega) = 0$ for $|\omega| \ge \omega_b$ and $\omega_c > \omega_b$ and $\phi = 0$ and $\omega_b < \omega_{co} < (2\omega_c - \omega_b)$, then the DSBAM signal $y(t) = x(t)\cos(\omega_c t)$ can be demodulated by the system in Fig. P-12.8. That is, for perfect adjustment of the demodulator frequency and phase, $v(t) = x(t)$. In the following parts, assume that the input signal $x(t)$ has a bandlimited Fourier transform represented by the following plot:

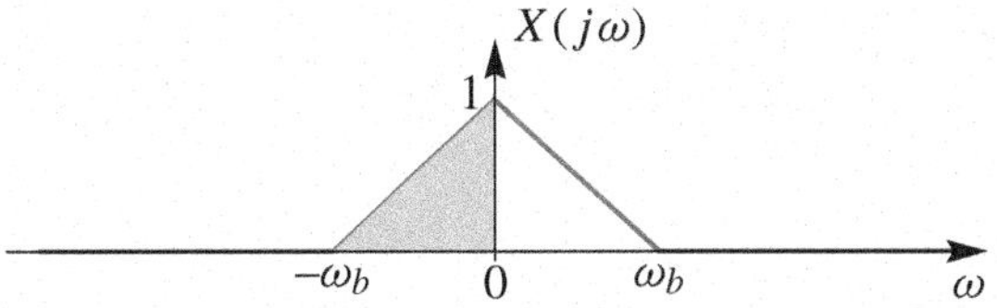

(a) Now suppose that $\phi \neq 0$. Use Euler's formula for the cosine to show that

$$\begin{aligned} w(t) &= y(t)\cos(\omega_c t + \phi) \\ &= x(t)\cos(\omega_c t)\cos(\omega_c t + \phi) \\ &= \tfrac{1}{2}x(t)\cos\phi + \tfrac{1}{2}x(t)\cos(2\omega_c t + \phi) \end{aligned}$$

(b) From this equation obtain an equation for $W(j\omega)$ in terms of $X(j\omega)$ and use this equation to make a plot of $W(j\omega)$ for the given $X(j\omega)$.

(c) From this plot, assuming $\omega_{co} = \omega_b$, determine a plot of $V(j\omega)$.

(d) From the plot of $V(j\omega)$, obtain an equation for $v(t)$ in terms of $x(t)$ and ϕ.

P-12.9 Using the same assumptions on $X(j\omega)$ and ω_c as in Prob. P-12.8, consider the case when the modulator and demodulator are in phase ($\phi = 0$) but the demodulator carrier frequency is mismatched by a small amount $\Delta \ll \omega_c$ as depicted in the block diagram in Fig. P-12.9.

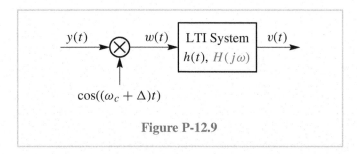

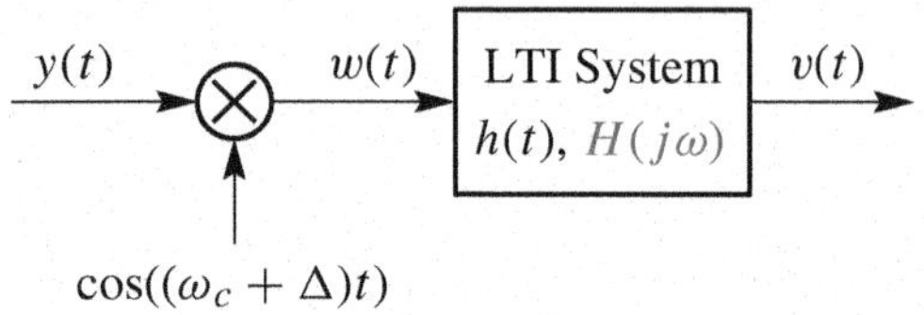

Figure P-12.9

(a) Make a plot of $W(j\omega)$ for the bandlimited $X(j\omega)$ given in Prob. P-12.8.

(b) Use this plot to help you determine an equation for $v(t)$ in terms of $x(t)$ and Δ. In this case, you will need to assume that the cutoff frequency of the ideal lowpass filter satisfies $(\omega_b + \Delta) < \omega_{co} < (2\omega_c + \Delta - \omega_b)$.

P-12.10 Figure P-12.10(a) depicts a system that is designed to detect signals of two different frequencies as in a frequency-shift keying (FSK) modem.

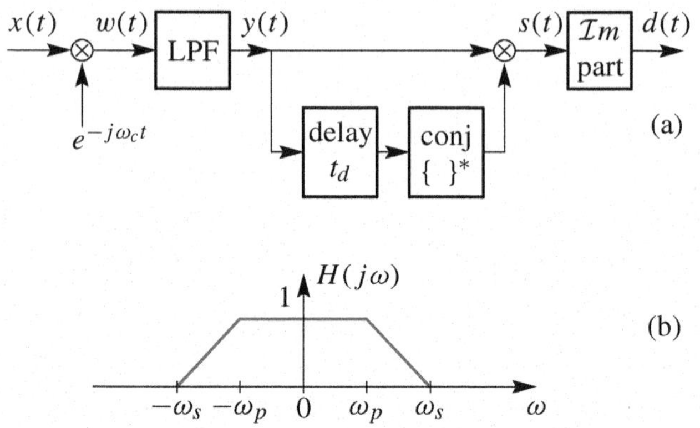

Figure P-12.10

(a) Suppose that the input signal to the system in Fig. P-12.10(a) is $x(t) = x_1(t) = \cos(\omega_1 t)$ where $\omega_1 = \omega_c - \omega_0$ with $\omega_0 > 0$. Using the frequency-shift property of Fourier transforms and the Fourier transform of the cosine signal, determine and plot the Fourier transform of the signal $w(t)$.

(b) Now suppose that the lowpass filter (LPF) in Fig. P-12.10(a) has frequency response as depicted in Fig. P-12.10(b). Determine the *smallest* value for ω_p and the *largest* value for ω_s such that the output of the filter is

$$y(t) = \tfrac{1}{2}e^{-j\omega_0 t}.$$

This will give the largest *transition region* between passband and stopband and therefore the simplest filter to implement.

(c) Show that for the passband and stopband frequencies found in part (b), the overall output is a constant; i.e., $d(t) = d_1(t) = -\frac{1}{4}\sin(\omega_0 t_d)$.

(d) Now suppose that $x(t) = x_2(t) = \cos(\omega_2 t)$ where $\omega_2 = \omega_c + \omega_0$ and the cutoff frequencies of the filter are the same as found in part (b). Show that the overall output is again a constant but in this case, $d(t) = d_2(t) = \frac{1}{4}\sin(\omega_0 t_d)$.

(e) Assume that the input signal can be either $x_1(t)$ or $x_2(t)$. Give a simple algorithm for determining which signal was used.

(f) Explain how you would implement the system of Fig. P-12.10(a) using digital signal processing. Draw a block diagram showing the signal $x(t)$ sampled with an ideal C-to-D converter followed by a discrete-time version of the system in Fig. P-12.10(a). For the input signals used in parts (a)–(e), what is the *minimum* sampling frequency $f_{\text{samp}} = 1/T_s$ that can be used? For this sampling rate, determine the frequency of the complex-exponential signal $e^{-j\hat{\omega}_c n}$ and the normalized cutoff frequencies of the discrete-time lowpass filter that will be needed.

P-12.11 Figure P-12.11(a) depicts a ***pulse amplitude modulation*** (PAM) system. In general, $p(t)$ is a periodic pulse signal with fundamental frequency $\omega_p = 2\pi/T_p$. Therefore, it can be represented as a Fourier series as shown in Fig. P-12.11(a). For this problem, we assume that $p(t)$ is the periodic square wave shown in Fig. P-12.11(b).

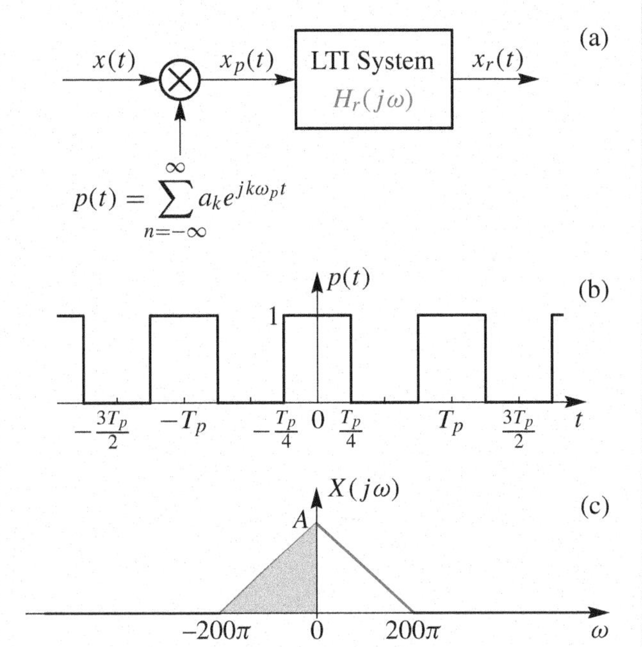

Figure P-12.11

(a) Assume that $x(t) = \cos(200\pi t)$. Make a plot over the interval $0 \le t \le 0.02$ of the signals $x(t)$, $p(t)$, and $x_p(t) = p(t)x(t)$. Assume that $\omega_p = 500\pi$ in making your plots of $p(t)$ and $x_p(t)$. Observe that multiplication of $x(t)$ by the periodic square wave has the effect of switching $x(t)$ on and off periodically with period $T_p = 1/250 = 0.004$ seconds.

(b) Show that the Fourier transform of $x_p(t) = p(t)x(t)$ is

$$X_p(j\omega) = \sum_{k=-\infty}^{\infty} a_k X(j(\omega - k\omega_p))$$

(c) For the input $x(t)$ with Fourier transform $X(j\omega)$ depicted in Fig. P-12.11(c), make a sketch of $X_p(j\omega)$ when $\omega_p = 2\pi/T_p = 500\pi$.

(d) If the frequency response of the LTI system is

$$H_r(j\omega) = \begin{cases} G & |\omega| \le \omega_{\text{co}} \\ 0 & |\omega| > \omega_{\text{co}} \end{cases}$$

and $\omega_p = 2\pi/T_p = 500\pi$, use the result of part (b) to determine the gain G and the cutoff frequency ω_{co} so that $x_r(t) = x(t)$.

(e) Determine the ***minimum*** value of ω_p and the corresponding values of G and ω_{co}, such that $x_r(t) = x(t)$.

(f) This system would work in essentially the same way if $p(t)$ was changed to almost any periodic signal with fundamental frequency ω_p. However, one important condition must be satisfied by the Fourier series coefficient a_0. What is that condition?

P-12.12 The derivation of the sampling theorem involves the operations of impulse train sampling and reconstruction as shown in Fig. P-12.12(a).

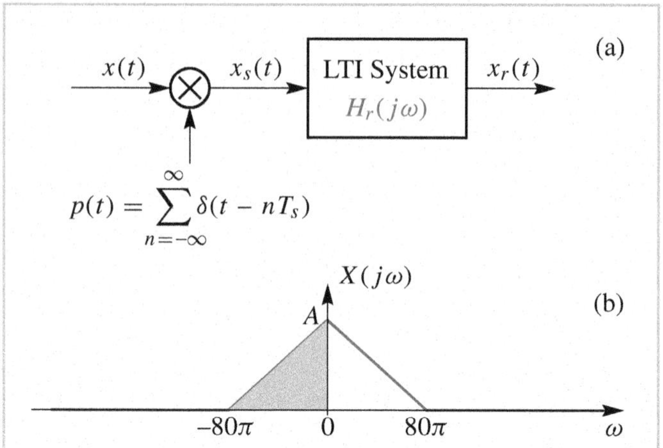

Figure P-12.12

(a) For the input with Fourier transform depicted in Fig. P-12.12(b), use the sampling theorem to choose the sampling rate $\omega_s = 2\pi/T_s$, so that $x_r(t) = x(t)$ when

$$H_r(j\omega) = \begin{cases} T_s & |\omega| \leq \pi/T_s \\ 0 & |\omega| > \pi/T_s \end{cases} \qquad (12.78)$$

Plot $X_s(j\omega)$ for the value of $\omega_s = 2\pi/T_s$ that is equal to the *Nyquist* rate.[5]

(b) If $\omega_s = 2\pi/T_s = 100\pi$ in the above system and $X(j\omega)$ is as depicted above, plot the Fourier transform $X_s(j\omega)$ and show that aliasing occurs. There will be an infinite number of shifted copies of $X(j\omega)$, so indicate the periodic pattern as a function of ω.

(c) For the conditions of part (b), determine and sketch the Fourier transform of the output $X_r(j\omega)$ if the frequency response of the LTI system is given by (12.78).

[5]Remember that the Nyquist rate is the *lowest* possible sampling rate that does not cause aliasing.

P-12.13 The input signal for the sampling and reconstruction system in Fig. P-12.12(a) is

$$x(t) = 2\cos(100\pi t - \pi/4) + \cos(300\pi t + \pi/3)$$

for $-\infty < t < \infty$. The frequency response of the lowpass reconstruction filter is

$$H_r(j\omega) = \begin{cases} T_s & |\omega| \leq \pi/T_s \\ 0 & |\omega| > \pi \end{cases}$$

where T_s is the sampling period.

(a) Determine the Fourier transform $X(j\omega)$ and plot the Fourier transform $X_s(j\omega)$ for $-2\pi/T_s < \omega < 2\pi/T_s$ for the case where $2\pi/T_s = 1000\pi$. Carefully label your sketch. What is the output $x_r(t)$ in this case?

(b) Now assume that $\omega_s = 2\pi/T_s = 500\pi$. Determine an equation for the output $x_r(t)$.

(c) Is it possible to choose the sampling rate so that

$$x_r(t) = A + 2\cos(100\pi t - \pi/4)$$

where A is a constant? If so, what is the value of T_s and what is the numerical value of A?

P-12.14 All parts of this problem are concerned with the system shown in Fig. P-12.14.

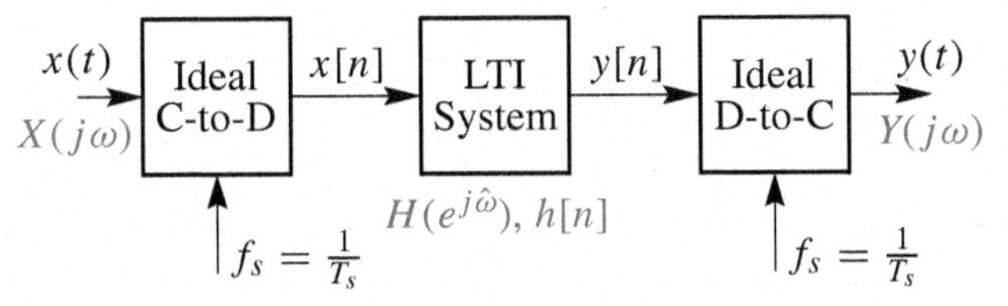

Figure P-12.14

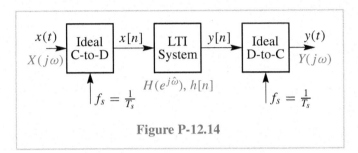

In all parts of this problem, assume that $X(j\omega) = 0$ for $|\omega| \geq 1000\pi$. In addition, assume that the C-to-D and D-to-C converters are ideal; i.e.,

$$x[n] = x(nT), \qquad y(t) = \sum_{n=-\infty}^{\infty} y[n] \frac{\sin \frac{\pi}{T_s}(t - nT_s)}{\frac{\pi}{T_s}(t - nT_s)}$$

(a) Suppose that the discrete-time system is defined by $y[n] = x[n]$. What is the *minimum* value of $2\pi/T_s$ such that $y(t) = x(t)$?

(b) Suppose that the LTI discrete-time system has system function $H(z) = z^{-10}$, and assume that the sampling rate satisfies the condition of part (a). Determine the overall effective frequency response $H_{\text{eff}}(j\omega)$ and from it determine a general relationship between $y(t)$ and $x(t)$.

(c) The input/output relation for the discrete-time system is

$$y[n] = \tfrac{1}{3}\left(x[n] + x[n-1] + x[n-2]\right)$$

For the value of T_s chosen in part (a), the input and output Fourier transforms are related by an equation of the form $Y(j\omega) = H_{\text{eff}}(j\omega)X(j\omega)$. Find an equation for the overall effective frequency response $H_{\text{eff}}(j\omega)$. Plot the magnitude and phase of $H_{\text{eff}}(j\omega)$. Use MATLAB to do this or sketch it by hand.

P-12.15 Consider the system for discrete-time filtering of a continuous-time signal shown in Fig. P-12.14. The Fourier transform of the input signal is shown in Fig. P-12.15(a), and the frequency response of the discrete-time system in Fig. P-12.14 is shown in Fig. P-12.15(b).

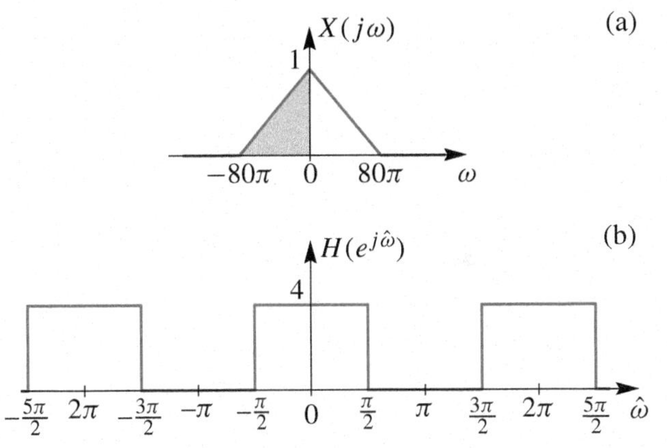

Figure P-12.15

(a) Assume that the input signal $x(t)$ has a bandlimited Fourier transform $X(j\omega)$ as depicted in Fig. P-12.15(a). For this input signal, what is the *smallest* value of the sampling frequency $f_s = 1/T_s$ such that the Fourier transforms of the input and output satisfy the relation $Y(j\omega) = H_{\text{eff}}(j\omega)X(j\omega)$?

(b) Assume that the discrete-time system is an ideal lowpass discrete-time filter with frequency response $H(e^{j\hat{\omega}})$ defined by the plot in Fig. P-12.15(b). Recall that $H(e^{j\hat{\omega}})$ is periodic with period 2π as shown.

Now, if $f_s = 100$ samples/sec, make a carefully labeled plot of $H_{\text{eff}}(j\omega)$, the effective frequency response of the overall system. Also plot $Y(j\omega)$, the Fourier transform of the output $y(t)$, when the input has Fourier transform $X(j\omega)$ as depicted in the graph Fig. P-12.15(a).

(c) For the input depicted in Fig. P-12.15(a) and the system defined in Fig. P-12.15(b), what is the *smallest sampling rate* such that the input signal passes through the lowpass filter unaltered; i.e., what is the minimum f_s such that $Y(j\omega) = X(j\omega)$?

CHAPTER 13

Computing the Spectrum

Throughout this text, as we studied sampling, linear filtering, and modulation, we have seen the value of frequency-domain analysis. The ***spectrum concept*** is fundamental to this way of thinking about signals and systems. Although we gained a great deal of insight from simple signals whose Fourier transforms (spectra) could be determined by evaluating the Fourier integral, most signals that arise in practice cannot be described by simple equations that can be looked up in a table or plugged into a Fourier integral. Therefore, if we want to apply the concepts that we have introduced in this text to real applications, we must address the problem of determining the spectrum of a signal that might arise in a practical setting.

Spectrum analysis via the fast Fourier transform (FFT) algorithm is the most likely situation in which an engineer or scientist would apply Fourier transform theory in practice. Nowadays, when signals are recorded from a system, the recording is almost always digital with an A-to-D converter digitizing the signal before placing it in computer storage. The process of computing the spectrum of the recorded signal generally amounts to ***running a Fourier transform program*** based upon the FFT. In effect, the Fourier transform integral is approximated or estimated by a computational algorithm. The FFT is particularly attractive not only because it has a well-understood interpretation as an approximation to the Fourier transform integral, but also because it runs significantly faster than other methods. For this reason, the FFT is used by experts and neophytes alike to extract the spectral content of signals, and gain understanding of their data. However, FFT spectrum analysis is not without some notable pitfalls. With a signal processing viewpoint, it is possible to understand how the FFT approximates the Fourier transform integral, and thus avoid embarrassing misinterpretation of data.

This chapter provides an introduction to these issues so that "computing the spectrum" leads to well-informed interpretation of data.

In addition to describing the technical details of spectrum computation, the primary goal of this chapter is to show how an FFT-computed spectrum relates to the "true" spectral content of a signal given by the Fourier transform integral. A secondary goal is to introduce the rudimentary ideas of what is commonly called time-frequency analysis. In one type of time-frequency analysis, we analyze a very long signal by doing many short FFTs, which are then assembled into a single grayscale image, called a ***spectrogram***. The resulting image shows the spectrum analyst how a localized frequency spectrum evolves with time. True understanding of the spectrogram requires study at an advanced level, but our discussion should help the reader to appreciate the limitations as well as the power of time-frequency spectrum analysis and perhaps provide motivation for further study.

13-1 Finite Fourier Sum

We have seen that the Fourier transform of a continuous-time signal $x_c(t)$ defined as[1]

The Continuous-Time Fourier Transform

$$X_c(j\omega) = \int_{-\infty}^{\infty} x_c(t)e^{-j\omega t}dt \tag{13.1}$$

is the formal mathematical embodiment of the spectrum concept introduced in Chapter 3 and used throughout this text. Because the spectrum is so useful, it is natural to want to be able to determine the spectrum of a signal even if it is not described by a simple mathematical formula. Indeed, given the power of digital computation, it would be handy to have a computer program that could calculate the spectrum from samples of a signal.

One simple approximation to (13.1) can be obtained by replacing the integral by a sum. This is justified by the fact that the Riemann integral is derived as the limit of a sum of samples of the integrand function. The Riemann sum becomes the Riemann integral as the spacing between samples approaches zero. With finite spacing between the samples, we get a sum (13.2) that approximates the integral in (13.1)

$$\hat{X}_c(j\omega) = \sum_{n=-\infty}^{\infty} x_c(nT_s)e^{-j\omega nT_s}T_s \tag{13.2}$$

where we have replaced t by nT_s, dt by the time increment between samples, T_s, and the integral by a sum over the nonzero samples of the integrand $x_c(nT_s)e^{-j\omega nT_s}$. As $T_s \to 0$, this sum becomes a very good approximation to the Fourier integral of a smooth[2] function, although there is always some approximation error when T_s is nonzero. There are two other major issues with (13.2). First, ω is a continuous variable, but we cannot compute (13.2) for all values of ω. We must be satisfied with a finite set of frequencies, denoted by ω_k. Second, the limits on the sum in (13.2) are infinite, which means that the sum is taken over all nonzero values of the integrand. An infinite duration signal $x_c(t)$ would require an infinite sum, but we have to limit the sum to a finite number of terms in order to compute it. For example, we could compute an approximation to the Fourier transform of a right-sided signal by using the equation

$$\hat{X}_c(j\omega_k) = T_s\sum_{n=0}^{L-1} x[n]e^{-j\omega_k nT_s} \tag{13.3}$$

[1]We will use the subscript $_c$ when we need to emphasize that $x_c(t)$ is a function of a continuous time variable.

[2]By virtue of the sampling involved in (13.2), we implicitly assume a bandlimited signal for $x_c(t)$, and such signals are very smooth.

The upper limit L would have to be large enough so that the signal samples $x[n] = x_c(nT_s)$ are either zero or negligibly small outside the interval $0 \leq nT_s < LT_s$.

We can choose a finite set of frequencies, $\{\omega_k\}$ for analysis, but which frequencies should be used to evaluate (13.3)? Because the approximation of (13.3) is based on sampled data $x[n] = x_c(nT_s)$, we know that there must be an underlying periodicity[3] to the spectrum $\hat{X}_c(j\omega_k)$. Therefore, it will be necessary and sufficient to limit our evaluation to a band of frequencies of width $\omega_s = 2\pi/T_s$. For reasons that will become apparent, it is common to choose that interval to be

$$0 \leq \omega_k < \frac{2\pi}{T_s}$$

and to evaluate (13.3) at the N equally spaced frequencies

$$\omega_k = \frac{2\pi k}{NT_s} \qquad k = 0, 1, \ldots, N-1 \tag{13.4}$$

Substituting (13.4) into (13.3) leads to

$$\frac{1}{T_s}\hat{X}_c\left(j\frac{2\pi k}{NT_s}\right) = \sum_{n=0}^{L-1} x[n]e^{-j(2\pi k/N)n} \tag{13.5}$$

for $k = 0, 1, \ldots, N-1$. Equation (13.5) is the finite "Fourier sum" that we have been seeking. If we define $X[k] = (1/T_s)\hat{X}_c\left(j\frac{2\pi k}{NT_s}\right)$, then $X[k]$ is an approximation to $(1/T_s)X(j\omega)$ at a discrete set of frequencies $\omega_k = 2\pi/(NT_s)$, and $X[k]$ can be obtained by a finite computation specified by the right-hand side of (13.5), which we write as

The Discrete Fourier Transform

$$X[k] = \sum_{n=0}^{L-1} x[n]e^{-j(2\pi/N)kn} \tag{13.6}$$
$$k = 0, 1, \ldots, N-1$$

[3]With sampling, we can expect that if the signal is not bandlimited, aliasing would also be an issue in determining the accuracy of the approximation in (13.3).

Note that (13.6) is a function only of k, which is the discrete frequency index.[4] Equation (13.6) is called the ***discrete Fourier transform*** or ***DFT*** in recognition of the fact that it is indeed a Fourier-like transformation, and it is discrete in both time and frequency.

Generally, we assume that $L = N$ in (13.6) because, when $L = N$, we have an exact ***inverse discrete Fourier transform*** and also efficient algorithms that exist for computing the N complex numbers specified by (13.6). These algorithms for implementing the efficient computation of the DFT are collectively known as the ***fast Fourier transform*** or ***FFT***. In MATLAB, the computation of (13.6) is done by the function called `fft()`. We will have more to say about the FFT in Sections 13-5.3 and 13-9 after learning more about the properties of the DFT and its use in spectrum analysis.

13-2 Too Many Fourier Transforms?

Often students protest with some justification that there are too many Fourier transforms to learn. Indeed, at this point, it may seem that there is a different Fourier transform for each type of signal. In a sense this is true, but it is really not as confusing as it may seem. We have just defined the DFT (discrete Fourier transform) in (13.6) and noted that it is a Fourier transform where both time and frequency are discrete. We arrived at the DFT by considering an approximation to the CTFT (continuous-time Fourier transform), which is the appropriate choice when both time and frequency are continuous variables. In Chapter 11, we showed that the continuous-time Fourier series can be considered a special case of the CTFT where time is continuous and frequency is discrete. Finally, in Chapter 12, we

[4]Because of the frequency-domain periodicity resulting from sampling of $x_c(t)$, the frequencies $\omega_k = 2\pi k/(NT_s)$ for $N/2 < k \leq N-1$ actually correspond to the negative frequencies $\omega_{k-N} = 2\pi(k-N)/(NT_s)$ of $X_c(j\omega)$. We will discuss this point further in Section 13-5.

Table 13-1: Types of Fourier transforms.

	Discrete-Time	Continuous-Time
Discrete Frequency	DFT, $X[k]$	Fourier Series $\{a_k\}$
Continuous Frequency	DTFT, $X(e^{j\hat{\omega}})$	CTFT, $X_c(j\omega)$

introduced the DTFT (discrete-time Fourier transform) for the situation where time is discrete and frequency is a continuous variable (see Table 13-1). While the definitions of these Fourier transforms are superficially different and while it is important to comprehend the subtle distinctions implicit in the terminology, the basic structure is the same for all, i.e., the signal is multiplied by a complex-exponential signal with variable frequency and the product is either summed or integrated depending on whether the time variable is discrete or continuous. This accounts for the fact that the basic properties of the different transforms are very similar; e.g., convolution in the time-domain always implies multiplication in the frequency-domain, and delay in time always implies multiplication of the Fourier transform by a complex exponential in the frequency-domain.

Not surprisingly, these different Fourier transforms are related through sampling. Indeed, our understanding of sampling and aliasing is heavily dependent on a frequency-domain point-of-view. So before we can understand how the Fourier spectrum of a signal can be computed using the DFT, we need to understand how the DFT, the DTFT and the CTFT are related. The key result has already been derived in Section 12-3.5, where it was shown that the DTFT of a sampled signal is equal to the sum of an infinite set of shifted and scaled copies of the CTFT of the continuous-time signal. (See (13.8) below.) This implies that *if the continuous-time signal is bandlimited, then each period of the DTFT of the corresponding sampled signal is a perfect (scaled) copy of the CTFT.* We will see that this is what allows us to estimate the Fourier spectrum of a continuous-time signal by using digital computation.

In this section, we will review and summarize the relationships among the different Fourier transforms that we have discussed. This will serve as the basis for a deeper understanding of spectrum analysis of sampled signals.

13-2.1 Relation of the DTFT to the CTFT

The Fourier transform of a discrete-time signal $x[n]$ is called the ***discrete-time Fourier transform (DTFT)***

The Discrete-Time Fourier Transform

$$X(e^{j\hat{\omega}}) = \sum_{n=-\infty}^{\infty} x[n]e^{-j\hat{\omega}n} \tag{13.7}$$

The DTFT is a function of a continuous frequency variable $\hat{\omega}$ and $X(e^{j\hat{\omega}})$ is always periodic with period 2π. If the sequence $x[n]$ was obtained by sampling a continuous-time signal $x_c(t)$ with sampling period T_s, then $x[n] = x_c(nT_s)$. We showed in Section 12-3.5, equation (12.61) on p. 376, that the DTFT of $x[n]$ and the CTFT of $x_c(t)$ are related by

$$X(e^{j\omega T_s}) = \frac{1}{T_s}\sum_{\ell=-\infty}^{\infty} X_c(j(\omega - \ell\omega_s)) \tag{13.8}$$

where $\omega_s = 2\pi/T_s$.

In the particular case of a bandlimited signal where $X_c(j\omega) = 0$ for $|\omega| \geq \pi/T_s$, the shifted copies $X_c(j(\omega - \ell\omega_s))$ do not overlap (i.e., no aliasing distortion occurs). As a result, it is true that

$$X(e^{j\omega T_s}) = \frac{1}{T_s}X_c(j\omega) \quad \text{for } |\omega| < \pi/T_s \tag{13.9}$$

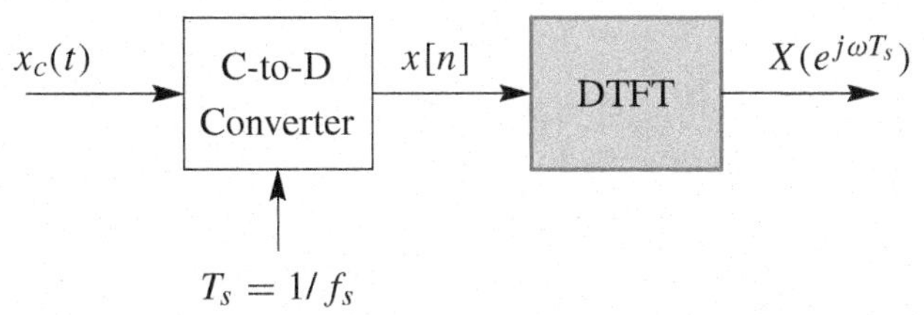

Figure 13-1: Computing the Fourier transform of a sampled signal by computing the DTFT. When the input $x_c(t)$ is bandlimited and the sampling rate is above the Nyquist rate, the DTFT is proportional to the CTFT; i.e., $X(e^{j\omega T_s}) = \frac{1}{T_s} X_c(j\omega)$.

This important result states that the CTFT of a bandlimited signal can be determined to within a constant multiplier by determining the DTFT of its samples, $x[n] = x_c(nT_s)$. This is the basis for using digital computation in spectrum analysis. Figure 13-1 shows how the DTFT would be used to compute the spectrum of a continuous-time bandlimited signal.

13-2.2 Relation of the DFT to the DTFT

The DFT is sampled in frequency, so how does the DFT fit into this picture? If we only consider finite-length sequences $x[n]$ such that $x[n] \neq 0$ within the interval $0 \leq n \leq L-1$, then the DTFT is

$$X(e^{j\hat{\omega}}) = \sum_{n=0}^{L-1} x[n] e^{-j\hat{\omega} n} \tag{13.10}$$

The DFT is obtained from the DTFT by evaluating (13.10) at a discrete set of equally spaced frequencies $\hat{\omega}_k = (2\pi/N)k$, for $k = 0, 1, \ldots, N-1$

$$X(e^{j(2\pi/N)k}) = \sum_{n=0}^{L-1} x[n] e^{-j(2\pi/N)kn} = X[k] \tag{13.11}$$

13-2.3 Relation of the DFT to the CTFT

To make the connection between the DFT and the CTFT, we combine (13.8) and (13.11) to obtain

$$\begin{aligned} X[k] &= X(e^{j(2\pi/N)k}) = X(e^{j\omega T_s})\Big|_{\omega=\frac{2\pi k}{NT_s}} \\ &= \frac{1}{T_s} \sum_{\ell=-\infty}^{\infty} X_c\left(j\left(\omega - \ell\frac{2\pi}{T_s}\right)\right)\Bigg|_{\omega=\frac{2\pi k}{NT_s}} \end{aligned} \tag{13.12}$$

The mathematical relationship (13.12) between the DFT of a finite segment of a sequence of samples and the CTFT of the corresponding continuous-time signal involves three issues:

- The DFT is a function of the frequency index k which can be related to a CTFT frequency, ω_k, via

$$k = \left(\frac{NT_s}{2\pi}\right)\omega_k \tag{13.13}$$

- The DFT uses only a finite-length segment of the sampled signal.
- The DFT evaluates the spectrum at only a discrete set of frequencies.

In the next section, we will address the finite-length issue, because it has the biggest impact on the appearance of the spectrum.

13-3 Time-Windowing

In order to use the DFT to determine the spectrum of a sampled signal, the length of the sequence must be finite. If the signal is infinitely long, we would have to select a finite number of samples for Fourier analysis, but we are free to choose any finite segment of the complete sequence. It is possible to represent this selection process as the multiplication of the complete sequence $x[n]$ by

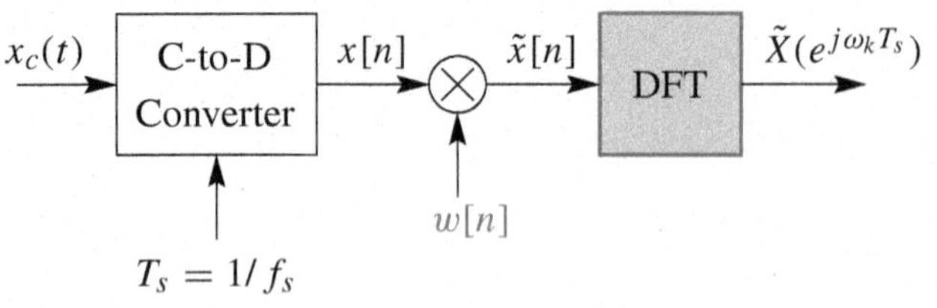

Figure 13-2: Discrete-time spectrum analysis using time-domain windowing and the DFT. The window $w[n]$ should be a finite-length sequence.

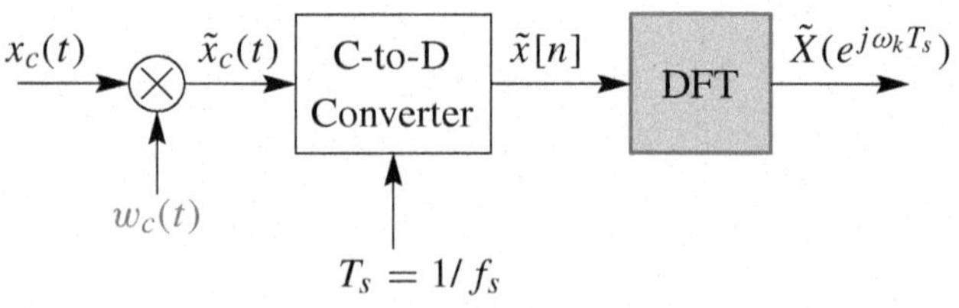

Figure 13-3: Equivalent system for discrete-time spectrum analysis using continuous-time windowing.

another sequence $w[n]$ that is generally called a *data window*, or simply a ***window***. For example, the window

$$w[n] = \begin{cases} 1 & 0 \le n \le L-1 \\ 0 & \text{otherwise} \end{cases} \tag{13.14}$$

is zero everywhere outside of a finite interval. The block diagram in Fig. 13-2 shows the signal $x_c(t)$ being sampled to produce the sequence $x[n]$, which is then truncated in time by the data window. If the window is $w[n]$ in (13.14), then $\tilde{x}[n]$ in Fig. 13-2 consists of L samples from $x[n]$.

The window in (13.14) is often called the ***rectangular window*** because of the shape of its plot. As defined in (13.14), the rectangular window is most useful for causal signals that start at $n = 0$, but shifted versions of the rectangular windows can also be used. Note that we have used L for the length of the window. Often $L = N$, the DFT length; however, there is no need to make this restriction unless an invertible DFT is needed, in which case L should satisfy $L \le N$.

Figure 13-2 makes it clear that the DFT computation is really the computation of "the DFT of the windowed signal" $\tilde{x}[n] = w[n]x[n]$, **not** the signal $x[n]$. Thus, understanding the spectrum computed by the DFT requires understanding how time-windowing changes the Fourier transform of the signal. To facilitate this, it is helpful to point out that the windowing operation can be applied either to the discrete-time signal as in Fig. 13-2, or to the original continuous-time signal as shown in Fig. 13-3. All that is required for Figs. 13-2 and 13-3 to be equivalent is that the window sequences be related by $w[n] = w_c(nT_s)$. If this is true, then it follows that

$$\begin{aligned} \tilde{x}[n] &= \tilde{x}_c(nT_s) \\ &= w_c(nT_s)x(nT_s) = w[n]x[n] \end{aligned} \tag{13.15}$$

making Figs. 13-2 and 13-3 equivalent.

For example, if the discrete-time window is the *rectangular window* in (13.14), then an equivalent continuous-time data window would be

$$w_c(t) = w_R(t) = \begin{cases} 1 & 0 \le t \le T \\ 0 & \text{otherwise} \end{cases} \tag{13.16}$$

where $(L-1)T_s < T < LT_s$.

We are now ready to explore the effect of time-windowing on the Fourier spectrum computation. To do this, we examine the DFT of the windowed sequence $\tilde{x}[n]$ as defined by

$$\tilde{X}(e^{j\omega_k T_s}) = \sum_{n=0}^{L-1} w[n]x[n]e^{-j\omega_k T_s n} \qquad 0 \le k \le N-1 \tag{13.17}$$

where the analysis frequencies are $\omega_k = 2\pi k/(NT_s)$. The window sequence $w[n]$ imposes finite limits on the sum so that (13.17) can be computed with the FFT.

EXERCISE 13.1: Show that the continuous-time windowing operation $\tilde{x}_c(t) = w_c(t)x_c(t)$, when viewed as a system with input $x_c(t)$ and output $\tilde{x}_c(t)$, is a *linear* system. Also show that the windowing system is *not time-invariant.*

13-4 Analysis of a Sum of Sinusoids

To illustrate and analyze the effects of time-windowing, we will consider a specific example from which we can draw some general conclusions. Let the signal be

$$x_c(t) = A_0 + A_1 \cos(\omega_1 t + \phi_1) \qquad (13.18)$$

for $-\infty < t < \infty$. This clearly is a bandlimited signal, and it is representative of all signals that can be written as a *finite* sum of sinusoids. Therefore, by choosing the sampling rate such that $\omega_s = 2\pi/T_s > 2\omega_1$ we can be assured that no aliasing distortion results from the initial sampling in Fig. 13-2. The Fourier transform (CTFT) of this signal is

$$X_c(j\omega) = 2\pi A_0 \delta(\omega) + \pi X_1 \delta(\omega - \omega_1) + \pi X_1^* \delta(\omega + \omega_1) \qquad (13.19)$$

where $X_1 = A_1 e^{j\phi_1}$ is the complex amplitude representation of the cosine at frequency ω_1. Note that in this case, the spectrum consists of three isolated spectral components represented by impulses at DC and at $\pm\omega_1$, the positive and negative frequencies of the sinusoid.

Now consider a new signal $\tilde{x}_c(t)$ formed by time-windowing $x_c(t)$

$$\tilde{x}_c(t) = w_c(t)x_c(t) \qquad (13.20)$$
$$= A_0 w_c(t) + \tfrac{1}{2} X_1 w_c(t) e^{j\omega_1 t} + \tfrac{1}{2} X_1^* w_c(t) e^{-j\omega_1 t}$$

where $X_1 = A_1 e^{j\phi_1}$. Using the frequency-shift property of the CTFT, it follows that the Fourier transform of $\tilde{x}_c(t)$ is

$$\tilde{X}_c(j\omega) = A_0 W_c(j\omega) + \tfrac{1}{2} X_1 W_c(j(\omega - \omega_1)) + \tfrac{1}{2} X_1^* W_c(j(\omega + \omega_1)) \qquad (13.21)$$

where $W_c(j\omega)$ is the CTFT of the window. From (13.21) we can see that the effect of time-windowing a sum of complex exponentials is to create a set of frequency-shifted copies of $W_c(j\omega)$, each scaled by the complex amplitude of the original complex-exponential term. If (13.18) contains more sinusoids at different frequencies, then (13.21) will contain more shifted and scaled copies of $W_c(j\omega)$.

EXERCISE 13.2: Suppose that the input to the spectrum analysis system of Fig. 13-2 is

$$x_c(t) = A_0 + \sum_{r=1}^{R} A_r \cos(\omega_r t + \phi_r) \qquad (13.22)$$

Determine an expression for the Fourier transform of the windowed signal $\tilde{x}_c(t) = w_c(t)x_c(t)$.

EXERCISE 13.3: Show that the CTFT of $w_R(t)$ in (13.16) is

$$W_R(j\omega) = \frac{\sin(\omega T/2)}{(\omega/2)} e^{-j\omega T/2} \qquad (13.23)$$

The rectangular window given by (13.16) is plotted in Fig. 13-4(a). The magnitude of its Fourier transform

$$\left|W_R(j\omega)\right| = \left|\frac{\sin(\omega T/2)}{(\omega/2)}\right| \qquad (13.24)$$

is plotted in Fig. 13-4(b). Note that the frequency width is inversely proportional to the duration of $w_R(t)$.

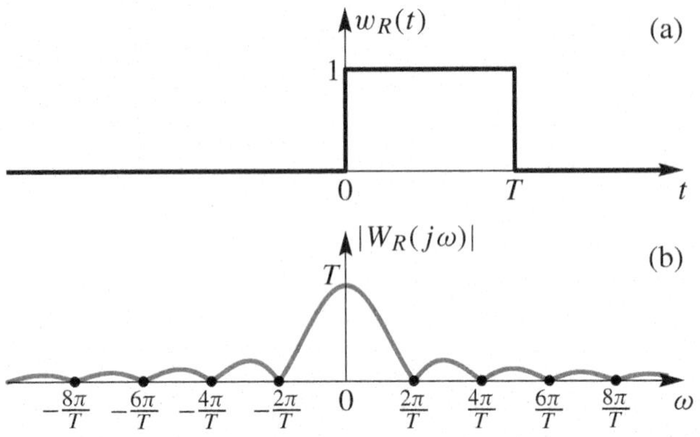

Figure 13-4: (a) Rectangular time window and (b) its Fourier transform. The part of the plot from $-2\pi/T$ to $2\pi/T$ is the *mainlobe*, while the *sidelobes* are evident for $|\omega| > 2\pi/T$.

A measure of the frequency width of $W_R(j\omega)$ around $\omega = 0$ is often taken to be the width of the ***mainlobe*** between the first zero crossings, which is $\Delta\omega = 4\pi/T$ in this case. We can see that as T increases, the time window becomes longer but $\Delta\omega$ decreases, so the Fourier transform becomes more concentrated around $\omega = 0$. Also note that the ***sidelobes*** have significant amplitude.

In the case of sinusoidal signals like (13.18) and (13.22), the effect of time windowing is to replace each impulse in the spectrum by a scaled and shifted copy of $W_R(j\omega)$ whose magnitude is shown in Fig. 13-4(b), thereby "blurring" and spreading the sharp spectral lines of the ideal bandlimited spectrum. This is illustrated by the following example.

Example 13-1: Numerical Example

Consider the special case of (13.18)

$$x_c(t) = 0.5 + 0.8\cos(\omega_1 t) \qquad (13.25)$$

That is, $A_0 = 0.5$, $A_1 = 0.8$, $\phi_1 = 0$, so that $X_1 = A_1 = 0.8$. The rectangular window is used with a window

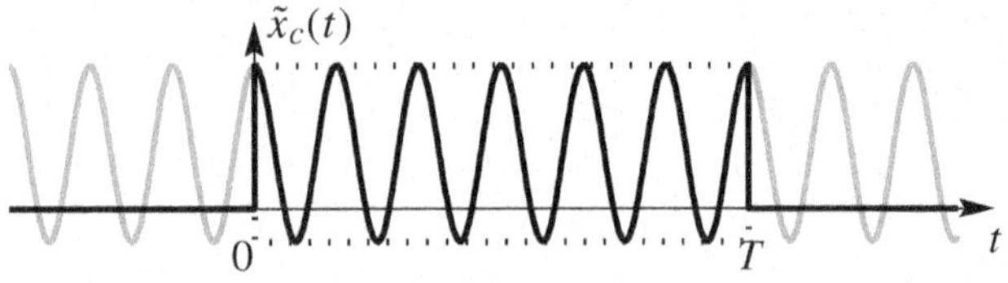

Figure 13-5: Waveform of sinusoidal signal $x_c(t) = 0.5 + 0.8\cos(\omega_1 t)$. Dark black line indicates portion selected by the rectangular window.

duration T chosen so that the frequency of the cosine component satisfies $\omega_1 = 3\Delta\omega = 12\pi/T$. The effect of time-windowing is shown in Figs. 13-5 and 13-6. Figure 13-5 shows the signal $x_c(t)$ defined in (13.25). The part of the signal with the black solid line is the part selected by the rectangular window in Fig. 13-4(a). The light gray part of the signal is the part of the ideal bandlimited cosine signal eliminated by the window.

Figure 13-6(a) shows the Fourier transform of the ideal bandlimited cosine signal, Fig. 13-6(b) shows the magnitudes of the three terms in (13.21), and Fig. 13-6(c) shows the magnitude of $\tilde{X}_c(j\omega)$, the (complex) sum of the three terms in (13.21). If we compare Fig. 13-6(a) to Fig. 13-6(c), we observe that the perfectly sharp spectral lines represented by the impulses in the ideal spectrum have been blurred by the windowing operation. Furthermore, since the Fourier transform of the window $W_c(j\omega)$ becomes more concentrated as the time window becomes longer, it follows that the way to improve the ***frequency resolution*** (reduce the blurring) is to use a longer time window. Indeed, since the width of the Fourier transform of the window is roughly $\Delta\omega = 4\pi/T$, it follows that the impulses in the ideal spectrum must be at least $\Delta\omega$ apart in frequency, or else the shifted copies of the Fourier transform of the window will overlap in forming $\tilde{X}_c(j\omega)$. Since in this case, $\omega_1 = 3\Delta\omega$, the DC component and the complex-

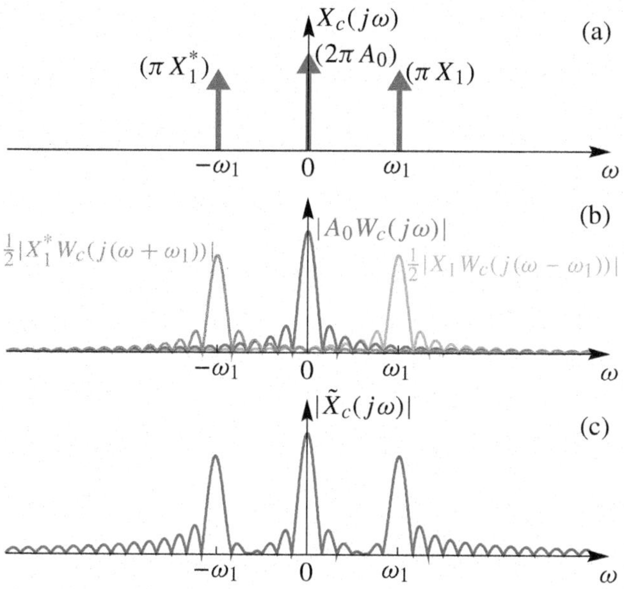

Figure 13-6: Spectrum of sinusoidal signal with rectangular window. (a) Fourier transform of infinite-duration cosine signal; (b) Fourier transform magnitudes of individual windowed complex-exponential terms; and (c) Fourier transform magnitude of windowed cosine signal.

exponential components at $\pm\omega_1$ are separated enough that we see three distinct peaks in the spectrum of the windowed signal. Each of these peaks is located close to one of the frequencies in the signal, and the height of each peak is approximately proportional to the amplitude of the corresponding complex-exponential component. ■

One of the major shortcomings of the rectangular window is that its sidelobes are relatively high and could be confused with low amplitude peaks due to other signals. The ***Hamming window***[5] (and other tapered shapes) eases this sidelobe problem at the expense of a

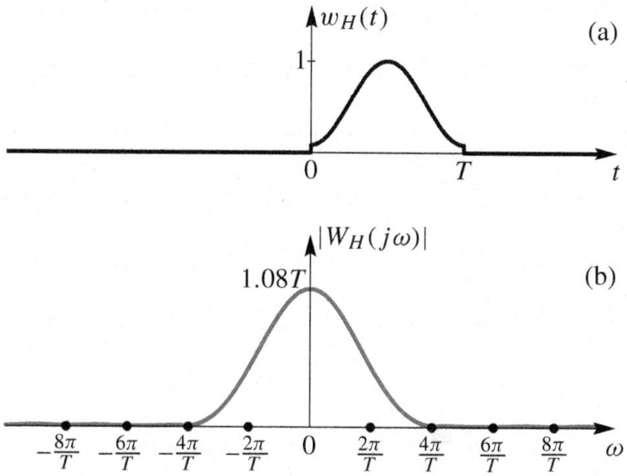

Figure 13-7: (a) Hamming time window and (b) its Fourier transform magnitude. The part of the plot from $-4\pi/T$ to $4\pi/T$ is the *mainlobe,* while the *sidelobes* are hardly visible over the range $|\omega| > 4\pi/T$.

wider mainlobe. The definition of the Hamming window shown in Fig. 13-7(a) is

$$w_H(t) = \begin{cases} 0.54 - 0.46\cos\left(\frac{2\pi}{T}t\right) & 0 \le t \le T \\ 0 & \text{otherwise} \end{cases} \tag{13.26}$$

Observe that the window tapers to near zero rather than dropping abruptly at the edges like the rectangular window. In the frequency-domain, Fig. 13-7(b), the mainlobe of $W_H(j\omega)$ is twice as wide ($\Delta\omega = 8\pi/T$) as that of $W_R(j\omega)$ for the rectangular window. On the other hand, the sidelobes are much lower—for the Hamming window they are not even visible in Fig. 13-7(b). Like the rectangular window, increasing the window's time duration T decreases the frequency width of its mainlobe; the sidelobes will remain at the same low level.

[5]Named for its inventor Richard W. Hamming

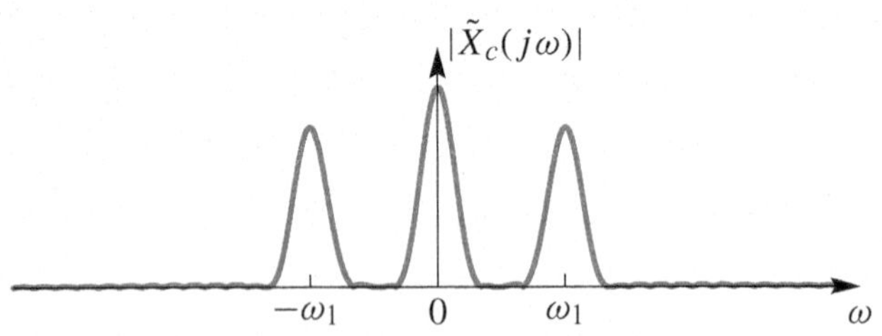

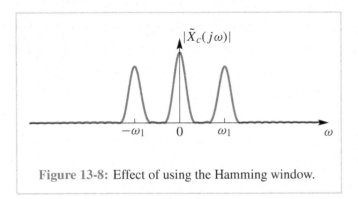

Figure 13-8: Effect of using the Hamming window.

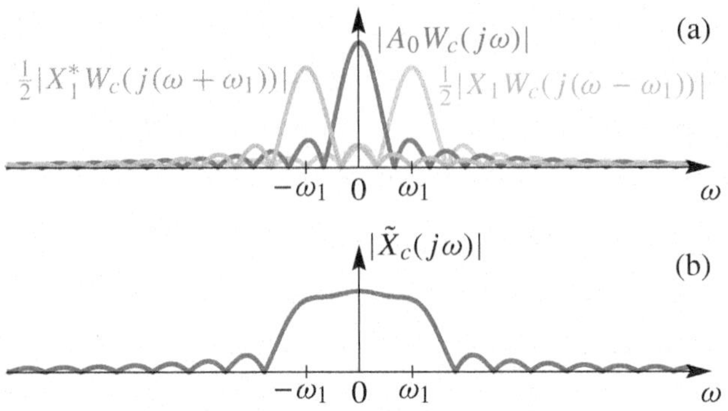

Figure 13-9: Spectrum of sinusoidal signal for short time window. (a) Fourier transform magnitudes of individual windowed complex-exponential terms and (b) Fourier transform magnitude of windowed cosine signal. (Scale enlarged compared to Fig. 13-6.)

The spectrum will look different depending upon which time window is used.

For the Hamming window, Fig. 13-8 shows the Fourier transform of $\tilde{x}_c(t) = w_H(t)x(t)$ when $x(t)$ is given by (13.25) with T and ω_1 the same as in Fig. 13-6. This plot has no confusing ripples caused by sidelobes, but a comparison to Fig 13-6(c) shows that the blurred spectrum lines are about twice as wide as those for the rectangular window.

No matter whether we use a rectangular or a Hamming window, if too much overlap occurs, we will no longer see distinct peaks around each of the original frequencies. This is illustrated in Fig. 13-9 where the three individual peaks merge into one wide spectral peak in Fig. 13-9(b). In this case, $\omega_1 = \frac{3}{4}\Delta\omega = 3\pi/T$ and the overlapping copies of $W_R(j\omega)$ fill in the entire region around $\omega = 0$. As a result, the complex-exponential components at $\pm\omega_1$ are not *resolved* from the DC component.

The phenomenon illustrated by Figs. 13-6 and 13-9 is often called the ***uncertainty principle*** of Fourier analysis. Specifically, in order to resolve all the frequencies in a sum of sinusoidal components, we need to take a window length T such that mainlobe width $\Delta\omega$ is less than the minimum spacing between adjacent sinusoidal frequencies. For the rectangular window, $\Delta\omega = 4\pi/T$, but if we use a Hamming window, the window length must be roughly twice the length of a rectangular window in order to have the same *frequency resolution* power.[6]

13-4.1 DTFT of a Windowed Sinusoid

Our main objective is to have a method for computing the spectrum, so we must use a system such as Fig. 13-2, where the windowing is applied *after* the C-to-D converter. The actual calculation will be frequency samples of the DTFT of the windowed discrete-time sequence

$$\tilde{X}(e^{j\omega_k T_s}) = \sum_{n=-\infty}^{\infty} w_r[n]x[n]e^{-j\omega_k T_s n} \qquad (13.27)$$

[6]The mainlobe width can be measured in different ways. For example, if it were defined as the frequency width at the half-amplitude level, the result would be $2.41\pi/T$ for $|W_R(j\omega)|$ and $3.63\pi/T$ for $|W_H(j\omega)|$, which implies that the Hamming window's duration must be 1.5 times that of the rectangular window to obtain comparable frequency resolution.

evaluated at N frequencies, $\omega_k = 2\pi k/NT_s$, $k = 0, 1, \ldots, N-1$. Normally, the DTFT is a function of the discrete-time frequency $\hat{\omega}$, but we have made the substitution $\hat{\omega} = \omega T_s$ so that the spectrum $\tilde{X}(e^{j\omega T_s})$ is a function of continuous-time frequency in rad/s. For the specific case of $x_c(t)$ in (13.18), the computed spectrum would be as shown in Fig. 13-10. The number of frequency samples is $N = 32$, and the sample values are indicated with dots over the region $0 \le \omega < \omega_s = 2\pi/T_s$. The smooth curve is the DTFT that would be obtained if $N \to \infty$. The dots are the DFT samples.

Since the spectrum in Fig. 13-10 differs significantly from the true spectrum of the bandlimited input signal, it is worthwhile to point out the major differences and the reasons for these differences. First, the spectrum in Fig. 13-10 is periodic versus ω. Such periodicity always occurs in the DTFT of sampled signals because the DTFT $X(e^{j\hat{\omega}})$ is the periodic repetition of the CTFT $X_c(j\omega)$

$$X(e^{j\omega T_s}) = \frac{1}{T_s} \sum_{\ell=-\infty}^{\infty} X_c(j(\omega - \ell\omega_s))$$

A plot of the DTFT $X(e^{j\omega T_s})$ would have copies of the spectrum centered at $\pm\omega_s$, $\pm 2\omega_s$, and so on. Second, the spectrum exhibits the effects of windowing. The DTFT of the time-windowed signal, $\tilde{x}[n] = w_r[n]x[n]$, can be obtained by using the frequency shift property

$$\tilde{X}(e^{j\omega T_s}) = A_0 W_r(e^{j\omega T_s}) + \tfrac{1}{2} X_1 W_r(e^{j(\omega-\omega_1)T_s}) + \tfrac{1}{2} X_1^* W_r(e^{j(\omega+\omega_1)T_s}) \tag{13.28}$$

where $W_r(e^{j\hat{\omega}})$ is the DTFT of the sampled rectangular window, $w_r[n]$ in (13.14). The shape of $|W_r(e^{j\hat{\omega}})|$ is a Dirichlet function versus $\hat{\omega}$, which looks very similar to the sinc function shape for $|W_R(j\omega)|$, the magnitude of the CTFT of the continuous-time window, $w_R(t)$. For large L the Dirichlet and the sinc are essentially identical in the mainlobe region.

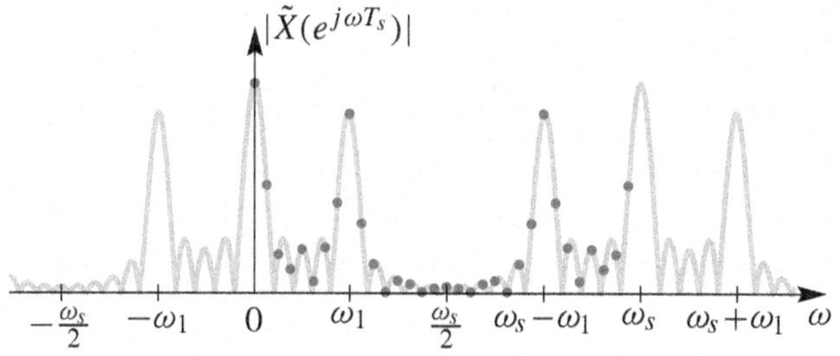

Figure 13-10: DTFT of windowed and sampled sinusoidal signal. The DTFT magnitude $|\tilde{X}(e^{j\omega_k T_s})|$ was evaluated at 32 frequencies $\omega_k = 2\pi/(32T_s)$, which are shown as dots on the smooth curve.

13-5 Discrete Fourier Transform

In (13.27) we showed that values of the DTFT of a finite segment of a sampled sequence can be computed by the equation

$$\tilde{X}(e^{j\omega_k T_s}) = \sum_{n=-\infty}^{\infty} \tilde{x}[n]e^{-j\omega_k T_s n} \qquad 0 \le k \le N-1 \tag{13.29}$$

where the analysis frequencies are $\omega_k = 2\pi k/(NT_s)$ and $\tilde{x}[n] = w[n]x[n]$ is the windowed time sequence. The window sequence $w[n]$ is a finite-length sequence such as the sampled rectangular window, or Hamming window. This equation can be simplified somewhat by assuming that the window length is N, and by substituting $\omega_k = 2\pi k/(NT_s)$ into (13.29) to obtain

$$\tilde{X}(e^{j(2\pi k/(NT_s))T_s}) = \sum_{n=0}^{N-1} \tilde{x}[n]e^{-j(2\pi k/(NT_s))T_s n} \qquad 0 \le k \le N-1$$

If we multiply the terms in the exponents, the T_s terms cancel and we get

$$\tilde{X}(e^{j(2\pi k/N)}) = \sum_{n=0}^{N-1} \tilde{x}[n]e^{-j(2\pi k/N)n} \qquad 0 \le k \le N-1 \tag{13.30}$$

which should be recognized as the ***discrete Fourier transform***, or ***DFT***, of the sequence $\tilde{x}[n]$, i.e.,

$$\tilde{X}[k] = \sum_{n=0}^{N-1} \tilde{x}[n]e^{-j(2\pi k/N)n} \qquad 0 \le k \le N-1 \tag{13.31}$$

The DFT is discrete (sampled) in both time and frequency. It takes n samples in the time-domain and produces N samples $X[k]$ in the frequency-domain.

13-5.1 The Inverse DFT

The DFT (13.31) has been introduced by showing that it results from computing the DTFT of a finite-length sequence at a finite set of frequencies. Equation (13.31) can in fact be applied to *any* finite-length sequence whether it resulted from time-windowing a sampled signal or not. Adding to the interest in the DFT is the fact that there exists an *inverse discrete Fourier transform* (or IDFT); i.e., a computation that converts $\tilde{X}[k]$ for $k = 0, 1, \ldots, N-1$ back into the sequence $\tilde{x}[n]$ for $n = 0, 1, \ldots, N-1$. The inverse DFT is

$$\tilde{x}[n] = \frac{1}{N}\sum_{k=0}^{N-1} \tilde{X}[k]e^{j(2\pi k/N)n} \qquad 0 \le n \le N-1 \tag{13.32}$$

Equations (13.32) and (13.31) are a unique relationship between a sequence $\tilde{x}[n]$ and its DFT $\tilde{X}[k]$. Following our earlier terminology for Fourier representations, the DFT defined by (13.31) is the *analysis* equation and IDFT defined by (13.32) is the *synthesis* equation. To prove that these equations are a consistent Fourier representation, we note that (13.32), because it is a finite, well-defined computation, would surely produce *some* sequence when evaluated for $n = 0, 1, \ldots, N-1$. So let us call that sequence $v[n]$ until we prove otherwise. Part of the proof is given by the following steps:

$$\begin{aligned} v[n] &= \frac{1}{N}\sum_{k=0}^{N-1} \tilde{X}[k]e^{j(2\pi k/N)n} \\ &= \frac{1}{N}\sum_{k=0}^{N-1}\left(\sum_{m=0}^{N-1} \tilde{x}[m]e^{-j(2\pi k/N)m}\right)e^{j(2\pi k/N)n} \\ &= \frac{1}{N}\sum_{m=0}^{N-1} \tilde{x}[m]\left(\sum_{k=0}^{N-1} e^{j(2\pi k/N)(n-m)}\right) \\ &= \tilde{x}[n] \end{aligned} \tag{13.33}$$

Several things happened in the manipulations leading up to the assertion of (13.33). On the second line, we substituted the right-hand side of (13.31) for $\tilde{X}[k]$ after changing the index of summation from n to m. We are allowed to make this change because this is a "dummy index" that can have any label, and we need to keep n reserved for the index of the sequence $v[n]$ that is synthesized by the IDFT. On the third line, the summations on k and m were interchanged. This is permissable since these are finite sums and can be done in any order. Now we need to consider the term in parenthesis in the third line of the proof. The following exercise states the result, which can be easily verified.

EXERCISE 13.4: Use the formula

$$\sum_{k=0}^{N-1} \alpha^k = \frac{1-\alpha^N}{1-\alpha}$$

to show that

$$\sum_{k=0}^{N-1} e^{j(2\pi k/N)(n-m)} = \frac{1-e^{j(2\pi)(n-m)}}{1-e^{j(2\pi/N)(n-m)}}$$

$$= \begin{cases} N & n-m=rN \\ 0 & \text{otherwise} \end{cases} \qquad (13.34)$$

where r is any positive or negative integer including $r = 0$.

If we substitute (13.34) into the third line of (13.33), we see that the only term in the sum on m that will be nonzero is the term corresponding to $m = n$ where $0 \le n \le N-1$. This means that $v[n] = \tilde{x}[n]$ for $0 \le n \le N-1$ as we wished to show.

13-5.2 Summary of the DFT Representation

The discussion so far has been concerned with the derivation of the discrete Fourier transform defined by (13.31) and its inverse defined by (13.32). These equations are rewritten together as follows:

$$X[k] = \sum_{n=0}^{N-1} x[n]e^{-j(2\pi/N)kn} \qquad k = 0, 1, \ldots, N-1 \qquad (13.35)$$

$$x[n] = \frac{1}{N}\sum_{k=0}^{N-1} X[k]e^{j(2\pi/N)kn} \qquad n = 0, 1, \ldots, N-1 \qquad (13.36)$$

Equation (13.35) is the (forward) DFT and (13.36) is the inverse DFT, or IDFT.[7] Note that we have removed the tildes from x and X in writing these equations again. This is just for notational convenience. The tildes were originally introduced in our discussion to denote the windowing that occurred in limiting the number of terms in the DTFT computation. Henceforth, we will always deal with finite-length sequences, so the tildes will usually be suppressed.

A final comment is in order about the limits of summation in (13.35). We have chosen to sum over N samples of $x[n]$ and to evaluate the DFT at N frequencies $\omega_k = 2\pi k/N$. Often, the sequence $x[n]$ has length L such that $L < N$. In such cases, we simply append $N-L$ zero samples to the nonzero samples of $x[n]$ in carrying out the computation. If, in fact, $x[n]$ is nonzero only in the interval $0 \le n \le L-1$, then appending zero samples is exactly the right thing to do.

Example 13-2: Short-Length DFT

In order to compute the 4-pt. DFT of the sequence $x[n] = \{1, 1, 0, 0\}$, we apply the definition for each value of $k = 0, 1, 2, 3$. To simplify the exponents in (13.35), we use $N = 4$, so that $2\pi/N = \pi/2$.

$$\begin{aligned} X[0] &= x[0]e^{-j0} + x[1]e^{-j0} + x[2]e^{-j0} + x[3]e^{-j0} \\ &= 1+1+0+0 = 2 \\ X[1] &= x[0]e^{-j0} + x[1]e^{-j\pi/2} + 0e^{-j\pi} + 0e^{-j3\pi/2} \\ &= 1+(-j)+0+0 = 1-j = \sqrt{2}e^{-j\pi/4} \\ X[2] &= x[0]e^{-j0} + x[1]e^{-j\pi} + 0e^{-j2\pi} + 0e^{-j3\pi} \\ &= 1+(-1)+0+0 = 0 \end{aligned}$$

[7] Note that we have used n as both the dummy summation index in (13.35) and the independent variable in (13.36), and vice versa for k, which is the dummy variable in (13.36). This usually causes no problem, but when necessary it is always possible to label the dummy indices of summation differently.

$$X[3] = x[0]e^{-j0} + x[1]e^{-j3\pi/2} + 0e^{-j3\pi} + 0e^{-j9\pi/2}$$
$$= 1 + (j) + 0 + 0 = 1 + j = \sqrt{2}e^{j\pi/4}$$

Thus we obtain the four DFT coefficients $X[k] = \{2,\ \sqrt{2}e^{-j\pi/4},\ 0,\ \sqrt{2}e^{j\pi/4}\}$. ■

13-5.3 The Fast Fourier Transform (FFT)

Both the DFT (13.35) and the IDFT summations (13.36) can be regarded simply as a computational method for taking N numbers in one domain and creating N (complex) numbers in the other domain. Equation (13.35) is really N separate summations, one for each value of k. To evaluate one of the $X[k]$ values,[8] we need N complex multiplications and $N-1$ complex additions. If we count up the arithmetic operations required to evaluate all of the $X[k]$ coefficients, the total is N^2 complex multiplications and $N^2 - N$ complex additions.

One of the most important discoveries[9] in the field of digital signal processing was the *fast Fourier transform*, or *FFT*, a set of algorithms that can evaluate (13.35) or (13.36) with a number of operations proportional to $N \log_2 N$ rather than N^2. When N is a power of two, the FFT algorithm computes the entire set of coefficients $X[k]$ with approximately $(N/2)\log_2 N$ complex operations. The $N \log_2 N$ behavior becomes increasingly significant for large N. For example, if $N = 1024$, the FFT will compute the set of coefficients $X[k]$ with $(N/2)\log_2 N = 5120$ complex multiplications rather than $N^2 = 1{,}048{,}576$, as required by direct evaluation of (13.35). The algorithm works best when the DFT length N is a power of two, but it also works efficiently if N has many small-integer factors. On the other hand, when N is a prime number, the standard FFT algorithm offers no savings over a direct evaluation of the DFT summation. FFT algorithms of many different variations are widely available in most computer languages and for almost any machine. In MATLAB, the command is simply `fft`, and most other spectral analysis functions in MATLAB call `fft` to do the bulk of their work.

More details on the FFT and its derivation can be found in Section 13-9 at the end of this chapter.

13-5.4 Negative Frequencies and the DFT

The signal defined in (13.36) has equally spaced normalized frequencies $\hat{\omega}_k = (2\pi/N)k$ over the positive frequency range, i.e.,

$$0 < (2\pi/N)k \le \pi \quad \text{for} \quad 0 < k \le N/2$$
$$\pi < (2\pi/N)k < 2\pi \quad \text{for} \quad N/2 < k < N-1$$

The index $k = N-1$ corresponds to the positive frequency $\hat{\omega}_{N-1} = 2\pi(N-1)/N$; however, since $\hat{\omega}_{N-1} = 2\pi(N-1)/N = 2\pi - 2\pi/N$, it follows that $\hat{\omega}_{N-1}$ is also the positive alias frequency of the negative frequency $\hat{\omega} = -(2\pi/N)$ because of the 2π periodicity in $\hat{\omega}$. Likewise, $k = N-2$ is the positive alias frequency of $\hat{\omega} = -(4\pi/N)$, etc.

When we have a real signal $x[n]$, there is conjugate symmetry in the spectrum, so the DFT coefficients satisfy the following constraints: $X[N-1] = X^*[1]$, $X[N-2] = X^*[2]$. In general, $X[N-k] = X^*[k]$ for $k = 0, 1, \ldots, N-1$.

Example 13-3: DFT Symmetry

In Example 13-2, the frequency index of the 4-pt. DFT corresponds to the four frequencies $\hat{\omega}_k = \{0,\ \pi/2,\ \pi,\ 3\pi/2\}$. Therefore, the DFT coefficients in Example 13-2 satisfy the conjugate-symmetric property, e.g., $X[1] = X^*[4-1] = X^*[3]$. ■

[8] We sometimes refer to these values as "coefficients" because they, in fact, are coefficients in (13.36) which looks like a Fourier series.

[9] J. W. Cooley and J. W. Tukey, "An Algorithm for the Machine Computation of Complex Fourier Series," *Mathematics of Computation,* vol. 19, pp. 297–301, April 1965. The basic idea of the FFT has been traced back as far as Gauss at the beginning of the 19th Century.

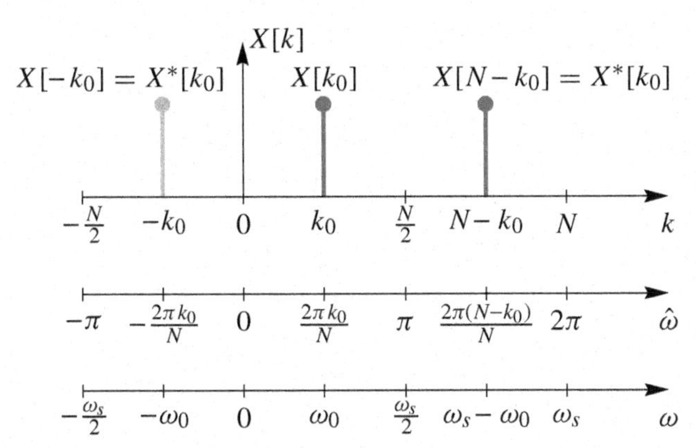

Figure 13-11: Illustration of the symmetry of the DFT coefficients showing that $X[N-k_0] = X^*[k_0]$.

Figure 13-11 shows a single DFT coefficient $X[k_0]$ corresponding to normalized frequency $\hat{\omega}_{k_0} = 2\pi k_0/N$ and the positive-frequency alias of the corresponding negative-frequency component (which is shown in gray). In addition, this graph has three "frequency scales." The top scale shows the DFT index k. The middle scale shows normalized frequency $\hat{\omega}$ for the discrete-time signal. The bottom scale shows the continuous-time frequency scale that would be appropriate if the sequence $x[n]$ had been obtained by sampling with a sampling frequency, $\omega_s = 2\pi/T_s$. Thus the DFT index k_0 corresponds to the analog frequency $\omega_0 = 2\pi k_0/(NT_s)$ rad/s.

EXERCISE 13.5: Create a MATLAB example that demonstrates the conjugate-symmetry property do `Xk=fft(1:8)`. List the values of the MATLAB vector `Xk` to verify that $X[N-k] = X^*[k]$ for $k = 0, 1, \ldots, 7$. In addition, list the value of $\hat{\omega}$ for each index k.

13-5.5 DFT Example

The DFT is used almost exclusively as a numerical operation that transforms a time-domain signal into its frequency-domain representation. However, in some important special cases, the finite sum can be "summed" to obtain a simple formula for the DFT coefficients. One such example is the DFT of a complex exponential.

Consider a general complex exponential whose frequency, $\hat{\omega}_0$, might or might not be an integer multiple of $2\pi/N$:

$$x_1[n] = e^{j(\hat{\omega}_0 n + \phi)} \qquad \text{for } n = 0, 1, 2, \ldots, N-1$$

The N-point DFT of $x_1[n]$ is

$$\begin{aligned}
X_1[k] &= \sum_{n=0}^{N-1} e^{j(\hat{\omega}_0 n+\phi)} e^{-j(2\pi k/N)n} \\
&= e^{j\phi} \sum_{n=0}^{N-1} e^{-j((2\pi k/N)-\hat{\omega}_0)n} \\
&= e^{j\phi}\Big(e^{-j(0)} + e^{-j((2\pi k/N)-\hat{\omega}_0)} \\
&\qquad + \cdots + e^{-j((2\pi k/N)-\hat{\omega}_0)(N-1)}\Big) \\
&= e^{j\phi}\frac{1-e^{-j((2\pi k/N)-\hat{\omega}_0)N}}{1-e^{-j((2\pi k/N)-\hat{\omega}_0)}} \qquad (13.37)
\end{aligned}$$

Notice that, in computing this result, we have not exploited any special information about the frequency $\hat{\omega}_0$. This would be the case if we had time-windowed a signal with a window of length N samples, but the window length did not contain an integer number of periods of the signal.

If we define $\theta = ((2\pi k/N) - \hat{\omega}_0)$ and substitute this into (13.37), then this last formula can be simplified somewhat by factoring out $e^{-j\theta N/2}$ from the numerator

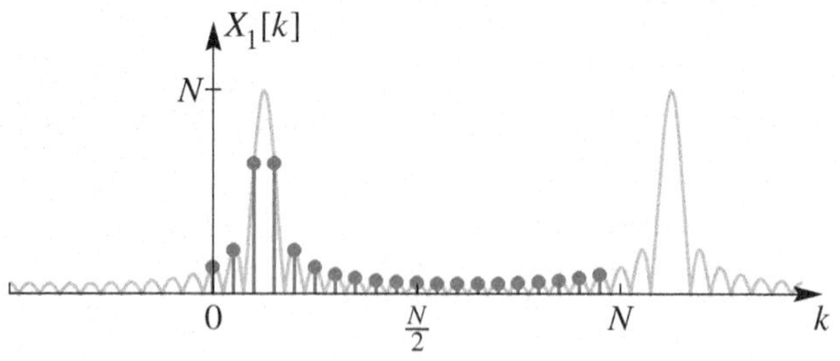

Figure 13-12: DFT of a complex exponential whose frequency is *not an integer multiple of* $2\pi/N$.

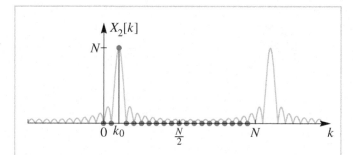

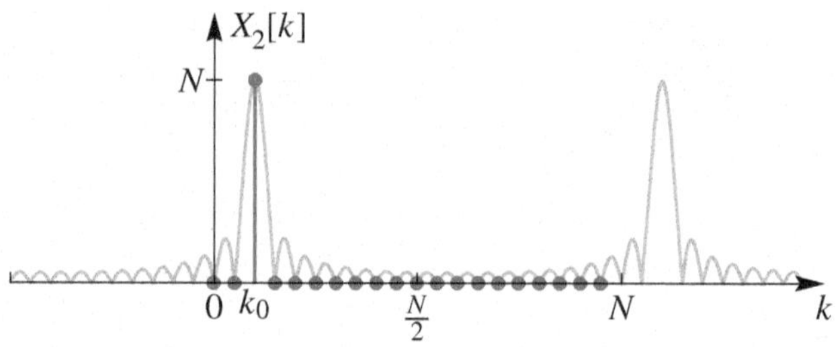

Figure 13-13: DFT of a complex exponential whose frequency *is* a multiple of $2\pi/N$; i.e., $k_0 = 2$.

and $e^{-j\theta/2}$ from the denominator. The final result, after a couple of algebraic steps, is

$$X_1[k] = D_N(e^{j((2\pi k/N)-\hat{\omega}_0)})e^{j\phi}\, e^{-j((2\pi k/N)-\hat{\omega}_0)(N-1)/2} \tag{13.38}$$

where $D_N(e^{j((2\pi k/N)-\hat{\omega}_0)})$ is a shifted and sampled version of the Dirichlet function[10]

$$D_N(e^{j\hat{\omega}}) = \frac{\sin(\hat{\omega}N/2)}{\sin(\hat{\omega}/2)}$$

Since the other terms in (13.38) only contribute to the phase, this result says that $|X_1[k]| = |D_N(e^{j((2\pi k/N)-\hat{\omega}_0)})|$ is composed of samples of a Dirichlet function, as shown in Fig. 13-12 for the case when $N = 20$ and $\hat{\omega}_0 = 5\pi/20 = 2\pi(2.5)/N$. The magnitude of the Dirichlet function has been plotted so that it is obvious where the samples are being taken. Notice that the peak of the envelope is at the non-integer value of 2.5.

When the frequency of the signal $x_1[n]$ is an integer multiple of $2\pi/N$, the resulting DFT is very simple. If we define $x_2[n] = e^{j(2\pi k_0/N)n}$, and then use $\hat{\omega}_0 = 2\pi k_0/N$ and $\phi = 0$ in (13.37), we obtain

$$X_2[k] = N\delta[k - k_0] \tag{13.39}$$

(i.e., a scaled discrete impulse at $k = k_0$ with all other DFT coefficients being zero). This result is confirmed in Fig. 13-13, where we can see that the DFT is obtained by sampling the Dirichlet function envelope exactly at the peak and at the zero crossings of the Dirichlet function. The peak value is N, and it is the only nonzero value in the DFT.

EXERCISE 13.6: Show that $X_2[k]$ in (13.39) can be obtained by substituting $\phi = 0$ and $\hat{\omega}_0 = 2\pi k_0/N$ in (13.38).

EXERCISE 13.7: Substitute (13.39) into the inverse DFT relation (13.36) to show that the corresponding time-domain sequence is

$$x_2[n] = e^{j(2\pi/N)k_0 n} \qquad n = 0, 1, \ldots, N-1$$

[10]The Dirichlet function was first defined in Section 6-7 on p. 145.

EXERCISE 13.8: Use Euler's relation to represent the signal $x_3[n] = \cos(2\pi k_0 n/N)$ as a sum of two complex exponentials. Use the fact that the DFT, like its cousins the z-transform and the DTFT, is a linear operation to show that

$$X_3[k] = \tfrac{1}{2}N\delta[k-k_0] + \tfrac{1}{2}N\delta[k+k_0]$$

or equivalently

$$X_3[k] = \tfrac{1}{2}N\delta[k-k_0] + \tfrac{1}{2}N\delta[k-(N-k_0)]$$

13-6 Spectrum Analysis of Finite-Length Signals

Many useful sampled signals have only a few hundred nonzero samples at most. These we call *finite-length* signals. One example would be a discrete-time signal obtained by sampling a very short continuous-time pulse. In other cases, sampling may not be involved at all. An example would be the impulse response of an FIR discrete-time system, where the sample values of $h[n]$ are usually computed by a filter design algorithm.

Suppose that a signal $x[n]$ is nonzero only over a finite interval $0 \le n \le L-1$. The length of this signal is L samples. Remember from Section 13-5 that the DFT of a finite-length sequence is identical to the DTFT of the sequence evaluated at frequencies $\hat{\omega}_k = 2\pi k/N$ with $k = 0, 1, \ldots, N-1$. That is,

$$X(e^{j(2\pi k/N)}) = \sum_{n=0}^{L-1} x[n]e^{-j(2\pi k/N)n} = X[k] \quad (13.40)$$

We recognized that $X[k] = X(e^{j(2\pi k/N)})$ even if $L < N$ since $N-L$ zero samples are effectively appended to the end of the sequence $x[n]$. This is sometimes called ***zero-padding*** the signal. In MATLAB, the DFT is computed by the function `fft`, which automatically zero-pads the sequence if the sequence length and the DFT length are different.

Example 13-4: Computation in MATLAB

If we create a vector `xx` with $L = 301$ elements, we can specify the FFT length to be different (e.g., $N = 512$) when we write the statement

```
XX = fft(xx, 512);
```

Since $L < 512$, the vector `xx` will be zero-padded automatically with 211 zeros. ■

One convenient application of FFT spectrum analysis is computing the frequency response of an FIR filter. If the impulse response $h[n]$ is of finite length L, the frequency response is simply the DTFT of $h[n]$, i.e.,

$$H(e^{j\hat{\omega}}) = \sum_{n=0}^{L-1} h[n]e^{-j\hat{\omega}n} \quad (13.41)$$

If we compute the N-point DFT of $h[n]$ with zero-padding, then we are evaluating (13.41) at N equally spaced frequencies in the interval $0 \le \hat{\omega} < 2\pi$, so we obtain

$$H[k] = \sum_{n=0}^{L-1} h[n]e^{-j(2\pi k/N)n} = H(e^{j2\pi k/N}) \qquad k = 0, 1, 2, \ldots, N-1 \quad (13.42)$$

If we use a large enough value for N, we can plot a smooth curve for the frequency response of the filter.

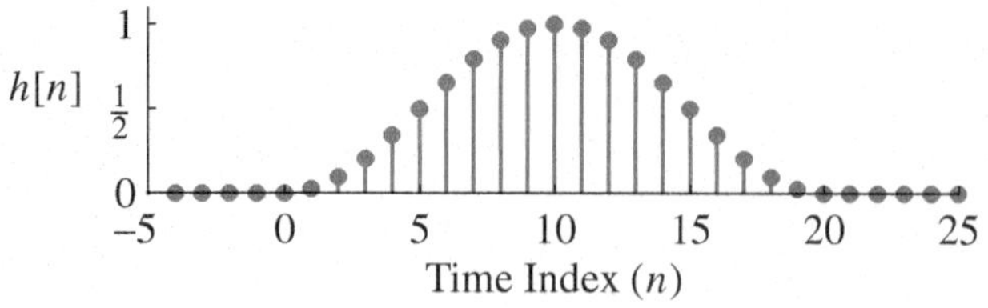

Figure 13-14: Impulse response of Hann filter with length $L = 20$.

Example 13-5: Frequency Response of a Hann Filter

Assume that the sequence

$$h[n] = \begin{cases} 0.5 - 0.5\cos\left(\frac{2\pi}{L}n\right) & 0 \le n \le L-1 \\ 0 & \text{otherwise} \end{cases} \tag{13.43}$$

is the impulse response of an FIR filter.[11] This impulse response is plotted in Fig. 13-14.

A filter with this impulse response is sometimes called a *Hann filter*.[12] Figure 13-15(a) shows plots of the magnitude of the frequency response for two Hann filters with lengths $L = 20$ and $L = 40$. The frequency responses were obtained using (13.42) with $N = 1024$. This large value of N results in a smooth curve when the points are connected by straight lines with the MATLAB plotting function. This figure illustrates that the Hann filter is a lowpass filter, because its passband is near $\hat{\omega} = 0$ where the frequency response is large, and its stopband covers the range $4\pi/L < \hat{\omega} < \pi$, where the frequency response values are relatively low.

[11]Note the similarity to the discrete-time Hamming window $w_h[n] = 0.54 - 0.46\cos\left(\frac{2\pi}{L}n\right)$, which is named for Richard Hamming.

[12]The same formula (13.43) can also be used to define a window, in which case we have the Hann window which is named for the Austrian scientist Julius von Hann. This window is often called the "hanning" window, a designation that seems to be derived from the Hamming window.

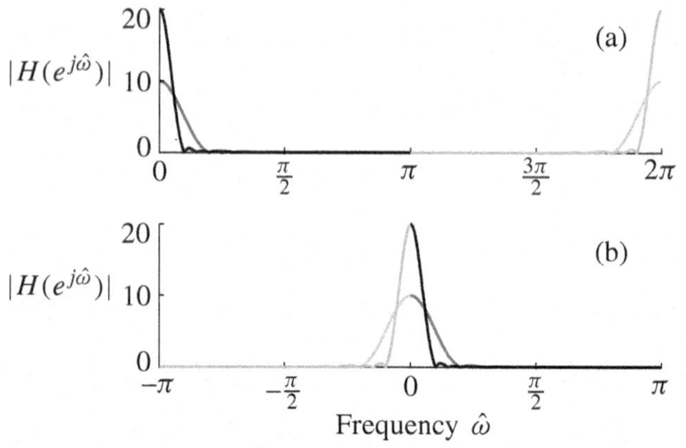

Figure 13-15: Frequency response of two Hann filters; length $L = 20$ (colored lines), and length $L = 40$ (black/gray lines) (a) as computed by 1024-point FFT and (b) as plotted with DFT values reordered.

Do not be confused by the region $\pi < \hat{\omega} < 2\pi$, which we have plotted in light colors. This is the normal DFT symmetry where the values of k for $N/2 < k \le N-1$ correspond to the normalized frequencies $\pi < \hat{\omega} < 2\pi$, which in turn are positive-frequency aliases of frequencies in the range $-\pi < \hat{\omega} < 0$. If we recognize this fact, we can construct the more conventional plot shown in Fig. 13-15(b), where the light colored plots at negative frequencies are the same as their counterparts in Fig. 13-15(a).

The essence of the MATLAB code for constructing the plot in Fig. 13-15(b) is as follows:

```
L = 20; N = 1024;
hh = 0.5*(1-cos(2*pi*(0:L-1)/L));
HH = fft(hh,N);
HHP = [HH(N/2+1:N), HH(1:N/2)];
omega = 2*pi*(-N/2+1:N/2)/N;
plot(omega,abs(HHP))
```

When we compare the two different impulse response lengths, we see that the frequency response becomes more concentrated around $\hat{\omega} = 0$ as we increase L. It can be shown that the first zero of $H(e^{j\hat{\omega}})$ occurs at $\hat{\omega} = 4\pi/L$ for the Hann filter. Thus, increasing L from 20 to 40 cuts the width of the passband in half (and also doubles its amplitude). Plotting the frequency response allows us to verify this result. ■

Example 13-5 shows how the DFT can be used to evaluate the frequency-domain representation of a finite-length sequence. In this example the sequence is an impulse response, so the resulting DFT values are samples of the frequency response of the FIR filter. In the next section, we will show how similar computations can be repeatedly applied to segments of a long signal to obtain a useful frequency-domain representation.

13-7 Spectrum Analysis of Periodic Signals

In this section, we will show how the DFT can be used to analyze periodic signals such as those that can be represented by a finite sum of harmonically related cosine signals. As a specific example, consider the periodic continuous-time signal

$$\begin{aligned} x_c(t) = 0.0472\cos(2\pi(200)t + 1.5077) \\ + 0.1362\cos(2\pi(400)t + 1.8769) \\ + 0.4884\cos(2\pi(500)t - 0.1852) \\ + 0.2942\cos(2\pi(1600)t - 1.4488) \\ + 0.1223\cos(2\pi(1700)t) \end{aligned} \qquad (13.44)$$

The fundamental frequency of this signal is $\omega_0 = 2\pi(100)$ rad/sec, and the signal consists of harmonics at 2, 4, 5, 16, and 17 times the fundamental. This signal is quite similar to the synthetic vowel studied in Section 3-3.1. If we sample this signal with sampling rate $f_s = 4000$ Hz, the resulting discrete-time signal will be periodic too, since 100 divides evenly into 4000. Specifically, the discrete-time signal $x[n] = x_c(n/4000)$ given by

$$\begin{aligned} x[n] = 0.0472\cos(0.1\pi n + 1.5077) \\ + 0.1362\cos(0.2\pi n + 1.8769) \\ + 0.4884\cos(0.25\pi n - 0.1852) \\ + 0.2942\cos(0.8\pi n - 1.4488) \\ + 0.1223\cos(0.85\pi n) \end{aligned} \qquad (13.45)$$

has frequencies that are all multiples of $\hat{\omega}_0 = 0.05\pi = 2\pi/40$. As in the continuous-time case, $x[n]$ in (13.45) has harmonics of the normalized fundamental frequency numbered 2, 4, 5, 16, and 17.

EXERCISE 13.9: Use the results of Section 13-5.5 to determine an equation for the 40-pt. DFT of $x[n]$ in (13.45). Verify that some of the nonzero DFT values are

$$X[k] = \begin{cases} 0.0236(40)e^{j1.5077} & k = 2 \\ 0.0681(40)e^{j1.8769} & k = 4 \\ 0.2442(40)e^{-j0.1852} & k = 5 \\ 0.1471(40)e^{-j1.4488} & k = 16 \\ 0.06115(40) & k = 17 \end{cases} \qquad (13.46)$$

What are the other nonzero values of $X[k]$? For which indices k are the DFT coefficients $X[k]$ equal to zero?

The discrete-time signal $x[n]$ defined in (13.45) is plotted in Fig. 13-16(a), from which we can see its periodicity $x[n] = x[n + 40]$. Now, if one cycle of $x[n]$ from Fig. 13-16(a) is used in a 40-point DFT (13.35), we obtain the 40 DFT coefficients $X[k]$ shown as magnitude and phase in Figs. 13-16(b) and (c), respectively. These 40 DFT coefficients represent the sequence $x[n]$ *exactly* through the synthesis formula of the IDFT (13.36).

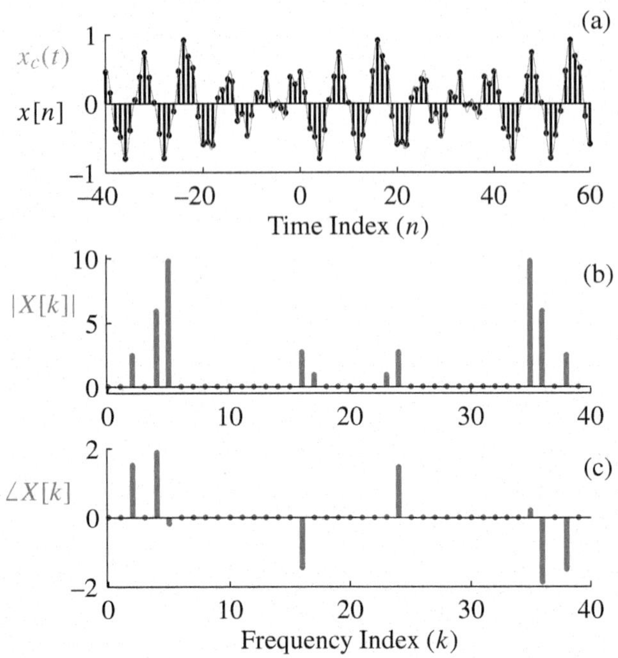

Figure 13-16: Periodic sequence and the corresponding 40-pt. DFT spectrum $X[k]$ of the sampled signal $x[n]$ in (13.45) with sampling rate $f_s = 4000$ samples/sec.

EXERCISE 13.10: Show that the IDFT of $X[k]$ defined in Exercise 13.9 gives $x[n]$ in (13.45). Furthermore, show that the signal $x[n]$ obtained via the IDFT is periodic with period N.

13-8 The Spectrogram

We have seen that the DFT can compute exact frequency-domain representations of both periodic and finite-length discrete-time signals. An equally important case is when the sampled signal is indefinitely long, but not necessarily periodic. In real-time processing of such signals, we might have to wait an inordinately long time just to acquire all the samples, and even having done so, we would then be faced with a huge DFT computation. Furthermore, the results of such a computation would often be of limited value because many of the temporal variations of the signal would not stand out in the computed spectrum. Good examples are audio signals such as speech or music. In both of these cases, it is the temporal variation of the frequency content that is of interest. Usually we have digital recordings that give a very long sequence obtained by sampling for many minutes or even hours. For example, when sampling an audio signal at a sampling rate of 44.1 kHz, one hour of stereophonic music would be represented by $44100 \times 2 \times 60 \times 60 = 317{,}520{,}000$ total samples. If we want to compute DFTs of approximately one hour of audio at 44.1 kHz, the closest power-of-two FFT needed would be $2^{28} = 268{,}435{,}456$ per channel. A better approach is to break the long signal into small segments and analyze each one with an FFT. This is reasonable because the very long recording probably contains a succession of short passages where the spectral content does not vary, so there may be a natural segment length. Indeed, we have already seen that music can be thought of in just this way.

To formalize this concept, assume that $x[m]$ is an indefinitely long sequence. We define the *time-dependent discrete Fourier transform* of this signal as

$$X[k, n] = \sum_{m=0}^{L-1} w[m]x[n+m]e^{-j(2\pi k/N)m} \qquad (13.47)$$

for $k = 0, 1, \ldots, N-1$. The interpretation of this equation is straightforward. First of all, the window $w[m]$ is a sequence that is nonzero only in the interval $m = 0, 1, \ldots, L-1$, where L is assumed to be much smaller than the total length of the sequence $x[m]$. The product $w[m]x[n+m]$ is also nonzero only for $m = 0, 1, \ldots, L-1$. In this way, the window selects a finite-length segment from the sequence $x[m]$. By adjusting the shift n, we can move any desired segment of $x[m]$ into the window. Now the right-hand side of (13.47) is easily recognized as the N-point DFT of the finite length sequence $w[m]x[n+m]$, so (13.47) can be evaluated for each choice of n by an FFT computation. This is illustrated by Fig. 13-17. Figure 13-17(a) shows a signal $x[m]$ plotted as a function of m. Also shown are two length-20 regions, one starting at $n = 10$ the other at $n = 50$. Figure 13-17(b) shows the left-shifted sequence $x[m+10]$ with the fixed window sequence $w[m]$.[13] The light colored samples in Figs. 13-17(a) and (b) are the samples that are multiplied by the window $w[m]$ when $n = 10$ and are thus selected for analysis. The DFT of $w[m]x[m+10]$ is $X[k, 10]$. Figure 13-17(c) shows the fixed window sequence along with the sequence $x[m+50]$, i.e., the sequence $x[m]$ shifted to the left by 50 samples. The gray samples in Figs. 13-17(a) and (c) are the samples that are multiplied by the window and are thus selected for DFT analysis to compute $X[k, 50]$.

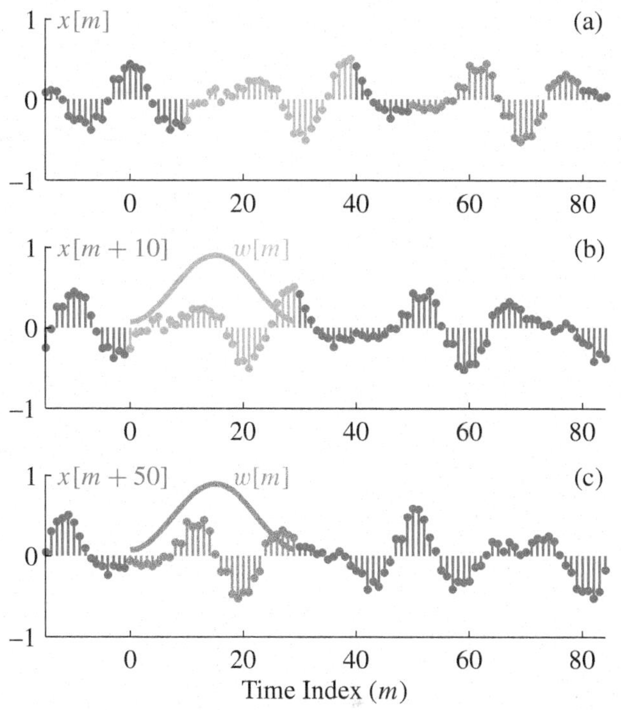

Figure 13-17: Time-dependent windowing of a signal. (a) The signal $x[m]$, (b) Fixed window with signal shifted by 10 samples, and (c) Fixed window with signal shifted by 50 samples.

13-8.1 Spectrogram Display

The spectrogram computation results in a two-dimensional sequence $X[k, n]$, where the k dimension represents frequency because $\hat{\omega}_k = (2\pi k/N)$ is the kth analysis frequency, and the n dimension represents time. Since the spectrogram is a function of both frequency and time, we cannot plot *the* spectrum, because there is a different *local spectrum* for each time. To deal with this additional complexity, a three-dimensional graphical display is needed. This is done by plotting $|X[k, n]|$ (or $\log|X[k, n]|$) as a function of both k and n using perspective plots, contour plots, or grayscale images. The preferred form is the ***spectrogram***, which is a grayscale (or pseudocolor) image where the gray level at point (k, n) is proportional to $|X[k, n]|$ or $\log|X[k, n]|$; large values are black, and small ones white. Examples of the spectrogram can be seen in Figs. 13-18, 13-20, 13-22, 13-23, 13-25, and 13-27. The horizontal axis is time and the vertical axis is frequency. In the spectrogram image for a real signal, only the range for $0 \le k \le N/2$, or $0 \le \hat{\omega} \le \pi$, need be shown. For the complex signal case, the frequency range is either $0 \le \hat{\omega} < 2\pi$, or $-\pi < \hat{\omega} \le \pi$, depending on whether or not the FFT values are reordered prior to being displayed.[14]

[13] The discrete-time window sequences are shown as continuous functions to distinguish them from the samples of $x[m]$.

[14] The *SP-First* toolbox function `spectgr`, which is equivalent to MATLAB's function `specgram`, uses the ordering $-\pi < \hat{\omega} \le \pi$ for complex signals; `specgram` uses $0 \le \hat{\omega} < 2\pi$.

13-8.2 Spectrograms in MATLAB

Since the spectrogram can be computed by doing many FFTs of windowed signal segments, MATLAB is an ideal environment for doing the calculation and displaying the image. Specifically, what MATLAB displays is

$$X[(2\pi k/N)f_s, rRT_s] = \sum_{m=0}^{L-1} w[m]x[rR+m]e^{-j(2\pi k/N)m} \quad (13.48)$$

$$k = 0, 1, \ldots, N/2$$

where $f_s = 1/T_s$ is the sampling rate associated with $x[n]$ and R determines spacing in time of the DFT computations.[15] The MATLAB command that invokes the computation of (13.48) is

```
[X,F,T] = specgram(xx, NFFT, ...
                   Fs, window, Noverlap)
```

where `X` is a two-dimensional array containing the complex spectrogram values, `F` is a vector of all the analysis frequencies, and `T` is a vector containing the times of the sliding window positions. The inputs are the signal `xx`, the FFT length `NFFT`, the sampling frequency `Fs`, the window coefficients `window`, and the number of points to overlap as the window slides `Noverlap`. Note that the window skip parameter R is `NFFT - Noverlap`. The overlap should be less than the window length, but choosing `Noverlap` equal to `length(window)-1` would generate a lot of needless computation, because the window skip would be one. It is common to pick the overlap to be somewhere between 50 percent and 80 percent of the window length, depending on how smooth the final image needs to be. See `help specgram` in MATLAB for more details.

The spectrogram image can be displayed by using

```
imagesc( T, F, abs(X) )
axis  xy, colormap(1-gray)
```

The color map of `(1-gray)` gives a negative gray scale that is useful for printing, but on a computer screen it is preferable to use color, e.g., `colormap(jet)`. Finally, it may be advantageous to use a logarithmic amplitude scale in `imagesc` in order to see small amplitude components.

13-8.3 Spectrogram of a Sampled Periodic Signal

A periodic sequence such as (13.45) is an example of a signal that can be represented exactly by a single DFT; it also is a continuing signal of indefinite length. Since we know its DFT exactly, it provides a useful example that has a simple spectrogram. If we use a rectangular window of length $L = 40$, and a DFT length of $N = 40$ in (13.47), we obtain

$$X[k,n] = \sum_{m=0}^{39} x[m+n]e^{-j(2\pi k/40)m} \qquad k = 0, 1, 2, \ldots, 39 \quad (13.49)$$

The resulting spectrogram for the signal in (13.45) is shown in Fig. 13-18. This grayscale image shows five constant horizontal lines of different grayscale intensity. There are no other spectral components. Why does the image look this way? First of all, since $N = L = 40$, each segment of the signal will be exactly one period of $x[n]$. Therefore, $|X[k,n]|$ will be equal to $|X[k]|$ in Fig. 13-16(b) for all n. In other words, the *magnitude* of the time-dependent Fourier transform will vary only in the frequency dimension, *not* in the time dimension. Furthermore, from our previous discussion, it follows that the only nonzero spectral components will be at $k = 2, 4, 5, 16, 17$.

The frequency axis in Fig. 13-18 is calibrated in terms of the continuous-time cyclic frequency variable, f in

[15] In other words, MATLAB's spectrogram function calibrates its frequency and time axes in continuous-time units.

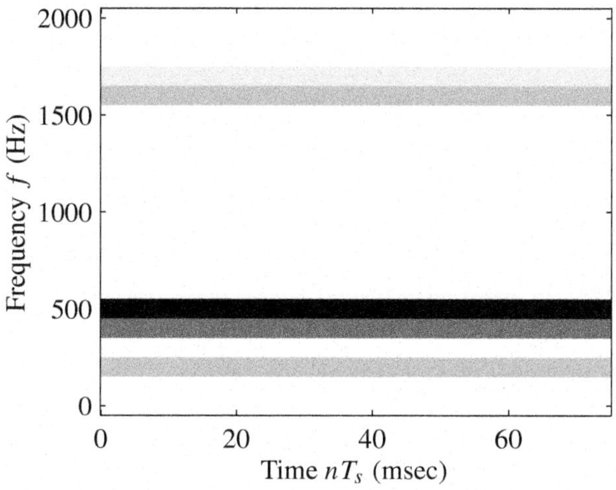

Figure 13-18: Spectrogram of the sampled periodic signal defined in (13.45). Only the second, fourth, fifth, sixteenth, and seventeenth harmonics are present.

Hz. In order to derive the relationship between f and the frequency index k, recall that the DFT frequencies are $\hat{\omega}_k = 2\pi k/N$ and (from the sampling theorem) $\hat{\omega}_k = 2\pi(f_k/f_s)$. Equating these two formulas, we get

$$f_k = (k/N)f_s \tag{13.50}$$

which allows us to label the frequency axis in Hz. Applying (13.50) to the frequency indices $0 \le k \le N/2$, we get the continuous-time frequencies $0 \le f_k \le f_s/2$, which is the usual range of frequencies displayed for a real signal. The negative-frequency spectral components can be inferred from the even symmetry of the magnitude spectrum $|X[k,n]|$ of a real signal. Therefore, in Fig. 13-18 there are only 21 discrete points on the vertical axis. The five frequency components are at $(k/40)4000 = 100k$ for $k = 2, 4, 5, 16, 17$. The width of the spectral lines plotted at these values of k is chosen by MATLAB's image-plotting function, which zooms to show a larger image with many more than 21 pixels vertically.[16] Since we chose $N = 40$, the spectral components for $k = 4$ and $k = 5$ are next to each other, as are components $k = 16$ and $k = 17$. Since MATLAB expands each point so that the 21 DFT values fill the vertical plotting range, we see no white space between these components.

13-8.4 Resolution of the Spectrogram

We have shown that calculation of the spectrum requires that the window length (L) be finite and that the number of analysis frequencies (N) be finite. The performance of a spectrogram analysis usually comes down to a statement about its resolution in either frequency or time. The key parameters that control resolution are the form of $w[n]$ and its length L.

The ability to *resolve* spectral components in the spectrogram is governed by the principles that were illustrated in Section 13-4. In that section, we showed that there is an inverse relationship between window length and the ability to resolve two adjacent frequencies. Specifically, we saw that the continuous-time rectangular window of duration T sec. has a Fourier transform mainlobe width of $\Delta\omega = 4\pi/T$. With a window length $L = T/T_s$ samples, the mainlobe width expressed in rad/sec is

$$\Delta\omega = \frac{4\pi}{L} f_s \quad \text{rad/sec} \tag{13.51}$$

The Hamming window defined in (13.26) has a Fourier transform main-lobe width that is twice as wide as that of the rectangular window. Thus, a window length twice as long is required to achieve the same resolution as that of the rectangular window.

DEMO: *Resolution of the Spectrogram*

[16] The gray scale uses white to represent zero magnitude and black to represent the highest magnitude. This scale was chosen to make the picture mostly white for printing. On a computer monitor, it may be better to invert the scale or to use pseudo-color.

There is a down side to increasing L; namely, the time resolution will suffer. When L is large, short-duration time events or rapid transitions will be lumped together with other parts of the signal by a long window. Therefore, we have again encountered the *uncertainty principle* of Fourier spectrum analysis:

> ***Uncertainty Principle:***
> *Time resolution and frequency resolution cannot be improved simultaneously in the spectrogram.*

13-8.4.1 Resolution Experiment

In order to test the resolving power of the spectrogram, we can perform a simple experiment. Our test will rely on analyzing a signal that is hard to "see" simultaneously in both time and frequency. The test signal consists of two components. The first is a constant 960-Hz tone that persists over the entire data set; the second is a short signal whose frequency is 1000 Hz, and whose beginning and ending times are 200 and 400 msec. The ideal form of such a spectrogram is drawn in Fig. 13-19. The goal in this test is to simultaneously determine the beginning and ending times of the second signal, as well as its individual frequency components. As we have just discussed, the frequency resolution is inversely proportional to the length the window. When a Hamming window of length L is used, this bandwidth is approximately $(8\pi/L)f_s$, so the frequency resolution in Hz is

$$\text{Frequency Resolution} \approx \frac{4}{L} f_s \quad \text{Hz} \tag{13.52}$$

The time resolution, on the other hand, is directly proportional to the length of the window.

To see the effect of the window on the ability to measure both time and frequency in the spectrogram, let us begin with the window positioned prior to the time marked 200 msec in Fig. 13-19. For values of n, such that the window only covers the signal prior to 200 msec, we will see a single horizontal line in the spectrogram centered on 960 Hz with a width that is inversely proportional to the window length L. For larger values of n, the window will straddle the starting time of the 1000 Hz component, so part of the signal under the window will contain both 960 Hz and 1000 Hz frequencies and part will only include a 960 Hz component. In this case, the entire window will *not* contain both frequencies. For these values of n, things will be somewhat muddled, and it will be difficult to distinguish exactly when the 1000 Hz component begins. Finally, when n is such that the entire window covers both frequency components, the issue switches back to that of frequency resolution. If the window is long enough so that $4f_s/L < 40$ Hz, then the two frequencies will be resolved and we will see two horizontal lines in the spectrogram. Otherwise, the two lines will blur together giving one broad line. Looking at Fig. 13-19, it is clear that the same sort of thing will happen as the window

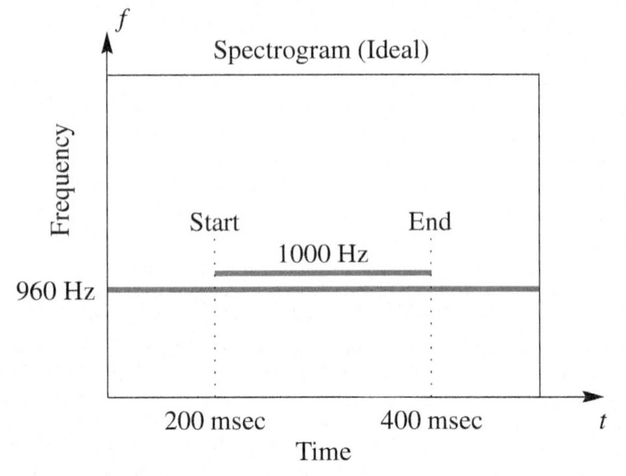

Figure 13-19: Resolution test for digital spectrum analysis. Idealized spectrum consisting of two sine waves with different durations and different frequencies.

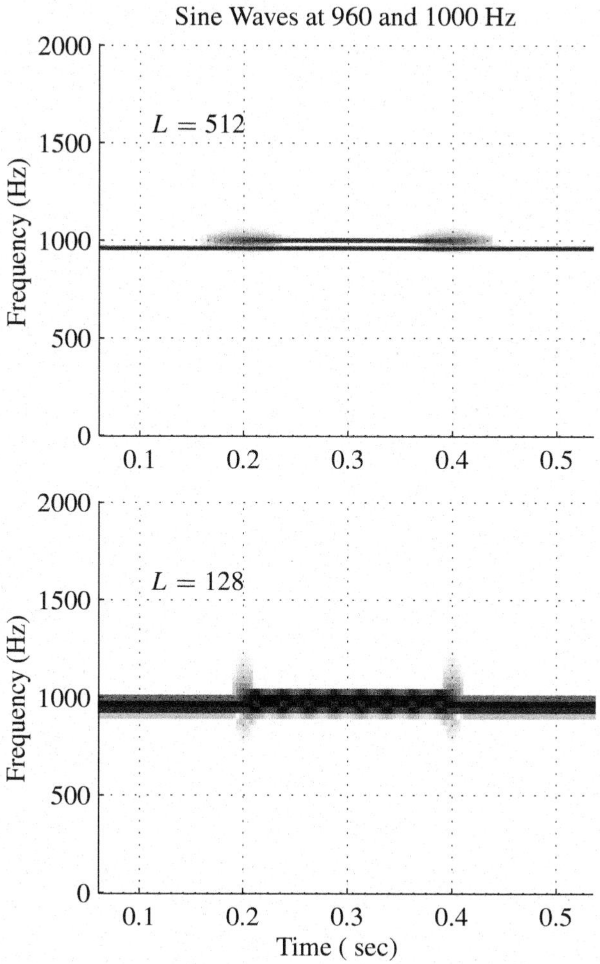

Figure 13-20: Resolution test for digital spectrum analysis using two different window lengths: (a) $L = 512$, which is 128 msec at $f_s = 4000$ Hz, and (b) $L = 128$ (or 32 msec). The FFT length N is equal to L.

approaches, then straddles, and finally clears the end of the 1000 Hz component.

Figure 13-20 shows the computed spectrograms for a discrete-time signal synthesized according to the idealized diagram in Fig. 13-19. The sampling rate was 4000 Hz. The frequency resolution of the longer window ($L = 512$) is adequate to resolve the two signals, but it also causes a blurring of the signal's endpoints, which inhibits an accurate calculation of the starting and ending times, as is evident in Fig. 13-20(a). Notice that the time duration of the 512-point window is $LT_s = 512/4000 = 0.128$ sec. The width of the blurred region around the ends of the 1000 Hz component is approximately half that of the window duration. This is because of the shape of the Hamming window, which tapers rapidly to zero on its ends. On the other hand, the shorter window ($L = 128$) in Fig. 13-20(b) does localize the ends of the 1000 Hz signal, but fails to resolve the two frequencies. This sort of time-frequency trade-off is always present in a spectrogram that relies on the DFT with its equally spaced analysis frequencies all having the same resolving power.[17]

DEMO: *Resolution of the Spectrogram*

13-8.5 Spectrogram of a Musical Scale

One case where the spectrogram matches our intuition comes from the analysis of musical instrument sounds, as we discussed in Chapter 3. A musical score (Fig. 13-21) employs a notation that corresponds to the "time-frequency" image found in the spectrogram. Each note specifies the frequency of the tone to be played, the duration of the tone, and the time at which it is to be played (in relation to other notes). Therefore, we can say that musical notation is an idealized spectrogram, although it does not use grayscale encoding to show amplitude.

As a simple example of a musical passage, we can synthesize a scale using pure tones (i.e., sinusoids). Eight notes played in succession make up a scale. If the scale

[17] One approach to the problem posed by this resolution example is called *wavelet analysis,* which uses non-uniform frequency spacing and resolution.

Figure 13-21: Musical score for Beethoven's *Für Elise.*

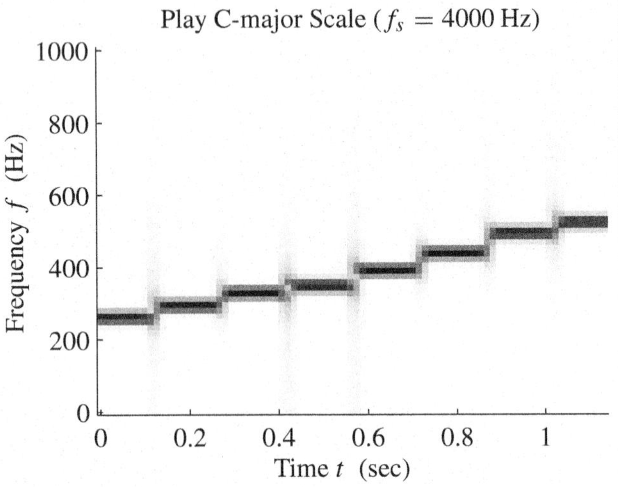

Figure 13-22: Spectrogram of an artificial piano scale composed of sine waves. Window was a Hann window of length $L = 256$, overlap was 200, and sampling rate $f_s = 4000$ Hz.

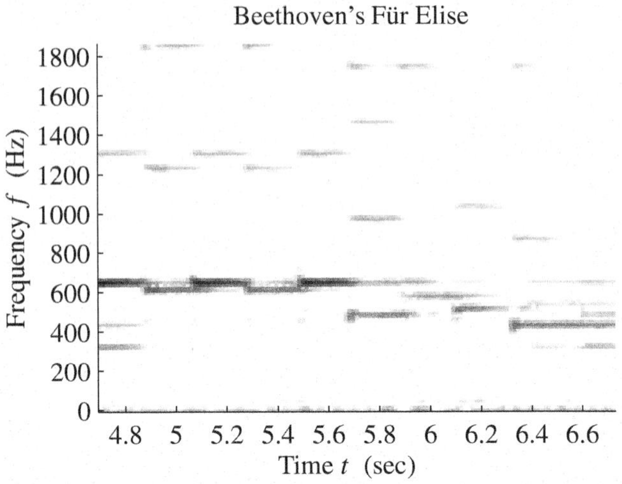

Figure 13-23: Spectrogram of *Für Elise* played on a piano. Notice the harmonics at twice the actual note frequency owing to the complex sounds made by the piano. Window was a Hann window of length $L = 256$, overlap was 200, and the FFT length was $N = 256$.

is C major, the notes are C, D, E, F, G, A, B, C, which have the frequencies given in the following table:

Middle C	D	E	F	G	A	B	C
262 Hz	294	330	349	392	440	494	523

The spectrogram of the synthetic scale is shown in Fig. 13-22. We can easily identify each note as it is played, although there is some blurring of the transitions between notes. This blurring is due to the fact that the window length of the spectrum analyzer is long enough to straddle two of the notes simultaneously. Shortening the window length would give sharper edges near the ends of each note, at the cost of fatter spectral lines for the tones.

A real piano plays notes that have a much more complex structure than the sine waves used for the C-major scale in Fig. 13-22. Most keys on a piano strike three strings when played, and the complex vibrations of these strings make the pleasing sound of the instrument. A spectrogram reveals the complex structure of the notes if the frequency spectrum has sufficient resolution. Figure 13-23 shows the spectrogram for the opening passage of Beethoven's *Für Elise.* The frequency resolution of the Hann window at the half-amplitude level is approximately

$$\frac{2}{L} f_s = 57.95 \text{ Hz}$$

which is sufficient to separate individual notes in Fig. 13-23. In the beginning of *Für Elise*, only one key is being played at any one time, but the spectrogram is certainly more complicated. In fact, we can identify the primary note being played, and we can also see second, and third, harmonics of that note, and sometimes even "undertones" at lower frequencies.

DEMO: *Music GUI*

These two examples suggest that the spectrogram might be a useful tool for making an automatic music-writing program. If we could analyze the spectrogram to find its large peaks, then the program could "read" the spectrogram and determine the frequency and duration of the notes being played.

13-8.6 Spectrogram of a Speech Signal

As another example, Fig. 13-24 shows a speech signal, which has been sampled at a rate of $f_s = 8000$ samples/sec. The time-domain plot is given in a strip format consisting of five waveform lines, each representing 800 samples, or 100 msec in time. The beginning of the second line is the sample after the last sample in the first line, etc. The plot is drawn as a continuous waveform, because 800 individual samples per line would be very close together on this plotting scale.

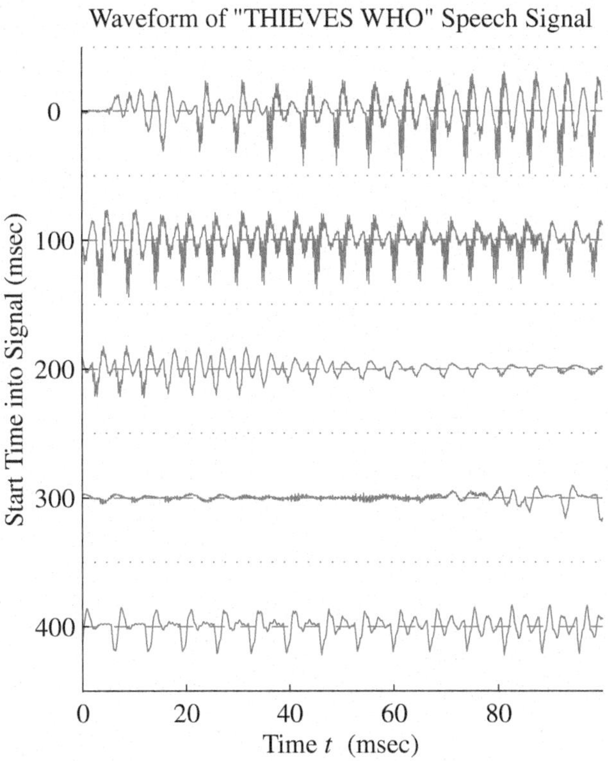

Figure 13-24: Waveform of a speech signal. Each line consists of 800 points.

We know that speech consists of a sequence of different sounds that alternate between voiced sounds (formed by vibrating vocal cords), such as vowels, and unvoiced sounds, such as "s", "sh" and "f". The waveform of Fig. 13-24 corresponds to the utterance "thieves who," so we can identify the voiced and unvoiced intervals. The vowel "ie" in "thieves" occupies the time region $0 \le t \le 200$ msec, while the unvoiced sound "s" can be found at $300 \le t \le 360$ msec. The vowel from "who" lies near the end, $450 \le t \le 500$ msec. The vowel signal is a loud, nearly periodic waveform; the unvoiced sound, on the other hand, is soft and seems to have a random structure. These sounds are the major events in the signal, but overall the waveform changes slowly with time, being relatively unchanged over intervals of 20 to 80 msec. Hence the spectral properties of the signal will also change slowly, and the spectrogram provides an invaluable tool for visualizing the changing character of the speech signal.

The spectrogram of the speech signal of Fig. 13-24 is shown in Fig. 13-25 for the case where $L = 250$. This image is a plot of $|X[(2\pi k/N)f_s, rRT_s]|$ defined in (13.48) on p. 410. In order for such a plot to be useful, we must understand why it looks the way that it does, and be able to interpret the various features of the image in terms of the time waveform and the concept of a time-varying frequency spectrum. Notice that in the first three lines of the waveform plot in Fig. 13-24, the vowel waveform is composed of pulses that occur somewhat evenly spaced in time. Indeed, if we limit our view to an interval of length 20 to 30 msec, we see that the waveform generally appears to be almost periodic over that interval. In computing the time-varying spectrum $|X[k,n]|$, the window length is $L = 250$, which corresponds to a time interval of $250/8000 = 31.25$ msec. As the window moves along the waveform, different segments of length 31.25 msec are analyzed. As the pulses change shape and their spacing changes with time, so do the spectral characteristics of the segments within the sliding window. We see this clearly in Fig. 13-25. The dark bars represent equally spaced frequency components whose spacing in the vertical (frequency) direction depends on the "fundamental frequency" at the corresponding time. The bars vary together; their spacing increases when the "fundamental frequency" increases (longer period), and vice versa. Both vowel regions, $0 \le t \le 200$ msec and $390 \le t \le 500$ msec, exhibit this regular structure, although the spacing of the bars is much closer in the last 110-msec interval.

DEMO: *Chapter 3 Spectrograms*

Returning to Fig. 13-24, notice that in the time interval 300 to 360 msec (the fourth line) the time waveform of the unvoiced sound decreases in amplitude and the waveform is no longer periodic. Correspondingly, in the spectrogram of Fig. 13-25, during the time interval 300 to 360 msec, the image fades, the regular structure

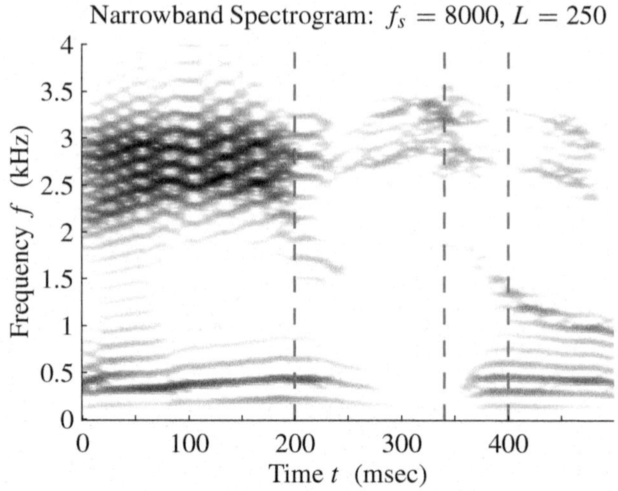

Figure 13-25: Spectrogram of speech signal; sampling frequency $f_s = 8000$ Hz, FFT length $N = 400$, and Hann window of length $L = 250$ (equivalent to 31.25 msec). Spectrum slices at times $nT_s = 200$, 340, and 400 msec are shown individually in Fig. 13-26.

disappears, and most of the spectral content lies at high frequencies near 3000 Hz.

The spectrogram image in Fig. 13-25 is just a sequence of spectral slices stacked side-by-side with the shade of gray representing the amplitude. In Fig. 13-25, the dotted vertical lines show the location of three "spectral slices" that have been extracted and plotted in Fig. 13-26. The two spectra at $t = 200$ and 400 msec have the general character of what we would expect for periodic signals, i.e., highly concentrated at equally spaced frequencies. Note that the peaks in the top panel ($t = 200$) are more widely spaced than those in the bottom panel ($t = 400$). Examination of the time waveform at 200 and 400 msec, respectively, shows that the fundamental period is significantly shorter at 200 msec than it is at 400 msec. Thus, the spectral peaks should be farther apart at 200 msec than they are at 400 msec. The spectrum in the middle

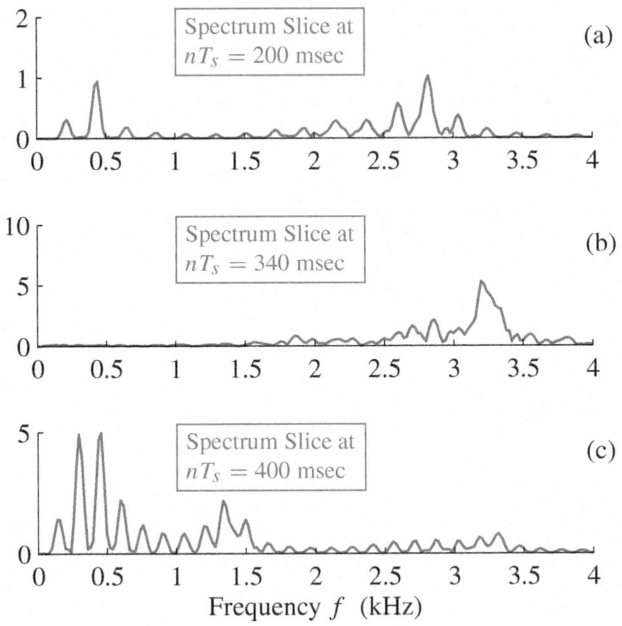

Figure 13-26: Spectrum slices ($L = 250$) at times $nT_s =$ 200, 340, and 400 msec taken from Fig. 13-25.

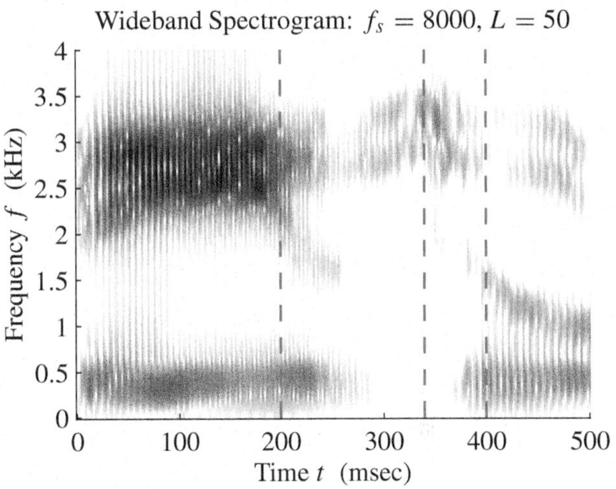

Figure 13-27: Spectrogram of speech signal; sampling frequency $f_s = 8000$ Hz, window length $L = 50$ (6.25 msec). Spectrum slices at times $nT_s = 200$, 340, and 400 msec are shown individually in Fig. 13-28.

panel ($t = 340$) is typical of unvoiced sounds which have their energy concentrated at relatively high frequencies.

An important point in our interpretation of the spectrogram of the speech signal is that, over the time interval of the window, the waveform sometimes "looks periodic." In other words, given only the waveform within the window, we cannot tell anything about the waveform outside the window. It could continue periodically, it could be zero outside the window, or it could change its character as the speech signal does. Thus, the length and shape of the window are important factors in the spectrum analysis of signals like speech that continue in time.

Suppose that the window is shorter than the local period of the signal. In this case, there is not enough signal within the window to measure the local period. Indeed, in this case, it would be more appropriate to think of the signal within the window interval as a finite-length signal. Thus, we would expect a significantly different spectrogram if the window is short. This is demonstrated in Fig. 13-27, which shows the spectrogram of the speech waveform of Fig. 13-24 for the case where $w[n]$ is a Hann window with $L = 50$ and the frequency spectrum is again evaluated with an FFT of length $N = 400$. Notice that the fine detail in the vertical direction in Fig. 13-25 is no longer present in Fig. 13-27. The thin wavy bars have been replaced by broader bars. In other words, the fine detail of the spectrum is no longer *resolved*. This point is made clear in the spectral slices taken at 200, 340, and 400 msec, as shown in Fig. 13-28. The frequency peaks in Fig. 13-28 are much wider than the corresponding ones in Fig. 13-26. Also note that the image of Fig. 13-27 seems to consist of vertical slices that alternate between dark and light. This is due to the fact that the window length is so

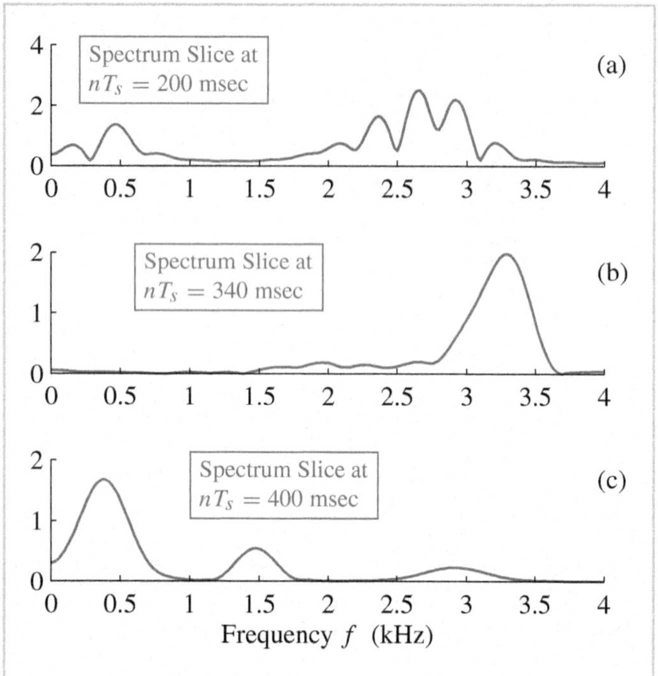

Figure 13-28: Spectrum slices ($L = 50$) at times $nT_s =$ 200, 340, and 400 msec taken from Fig. 13-27.

short that sometimes it covers the high-amplitude portion of one of the pulses and later covers the low-amplitude part. Thus, the spectrum appears to fade in and out along the time dimension.

To conclude this discussion, we answer the question, "What would the spectrum of the entire 4000 samples of the speech signal be like?" Since an FFT length of $N = 4000$ is not really a problem for modern computers, we can simply use the entire sequence as the input and evaluate (13.35) with $N = 4000$. The result is shown in Fig. 13-29. In this case, we have a single computed spectrum. Observe that Fig. 13-29 shows 2001 distinct discrete-time frequencies $f_k = 2k$ Hz, for $k = 0, 1, \ldots, 2000$, corresponding to the continuous-time frequencies in the range $0 \le f \le 4000$ Hz. Notice that Fig. 13-29 is similar in some respects to the spectral slices in Figs. 13-26 and 13-28, but it has much more detail and does not have the regularly spaced peaks that characterize a periodic signal. This is because the signal is *not* periodic over the entire 4000-sample segment. The large broad peaks give an indication of the concentration of frequencies in the signal, but that is about all that can be inferred from this plot. Thus we have demonstrated that we can have either a long-term or a short-term spectrum for the speech signal. The choice will be determined by many factors, but it is clear that the short-term, time-dependent spectrum portrays a great deal about the speech signal that would be hidden in the long-term spectrum.

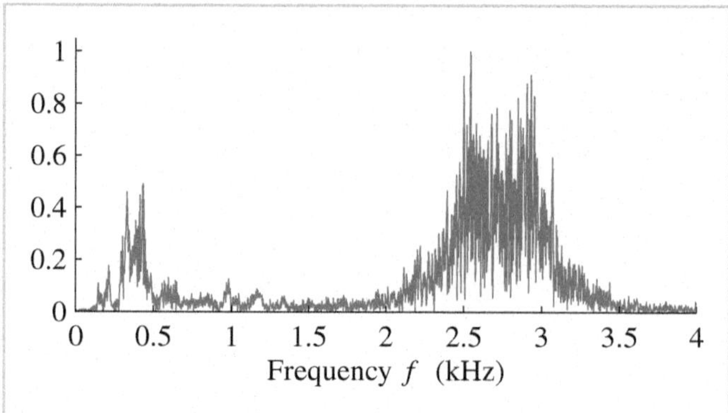

Figure 13-29: Spectrum of the entire sequence of 4000 samples in Fig. 13-24.

13-8.7 Filtered Speech

As a final example, we show that lowpass filtering applied to the speech signal will alter the spectrogram by removing all high-frequency components. For the lowpass filter, we use the LTI system whose impulse response is plotted in the top panel of Fig. 13-30 and whose frequency response (magnitude) is shown in the bottom panel of the figure. If the speech signal of Fig. 13-24 is the input to this filter, we would expect that "the high frequencies would be removed by the filter."

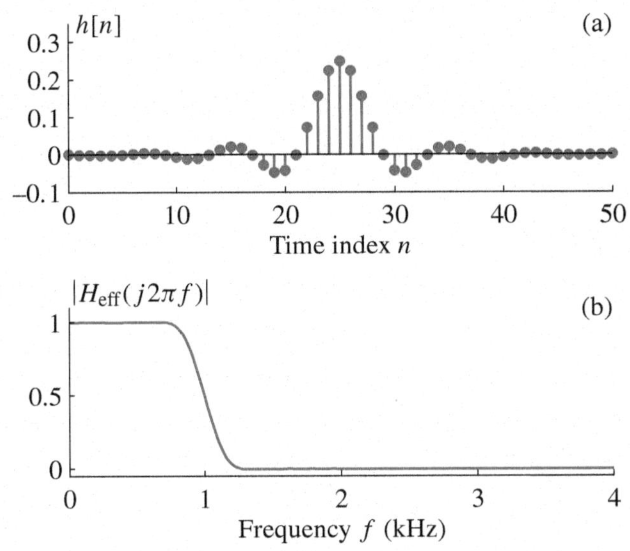

Figure 13-30: Impulse response and effective frequency response of a lowpass discrete-time length-51 FIR filter.

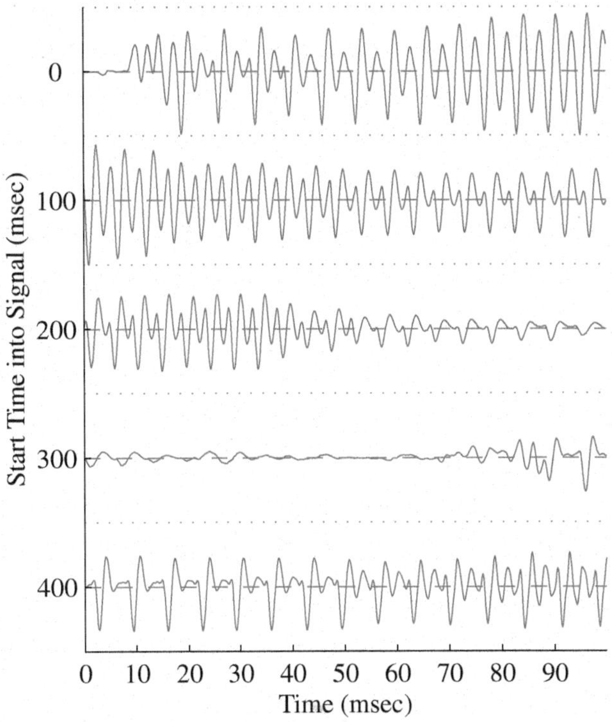

Figure 13-31: Waveform of lowpass filtered speech signal. Each row is 100 msec long.

Figure 13-31 shows the output signal when the input is the speech signal of Fig. 13-24. A careful comparison of the two waveforms shows that (1) the output is delayed relative to the input, and (2) the output is "smoother" than the input. Both of these results are expected, since the filter has a delay of 25 samples, and high frequencies are required to make rapid changes in the waveform. Clearly, the filter removes the high frequencies.

This is further demonstrated by the spectrogram in Fig. 13-32, where all components above 1000 Hz have been removed from the spectrum, while below 1000 Hz the spectrograms of Figs. 13-25 and 13-32 appear to be identical. This is entirely consistent with our understanding of lowpass filtering, since the "cut-off frequency" of the discrete-time filter is approximately $\hat{\omega}_c = \pi/4$, which corresponds to continuous-time frequency $\omega_c = \hat{\omega}_c f_s = (\pi/4)8000 = 2\pi(1000)$. The detailed spectral slices at times $t = 200$ and 400 msec shown in Fig. 13-33 also confirm our conclusions. Note that the solid lines show the local spectrum at those times, and the dotted lines are a plot of the magnitude of the frequency response superimposed to show the location of the passband (plotted versus continuous-time cyclic frequency). The spectral slice at $t = 340$ msec corresponding to the middle panel in Fig. 13-26 is not shown since it is effectively zero because the speech signal has virtually no energy below 1000 Hz for the unvoiced sound.

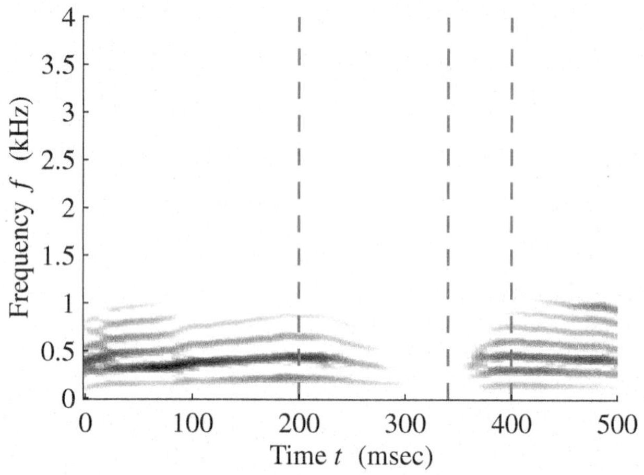

Figure 13-32: Spectrogram of lowpass filtered speech signal. Window length $L = 250$; sampling frequency $f_s = 8000$ Hz.

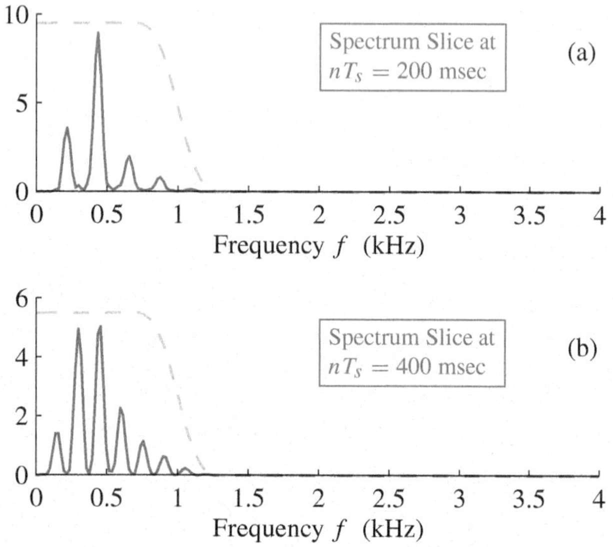

Figure 13-33: Spectrum slices at times (a) $nT_s = 200$ and (b) $nT_s = 400$ msec taken from Fig. 13-32.

13-9 The Fast Fourier Transform (FFT)

The material in this section is optional reading. It is included for completeness, since the FFT is the most important algorithm and computer program for doing spectrum analysis.

13-9.1 Derivation of the FFT

In Section 13-5.3 we discussed the FFT as an efficient algorithm for computing the DFT. In this section, we will give the basic divide-and-conquer method that leads to the FFT. From this derivation, it should be possible to write an FFT program that runs in time proportional to $(N/2)\log_2 N$ time. We need to assume that N is a power of two, so that the decomposition can be carried out recursively. Such algorithms are called *radix-2* algorithms.

The DFT summation (13.35) and the IDFT summation (13.36) are essentially the same, except for a minus sign in the exponent of the DFT and a factor of $1/N$ in the inverse DFT. Therefore, we will concentrate on the DFT calculation, knowing that a program written for the DFT could be modified to do the IDFT by changing the sign of the complex exponentials and multiplying the final values by $1/N$. The DFT summation can be broken into two sets, one sum over the even-indexed points of $x[n]$ and another sum over the odd-indexed points.

$$X[k] = \text{DFT}_N\{x[n]\} \tag{13.53}$$

$$= \sum_{n=0}^{N-1} x[n]e^{-j(2\pi/N)kn} \tag{13.54}$$

$$\begin{aligned} &= \big(x[0]e^{-j0} + x[2]e^{-j(2\pi/N)2k} \\ &\quad + \cdots + x[N-2]e^{-j(2\pi/N)k(N-2)}\big) \\ &\quad + \big(x[1]e^{-j(2\pi/N)k} + x[3]e^{-j(2\pi/N)3k} \\ &\quad + \cdots + x[N-1]e^{-j(2\pi/N)k(N-1)}\big) \end{aligned} \tag{13.55}$$

$$X[k] = \sum_{\ell=0}^{N/2-1} x[2\ell]e^{-j(2\pi/N)k(2\ell)} + \sum_{\ell=0}^{N/2-1} x[2\ell+1]e^{-j(2\pi/N)k(2\ell+1)} \quad (13.56)$$

At this point, two clever steps are needed: First, the exponent in the second sum must be broken into the product of two exponents, so we can factor out the one that does not depend on ℓ. Second, the factor of two in the exponents (2ℓ) can be associated with the N in the denominator of $2\pi/N$.

$$X[k] = \sum_{\ell=0}^{N/2-1} x[2\ell]e^{-j(2\pi k/N)(2\ell)} + e^{-j(2\pi k/N)} \sum_{\ell=0}^{N/2-1} x[2\ell+1]e^{-j(2\pi k/N)(2\ell)}$$

$$X[k] = \sum_{\ell=0}^{N/2-1} x[2\ell]e^{-j(2\pi k/(N/2))\ell} + e^{-j(2\pi k/N)} \sum_{\ell=0}^{N/2-1} x[2\ell+1]e^{-j(2\pi k/(N/2))\ell}$$

Now we have the correct form. Each of the summations is a DFT of length $N/2$, so we can write

$$X[k] = \text{DFT}_{N/2}\{x[2\ell]\} + e^{-j(2\pi/N)k}\,\text{DFT}_{N/2}\{x[2\ell+1]\} \quad (13.57)$$

The formula (13.57) for reconstructing $X[k]$ from the two smaller DFTs has one hidden feature: It must be evaluated for $k = 0, 1, 2, \ldots, N-1$. The $N/2$-point DFTs give output vectors that contain $N/2$ elements; e.g., the DFT of the odd-indexed points would be

$$X^o_{N/2}[k] = \text{DFT}_{N/2}\{x[2\ell+1]\}$$

for $k = 0, 1, 2, \ldots, N/2-1$. Thus we need an extra bit of information to calculate $X[k]$, for $k \geq N/2$. It is easy to verify that

$$X^o_{N/2}[k + N/2] = X^o_{N/2}[k]$$

and likewise for the DFT of the even-indexed points, so we need merely to periodically extend the results of the $N/2$-point DFTs before doing the sum in (13.57). This requires no additional computation.

The decomposition in (13.57) is enough to specify the entire FFT algorithm: Compute two smaller DFTs and then multiply the outputs of the DFT over the odd indices by the exponential factor $e^{-j(2\pi/N)k}$. Refer to Fig. 13-34, where three levels of the recursive decomposition can be seen. If a recursive structure is adopted, the two $N/2$ DFTs can be decomposed into four $N/4$-point DFTs, and those into eight $N/8$-point DFTs, etc. If N is a power of two, this decomposition will continue $(\log_2 N-1)$ times and then eventually reach the point where the DFT lengths are equal to two. For two-point DFTs, the computation is trivial:

$$X_2[0] = x_2[0] + x_2[1]$$
$$X_2[1] = x_2[0] + e^{-j2\pi/2}x_2[1] = x_2[0] - x_2[1]$$

The two outputs of the two-point DFT are the sum and the difference of the inputs. The last stage of computation would require $N/2$ two-point DFTs.

13-9.1.1 FFT Operation Count

The foregoing derivation is a bit sketchy, but the basic idea for writing an FFT program using the two-point DFT and the complex exponential as basic operators has been covered. However, the important point about the FFT is not how to write the program, but rather the number

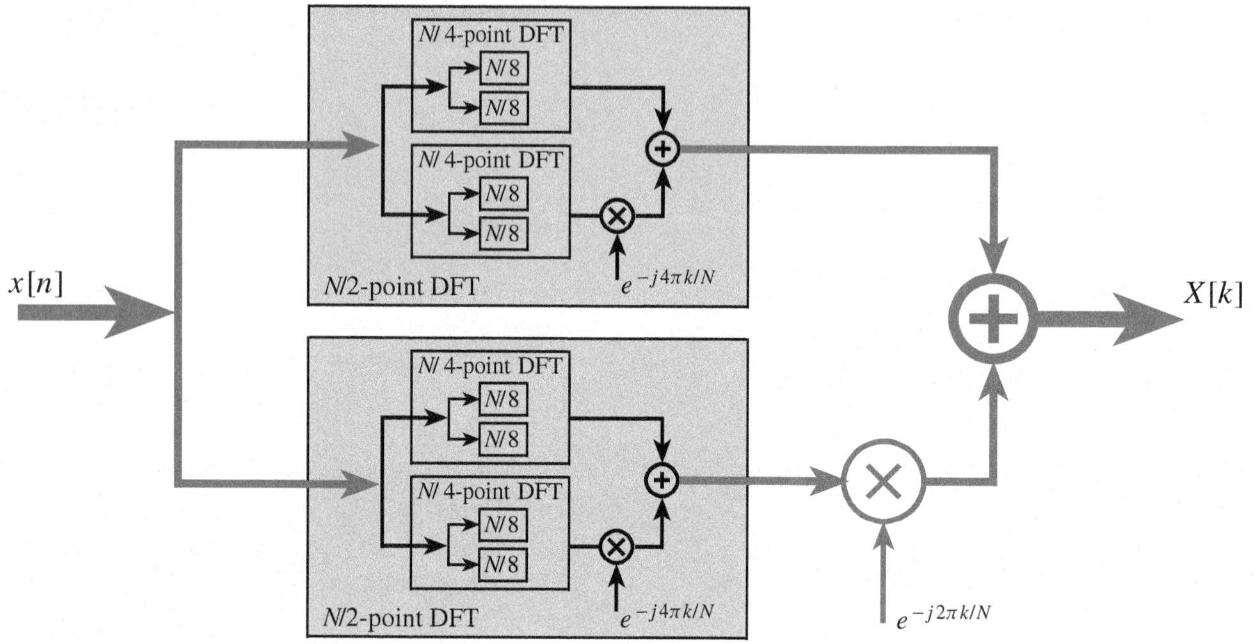

Figure 13-34: Block diagram of a radix-2 FFT algorithm for $N = 2^\nu$. The width of the lines is proportional to the amount of data being processed. For example, each $N/4$-point DFT must transform a data vector containing $N/4$ elements.

of operations needed to complete the calculation. When it was first published, the FFT made a huge impact on how people thought about problems, because it made the frequency-domain accessible numerically. Spectrum analysis became a routine calculation, even for very long signals. Operations such as filtering, which seem to be natural for the time-domain, could be done more efficiently in the frequency-domain for very long FIR filters.

The number of operations needed to compute the FFT can be expressed in a simple formula. We have said enough about the structure of the algorithm to count the number of operations. The count goes as follows: the N point DFT can be done with two $N/2$ point DFTs followed by N complex multiplications and N complex additions, as we can see in (13.57).[18] Thus, we have

$$\mu_c(N) = 2\mu_c(N/2) + N$$
$$\alpha_c(N) = 2\alpha_c(N/2) + N$$

where $\mu_c(N)$ is the number of complex multiplications for a length-N DFT, and $\alpha_c(N)$ is the number of complex additions. This equation can be evaluated successively for $N = 2, 4, 8, \ldots,$ because we know that $\mu_c(2) = 0$ and $\alpha_c(2) = 2$. Table 13-2 lists the number of operations for some transform lengths that are powers of two.

[18] Actually, the number of complex multiplications can be reduced to $N/2$, because $e^{-j2\pi(N/2)/N} = -1$.

Table 13-2: Number of operations for radix-2 FFT when N is a power of two. Notice how much smaller $\mu_c(N)$ is than $4N^2$.

N	$\mu_c(N)$	$\alpha_c(N)$	$\mu_r(N)$	$\alpha_r(N)$	$4N^2$
2	0	2	0	4	16
4	4	8	16	16	64
8	16	24	64	48	256
16	48	64	192	128	1024
32	128	160	512	320	4096
64	320	384	1280	768	16384
128	768	896	3072	1792	65536
256	1792	2048	7168	4096	262144
⋮	⋮	⋮	⋮	⋮	⋮

The formula for each can be derived by matching the table:

$$\mu_c(N) = N(\log_2 N - 1)$$
$$\alpha_c(N) = N \log_2 N$$

Since complex number operations ultimately must be done as multiplies and adds between real numbers, it is useful to convert the number of operations to real adds and real multiplies. Each complex addition requires two real additions, but each complex multiplication is equivalent to four real multiplies and two real adds. Therefore, we can put two more columns in Table 13-2 with these counts.

The bottom line for operation counts is that the total count is something proportional to $N \log_2 N$. The exact formulas from Table 13-2 are

$$\mu_r(N) = 4N(\log_2 N - 1)$$
$$\alpha_r(N) = 2N(\log_2 N - 1) + 2N = 2N \log_2 N$$

for the number of real multiplications and additions, respectively. Even these counts are a bit high because certain symmetries in the complex exponentials can be exploited to further reduce the computations.

13-10 Summary and Links

In this chapter, we have tried to establish a basic understanding of spectrum analysis and an appreciation for the concept of time-frequency representations of a discrete-time signal. We derived the DFT which plays a crucial role because it is efficiently computable using special (FFT) algorithms. We examined the DFT for finite-length signals, as well as continuing periodic signals. Also, we defined the spectrogram and investigated its resolution which is controlled by the window used prior to the FFT. The windowing property of the Fourier transform plays a central role in explaining the shapes that are commonly seen in spectrum analysis plots. Several examples were shown of spectrograms for speech and music to illustrate the resolution trade-off inherent in spectrum analysis.

The filtered speech example of Section 13-8.7 is ample evidence that the concept of frequency spectrum has wide-ranging validity. It means that we can think of the effect of an LTI system on *any* signal as being represented by its frequency response. We can say things like "The signal has a bandwidth of 4 kHz," or "The high frequencies were emphasized by the filter." This should not be a surprising conclusion, since even without this semiformal discussion, most of us, engineers or not, are comfortable with thinking about frequency responses of hi-fi sets, loudspeakers, headphones, and even our ears. What makes the frequency response such a useful concept is that any signal can be thought of as a sum of (perhaps infinitely many) complex-exponential signals. This chapter has tried to demonstrate this with many examples and some equations. There is much more to learn about Fourier analysis and the concept of the frequency spectrum, but the essence of this infinitely interesting subject is contained here.

The demonstrations and projects for this chapter have much in common with those from Chapter 3, so the reader should consult those earlier demonstrations. Two additional laboratory projects available on the CD-ROM are Lab #19 which explores the FFT, and Lab #20 which requires students to develop a music-analysis program that will write the music for a song from a recording. In Lab #20, the spectrogram is used to get a time-frequency representation of the music, and then a peak-picking algorithm and editing program must be written to find the spectral peaks that correspond to actual notes. This project can be considered as the inverse to Lab #4, the music synthesis lab. Another similar lab is Lab #9 on the CD which also appears in Section C-2 of Appendix C. Lab #9 involves sinusoidal signals generated by a touch-tone phone. A spectrogram analysis of these signals reveals their simple structure.

The CD-ROM also contains three demonstrations related to the spectrogram:

(a) Computation of the spectrogram is illustrated by a movie showing a windowed FFT moving through a signal.

DEMO: *Resolution of the Spectrogram*

(b) Spectrograms of chirp signals show how the window length of the FFT affects the spectrogram image that you see. For the chirp case, the rate of change of the frequency affects the spectrogram image depending on how fast the frequency changes within the window duration. It also affects the sound your hear, because the human auditory system can be modeled as a spectrum analyzer with variable (but known) frequency resolution.

DEMO: *Wideband FM Signals*

(c) A MATLAB graphical user interface that presents the connection between musical notation and the spectrogram.

DEMO: *Music GUI*

Finally, the reader is reminded that there are many solved problems that are available for review and practice.

NOTE: *Hundreds of Solved Problems*

13-11 Problems

P-13.1 Suppose that a discrete-time signal $x[n]$ is a sum of complex-exponential signals

$$x[n] = 3 + 2e^{j0.2\pi n} + 2e^{-j0.2\pi n} - 7je^{j0.7\pi n} + 7je^{-j0.7\pi n}$$

(a) Make a plot of the DTFT for $x[n]$ using only positive frequencies, i.e., $0 \le \hat{\omega} < 2\pi$.

(b) Suppose that $x_1[n] = x[n]e^{j0.4\pi n}$. Make a plot of the DTFT for $x_1[n]$ using only positive frequencies.

(c) Suppose that $x_2[n] = (-1)^n x[n]$. Make a plot of the DTFT for $x_2[n]$ using only positive frequencies.

P-13.2 Consider the continuous-time rectangular window $w_R(t)$, defined by (13.16) and the Hamming window, $w_H(t)$, defined by (13.26).

(a) Show that the continuous-time Fourier transform of the rectangular window $w_R(t)$ is as given in (13.23).

(b) Use the fact that

$$w_H(t) = \left[0.54 - 0.46\cos\left(\frac{2\pi}{T}t\right)\right] w_R(t)$$

to show that

$$W_H(j\omega) = 0.54 W_R(j\omega) - 0.23 W_R(j(\omega - 2\pi/T)) - 0.23 W_R(j(\omega + 2\pi/T))$$

(c) Use the result of (b) to construct a sketch of $|W_H(j\omega)|$, and from this sketch argue that the width of the main-lobe of $|W_H(j\omega)|$ is $\Delta\omega_H = 8\pi/T$.

P-13.3 A continuous-time signal $x_c(t)$ is bandlimited such that $X_c(j\omega) = 0$ for $|\omega| \geq 2\pi(1000)$. This signal is sampled with sampling rate $f_s = 1/T_s$ producing the sequence $x[n] = x_c(nT_s)$ as in Fig. 13-2. Assume that the window is a rectangular window whose length L is equal to the DFT length, N. Furthermore, for efficiency in computation, assume that N is a power of two. Both f_s and N can be chosen at will subject to the constraints that aliasing be avoided and $N = 2^\nu$. Determine the *minimum* value of N and the range of sampling rates $f_s^{\min} < f_s < f_s^{\max}$ so that the *effective* spacing between DFT frequencies is less than or equal to 5 Hz. Give numerical values for N, $f_s^{\min}$ and $f_s^{\max}$.

P-13.4 Determine the 10-point DFT of the following:

(a) $x_0[n] = \begin{cases} 1 & n = 0 \\ 0 & n = 1, 2, \ldots, 9 \end{cases}$

(b) $x_1[n] = 1 \qquad \text{for } n = 0, 1, 2, \ldots, 9$

(c) $x_2[n] = \begin{cases} 1 & n = 4 \\ 0 & n \neq 4 \end{cases}$

(d) $x_3[n] = e^{j2\pi n/5} \qquad \text{for } n = 0, 1, 2, \ldots, 9$

P-13.5 Determine the 10-point inverse DFT (IDFT) of the following:

(a) $X_a[k] = \begin{cases} 1 & k = 0 \\ 0 & k = 1, 2, \ldots, 9 \end{cases}$

(b) $X_b[k] = 1 \qquad \text{for } k = 0, 1, 2, \ldots, 9$

(c) $X_c[k] = \begin{cases} 1 & k = 3, 7 \\ 0 & k = 0, 1, 2, 4, 5, 6, 8, 9 \end{cases}$

(d) $X_d[k] = \cos(2\pi k/5) \qquad \text{for } k = 0, 1, 2, \ldots, 9$

P-13.6 Determine the 12-point DFT of the following:

(a) $y_0[n] = \begin{cases} 1 & n = 0, 1, 2, 3 \\ 0 & n = 4, 5, \ldots, 11 \end{cases}$

(b) $y_1[n] = \begin{cases} 1 & n = 0, 2, 4, 6, 8, 10 \\ 0 & n = 1, 3, 5, 7, 9, 11 \end{cases}$

P-13.7 The 8-point running-sum filter is an FIR filter with coefficients $\{b_k\} = \{1,\ 1,\ 1,\ 1,\ 1,\ 1,\ 1,\ 1,\}$.

(a) Use FFT spectrum analysis to plot the frequency response of the 8-pt. running-sum filter.

(b) Suppose that the input to the 8-pt. running-sum filter is $x[n]$ whose DTFT has nonzero components only at $\hat{\omega}_1 = 0.25\pi$, $\hat{\omega}_2 = 0.5\pi$, and $\hat{\omega}_3 = 0.75\pi$, and a DC value of $X_0 = 3$. Explain why the output of the 8-pt. running-sum filter is then $y[n] = 24$ for all n. Use values from the plot of $|H(e^{j\hat{\omega}})|$ versus $\hat{\omega}$ to support your explanation.

P-13.8 Suppose that a continuous-time signal $x(t)$ consists of several sinusoidal sections

$$x(t) = \begin{cases} \cos(2\pi(600)t) & 0 \leq t < 0.5 \\ \sin(2\pi(1100)t) & 0.3 \leq t < 0.7 \\ \cos(2\pi(500)t) & 0.4 \leq t < 1.2 \\ \cos(2\pi(700)t - \pi/4) & 0.4 \leq t < 0.45 \\ \sin(2\pi(800)t) & 0.35 \leq t < 1.0 \end{cases}$$

(a) If the signal is sampled with a sampling frequency of $f_s = 8000$ Hz, make a sketch of the *ideal* spectrogram that corresponds to the signal definition.

(b) Make a sketch of the *actual* spectrogram that would be obtained with an FFT length of $N = 256$ and a Hann window length of $L = 256$. Make approximations in order to do the sketch without actually calculating the spectrogram in MATLAB.

P-13.9 The C-major scale was analyzed in the spectrogram of Fig. 13-22 with a window length of $L = 256$. Make a sketch of the spectrogram that would be obtained if the window length were $L = 100$. Explain how your sketch would differ from Fig. 13-22.

P-13.10 In the spectrogram of the C-major scale in Fig. 13-22, estimate the time duration of each note.

P-13.11 The spectrogram of Fig. 13-23 shows the period of time corresponding to the first nine notes in the treble clef in Fig. 13-21. Compute the theoretical frequencies (based on A440) of these nine notes and compare your result to what you measure on the spectrogram of Fig. 13-23.

P-13.12 Assume that a speech signal has been sampled at 8000 Hz and then analyzed with MATLAB's `specgram` function using the following parameters: Hamming window with length $L = 100$, FFT length of $N = 256$, and overlap of 80 points. Determine the resolution of the resulting spectrogram image.

(a) Determine the frequency resolution (in Hz)

(b) Determine the time resolution (in sec)

P-13.13 The resolution of a spectrum analysis system is determined by the frequency response of the time window. Suppose that we sample a continuous-time signal at $f_s = 10{,}000$ Hz and we want to have frequency resolution of 250 Hz.

(a) If we use a Hamming window for $w[n]$, what window length L would be needed? Estimate a minimum value for L.

(b) Use the specific value for L determined in part (a), and demonstrate that the resulting Hamming window will have sufficient frequency resolution by plotting its frequency response (magnitude).

APPENDIX A

Complex Numbers

The basic manipulations of complex numbers are reviewed in this appendix. The algebraic rules for combining complex numbers are reviewed, and then a geometric viewpoint is taken to visualize these operations by drawing vector diagrams. This geometric view is a key to understanding how complex numbers can be used to represent signals. Specifically, the following three significant ideas about complex numbers are treated in this appendix:

- *Simple Algebraic Rules:* Operations on complex numbers ($z = x + jy$) follow exactly the same rules as real numbers, with j^2 replaced everywhere by -1.[1]
- *Elimination of Trigonometry:* Euler's formula for the complex exponential $z = re^{j\theta} = r\cos\theta + jr\sin\theta$ provides a connection between trigonometric identities and simple algebraic operations on complex numbers.
- *Representation by Vectors:* A vector drawn from the origin to a point (x, y) in a two-dimensional plane is equivalent to $z = x + jy$. The algebraic rules for z are, in effect, the basic rules of vector operations. More important, however, is the *visualization* gained from the vector diagrams.

The first two ideas concern the algebraic nature of $z = x + jy$, the other its role as a ***representer*** of signals. Skill in algebraic manipulations is important, but the use of complex numbers in representation is more important in the long run. Complex numbers in electrical engineering are used as a convenience because

[1] Mathematicians and physicists use the symbol i for $\sqrt{-1}$; electrical engineers prefer to reserve the symbol i for current in electric circuits.

when they stand for the sinusoidal signals they can simplify manipulations of the signals. Thus, a sinusoidal problem (such as the solution to a differential equation) is converted into a complex number problem that can be (1) solved by the simple rules of algebra and (2) visualized through vector geometry. The key to all this is the higher-order thinking that permits abstraction of the problem into the world of complex numbers. Ultimately, we are led to the notion of a "transform" such as the Fourier or Laplace transform to reduce many other sophisticated problems to algebra. Because complex numbers are so crucial to almost everything we do in the study of signals and systems, a careful study of this appendix is advised.

Once such insight is gained, it still will be necessary to return occasionally to the lower-level drudgery of calculations. When you have to manipulate complex numbers, a calculator will be most useful, especially one with built-in complex arithmetic capability. It is worthwhile to learn how to use this feature on your calculator. However, it is also important to do some calculations by hand, so that you will *understand* what your calculator is doing!

Finally, it's too bad that complex numbers are called "complex." Most students, therefore, think of them as complicated. However, their elegant mathematical properties usually simplify calculations quite a bit.

A-1 Introduction

A complex number system is an extension of the real number system. Complex numbers are necessary to solve equations such as

$$z^2 = -1 \tag{A.1}$$

The symbol j is introduced to stand for $\sqrt{-1}$, so the previous equation (A.1) has the two solutions $z = \pm j$. More generally, complex numbers are needed to solve for the two roots of a quadratic equation

$$az^2 + bz + c = 0$$

which, according to the quadratic formula, has the two solutions:

$$z = \frac{-b \pm \sqrt{b^2 - 4ac}}{2a}$$

Whenever the discriminant $(b^2 - 4ac)$ is negative, the solution must be expressed as a complex number. For example, the roots of

$$z^2 + 6z + 25 = 0$$

are $z = -3 \pm j4$, because

$$\begin{aligned}(1/2a)\sqrt{b^2 - 4ac} &= \tfrac{1}{2}\sqrt{36 - 4(25)} \\ &= \tfrac{1}{2}\sqrt{-64} = \pm j4.\end{aligned}$$

A-2 Notation for Complex Numbers

Several different mathematical notations can be used to represent complex numbers. The two basic types are polar form and rectangular form. Converting between them quickly and easily is an important skill.

A-2.1 Rectangular Form

In ***rectangular form***, all of the following notations define the same complex number.

$$\begin{aligned} z &= (x, y) \\ &= x + jy \\ &= \Re e\{z\} + j\Im m\{z\} \end{aligned}$$

The ordered pair (x, y) can be interpreted as a point in the two-dimensional plane.[2]

[2]This is also the notation used on some calculators when entering complex numbers.

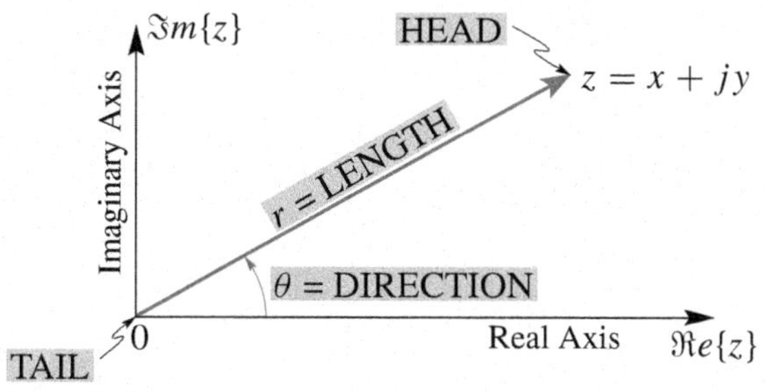

Figure A-1: Complex number represented as a vector from the origin to (x, y).

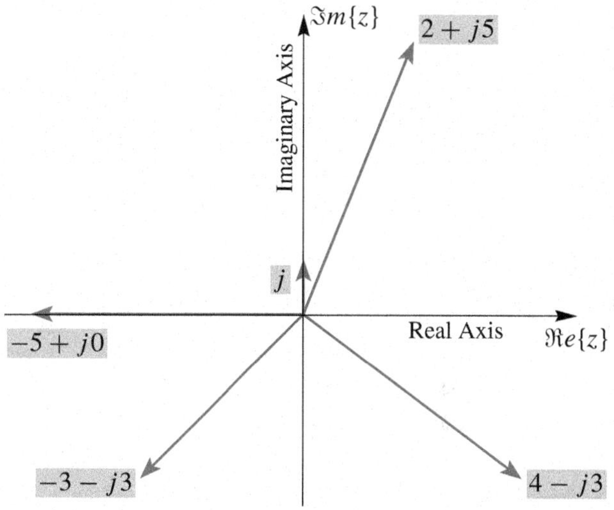

Figure A-2: Complex numbers plotted as vectors in the two-dimensional "complex plane." Each $z = x + jy$ is represented by a vector from the origin to the point with coordinates (x, y) in the complex plane.

A complex number can also be drawn as a vector whose tail is at the origin and whose head is at the point (x, y), in which case x is the horizontal coordinate of the vector and y the vertical coordinate (see Fig. A-1).

Figure A-2 shows some numerical examples. The complex number $z = 2 + j5$ is represented by the point $(2, 5)$, which lies in the first quadrant of the two-dimensional plane; likewise, $z = 4 - j3$ lies in the fourth quadrant at the location $(4, -3)$. Since the complex number notation $z = x + jy$ represents the point (x, y) in the two-dimensional plane, the number j represents $(0, 1)$, which is drawn as a vertical vector from the origin up to $(0, 1)$, as in Fig. A-2. Thus multiplying a real number, such as 5, by j changes it from pointing along the horizontal axis to pointing vertically; i.e., $j(5 + j0) = 0 + j5$.

Rectangular form is also referred to as ***Cartesian form***. The horizontal coordinate x is called the ***real part***, and the vertical coordinate y the ***imaginary part***. The operators $\Re e\{z\}$ and $\Im m\{z\}$ are provided to extract the real and imaginary parts of $z = x + jy$:

$$x = \Re e\{z\} \quad \text{(A.2a)}$$

$$y = \Im m\{z\} \quad \text{(A.2b)}$$

A-2.2 Polar Form

In ***polar form***, the vector is defined by its length (r) and its direction (θ) as in Fig. A-1. Therefore, we use the following descriptive notation sometimes:

$$z \longleftrightarrow r\angle\theta$$

Some examples are shown in Fig. A-3, where the direction θ is given in degrees. The angle is always measured from the positive x-axis and may be either positive (counterclockwise) or negative (clockwise). However, we generally specify the ***principal value*** of the angle so that $-180^\circ < \theta \leq 180^\circ$. This requires that integer multiples of 360° be subtracted from or added to the angle until the result is between -180° and $+180^\circ$. Thus, the vector $3\angle{-80^\circ}$ is the principal value of $3\angle 280^\circ$.

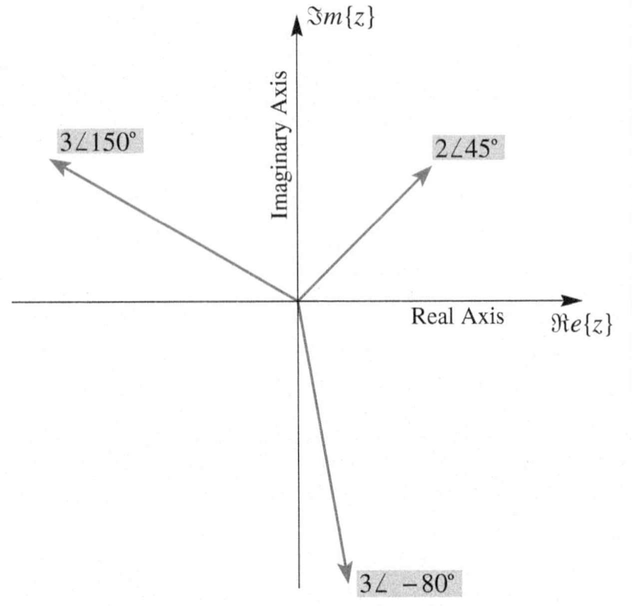

Figure A-3: Several complex numbers plotted in terms of length (r) and direction (θ) of their vector representation. The angle is always measured with respect to the positive real axis; its units are usually radians, but are shown as degrees in this case.

A-2.3 Conversion: Rectangular and Polar

Both the polar and rectangular forms are commonly used to represent complex numbers. The prevalence of the polar form, for sinusoidal signal representation, makes it necessary to convert quickly and accurately between the two representations.

DEMO: ***Zdrill***

From Fig. A-4, we see that the x and y coordinates of the vector are given by

$$x = r\cos\theta \tag{A.3a}$$

$$y = r\sin\theta \tag{A.3b}$$

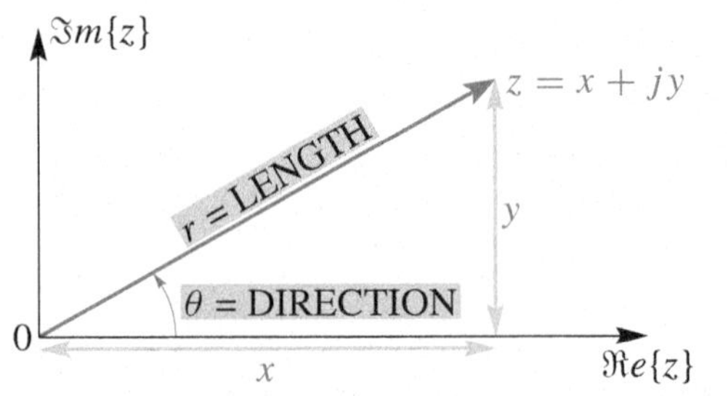

Figure A-4: Basic layout for relating Cartesian coordinates (x, y) to polar coordinates r and θ.

Therefore, a valid formula for z is

$$z = r\cos\theta + j\,r\sin\theta \tag{A.4}$$

Example A-1: Polar to Rectangular

In Fig. A-3, the three complex numbers are

$$2\angle 45° \quad\longleftrightarrow\quad z = \sqrt{2} + j\sqrt{2}$$

$$3\angle 150° \quad\longleftrightarrow\quad z = -\frac{3\sqrt{3}}{2} + j\frac{3}{2}$$

$$3\angle -80° \quad\longleftrightarrow\quad z = 0.521 - j2.954$$

The conversion from polar to rectangular was done via (A.3). ■

The conversion from (x, y) to $r\angle\theta$ is a bit trickier. From Fig. A-4, the formulas are

$$r = \sqrt{x^2 + y^2} \qquad \text{(Length)} \tag{A.5a}$$

$$\theta = \arctan(y/x) \qquad \text{(Direction)} \tag{A.5b}$$

The arctangent must give a four-quadrant answer, and the direction θ is usually given in radians rather than degrees.

EXERCISE A.1: At this point, the reader should convert to polar form the five complex numbers shown in Fig. A-2. The answers, in a random order, are: $1\angle 90°$, $5\angle -36.87°$, $4.243\angle -135°$, $5.385\angle 68.2°$, and $5\angle 180°$.

In Section A-3, we will introduce two other polar notations:

$$z = re^{j\theta} = |z|e^{j \arg z}$$

where $|z| = r = \sqrt{x^2 + y^2}$ is called the ***magnitude*** of the vector and $\arg z = \theta = \arctan(y/x)$ is its ***phase*** in radians (not degrees). This exponential notation, which relies on Euler's formula, has the advantage that when it is used in algebraic expressions, the standard laws of exponents apply.

A-2.4 Difficulty in Second or Third Quadrant

The formula (A.5b) for the angle θ as the $\arctan(y/x)$ must be used with care, especially when the real part is negative (see Fig. A-5). For example, the complex number $z = -3 + j2$ would require that we evaluate $\arctan(-2/3)$ to get the angle; the same calculation would be needed if $z = 3 - j2$. The arctangent of $-2/3$ is -0.588 rad, or about $-33.7°$, which is the correct angle for $z = 3 - j2$. However, for $z = -3 + j2$, the vector lies in the second quadrant and the angle must satisfy $90° \leq \theta \leq 180°$. We get the correct angle by adding $180°$. In this case, the correct angle is $\pi - 0.588 = 2.55$ rad, or about $180° - 33.7° = 146.3°$. The general method for getting the angle of z in polar form is

$$\theta = \begin{cases} \arctan(y/x) & \text{if } x \geq 0 \\ \pm 180° + \arctan(y/x) & \text{if } x < 0 \end{cases} \tag{A.6}$$

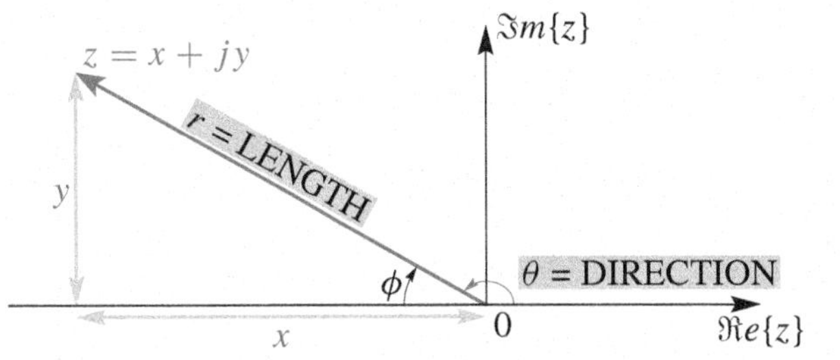

Figure A-5: In the second quadrant, the interior angle ϕ is easily calculated from x and y, but is not the correct angle for the polar form, which requires the exterior angle θ with respect to the positive real axis.

Since $180°$ and $-180°$ are both angles to $z = -1$, when $x < 0$ we should pick the appropriate one to make the final answer satisfy $|\theta| \leq 180°$.

A-3 Euler's Formula

The conversion from polar to rectangular form (A.3) suggests the following formula:

$$e^{j\theta} = \cos\theta + j\sin\theta \tag{A.7}$$

Equation (A.7) defines the ***complex exponential*** $e^{j\theta}$, which is equivalent to $1\angle\theta$ (i.e., a vector of length 1 at angle θ). A proof of Euler's formula based on power series is outlined in Problem 2.4 of Chapter 2.

The amazing discovery was that the laws of exponents apply to $e^{j\theta}$. Euler's formula (A.7) is so important that it must be instantly recalled; likewise, the inverse Euler formulas (A.8) and (A.9) should also be committed to memory.

Example A-2: Euler's Formula

Here are some examples:

$$(\mathbf{90^\circ}):\quad e^{j\pi/2} = \cos(\pi/2) + j\sin(\pi/2)$$
$$= 0 + j1 = j \quad\longleftrightarrow\quad 1\angle\pi/2$$

$$(\mathbf{180^\circ}):\quad e^{j\pi} = \cos(\pi) + j\sin(\pi)$$
$$= -1 + j0 = -1 \quad\longleftrightarrow\quad 1\angle\pi$$

$$(\mathbf{45^\circ}):\quad e^{j\pi/4} = \cos(\pi/4) + j\sin(\pi/4)$$
$$= \frac{1}{\sqrt{2}} + j\frac{1}{\sqrt{2}} \quad\longleftrightarrow\quad 1\angle\pi/4$$

$$(\mathbf{60^\circ}):\quad e^{j\pi/3} = \cos(\pi/3) + j\sin(\pi/3)$$
$$= \tfrac{1}{2} + j\tfrac{1}{2}\sqrt{3} \quad\longleftrightarrow\quad 1\angle\pi/3$$

■

Example A-3: Degrees to Radians

Referring back to Fig. A-3, the three complex numbers can be rewritten as:

$$2\angle 45^\circ \quad\longleftrightarrow\quad z = 2e^{j\pi/4}$$
$$3\angle 150^\circ \quad\longleftrightarrow\quad z = 3e^{j5\pi/6}$$
$$3\angle -80^\circ \quad\longleftrightarrow\quad z = 3e^{-j4\pi/9} = 3e^{-j1.396}$$

■

Numbers like -1.396 rad are difficult to visualize, because we are more used to thinking of angles in degrees. It may be helpful to express the angles used in the exponents as a fraction of π rad; i.e., $-1.396 = -(1.396/\pi)\pi = -0.444\pi$ rad. This is a good habit to adopt, because it simplifies the conversion between degrees and radians. If θ is given in radians, the conversion is

$$\theta \times \left(\frac{180}{\pi}\right) = \text{direction in degrees}$$

A-3.1 Inverse Euler Formulas

Euler's formula (A.7) can be solved separately for the cosine and sine parts. The result will be called the ***inverse Euler relations***

$$\cos\theta = \frac{e^{j\theta} + e^{-j\theta}}{2} \tag{A.8}$$

$$\sin\theta = \frac{e^{j\theta} - e^{-j\theta}}{2j} \tag{A.9}$$

Recalling that $\cos(-\theta) = \cos\theta$ and $\sin(-\theta) = -\sin\theta$, the proof of the $\sin\theta$ version is as follows:

$$e^{-j\theta} = \cos(-\theta) + j\sin(-\theta)$$
$$= \cos\theta - j\sin\theta$$
$$e^{+j\theta} = \cos\theta + j\sin\theta$$
$$\Longrightarrow e^{j\theta} - e^{-j\theta} = 2j\sin\theta$$
$$\Longrightarrow \sin\theta = \frac{e^{j\theta} - e^{-j\theta}}{2j}$$

A-4 Algebraic Rules for Complex Numbers

The basic arithmetic operators for complex numbers follow the usual rules of algebra as long as the symbol j is treated as a special token that satisfies $j^2 = -1$. In rectangular form, all of these rules are relatively straightforward. The five fundamental rules are the following:

Addition:

$$z_1 + z_2 = (x_1 + jy_1) + (x_2 + jy_2)$$
$$= (x_1 + x_2) + j(y_1 + y_2)$$

Subtraction:

$$z_1 - z_2 = (x_1 + jy_1) - (x_2 + jy_2)$$
$$= (x_1 - x_2) + j(y_1 - y_2)$$

Multiplication:

$$z_1 \times z_2 = (x_1 + jy_1) \times (x_2 + jy_2)$$

$$\begin{aligned} z_1 z_2 &= x_1x_2 + j^2 y_1 y_2 + jx_1y_2 + jx_2y_1 \\ &= (x_1x_2 - y_1y_2) + j(x_1y_2 + x_2y_1) \end{aligned}$$

Division:

$$z_1 \div z_2 = (x_1 + jy_1)/(x_2 + jy_2)$$

$$\begin{aligned} \frac{z_1}{z_2} &= \frac{z_1 z_2^*}{z_2 z_2^*} = \frac{z_1 z_2^*}{|z_2|^2} \\ &= \frac{(x_1 + jy_1)(x_2 - jy_2)}{x_2^2 + y_2^2} \\ &= \frac{(x_1x_2 + y_1y_2) + j(x_2y_1 - x_1y_2)}{x_2^2 + y_2^2} \end{aligned}$$

Conjugate:

$$\begin{aligned} z_1^* &= (x_1 + jy_1)^* \\ &= x_1 - jy_1 \end{aligned}$$

Addition and subtraction are straightforward because we need only add or subtract the real and imaginary parts. On the other hand, addition (or subtraction) in polar form cannot be carried out directly on r and θ; instead, an intermediate conversion to rectangular form is required. In contrast, the operations of multiplication and division, which are rather messy in rectangular form, reduce to simple manipulations in polar form. For multiplication, multiply the magnitudes and add the angles; for division, divide the magnitudes and subtract the angles. The conjugate in polar form requires only a change of sign of the angle.

Multiplication:

$$\begin{aligned} z_1 \times z_2 &= r_1 e^{j\theta_1} \times r_2 e^{j\theta_2} \\ &= (r_1 r_2)\, e^{j(\theta_1 + \theta_2)} \end{aligned}$$

Division:

$$\begin{aligned} z_1 \div z_2 &= \frac{r_1 e^{j\theta_1}}{r_2 e^{j\theta_2}} \\ &= \frac{r_1}{r_2} e^{j(\theta_1 - \theta_2)} \end{aligned}$$

Conjugate:

$$\begin{aligned} z_1^* &= (r_1 e^{j\theta_1})^* \\ &= r_1 e^{-j\theta_1} \end{aligned}$$

EXERCISE A.2: The inverse or reciprocal of a complex number z is the number z^{-1} such that

$$z^{-1} z = 1.$$

A common mistake with the inverse is to invert $z = x + jy$ by taking the inverse of x and y separately. To show that this is wrong, take the specific case where $z = 4 + j3$ and $w = \frac{1}{4} + j\frac{1}{3}$. Show that w is not the inverse of z, because $wz \neq 1$. Determine the correct inverse of z.

Polar form presents difficulties when adding (or subtracting) two complex numbers and expressing the final answer in polar form. An intermediate conversion to rectangular form must be done. Here is the recipe for adding complex numbers in polar form.

1. Starting in polar form, we have

$$z_3 = z_1 \pm z_2 = r_1 e^{j\theta_1} \pm r_2 e^{j\theta_2}$$

2. Convert both z_1 and z_2 to Cartesian form:

$$z_3 = (r_1 \cos\theta_1 + j\, r_1 \sin\theta_1) \pm (r_2 \cos\theta_2 + j\, r_2 \sin\theta_2)$$

3. Perform the addition in Cartesian form:

$$z_3 = (r_1 \cos\theta_1 \pm r_2 \cos\theta_2) + j\,(r_1 \sin\theta_1 \pm r_2 \sin\theta_2)$$

4. Identify the real and imaginary parts of z_3:

$$\begin{aligned} x_3 &= \Re e\{z_3\} = r_1 \cos\theta_1 \pm r_2 \cos\theta_2 \\ y_3 &= \Im m\{z_3\} = r_1 \sin\theta_1 \pm r_2 \sin\theta_2 \end{aligned}$$

5. Convert back to polar form using (A.5a) and (A.5b):

$$z_3 = x_3 + jy_3 \quad \longleftrightarrow \quad z_3 = r_3 e^{j\theta_3}$$

If you have a calculator that converts between polar and rectangular form, learn how to use it; it will save many hours of hand calculation and also be more accurate. Most "scientific" calculators even have the capability to use both notations, so the conversion is transparent to the user.

Example A-4: Adding Polar Forms

Here is a numerical example of adding two complex numbers given in polar form:

$$z_3 = 7e^{j4\pi/7} + 5e^{-j5\pi/11}$$

$$z_3 = (-1.558 + j6.824) + (0.712 - j4.949)$$

$$z_3 = -0.846 + j1.875$$

$$z_3 = 2.057e^{j1.995} = 2.057e^{j0.635\pi} = 2.057\angle 114.3^\circ$$

Remember: When the angle appears in the exponent, its units must be in radians. ■

A-4.1 Complex Number Exercises

To practice computations for complex numbers, try the following exercises.

EXERCISE A.3: Add and multiply the following, then plot the results:

$$z_4 = 5e^{j4\pi/5} + 7e^{-j5\pi/7}$$

$$z_5 = 5e^{j4\pi/5} \times 7e^{-j5\pi/7}$$

The answers are $z_4 = -8.41 - j2.534 = 8.783e^{-j0.907\pi}$ and $z_5 = 35e^{j3\pi/35}$.

EXERCISE A.4: For the conjugate the simple rule is to change the sign of all the j terms. Work the following:

$$(3 - j4)^* = \ ?$$

$$(j(1 - j))^* = \ ?$$

$$(e^{j\pi/2})^* = \ ?$$

EXERCISE A.5: Prove that the following identities are true:

$$\Re e\{z\} = (z + z^*)/2$$

$$\Im m\{z\} = (z - z^*)/2j$$

$$|z|^2 = zz^*$$

More drill problems can be generated by using the MATLAB program `zdrill.m`, which presents a GUI (graphical user interface) that asks questions for each of the complex operations, and also plots the vectors that represent the solutions. The `zdrill` GUI has both a novice and an advanced level. A screen shot of the GUI is shown in Fig. A-6.

DEMO: *Zdrill*

A-5 Geometric Views of Complex Operations

It is important to develop an ability to visualize complex number operations. This is done by plotting the vectors that represent the numbers in the (x, y) plane, where $x = \Re e\{z\}$ and $y = \Im m\{z\}$. The key to this is to recall that, as shown in Fig. A-1, the complex number $z = x + jy$ is a vector whose tail is at the origin and whose head is at (x, y).

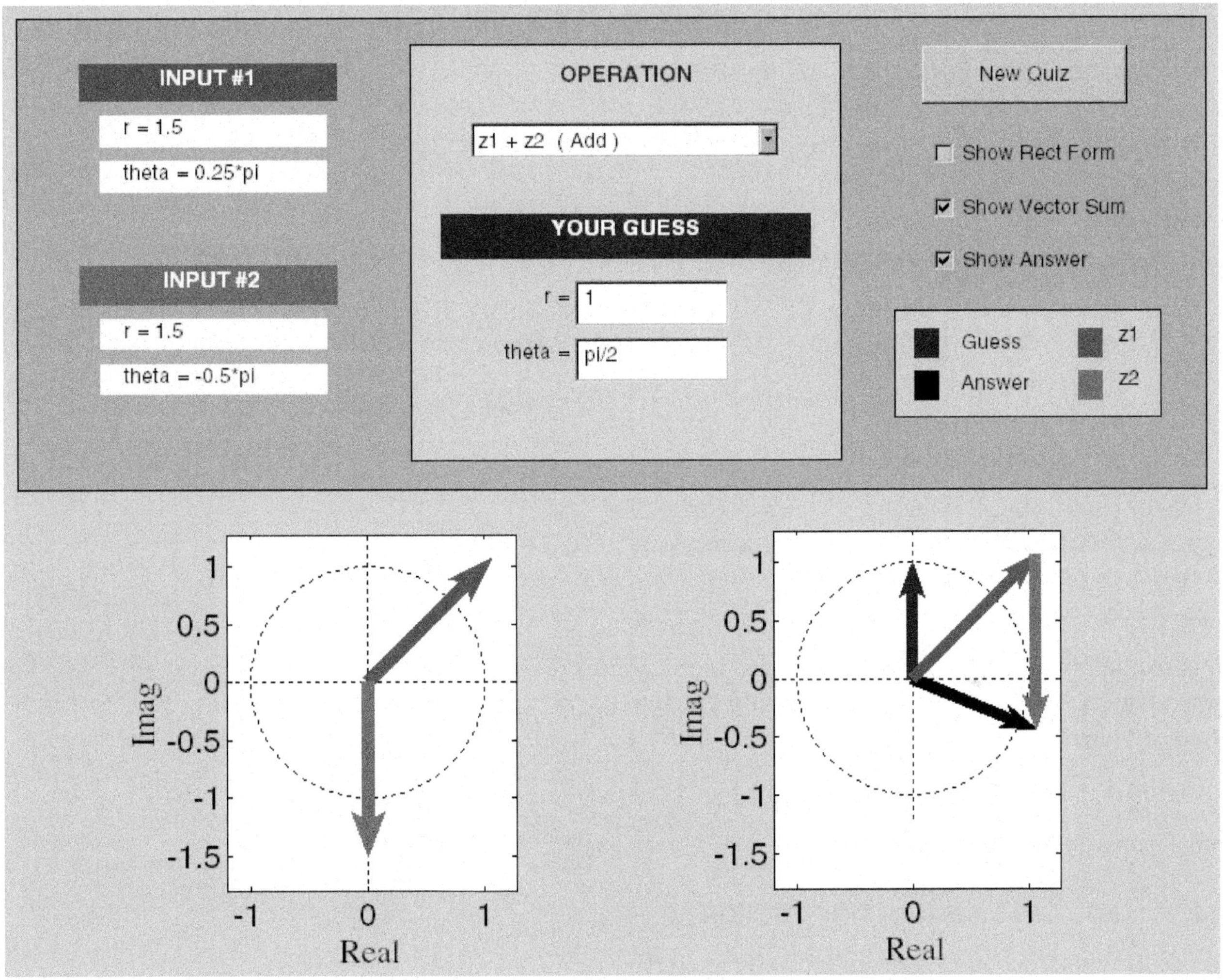

Figure A-6: The MATLAB GUI `zdrill` for practicing complex number operations.

A-5.1 Geometric View of Addition

For complex addition, $z_3 = z_1 + z_2$, both z_1 and z_2 are viewed as vectors with their tails at the origin. The sum z_3 is the result of vector addition, and is constructed as follows (see Fig. A-7):

1. Draw a copy of z_1 with its tail at the head of z_2. Call this displaced vector $\hat{z}_1$.

2. Draw the vector from the origin to the head of $\hat{z}_1$. This vector is the sum z_3.

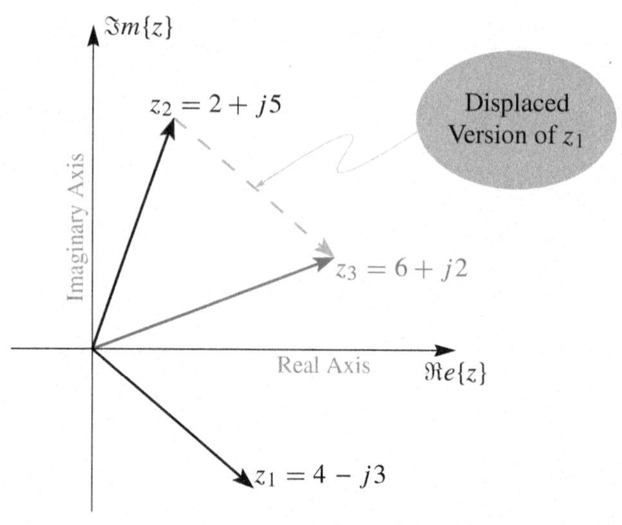

Figure A-7: Graphical construction of complex number addition $z_3 = z_1 + z_2$.

This method of addition can be generalized to many vectors. Figure A-8 shows the result of adding the four complex numbers

$$(1 + j) + (-1 + j) + (-1 - j) + (1 - j)$$

where the answer happens to be zero.

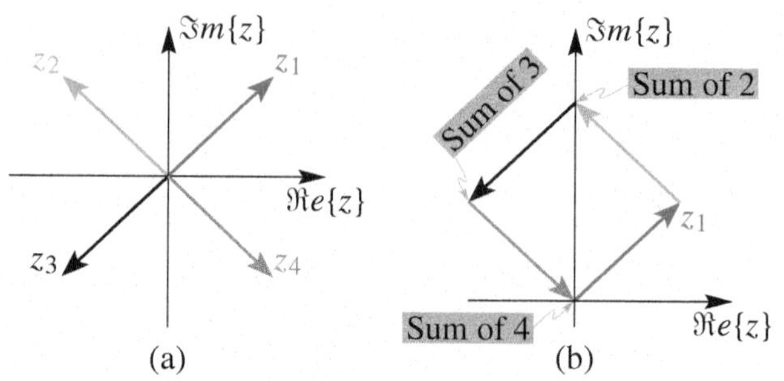

Figure A-8: Result of adding the four vectors $\{1+j,\ -1+j,\ -1-j,\ 1-j\}$ shown in (a) is zero. The "head-to-tail" graphical method is illustrated in (b).

A-5.2 Geometric View of Subtraction

The visualization of subtraction requires that we interpret the operation as addition:

$$z_3 = z_2 - z_1 = z_2 + (-z_1)$$

Thus we must flip z_1 to $-z_1$ and then add the result to z_2. This method is shown in Fig. A-9 and is equivalent to the visualization of addition, as in Fig. A-7. There are two additional comments to be made about subtraction:

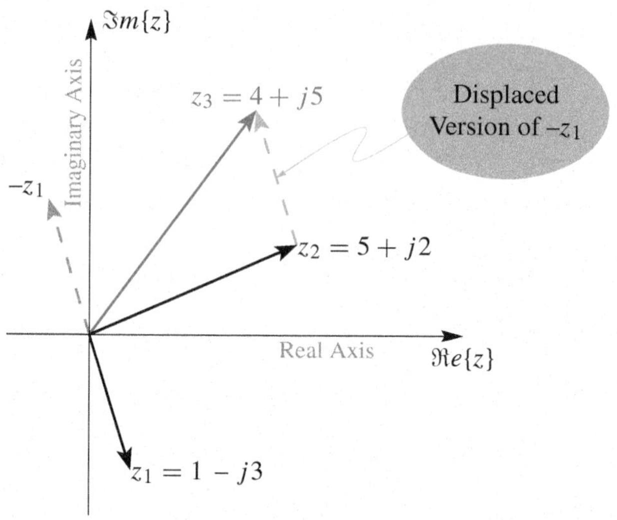

Figure A-9: Geometric view of subtraction. The vector z_1 is flipped to make $(-z_1)$ and then a displaced version is added to z_2 to obtain $z_3 = z_2 + (-z_1)$.

1. Since $z_2 = z_1 + z_3$, a displaced version of the difference vector could be drawn with its tail at the head of z_1 and its head at the head of z_2.
2. The triangle with vertices at the three points 0, z_1, and z_2 has sides equal to z_1, z_2, and $z_2 - z_1$.

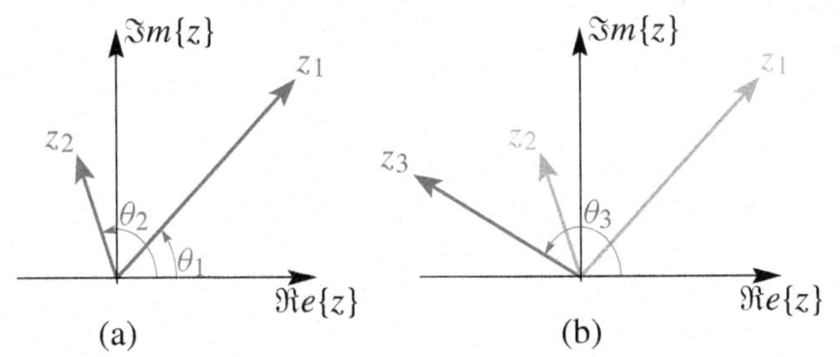

Figure A-10: Geometric view of complex multiplication $z_3 = z_1 z_2$.

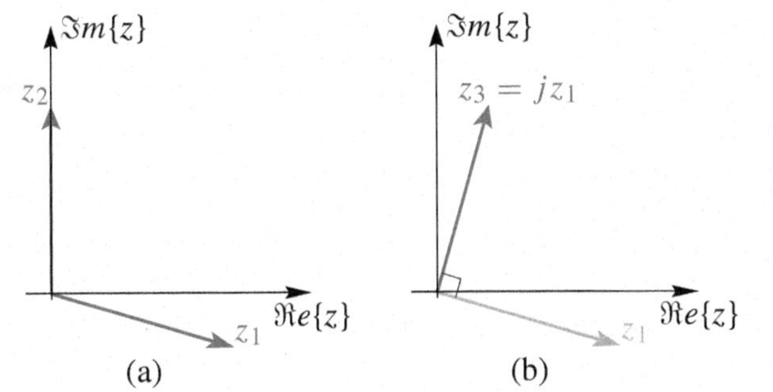

Figure A-11: Complex multiplication becomes a rotation when $|z_2| = 1$. The case where $z_2 = j$ gives a rotation by 90°.

EXERCISE A.6: Define a vector w to have its tail at z_1 and its head at z_2. Prove that w has the same length and direction as $z_2 - z_1$.

EXERCISE A.7: Prove the triangle inequality:

$$|z_2 - z_1| \leq |z_1| + |z_2|$$

Use either an algebraic method by squaring both sides, or a geometric argument based on the intuitive idea that "the shortest distance between two points is a straight line."

A-5.3 Geometric View of Multiplication

It is best to view multiplication in terms of polar form where we multiply the magnitudes, and add the angles. In order to draw the product vector z_3, we must decide whether or not $|z_1|$ and/or $|z_2|$ are greater than 1. In Fig. A-10, it is assumed that only $|z_1|$ is larger than 1.

A special case occurs when $|z_2| = 1$, because then there is no scaling, and the multiplication by $z_2 = e^{j\theta_2}$ becomes a rotation. Figure A-11 shows the case where $z_2 = j$, which gives a rotation by $\pi/2$ or 90°, because $j = e^{j\pi/2}$.

A-5.4 Geometric View of Division

Division is very similar to the visualization of multiplication, except that we must now subtract the angles and divide the magnitudes (Fig. A-12).

$$z_3 = \frac{z_1}{z_2} = \frac{r_1}{r_2} e^{j(\theta_1 - \theta_2)}$$

EXERCISE A.8: Given two complex numbers z_4 and z_2, as in Fig. A-12, where the angle between them is 90° and the magnitude of z_4 is twice that of z_2. Evaluate z_4/z_2.

A-5.5 Geometric View of the Inverse, z^{-1}

This is a special case of division where $z_1 = 1$, so we just negate the angle and take the reciprocal of the magnitude.

$$z^{-1} = \frac{1}{z} = \frac{1}{r} e^{-j\theta}$$

Refer to Fig. A-13 for examples of the inverse.

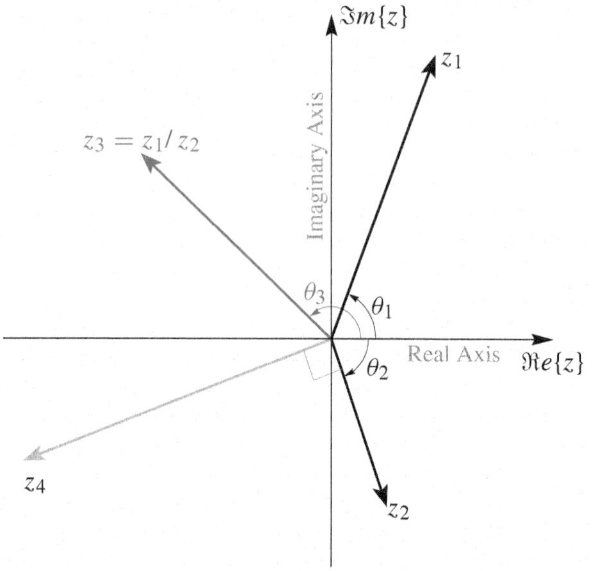

Figure A-12: Graphical visualization of complex number division $z_3 = z_1/z_2$. Notice that the angles subtract; i.e., $\theta_3 = \theta_1 - \theta_2$. Since $|z_2| > 1$, the vector z_3 is shorter than z_1.

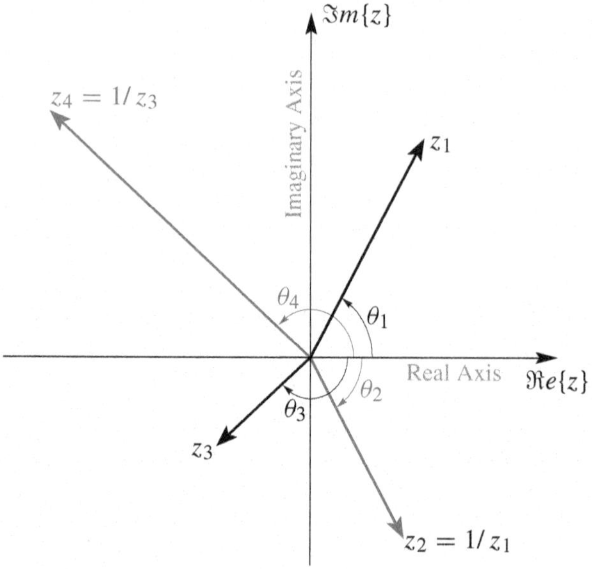

Figure A-13: Graphical construction of complex number inverse $1/z$. For the vectors shown, $|z_1| > 1$ and $|z_3| < 1$.

A-5.6 Geometric View of the Conjugate, z^*

In this case, we negate the angle, which has the effect of flipping the vector about the horizontal axis. The length of the vector remains the same.

$$z^* = x - jy = re^{-j\theta}$$

Figure A-14 shows two examples of the geometric interpretation of complex conjugation.

EXERCISE A.9: Prove the following fact:

$$\frac{1}{z^*} = \frac{1}{r}\, e^{j\theta}$$

Plot an example for $z = 1 + j$; also plot $1/z$ and z^*.

A-6 Powers and Roots

Integer powers of a complex number can be defined in the following manner:

$$z^N = \left(re^{j\theta}\right)^N = r^N e^{jN\theta}$$

In other words, the rules of exponents still apply, so the angle θ is multiplied by N and the magnitude is raised to the Nth power. Figure A-15 shows a sequence of these:

$$\{z^\ell\} = \{z^0,\ z^1,\ z^2,\ z^3, \ldots\}$$

where the angle steps by a constant amount; in this case, exactly $\pi/6$ rad. The magnitude of z is less than one, so the successive powers spiral in toward the origin. If $|z| > 1$, the points would spiral outward; if $|z| = 1$, all

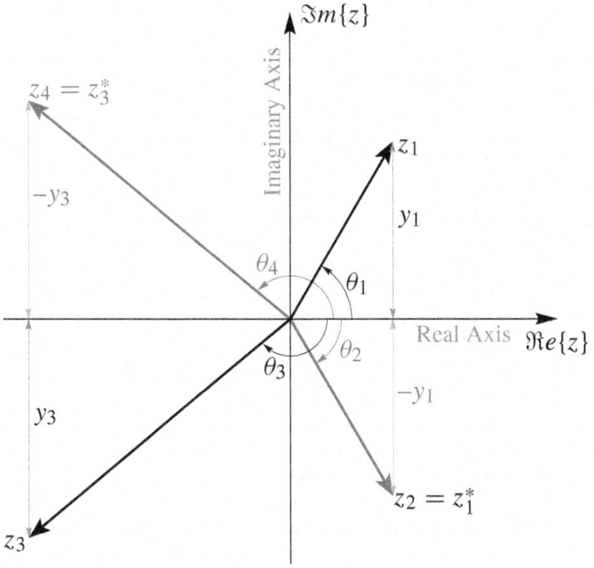

Figure A-14: Graphical construction of the complex conjugate z^*; only the imaginary part changes sign. The vectors flip about the real axis: z_1 flips down, and z_3 flips up.

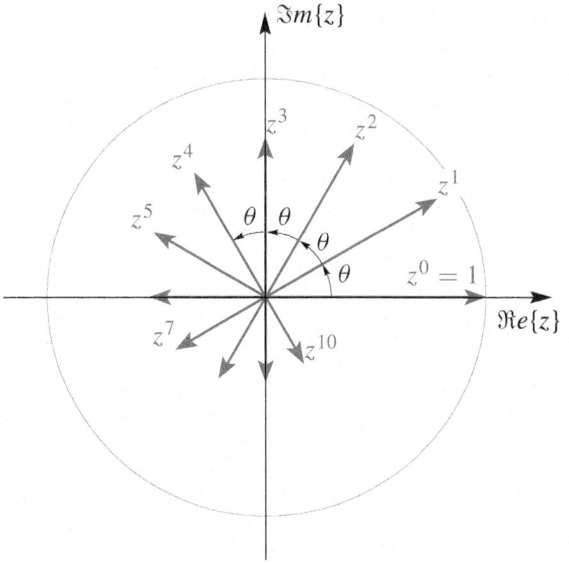

Figure A-15: A sequence of powers z^ℓ for $\ell = 0, 1, 2, \ldots, 10$. Since $|z| = 0.9 < 1$, the vectors spiral in toward the origin. The angular change between successive powers is constant: $\theta = \arg z = \pi/6$.

the powers z^N would lie on the ***unit circle*** (a circle of radius 1). One famous identity is DeMoivre's formula:

$$(\cos\theta + j\sin\theta)^N = \cos N\theta + j\sin N\theta$$

The proof of this seemingly difficult trigonometric identity is actually trivial if we invoke Euler's formula (A.7) for $e^{jN\theta}$.

EXERCISE A.10: Let z^{N-1}, z^N, and z^{N+1} be three consecutive members of a sequence such as shown in Fig. A-15. If $z = 0.98e^{-j\pi/6}$ and $N = 11$, plot the three numbers.

A-6.1 Roots of Unity

In a surprising number of cases, the following equation must be solved:

$$z^N = 1 \tag{A.10}$$

In this equation, N is an integer. One solution is $z = 1$, but there are many others, because (A.10) is equivalent to finding all the roots of the N^{rmth}-degree polynomial $z^N - 1$, which must have N roots. It turns out that all the solutions are given by

$$z = e^{j2\pi\ell/N} \qquad \text{for } \ell = 0, 1, 2, \ldots N-1$$

which are called the ***N^{th} roots of unity***. As shown in Fig. A-16, these N solutions are complex numbers equally spaced around the unit circle. The angular spacing between them is $2\pi/N$.

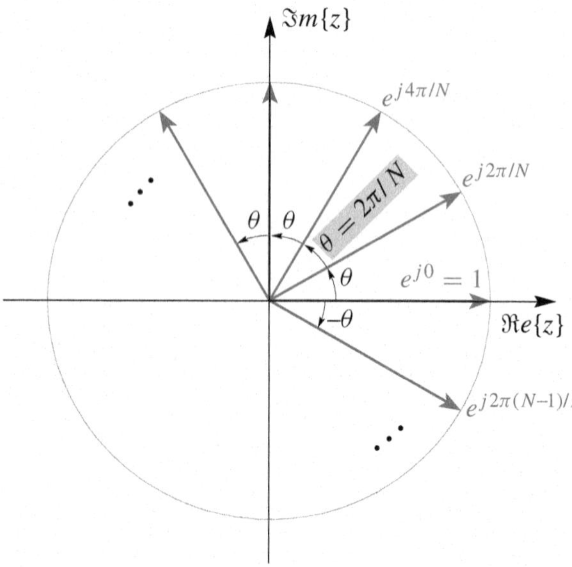

Figure A-16: Graphical display of the Nth roots of unity ($N = 12$). These are the solutions of $z^N = 1$. Notice that there are only N distinct roots.

A-6.1.1 Procedure for Finding Multiple Roots

Once we know that the solutions to (A.10) are the Nth roots of unity, we can describe a structured approach to solving equations with multiple roots. In order to get a general result, we consider a slightly more general situation

$$z^N = c$$

where c is a complex constant, $c = |c|e^{j\phi}$. The procedure involves the following six steps:

1. Write z^N as $r^N e^{jN\theta}$.
2. Write c as $|c|e^{j\phi}e^{j2\pi\ell}$ where ℓ is an integer. Note that when $c = 1$ we would write the number 1 as
$$1 = e^{j2\pi\ell} \qquad \text{for } \ell = 0, \pm1, \pm2, \ldots$$
3. Equate the two sides and solve for the magnitudes and angles separately.
$$r^N e^{jN\theta} = |c|e^{j\phi}e^{j2\pi\ell}$$
4. The magnitude is the positive N^{th} root of a positive number $|c|$:
$$r = |c|^{1/N}$$
5. The angle contains the interesting information, because there are N different solutions:
$$N\theta = \phi + 2\pi\ell \qquad \ell = 0, 1, \ldots N-1$$
$$\Longrightarrow \quad \theta = \frac{\phi + 2\pi\ell}{N}$$
$$\theta = \frac{\phi}{N} + \frac{2\pi\ell}{N}$$
6. Thus, the N different solutions all have the same magnitude, but their angles are equally spaced with a difference of $2\pi/N$ between each one.

Example A-5: 7-th Roots of Unity

Solve $z^7 = 1$, using the procedure above.

$$r^7 e^{j7\theta} = e^{j2\pi\ell}$$
$$\Longrightarrow r = 1$$
$$\Longrightarrow 7\theta = 2\pi\ell$$
$$\theta = \frac{2\pi}{7}\ell \qquad \ell = 0, 1, 2, 3, 4, 5, 6$$

Therefore, these solutions are equally spaced around the unit circle, as shown in Fig. A-17. In this case, the solutions are called the ***seventh roots of unity***. ■

EXERCISE A.11: Solve the following equation:

$$z^5 = -1$$

Use the fact that $-1 = e^{j\pi}$. Plot all the solutions.

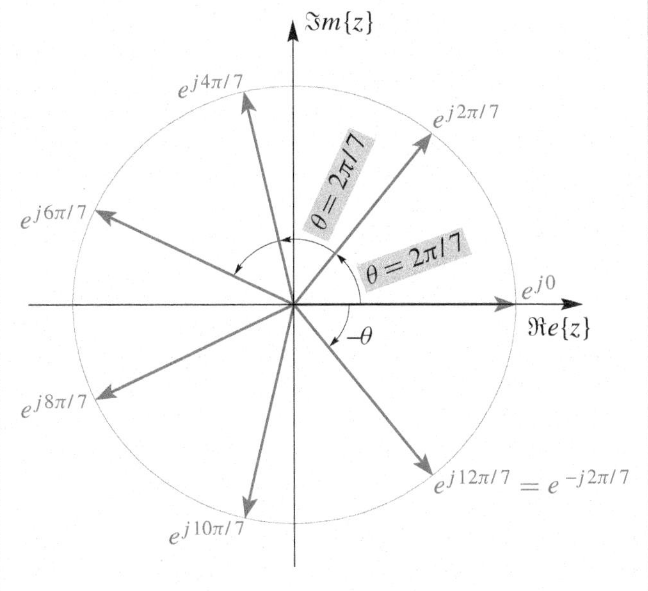

Figure A-17: Graphical display of the 7th roots of unity. Notice that the sequence $e^{j2\pi\ell/7}$ repeats with a period equal to 7.

A-7 Summary and Links

This appendix has presented a brief review of complex numbers and their visualization as vectors in the two-dimensional complex plane. Although this material should have been seen before by most students during high-school algebra, our intense use of complex notation demands much greater familiarity. The labs in Chapter 2 deal with various aspects of complex numbers, and also introduce MATLAB. In Lab #2, we also have included a number of MATLAB functions for plotting vectors from complex numbers (`zvect`, `zcat`) and for changing between Cartesian and polar forms (`zprint`).

LAB: *#2 Introduction to Complex Exponentials*

The complex numbers via MATLAB demo is a quick reference to these routines.

DEMO: *Complex Numbers via MATLAB*

In addition to the labs, we have written a MATLAB GUI (graphical user interface) that will generate drill problems for each of the complex operations studied here: addition, subtraction, multiplication, division, inverse, and conjugate. A screen shot of the GUI is shown in Fig. A-6.

DEMO: *Zdrill*

A-8 Problems

P-A.1 Convert the following to polar form:

(a) $z = 0 + j2$

(b) $z = (-1, 1)$

(c) $z = -3 - j4$

(d) $z = (0, -1)$

P-A.2 Convert the following to rectangular form:

(a) $z = \sqrt{2}\, e^{j(3\pi/4)}$

(b) $z = 1.6\angle(\pi/6)$

(c) $z = 3e^{-j(\pi/2)}$

(d) $z = 7\angle(7\pi)$

P-A.3 Evaluate the following by reducing the answer to rectangular form:

(a) j^3

(b) $e^{j(\pi+2\pi m)}$ (m an integer)

(c) j^{2n} (n an integer)

(d) $j^{1/2}$ (find two answers)

P-A.4 Simplify the following complex-valued expressions:

(a) $3e^{j2\pi/3} - 4e^{-j\pi/6}$

(b) $(\sqrt{2} - j2)^8$

(c) $(\sqrt{2} - j2)^{-1}$

(d) $(\sqrt{2} - j2)^{1/2}$

(e) $\Im m\{je^{-j\pi/3}\}$

Give the answers in both Cartesian and polar form.

P-A.5 Evaluate each expression and give the answer in both rectangular and polar form. In all cases, assume that $z_1 = -4 + j3$ and $z_2 = 1 - j$.

(a) z_1^*

(b) z_2^2

(c) $z_1 + z_2^*$

(d) jz_2

(e) $z_1^{-1} = 1/z_1$

(f) z_1/z_2

(g) e^{z_2}

(h) $z_1 z_1^*$

(i) $z_1 z_2$

P-A.6 Simplify the following complex-valued sum:

$$z = e^{j9\pi/3} + e^{-j5\pi/8} + e^{j13\pi/8}$$

Give the numerical answer for z in polar form. Draw a vector diagram for the three vectors and their sum (z).

P-A.7 Simplify the following complex-valued expressions. Give your answers in polar form. Reduce the answers to a simple numerical form.

(a) For $z = -3 + j4$, evaluate $1/z$.

(b) For $z = -2 + j2$, evaluate z^5.

(c) For $z = -5 + j13$, evaluate $|z|^2$.

(d) For $z = -2 + j5$, evaluate $\Re e\{ze^{-j\pi/2}\}$

P-A.8 Solve the following equation for z:

$$z^4 = j$$

Be sure to find all possible answers, and express your answer(s) in polar form.

P-A.9 Let $z_0 = e^{j2\pi/N}$. Prove that $z_0^{N-1} = 1/z_0$.

P-A.10 Evaluate $(-j)^{1/2}$ and plot the result(s).

APPENDIX B

Programming in MATLAB

MATLAB will be used extensively in the laboratory exercises on the CD-ROM and in Appendix C. This appendix provides an overview of MATLAB and some of its capabilities. We focus on programming issues since a wealth of information is already available on syntax and basic commands.[1] MATLAB has an extensive on-line help system, which can be used to answer any questions not answered in this brief presentation. In fact, an ideal way to read this appendix would be to have MATLAB running so that `help` can be used whenever necessary, and the examples can be run and modified.

MATLAB (short for Matrix Laboratory) is an environment for numerical analysis and computing. It originated as an interface to the collections of numerical routines from the LINPACK and EISPACK projects, but it is now a commercial product of The Mathworks, Inc. MATLAB has evolved into a powerful programming environment containing many built-in functions for doing signal processing, linear algebra, and other mathematical calculations. The language has also been extended by means of toolboxes containing additional functions for MATLAB. For example, the CD-ROM that accompanies this book contains a toolbox of functions needed for the laboratory exercises. The toolboxes are installed as separate directories within the MATLAB directory.

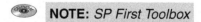

NOTE: *SP First Toolbox*

Please follow the instructions on the CD-ROM to install the ***SP First toolbox*** before doing the laboratory exercises.

[1]Two useful reference books are: D. Hanselman and B. Littlefield, *Mastering MATLAB 6: A Comprehensive Tutorial and Reference*, Prentice Hall, Upper Saddle River, NJ, 2000, ISBN: 0130194689. E. B. Magrab, S. Azarm, B. Balachandran, and J. Duncan, *An Engineer's Guide to MATLAB*, Prentice Hall, Upper Saddle River, NJ, 2000, ISBN: 0130113352.

Since MATLAB is extensible, users find it convenient to write new functions whenever the built-in functions fail to do a vital task. The programming necessary to create new functions and scripts is not too difficult if the user has some experience with C, PASCAL, or FORTRAN. This appendix gives a brief overview of MATLAB for the purpose of programming.

Version 5.2.1 or greater is needed for ***SP First.***

B-1 MATLAB Help

MATLAB provides an on-line help system accessible by using the `help` command. For example, to get information about the function `filter`, enter the following at the command prompt:

```
>> help filter
```

The command prompt is indicated by » in the command window. The `help` command will return text information in the command window. Help is also available for categories; for example, `help punct` summarizes punctuation as used in MATLAB's syntax. The help system now has a Web-browser interface, so the commands `helpdesk`, `helpwin` and `doc` bring up this interface.

A useful command for getting started is `intro`, which covers the basic concepts in the MATLAB language. Also, there are many demonstration programs that illustrate the various capabilities of MATLAB; these can be started with the command `demo`.

Finally, if you are searching for other tutorials, some are freely available on the Web.

When unsure about a command, use `help`.

B-2 Matrix Operations and Variables

The basic variable type in MATLAB is a matrix.[2] To declare a variable, simply assign it a value at the MATLAB prompt. For example,

```
>> M = [1 2 6; 5 2 1]
   M =
         1   2   6
         5   2   1
```

When the definition of a matrix involves a long formula or many entries, then a very long MATLAB command can be broken onto two (or more) lines by placing an ellipses (`...`) at the end of the line to be continued. For example,

```
P = [ 1, 2, 4, 6, 8 ] + ...
      [ pi, 4, exp(1), 0, -1 ] + ...
       [ cos(0.1*pi), sin(pi/3), ...
        tan(3), atan(2), sqrt(pi) ];
```

If an expression is followed by a semicolon (`;`), then the result is not echoed to the screen. This is very useful when dealing with large matrices. Also note that ; is used to separate rows in a matrix definition. Likewise, spaces or commas (,) are used to separate column entries.

The size of the matrix can always be extracted with the `size` operator:

```
>> Msize = size(M)
   Msize =
           2   3
```

Therefore, it becomes unnecessary to assign separate variables to track the number of rows and number of columns. Two special types of matrix variables are worthy of mention: *scalars* and *vectors*. A scalar is a matrix with only one element; its size is 1×1. A vector is a matrix that has only one row or one column. In the

[2]In Versions 5 and later, MATLAB offers many other data types found in conventional programming languages. This appendix will only discuss matrices.

SP First laboratory exercises, signals will often be stored as vectors.

Individual elements of a matrix variable may be accessed by giving the row index and the column index, for example:

```
>> M13 = M(1,3)
   M13 =
         6
```

Submatrices may be accessed in a similar manner by using the colon operator as explained in Section B-2.1.

B-2.1 The Colon Operator

The colon operator `(:)` is useful for creating index arrays, creating vectors of evenly spaced values, and accessing submatrices. Use `help colon` for a detailed description of its capabilities.

The colon notation is based on the idea that an index range can be generated by giving a start, a skip, and then the end. Therefore, a regularly spaced vector of numbers is obtained by means of

```
iii = start:skip:end
```

Without the `skip` parameter, the default increment is 1. This sort of counting is similar to the notation used in FORTRAN DO loops. However, MATLAB takes it one step further by combining it with matrix indexing. For a 9×8 matrix `A`, `A(2,3)` is the scalar element located at the 2nd row and 3rd column of `A`, so a 4×3 submatrix can be extracted with `A(2:5,1:3)`. The colon also serves as a wild card; i.e., `A(2,:)` is a row vector consisting of the entire 2nd row of the matrix `A`. Indexing backwards flips a vector, e.g., `x(9:-1:1)` for a length-9 vector. Finally, it is sometimes necessary to work with a list of all the values in a matrix, so `A(:)` gives a 72×1 column vector that is merely the columns of `A` concatenated together. This is an example of ***reshaping*** the matrix. More general reshaping of the matrix `A` can be accomplished with the `reshape(A,M,N)` function. For example, the 9×8 matrix `A` can be reshaped into a 12×6 matrix with: `Anew = reshape(A,12,6)`.

B-2.2 Matrix and Array Operations

The default operations in MATLAB are matrix operations. Thus `A*B` means matrix multiplication, which is defined and reviewed next.

B-2.2.1 A Review of Matrix Multiplication

The operation of matrix multiplication AB can be carried out only if the two matrices have compatible dimensions, i.e., the number of columns in A must equal the number of rows in B. For example, a 5×8 matrix can multiply an 8×3 matrix to give a result AB that is 5×3. In general, if A is $m \times n$, then B must be $n \times p$, and the product AB would be $m \times p$. Usually matrix multiplication is *not* commutative, i.e., $AB \neq BA$. If $p \neq m$, then the product BA cannot be defined, but even when BA is defined, we find that the commutative property applies only in special cases.

Each element in the product matrix is calculated with an inner product. To generate the first element in the product matrix, $C = AB$, simply take the first row of A and multiply *point by point* with the first column of B, and then sum. For example, if

$$A = \begin{bmatrix} a_{1,1} & a_{1,2} \\ a_{2,1} & a_{2,2} \\ a_{3,1} & a_{3,2} \end{bmatrix} \quad \text{and} \quad B = \begin{bmatrix} b_{1,1} & b_{1,2} \\ b_{2,1} & b_{2,2} \end{bmatrix}$$

then the first element of $C = AB$ is

$$c_{1,1} = a_{1,1}b_{1,1} + a_{1,2}b_{2,1}$$

which is, in fact, the inner product between the first row of A and the first column of B. Likewise, $c_{2,1}$ is found by taking the inner product between the *second* row of

A and the *first* column of B, and so on for $c_{1,2}$ and $c_{2,2}$. The final result would be

$$C = \begin{bmatrix} c_{1,1} & c_{1,2} \\ c_{2,1} & c_{2,2} \\ c_{3,1} & c_{3,2} \end{bmatrix}$$

$$= \begin{bmatrix} a_{1,1}b_{1,1} + a_{1,2}b_{2,1} & a_{1,1}b_{1,2} + a_{1,2}b_{2,2} \\ a_{2,1}b_{1,1} + a_{2,2}b_{2,1} & a_{2,1}b_{1,2} + a_{2,2}b_{2,2} \\ a_{3,1}b_{1,1} + a_{3,2}b_{2,1} & a_{3,1}b_{1,2} + a_{3,2}b_{2,2} \end{bmatrix} \quad \text{(B.1)}$$

Some special cases of matrix multiplication are the *outer product* and the *inner product*. In the ***outer product***, a column vector multiplies a row vector to give a matrix. If we let one of the vectors be all 1s, then we can get a repeating result:

$$\begin{bmatrix} a_1 \\ a_2 \\ a_3 \end{bmatrix} \begin{bmatrix} 1\,1\,1\,1 \end{bmatrix} = \begin{bmatrix} a_1 & a_1 & a_1 & a_1 \\ a_2 & a_2 & a_2 & a_2 \\ a_3 & a_3 & a_3 & a_3 \end{bmatrix}$$

With all 1s in the row vector, we end up repeating the column vector four times.

For the ***inner product***, a row vector multiplies a column vector, so the result is a scalar. If we let one of the vectors be all 1s, then we will sum the elements in the other vector:

$$\begin{bmatrix} a_1\, a_2\, a_3\, a_4 \end{bmatrix} \begin{bmatrix} 1 \\ 1 \\ 1 \\ 1 \end{bmatrix} = a_1 + a_2 + a_3 + a_4$$

B-2.2.2 Pointwise Array Operations

If we want to do a pointwise multiplication between two arrays, some confusion can arise. In the pointwise case, we want to multiply the matrices together element-by-element, so they must have exactly the same size in both dimensions. For example, two 5×8 matrices can be multiplied pointwise, although we cannot do matrix multiplication between two 5×8 matrices. To obtain pointwise multiplication in MATLAB, we use the "point-star" operator `A .* B`. For example, if A and B are both 3×2, then

$$\texttt{D} = \texttt{A.*B} = \begin{bmatrix} d_{1,1} & d_{1,2} \\ d_{2,1} & d_{2,2} \\ d_{3,1} & d_{3,2} \end{bmatrix} = \begin{bmatrix} a_{1,1}\, b_{1,1} & a_{1,2}\, b_{1,2} \\ a_{2,1}\, b_{2,1} & a_{2,2}\, b_{2,2} \\ a_{3,1}\, b_{3,1} & a_{3,2}\, b_{3,2} \end{bmatrix}$$

where $d_{i,j} = a_{i,j}\, b_{i,j}$. We will refer to this type of multiplication as ***array multiplication***. Notice that array multiplication is commutative because we would get the same result if we computed `D = B.*A`.

A general rule in MATLAB is that when "point" is used with another arithmetic operator, it modifies that operator's usual matrix definition to a pointwise one. Thus we have `./` for pointwise division and `.^` for pointwise exponentiation. For example,

```
xx = (0.9) .^ (0:49)
```

generates a vector whose values are equal to $(0.9)^n$, for $n = 0, 1, 2, \ldots, 49$.

B-3 Plots and Graphics

MATLAB is capable of producing two-dimensional x-y plots and three-dimensional plots, displaying images, and even creating and playing movies. The two most common plotting functions that will be used in the *SP First* laboratory exercises are `plot` and `stem`. The calling syntax for both `plot` and `stem` takes two vectors, one for the x-axis points, and the other for the y-axis.[3] The statement `plot(x,y)` produces a connected plot with straight lines between the data points

```
{(x(1),y(1)),(x(2),y(2)),...,(x(N),y(N))}
```

[3] If only one argument is given, `plot(y)` uses the single argument as the y-axis, and uses `[1:length(y)]` for the x-axis. These functions also have optional arguments. To learn about them type `help plot`.

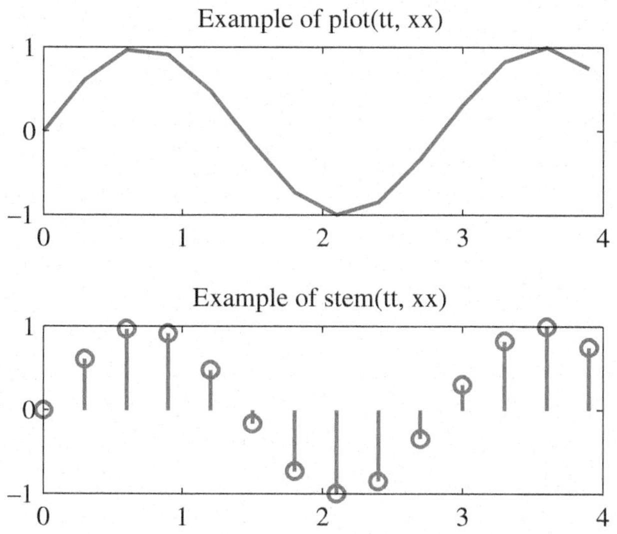

Figure B-1: Example of two different plotting formats, `plot` and `stem`.

as shown in the top panel of Fig. B-1. The same call with `stem(x,y)` produces the "lollipop" presentation of the same data in the bottom panel of Fig. B-1. MATLAB has numerous plotting options that can be studied by using `help plotxy`, `help plotxyz`, or by using `help graph2d`, `help graph3d`, or `help specgraph`.

B-3.1 Figure Windows

Whenever MATLAB makes a plot, it writes the graphics to a *figure window*. You can have multiple figure windows open, but only one of the them is considered the *active* window. Any plot command executed in the command window will direct its graphical output to the active window. The command `figure(n)` will pop up a new figure window that can be refered to by the number `n`, or makes it active if it already exists. Control over many of the window attributes (size, location, color, etc.) is also possible with the `figure` command, which does initialization of the plot window.

B-3.2 Multiple Plots

Multiple plots per window can be done with the ***`subplot`*** function. This function does not do the actual plotting; it merely divides the window into tiles. To set up a 3×2 tiling of the figure window, use `subplot(3,2,tile_number)`. For example, `subplot(3,2,3)` will direct the next `plot` to the third tile, which is in the second row, left side. The graphs in Fig. B-1 were done with `subplot(2,1,1)` and `subplot(2,1,2)`. The program that creates Fig. B-1 is given in Section B-5.

B-3.3 Printing and Saving Graphics

Plots and graphics may be printed to a printer or saved to a file using the ***`print`*** command. To send the current figure window to the default printer, simply type `print` without arguments. To save the plot to a file, a device format and filename must be specified. The device format specifies which language will be used to store the graphics commands. For example, a useful format for including the file in a document is encapsulated PostScript (EPS), which can be produced as follows:

```
>> print -deps myplot
```

The postscript format is also convenient when the plots are kept for printing at a later time. For a complete list of available file formats, supported printers, and other options, see `help print`.

B-4 Programming Constructs

MATLAB supports the paradigm of "functional programming" in which it is possible to nest a sequence of function calls. Consider the following equation, which can be implemented with one line of MATLAB code.

$$\sum_{n=1}^{L} \log(|x_n|)$$

Here is the MATLAB equivalent:

```
sum( log( abs(x) ) )
```

where x is a vector containing the elements x_n. This example illustrates MATLAB in its most efficient form, where individual functions are combined to get the output. Writing efficient MATLAB code requires a programming style that generates small functions that are vectorized. Loops should be avoided. The primary way to avoid loops is to use calls to toolbox functions as much as possible.

B-4.1 MATLAB Built-in Functions

Many MATLAB functions operate on arrays just as easily as they operate on scalars. For example, if x is an array, then cos(x) returns an array of the same size as x containing the cosine of each element of x.

$$\texttt{cos(x)} = \begin{bmatrix} \cos(x_{1,1}) & \cos(x_{1,2}) & \cdots & \cos(x_{1,n}) \\ \cos(x_{2,1}) & \cos(x_{2,2}) & \cdots & \cos(x_{2,n}) \\ \vdots & \vdots & \vdots & \vdots \\ \cos(x_{m,1}) & \cos(x_{m,2}) & \cdots & \cos(x_{m,n}) \end{bmatrix}$$

Notice that no loop is needed, even though cos(x) does apply the cosine function to every array element. Most transcendental functions follow this pointwise rule. In some cases, it is crucial to make this distinction, such as the matrix exponential (expm) versus the pointwise exponential (exp):

$$\texttt{exp(A)} = \begin{bmatrix} \exp(a_{1,1}) & \exp(a_{1,2}) & \cdots & \exp(a_{1,n}) \\ \exp(a_{2,1}) & \exp(a_{2,2}) & \cdots & \exp(a_{2,n}) \\ \vdots & \vdots & \vdots & \vdots \\ \exp(a_{m,1}) & \exp(a_{m,2}) & \cdots & \exp(a_{m,n}) \end{bmatrix}$$

B-4.2 Program Flow

Program flow can be controlled in MATLAB using if statements, while loops, and for loops. There is also a switch statement. These are similar to any high-level language. Descriptions and examples for each of these program constructs can be viewed by using the MATLAB help command.

B-5 MATLAB Scripts

Any expression that can be entered at the MATLAB prompt can also be stored in a text file and executed as a script. The text file can be created with any plain ASCII editor such as notepad on Windows, emacs or vi on UNIX or Linux, and the built-in MATLAB editor on a Macintosh or Windows platform. The file extension must be .m, and the script is executed in MATLAB simply by typing the filename (without the extension). These programs are usually called *M-files*. Here is an example:

```
tt = 0:0.3:4;
xx = sin(0.7*pi*tt);
subplot(2,1,1)
plot( tt, xx )
title('Example of plot( tt, xx)')
subplot(2,1,2)
stem( tt, xx )
title('Example of stem( tt, xx)')
```

If these commands are saved in a file named plotstem.m, then typing plotstem at the command prompt will run the file, and all eight commands will be executed as if they had been typed in at the command prompt. The result is the two plots that were shown in Fig. B-1.

B-6 Writing a MATLAB Function

You can write your own functions and add them to the MATLAB environment. These functions are another type of M-file, and are created as an ASCII file with a text editor. The first word in the M-file must be the keyword `function` to tell MATLAB that this file is to be treated as a function with arguments. On the same line as the word `function` is the calling template that specifies the input and output arguments of the function. The filename for the M-file must end in `.m`, and the filename will become the name of the new command for MATLAB. For example, consider the following file, which extracts the last L elements from a vector:

```
function  y = foo( x, L )
%FOO    get last L points of x
%  usage:
%        y = foo( x, L )
%  where:
%
x = input vector
%        L = number of points to get
%        y = output vector
N = length(x);
if( L > N )
  error('input vector too short')
end
y = x((N-L+1):N);
```

If this file is called `foo.m`, the operation may be invoked from the MATLAB command line by typing

```
aa = foo( (1:2:37), 7 );
```

The output will be the last seven elements of the vector `(1:2:37)`, i.e.,

```
aa = [ 25 27 29 31 33 35 37 ]
```

B-6.1 Creating A Clip Function

Most functions can be written according to a standard format. Consider a clip *function M-file* that takes two input arguments (a signal vector and a scalar threshold) and returns an output signal vector. You can use an editor to create an ASCII file `clip.m` that contains the statements in Fig. B-2. We can break down the M-file `clip.m` into four elements:

Definition of Input–Output: Function M-files must have the word `function` as the very first item in the file. The information that follows `function` on the same line is a declaration of how the function is to be called and what arguments are to be passed. The name of the function should match the name of the M-file; if there is a conflict, it is the name of the M-file on the disk that is known to the MATLAB command environment.

Input arguments are listed inside the parentheses following the function name. Each input is a matrix. The output argument (also a matrix) is on the left side of the equals sign. Multiple output arguments are also possible if square brackets surround the list of output arguments; e.g., notice how the `size(x)` function in line 11 returns the number of rows and number of columns into separate output variables. Finally, observe that there is no explicit command for returning the outputs; instead, MATLAB returns whatever value is contained in the output matrix when the function completes. For `clip` the last line of the function assigns the clipped vector to `y`, so that the clipped vector is returned. MATLAB does have a command called `return`, but it just exits the function, it does not take an argument.

The essential difference between the function M-file and the script M-file is dummy variables versus permanent variables. MATLAB uses ***call by value*** so that the function makes local copies of its arguments. These local variables disappear after the function completes. For example, the following statement creates a clipped vector `wwclipped` from the input vector `ww`.

```
>> wwclipped = clip(ww, 0.9999);
```

The arrays `ww` and `wwclipped` are permanent variables in the MATLAB workspace. The temporary arrays created inside `clip` (i.e., `y`, `nrows`, `ncols`,

The eight lines of comments at the beginning of the function will be the response to `help clip`.

First step is to figure out the matrix dimensions of `x`.

Input could be a row or column vector.

Since `x` is local, we can change without affecting the workspace.

Create output vector.

```
 1: function   y = clip( x, Limit )
 2: %CLIP    saturate mag of x[n] at Limit
 3: %      when |x[n]| > Limit, make |x[n]| = Limit
 4: %
 5: %  usage:  Y = clip( X, Limit )
 6: %
 7: %       X  - input signal vector
 8: %  Limit  - limiting value
 9: %       Y  - output vector after clipping
10:
11: [nrows ncols] = size(x);
12:
13: if( ncols ~= 1 & nrows ~= 1 )       %-- NEITHER
14:    error('CLIP: input not a vector')
15: end
16: Lx = max([nrows ncols]);          %-- Length
17:
18: for n=1:Lx             %-- Loop over entire vector
19:    if( abs(x(n)) > Limit )
20:       x(n) = sign(x(n))*Limit;     %-- saturate
21:    end
22: end
23: y = x;                 %-- copy to output vector
```

Figure B-2: Illustration of a MATLAB function called `clip.m`.

`Lx` and `i`) exist only while `clip` runs; then they are deleted. Furthermore, these variable names are local to `clip.m`, so the name `x` may also be used in the workspace as a permanent name. These ideas should be familiar to anyone experienced with a high-level computer language like C, Java, FORTRAN, or PASCAL.

Self-Documentation: A line beginning with the `%` sign is a comment line. The first group of these in a function is used by MATLAB's help facility to make M-files automatically self-documenting. That is, you can now type `help clip` and the comment lines from *your* M-file will appear on the screen as help information. The format suggested in `clip.m` (lines 2–9) follows the convention of giving the function name, its calling sequence, a brief explanation, and then the definitions of the input and output arguments.

Size and Error Checking: The function should determine the size of each vector or matrix that it will operate on. This information does not have to be passed as a separate input argument, but can be extracted with the `size` function (line 11). In the case of the `clip` function, we want to restrict the function to operating on vectors, but we would like to permit either a row ($1 \times L$) or a column ($L \times 1$). Therefore, one of the variables `nrows` or `ncols` must be equal to 1; if not, we terminate the function with the bail-out function `error`, which prints a message to the command line and quits the function (lines 13–15).

Actual Function Operations: In the case of the `clip` function, the actual clipping is done by a `for` loop (lines 18–22), which examines each element of the `x` vector for its size compared to the threshold `Limit`. In the case of negative numbers, the clipped value must be set to `-Limit`, hence the multiplication by `sign(x(n))`. This assumes that `Limit` is passed in as a positive number, a fact that might also be tested in the error-checking phase.

This particular implementation of `clip` is very inefficient owing to the `for` loop. In Section B-7.1, we will show how to vectorize this program for speed.

B-6.2 Debugging a MATLAB M-file

Since MATLAB is an interactive environment, debugging can be done by examining variables in the workspace. MATLAB contains a symbolic debugger that is now integrated with the text editor. Since different functions can use the same variable names, it is important to keep track of the local context when setting break points and examining variables. Several useful debugging commands are listed here, others can be found with `help debug`.

dbstop is used to set a breakpoint in an M-file. It can also be used to give you a prompt when an error occurs by typing `dbstop if error` before executing the M-file. This allows you to examine variables within functions and also the calling workspace (by typing `dbup`).

dbstep incrementally steps through your M-file, returning you to a prompt after each line is executed.

dbcont causes normal program execution to resume, or, if there was an error, returns you to the MATLAB command prompt.

dbquit quits the debug mode and returns you to the MATLAB command prompt.

keyboard can be inserted into the M-file to cause program execution to pause, giving you a MATLAB prompt of the form `K>` to indicate that it is not the command-line prompt.

B-7 Programming Tips

This section presents a few programming tips that should improve the speed of your MATLAB programs. For more ideas and tips, list some of the function M-files in the toolboxes of MATLAB by using the `type` command. For example,

```
type angle
type conv
type trapz
```

Copying the style of other (good) programmers is always an efficient way to improve your own knowledge of a computer language. In the following hints, we discuss some of the most important points involved in writing good MATLAB code. These comments will become increasingly useful as you develop more experience in MATLAB.

B-7.1 Avoiding Loops

Since MATLAB is an interpreted language, certain common programming habits are intrinsically inefficient. The primary one is the use of `for` loops to perform simple operations over an entire matrix or vector. *Whenever possible*, you should try to find a vector function (or the nested composition of a few vector functions) that will accomplish the desired result, rather than writing a loop. For example, if the operation is summing all the elements in a matrix, the difference between calling `sum` and writing a loop that looks like FORTRAN or C code can be astounding; the loop is unbelievably slow owing to the interpreted nature of MATLAB. Consider the three methods for matrix summation in Fig. B-3.

The first method in Fig. B-3(a) is the MATLAB equivalent of conventional programming. The last two methods rely on the built-in function `sum`, which has different characteristics depending on whether its argument is a matrix or a vector (called "operator overloading"). When acting on a matrix as in Fig. B-3(b), `sum` returns a row vector containing the column sums; when acting on a row (or column) vector as in Fig. B-3(c), the sum is a scalar. To get the third (and most efficient) method, the matrix `x` is converted to a column vector with the colon operator. Then one call to `sum` will suffice.

Double loop is needed to index all of the matrix xx.

```
function ss = mysum1(xx)
%
[Nrows,Ncols] = size(xx);
ss = 0;
for jj=1:Nrows
  for kk=1:Ncols
    ss = ss + xx(jj,kk);
  end
end
```

(a)

sum acts on a matrix to give the sum down each column.

```
function ss = mysum2(xx)
%
xx = sum( sum(xx) );
```

(b)

x(:) is a column vector of all elements in the matrix.

```
function ss = mysum3(xx)
%
xx = sum(xx(:));
```

(c)

Figure B-3: Three ways to add all of the elements of a matrix.

B-7.2 Repeating Rows or Columns

Often it is necessary to form a matrix from a vector by replicating the vector in the rows or columns of the matrix. If the matrix is to have all the same values, then functions such as `ones(M,N)` and `zeros(M,N)` can be used. But to replicate a column vector `x` to create a matrix that has identical columns, a loop can be avoided by using the outer-product matrix multiply operation discussed in Section B-2.2. The following MATLAB code fragment will do the job for eleven columns:

```
x = (12:-2:0)';
% prime indicates conjugate transpose
X = x * ones(1,11)
```

If `x` is a length L column vector, then the matrix `X` formed by the outer product is $L \times 11$. In this example, $L = 7$. Note that MATLAB is case-sensitive, so the variables `x` and `X` are different. We have used capital `X` to indicate a matrix, as would be done in mathematics.

B-7.3 Vectorizing Logical Operations

It is also possible to vectorize programs that contain `if`, `else` conditionals. The `clip` function (Fig. B-2) offers an excellent opportunity to demonstrate this type of vectorization. The `for` loop in that function contains a logical test and might not seem like a candidate for vector operations. However, the relational and logical operators in MATLAB, such as greater than, apply to matrices. For example, a greater than test applied to a 3×3 matrix returns a 3×3 matrix of ones and zeros.

```
>> %-- create a test matrix
>> x = [ 1  2 -3; 3 -2  1; 4  0  -1]
   x = [ 1  2 -3
         3 -2  1
         4  0 -1 ]
>> %-- check the greater than condition
>> mx = x > 0
   mx = [ 1  1  0
          1  0  1
          1  0  0 ]
>> %-- pointwise multiply by mask, mx
>> y = mx .* x
   y = [ 1  2  0
         3  0  1
         4  0  0 ]
```

The zeros mark where the condition was false; the ones denote where the condition was true. Thus, when we do the pointwise multiply of `x` by the masking matrix `mx`, we get a result that has all negative elements set to zero. Note that these last two statements process the entire matrix without ever using a `for` loop.

Since the saturation done in `clip.m` requires that we change the large values in `x`, we can implement the entire `for` loop with three array multiplications. This leads to a vectorized saturation operator that works for matrices as well as vectors:

```
y = Limit*(x > Limit) ...
     - Limit*(x < -Limit) ...
      + x.*(abs(x) <= Limit);
```

Three different masking matrices are needed to represent the three cases of positive saturation, negative saturation, and no action. The additions correspond to the logical OR of these cases. The number of arithmetic operations needed to carry out this statement is $3N$ multiplications and $2N$ additions, where N is the total number of elements in `x`. This is actually more work than the loop in `clip.m` if we counted only arithmetic operations. However, the cost of code interpretation is high. This vectorized statement is interpreted only once, whereas the three statements inside the `for` loop must be reinterpreted N times. If the two implementations are timed with `etime`, the vectorized version will be much faster for long vectors.

B-7.4 Creating an Impulse

Another simple example of use of logical operators is given by the following trick for creating an impulse signal vector:

```
nn = [-10:25];
impulse = (nn==0);
```

This result may be plotted with `stem(nn, impulse)`. In a sense, this code fragment is perfect because it captures the essence of the mathematical formula that defines the impulse as existing only when $n = 0$.

$$\delta[n] = \begin{cases} 1 & n = 0 \\ 0 & n \neq 0 \end{cases}$$

B-7.5 The Find Function

An alternative to masking is to use the `find` function. This is not necessarily more efficient; it just gives a different approach. The `find` function will determine the list of indices in a vector where a condition is true. For example, `find( x > Limit )` will return the list of indices where the vector is greater than the `Limit` value. Thus we can do saturation as follows:

```
y = x;
jkl = find(y > Limit);
y( jkl ) =  Limit;
jkl = find(y < -Limit);
y( jkl ) = -Limit;
```

B-7.6 Seek to Vectorize

The dictum to "avoid `for` loops" is not always easy to obey, because it means the algorithm must be cast in a vector form. If matrix-vector notation is incorporated into MATLAB programs, the resulting code will run much faster. Even loops with logical tests can be ***vectorized*** if masks are created for all possible conditions. Thus, a reasonable goal is the following:

> *Eliminate all `for` loops.*

B-7.7 Programming Style

If there were a proverb to summarize good programming style, it would probably read something like

> *May your functions be short*
> *and your variable names long.*
> –Anon

This is certainly true for MATLAB. Each function should have a single purpose. This will lead to short, simple modules that can be linked together by functional composition to produce more complex operations. Avoid the temptation to build super functions with many options and a plethora of outputs.

MATLAB supports long variable names, so you can take advantage of this feature to give variables descriptive names. In this way, the number of comments littering the code can be drastically reduced. Comments should be limited to `help` information and the documentation of tricks used in the code.

APPENDIX C

Laboratory Projects

This appendix contains three sample laboratories out of the 20 that are available on the CD-ROM. Each of the 20 labs corresponds to material primarily from an individual chapter, although many of the later labs use concepts from several chapters. Table C-1 summarizes all of the labs available on the CD-ROM, with a cross reference to the primary chapter(s) involved in each lab. The structure of the labs is as follows:

Overview: Each lab starts with a brief review of the key concepts to be studied and implemented.

Pre-Lab: The Pre-Lab section consists of a few exercises that introduce some simple MATLAB commands that will be needed in the lab. These should be easy enough that students can finish them successfully without any outside help.

Warm-up: The warm-up section consists of more simple exercises that introduce the MATLAB functionality needed for the Exercises or Project of that lab. The warm-up is intended to be completed during a *supervised lab time,* so that students can ask questions of an expert. In our own use of these experiments, we have implemented a system in which the laboratory instructor must verify the appropriate steps by initialing an ***Instructor Verification*** sheet for a few key steps. Examples of this verification sheet are included at the end of sample labs in this appendix.

Exercises: The bulk of the work in each lab consists of exercises that require some MATLAB programming and plotting. All of the exercises are designed to illustrate theoretical ideas presented in the text. Furthermore, we have included numerous processing examples that involve real signals such as speech, music, and images.

Table C-1: Laboratory material on *SP-First* CD-ROM.

Lab	*Subject*	*Cross-Reference*
Lab #1	Introduction to MATLAB	Ch. 2
Lab #2a	Introduction to Complex Exponentials—Multipath	Ch. 2
Lab #2b	Introduction to Complex Exponentials—Direction Finding	Ch. 2
Lab #3	AM and FM Sinusoidal Signals	Ch. 3
Lab #4	Synthesis of Sinusoidal Signals	Ch. 3
Lab #5	FM Synthesis for Musical Instruments	Ch. 3
Lab #6	Digital Images: A/D and D/A	Ch. 4
Lab #7	Sampling, Convolution, and FIR Filtering	Ch. 5
Lab #8	Frequency Response: Bandpass & Nulling Filters	Ch. 6
Lab #9	Encoding and Decoding Touch-Tone Signals	Ch. 6 and 7
Lab #10	Octave Band Filtering	Ch. 7
Lab #11	PeZ—The z, n, and $\hat{\omega}$ Domains	Ch. 8
Lab #12	Two Convolution GUIs	Ch. 9
Lab #13	Numerical Evaluation of Fourier Series	Ch. 10
Lab #14a	Design with Fourier Series—Power Supply	Ch. 12
Lab #14b	Design with Fourier Series—Distortion	Ch. 12
Lab #15	Fourier Series	Ch. 12
Lab #16	AM Communication System	Ch. 12
Lab #17	Digital Communication: FSK Modem (Encoding)	Ch. 13
Lab #18	Digital Communication: FSK Modem (Decoding)	Ch. 13
Lab #19	The Fast Fourier Transform	Ch. 13
Lab #20	Extracting Frequencies of Musical Tones	Ch. 13

Projects: In some cases, the labs require implementations that are so large and complicated that it is no longer fair to call them exercises. In labs such as Lab #4 on Synthesis of Sinusoidal Signals (Music Synthesis), Lab #9 on Touch-Tone Decoding, and others, the problem statement is more like that of a design project. In these cases, students should be given some flexibility in creating an implementation that satisfies a general objective. In addition, many individual parts must be completed to make the whole project function correctly.

These laboratory exercises can be done with version 5.2.1 (or later) of MATLAB. Version 4 would suffice in most cases, except for those labs where GUIs are used. Student versions of MATLAB have been available for a number of years, with varying capabilities. Since 2001, the student version has been made equivalent to the professional version of MATLAB 5.3 and later, so it is more than adequate to do all the labs. Student versions prior to 2001 will handle some of the labs, but not when the processing involves long signals or large images. Since a few of the labs require M-files that are not part of the standard MATLAB release, we have also provided a package of *SP First* M-files containing functions developed for this book. Lab #2 (either version) gives instructions on how to *install the SP First toolbox.*

NOTE: *Install SP-First Toolbox for* MATLAB

In the following sections, three labs from the entire set of 20 are presented in order to convey the type of lab exercises that are possible with MATLAB. The first one is Lab C-1 which is the introductory lab. The second is Lab #9, the Touch-Tone dialing and decoding lab. The third one (Lab #12) deals with convolution by having students use the two GUIs developed for visualizing convolution.

LAB: *All 20 Labs are on the CD-ROM*

C-1 Introduction to MATLAB

LAB: *This is Lab #1 on the CD-ROM.*

The *Pre-Lab* exercises are intended to be completed while preparing for the actual lab.

The *Warm-up* section of this lab would normally be completed *during supervised lab time,* and the steps marked *Instructor Verification* would be signed off by the lab supervisor.

The *Laboratory Exercises* should be done individually and summarized with a lab report. A good report should include graphs and diagrams along with explanations and calculations to verify how the theory relates to the MATLAB programs. All MATLAB plots should have titles and their axes should be labeled.

C-1.1 Pre-Lab

In this introductory lab, the Pre-Lab will be extremely short and very easy. Make sure that you read through the entire lab prior to working the various exercises.

C-1.1.1 Overview

MATLAB will be used extensively in all the labs. The primary goal of the present lab is to become familiar with using MATLAB. Please read Appendix B: *Programming in* MATLAB for a quick overview. Here are three specific goals for this lab:

1. Learn basic MATLAB commands and syntax, including the `help` system.

2. Write and edit your own script files in MATLAB, and run them as commands.

3. Learn a little about advanced programming techniques for MATLAB, such as vectorization.

C-1.1.2 Movies: MATLAB Tutorials

In addition to the MATLAB review in Appendix B, there are movies on the CD-ROM covering the basic topics in MATLAB(e.g., colon operator, indexing, functions, etc).

C-1.1.3 Getting Started

From the desktop of the OS, you can start MATLAB by double-clicking on a MATLAB icon, typing `matlab` in a terminal window, or by selecting MATLAB from a menu such as the `START` menu under Windows-95/98/NT/2000. The following steps will introduce you to MATLAB.

(a) View the MATLAB introduction by typing `intro` at the MATLAB prompt. This short introduction will demonstrate some of the basics of using MATLAB.

(b) Run the MATLAB help desk by typing `helpdesk`. The help desk provides a hypertext interface to the MATLAB documentation. The MATLAB preferences can be set to use Netscape Navigator or Internet Explorer as the browser for help. Two links of interest are **Getting Help** (at the bottom of the right-hand frame), and **Getting Started** which is under MATLAB in the left-hand frame.

(c) Explore the MATLAB help capability available at the command line. Try the following:

```
help
help edit
help plot
help colon   %<-- VERY IMPORTANT syntax
help ops
help zeros
help ones
lookfor filter  %<-- keyword search
```

Note: It is possible to force MATLAB to display only one screen-full of information at once by issuing the command `more on`.

(d) Run the MATLAB demos: type `demo` and explore a variety of basic MATLAB commands and plots.

(e) Use MATLAB as a calculator. Try the following:

```
pi*pi - 10
sin(pi/4)
ans^2 %<--- "ans" holds the last result
```

(f) Do variable name assignment in MATLAB. Try the following:

```
x = sin( pi/5 );
cos( pi/5 )  %<--- assigned to what?
y = sqrt( 1 - x*x )
ans
```

Notes: The semicolon at the end of a statement will suppress the echo to the screen. The text following the percent sign `%` is a comment.

When unsure about a command, use `help`.

(g) Complex numbers are natural in MATLAB. The basic operations are supported. Try the following:

```
z = 3 + 4i, w = -3 + 4j
real(z),  imag(z)
abs([z,w])   %<--- Vector constructor
conj(z+w)
angle(z)
exp( j*pi )
exp(j*[ pi/4, 0, -pi/4,...
           pi/2, pi, 2*pi, 3*pi ])
```

The comma in the first or second line allows multiple MATLAB statements on one line. The three dots at the end of the seventh line indicate continuation of the MATLAB statement to the next line.

C-1.2 Warm-up

This warm-up section covers many of the basic MATLAB commands and functions that you will need to complete the *Laboratory Exercises* in Section C-1.3.

C-1.2.1 MATLAB Array Indexing

(a) Make sure that you understand the *colon notation*. In particular, explain in words what the following MATLAB code will produce

```
jkl =  0 : 6
jkl =  2 : 4 : 17
jkl =  99 : -1 : 88
ttt =  2 : (1/9) : 4
tpi = pi * [ 0:0.1:2 ];
```

(b) Extracting and/or inserting numbers in a vector is very easy to do. Consider the following operations on the vector xx as a model for doing vector replacement.

```
xx = [zeros(1,3),...
        linspace(0,1,5),ones(1,4)]
xx(4:6)
size(xx)
length(xx)
xx(2:2:length(xx))
```

Explain the results echoed from the last four lines of the above code.

(c) Observe the result of the following assignments:

```
yy = xx;   yy(4:6) = pi*(1:3)
```

Now write a statement that will take the vector xx defined in part (b) and replace the even indexed elements (i.e., xx(2), xx(4), etc) with the constant π^{π}. *Use a vector replacement, not a loop.*

Instructor Verification (separate page)

C-1.2.2 MATLAB Script Files

(a) Experiment with vectors in MATLAB. Think of the vector as a set of numbers. The following code should compute a vector of cosine values:

```
xk = cos( pi*(0:11)/4 )
```

Explain how the different values of cosine are stored in the vector xk. What is xk(1)? Is xk(0) defined?

(b) (A taste of vectorization) Loops can be written in MATLAB, but they are *not* the most efficient way to get things done. It's better to ***always avoid loops*** and use the colon notation instead. The following code has a loop that computes values of the cosine function. The index of yy() must start at 1. Rewrite this computation without using the loop by using colon notation for array indexing.

```
yy = [ ];   %<-- initialize yy as empty
for k=-5:5
   yy(k+6) = cos( k*pi/3 )
end
yy
```

Explain why it is necessary to write yy(k+6). What happens if you use yy(k) instead?

Instructor Verification (separate page)

(c) Plotting is easy in MATLAB for both real and complex numbers. The basic plot command will plot a vector y versus a vector x. Try the following:

```
x = [-3 -1 0 1 3 ];
y = x.*x - 3*x;
plot( x, y )
%-- complex number example:
z = x + y*sqrt(-1)
plot( z )   %<-- will plot imag vs. real
```

Use `help arith` to learn how the operation xx.*xx works when xx is a vector; compare to matrix multiply.

When unsure about a command, use `help`.

(d) Use the built-in MATLAB editor, or an external one such as EMACS on UNIX/LINUX, to create a script file called `mylab1.m` containing the following lines:

```
tt = -1 : 0.01 : 1;
xx = cos( 5*pi*tt );
zz = 1.4*exp(j*pi/2)*exp(j*5*pi*tt);
%--- plot a sinusoid
plot( tt,xx,'b-', tt,real(zz),'r--' )
grid on
title('TEST PLOT of a SINUSOID')
xlabel('TIME (sec)')
```

Explain why the plot of `real(zz)` is a sinusoid. What is its phase and amplitude? Make a calculation of the phase from a time-shift measured on the plot.

Instructor Verification (separate page)

(e) Run your script from MATLAB. To run the file `mylab1` that you created previously, try

```
mylab1       %<-run commands in the file
type mylab1 %<-type out  contents of
            %   mylab1.m to the screen
```

C-1.2.3 MATLAB Sound (optional)

The exercises in this section involve sound signals, so you need headphones for listening.

(a) Run the MATLAB sound demo by typing `xpsound` at the MATLAB prompt. If you are unable to hear sounds in the MATLAB demo then ask an instructor for help.

(b) Now generate a tone (i.e., a sinusoid) in MATLAB and listen to it with the `soundsc()` command.[1] Refer back to part C-1.2.2(d) for some code that creates values of a sinusoid. The frequency of your sinusoidal tone should be 1300 Hz and its duration should be 0.9 sec. Use a sampling rate `(fs)` equal to 11025 samples/sec. The sampling rate dictates the time interval between time points, so the time vector should be defined as follows:

```
tt = 0:(1/fs):dur;
```

where `fs` is the desired sampling rate and `dur` is the desired duration (in seconds). Read the online `help` for both `sound()` and `soundsc()` to get more information on using this command. What is the length of your `tt` vector?

Instructor Verification (separate page)

C-1.3 Laboratory: Manipulating Sinusoids with MATLAB

Now you are on your own. As you work this section, write a short summary of each part including plots in order to create a Lab report. It would be efficient to write one MATLAB script file to do steps (a) through (d) below. Then you can include a listing of the script file with your report to document your work.

(a) Generate a time vector `(tt)` to cover a range of t that will exhibit approximately two cycles of the 4000 Hz sinusoids defined in the next part, part (b). Use a definition for `tt` similar to Section C-1.2.2(d). If we use T to denote the period of the sinusoids, define the starting time of the vector `tt` to be equal to $-T$, and the ending time as $+T$. Then the two cycles will include $t = 0$. *Finally, make sure that you have* ***at least 25 samples per period*** *of the sinusoidal wave.* In other words, when you use the colon operator to define the time vector, make the increment small enough to generate 25 samples per period.

[1] The `soundsc(xx,fs)` function requires **two** arguments: the first one `(xx)` contains the vector of data to be played and the second argument `(fs)` is the rate for playing the samples. In addition, `soundsc(xx,fs)` does automatic scaling and then calls `sound(xx,fs)` to actually play the signal.

(b) Generate two 4000 Hz sinusoids with arbitrary amplitude and time-shift.

$$x_1(t) = A_1 \cos(2\pi(4000)(t - t_{m_1}))$$
$$x_2(t) = A_2 \cos(2\pi(4000)(t - t_{m_2}))$$

Select the value of the amplitudes and time-shifts as follows: Let A_1 be equal to your age and set $A_2 = 1.2A_1$. For the time-shifts, set $t_{m_1} = (37.2/M)T$ and $t_{m_2} = -(41.3/D)T$ where D and M are the day and month of your birthday, and T is the period.

Make a plot of both signals over the range of $-T \leq t \leq T$. For your final printed output in part (d) below, use `subplot(3,1,1)` and `subplot(3,1,2)` to make a three-panel subplot that puts both of these plots in the same figure window. See `help subplot`.

(c) Create a third sinusoid as the sum

$$x_3(t) = x_1(t) + x_2(t)$$

In MATLAB this amounts to summing the vectors that hold the values of each sinusoid. Make a plot of $x_3(t)$ over the same range of time as used in the plots of part (b). Include this as the third panel in the plot by using `subplot(3,1,3)`.

(d) Before printing the three plots, put a title on each subplot, and include your name in one of the titles. See `help title`, `help print` and `help orient`, especially `orient tall`.

C-1.3.1 Theoretical Calculations

Remember that the phase of a sinusoid can be calculated after measuring the time location of a positive peak,[2] if we know the frequency or the period.

[2]Usually we say time delay or time shift instead of the "time location of a positive peak."

(a) Make measurements of the "time location of a positive peak" and the amplitude from the plots of $x_1(t)$ and $x_2(t)$, and write those values for A_i and t_{m_i} directly on the plots. Then calculate by hand the phases of the two signals, $x_1(t)$ and $x_2(t)$, by converting each time shift t_{m_i} to phase. Write the calculated phases ϕ_i directly on the plots.
Note: When doing computations, express phase angles in radians, not degrees!

(b) Measure the amplitude and time-shift of $x_3(t)$ directly from the plot and then calculate the phase (ϕ_3) by hand. Write these values directly on the plot to show how the amplitude and time shift were measured, and how the phase was calculated.

(c) Now use the phasor addition theorem. Carry out a phasor addition of complex amplitudes for $x_1(t)$ and $x_2(t)$ to determine the complex amplitude for $x_3(t)$. Use the complex amplitude for $x_3(t)$ to verify that your previous calculations of A_3 and ϕ_3 in part (b) were correct.

C-1.3.2 Complex Amplitude

Write one line of MATLAB code that will generate values of the sinusoid $x_1(t)$ above by using the complex-amplitude representation

$$x_1(t) = \Re e\{Xe^{j\omega t}\}$$

Use constants for X and ω.

C-1.4 Lab Review Questions

In general, your lab write-up should indicate that you have acquired a new understanding of the topics treated by the laboratory assignment. Answer the questions below in your lab report as an assessment of your understanding of this lab's objective; i.e., obtaining a working knowledge of the basics of MATLAB. If you do not know the answers to these questions go back to the lab and try to figure them out in MATLAB (remember the commands `help` and `lookfor`). Also, consult Appendix B as a reference source.

1. You saw how easy it is for MATLAB to generate and manipulate vectors (i.e., 1-dimensional arrays of numbers). For example, consider the following:

```
nn = 0*(0:44);
mm = zeros(1,44);
kk = 0:pi/44:pi;
```

   (a) Is the length of `kk` equal to 44 or 45? Explain.

   (b) Which one of the lines above will produce 44 zeros?

   (c) How would you modify one of the above lines of MATLAB code to create a vector containing 45 sevens?

2. You also learned that MATLAB has no problem handling complex numbers. Consider the following line of code:

```
yy = sqrt(3) - j;
```

   (a) How do you get MATLAB to return the magnitude of the complex number `yy`?

   (b) How do you get MATLAB to return the phase of the complex number `yy`? What are the units of the phase?

   (c) Use the relationship $|z|^2 = (z^*)z$ to write a line of MATLAB code that returns the magnitude-squared of the complex number `yy`.

Lab #1
INSTRUCTOR VERIFICATION

SAMPLE

For each verification, be prepared to explain your answer and respond to other related questions that the lab instructor might ask.

Name: ______________________________

Date of Lab: __________

Part C-1.2.1 Vector replacement using the colon operator.

Instructor Verification ____________________

Part C-1.2.2(b) Explain why it is necessary to write `yy(k+6)`. What happens if you use `yy(k)` instead?

Instructor Verification ____________________

Part C-1.2.2(d) Explain why the plot of `real(zz)` is a sinusoid. What is its amplitude and phase? In the space below, make a calculation of the phase from time shift.

Instructor Verification ____________________

(optional) Part C-1.2.3 Use `soundsc()` to play a 1300 Hz tone in MATLAB:

Instructor Verification ____________________

C-2 Encoding and Decoding Touch-Tone Signals

LAB: *This is Lab #9 on the CD-ROM.*

The *Pre-Lab* exercises are intended to be completed while preparing for the actual lab.

The *Warm-up* section of this lab would normally be completed *during supervised lab time,* and the steps marked *Instructor Verification* would be signed off by the lab supervisor.

The *Laboratory Exercises* should be done individually and summarized with a lab report.

C-2.1 Introduction

This lab introduces a practical application where sinusoidal signals are used to transmit information: a Touch-Tone[3] phone dialer. Bandpass FIR filters can be used to extract the information encoded in the waveforms. The goal of this lab is to design and implement bandpass FIR filters in MATLAB, and use them to do the decoding automatically. In the experiments of this lab, you will use `firfilt()`, or `conv()`, to implement filters and `freqz()` to obtain the filter's frequency response.[4] As a result, you should learn how to characterize a filter by knowing how it reacts to different frequency components in the input.

C-2.1.1 Review

In other labs, you have learned about FIR filters such as L-point averagers and nulling filters. Another very important FIR filter is known as the bandpass filter (BPF). For the rest of the lab, you will learn how to design these filters and how to use them to do certain tasks. One practical example is the decoding of *Dual-Tone Multi-Frequency* (DTMF) signals used to dial a telephone. The following *Background* section explains the encoding scheme for DTMF signals.

Table C-2: Extended DTMF encoding table for Touch-Tone dialing. When any key is pressed, the tones of the corresponding column and row are generated and summed. Keys A–D (in the fourth column) are not implemented on commercial and household telephone sets, but are used in some military and other signaling applications.

FREQS	1209 Hz	1336 Hz	1477 Hz	1633 Hz
697 Hz	**1**	**2**	**3**	**A**
770 Hz	**4**	**5**	**6**	**B**
852 Hz	**7**	**8**	**9**	**C**
941 Hz	*	**0**	**#**	**D**

C-2.1.2 Background: Telephone Touch-Tone Dialing

Telephone key pads generate *Dual-Tone Multi-Frequency* (DTMF) signals to dial a telephone number. When any key is pressed, the sinusoids of the corresponding row and column frequencies (in Table C-2) are generated and summed producing two simultaneous or dual tones. As an example, pressing the **5** key generates a signal containing the sum of the two tones at 770 Hz and 1336 Hz.

The frequencies in Table C-2 were chosen by the design engineers to avoid harmonics. No frequency is an integer multiple of another, the difference between any two frequencies does not equal any of the frequencies, and the sum of any two frequencies does not equal any of the frequencies.[5] This makes it easier to detect exactly

[3] Touch-Tone is a registered trademark of AT&T.

[4] If you do not have access to the function `freqz.m`, there is a substitute available in the *SP First* toolbox called `freekz.m`.

[5] More information can be found by searching for "DTMF" or "Touch-Tone" on the Internet.

which tones are present in the dialed signal in the presence of nonlinear line distortions.

C-2.1.3 DTMF Decoding

There are several steps in decoding a DTMF signal:

1. Divide the time signal into short time segments representing individual key presses.

2. Filter the individual segments to extract the possible frequency components. Bandpass filters can be used to isolate the sinusoidal components.

3. Determine which two frequency components are present in each time segment by measuring the size of the output signal from all of the bandpass filters.

4. Decode which key was pressed, **0–9**, **A–D**, *, or # by converting frequency pairs back into key names according to Table C-2.

It is possible to decode DTMF signals using a simple FIR filter bank. The filter bank in Fig. C-1 consists of eight bandpass filters, each of which passes only one of the eight possible DTMF frequencies. The input signal for all the filters is the same DTMF signal.

Here is how the system should work: When the input to the filter bank is a DTMF signal, the outputs from two of the bandpass filters (BPFs) should be larger than the rest. If we detect (or measure) which two outputs are the large ones, then we know the two corresponding frequencies. These frequencies are then used as row and column pointers to determine the key from the DTMF code. A good measure of the output levels is the *peak value* at the filter outputs, because when the BPF is working properly it should pass only one sinusoidal signal and the peak value would be the amplitude of the sinusoid passed by the filter. More discussion of the detection problem can be found in Section C-2.4.

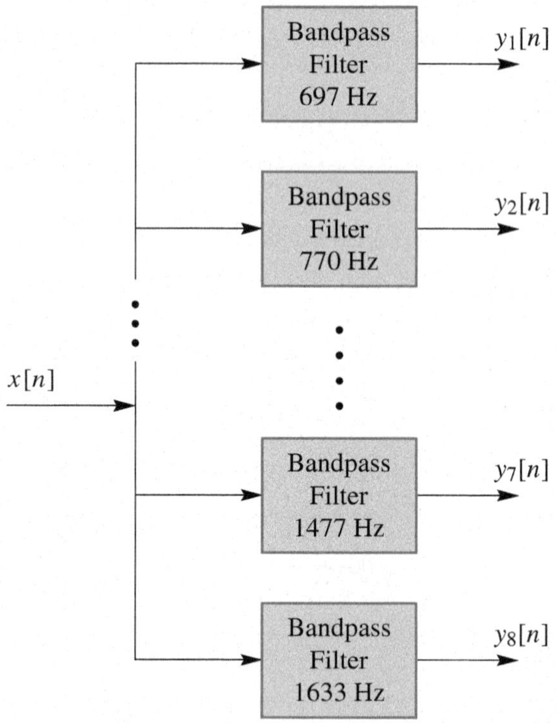

Figure C-1: Filter bank consisting of bandpass filters (BPFs) that pass frequencies corresponding to the eight DTMF component frequencies listed in Table C-2. The number in each box is the *center frequency* of the BPF.

C-2.2 Pre-Lab

C-2.2.1 Signal Concatenation

In Lab #4 on Sinusoidal Synthesis, a very long music signal was created by joining together many sinusoids. When two signals were played one after the other, the composite signal was created by the operation of *concatenation.* In MATLAB, this can be done by making each signal a row vector, and then using the matrix building notation as follows:

```
xx = [ xx, xxnew ];
```

where `xxnew` is the sub-signal being appended. The length of the new signal is equal to the sum of the lengths of the two signals `xx` and `xxnew`. A third signal could be added later on by concatenating it to `xx`.

C-2.2.2 Comment on Efficiency

In MATLAB the concatenation method, `xx = [ xx, xxnew ]`, would append the signal vector `xxnew` to the existing signal `xx`. However, this becomes an *inefficient* procedure if the signal length gets to be very large. The reason is that MATLAB must re-allocate the memory space for `xx` every time a new sub-signal is appended via concatenation. If the length `xx` were being extended from 400,000 to 401,000, then a clean section of memory consisting of 401,000 elements would have to be allocated followed by a copy of the existing 400,000 signal elements and finally the append would be done. This is clearly inefficient, but would not be noticed for short signals. An alternative is to pre-allocate storage for the complete signal vector, but this can only be done if the final signal length is known ahead of time.

C-2.2.3 Encoding from a Table

Explain how the MATLAB program in Fig. C-2 uses frequency information stored in a table to generate a long signal via concatenation. Determine the size of the table and all of its entries, and then state the playing order of the frequencies. Determine the total length of the signal played by the `soundsc` function. How many samples and how many seconds?

C-2.2.4 Overlay Plotting

Sometimes it is convenient to overlay information onto an existing MATLAB plot. The MATLAB command `hold on` will inhibit the "figure erase" that is usually done just before a new plot. Demonstrate that you can do an overlay by following these instructions:

```
ftable = [1;2;3;4;5]*[80,110]
fs = 8000;
xx = [ ];
disp('-Here we go through the Loop-')
keys = rem(3:12,10) + 1;
for ii = 1:length(keys)
  kk = keys(ii);
  xx = [xx,zeros(1,400)];
  krow = ceil(kk/2);
  kcol = rem(kk-1,2) + 1;
  freq = ftable(krow,kcol);
  xx = [xx,...
           cos(2*pi*freq*(0:1199)/fs)];
end
soundsc(xx,fs);
```

Figure C-2: Generate a table of frequencies and synthesize a signal from it.

(a) Plot the magnitude response of the 5-point averager, created from

```
HH = freqz(ones(1,5)/5,1,ww)
```

Make sure that the horizontal frequency axis extends from $-\pi$ to $+\pi$.

(b) Use the `stem` function to place vertical markers at the zeros of the frequency response.

```
hold on
stem(.4*pi*[-2,-1,1,2],.3*ones(1,4),'r.')
hold off
```

C-2.3 Warm-up: DTMF Synthesis

To prepare for the DTMF design problem, two tasks are addressed in the warm-up. First, a DTMF dialing program is created. If the output of `dtmfdial` is played as a sound, it should be able to dial an actual phone.

```
function xx = dtmfdial(keyNames,fs)
%DTMFDIAL  Create signal vector of
%         DTMF tones that will dial
%         a Touch Tone telephone.
%
% usage:  xx = dtmfdial(keyNames,fs)
%
%  keyNames = character string with
%               valid key names
%        fs = sampling frequency
%    xx = signal vector that is the
%         concatenation of DTMF tones
%
dtmf.keys = ...
   ['1','2','3','A';
    '4','5','6','B';
    '7','8','9','C';
    '*','0','#','D'];
ff_cols = [1209,1336,1477,1633];
ff_rows = [ 697; 770; 852; 941];
dtmf.colTones = ones(4,1)*ff_cols;
dtmf.rowTones = ff_rows*ones(1,4);
```

Figure C-3: Skeleton of `dtmfdial.m`, a DTMF phone dialer. Complete this function with additional lines of code.

Second, a method for bandpass filter design is presented. This will be useful in producing the BPFs needed in the filter bank of Fig. C-1.

C-2.3.1 DTMF Dial Function

Write a function, `dtmfdial.m`, to implement a DTMF dialer based on the frequency table defined in Table C-2. A skeleton of `dtmfdial.m` is given in Fig. C-3. In this warm-up, you must complete the dialing code so that it implements the following:

(a) The input to the function is a vector of characters, each one being equal to one of the key names on the telephone. In MATLAB a character string is the same as a vector of characters. The MATLAB structure[6] called `dtmf` contains the key names in the field `dtmf.keys` which is a 4×4 matrix that corresponds exactly to the keyboard layout in Table C-2.

(b) The output should be a vector of samples at sampling rate $f_s = 8000$ Hz containing the DTMF tones, one tone pair per key. Remember that each DTMF signal is the sum of a pair of equal amplitude sinusoidal signals. The duration of each tone pair should be exactly 0.20 sec, and a silence, about 0.05 sec long, should separate the DTMF tone pairs. These times can be declared as fixed parameters in the code of `dtmfdial`. (You do not need to make them variable in your function.)

(c) The frequency information is given as two 4×4 matrices (`dtmf.colTones` and `dtmf.rowTones`): one contains the column frequencies, the other has the row frequencies. You can translate a key such as the **6** key into the correct location in these 4×4 matrices by using MATLAB's `find` function. For example, the key **6** is in row 2 and column 3, so we would generate sinusoids with frequencies equal to `dtmf.colTones(2,3)` and `dtmf.rowTones(2,3)`.

To convert a key name to its corresponding row–column indices, consider the following example:

```
[ii,jj] = find('3'==dtmf.keys)
```

Also, consult the MATLAB code in Fig. C-3 and modify it for the 4×4 tables in `dtmfdial.m`. You should implement error checking so that an illegitimate key name is rejected.

(d) Your function should create the appropriate tone sequence to dial an arbitrary phone number. When played

[6] MATLAB structures are collections of different kinds of data organized by named fields. You can find out more about structures by consulting MATLAB help "Programming and Data Types: Structures and Cell Arrays."

through a telephone handset, the output of your function will be able to dial the phone. You could use `specgram` to check your work.[7]

Instructor Verification (separate page)

C-2.3.2 Simple Bandpass Filter Design

The L-point averaging filter is a lowpass filter. Its passband width is inversely proportional to L. It is also possible to create a filter whose passband is centered around some frequency other than zero. One simple way to do this is to define the impulse response of an L-point FIR filter as

$$h[n] = \beta \cos\left(\hat{\omega}_c n\right) \qquad 0 \le n < L$$

where L is the filter length, and $\hat{\omega}_c$ is the center frequency that defines the frequency location of the passband and β is used to adjust the gain in the passband. For example, we pick $\hat{\omega}_c = 0.2\pi$ if we want the peak of the filter's passband to be centered at 0.2π. Also, it is possible to choose β so that the maximum value of the frequency response magnitude will be one. The bandwidth of the bandpass filter is controlled by L; the larger the value of L, the narrower the bandwidth. This particular filter is also discussed in the section on useful filters in Chapter 7. Also, you might have designed some BPFs based on this idea in other labs.

(a) Generate a bandpass filter that will pass a frequency component at $\hat{\omega} = 0.2\pi$. Make the filter length (L) equal to 51. Figure out the value of β so that the maximum value of the frequency response magnitude will be one. Make a plot of the frequency response magnitude and phase.
Hint: Use MATLAB's `freqz()` function or the `freekz()` function in *SP First* to calculate these values.

[7] In MATLAB, the demo called `phone` also shows the waveforms and spectra generated in a DTMF system.

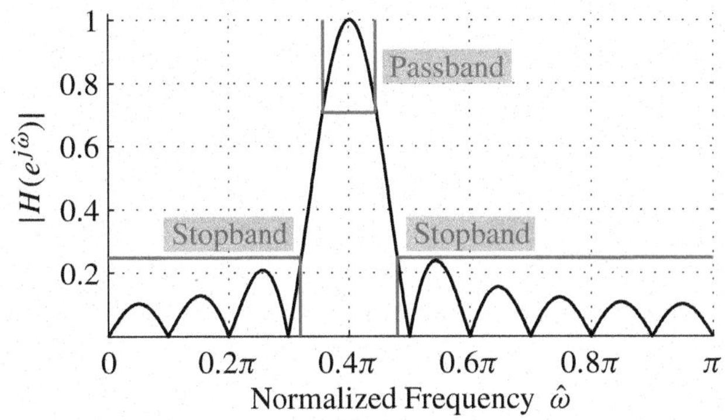

Figure C-4: The frequency response (magnitude) of an FIR bandpass filter is shown with its passband and stopband regions.

(b) The *passband* of the BPF filter is defined by the region of the frequency response where $|H(e^{j\hat{\omega}})|$ is close to its maximum value of one. Typically, the passband width is defined as the length of the frequency region where $|H(e^{j\hat{\omega}})|$ is greater than $1/\sqrt{2} = 0.707$. *Note*: You can use MATLAB's `find` function to locate those frequencies where the magnitude satisfies $|H(e^{j\hat{\omega}})| \ge 0.707$. Plot markers at the passband and stopband edges (similar to Fig. C-4).

Use the plot of the frequency response for the length-51 bandpass filter from part (a) to determine the passband width.

Instructor Verification (separate page)

(c) If the sampling rate is $f_s = 8000$ Hz, determine the analog frequency components that will be passed by this bandpass filter. Use the passband width and also the center frequency of the BPF to make this calculation.

Instructor Verification (separate page)

C-2.4 Lab: DTMF Decoding

A DTMF decoding system needs two sub-systems: a set of bandpass filters (BPF) to isolate individual frequency components, and a detector to determine whether or not a given component is present. The detector must "score" each BPF output and determine which two frequencies are most likely to be contained in the DTMF tone. In a practical system where noise and interference are also present, this scoring process is a crucial part of the system design, but we will only work with noise-free signals to understand the basic functionality in the decoding system.

To make the whole system work, you will have to write three M-files: `dtmfrun`, `dtmfscore` and `dtmfdesign`. An additional M-file called `dtmfcut` is available on the *SP First* CD-ROM. The main M-file should be named `dtmfrun.m`. It will call `dtmfdesign.m`, `dtmfcut.m`, and `dtmfscore.m`. The following sections discuss how to create or complete these functions.

C-2.4.1 Filter Bank Design: `dtmfdesign.m`

The FIR filters that will be used in the filter bank of Fig. C-1 are a simple type constructed with sinusoidal impulse responses, as already shown in the Warm-up. In the section on useful filters in Chapter 7, a *simple* bandpass filter design method is presented in which the impulse response of the FIR filter is simply a finite-length cosine of the form

$$h[n] = \beta \cos\left(\frac{2\pi f_b n}{f_s}\right), \qquad 0 \le n \le L-1$$

where L is the filter length, and f_s is the sampling frequency. The constant β gives flexibility for scaling the filter's gain to meet a constraint such as making the maximum value of the frequency response equal to one. The parameter f_b defines the frequency location of the passband; e.g., we pick $f_b = 852$ if we want to isolate the 852 Hz component. The bandwidth of the bandpass filter is controlled by L; the larger the value of L, the narrower the bandwidth.

```
function  hh = dtmfdesign(fcent,L,fs)
%DTMFDESIGN
%    hh = dtmfdesign(fcent, L, fs)
%     returns L by 8 matrix
%     where each column contains the
%     impulse response of a BPF,
%     one for each frequency in fcent
% fcent = vector of center freqs
%       L = length of bandpass filters
%     fs = sampling frequency
%
%    Each BPF must be scaled so that
%     its frequency response has a
%     maximum magnitude equal to one.
```

Figure C-5: Skeleton of the `dtmfdesign.m` function. Complete this function with additional lines of code.

(a) Devise a MATLAB strategy for picking the constant β so that the maximum value of the frequency response will be equal to one. Write one or two lines of MATLAB code that will do this scaling operation in general. There are two approaches here:

> *Mathematical:* Derive an approximate formula for β from the formula for the frequency response of the BPF. Then use MATLAB to evaluate this closed-form expression for β.
>
> *Numerical:* Let MATLAB measure the peak value of the unscaled frequency response, and then have MATLAB compute β to scale the peak to be one.

(b) Complete the M-file `dtmfdesign.m` which is described in Fig. C-5. This function should produce all eight bandpass filters needed for the DTMF filter bank system. Store the filters in the columns of the matrix `hh` whose size is $L \times 8$.

(c) The rest of this section describes how you can exhibit that you have designed a correct set of BPFs. In particular, you should justify how to choose L, the length of the filters. When you have completed your filter design function, you should run the $L = 40$ and $L = 80$ cases, and then you should determine empirically the minimum length L so that the frequency response will satisfy the specifications on passband width and stopband rejection given in part (f) below.

(d) Generate the eight (scaled) bandpass filters with $L = 40$ and $f_s = 8000$. Plot the magnitude of the frequency responses all together on one plot (the range $0 \leq \hat{\omega} \leq \pi$ is sufficient because $|H(e^{j\hat{\omega}})|$ is symmetric). Indicate the locations of each of the eight DTMF frequencies (697, 770, 852, 941, 1209, 1336, 1477, and 1633 Hz) on this plot to illustrate whether or not the passbands are narrow enough to separate the DTMF frequency components.[8] *Hint*: use the `hold on` command and markers as you did in the warm-up.

(e) Repeat the previous part with $L = 80$ and $f_s = 8000$. The width of the passband is supposed to vary inversely with the filter length L. Explain whether or not that is true by comparing the length 80 and length 40 cases.

(f) As help for the previous parts, recall the following definitions: The *passband* of the BPF filter is defined by the region of $\hat{\omega}$ where $|H(e^{j\hat{\omega}})|$ is close to one. Typically, the passband width is defined as the length of the frequency region where $|H(e^{j\hat{\omega}})|$ is greater than $1/\sqrt{2} = 0.707$.

The *stopband* of the BPF filter is defined by the region of $\hat{\omega}$ where $|H(e^{j\hat{\omega}})|$ is close to zero. In this case, it is reasonable to define the stopband as the region where $|H(e^{j\hat{\omega}})|$ is less than 0.25.

Filter Design Specifications: For each of the eight BPFs, choose L so that only one frequency lies within the passband of the BPF and all other DTMF frequencies lie in the stopband.

Use the `zoom on` command to show the frequency response over the frequency interval where the DTMF frequencies lie. Comment on the selectivity of the bandpass filters; i.e., use the frequency response to explain how the filter passes one component while rejecting the others. Is each filter's passband narrow enough so that only one frequency component lies in the passband and the others are in the stopband? Since the same value of L is used for all the filters, which filter drives the problem? In other words, for which center frequency is it hardest to meet the specifications for the chosen value of L?

[8] Remember that normalized discrete-time frequency is related to continuous-time frequency by the relation $\hat{\omega} = 2\pi f/f_s$.

C-2.4.2 A Scoring Function: `dtmfscore.m`

The final objective is decoding—a process that requires a binary decision on the presence or absence of the individual tones. In order to make the signal detection an automated process, we need a *score* function that rates the different possibilities.

(a) Complete the `dtmfscore` function based on the skeleton given in Fig. C-6. The input signal `xx` to the `dtmfscore` function must be a short segment from the DTMF signal. The task of breaking up the signal so that each short segment corresponds to one key is done by the function `dtmfcut` prior to calling `dtmfscore`.

(b) Use the following rule for scoring: If the signal $y_i[n]$ is the output of the ith BPF, the score equals one when $\max_n |y_i[n]| \geq 0.59$; otherwise, it is zero.

```
function  sc = dtmfscore(xx, hh)
%DTMFSCORE
%  usage:   sc = dtmfscore(xx, hh)
%  returns score based on the max
%  amplitude of the filtered output
%   xx = input DTMF tone
%   hh = impulse response of ONE BPF
%
%  Signal detection after filtering
%    ss with a length-L BPF, hh,
%    and then finding the maximum
%    amplitude of the output.
%  The score is either 1 or 0.
%     sc = 1 if max(|y[n]|) >= 0.59
%     sc = 0 if max(|y[n]|) < 0.59
%
%-- Scale the input signal x[n]
%-- to the range [-2,+2]
xx = xx*(2/max(abs(xx)));
```

Figure C-6: Skeleton of the dtmfscore.m function. Complete this function with additional lines of code.

(c) Prior to filtering and scoring, make sure that the input signal $x[n]$ is normalized to the range $[-2, +2]$. With this scaling, the two sinusoids that make up $x[n]$ should each have amplitudes of approximately 1.0.[9] Therefore, the scoring threshold of 0.59 corresponds to a 59% level for detecting the presence of one sinusoid.

(d) The scoring rule above depends on proper scaling of the frequency response of the bandpass filters. Explain why the maximum value of the magnitude for $H(e^{j\hat{\omega}})$ must be equal to one. Consider the fact that both sinusoids in the DTMF tone will experience a known gain (or attenuation) through the bandpass filter, so the amplitude of the output can be predicted if we control both the frequency response and the amplitude of the input.

[9]The two sinusoids in a DTMF tone have frequencies that are not harmonics. When plotted versus time, the peaks of the two sinusoids will eventually line up.

(e) When debugging your program it might be useful to have a plot command inside the `dtmfscore.m` function. If you plot the first 200–500 points of the filtered output, you should be able to see two cases: either $y[n]$ is a strong sinusoid with an amplitude close to one (when the filter is matched to one of the component frequencies), or $y[n]$ is relatively small when the filter passband and input signal frequency are mismatched.

C-2.4.3 DTMF Decode Function: `dtmfrun.m`

The DTMF decoding function, `dtmfrun` must use information from `dtmfscore` to determine which key was pressed based on an input DTMF tone. The skeleton of this function in Fig. C-7 includes the help comments.

The function `dtmfrun` works as follows: first, it designs the eight bandpass filters that are needed, then it breaks the input signal down into individual segments (using `dtmfcut` provided). For each segment, it will have to call the user-written `dtmfscore` function to score the different BPF outputs and then determine the key for that segment. The final output is the list of decoded keys. You must add the logic to decide which key is present.

The input signal to the `dtmfscore` function must be a short segment from the DTMF signal. The task of breaking up the signal so that each segment corresponds to one key is done with the `dtmfcut` function which is called from `dtmfrun`. The score returned from `dtmfscore` *must be* either a 1 or a 0 for each frequency. Then the decoding works as follows: If exactly one row frequency and one column frequency are scored as 1's, then a unique key is identified and the decoding is probably successful. In this case, you can determine the key by using the row and column index. It is possible that there might be an error in scoring if too many or too few frequencies are scored as 1's. In this case, you

```
function keys = dtmfrun(xx,L,fs)
%DTMFRUN    keys = dtmfrun(xx,L,fs)
% returns list of key names in xx
%  keys = array of characters,
%         i.e., the decoded key names
%    xx = DTMF waveform
%     L = filter length
%    fs = sampling freq
%
dtmf.keys = ...
  ['1','2','3','A';
   '4','5','6','B';
   '7','8','9','C';
   '*','0','#','D'];
ff_cols = [1209,1336,1477,1633];
ff_rows = [ 697; 770; 852; 941];
dtmf.colTones = ones(4,1)*ff_cols;
dtmf.rowTones = ff_rows*ones(1,4);
                  %<====
center_freqs =%<=FILL IN THE CODE HERE
                  %<====
hh = dtmfdesign( center_freqs,L,fs );
% hh = L x 8 MATRIX of all the filters
%      Each column contains the impulse
%      response of one bandpass filter
%
%-----Find the tone bursts:
[nstart,nstop] = dtmfcut(xx,fs);
keys = [];
for kk=1:length(nstart)
   %----Extract one DTMF tone
   x_seg = xx(nstart(kk):nstop(kk));
%<==========FILL IN THE CODE HERE
end
```

Figure C-7: Skeleton of `dtmfrun.m`. Complete the `for` loop in this function with additional lines of code.

should return an error indicator (perhaps by setting the key equal to -1). There are several ways to write the `dtmfrun` function, but you should avoid excessive use of "`if`" statements to test all 16 cases.

Hint: Use MATLAB's logicals (e.g., `help find`) to implement the tests in a few statements.

```
fs = 8000;  %<--sampling rate
            %  for all functions
tk = ['A','B','C','D','*','#',...
      '0','1','2','3','4','5',...
      '6','7','8','9'];
xx = dtmfdial( tk, fs );
soundsc(xx, fs)
L = ...     %<--use your value of L
dtmfrun(xx, L, fs)
%-- The answer should be
%     ans = ABCD*#0123456789
```

Figure C-8: Testing the DTMF system.

C-2.4.4 Testing

Once you get your system working, there should be no errors with a large value of L, but when you try to reduce the filter length, the error indicator (key equal to -1) would tell you that the filter length is getting too small. Run tests to find the minimum value for L that gives reliable operation. In your lab report, describe how you tested the system to get this minimum value for L.

C-2.4.5 Telephone Numbers

The functions `dtmfdial.m` and `dtmfrun.m` can be used to test the entire DTMF system as shown in Fig. C-8. For the `dtmfrun` function to work correctly, all the M-files must be on the MATLAB path. It is also essential to have short pauses between the tone pairs so that `dtmfcut` can parse out the individual signal segments.

If you are presenting this project in a lab report, demonstrate a working version of your programs by running it on the following phone number:

`407*89132#BADC`

In addition, make a spectrogram of the signal from `dtmfdial` to illustrate the presence of the dual tones.

C-2.4.6 Demo

When you submit your lab report, you should demonstrate your working system to your lab supervisor. You would call `dtmfrun` for a signal `xx` provided at random. The output should be the decoded telephone number. The evaluation criteria are shown below.

DTMF Decoding Evaluation

Does the designed DTMF decoder decode the telephone numbers correctly for the following values of L?

For $L = 111$

All Numbers _____ Most _____ None _____

For L =Student's Minimum Length

All Numbers _____ Most _____ None _____

Lab #9
INSTRUCTOR VERIFICATION

For each verification, be prepared to explain your answer and respond to other related questions that the lab instructors might ask.

Name: ______________________________

Date of Lab: __________

Part C-2.3.1: Complete the dialing function `dtmfdial.m`:

Instructor Verification ______________________

Part C-2.3.2(b): Measure the passband width of the BPF

Instructor Verification ______________________

Part C-2.3.2(c): Determine the analog frequency components passed by the BPF when $f_s = 8000$ Hz. Give the range of frequencies.

Instructor Verification ______________________

C-3 Two Convolution GUIs

LAB: *This is Lab #12 on the CD-ROM.*

The *Pre-Lab* exercises are intended to be completed while preparing for the actual lab.

The *Warm-up* section of this lab would normally be completed *during supervised lab time,* and the steps marked *Instructor Verification* would be signed off by the lab supervisor.

The *Laboratory Exercises* should be done on your own and summarized with a lab report.

C-3.1 Introduction

This lab concentrates on the use of two MATLAB GUIs for convolution.

1. `dconvdemo:` GUI for discrete-time convolution. This is exactly the same as the MATLAB functions `conv()` and `firfilt()` used to implement FIR filters.

   DEMO: *Chap. 5: Discrete Convolution*

2. `cconvdemo:` GUI for continuous-time convolution.

   DEMO: *Chap. 9: Continuous Convolution*

Each one of these demos illustrates an important point about the behavior of a linear, time-invariant (LTI) system. They also provide a convenient way to visualize the output of an LTI system. Both of these demos are on the *SP First CD-ROM* .

C-3.2 Pre-Lab: Run the GUIs

The first objective of this lab is to demonstrate usage of two convolution GUIs, so you must install the GUIs and practice using them.

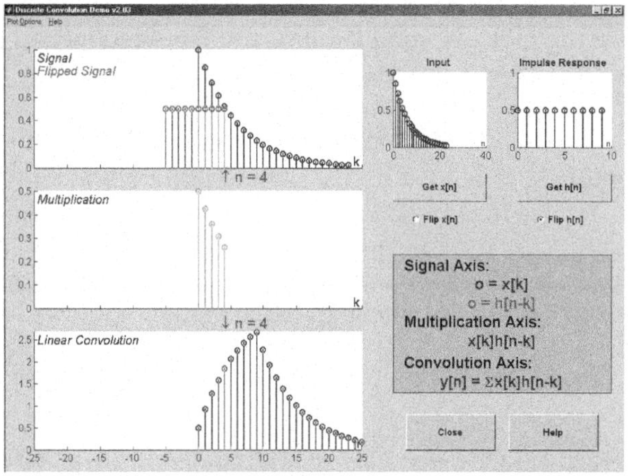

Figure C-9: Interface for discrete-time convolution GUI.

C-3.2.1 Discrete-Time Convolution Demo

In the `dconvdemo` GUI, you can select an input signal $x[n]$, as well as the impulse response of the filter $h[n]$. Then the demo shows the "flipping and shifting" used when a convolution is computed. This corresponds to the sliding window of the FIR filter. Figure C-9 shows the interface for the `dconvdemo` GUI.

In the Pre-Lab, you should perform the following steps with the `dconvdemo` GUI.

(a) Set the input to a finite-length pulse

$$x[n] = 2\{u[n] - u[n-10]\}$$

(b) Set the filter's impulse response to obtain a 5-point averager.

(c) Use the GUI to produce the output signal.

(d) When you move the mouse pointer over the index "n" below the signal plot and do a click-hold, you will get a *hand tool* that allows you to move the "n"-pointer. By moving the pointer horizontally you can observe the sliding window action of convolution. You can even move the index beyond the limits of the window and the plot will scroll over to align with "n."

(e) Set the filter's impulse response to a length-10 averager,

$$h[n] = \tfrac{1}{10}\{u[n] - u[n-10]\}$$

Use the GUI to produce the output signal.

(f) Set the filter's impulse response to a shifted impulse,

$$h[n] = \delta[n-3]$$

Use the GUI to produce the output signal.

(g) Compare the outputs from parts (c), (e) and (f). Notice the different shapes (triangle, rectangle or trapezoid), the maximum values, and the different lengths of the outputs.

C-3.2.2 Continuous-Time Convolution Demo

In the `cconvdemo` GUI, you can select an input signal $x(t)$, as well as the impulse response of an *analog* filter $h(t)$. Then the demo shows the "flipping and shifting" used when a convolution integral is performed. Figure C-10 shows the interface for the `cconvdemo` GUI. A more expanded view of the interface for this GUI can be found in Chapter 9 on p. 266.
In the Pre-Lab, you should perform the following steps with the `cconvdemo` GUI.

(a) Set the input to a four-second pulse

$$x(t) = u(t) - u(t-4)$$

(b) Set the filter's impulse response to a two-second pulse with amplitude $\frac{1}{2}$,

$$h(t) = \tfrac{1}{2}\{u(t) - u(t-2)\}$$

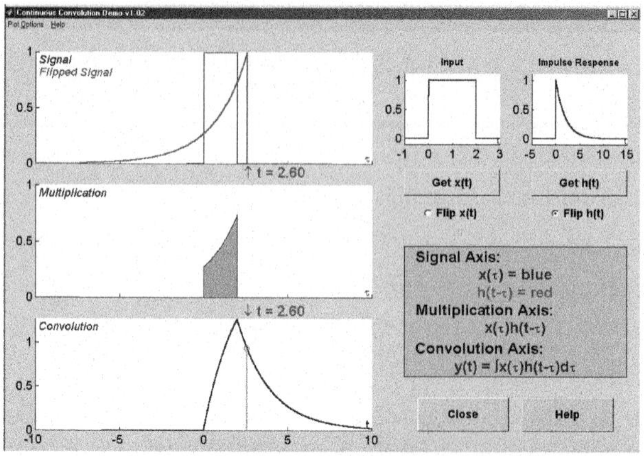

Figure C-10: Interface for continuous-time convolution GUI.

(c) Use the GUI to produce the output signal. Use the *sliding hand tool* to grab the time marker and move it to produce the flip-and-slide effect of convolution.

(d) Set the filter's impulse response to a four-second pulse with amplitude $\frac{1}{4}$,

$$h(t) = \tfrac{1}{4}\{u(t) - u(t-4)\}$$

Use the GUI to produce the output signal.

(e) Set the filter's impulse response to a shifted impulse,

$$h(t) = \delta(t-3)$$

Use the GUI to produce the output signal.

(f) Compare the outputs from parts (c), (d) and (e). Notice the different shapes (triangle, rectangle or trapezoid), the maximum values, and the different lengths of the outputs.

C-3.3 Warm-up: Run the GUIs

The objective of the warm-up in this lab is to use the two convolution GUIs to solve problems (some of which may be homework problems). If you are working on your own machine, you must install the SP First Toolbox, which contains `dconvdemo` and `cconvdeno`.

C-3.3.1 Continuous-Time Convolution GUI

In the warm-up, you should perform the following steps with the `cconvdemo` GUI.

(a) Set the input to an exponential signal

$$x(t) = e^{-0.25t}\{u(t) - u(t-4)\}$$

(b) Set the filter's impulse response to a different exponential

$$h(t) = e^{-t}\{u(t-1) - u(t-5)\}$$

(c) Use the GUI to produce a plot of the output signal. Use the ***sliding hand tool*** to grab the time marker and move it to produce the flip-and-slide effect of convolution.
Note: If you move the hand tool past the end of the plot, the plot will automatically scroll in that direction.

(d) Usually, the convolution integral must be evaluated in *five* different regions: no overlap (on the left side), partial overlap (on the left side), complete overlap, partial overlap (on the right side), and no overlap (on the right side). In this case, there are only *four* regions. Why?

(e) Set up the convolution integral for the second region. This is the case of partial overlap (on the left side). In addition, determine the boundaries (in secs) of the second region (i.e., the starting and ending times of region #2). Use the ***flip-and-slide*** interpretation of convolution along with the GUI to help answer this question.

Instructor Verification (separate page)

C-3.3.2 Discrete Convolution GUI

In the warm-up, you should perform the following steps with the `dconvdemo` GUI.

(a) Set the input to be a finite-length sinusoid

$$x[n] = 2\cos(2\pi n/3)\,(u[n] - u[n-10])$$

(b) Set the filter's impulse response to obtain a 3-point averager.

(c) Use the GUI to produce the output signal.

(d) Explain why the output has five different regions and why the output is zero in three of the five.

Instructor Verification (separate page)

C-3.4 Lab Exercises

In each of the following exercises, you should make a screen shot of the final picture produced by the GUI to validate that you were able to do the implementation. In all cases, you will have to do some mathematical calculations to verify that the MATLAB GUI result is correct.

C-3.4.1 Continuous-Time Convolution

In this section, use the continuous-time convolution GUI, `cconvdemo`, to do the following:

(a) Set the input to an exponential signal

$$x(t) = e^{-0.25t}\{u(t) - u(t-6)\}$$

(b) Set the filter's impulse response to a different exponential

$$h(t) = e^{-t}\{u(t+1) - u(t-5)\}$$

(c) Use the GUI to produce a plot of the output signal.

(d) Usually the convolution integral must be evaluated in five different regions: no overlap (on the left side), partial overlap (on the left side), complete overlap, partial overlap (on the right side), and no overlap (on the right side). In this case, there are only four regions. Determine the boundaries of each region (i.e., the starting and ending times in secs).

(e) Determine the mathematical formula for the convolution in each of the four regions. Use the GUI to help in setting up the integrals, but carry out the mathematics of the integrals by hand.

C-3.4.2 Continuous-Time Convolution Again

(a) Find the output of an analog filter whose impulse response is

$$h(t) = u(t+3) - u(t)$$

when the input is

$$x(t) = 2\cos(2\pi(t-2)/3)\,\{u(t-2) - u(t-12)\}$$

(b) Use the GUI to determine the length of the output signal and the boundaries of the five regions of the convolution.
Note: The regions of partial overlap would be called *transient regions,* while the region of complete overlap would be the *steady-state region.*

(c) Perform the mathematics of the convolution integral to get the exact analytic form of the output signal and verify that the GUI is correct. Also, verify that the duration of the output signal is correct.

C-3.4.3 Discrete-Time Convolution

Use the discrete-time convolution GUI, `dconvdemo`, to do the following:

(a) Find the output of a digital filter whose impulse response is

$$h[n] = u[n+3] - u[n]$$

when the input is

$$x[n] = 2\cos(2\pi(n-2)/3)\{u[n-2] - u[n-12]\}$$

(b) Use the GUI to determine the length of the output signal. Notice that you can see five separate regions just like for the continuous-time convolution in Section C-3.4.2. Identify the boundaries of these regions. *Note*: The regions of partial overlap would be called ***transient regions*** while the region of complete overlap would be the ***steady-state region***.

(c) Use numerical convolution to get the exact values of the output signal for each of the five regions. Thus, you will verify that the GUI is correct. Also verify that the duration of the output signal is correct.

(d) Discuss the relationship between this output and the continuous-time output signal in Section C-3.4.2. Point out similarities and differences.

Lab #12
INSTRUCTOR VERIFICATION

For each verification, be prepared to explain your answer and respond to other related questions that the lab instructors might ask.

Name: ______________________________

Date of Lab: __________

Part C-3.3.1: Demonstrate that you can run the continuous-time convolution demo. Explain how to find the FOUR regions for this convolution integral. In the space below, write the specific convolution integral that would be performed for these specific exponential signals in REGION #2. Make sure that the limits of integration are correct.

Instructor Verification ______________________

$y(t) = \int$ for $\leq t \leq$

Part C-3.3.2: Demonstrate that you can run the discrete-time convolution demo. Explain why the output is zero in three of the five regions identified for the output signal.

Instructor Verification ______________________

APPENDIX D

CD-ROM Demos

This appendix lists all of the demos available on the *Signal Processing First* CD-ROM and provides a brief description of each. Some of the demos are applicable to more than one chapter.

Chapter 2

Sinusoids: An introduction to plotting sinusoids (both sine and cosine waves) from equations. In addition, the tutorial reviews how to write the equation given a plot of the sinusoid. Also, it contains links to the *Sine Drill* MATLAB GUI.

Rotating Phasors: Here are four movies showing rotating phasors and how the real part of the phasor $e^{j\omega t}$ traces out a sinusoid versus time. Two of the movies show how rotating phasors of different frequencies interact to produce complicated waveforms such as beat signals.

Sine Drill: MATLAB GUI that tests the user's ability to determine basic parameters of a sinusoid. After a plot of a sinusoid is displayed, the user must correctly determine its amplitude, frequency, and phase

Tuning Fork: This demo shows how the size and stiffness of a tuning fork affect the waveform and tone produced by three different tuning forks.

Clay Whistle: Sounds and waveforms recorded from two clay whistles. These real sounds are nearly sinusoidal.

ZDrill: GUI that tests your ability to perform a variety of complex arithmetic operations. The program emphasizes the vector (geometric) view of a complex number. Problems are presented and the user must give the correct answer. See Appendix A also.

Chapter 3

FM Synthesis: Brief summary of the mathematical theory that underlies music synthesis using the principles of frequency modulation. Parameters for brass, clarinet, bell and knocking sounds are given.

Rotating Phasors: Same as Chapter 2.

Fourier Series: MATLAB GUI that shows Fourier synthesis of various waveforms with a finite number of Fourier coefficients. Examples include square, sawtooth and triangle waveforms.

Spectrograms: Plots of the spectrogram of various sounds along with the waveform and sound in order to illustrate the connection between the time- and frequency-domains. Examples include sinusoids, square waves, real and synthesized music for a piano, and a number of chirps.

Vowel Sound: Synthesis of a vowel sound from its Fourier series components demonstrates the idea of harmonic sinusoids.

Chapter 4

Aliasing and Folding: Quicktime Movies showing a sinusoid and its spectrum when sampled at rates below the Nyquist rate.

Continuous-to-Discrete Sampling (`con2dis`): MATLAB GUI that shows sampling of a sinusoid in the time- and frequency-domains. The user can interactively change the sampling rate and the input frequency, and then observe the discrete-time signal, as well as the reconstructed output.

Reconstruction: Quicktime movies that show the D-to-C conversion process as interpolation with a pulse shape. Pulse shapes include square, triangular and truncated sincs.

Sampling Theory Tutorial: Short tutorial movies that show how to use MATLAB to create sampled sinusoids. The basic ideas of sampling rate and sampling period are explained, as well as how to use MATLAB to simulate a sinusoid of a given frequency and duration.

Strobe Movies: Quicktime movies of an actual rotating disk taken by a camcorder. The frame rate of the camera (30 frames per second) does the sampling so that real aliasing can be observed.

Synthetic Strobe Movies: Quicktime movies produced in MATLAB simulate the strobe-like sampling of a rotating disk. Twelve different combinations of sampling rate and rotation rate are shown.

Chapter 5

Discrete Convolution Demo (`dconvdemo`): MATLAB GUI that illustrates the *flip-and-slide* view of convolution. The user can interactively choose different signals, and also slide the flipped signal along the horizontal axis with the mouse pointer.

Chapter 6

Cascading FIR Filters: An image is processed through a first-difference filter and a five-point averager. The final output is shown to be independent of the processing order. In addition, blurring and sharpening of the image are illustrated by the lowpass and highpass filters.

DLTI Demo: MATLAB GUI that illustrates the *sinusoid-in gives sinusoid-out* property of LTI systems. The user can pick the input frequency and the filter's frequency response, and then observe that the output is a sinusoid with different amplitude and phase.

Introduction to FIR filters: The waveform, sound and spectrogram of a signal are shown before and after lowpass filtering. The signal is a speech signal between two chirps, so the example illustrates how lowpass filtering can change the character of signals.

Chapter 7

Three-Domains FIR: The connection between the z-transform domain of poles and zeros and the time-domain, and also the frequency-domain is illustrated with several movies where individual zeros or zero pairs are moved continuously.

PeZ: A MATLAB GUI for pole-zero manipulation. Poles and zeros can be placed anywhere on a map of the z-plane. The corresponding time-domain (n) and frequency-domain ($\hat{\omega}$) plots will be displayed. When a zero pair (or pole pair) is dragged, the impulse response and frequency response plots will be updated in real time.

PeZ Tutorial: These movies describe how to use the `PeZ` graphical user interface to place/move poles and zeros. They also show how to display the associated impulse and frequency response.

Z to Freq: A movie that illustrates the connection between the complex z-plane and the frequency response of a system. The frequency response is obtained by evaluating $H(z)$ on the unit circle in the complex z-plane.

Chapter 8

Three-Domains IIR: The connection between the z-transform domain of poles and zeros and the time-domain, and also the frequency-domain is illustrated with several movies where individual poles, or zeros or pole pairs of IIR filters are moved continuously.

PeZ: Same as Chapter 7, but note that it can do IIR filters, as well as FIR filters.

PeZ Tutorial: Same as Chapter 7.

IIR Filtering: A short tutorial on first- and second-order IIR (infinite-length impulse response) filters. This demo shows plots in the three domains for a variety of IIR filters with different filter coefficients.

Z to Freq: Same as Chapter 7.

Chapter 9

Continuous Convolution Demo (`cconvdemo`): The Continuous Convolution Demo is a program that helps visualize the process of continuous-time convolution. Features: Users can choose from a variety of different signals, signals can be dragged around with the mouse with results displayed in real-time, tutorial mode lets students hide convolution result until requested, and various plot options enable the tool to be effectively used as a lecture aid in a classroom environment.

Discrete Convolution Demo (`dconvdemo`): Same as Chapter 5.

Chapter 10

CLTI Demo: A MATLAB GUI that illustrates the sinusoidal response of LTI systems.

Fourier Series: This MATLAB demo reconstructs a square, triangle, or sawtooth waveform, using a given number of Fourier series coefficients. Same as Chapter 3.

Chapter 11

CLTI Demo: Same as Chapter 10.

Fourier Series: Same as Chapters 3, 10 and 12.

Chapter 12

Fourier Series: Same as Chapters 3, 10 and 11.

Chapter 13

Music GUI: A MATLAB GUI for composing music, listening to sinusoidal synthesis and viewing the spectrogram of the song.

Resolution of the Spectrogram: These Quicktime movies illustrate the inherent trade-offs between time and frequency resolution of the spectrogram when it is computed by a sliding-window FFT.

Appendix A

Complex Numbers via MATLAB*:* Examples of how complex numbers and complex exponentials can be handled by MATLAB.

Phasor Races: This MATLAB GUI is a speed drill for testing complex addition. It includes many other related operations that can be tested in a "drill" scenario: adding sinusoids, z-transforms, etc. A timer starts as soon as the problem is posed, so that a student can try to solve questions quickly and accurately.

ZDrill: GUI that tests the user's ability to calculate the result of simple operations on complex numbers. The program emphasizes the geometric (vector) view of a complex number. The following six operations are supported: Add, Subtract, Multiply, Divide, Inverse, and Conjugate.

Appendix B

MATLAB Tutorial Movies: A set of movies illustrating MATLAB programming is available on the CD-ROM.

Index

YOU SHOULD CAREFULLY READ THE TERMS AND CONDITIONS BEFORE USING THE SOFTWARE. USING THIS SOFTWARE INDICATES YOUR ACCEPTANCE OF THESE TERMS AND CONDITIONS.

Pearson Education, Inc. provides this program and licenses its use. You assume responsibility for the selection of the program to achieve your intended results, and for the installation, use, and results obtained from the program. This license extends only to use of the program in the United States or countries in which the program is marketed by authorized distributors.

LICENSE GRANT

You hereby accept a nonexclusive, nontransferable, permanent license to install and use the program ON A SINGLE COMPUTER at any given time. You may copy the program solely for backup or archival purposes in support of your use of the program on the single computer. You may not modify, translate, disassemble, decompile, or reverse engineer the program, in whole or in part.

TERM

The License is effective until terminated. Pearson Education, Inc. reserves the right to terminate this License automatically if any provision of the License is violated. You may terminate the License at any time. To terminate this License, you must return the program, including documentation, along with a written warranty stating that all copies in your possession have been returned or destroyed.

LIMITED WARRANTY

THE PROGRAM IS PROVIDED "AS IS" WITHOUT WARRANTY OF ANY KIND, EITHER EXPRESSED OR IMPLIED, INCLUDING, BUT NOT LIMITED TO, THE IMPLIED WARRANTIES OR MERCHANTABILITY AND FITNESS FOR A PARTICULAR PURPOSE. THE ENTIRE RISK AS TO THE QUALITY AND PERFORMANCE OF THE PROGRAM IS WITH YOU. SHOULD THE PROGRAM PROVE DEFECTIVE, YOU (AND NOT PEARSON EDUCATION, INC. OR ANY AUTHORIZED DEALER) ASSUME THE ENTIRE COST OF ALL NECESSARY SERVICING, REPAIR, OR CORRECTION. NO ORAL OR WRITTEN INFORMATION OR ADVICE GIVEN BY PEARSON EDUCATION, INC., ITS DEALERS, DISTRIBUTORS, OR AGENTS SHALL CREATE A WARRANTY OR INCREASE THE SCOPE OF THIS WARRANTY. SOME STATES DO NOT ALLOW THE EXCLUSION OF IMPLIED WARRANTIES, SO THE ABOVE EXCLUSION MAY NOT APPLY TO YOU. THIS WARRANTY GIVES YOU SPECIFIC LEGAL RIGHTS AND YOU MAY ALSO HAVE OTHER LEGAL RIGHTS THAT VARY FROM STATE TO STATE.

Pearson Education, Inc. does not warrant that the functions contained in the program will meet your requirements or that the operation of the program will be uninterrupted or error-free. However, Pearson Education, Inc. warrants the disc(s) on which the program is furnished to be free from defects in material and workmanship under normal use for a period of ninety (90) days from the date of delivery to you as evidenced by a copy of your receipt. The program should not be relied on as the sole basis to solve a problem whose incorrect solution could result in injury to person or property. If the program is employed in such a manner, it is at the user's own risk and Pearson Education, Inc.. explicitly disclaims all liability for such misuse.

LIMITATION OF REMEDIES

Pearson Education, Inc.'s entire liability and your exclusive remedy shall be:

1. the replacement of any disc not meeting Pearson Education, Inc.'s "LIMITED WARRANTY" and that is returned to Pearson Education, Inc., or

2. if Pearson Education, Inc. is unable to deliver a replacement disc that is free of defects in materials or workmanship, you may terminate this agreement by returning the program

IN NO EVENT WILL PEARSON EDUCATION, INC. BE LIABLE TO YOU FOR ANY DAMAGES, INCLUDING ANY LOST PROFITS, LOST SAVINGS, OR OTHER INCIDENTAL OR CONSEQUENTIAL DAMAGES ARISING OUT OF THE USE OR INABILITY TO USE SUCH PROGRAM EVEN IF PEARSON EDUCATION, INC. OR AN AUTHORIZED DISTRIBUTOR HAS BEEN ADVISED OF THE POSSIBILITY OF SUCH DAMAGES, OR FOR ANY CLAIM BY ANY OTHER PARTY.SOME STATES DO NOT ALLOW FOR THE LIMITATION OR EXCLUSION OF LIABILITY FOR INCIDENTAL OR CONSEQUENTIAL DAMAGES, SO THE ABOVE LIMITATION OR EXCLUSION MAY NOT APPLY TO YOU.

GENERAL

You may not sublicense, assign, or transfer the license of the program. Any attempt to sublicense, assign or transfer any of the rights, duties, or obligations hereunder is void. This Agreement will be governed by the laws of the State of New York. Should you have any questions concerning this Agreement, you may contact Pearson Education, Inc. by writing to: Electrical Engineering Editor, Higher Education Division, Pearson Education, Inc., 1 Lake Street, Upper Saddle River, NJ 07458Should you have any questions concerning technical support, you may write to: Engineering Media Production, Higher Education Division, Pearson Education, Inc., 1 Lake Street, Upper Saddle River, NJ 07458

YOU ACKNOWLEDGE THAT YOU HAVE READ THIS AGREEMENT, UNDERSTAND IT, AND AGREE TO BE BOUND BY ITS TERMS AND CONDITIONS. YOU FURTHER AGREE THAT IT IS THE COMPLETE AND EXCLUSIVE STATEMENT OF THE AGREEMENT BETWEEN US THAT SUPERSEDES ANY PROPOSAL OR PRIOR AGREEMENT, ORAL OR WRITTEN, AND ANY OTHER COMMUNICATIONS BETWEEN US RELATING TO THE SUBJECT MATTER OF THIS AGREEMENT.

McClellan, Schafer, and Yoder, Signal Processing First, ISBN 0-13-065562-7. Prentice Hall, Upper Saddle River, NJ 07458. © 2003 Pearson Education, Inc.